73.50
100K

ADVANCES IN
LASER SCIENCE–II

AMERICAN INSTITUTE OF PHYSICS
CONFERENCE PROCEEDINGS NO. **160**
NEW YORK 1987

OPTICAL SCIENCE AND ENGINEERING SERIES 8

SERIES EDITOR: RITA G. LERNER

ADVANCES IN LASER SCIENCE–II

PROCEEDINGS OF THE SECOND INTERNATIONAL
LASER SCIENCE CONFERENCE

SEATTLE, WA 1986

EDITORS:

MARSHALL LAPP
SANDIA NATIONAL LABORATORIES

WILLIAM C. STWALLEY
UNIVERSITY OF IOWA

GERALDINE A. KENNEY-WALLACE
UNIVERSITY OF TORONTO

Authorization to photocopy items for internal or personal use, beyond the free copying permitted under the 1978 US Copyright Law (see statement below), is granted by the American Institute of Physics for users registered with the Copyright Clearance Center (CCC) Transactional Reporting Service, provided that the base fee of $3.00 per copy is paid directly to CCC, 27 Congress St., Salem, MA 01970. For those organizations that have been granted a photocopy license by CCC, a separate system of payment has been arranged. The fee code for users of the Transactional Reporting Service is: 0094-243X/87 $3.00.

Copyright 1987 American Institute of Physics

Individual readers of this volume and non-profit libraries, acting for them, are permitted to make fair use of the material in it, such as copying an article for use in teaching or research. Permission is granted to quote from this volume in scientific work with the customary acknowledgment of the source. To reprint a figure, table or other excerpt requires the consent of one of the original authors and notification to AIP. Republication or systematic or multiple reproduction of any material in this volume is permitted only under license from AIP. Address inquiries to Series Editor, AIP Conference Proceedings, AIP, 335 E. 45th St., New York, NY 10017.

L.C. Catalog Card No. 87-71962
ISBN 0-88318-360-9
DOE CONF-861038

Printed in the United States of America

PREFACE

When the preface to *Advances in Laser Science–I* was written a year ago, we were pleased in the sense one feels when the first data from a new experiment are viewed as yielding positive results. The International Laser Science Conference (ILS) was established as a Topical Conference of the American Physical Society (APS) in order to provide an international forum for discussion of core research areas related to laser science, focusing on fundamental areas relating to laser mechanisms and properties, laser interactions with materials and processes, and the underlying spectroscopy that forms the basis for understanding these laser-related issues. This endeavor was facilitated by being designated the annual meeting of the newly-established APS Topical Group on Laser Science.

Contrary to one's concerns for "yet another new meeting", the response to the announcement of the first ILS Conference (ILS–I) was outstanding, as was the high level of scientific presentations, made all the more successful by the conducive atmosphere provided by our hosts, the University of Texas at Dallas, in November 1985. Therefore, we embarked on the next step in our growth process in much the same way one builds on a positive first experiment, aided now by co-sponsorship by the Optical Society of America (OSA).

Thus, ILS–II, was held in Seattle from October 20–24 concurrently with the 1986 OSA Annual Meeting, the Topical Meeting on Multiple Excitation of Atoms, and the Workshop on Optical Fabrication and Testing—all grouped together at the Seattle Center under the banner of the Coordinated Conference on Optical and Laser Science and Technology. Through the cooperation of the other program organizers, seven of the 32 ILS symposia were co-sponsored with the OSA and one with the Topical Meeting on Multiple Excitation of Atoms, providing useful ties between the respective programs. The ILS symposia, usually of half-day length, were distributed in up to four parallel sessions, with six outstanding plenary lectures interspersed. The Conference was initiated the evening before the first symposia with a wide-ranging panel discussion organized by Arthur H. Guenther on Limits to Laser Advancements—Fundamental and Otherwise, and treated to a probing review by C. Kumar N. Patel and other experts in a subsequent evening panel on the methodology of the APS Study on the Scientific and Technological Aspects of the Strategic Defense Initiative (SDI). Attendance at the various ILS–II symposia was well within our desired goals (with, of course, the occasional glitches caused by the expected unexpected, such as fogged-in airports). More than 2000 people attended the Coordinated Conferences, with typically 40% attending ILS sessions. Over a quarter of the papers involved foreign authors. We are indebted to the OSA staff and leadership for their efforts on our behalf in this undertaking, and especially to Jarus Quinn, the Executive Secretary, and Barbara Hicks, the Meetings Director.

The Conference Chair of ILS–II was Richard C. Powell (Oklahoma State University), who provided overall guidance and oversight as well as expert financial control. William C. Stwalley (University of Iowa) was Conference Co-Chair, Marshall Lapp (Sandia National Laboratories) was Program Chair, and Geraldine A. Kenney-Wallace (University of Toronto) was Program Co-Chair. Rolf W. F. Gross (Aerospace Corporation) was International Co-Chair, a position of key importance in facilitating the attendance of our overseas colleagues, which Rolf executed with expertise and dedication.

The Conference program was originated and assembled by the Program Chair and Co-Chair together with inputs from the Conference Co-Chair. Substantial contributions were made by the various Session Organizers and several of the ILS–II Program Advisors. A few of the Session Organizers, in fact, informally assumed the role of Area Chairs, taking responsibility for coordinating several related symposia—an approach used formally in the present organization of ILS–III, to be held in Atlantic City, November 1-5, 1987.

We note that this Proceedings volume is not intended to replace the publication of new and complete research results in the technical journals that cover our fields. Instead, the brief accounts of research from the contributed and some of the invited papers are intended to be focused on providing entry-level information about the research topic chosen, with just enough description aided by a listing of the key references to enable the reader to (1) understand the basis of what has been accomplished, and (2) know where to look in order to find more detailed information. Others of the invited papers, the plenary lectures, and an accounting of the initial panel discussion are directed more at providing reviews of technical areas as well as original results, and in some instances include historical perspectives and thoughts on future directions.

One advantage of providing this information in a published proceedings volume is that it is organized in a more useful fashion for subsequent use than the technical digest provided at many conferences. Thus, the organization of this volume follows that of the general areas for which papers were solicited, rather than in the order of presentation at the meeting. (For the convenience of the reader, a listing of these areas and the corresponding conference topics follows this preface. The reader will note, however, that some poster and postdeadline papers have been, inevitably, assigned with less than perfect precision.) Another advantage is that all papers in this volume have been reviewed by at least one expert in the field. Thus, quoting results from these papers is appropriate, although we strongly encourage the reader to explore the subsequent detailed journal publications resulting from this work for more definitive citations. For their reviews, we thank those session organizers and other colleagues who provided this valuable assistance.

Special thanks are extended to Lynn Borders, the ILS Conference Administrative Assistant, and to the secretarial staff at the University of Iowa for their tireless devotion to the countless administrative tasks for which they took responsibility. Indeed, Lynn often provided the conscience and corporate memory of the Conference, never permitting the Organizing Committee to lose track of important issues. The extensive work involved in interfacing with the authors of the papers in this volume and with the reviewers, as well as in providing manuscript correction capabilities and assembly of the volume was provided by Lynn.

We gratefully acknowledge the organizations that provided generous support to ILS–II—the Air Force Office of Scientific Research, the Army Research Office, the Office of Naval Research, Lawrence Livermore National Laboratory, the National Science Foundation, and the University of Iowa—and trust that they will find this record of scientific results presented at the Conference to be of value.

The Editors

Marshall Lapp, Sandia National Laboratories, Livermore (Program Chair)
William C. Stwalley, University of Iowa (Conference Co-Chair)
Geraldine A. Kenney-Wallace, University of Toronto (Program Co-Chair)

Contents

I. Limits to Laser Advancements—Fundamental and Otherwise (Panel Discussion)

Limits to Laser Advancements—Fundamental and Otherwise 2
    A. H. Guenther (Moderator)

II. Advanced Lasers and Coherent Sources

II. A. High Energy Laser Overviews

Advances in Chemical Lasers (*invited*) 10
    J. Miller

A Historical Perspective of Early High Power Gas Laser Research (*invited*) 23
    A. Hertzberg

Injection Controlled Operation of a Broadband Excimer Laser 30
    Y. Zhu, R. A. Sauerbrey, F. K. Tittel, and W. L. Wilson, Jr.

High-Performance DF-CO_2 Chain-Reaction Laser 33
    S. T. Amimoto, J. S. Whittier, G. N. Harper, R. Hofland, Jr., J. M. Walters, Jr., T. A. Barr, Jr., R. L. Kerber, and W. K. Jaul

Investigations of a Transverse-Excited High-Power CO_2 Laser 39
    J. Stanco, G. Sliwinski, J. Konefal, P. Kukiello, G. Rabczuk, Z. Rozkwitalski, and R. Zaremba

The KMSF Low Preheat Implosion Experiments 42
    J. T. Larsen and R. R. Johnson

II. B. Gamma Ray Lasers

Prospects for a Gamma Ray Laser Based upon Upconversion (*plenary*) 45
    C. B. Collins

The Multi-Beam Borrmann Effect in Crystalline Gamma-Ray Lasers (*invited*) 55
    J. T. Hutton, G. T. Trammell, and J. P. Hannon

Progress in the Gamma Ray Laser Program at Texas 1: Flash X-ray Techniques for Pumping Nuclear Materials 63
    F. Davanloo, T. S. Bowen, and C. B. Collins

Progress in the Gamma Ray Laser Program at Texas 2: Coherent Techniques for Pumping a Gamma Ray Laser 66
    S. Wagal, P. Reittinger, E. Juengerman, C. Eberhard, and C. B. Collins

Classical and Semiclassical Calculation of Nuclear-Electron Coupling 69
    D. W. Noid, F. X. Hartmann, and M. L. Koszykowski

Nuclear Interlevel Transfer Driven by Collective Outer Shell Electron Oscillations (*invited*) 75
    G. A. Rinker, J. C. Solem, and L. C. Biedenharn

Nuclear Transitions Induced by Atomic Excitations 87
    J. A. Bounds, P. Dyer, and R. C. Haight

Why Gamma Rays Are Different. Some Notes for Pedestrians (*invited*) 90
    H. J. Lipkin

A Nanosecond Flash X-ray Device of Subangstrom Excitation—A Table Top Alternative to Synchrotron Radiation 98
    T. S. Bowen, J. J. Coogan, F. Davanloo, and C. B. Collins
Nuclear Raman Spectrometer 101
    P. Reittinger, S. Wagal, and C. B. Collins

II. C. Novel Solid-State Lasers

Spectroscopy of New Chromium/Neodymium Doped Oxide Laser Materials: Garnets and Hexaaluminates (*invited*) 104
    G. Boulon, C. Garapon, and A. Monteil
An Extended Foerster-Dexter Model for Correlated Donor-Acceptor Placement in Solid State Materials 114
    S. R. Rotman and F. X. Hartmann
Large Nd,Cr:GSGG Boule Growth and Quality 118
    S. E. Stokowski, M. D. Shinn, M. Randles, and D. Dawes
Single Crystal $Na_3Ga_2Li_3F_{12}$:Cr^{3+} Laser Pumped Laser Experiments 120
    J. A. Caird, P. R. Staver, M. D. Shinn, H. J. Guggenheim, and D. Bahnck
Nonlinear Refractive Index Measurements of Glasses and Crystals 124
    R. Adair, L. L. Chase, and S. A. Payne
High Efficiency Laser-Pumped Emerald Lasers 128
    S. T. Lai
New Generation of High Power Laser Systems Based on Multiple-Pass Amplifiers 130
    S. Jackel, R. Lalluz, E. Yarkoni, M. Givon, B. Arad, S. Eliezer, and A. Zigler
Intracavity Frequency-Doubling of Quasi-CW Pumped YAG Laser 133
    J. Q. Yao, Y. Li, and Y. M. Liu
Thermo-Optical Modeling of Flashlamp-Pumped Zig-Zag Slabs 136
    R. J. Gelinas, S. K. Doss, and S. S. Murty
Near-Field Phase Measurements of Diode Laser Arrays 139
    G. C. Dente, K. A. Wilson, D. Depatie, and J. Querns

II. D. Enhancement of Free-Electron Laser Performance

Prevention of Sideband-Induced Detrapping in Tapered-Undulator Free-Electron Lasers (*invited*) 142
    D. C. Quimby

II. E. Ultrafast Short-Wavelength Light Sources and Techniques

X-ray Characterization of Picosecond Laser Plasmas (*invited*) 157
    O. L. Landen, E. M. Campbell, and M. D. Perry
Soft X-ray Population Inversion of Na XI Levels by Intercombination Line Resonant Photoexcitation 163
    Z. Q. Zhang, R. X. Lu, and G. Y. Yin

II. F. Applications of Ultrashort Wavelength Lasers

Short Wavelength Limitations of Four Wave Mixing in Gases (*invited*) 164
    H. Scheingraber and C. R. Vidal
Photofragment Spectroscopy with Coherent VUV: Product Correlations and Alignment (*invited*) 170
    G. E. Hall, N. Sivakumar, G. Chawla, P. L. Houston, I. Burak, I. M. Waller, H. F. Davis, and J. W. Hepburn

Measurements of Subpicosecond Laser-Produced Plasmas 179
    H. Milchberg, R. R. Freeman, and S. C. Davey
A Laser Produced Plasma Light Source for High Resolution Spectroscopy and Soft X-ray Lithography 182
    M. L. Ginter and T. J. McIlrath
XUV Laser Stark Spectroscopy of Xe Autoionizing Rydberg States 185
    W. E. Ernst, T. P. Softley, L. Tashiro, and R. N. Zare

III. NONLINEAR OPTICAL PHENOMENA AND APPLICATIONS

III. A. Fluctuations, Noise, and Chaos in Laser Systems

Optical Bistability Switching with External Noise (*invited*) 190
    E. Arimondo, D. Dangoisse, L. Fronzoni, O. Incani, and N. K. Rahman
Bi-directional Oscillation and Bichromatic Emission in a Ring Dye Laser: Evidence for a New Mode Structure Hierarchy (*invited*) 196
    N. M. Lawandy

III. B. Stochastic Fluctuation Effects in Nonlinear Optics

Role of Fluctuations in Nonlinear Optical Absorption Processes (*invited*) 202
    S. J. Smith

III. C. Nonlinear Optical Techniques for Gas-Phase Measurements

Parametric Processes and Gain Saturation in Resonantly Enhanced Optical Phase Conjugation in Na Vapor Near a Two-Photon Resonance 208
    R. K. Wunderlich, W. R. Garrett, and M. G. Payne
Multiwave Mixing and Multiphoton Ionization in Strontium Vapor 211
    K. Böhmer, J. Reif, and E. Matthias
Angular Dependence of the Vibrational Raman Linewidths for Stimulated Raman Scattering in H_2 214
    G. C. Herring, M. J. Dyer, and W. K. Bischel
Observation of Long-Lived Collision-Induced Coherences and Ground-State-Spin Gratings in a Flame 217
    R. Trebino and L. A. Rahn
Generation of Solitons in Transient Stimulated Raman Scattering by Optical Phase Shifts 220
    D. C. MacPherson and J. L. Carlsten

III. D. Lasers and Nonlinear Optics

High Efficiency Distributed Feedback Lasers Optimally Designed for Stable Single Mode Operation 223
    H. Ishikawa, H. Soda, Y. Kotaki, K. Kihara, and H. Imai
Longitudinal Mode Width in Excimer Lasers 226
    G. G. Lombardi and W. H. Long, Jr.
High Power He-Cd^+ White Light Laser 229
    A. Fuke, K. Masuda, and Y. Tokita
Ultrashort Pulse Chirp Parameter Determination by Interferometric Methods 232
    R. Fischer, C. Rempel, J. Gauger, and J. Tilgner

Phase Conjugation of 2.91 μm HF Laser Radiation Via Stimulated Brillouin Scattering 235
    M. T. Duignan, B. J. Feldman, and W. T. Whitney

Efficient Harmonic Generation of CO_2 Laser Radiation in Thallium Arsenic Selenide 238
    R. C. Y. Auyeung, D. M. Zielke, and B. J. Feldman

Radiative Trapping Effects in Ruby: 77 °K to 300 °K 240
    M. Birnbaum, C. L. Fincher, J. Machan, and M. Bass

A Study of Multiphoton Resonances in Kr and Ar under Intense Laser Field 243
    J. J. Tiee, M. J. Ferris, and G. K. Anderson

Simultons versus Raman Solitons: Diffraction and Pulse Shape Effects from a Numerical Experiment Point of View 247
    F. P. Mattar, J. DeLettrez, J. P. Babuel-Peyrissac, J. P. Marinier, and C. Bardin

Direct Measurement of Nonlinear Energy Deposition from an Intense 532 nm Photon Field into Alkali Halides 254
    S. C. Jones, X.-A. Shen, P. Braunlich, and P. Kelly

Degenerate Stimulated Parametric Scattering in $LiNbO_3$:Fe 257
    G. Y. Zhang, S. M. Liu, Z. K. Wu, Q. X. Li, P. P. Ho, and R. R. Alfano

Third Harmonic Generation with High Efficiency 260
    R. Fischer

Intensity Dependence of the Polarization Tensor in a Quantum Treatment 263
    N. Chencinski and A. N. Weiszmann

Transformation of Coherent States of Electromagnetic Field to Quasi-Fock States 266
    V. I. Zakharov

Application of the Split Operator Fourier Transform Method to the Solution of the Nonlinear Schrödinger Equation 269
    P. L. DeVries

The Effects of Diffraction, Dispersion, and Transient Propagation on Optical Bistability in a One-Directional Ring Cavity 272
    F. P. Mattar, J. Teichmann, Y. Claude, and C. Goutier

Comparison of Laser Phase and Suppression of Noise 275
    X. M. Yang, M. Yi, and Z. H. Huan

Observation of a Cone Radiation in Sodium Vapor 278
    X. F. Han, Z. G. Lu, and Z. G. Ma

IV. Atomic, Molecular, and Ionic Spectroscopy

IV. A. Chaos in Molecular Systems

Quasiperiodic and Chaotic Motions in Intense Field Multiphoton Processes (*invited*) 282
    S.-I. Chu

Dissipative Molecular Dynamics: Quantal versus Classical Treatment (*invited*) 289
    M. W. Tung, J. M. Yuan, and J. F. Heagy

IV. B. Orientational and Nonadiabatic Effects in Collision Dynamics

Inelastic Collisions in Laser Excited Alkali Atoms (*invited*) 295
    M. Allegrini, S. Gozzini, and L. Moi

Isotope Dependence of Fine-Structure Branching in He-Na Optical Collisions   301
    L. L. Vahala and P. S. Julienne

Near Resonant Energy Transfer Theory Applied to Non-Resonant Processes   304
    D. E. Godar, K. L. McNesby, and R. D. Bates, Jr.

IV. C. Laser Cooling and Trapping

Single Atomic Particle at Rest in Free Space: New Value for Electron Radius (*invited*) 307
    H. Dehmelt

Diffraction of Atomic Waves by Non-Orthogonal Standing Waves 315
    P. J. Martin, P. L. Gould, B. Oldaker, and D. E. Pritchard

Theory of Atomic Motion in Laser Light (*invited*) (*abstract only*) 318
    C. Cohen-Tannoudji

Laser Cooling and Trapping of Atoms (*invited*) 319
    J. E. Bjorkholm, S. Chu, A. Ashkin, and A. Cable

Trapping Atoms with Radiation Pressure 329
    D. E. Pritchard and E. L. Raab

Magnetic Trapping of Neutral Atoms 332
    T. Bergeman and H. Metcalf

Progress Toward an Alexandrite Laser Trap for Potassium Atoms 335
    K.-H. Yang, X.-Z. Zeng, and W. C. Stwalley

'Quantum Jumps' Observed in the Fluorescence of a Single Ion 337
    Th. Sauter, R. Blatt, W. Neuhauser, and P. E. Toschek

The 'Containerless' Condensation of 'Mirror' Matter 344
    J. T. Bahns

IV. D. Molecular Ion Spectroscopy and Applications

Autodetachment Spectroscopy of Negative Ions (*invited*) 347
    D. M. Neumark, K.R. Lykke, T. Andersen, and W. C. Lineberger

High-Resolution Measurement of the Infrared Rotation-Vibration Spectrum of NH^- 354
    H. C. Miller, M. Al-Za'al, and J. W. Farley

SO^+ Emission in a Supersonic Jet 357
    I. W. Milkman, J. C. Choi, J. L. Hardwick, and J. T. Moseley

IV. E. Molecular Spectroscopy

Double Resonance Techniques for the High Resolution Spectroscopy of Unstable Molecules (*invited*) 359
    W. E. Ernst

Reduction of 1+1 Rempi Spectra to Population Distributions: Saturation and Intermediate State Alignment Effects (*invited*) 364
    D. C. Jacobs, R. J. Madix, and R. N. Zare

Transient Excited Singlet State Absorption in the Laser Dye alpha-NPO 371
    P. Venkateswarlu, M. C. George, Y. V. Rao, H. Jagannath and G. Chakrapani

Four-Atomic Rare Gas Halide Exciplexes and Their Impact on High-Power Laser Kinetics 373
    R. Sauerbrey, F. K. Tittel, Y. Zhu, and W. L. Wilson, Jr.

Experimental Determination of the Spin-Rotation Coupling in NaXe
Molecules 376
    M. Y. Hou, B. H. Feng, J. Zhang, Y. B. She, Z. L. Mi, and T. J. Chen

Hyperfine Interaction of the Rydberg Triplet States of Na_2 378
    L. Li, R. W. Field, and Q. S. Zhu

Studies of the Diffuse Bands of K_2, RB_2 and Cs_2 381
    W.-T. Luh, J. T. Bahns, K. M. Sando, A. M. Lyyra, P. D. Kleiber, and
    W. C. Stwalley

The Li_2 $b^3\Sigma_g^+ - X^3\Sigma_u^+$ Transition and A New UV Absorption Band 386
    F. Jin, C. Zhang, and B. Y. Tang

Precise Multiphoton Spectroscopy of Excited States of H_2 388
    E. E. Eyler, J. M. Gilligan, and E. McCormack

Kinetic Spectroscopy Using a Color Center Laser 391
    J. W. Stephens, J. L. Hall, W. B. Yan, H. Solka, M. L. Richnow, R. F. Curl, Jr.,
    G. P. Glass, and F. K. Tittel

$Fe(CO)_5$ Multiphoton Ionization Mass Spectra 394
    S. T. Li, J. C. Han, J. L. Shi, H. X. Liu, F. L. Li, J. P. Gu, and C. K. Wu

IV. F. Multiple Excitation of Atoms—Summary Session (published in the
Technical Digest (Optical Society of America, Washington, DC, 1986), pp. 75–83.

IV. G. Atomic Spectroscopy

Light-Induced Drift in a Discharge Tube 396
    F. C. Lin, S. F. Li, Q. Q. Hu, and Z. J. Huang

ns' Autoionizing Rydberg Series Lines of Xe 399
    K. Ueda

Remeasurement of the Rydberg Constant by a Crossed-Beam Laser 402
    P. Zhao, W. Lichten, H. Layer, and J. Bergquist

Laser Absorption and Fluorescence Studies of the Lithium 2S-3D Transition ... 404
    G. C. Tisone and P. J. Hargis, Jr.

V. CONDENSED MATTER, SURFACE, AND PARTICLE SPECTROSCOPY

V. A. Laser Raman Spectroscopy 1: Coherent Phenomena

Preliminary Observation of Nonrelaxational, Inertial Motion in CS_2 Liquid by
Femtosecond Time-Resolved Impulsive Stimulated Scattering (*invited*) 408
    L. R. Williams, S. Ruhman, A. G. Joly, B. Kohler, and K. A. Nelson

Development of Rike Techniques using Picosecond Lasers 417
    M. W. Schauer, M. J. Pellin, B. M. Biwer, and D. M. Gruen

Steady State Light Pulses in Stimulated Backward Scattering 420
    D. N. Ghosh Roy and D. V. G. L. N. Rao

The Raman Solitons 422
    F. P. Mattar

Techniques for Far Ultraviolet Resonance Raman Spectroscopy 430
    P. B. Kelly, S. Li, G. D. Strahan, and B. Hudson

Applications of Ultraviolet Resonance Raman Scattering in Molecular
Electronic Spectroscopy 433
    P. B. Kelly, S. J. Li, and B. Hudson

Stimulated Electronic Raman Scattering in Sodium Vapor 436
    X. F. Han, Z. G. Lu, Z. G. Ma, and Y. K. Cheng

V. B. Clusters 2: Metals

Quantum Level Probes of Small Metal Clusters and Their Oxidation (*invited*) 439
    J. L. Gole

Silver Clusters as the Active Sites for Surface Enhanced Raman Scattering ... 452
    T. E. Furtak and D. Roy

Optical Properties of Fractal Clusters 455
    Z. Chen, P. Sheng, D. A. Weitz, H. M. Lindsay, M. Y. Lin, and P. Meakin

V. C. Clusters 1: Non-Metals

Dynamics of Cluster Dissociation (*invited*) 458
    R. G. Keesee and A. W. Castleman, Jr.

The Chemistry of Size-Selected Silicon Clusters as Studied by Fourier Transform Mass Spectrometry (*invited*) 465
    W. D. Reents, Jr., M. L. Mandich, and V. E. Bondybey

Photodetachment and Photodissociation Studies of Semiconductor Cluster Ions 472
    Y. Liu, Q. Zhang, S. C. O'Brien, J. R. Heath, R. F. Curl, F. K. Tittel, and R. E. Smalley

V. D. Time-Resolved Laser Probes of Surface Dynamics

Studies of Surface Dynamics Using Second-Harmonic Generation (*invited*) .. 475
    H. W. K. Tom

Infrared Laser-Induced Desorption of NO and CO from Alumina Substrates 484
    W. H. Weber and B. D. Poindexter

Second Harmonic Generation and Differential Capacitance Studies of Smooth Silver Electrode-Aqueous Electrolytes 487
    H. M. Rojhantalab and G. L. Richmond

Nonlinear Optical Studies of Semiconductor Interfacial Properties 490
    J. M. Robinson and G. L. Richmond

Vibrational Excitation of an Adbond by a Short-Pulsed Laser 493
    S. van Smaalen and T. F. George

V. E. Atomic and Molecular Surface Imaging

Atomic and Electronic Imaging of Semiconductor Surfaces with Scanning Tunneling Microscopy (*invited*) 496
    J. E. Demuth, R. J. Hamers, and R. M. Tromp

Second Harmonic and Sum Frequency Generation on Dye-Coated Surfaces using Collinear and Non-Collinear Excitation Geometries 506
    R. E. Muenchausen, D. C. Nguyen, R. A. Keller, and N. S. Nogar

V. F. Laser Particle Interactions

Micrometer-Size Droplets as Optical Cavities: Lasing and Other Nonlinear Effects (*plenary*) 509
    R. K. Chang

Raman-Mie Scattering from Optically Levitated Single Particles (*invited*) ... 516
    W. Kiefer

Resonance Light Scattering from a Suspension of Microspheres 523
    T. R. Lettieri and E. Marx

Two-Wave Mixing in Liquid Suspensions of Microparticles 526
    R. McGraw and D. Rogovin

Laser Doppler Velocimetry for Sub-Micrometer Particle Size Determination 529
    N. J. Dovichi and F. Zarrin
Diffusive and Convective Evaporation of Irradiated Droplets 532
    R. L. Armstrong and A. Zardecki
A Self-Similar Approach to the Explosion of Droplets by a High Energy Laser Beam 534
    S. M. Chitanvis
Near-Field Light Scattering by Parallel Glass Fibers 537
    D. S. Benincasa, T.-G. Tsuei, and P. W. Barber
Optical Bistable Interaction of Laser Radiation with Microparticles 540
    K. M. Leung

V. G. Ultrahigh-Speed Photodetectors

High Gain-Bandwidth-Product Avalanche Photodiodes for Multigigabit Data Rates (*invited*) 543
    J. C. Campbell

VI. LASER PHOTOCHEMISTRY AND PHOTOPHYSICS

VI. A. Laser Challenges in Photochemistry and Photophysics

Using Incoherent Light to Generate Coherent Excitations (*invited*) 556
    S. R. Hartmann
A Theory of Coherent Multi-Color Laser Excitation of Localized States 563
    J. S. Hutchinson
Application of Semiconductor Diode Lasers to Probe Photodissociation Dynamics 566
    H. K. Haugen, W. P. Hess, and S. R. Leone
Selective IR Photoisomerizations in Solids 569
    J. S. Shirk and C. L. Marquardt

VI. B. Laser Raman Spectroscopy 2: Chemical and Biochemical Reactions

Interferences in the Raman Excitation Profile for the Intensity of Normal Modes of Aggregated Chlorophyll a (*invited*) 571
    L. V. Haley, T. L. Collier, T. A. Mattioli, D. L. Thibodeau and
    J. A. Koningstein
Studies of Enzymes by Resonance Raman Spectroscopy (*invited*) 574
    P. R. Carey and A. C. Storer
Laser Raman Phonon Spectroscopy of Solid State Photoreaction: Photodimerization of O-Methoxy Trans Cinnamic Acid 580
    U. Ghosh and T. N. Misra
Applications of Ultraviolet Resonance Raman Spectroscopy to Protein Structure 583
    L. Mayne and B. Hudson

VI. C. Laser-Induced Surface Reactions

Laser-Induced Etching (*invited*) 586
    C. I. H. Ashby
UV Laser Induced Thin Film Deposition (*invited*) 594
    R. Solanki

Laser-Induced Deposition of Gold 600
    T. H. Baum
Laser-Induced Desorption from the (111) Surface of BaF_2 602
    E. Matthias, H. B. Nielsen, J. Reif, A. Rosén, and E. Westin
Laser Excitation Spectroscopic Studies of Metal Ion Binding in Polymers 605
    E. K. L. Wong and G. L. Richmond

VI. D. Laser Photophysics

Picosecond Reorientational Dynamics in Polymer Solutions 608
    E. L. Quitevis, K.G. Casey, and T. W. Sinor

VII. DIAGNOSTIC AND ANALYTICAL APPLICATIONS OF LASERS

VII. A. Laser Diagnostics for Large Molecules

Supercritical Fluid Injection of Nonvolatiles with Resonant Two Photon
Ionization Detection in Supersonic Beam Mass Spectrometry 612
    D. M. Lubman, C. H. Sin, and H. M. Pang
Laser-Based Circular Dichroism Detection of Molecules in Flowing Liquid
Systems using High Frequency Polarization Modulation 615
    R. E. Synovec and E. S. Yeung
Laser Spectroscopy of Jet-Cooled Chlorinated Aromatic Hydrocarbons 618
    E. A. Rohlfing and D. W. Chandler
Photo-Thermal Deflection Velocimetry in Laminar and Turbulent Flows Using
a Transient Grating 621
    C. J. Dasch and J. A. Sell

VII. B. Laser Microprobes and Microscopy

Laser Raman and Fluorescence Microprobing Techniques (*invited*) 624
    F. Wallart and P. Dhamelincourt
A Review of the NRL CARS Microscope (*invited*) 631
    M. D. Duncan, J. Reintjes and T. J. Manuccia
Crossed-Beam Thermal Lens as a Scanning Laser Microscope (*invited*) 638
    D. S. Burgi and N. J. Dovichi

VII. C. Optical Aspects of Laser Diagnostics

Folk Wisdom in Optical Design (*invited*) 644
    A. E. Smart
Planning and Implementing a Formal Test Program for Spaceflight
Instruments (*invited*) 650
    L. E. Mauldin III
Optical Ray Tracing for Crossed Beam Photothermal Deflection Spectroscopy 658
    J. A. Sell
Classical and Holographic Interferometry Systems Combined to Document
Inertial Fusion Experiments 661
    J. S. Ankney and G. E. Busch
Visualization of RF Acoustic Wavefront by Laser Correlation Theory 664
    X. M. Yang and M. Yi
Cross-Correlation Theory of Gratings and Its Use for Measuring the 2-D
Microvibration by Laser Beam 667
    X. M. Yang, M. Yi, and H. Pan

A New Technique for Pattern Recognition Using Fresnel Hologram and
Extended Source 670
    G. G. Mu, Z. Q. Wang, and D. Q. Chen
Computerization of an Infrared Diode Laser Spectrometer 673
    C. B. Dane, D. R. Lander, R. F. Curl, Jr., J. V. V. Kasper, F. K. Tittel, and
    R. Brüggemann

VII. D. Imaging in Multiphase Media

Application of a Laser-Induced Breakdown Time-of-Flight Technique as a
Flow Diagnostic in a CO_2 Free-Jet Expansion 676
    P. J. Wantuck and D. E. Hof
Multispot Laser Vibrometry for Materials and Structure Evaluations 680
    G. L. Fitzpatrick, R. L. Skaugset, and T. J. Davis

VII. E. Atomic and Molecular Diagnostics

Detection of Transient Fluorine Atoms 683
    G. W. Loge, N. Nereson, and H. A. Fry
Laser Spectroscopic Detection of OH in Catalytic Reactions on Platinum 687
    S. Ljungström, A. Rosén, T. Wahnström, and B. Kasemo
IR Laser Absorption Eddy Correlation Measurement Devices for Trace
Atmospheric Gases 690
    M. S. Zahniser, P. L. Kebabian, S. Anderson, A. Freedman, and C. E. Kolb
Propellant Combustion Study by Coherent Anti-Stokes Raman Scattering 693
    T. H. Vu and R. Field

VIII. LASER RESEARCH AND TECHNIQUES IN MEDICINE AND BIOLOGY

VIII. A. Interactions of Laser Radiation with Biological Tissue

Interactions of Excimer Lasers with Polymers (*invited*) 698
    Y. S. Liu, H. S. Cole, and H. R. Philipp
Dynamics of the Ultraviolet Laser Ablation of Corneal Tissue (*invited*) 703
    R. Srinivasan
Internal Biological Tissue Temperature Measurements Using Zirconium
Fluoride Fiber (*invited*) 710
    E. Sinofsky and G. Gofstein

VIII. B. Biomedical Laser Applications

Tissue Diagnostics Using Laser-Induced Fluorescence Techniques (*invited*) .. 715
    P. S. Andersson, J. Ankerst, E. Kjellén, S. Montán, K. Svanberg, and S.
    Svanberg
Mechanistic and Diagnostic Aspects of Photodynamic Enhancement and Stone
Fragmentation (*invited*) 722
    D. I. Rosen, S. J. Davis, A. A. Boni and J. P. Campbell
Arterial Aneurysm Model Using Laser Energy (*invited*) 735
    J. LoCicero III, R. S. Hartz, W. J. McCarthy, and S.-R. Shih

LISTING OF MAJOR AREAS, CONSTITUENT
SYMPOSIA, AND SYMPOSIUM ORGANIZERS FOR ILS–III

ADVANCED LASERS AND COHERENT SOURCES

High Energy Laser Overviews	Edward T. Gerry
Applications of Ultrashort Wavelength Lasers	Stephen C. Wallace
Gamma-Ray Lasers	Leslie Cohen
	Bohdan Balko
Novel Solid-State Lasers	William F. Krupke
Optical Techniques for the Enhancement of Free-Electron Laser Performance*	John M. J. Madey (OSA)
	Cha-Mei Tang (ILS)
Ultrafast Short-Wavelength Light Sources and Techniques*	Roger W. Falcone (OSA)
	Herschel S. Pilloff (ILS)

NONLINEAR OPTICAL PHENOMENA AND APPLICATIONS

Fluctuations, Noise, and Chaos in Laser Systems	Lorenzo M. Narducci
Stochastic Fluctuation Effects in Nonlinear Optics	Yehiam Prior
Nonlinear Optical Techniques for Gas-Phase Measurements	Lawrence A. Rahn
	Roger L. Farrow
Lasers and Nonlinear Optics	Geraldine A. Kenney-Wallace

ATOMIC, MOLECULAR AND IONIC SPECTROSCOPY

Chaos in Molecular Systems	Jian-Min Yuan
	Eric J. Heller
Laser Cooling and Trapping	William D. Phillips
Orientational and Nonadiabatic Effects in Collision Dynamics	Michael G. Raymer
	John Weiner
Molecular Ion Spectroscopy and Applications*	John T. Moseley (OSA)
	Richard J. Saykally (ILS)
Multiple Excitations of Atoms: Summary Session**	William E. Cooke (Topical Meeting)
	William C. Stwalley (ILS)
Molecular Spectroscopy	William C. Stwalley

CONDENSED MATTER, SURFACE, AND PARTICLE SPECTROSCOPY

Laser Raman Spectroscopy 1: Coherent Phenomena	George H. Atkinson
	J. A. Koningstein
Laser–Particle Interactions	Peter W. Barber
Atomic and Molecular Surface Imaging	Richard P. Van Duyne
	Thomas E. Furtak
Clusters I: Metals* and	Vladimir E. Bondybey (OSA)
Clusters II: Nonmetals*	William C. Stwalley (ILS)
Time-Resolved Laser Probes of Surface Dynamics*	Jeffrey Bokor (OSA)
	Geraldine L. Richmond (ILS)
Ultra-High Speed Photodetectors*	Federico Capasso (OSA)
	Lawrence S. Goldberg (ILS)

LASER PHOTOCHEMISTRY AND PHOTOPHYSICS

Laser Challenges in Photochemistry and Photophysics — Paul Brumer

Laser-Induced Surface Reactions — Susan D. Allen
Jeffrey I. Steinfeld

Laser Raman Spectroscopy 2: Chemical and Biochemical Reactions — J. A. Koningstein
George H. Atkinson

DIAGNOSTIC AND ANALYTICAL APPLICATIONS OF LASERS

Laser Diagnostics for Large Molecules — Normand Laurendeau
David M. Lubman

Imaging in Multiphase Media — Marshall B. Long
Julian M. Tishkoff

Laser Microprobes and Microscopy — Thomas J. Mannuccia
Marshall Lapp

Optical Aspects of Laser Diagnostics — James D. Trolinger
Anthony E. Smart

LASER RESEARCH AND TECHNIQUES IN MEDICINE AND BIOLOGY

Interactions of Laser Radiation with Biological Tissue — R. Srinivasan

Biomedical Laser Applications — Joseph LoCicero III
Warren S. Grundfest

*Symposia sponsored jointly with the OSA Annual Meeting
**Symposium sponsored jointly with the OSA Topical Meeting on Multiple Excitation of Atoms

I. Limits to Laser Advancements—Fundamental and Otherwise
(Panel Discussion)

LIMITS TO LASER ADVANCEMENT: FUNDAMENTAL AND OTHERWISE

PANEL DISCUSSION HELD AT THE SECOND INTERNATIONAL
LASER SCIENCE CONFERENCE

OCTOBER 20, 1986
SEATTLE, WASHINGTON

Arthur H. Guenther, Moderator
Participants: Robert Byer, Michael Bass, Elsa Garmire, and David Sliney.

This panel addressed four constraints to the further development of lasers. They were primarily of two kinds -- natural and man-made. Of the former, one is obviously related to materials. This aspect was discussed by Dr Robert L. Byer, Professor of Applied Physics and Associate Dean, Humanities and Sciences, Stanford University. Because we are dealing with lasers, a coherent light source capable of extreme intensities, one must of necessity be concerned as well with the interaction of such radiation and matter, even if it is nonlinear or catastrophic in character. This aspect was addressed by Professor Michael Bass, Professor and Chairman of Electrical Engineering and Electrophysics at the University of Southern California.

The more subtle limits, those that we place on ourselves, were next elucidated; first that aspect which is related to impeded communication or scientific secrecy was discussed by Professor Elsa Garmire, Director of the Center for Laser Studies and Professor of Electrical Engineering and Physics at the University of Southern California. Finally, Dave Sliney, Chief of the Laser Branch, Laser Microwave Division of the US Army Environmental Hygiene Agency, concluded the panel's deliberations by presenting the impact of regulatory constraints, such as those relating to safety, standards, etc., as issues affecting the advancement of laser technology and applications. All panelists addressed the significant and beneficial role education can play in overcoming or at least reducing the negative factors surfacing during the evening in a lively interplay with the audience which followed each of their presentations. The panel was assembled and discussion led by Dr. Arthur H. Guenther, Chief Scientist, Air Force Weapons Laboratory, Kirtland AFB, NM.

Professor Byer's discussion of the influence of materials as they limit the advancement of lasers centered on thermal considerations. Issues such as overall efficiency, removal of heat, pumping, energy storage, and extraction were considered in light of various material properties and related to different material classes, whether they be solid (glassy or crystalline) liquid, or even gaseous. Professor Byer also included a discussion of the potential of free electron lasers as a major laser candidate in the future, but his conclusions indicated that, based upon our long history and the knowledge gained therein, glass offered the greatest potential from an active storage medium standpoint, for long life, efficient high average power systems. He felt that much could be accomplished through materials engineering to alleviate what many believe to be present-day limitations in these classes of materials. The figure of merit that he applied which lent credence to his conjecture was centered upon the architecture of the eventual laser system exemplified by modern advances such as diode pumping, etc. Mass production of these types of solid state semiconductor materials will eventually lead to significant cost reductions per unit weight resulting in reduced size and increased efficiency, greatly enhancing the lasers range of application and utility. He concluded his presentation by espousing the well-known KISS principle as limits to materials and felt that certainly for the foreseeable future the greatest advances could be made by applying what we know to those materials with which we have the greatest experience. This is not to say that he ruled out the development of other types of lasers for very specific and unique applications such as chemical lasers for defense, etc.

The second panelist, Professor Michael Bass of the University of Southern California, discussing the interaction of optical radiation with material, of course touched upon both linear and nonlinear aspects. Again, he of necessity discussed thermal issues early in his presentation, but tried to draw the distinction within an application between experimentally desirable parameters and those that may be made manifest by material properties, with an awareness that in some cases this tradeoff must as well address operator influences. As might be expected he was quick to point out that the very nature of coherent radiation and the interaction with specific material classes can lead to situations which can overcome or at least circumvent to some degree what otherwise might be limitations in the advancement of laser technology -- the most obvious to come to mind is the correction of wavefront errors by the nonlinear optical process of phase conjugation. He addressed many other comparisons between true constraints on one hand and opportunities for advancements on the other by clever application of physical principles. Of course Professor Bass' strong background in laser induced damage studies brought him back on many occasions to the troublesome problem of the power handling limits of optical materials in laser systems. So, it is no surprise that the issue that surfaced as that of greatest concern was optical coatings since they have not only the greatest diversity of applications but are an area where one can realize the greatest improvement in damage resistance. Professor Bass concluded his discussion by a common sense view of addressing other areas which could stand improvement to enhance the further development of lasers and their applications in such fields as manufacturing, medicine, materials processing, etc. Here he called for the development of high quality fixtures, stage and mounts, power supplies, coating systems, and the necessary architecture to aid in efficient cooling.

Much of Professor Byer and Bass' comments, from a physical limitation standpoint, centered on materials interaction and thermal problems. They both concluded that this was an issue that if properly addressed can greatly enhance the reliability, utility, and acceptability of lasers in today's society. They specifically suggested, for further reference, a document entitled "Thermal Loading Limits and Scaling Possibilities," by John Emmett, William Krupke, and Walt Sooy, UCRL Document No. 53571, Lawrence Livermore National Laboratory.

After addressing the fundamental scientific limit to laser advances, we delved into the more difficult to define areas imposed by men and women themselves.

Our first speaker in this area was Professor Elsa Garmire, who gave a somewhat impassioned deliberation on the problematical subject of impeded communications. Professor Garmire was quick to point out early in her discussion the appropriateness of the need for tightly holding certain information in the sense of proprietariness from a commercial standpoint, a well accepted safeguard to commercial success in a competitive world. She applied this same acceptance of protected information to sensitive technology from a manufacturing standpoint or to engineering designs as might be applied to defensive systems, but where the line was drawn was at the basic sciences level. The distinction here was between basic and applied sciences akin to the 6.1 and 6.2 designations as used in the government. The fundamental question is who should do the defining or where the line should be drawn. Some feel that this line should be drawn by those individuals who pay to have the research performed. It was generally agreed however, that the US technical community can better (and I mean better in the sense of more rapidly) assimilate, digest and build from ideas through its high speed communications network of computer professionals, meeting openness, etc.; as such it can advance further than others in the world which tend to be more secretive and hampered by bureaucracy. In fact, in her words, in the United States, it is the interaction that is the advantage to our science. The interaction, publication and peer review, and open presentation all help us recruit, professionally advance, avoid publishing bad work, and engender more ideas through this open flow of information. We all know the way to advance is to sow good ideas to each other.

There was much discussion following Professor Garmire's plea for openness in basic science. Comments from the audience contrasted our system with that of others in the world. It was generally felt that these more inhibited systems generally shackle the growth of technology and initiative. This is certainly an observation for which there was wide acceptance in the audience.

Finally, Dave Sliney of the Army Environmental Hygiene Agency talked of the regulatory limits on laser advancement and he also was quick to point out areas in which the laser may in fact be the least hazardous item in an otherwise hazardous environment. I am sure most of us who have worked with lasers are more aware and concerned, as we should be, of high voltage and high current electrical hazards of laser systems and know of many more adverse incidents involving that aspect than with optical radiation. The concern for regulation of the laser, etc., is at once driven because the laser is new and, relatively speaking, unfamiliar. There is a solution to this problem of "regulatory" impediment and that is one of education at all levels. Dave Sliney pointed out that a lot of paper (in the sense of regulations and other ideas) just doesn't make the laser any safer. It was interesting that he noted that much of the regulations come through the Food and Drug Administration, which was a surprise to some participants.

When one is not aware or adequately knowledgeable of an item such as the laser because of a lack of appreciation or familiarity or knowledge, it becomes more of a theoretical hazard than a real or practical hazard. Dave Sliney indicated that one way of overcoming much of the difficulty in what some might consider overregulation would be to consider the adequacy of standards rather than the speed of their implementation. This stems from the fact that most standards are "concensus" standards, standards which are established to enhance commercialization. They include members of the commercial sector (involving both the vendors and the buyers together) with objective involvement by government agencies such as the National Bureau of Standards. Standards then, through agreements on definitions and procedures will lead to a more educated community able to communicate more clearly. thus, a plea was made for people to get involved. His talk was highlighted by a most interesting slide presentation depicting safety issues, their importance and their relevance to the subject of the panel's deliberation.

In summary, the subject of the panel's discussion, Limits to Laser Advancement: Fundamental and Otherwise, was deemed most appropriate since a similar panel convened by Dr. Alex Glass of KMS Fusion at last year's meeting concluded that all the best lasers had not been discovered, so it was correct to discuss what might impede their development or more properly their application. Last year's panel concluded that lasers are not so much "discovered" as defined or invented. From that viewpoint the subject and presenters at this year's panel were optimum choices. Our panel of experts was just that, both well informed and highly regarded in that subject which embraced the panel's central theme.

I would be remiss if I did not make a few personal observations. I have great confidence that our technology will push the limits of the laser's physical barriers by employing sound basic science as well as engineering principles, but I hope most sincerely that we do not impede the benefit of mankind from the progress of laser development by overzealously applying shackles to that development. There is much talk about our safeguarding against technology transfer, the bad kind; one can't tell these days if that means political or economic, but I am afraid that as unrealistic, but well intentioned, constraints are passed on concerning the release of basic information, the losers will be us, all of us. I have great confidence that we can assimilate information and build upon it better and faster than any. Let us all work to overcome this potentially disastrous impediment through education of those who will work in the field, those who will benefit from the field, and mostly those who will somehow control the field.

II. Advanced Lasers and Coherent Sources
   A. High Energy Laser Overviews 10
   B. Gamma Ray Lasers 45
   C. Novel Solid-State Lasers 104
   D. Enhancement of Free-Electron Laser Performance 142
   E. Ultrafast Short-Wavelength Light Sources and
      Techniques 157
   F. Applications of Ultrashort Wavelength Lasers 164

ADVANCES IN CHEMICAL LASERS

Dr. Joseph Miller
TRW S&TG, One Space Park, Redondo Beach, CA. 90278

ABSTRACT

High-power chemical lasers thrive in an array of special environments and present many fascinating associated subjects ripe for developmental research. Included are processes to produce the source reactants; supersonic mixing and reacting flow fields; the production and dissipation of multiple vibrational-rotational molecular states; optical gain extraction in complex geometries; media inhomogeneity effects, and waste energy and reaction products removal. Some configurations require wavelength selectivity, special optical components, and coherent cavity or beam combining. In recent years, progress has been made in these areas on behalf of continuous-wave and repetitively pulsed hydrogen fluoride and deuterium fluoride lasers, subsonic and supersonic oxygen-iodine lasers, and potential shorter wavelength chemical lasers based on chemically excited higher electronic states. This paper presents a brief review of the technical approach of some of the technology areas, and the status in achieving practical, integrated high-power chemical lasers.

HISTORY OF CHEMICAL LASERS

Interest in chemical lasers dates from shortly after the birth of lasers themselves. In 1961, Polanyi predicted infrared lasers from chemical reactions. He discussed both total and partial vibrational state population inversions and suggested the pulsed discharge chemical laser as well as the use of low temperature, low pressure and fast flow to minimize collisional deactivation effects in a continuous-wave chemical laser.[1] A special issue of Applied Optics on "Chemical Lasers" in 1965 contained comprehensive review papers and conference proceedings predating the operation of the first chemical laser.[2] Kasper and Pimental reported the first chemical laser, HCl, also in 1965.[3] By 1969, T. A. Cool, R. R. Stephens, and T. J. Falk at Cornell[4] and D. J. Spencer, T. A. Jacobs, H. Mirels and R. W. F. Gross at Aerospace Corporation were working on the first continuous-wave hydrogen flouride (HF) laser.[5] Shortly thereafter, in 1970, R. A. Meinzer at United Technologies Research Center[6] and people at TRW were operating combustion-driven versions of the Aerospace scheme. W. R. Warren in 1975[7] and L. E. Wilson in 1980[8] reviewed the field. In 1976, sufficient body and intensity of activity existed that a Handbook of Chemical Lasers was published.[9] Since 1970, continuous-wave HF lasers have been scaled to very high powers and used to study high power beam control, propagation through the atmosphere, and material interaction effects.[10]

HIGH-POWER CONTINUOUS-WAVE CHEMICAL LASERS

High-power continuous-wave HF chemical lasers have many similarities to CO_2 gas dynamic lasers (GDLs), whose heyday was the late '60s and early '70s. In both, rapid expansion into supersonic flow establishes desirable temperature, pressure, and nonequilibrium vibrational-rotational molecular state distributions. The environment provides low collisional deactivation rates and rapid removal of both ground state molecules and excess thermal energy. Here, however, the similarity ends. In the chemical laser cavity, intimate mixing, chemical reaction, and attendant energy release occur, all of which complicate matters. We might say chemical lasers are born in the wake of the CO_2 gas dynamic lasers, which makes a good pun but would be unkind. In fact, CO_2 GDLs provided background for the development of combustion hardware, optical components, and optical configurations which in turn allowed the very rapid development of the technology and high-power scaling for the chemical lasers. Also, the pioneers of the GDL efforts, including Abe Hertzberg and Ed Gerry, who headed DARPA's Strategic Technology Office in the early '70s, recognized the potential of the chemical laser for higher efficiency and more desirable wavelengths, and consequently steered research in this direction.

The most common high power continuous-wave HF lasers use the chemistry

$$F + H_2 \longrightarrow HF^* + H \quad 32 \text{ kcal/mole} \quad (1)$$

This reaction has the property that the energy release is found in the first through third excited vibrational states, not in the translation --an inherent, nascent population inversion from which to accomplish stimulated emission. To establish this reaction in a laser optical cavity, two stages of combustion are used (Figure 1). The first dissociates a fluorine compound to make F; the second injects and mixes H_2 in the laser optical cavity. As indicated above, this second

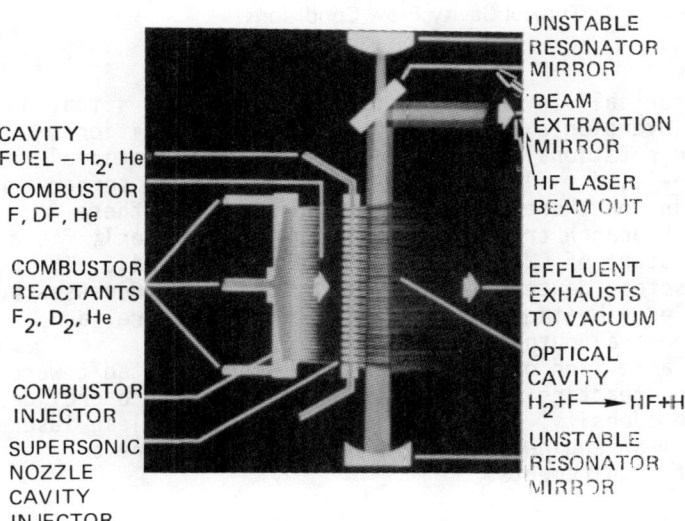

Figure 1. CW HF Chemical Laser

combustion stage occurs in the midst of a mixing, supersonic flow field. Figure 2 shows typical flow conditions for an efficient laser. As the mixing, reacting flow field progresses, new, excited state molecules are born; the low pressure, low temperature and continued flow expansion mitigates the depopulation, heating and pressure rise that ensue. The result is a modest temperature and pressure rise through the cavity as the excited states are depopulated by collisional deactivation, in competition with stimulated emission. First growth and then decay of excited vibrational-rotational states occur along the flow direction. Interestingly enough, the bulk of the cavity may not contain a total vibrational state population inversion but still allows stimulated

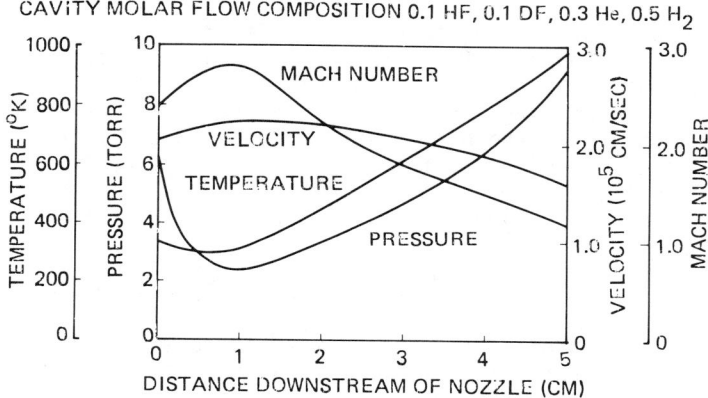

Figure 2. Typical Cavity Flow Conditions

emission from a "partial" inversion. Figure 3 shows this situation, where population inversion exists only for P-Branch transitions, which is where the rotational state increases as the vibrational state drops ($\Delta V = -1, \Delta j = +1$). The lower the rotational temperature the higher the gain for these partial inversions. It is these partial inversion P-Branch transitions that are most commonly found in the multiline output of CW/HF chemical Lasers (Figure 4). The use of the hydrogen isotope deuterium produces, by the same processes but a different detailed vibrational-rotational state structure, a different laser spectral output, in the range 3.6-4.1μ.

Much of the attention of researchers has been directed toward producing a mixing, reacting supersonic flow environment to provide scalable high-power density, high-efficiency, optically-clean lasers. Unfortunately, from an engineering point of view, these are a conflicting set of requirements.

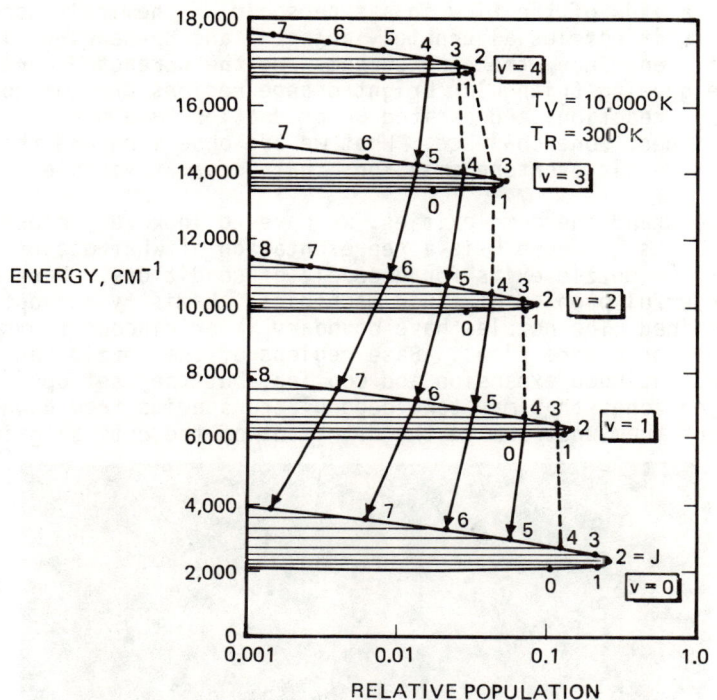

Figure 3. Energy-Level Diagram for Excited HF Molecules

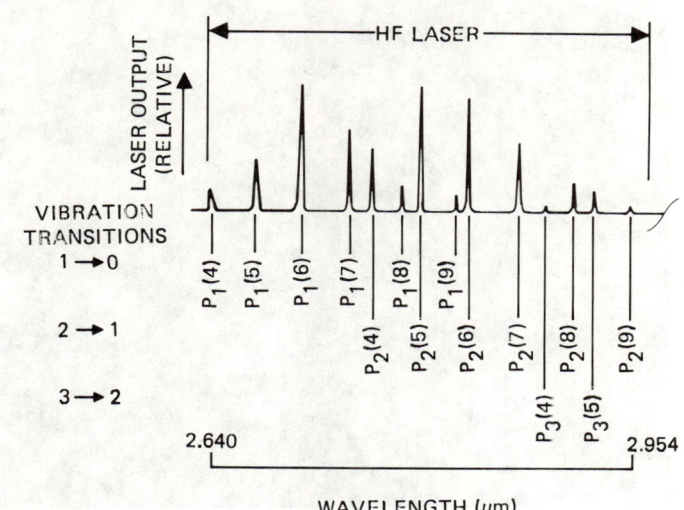

Figure 4. Typical Continuous-Wave HF Laser Beam Spectral Content

Figure 5 is a view of the flow in a supersonic HF chemical laser. In this laser, slit nozzles alternately inject F- and H_2-bearing flows. At the cavity entrance, the dark streams are the unreacted core flows. The growing triangular bright orange regions are the zones in which mixing, reaction, and excited HF production take place; eventually, these zone coalesce. What we are observing are the excited HF molecule overtone emissions that occur at visible wavelengths.

To understand the complexities, we have to look very closely at the nozzle exits. Figure 6 is a representation of what occurs in the vicinity of the nozzle exits for a couple of conditions. Here we see a truly wonderful problem of fluid mechanics, chemistry and optical physics combined. The nozzles have boundary layer viscous flows, as well as supersonic core flows. Base regions of the nozzle can provide for continued expansion and cooling, but they set up recirculation zones that may take depopulated species from downstream and transport them upstream to the detriment of the optical gain.

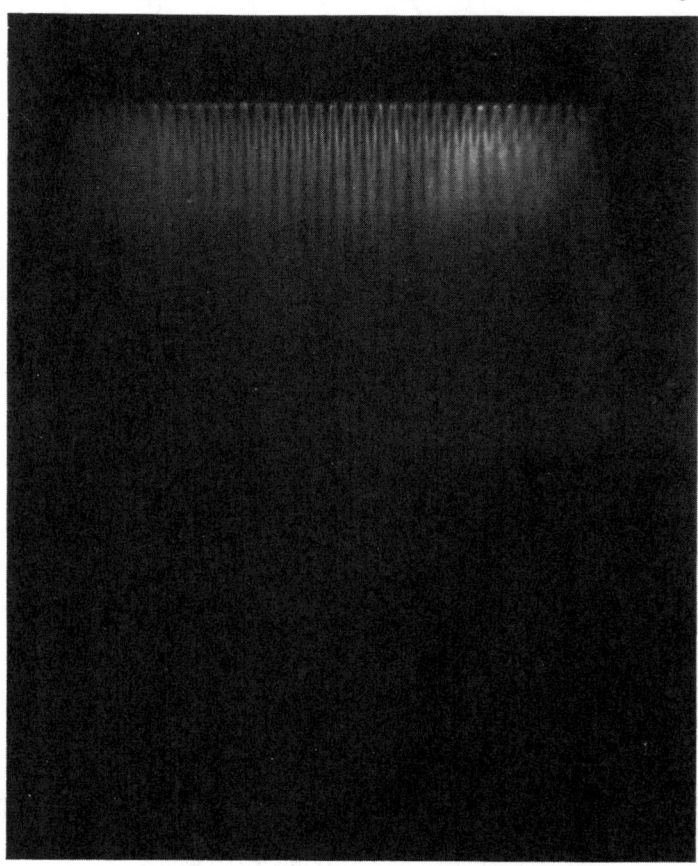

Figure 5. Supersonic, Mixing, Reacting Chemical Laser Flow Field

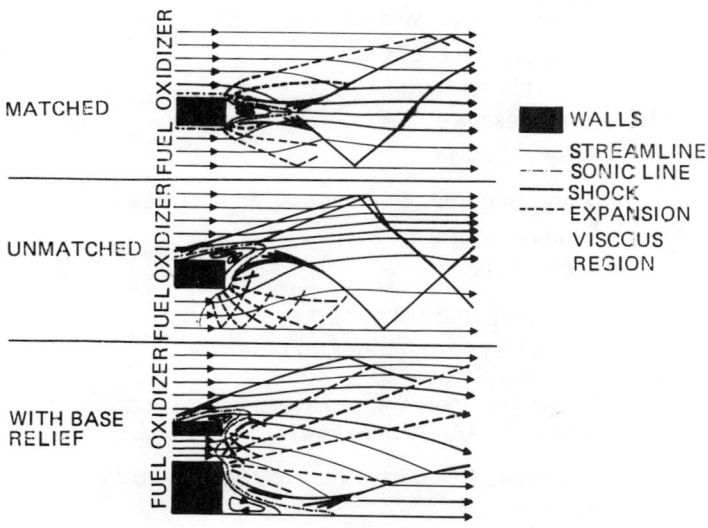

Figure 6. Supersonic Nozzle Exit Flows

The nozzle boundary layer viscous flows persist and become wakes capable of disturbing optical homogeneity. As the supersonic flows enter the cavity, mix with adjacent flows, and encounter the chemical energy release, supersonic shocks occur, another source of local optical path differences potentially detrimental to beam quality. To design a high-power laser, one has to understand this flow field, the growth and decay of each vibrational-rotational state population, and the optical disturbance features, in the presence of an interacting optical field established by the optical resonator with all of its physical optics diffraction properties and characteristics.

As Figure 7 indicates, high efficiency requires good utilization of available reactants and a high rate of production of the excited state species (rapid mixing and chemical reaction), a high rate of stimulated emission (high optical-field strength), a low rate of deactivation (low temperature, low pressure), and rapid removal of depopulated species (high flow-field velocity). These tend to create conflicting requirements. Rapid mixing and chemical reaction argues for a fine mixing scale; however, small nozzles fill with boundary layers which give higher temperature, higher pressure, and lower Mach number. If one tries to promote rapid mixing with boundary layer trip flows or transverse injections, the resulting shocks create temperature and pressure increases, and optical path disturbances. Furthermore, for applications in the earth's atmosphere, we would like to minimize the size of the exhaust pump; this argues for high-pressure, high-Mach-Number (high-pressure recovery) flow--another partially conflicting requirement.

HIGH-EFFICIENCY CAVITY CONDITIONS
- RAPID, COMPLETE MIXING
- LOW TEMPERATURE
- LOW PRESSURE
- HIGH VELOCITY

HIGH-POWER DENSITY AND PRESSURE RECOVERY
- HIGH PRESSURE
- HIGH VELOCITY

HIGH OPTICAL MEDIUM QUALITY
- LOW CAVITY PRESSURE AND INDEX OF REFRACTION
- MINIMIZED WAKES, SHOCKS AND OTHER FLOW DENSITY VARIATION EFFECTS

HIGH OPTICAL EXTRACTION EFFICIENCY AND QUALITY
- SATURATION
- MODE MATCHING
- OPTICAL SYSTEM QUALITY

Figure 7. Chemical Laser Design

The trend through the 70's was toward finer and finer nozzle scales, mixing promotion, and the heavy use of diluent heat capacity. Wilson[8] reports slit nozzle structures with nozzle throats on the order of 0.1 mm, nozzle exits varying from 1.0 to 2.0 mm, and boundary layer trip injection holes along the exit edges of the nozzle blades. Further, intricate internal passages and feeds exist if one attempts to provide regenerative cooling using the source reactants. These fine-mixing laser configurations provide high density, high-pressure recovery flows and allow large-scale HF and DF lasers with good pressure recovery (lower exhaust pumping requirements). Nozzles of this character are used in the Mid-Infrared Advanced Chemical Laser (MIRACL) (Figure 8) currently located at the National High Energy Laser Systems Test Facility at the White Sands Missile Range. In this laser, the desire for high-power, high power-density, and high-pressure recovery has compromised efficiency and, somewhat, optical quality. Concerning the optical quality, one can see that the gain path length is several meters. The optics form a simple, confocal unstable resonator as shown in Figure 1. The optial axis and the collimated output are aligned perpendicular to the flow direction, parallel to the face of the nozzle array. In MIRACL, the small, local flow wakes created by the structures that form the small blades are not aligned along the output pass. If they were, they would reinforce and affect the the output beam quality. By tilting the nozzle banks slightly, so as to misalign these disturbances and thus smear rather than

INITIAL OPERATION 1980

SUCCESSFUL POWER SCALING DEMONSTRATION

RELOCATED TO WHITE SANDS MISSILE RANGE, NEW MEXICO

Figure 8. Mid-IR Advanced Chemical Laser (MIRACL)

reinforce their effect, a significant increase in beam quality is realized. The MIRACL laser, a successful scaling experience, has provided dramatic and important data in behalf of the Strategic Defense Initiative (SDI).

As HF/DF laser technology has advanced, particularly on behalf of space-based applications, it has become apparent that the fine-mixing scale nozzle is inappropriate. Fine nozzles are filled with boundary layers to the extent that low-pressure, high-Mach-number conditions are difficult to achieve. Army technology programs have encountered this same problem where N_2 has been substituted for He as diluent. Since space lasers have their own natural exhaust pump and need no pressure recovery, they can operate at the high-efficiency, low-pressure, low-temperature conditions provided by large nozzles. Thus the technology has turned in the direction of large primary fluorine nozzles and injection of the H_2 or D_2 by hypersonic wedges located across the nozzle exit plane (but misaligned with respect to the optical axis and/or output pass). This method provides a well-established supersonic flow in the primary nozzle, with relatively little boundary layer growth along the hypersonic wedges. Care has to be taken to assure that the shocks created by the hypersonic wedges do not block the primary nozzle or create undesirable optical path differences (OPDs). Large nozzle, hypersonic wedge injection technology is being incorporated in the SDI Project ALPHA.

It is apparent that the design problem for a high-power, efficient, high quality, HF laser is technologically complex. However, it is a matter of record that the complex set of interactions has yielded to scientific understanding and engineering process combined, resulting in efficient, high-power, high-beam-quality HF lasers. The engineering process consists of constructing on a modest scale a module of the combustor/gain generator configuration. Figure 9 shows the ALPHA verification module. Such a module verifies structural, thermal, and flow engineering predictions. Further, precise measurements are made using absorption spectroscopy, optical interferometry, and power extraction techniques. These data verify the gain models for unsaturated and saturated growth and decay of each of the excited state populations in the physical-optics-resonator-design computer codes. The measurements also provide verification or input for the laser medium OPD models in these design tools. Thus, a rather explicit set of measurements is achieved early on, assuring that the large-scale laser design tools and processes are sound. The ALPHA verification module has been used in this way, through over 10,000 seconds of experimentation, to lay the foundation for the ALPHA laser shown in Figure 10. ALPHA, an SDI/Air Force project at TRW, is currently well underway and will provide the technology for a high-power, efficient, lightweight, high-beam-quality HF laser.

REPETITIVELY PULSED CHEMICAL LASERS

Let us now turn to a different form of HF/DF lasers, the repetitively pulsed version. It is a matter of interest that the previously described HF lasers use a relatively inferior reaction. When premixed H_2 and F_2 are ignited, the reaction proceeds by way of a chain

$$H_2 + F \longrightarrow H + HF \quad 32 \text{ kcal/mole} \quad (2)$$

$$H + F_2 \longrightarrow F + HF \quad 97 \text{ kcal/mole} \quad (3)$$

Each reaction provides the atomic specie that feeds its sister reaction. The second reaction is much more energetic. We don't use it in the supersonic flowing continuous-wave lasers because there is so much energy release that in the presence of a mixing-controlled production rate, the heating and depopulation effects dominate and efficiencies are low. However, in a premixed laser, we can use the full chain. To create such a laser one mixes H_2 and F_2 (amateur chemists be careful--without any H or F present); then an E-Beam or photolytic dissociation of a small percentage of the F_2 lets the reaction proceed. At atmospheric pressure the mixture reacts in about a microsecond, and in the presence of sufficient diluents, hits a 5 atmosphere overpressure. By flowing the cavity gases (subsonic

Figure 9. **ALPHA** VM Module

ALPHA EXPERIMENT

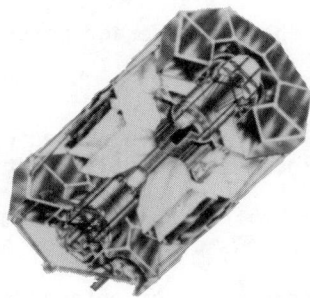

GROUND-BASED EXPERIMENT
OF KEY TECHNOLOGIES
- HIGH POWER
- HIGH EFFICIENCY
- HIGH BEAM QUALITY
- LIGHTWEIGHT HARDWARE

FACILITIES PROVIDE
- REACTANT STORAGE AND DELIVERY
- EXHAUST GAS PUMPING
- DIAGNOSTIC AND DATA SERVICES

Figure 10. Project ALPHA

in this case), one can establish a repetitively-pulsed chemical laser, HF or DF. Figure 11 shows a repetitively pulsed HF/DF laser at TRW. The advantages of such a laser are that it operates above atmospheric pressure (no exhaust pump required) and the pulse energy form is desirable 1) for mitigating atmospheric thermal blooming effects when atmospheric exchange is present (winds or beam slewing), and/or 2) for providing enhanced material interaction effects because the local µsec pulse power is on the order of 10^3 to 10^4 times higher than the average power.

Figure 11. Repetitively-Pulsed Chemical Laser

FUTURE OF CHEMICAL LASER TECHNOLOGY

This latter property of very high local power in pulsed chemical lasers has led to some fascinating developments recently. Early in 1986, W. T. Whitney, M. T. Duignan, and B. J. Feldman at NRL reported using a pulsed HF laser to achieve threshold for Stimulated Brillouin Scattering.[11] This discovery allows consideration of phase conjugation techniques in HF lasers.

Attendant possibilities exist for correction of optical and medium aberrations, for beam combining and array phasing, and for jitter stabilization of very-high-power HF lasers. One can look forward to higher power, higher quality chemical laser systems simultaneous with relaxed requirements on optical components, laser media homogeneity, and jitter stabilization systems. This is particularly important for the pulsed chemical laser because the cavity is at such high pressure that it experiences an acoustic shock and must be contained with transmissive optics. These systems may even be a possibility for the very-high-power continuous-wave lasers. One can foresee continuing research in this direction.

The other research direction that continues is the search for yet shorter wavelength chemical lasers, of high efficiency, and suitable for scaling to high powers. So far, only the $O_2^* - I$ chemical laser, which operates at 1.3μ, has shown promise. A multi-kilowatt version of this laser is shown in Figure 12. The oxygen-iodine uses a peroxide-chlorine reaction to create excited singlet-delta oxygen. Subsequent mixing with iodine provides dissociation to atomic iodine and excitation from which stimulated emission can be realized. This and other potential new chemical systems are the subject of other sessions at this conference.

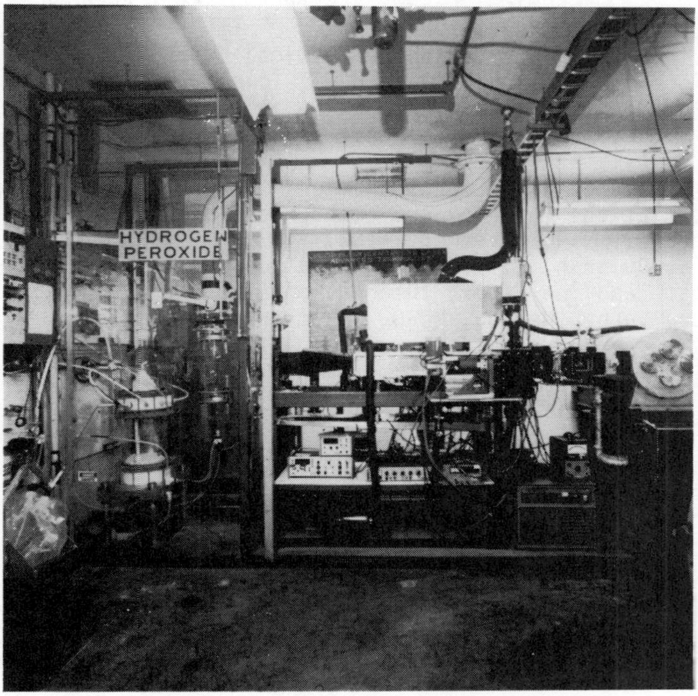

In summary, chemical lasers have come quite a way in the high-power domain. Significant technology developments currently underway are Project ALPHA for HF lasers, advanced technology directions including non-linear optics application, and shorter wavelength lasers.

Figure 12. Chemical Oxygen-Iodine Laser

REFERENCES

1. Polanyi, J. C., "Proposal for an Infrared Maser Dependent on Vibrational Excitation," J. Chem. Phys., 34(1), 347-348, 1961.
2. Howard, J. N., ed., Applied Optics, "Supplement 2: Chemical Lasers," (publication of the Optical Society of America, Inc.), 1965.
3. Kasper, J. V. V., and Pimentel, G. C., "HCl Chemical Laser," Phys. Rev. Letters, 14(10), 352-354, 1965.

4. Cool, T. A, Stephens, R.R., and Falk, T. J., "A Continuous-Wave Chemically Excited CO_2 Laser," International J. Chem. Kinetics, 1 (5), 495-497, 1969.
5. Spencer, D. J., Jacobs, T. A., Mirels, H., and Gross, R. W. F., "Continuous-Wave Chemical Laser," International J. Chem. Kinetics, 1 (5), 493-494, 1969.
6. Meinzer, R. A., "Communication--A Continuous-wave Combustion Laser," International J. Chem. Kinetics, 2, 335, 1970.
7. Warren, Jr., W. R., "Chemical Lasers," Astronautics & Aeronautics, Vol. 13, No. 4, 36-49, 1975.
8. Wilson, L. E., "Deuterium Fluoride CW Chemical Lasers," J. De Physique, Colloque 09, Supplement au n 11, Tome 41, C9-1-C9-8, Novembre 1980.
9. Gross, R. W. F., and Bott, J. F., eds., Handbook of Chemical Lasers, John Wiley & Sons, Inc., 1976.
10. Miller, J., "Status of High Energy Chemical Lasers," Invited paper presented at the International Conference on LASERS '85, Las Vegas, Nevada, TB.1, December, 1985.
11. Whitney, W. T., Duignan, M. T., and Feldman, B. J., "SBS With An HF Laser," Post Deadline Paper, International Conference on LASERS '85, Las Vegas, Nevada, MD-8, December, 1985.

A HISTORICAL PERSPECTIVE OF EARLY HIGH POWER GAS LASER RESEARCH

A. Hertzberg
Aerospace and Energetics Research Program
University of Washington
Seattle, WA 98195

ABSTRACT

This paper traces the author's involvement in the early development of high power lasers. Over a brief period in the early 1960's the concepts of the gasdynamic laser, the mixing laser, and laser-induced implosions were spawned at Cornell Aeronautical Laboratories.

INTRODUCTION

The invention of the gas laser in 1959 attracted the attention of many engineers and scientists working in the field of high energy gasdynamics and reentry physics, as well as scientists working in chemical and optical physics. Much of the science base created by gasdynamic research was applicable to laser physics and it was recognized that fast flow could be used to create the population inversion required for lasing or it could be used to improve laser operation. At the present time, most lasers which are capable of continuous high average power employ flow as an indispensable aspect of their operation. I was fortunate enough to play a small role in the development of high power gas lasers, and it is the purpose of this paper to recall some of the excitement and pure fun experienced by myself and my colleagues who had an opportunity to be involved.

While there are, of course, a great number of research scientists and engineers in the field of high energy gasdynamics, chemical kinetics and laser technology who were responsible for critical and important contributions, it is beyond the scope of this paper to give full credit and accounting to all involved and a comprehensive review is not intended. The author is also circumscribed by the information available to him and the nature of his own involvement. The author is furthermore aware of his own humanity and presumes that there will be self-serving statements contained in this paper, which is after all only intended as a subjective historical review.

EARLY GAS LASER RESEARCH

In 1960, "laser" was only a word to me. I was at that time the director of a small basic research group working in hypersonics and other aspects of flow physics at Cornell Aeronautical Laboratory. This field was very fashionable at the time as part of the so-called "reentry problem and reentry signature problem" which brought together a great many fluid dynamicists, physicists and chemists. We were conditioned to think about non-equilibrium flows and it was between 1950 and 1960 that this branch of aero physics reached maturity. My interest in lasers probably resulted from boredom with a mature field which no longer encouraged innovative science. I can still remember the excitement long ago of finding out what a laser was. I made the mistake at lunch of passing on a few remarks about the laser that I had picked up from the popular press to some of my more sarcastic colleagues. They accused me of not even knowing what a laser was. Their comments were perfectly true, but I found out in the next few months that they did not know what a laser was either.

I went to the library for Ali Javan's first articles in Physical Review Letters.[1,2] I believe this was in the late 1960 time period. I selected Javan's papers since they involved gas lasers and were therefore more closely allied to my background than solid state lasers. The papers proved exciting and I stayed up much of the night studying and reading his short articles as the ramifications began to sink in. The possibility of a connection between the laser physics of Javan and the flow physics that we were dealing with at the time was tantalizing. The following day I probably drove my colleagues crazy explaining the operation of lasers to them. I remember this happy period very well since many of the ideas that I worked on for the next five years, particularly in the areas of gasdynamic lasers and flowing gas lasers, were then born.

There was no particular job-oriented motive that drove me towards the development of laser concepts; rather, it was the sense of adventure which led to the direct connections between the work of non-equilibrium fluid expansion that my group was currently studying and the possibility of producing population inversion. At the time, I had had no contact with the still small laser community, and all of our work occurred within my particular basic research group at Cornell Aeronautical Laboratory. However, at this time it was possible for a single individual to track the entire laser literature and we quickly assembled a very active file of most of the important publications in the field.

I did believe my ideas to be quite original and was somewhat reluctant in the first few months to expose any of my work to the outside community until we had a better grasp of the problem and the possibilities. In particular, we realized that a flowing gas system contained the implication of very high continuous power and we were able to sketch out the basic configuration of a gasdynamic laser without even knowing what the possible lasing fluids might be (see Fig. 1). The technology for creating such a system was at hand. For example, the concept of a grid nozzle was an idea available in aerodynamic research since 1945.

It is amusing, perhaps even cynical on my part, to note that by and large my colleagues at the lab and even the co-authors on my papers were not particularly encouraging. They tended to think that this was just a nutty idea and they had to humor the boss.

EARLY GDL EFFORTS

It was in the late 1960/early 1961 time period that we developed the concept of the flowing gas GDL. Our concept naturally evolved from our understanding of the capabilities of fluid mechanical expansions to obtain rapid cooling. In some of my earlier work on NO fixation back in the 50's we had devised several nitrogen fixation schemes built around this principle so that the idea of non-equilibrium freezing was integral to our work. We were also quite familiar with the freezing of the nitrogen vibration and dissociation levels in rapid expansion nozzles due to our contractual commitments to the Air Force and NASA to study this phenomenon in relationship to reentry physics.[3]

While I hardly considered myself an expert in spectroscopy, I did know enough about rate data to estimate that if we could cool a gas fast enough we might invert the electronic levels of a monoatomic gas such as Xenon, which could be readily heated in a shock tube. Depletion of the lower level through radiative decay can in principle result in population inversion. The idea is not dissimilar to that proposed later by Basov and his colleagues,[4] in which they also proposed creating population inversions by rapid cooling. However, they did not suggest a method in their papers for obtaining rapid cooling, and were not even aware of Basov's work until we submitted our own work for archival publication in 1964.[5]

As department head, I was able to requisition old shock tubes and shanghai some of my colleagues into carrying out an experiment with shock-heated Xenon expanding through a slit nozzle, a flow system we understood how to characterize. The only archival publications to come out of this work involved our fluid mechanical work on slit nozzle expansions, which in the late 60's found its way into use in relationship to various chemical laser research programs. As an experiment to demonstrate population inversions, the results were disappointing.

There were several sidelines in our research program which we only briefly pursued. One of these was an early concept of a mixing laser, in which we proposed to manufacture helium metastables in a separate container and mix them with neon to produce a more efficient helium neon laser.[5] Sometime after this, Terry Cool at Cornell University carried out several experiments in this area and had some success, but the complexity of the system did not warrant the level of improvement. In concept, the idea is somewhat similar to Patel's later concept of externally exciting the nitrogen and mixing it with the CO_2.

INTERACTION WITH FUNDING AGENCIES

By then we rather freely discussed our work with all visitors - and in those days we had many, mainly from the reentry flow community, since we were active in that field. I attempted to get extra funding from the various contracting agencies I was dealing with such as AFOSR and NASA and even ARPA. Except for my interview at ARPA, where it was carefully explained to me that an expansion laser couldn't possibly work, I received a rather sympathetic if not financially successful hearing. My contractors quite correctly pointed out that they gave me wide latitude in my research and I was free to drop any aspect of it to undertake this new work. They did not, however, offer to increase the funding. I secretly agreed with them, and being quite human I was very reluctant to face my unenthusiastic superiors and change the direction of what they considered a rather profitable hypersonic program. The lack of increased funding combined with management attitudes did require me to slow down on our efforts since there were priority demands for various pieces of equipment that were borrowed or stolen to get this work going.

LASER-INDUCED IMPLOSIONS

The year 1961-62 was quite an exciting time in the nature of the research but contained no material rewards or technical success. It was a time for the fermentation of ideas and nearly everything I have done in lasers since has been influenced by our work at that time. For example, in studying the properties of lasers in 1962 I began to realize the enormous power concentration ability of coherent radiation and we started our work on laser-driven implosions. Here again, the precursor structure of ideas came from my background in gasdynamics, to which we added the capability of the rapidly developing ruby and solid state laser technology. The fascinating capability of being able to maneuver light to create a light implosion to drive a gasdynamic implosion was appealing and supplemented ideas that we had already been working on in attempting to obtain electric shock tube driven implosions.

In this case I was able to get limited funding (which later increased) due to the argument that the laser offered energy density enough to heat gases to temperatures far beyond anything a shock tube could. I must admit we didn't have very clear ideas about how to couple the radiation to the gas and glibly told our sponsors (in this case the Aeronautical Research Laboratory at Wright Field) about its ultimate potential. The contract monitors there were surprisingly far-sighted and gave us free reign to explore this possibility.

We had few tools comparable to those available at the national labs such as Los Alamos or Livermore, but we were able to make a few gross estimates of the energies required to produce breakeven.[6-8] When I look back at our work I note with some amusement that, as naive as our calculations were, we were closer than the national labs in estimating the breakeven energy. (I believe the current estimate is about 10^6-10^7 Joules, very close to the numbers we produced in 1962-63.) I did have to withstand a certain amount of personal ridicule, not only from my own colleagues at Cornell Aero Lab but also from the general scientific community, with the exception of my friends such as Keith Boyer at Los Alamos and Russell Duff at Livermore, who provided my first contacts with the national labs. There, as I remember it, I was heard very fairly and not judged too extreme. I presented seminars on our work at these labs in 1963 and came into contact with scientists such as Ray Kidder, John Dawson and others, who were of immense encouragement and help during a very difficult period in my life.

To summarize, I believe in 1960-61 we conceived of the first ideas relating to mixing lasers, gasdynamic lasers, and laser-driven implosions. Our first publication was a Cornell Aero Lab report dated December 1962[9] in which we summarized some of our ideas on expansion-driven gasdynamic lasers and basically conceived almost in its entirety the final configuration of a gasdynamic laser. This document had a moderate distribution of about 400 or 500 copies that were sent to other labs and universities. A copy was sent to the AVCO Everett Research Lab, where some of my friends have led me to believe it had an impact on their thinking and was one of the initiators, if not the initiator, of their own rather remarkable program on gasdynamic expansion lasers.

CORNELL AERO LAB AND THE NITROGEN CO_2 GASDYNAMIC LASER

While it is my belief that my group at Cornell Aero Lab conceived of the configuration of the gasdynamic laser sometime between 1960 and 1961, we really missed the boat in connecting this with the later work of Patel on the nitrogen CO_2 laser. That was left to the very active group at AVCO, the AVCO Everett Research Lab. I should point out that it is rather galling that we let the chance slip by, since after we published our Cornell Aero Lab report in 1962 we received a visitor from AVCO Everett, who discussed with us in an open seminar the resonance between the nitrogen v=1 state and the upper lasing level of the nitrogen CO_2 laser. It is my memory, corroborated from that period, that the visitor did suggest in that open meeting that we make a laser out of this system. After he left, my group talked about it briefly but decided that, due to our competitive relations with AVCO Everett, they would hardly give us an idea that was worth following up. Also, we ourselves had little confidence in it. While we did nothing, with the equipment that we had available at that time we could have easily produced both the electrical and the gasdynamic nitrogen CO_2 laser systems in 1962. It is interesting to note that the AVCO Everett Research Lab also did nothing until Patel's work was published.[10] It was left to AVCO, Bosov's group[11] and Prokhorov's group[12] to connect Patel's idea with the gasdynamic configuration.

THE CHEMICAL LASER AND THE GASDYNAMIC LASER

The chemical laser, I believe, started out quite independently of any thoughts about flow. It was only the activities at the Aerospace Research Corporation led by the pioneering work of Jacobs and Gross which combined these two ideas. They developed the early successful concepts, particularly with the help of Hal Mirels and Wally Warren. I did act as a consultant to them in 1966 when they first observed the lasing phenomena. I must say that my own feelings were somewhat negative and I acted only as an advisor on aerodynamic problems, though I do believe my position there as a consultant helped stimulate them in regard to the grid nozzle approach.[13]

STANCE OF LABORATORY MANAGEMENT

It should be remarked that the management at Cornell Aero Lab at that period of time was shifting from a research-oriented group to a business-oriented, systems-type management organization. They were interested in the science my group provided as it supported larger systems contracts. Basic research in those days at Cornell Aero Lab was tolerated, not encouraged. Indeed, it was my disagreement and rather strenuous opposition to this policy which later led me to leave. I would hasten to point out, though, that my dismissal, far from being a career disaster, was a blessing in that it forced me to make a decision to leave industry and join a university.

MY WORK 1966-1978

When I joined the University of Washington in 1966, the CO_2 gasdynamic laser (GDL) was well established and there were many people kind enough to associate my name with the creation of this concept. This proved of great value in helping me start our laser program here at the University of Washington. By then the idea of the GDL was far more acceptable to the community and significant amounts of funding were becoming available. We worked hard at that time to improve the efficiency of the GDL and indeed had quite a bit of fun developing new concepts for various closed flow cycle systems. This led to a paper on "photon engines," which established some of the first theoretical limits to the performance capabilities of gas lasers and lasers in general.[14] In addition, we continue to study the use of lasers in fusion.[15]

CONCLUDING REMARKS

The ten years between 1960 and 1970 were indeed an unusual and creative time in all areas of laser research. It was, however, particularly exciting to me in that the concepts which later flowered into the current technology of high power flowing gas lasers was born. It was a period where almost everything one did was original by definition. Even our mistakes were original and there existed a level of camaraderie among the various research groups that seemed unusual by current standards. While I would hesitate to say that laser research has matured, it certainly has increased in size by many orders of magnitude. I can only hope that the young people now going into laser research will be able to capture some of the same excitement and fun that we shared during those early years. Many of the ideas which were discussed at this meeting, ranging from laser material working to laser fusion, were conceived during this exciting period.

REFERENCES

1. A. Javan, Phys. Rev. Lett. $\underline{3}$, 87 (1959).
2. A. Javan, W.R. Bennett, Jr. and R. Herriott, Phys. Rev. Lett. $\underline{6}$, 106 (1961).
3. J.D. Buckmaster, AIAA J $\underline{2}$, 1649 (1964).
4. N.G. Basov and A.N. Oraevskii, Sov. Phys. JETP $\underline{17}$, 1171 (1963).
5. I.R. Hurle and A. Hertzberg, Phys. Fluids $\underline{8}$, $16\underline{01}$ (1965).
6. J.W. Daiber, C.E. Wittliff and A. Hertzberg, Bull. Amer. Phys. Soc. $\underline{10}$, 225 (1965).
7. J.W. Daiber, A. Hertzberg and C.E. Wittliff, Phys. Fluids $\underline{9}$, 617 (1966).
8. U.S. Patent No. 3,489,645, Jan. 1970.
9. I.R. Hurle, A. Hertzberg and J.D. Bruckmaster, Cornell Aeronautical Laboratory, Inc., Buffalo, NY, CAL Rept. RH-1670-A-1, Dec. 1962.
10. C.K.N. Patel, Phys. Rev. Lett. 13, 617 (1964).
11. N.G. Basov, V.G. Mikhailov, A.N. Oraevskii and V.A. Shcheglov, Soviet Physics-Technical Physics $\underline{13}$, 1630 (1969).
12. V.K. Konyukhov, I.V. Matrosov, A.M. Prokhorov, D.T. Shalunov and N.N. Shirokov, J. Exper. Theoret. Phys. $\underline{10}$, 53 (1969).
13. D.J. Spencer, T.A. Jacobs, H. Mirels and R.W.F. Gross, Int. J. Chem. Kinetics 1, 493 1969).
14. A. Hertzberg, W.H. Christiansen, E.W. Johnston and H.G. Ahlstrom, AIAA J $\underline{10}$, 394 (1972).
15. J.M. Dawson, A. Hertzberg, R. Kidder, G.C. Vlases, H. Ahlstrom and L.C. Steinhauer, Proc. Int. Conf. on Plasma Physics and Controlled Nuclear Fusion Research, Vol. 1, International Atomic Energy Agency, Vienna, 1971, p. 673.

INJECTION CONTROLLED OPERATION OF A BROADBAND EXCIMER LASER*

Y. Zhu, R.A. Sauerbrey, F.K. Tittel, and W.L. Wilson, Jr.
Department of Electrical and Computer Engineering, Rice University, Houston, TX 77251

W.L. Nighan
United Technologies Research Center, East Hartford, CT 06108

ABSTRACT

This paper reports on a semi-empirical model of an injection-controlled $XeF(C \to A)$ excimer laser. Calculated results are compared with experimental data on injection timing, injection gain, and cavity magnification for an unstable resonator geometry.

In an effort to achieve optimized performance of a broadband $XeF(C \to A)$ excimer laser, injection control of such a laser by means of a dye laser has been investigated.[1] In any low gain, short-pulse laser system, rapid build-up of the optical field within the resonator is critical to good laser performance and extraction efficiency. Simultaneous injection of a "seed" signal into a laser cavity with laser pump excitation can create a much faster increase in the intensity of the optical fields, compared to one that results from a build-up of spontaneous emission. Efficient, ultranarrow spectral output from an electron-beam excited $XeF(C \to A)$ excimer laser medium has recently been observed by injection control. Several different confocal unstable resonator geometries were investigated. A maximum $XeF(C \to A)$ laser output of ~150 mJ was measured at 482.5 nm for a cavity with magnification $M = 1.1$, which corresponds to an energy density and intrinsic efficiency of ~8 J/liter and ~6%, respectively. These values are comparable to those of UV rare gas-halide lasers.

In order to gain a better understanding of the injection control process, an analytical model was established using pulsed regenerative amplifier approach. The amplifier was described by a set of coupled rate equations:

$$\frac{dq(x,t)}{dx} = G(t)\, q(x,t) - L(t)\, q(x,t) - \frac{\ln M}{d} q(x,t) + \frac{N_C(t)}{\tau_{CA}} \Omega \int g(v)\, dv \qquad (1)$$

$$\frac{dG(t)}{dt} = P_C(t)\, \sigma_{SE} - \frac{G(t)}{\tau_C} - G(t)\, \sigma_{SE}\, q(x,t) \qquad (2)$$

$$\frac{dL(t)}{dt} = P_a(t)\, \sigma_a - \frac{L(t)}{\tau_a} - L(t)\, \sigma_a\, q(x,t) \qquad (3)$$

where $q(x,t)$ is photon flux; $G(t), L(t)$ are gain and loss per unit length at 488 nm with optimized gas conditions; d is the cavity length; M stands for the cavity magnification; $N_C(t)$ is the population density of $XeF(C)$ states; $\tau_{CA}, \tau_C, \tau_a$ are the radiative decay time of the $C \to A$ transition, effective decay times of the C state and absorbers, respectively; Ω is the solid angle into which the spontaneous radiation has to be emitted in order to contribute to the build-up of a free running laser signal; $g(v)$ is the line shape function; $P_C(t), P_a(t)$ denote the production rates for C states and absorbers derived from the measured small signal gain and loss, σ_{SE}, σ_a represent the stimulated emission and absorption cross-section.

© American Institute of Physics 1987

On the right side of Equation (1), the first and second terms describe the gain and loss, respectively, due to the active medium. The third term $(-\frac{\ln M}{d} q)$ is the loss of photon flux per unit length due to the expansion of the beam in the unstable resonator. The fourth term is due to the spontaneous emission contribution. On the right side of both Equations (2) and (3), the first terms are the production terms. The second terms are the decay terms, due to radiation and nonradiative collision processes; and the third terms describe the losses due to stimulated transitions induced by the radiation field and lead to saturation. The injected dye laser flux was coupled into the rate equations by a boundary condition $q(0,t)$ at the injection hole position ($x = 0$). The unstable cavity was treated like a folded amplifier. The influence of the unpumped region (~ 2.5 cm) is neglected compared to the pumped region (~ 10 cm). In order to solve the rate equations, an analytical expression for the small signal gain and loss must be obtained. Therefore, the generally unknown production rates $P_C(t)$ and $P_a(t)$ can then be derived from Equations (2) and (3) in the small signal approximation ($q = 0$) by fitting the solutions of (2) and (3) to experimentally determined gain and absorption signals. σ_a was $\sim 1 \times 10^{-17}$ cm^2. τ_C and τ_a were derived from earlier experiments and had values ~ 10 ns and ~ 8.3 ns, respectively.

A computer code was developed for numerical integration of the coupled rate equations.[2] Of course, a rate equation approximation of the photons flux neglects all wave effects, in particular diffraction. Thus, it is the purpose of this semiempirical model of an injection controlled laser to describe its performance parameters based on experimentally determined properties of the gain medium, using a minimum of free parameters.

Experimental and analytical results of the temporal relation of the injection controlled laser pulse to the injected dye laser pulse are shown in Fig. 1. Generally good agreement exists between calculated results and experimental data. Only the calculated duration of injection controlled output is shorter than the measured output pulse due to the pulsed regenerative amplifier model.[2] Another observation is the injection gain and saturation behavior as shown in Fig. 2. For cavity magnification $M = 1.23$, the injection controlled output was saturated for injection energy over 1 mJ. The same behavior is predicted by the analytical model. Fig. 3 shows the dependence of the amplified output pulse on cavity magnification with an optimum timing relation between dye injection and e-beam pumping. The output power was optimized at about $M = 1.1$ for both the experiment and analytical model with a 12.5 cm long cavity. This analytical model can be used to assist the design of the cavity optics for a future experiment where a cavity length of up to 25 cm will be employed and the optimized magnification predicted to be around $M = 1.3$ must be verified by experiment.

*Supported in part by the Office of Naval Research, the National Science Foundation, and the Robert Welch Foundation.

REFERENCES

1. F.K. Tittel, G. Marowsky, W.L. Nighan, Y. Zhu, R.A. Sauerbrey, and W.L. Wilson, Jr., "Injection controlled tuning of an electron-beam excited $XeF(C \rightarrow A)$ laser," IEEE J. Quan. Elec., vol. QE-22, pp. 2168-2173, 1986.

2. A. Icsevgi and W.E. Lamb, Jr., "Propagation of light pulses in a laser amplifier," Phys. Rev., vol. 185, pp. 517-545, 1969.

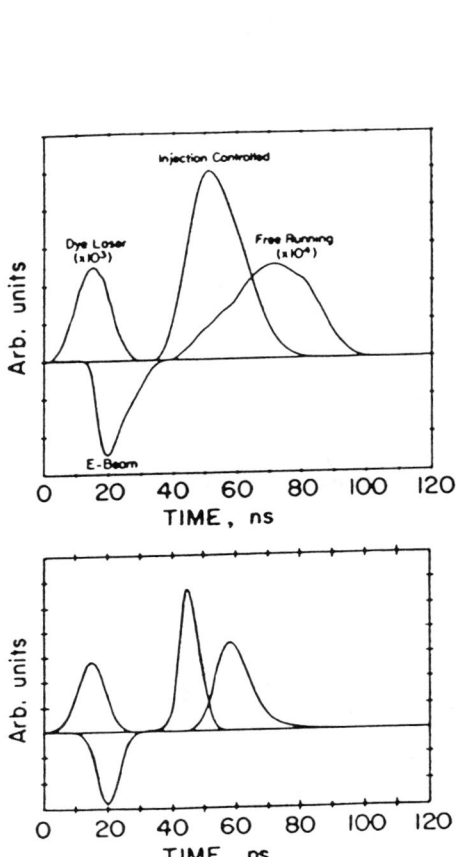

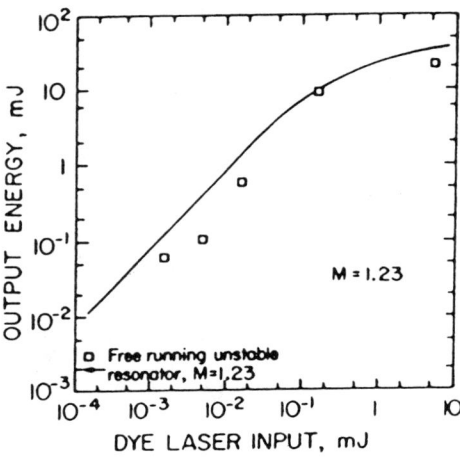

Fig. 2: $XeF(C \to A)$ output pulse energy dependence on the injected dye laser energy. The squares are the experimental results for cavity magnification of M = 1.23.

Fig. 1: Temporal relationships of the dye laser, the e-beam excitation pulse, the amplified $XeF(C \to A)$ output, and the broadband $XeF(C \to A)$ output with the system operating as a free running oscillator. Calculated results (bottom view) are compared with experimental data (top view).

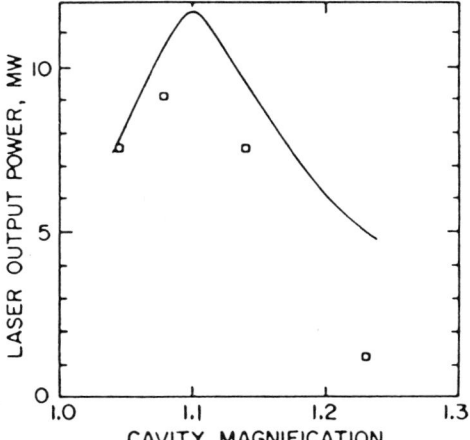

Fig. 3: Dependence of the amplified laser output power on cavity magnification for optimum timing from the analytical model. The squares are the experimental data.

HIGH PERFORMANCE DF-CO_2 CHAIN-REACTION LASER

S. T. Amimoto, J. S. Whittier, G. N. Harper
and R. Hofland, Jr.
Aerophysics Laboratory, Laboratory Operations
The Aerospace Corporation, El Segundo, Calif. 90245

J. M. Walters, Jr. and T. A. Barr, Jr.
U. S. Army Missile Command, Redstone Arsenal,
Ala. 35898

and

R. L. Kerber and W. K. Jaul
Michigan State University, East Lansing, Mich. 48824

ABSTRACT

Performance of a pulsed DF-CO_2 transfer laser was investigated both for photolytic initiation and for transverse initiation by a magnetically-confined electron beam. Laser output is presented as a function of e-beam charge fluence, O_2 concentration, F_2 and D_2 concentration, diluent gas, initiation level and total mixture pressure. The experimental results are in agreement with a comprehensive DF-CO_2 laser model. The DF-CO_2 laser performance is found to be comparable to that of a DF chain laser using mixtures of comparable chemical energy content.

INTRODUCTION

The problem of photolysis initiation of pulsed DF-CO_2 transfer lasers has been studied by a number of investigators.[1-3] Interest in pumping a CO_2 laser by use of the D_2-F_2 chain reaction is motivated by the simplicity of photolytic initiation compared with e-beam controlled excitation and by the possibility of high energy density output. In a recent demonstration, Basov et al.,[1] have reported output energies of 150 J/liter and chemical efficiencies of 7% from a DF-CO_2 amplifier system operating with nondilute mixtures. Performance of an electron-beam initiated DFPCO_2 laser has been briefly examined by Bashkin et al.;[4] electron-beam initiation has received little attention in comparison with the photolytically initiated case. The present investigation extends the work of Reference 4 to longer e-beam pulse lengths and lower densities. Photolytic transfer laser initiation was included as a part of the present study to permit performance comparisons with electron-beam transfer laser initiation. The apparatus of this study has been previously[5,6] employed to investigate the performance of HF(DF) chain lasers. The extension to the DF(HF)-CO_2 transfer laser case reported herein was intended to establish scale-size and mixture composition for CO_2 wavelengths and, hence, to establish scaling limitations for multicolor laser operation in general.

© American Institute of Physics 1987

emission from a pulsed HF/DF chain laser.[5,6] Further confirmation of CO_2 lasing was obtained by insertion of narrow-bandpass filters (3.5 – 4.5 m or 9 – 11 m) in the laser beam.

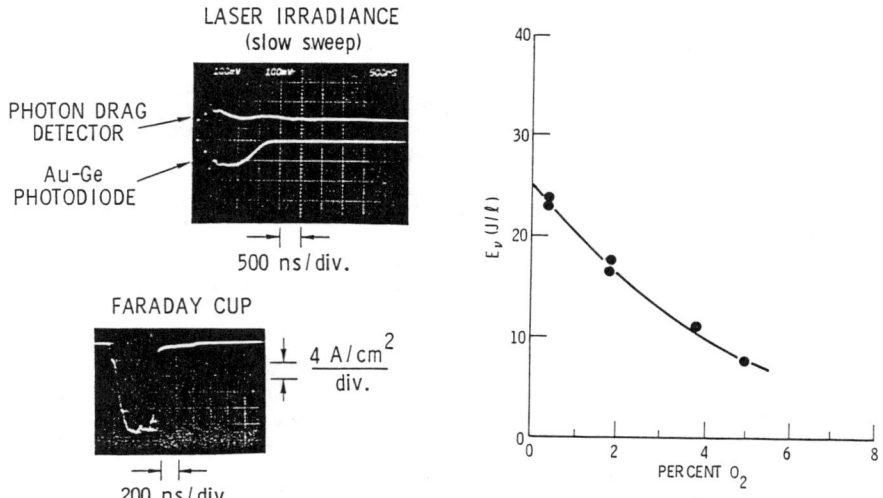

Fig. 1 Oscillograms of laser output and e-beam current (Faraday Cup).

Fig. 2 Effect of O_2 on e-beam initiated DF-CO_2 laser.

Dependence of DF(HF)-CO_2 laser performance on several parameters has been studied. Sensitivity of output energy to O_2 concentration in the laser mixture was studied first (Figure 2). Best performance for the DF-CO_2 transfer laser was found to occur at an oxygen concentration of 0.3%. An approximate expression for the fractional laser-energy degradation due chain termination by of O_2 can be written as[8]

$$E(O_2)/E(O_2 = 0) = (1 + \tau NO_2 \sum_i k_i n_i)^{-1} \quad (1)$$

where τ is the laser pulse duraction, n_i is the number density of species i, and k_i is the rate coefficient for the chain termination reaction

$$H + O_2 + M_i \rightarrow HO_2 + M_i \quad (2)$$

Eq. (1) has been fit to the data of Figure 2 to obtain an estimate for the unknown rate coefficient for Eq. (2) with CO_2 as a collision partner. If one assumed[8] $k_D = 1.4 \times 10^{-32}$ exp $[500/T (°K)]$ cm^6/s and $k_{F2} = k_{HE} = 4.1 \times 10^{-33}$ exp $[500/T(°K)]$ cm^6/s, then the data of Figure 2 imply that $k_{CO} = 4.7 \times 10^{-32}$ exp $[500/T (°K)]$ cm^6/s.

The effect of initiation strength (F/F_2) on CO_2 output was studied by holding the e-beam current density constant and varying the e-beam pulse duration and hence, the incident charge fluence. Laser output energy density is plotted as a function of e-beam charge fluence in Figure 3. For the HF-CO_2 laser case, specific

EXPERIMENTAL TECHNIQUE

The DF(HF)-CO_2 electron-beam tansfer laser was initiated by means of a pulsed 4-stage Marx-bank accelerator system. Magnetic confinement for transverse initiation geometry was provided by a pair of Helmoltz coils capable of generating magnetic fields up to 1.3 kG with good uniformity. At the exit of the anode window, a uniform, (±10%) beam of 20A/cm^2 current density and 175 keV energy was transmitted across the laser cavity. A low-inductance, a crowbar switch located at the output of the Marx generator was used to obtain continuously variable e-beam pulse lengths in the range 0.2 - 1.2 μs.

Performance of the DF(HF)-CO_2 transfer laser was also studied by means of a portable photolysis initiation device.[6] The photolysis laser consisted of a 9-cm-diameter × 1-m-long quartz reactor vessel surrounded by four 1-m-long flashlamps that provided ultraviolet (UV) radiation to initiate the chain reaction between F_2 and $D_2(H_2)$, and four cusp-shaped reflectors behind each flashlamp. The flashlamps were energized by a 2.8-F capacitor (Maxwell Laboratory), a single spark gap (Physics International 675) that was used to initiate the discharge, and twelve RG-213 cables (three per lamp) that connected the flashlamps to the output end of the high voltage switch. UV conversion efficiency of the Aerospace xenon flashlamps was measured based on the absolute radiant energy emission into the molecular fluorine absorption continuum (240-360 nm) and the stored capacitor energy. Lamp efficiencies were observed to vary from 11% to 15% as the lamp pressure was increased from 20 to 100 Torr. The FWHM pulsewidth was also seen to increase somewhat above 2 μsec with increasing fill pressure.

The optical extraction volume of the photolysis laser was 5.8 liters, while that of the e-beam laser was 2.0 liters, as verified by near-field burn patterns. Commercial grade gases were metered and mixed in an aluminum mixer block prior to their introduction into the laser cavity.

RESULTS AND DISCUSSION

Data obtained during a typical laser experiment are presented in Figure 1. A mixture consisting of 11% F_2: 11% D_2: 51% CO_2: 0.3% O_2: 26.7% He by molar volume was irradiated at an initial cavity pressure of 400 Torr. The measured CO_2 pulse energy of 50 J corresponds to an output energy density of 25 J/liter (48 J/liter-atm). The Faraday cup record of Figure 1 shows that the peak e-beam current density for this case was 22 A/cm^2 and the FWHM pulse duration was 455 ns for peak cathode voltage of 190 kV. Time history of the laser irradiance is illustrated in oscillograms in the upper half of Figure 1. The upper oscillogram trace is the output of the photon drag detector, while the lower trace is the gold-doped germanium detector output. The overall laser FWHM pulsewidth is observd to be 1.4 μs. Each detector shows a 50-ns-wide gain spike and the presence of mode oscillations of 18 ns; both features are characteristic of CO_2 lasers and are absent during

energy is seen to be extremely low and nearly independent of charge fluence due to the large mismatch in energy levels between HF(v) and CO_2. For the $DF-CO_2$ transfer laser case, a five-fold increase in charge fluence produced a 50% increase in output transfer energy density (from 17 to 25 J/liter), i.e., $(J/liter)_{DF-CO} \propto (Cb/cm^2)^{1/4}$.

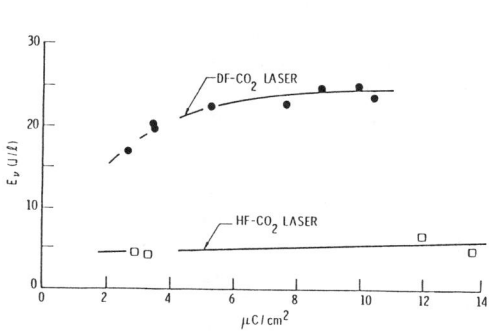

Fig. 3. Effect of charge fluence on laser performance.

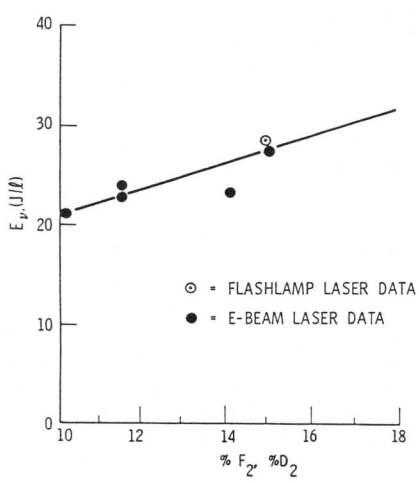

Fig. 4 Effect of F_2 (D_2) content on laser performance.

The dependence of specific output energy on reagent concentration, F_2 and D_2, is illustrated in Figure 4 at a cavity pressure of 400 Torr. In this parametric study, fuel and oxidizer concentrations were increased at the expense of CO_2 concentration such that $F_2 = -CO_2/2$ and $F_2 = D_2$. The observed increase in CO_2 output energy density is believed to be caused by the increase in energy content of the mixture. Both photolysis and e-beam laser data are included in the parametric study of Figure 4. Because of the weak dependence of laser output energy on initiation level, the flashlamp and e-beam laser results were found to be comparable. Unfortunately, crazing of the photolysis laser windows limited further study to a single photolysis laser data point. Were the data of Figure 4 extrapolated from 10-15% F_2 to 20% F_2 mixtures, output energy densities of 32 J/liter (60 J/liter-atm) would be predicted. Assuming (F/F_2) is approximately unchanged for the data of Figure 4, the observed dependence is J/liter $\propto$ (F_2). Taken with the data of Figure 3, a scaling law for a $DF-CO_2$ laser of the form $(J/liter) \propto (F/F_2)^{1/4}$ (F_2) appears to be valid in the present operating regime.

Laser performance data obtained at 200, 400, and 800 Torr are plotted as a function of cavity pressure in Figure 5. Except for the two photolysis laser data points at 200 Torr, all data plotted in Figure 5 were obtained using e-beam laser initiation. Substitution of nitrogen gas for helium gas to buffer the laser mixture is seen to have had little effect on laser performance. The

800-Torr performance is, nevertheless, comparable to that of a DF chain laser using mixtures of comparable chemical energy content and level of initiation.

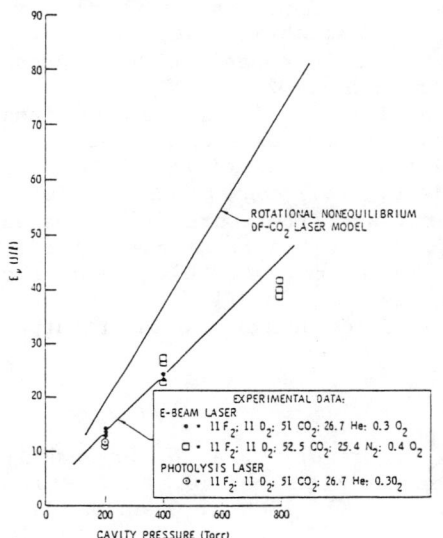

Fig. 5. Pressure scaling of laser performance.

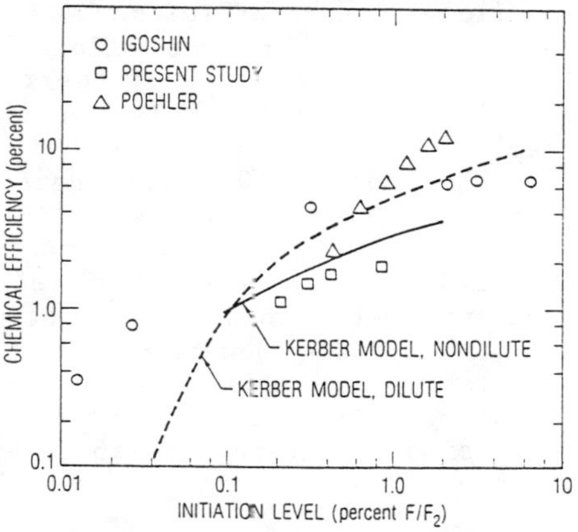

Fig. 6. Dependence of chemical efficiency on initiation level.

The predictions of Kerber's DF-CO_2 laser model[3,9] are included in Figure 5. The initiation level that is required as initial data in the calculations was obtained by matching the model pulse lengths to the experimentally-determined pulse lengths. Using this procedure it was found that $(F/F_2) = 0.006 - 0.008$ for the e-beam laser data presented herein. We find that all the trends of the experimental data are correctly predicted by the present code; unfortunately, the model overpredicts the measured energy densities by approximately 50 to 100%, as indicated in Figure 5. The cause for this disagreement is unknown at present.

The present data are compared with the experimental results of Igoshin, et al.[3] and Poehler, et al.[10] in Figure 6. The Igoshin data were obtained for a 1 atm mixture containing 9.1% F_2: 9.1% D_2: 36.4% CO_2: 45.4% He by molar percent. The Poehler data were obtained with a 250 Torr mixture of 3.7% F_2: 3.7% D_2; 22.2% He by molar percent. Laser chemical efficiency, defined here to be the laser output energy density divided by the available chemical energy density, is plotted in Figure 6 as a function of level of initiation, F/F_2. Included in Figure 6 are the theoretical predictions of Kerber[9] for low-pressure dilute mixtures (2% F_2: 2% D_2: 16% CO_2: 80% He at 50 Torr total pressure) and for nondilute high-pressure mixtures (11% F_2: 11% D_2: 51% CO_2; 27% He at 400 Torr total pressure). As anticipated on theoretical grounds, the more dilute laser mixtures generally yield substantially higher chemical efficiencies than laser mixtures that contain high reagent concentrations. The model predictions are again seen to overpredict laser performance, except at very low levels of initiation. The nondilute laser efficiency is observed to follow an $(F/F_2)^{1/4}$ dependence, whereas the dilute laser efficiency is found to vary linearly with F/F_2. Over the range of initiation levels from 0.2% to 1%, $DF-CO_2$ transfer lasers are seen to be capable of generating chemical efficiencies in the neighborhood of 1-5%.

The best $DF-CO_2$ laser performance of 41 J/liter was obtained with an electron-beam energy incident on the extraction volume of about 350 J; chemical efficiency relative to the initial D_2 content was 1.8%. The electrical efficiency for conversion of total incident energy to laser output energy was, therefore, 23%. Based on the stopping power data of Ref. 11, the intrinsic electrical efficiency for this case is calculated to be 200%. For the flashlamp initiated laser performance of 28 J/1 at 400 Torr, conversion efficiency of stored energy on capacitors to laser output was 4.6%.

REFERENCES

1. N. G. Basov, A. S. Bashkin, P. G. Grigor'ev, A. N. Oraevskii, and O. E. Porodinkov, Sov. J. Quantum Electron. 10, 135 (1980).
2. R. L. Kerber, N. Cohen, and G. Emanuel, IEEE J. Quantum Electron. QE-9, 94 (1973).
3. V. I. Igoshin, V. Yu. Nikitin, and A. N. Oraevskii, Sov. J. Quantum Electron. 10, 828 (1980).
4. A. S. Bashkin, A. N. Oraevskii, V., Tomashov, and N. N. Yuryshev, Sov. J. Quantum Electron. 10, 781 (1980).
5. S. T. Amimoto, J. W. Whittier, M. L. Lundquist, F. G. Ronkowski, P. J. Ortwerth, and R. Hofland, Appl. Phys. Lett. 40, 20 (1982).
6. S. T. Amimoto, R. W. F. Gross, D. A. Durran, G. N. Harper, L. S. Azevedo, L. E. Schneider, J. M. Walters, and R. Hofland, Jr., "Polyatomic-Buffered Pulsed DF/HF Laser Compatible with Solid-Reagent Generator and Chemical Pump," accepted for publication, J. Appl. Phys.
7. R. L. Taylor, P. F. Lewis, and J. Cronin, J. Chem. Phys. 73, 2218 (1980).
8. D. L. Baulch, D. D. Drysdale, D. J. Horne, A. C. Lloyd, Evaluated Kinetic Data for High Temperature Reactions, 377 (1972), CCR Press, Cleveland, Ohio.
9. R. L. Kerber, Appl. Opt. 12, 1157 (1973).
10. T. O. Poehler, J. C. Pirkle, Jr., and R. E. Walter, IEEE J. Quantum Electron. QE-9, 83 (1973).
11. L. Pages, E. Bertel, H. Jaffre, and L. Sklavenitis, Atomic Data. 4, 1-127 (1972).

ACKNOWLEDGEMENTS

This research was supported by the U. S. Army Missile Command under U. S. Air Force Space Division Contract F04701-80-C-0081.

INVESTIGATIONS OF A TRANSVERSE-EXCITED HIGH-POWER CO_2 LASER

J.Stańco, G.Śliwiński, J.Konefał, P.Kukiełło, G.Rabczuk,
Z.Rozkwitalski, R.Zaremba
Institute of Fluid-Flow Machines, Polish Academy of Sciences,
80-952 Gdańsk, Poland

ABSTRACT

Investigations of a high-power, transverse-flow, transverse-discharge, closed-cycle cw CO_2 laser are reported. The laser has been designed as a laboratory facility, primarily for the materials processing research.

INTRODUCTION

A high-power transverse-flow, transverse-discharge laser has been built to serve as a radiation beam source for investigations of interaction of radiation with matter, including especially research into the laser processing of materials[1]. In the laser, a mixture of CO_2, N_2 and He serving as the working medium is excited in a self--sustained normal glow discharge between an array of individually ballasted pin cathodes (tungsten wire of 1.5 mm diameter) and a plate anode (copper). In most experiments 816 cathodes, 1 per 0.9 cm^2, were used. The distance between the cathode tips and the anode having been 63 mm and the first row of pins protruded deeper into the flow. The discharge chamber forms a box 0.07 (h) x 1.25 (w) x x 0.35 (l) m, the cathodes being distributed over the upstream, 0.08-m-long part of the cathode board.
Circulation of gas, ensuring forced convection cooling, occurs in a loop of about 6 m^3 volume comprising the discharge chamber/optical cavity, a heat exchanger, a controllable-speed centrifugal compressor, and connecting ducts in the circuit. Typical values of the operating pressure and gas velocity at the discharge chamber inlet are 80 kPa and 100 m/s, respectively. For the $CO_2:N_2:He$ = = 1:2:7 mixture at a pressure of 80 kPa the electric power deposited in the gas reached 60 kW, with an average current density of about 22 mA/cm^2, a specific power per unit volume of about 12 W/cm^3, and a specific power per unit mass flow equal to about 220 $kW/kg \cdot s^{-1}$.

EXPERIMENTS

Measurements aimed at optimizing the laser parameters showed advantages of using a baffle array upstream of the electrode system. It was shown that adjustment of the baffles improves significantly the discharge uniformity and stability, allowing a considerable increase in the electric input power.[2]
The laser has been designed for operation with an unstable multipass optical resonator, enabling one to fully utilize the optically active medium.[3] For the relatively narrow discharge region used,

© American Institute of Physics 1987

single- or two-pass resonator, with an additional amplifying pass, was assumed to provide adequate coverage of the gain region. The laser beam was coupled out of the resonator with the aid of an annular coupling mirror and a KCl, NaCl or KRS-5 window.

Examples of the laser output characteristics are shown in Fig. 1. It was found, when optimizing the working medium composition, that the $CO_2:N_2:He = 1:3:5.7$ mixture ensured the most efficient electric-to-optical energy conversion[1]. For such a mixture a maximum output power of 4.4 kW was obtained with a two-pass (M = 1.95) resonator; the electric-to-optical efficiency amounted to 10%.

Measurements of the small-signal gain[4] confirmed the anticipated narrowness of the gain region. The combined influence of discharge nonuniformity and the dynamics of molecular excitation and relaxation processes in the flowing gas of moderate pressure resulted in a nearly triangular gain distribution with the maximum (of up to 1.0 m^{-1}) located just downstream of the of cathode pin array.

The results of measurements were used for adjustment of the optical resonator position, aimed at maximizing the output power. The improvement was clearly evident for a one-pass cavity (Fig. 1), while the interplay between the oscillator and amplifier parts of the cavity masked the effect in a two-pass resonator.

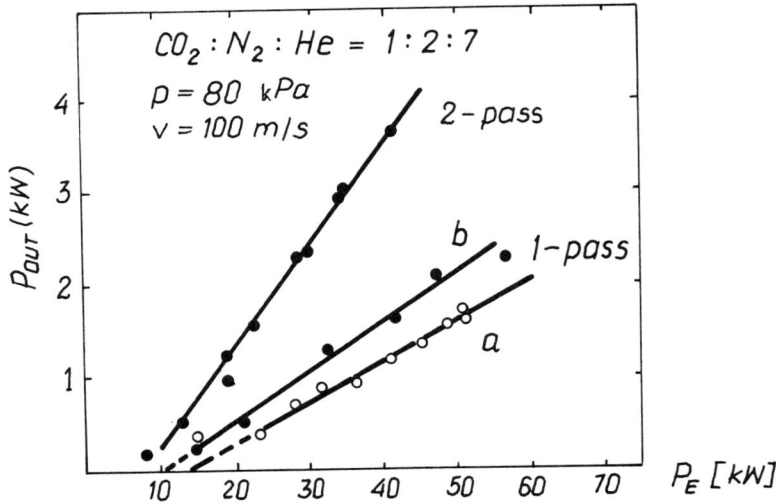

Fig. 1. Laser output vs electric power dissipated in the discharge for the laser with different optical resonators (2-pass: magnification M = 1.95, single-mirror loss T_L = = 1.5 %; 1-pass: M = 1.6, T_L = 12 %, resonator axis (a) 3.5 cm and (b) 7.5 cm downstream of the first row of pin cathodes)

To determine the laser beam divergence the focal spot at the focus of a mirror of large focal length (f/40 optics) was measured using burn patterns in plexiglass. For the single-pass resonator the divergence amounted to about 1.7 mrad (half-angle) while the calculated diffraction-limited value was 0.75 mrad for the central lobe plus diffraction rings containing together 90 % of the laser output power.

First tests of laser welding and cutting steel sheet up to 3 mm thickness were carried out at power levels of about 1.5 kW. Reflective f/3 optics allowed a radiation intensity of 1.6×10^6 W/cm^3 in a focal spot of 400 μm diameter. Helium was used as the shielding gas. Conclusions drawn from the tests have been utilized in the designing of a laser head for metals processing. Also a stand has been designed for investigations of continuous electrical discharge in gases.

ACKNOWLEDGMENTS

This work has been sponsored by the shipbuilding industry. The laser facility was built in cooperation with the PROMOR Shipbuilding Technology Centre in Gdańsk, the Gdańsk Shipyard, and the Institute of Quantum Electronics in Warsaw. Contributions of many colleagues in various stages of laser construction and investigation is greatly appreciated.

REFERENCES

1. J. Stańco, E. Antropik, P. Grodecki, M. Irczuk, P. Kozyro, J. Konefał, P. Kukiełło, W. Mikienko, M. Piskulski, G. Rabczuk, Z. Rozkwitalski, T. Stelter, G. Śliwiński, R. Werdon, R. Zaremba, Lasers and Applications, Proc. 2nd Intern. Conf. Bucharest 1985 (1986), p. 551.
2. G. Śliwiński, J. Stańco, E. Antropik, M. Irczuk, J. Konefał, P. Kukiełło, W. Mikienko, G. Rabczuk, Z. Rozkwitalski, R. Werdon, R. Zaremba, Bull. Pol. Ac.: Tech. 34 (1986)(in print).
3. G. Rabczuk, Proc. 5th GCL Symp. (Oxford 1984), Int. Phys. Conf. Ser. 72 (1985), p. 391.
4. J. Stańco, Z. Rozkwitalski, G. Śliwiński, J. Konefał, P. Kukiełło, G. Rabczuk, R. Zaremba, Appl. Phys. B41 (1986)(in print).

THE KMSF LOW PREHEAT IMPLOSION EXPERIMENTS

Jon T. Larsen and Roy R. Johnson
KMS Fusion, Inc., Ann Arbor, MI 48106

ABSTRACT

Laser-driven inertial confinement fusion (ICF) experiments are being performed at several laboratories throughout the world. Nearly all are seeking to demonstrate compressed fuel densities of several tens g/cm^3. For ICF to be an efficient process, it is desirable to compress the fuel along a near-Fermi degenerate adiabat to densities of several hundred g/cm^3. Addition of thermal energy prior to achieving the compressed state is detrimental in that a significantly larger driver would be required to achieve ignition. Preheating of the fuel prior to compression may result from radiation or thermal electron conduction, or non-isentropic processes such as shock heating. Recent experiments at KMSF sought to control these by a combination of target and laser parameters. Using a spherical illumination system, the target is irradiated with a carefully prescribed temporal pulse of frequency doubled neodymium glass laser light. Low preheat targets are designed with thick polymer (poly vinyl alcohol) shells and cryogenic fuel layers. The implosion process is recorded by a number of optical and x-ray diagnostics. A four-frame holographic interferometer records the temporal evolution of the ablated plasma while an x-ray pinhole camera records the symmetry of the compressed core. The most important diagnostic is the combined x-ray backlighter/streak camera combination which records the implosion trajectory in one dimension. A comparison of the data from these three major instruments with the results from the hydrodynamic simulation code allows one to interpret the final state the fuel reached and thus, the preheat level. Assuming spherical symmetry, record fuel densities in excess of 35 g/cm^3 have been achieved.

DISCUSSION

In 1976, KMSF began to experiment with implosions driven with frequency doubled 1.06 μm laser light, polymer fuel capsules, and cryogenic fuel configurations. Several series of experiments during the past ten years perfected each of these ingredients independently. In 1982, we demonstrated an initial success in combining all these features.

The purpose of the recent experiment series was to demonstrate improved fuel compression by carefully controlling the temperature the fuel was brought to prior to performing PdV work. It is energetically advantageous to keep the fuel near its Fermi degenerate temperature. It is also advantageous to start with as high a fuel density as possible, which is accomplished by forming a thin frozen layer on the inside of the shell. The integrity of this layer is maintained only if it can be kept relatively cool and free of excessive preheat. Numerical simulations and analytic model calculations show that a 1 to 3 eV temperature is sufficient to achieve compressed fuel densities several hundred times that of liquid DT with only a few hundred joules of laser light.

Uniform spherical illumination was achieved with a new set of ellipsoidal mirrors designed to be more resistant to damage than the previous TBIS system. This new system uses a double-bounce (DBIS) configuration from the mirrors in conjunction with aspheric focusing lenses. Figure 1 shows the trajectory of the marginal rays through the system. Careful positioning of the two mirrors and lenses relative to the target position allow the marginal rays to slightly overlap, thus providing uniform illumination over the whole of the target's surface. DBIS increases the first bounce area on the mirror, relative to that of the TBIS, so that the energy density is less. In addition, AR coatings were optimized for 527 nm rather than the dual wavelength coatings chosen for TBIS. Care was taken to minimize the debris from the target and support stalk from reaching the mirror surfaces.

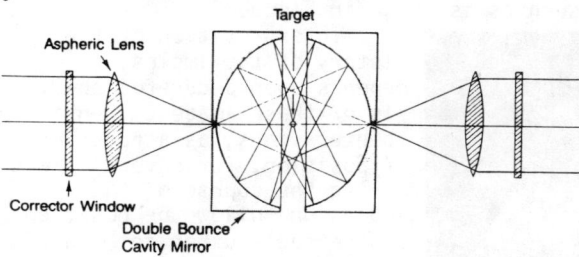

Fig. 1. Uniform spherical illumination is achieved from two laser beams with a double bounce configuration. Each 18.6 cm laser beam passes through a f/1.39 lens and into a two mirror system. A reflection from each surface puts the marginal ray at an angle of 70° to the axis of rotation. A slight (about 20 μm) shift of the focal point behind the target's center provides uniform coverage. All diagnostic instruments and target handling access is through the 5 cm annular gap between the two mirrors.

The large solid angle occupied by the DBIS mirrors severely restricts the diagnostic instrument access to the target. All instruments and target handling equipment must pass through the annular gap between the two mirrors. The cryogenic target positioner, cryogenic shroud, heating laser, and characterization interferometer take up a significant fraction of the 2" gap. The x-ray backlighter diagnostic consists of the backlighter target and positioner, alignment telescope, and x-ray streak camera. Other instruments competing for position include pinhole cameras, holographic interferometer, and plasma calorimeters.

A measure of the uniformity of illumination is provided by the two-dimensional reconstructed image from the holographic interferometer. Each frame, made with a 20 ps exposure, is digitized and mathematically (Abel) inverted to provide a radial plasma density profile.

One of the key ingredients for a low preheat implosion is the control of the shock wave generated by the laser pulse interacting with the target. A laser pulse shaped in time, roughly as t^2, accomplishes this. The other major source of preheat is from hot electron production which is minimized by the use of 527 nm light. Approximately 160 to 180 joules of light was incident on the target.

Low atomic number shells are preferred for these low preheat implosions, primarily for the compatibility with the x-ray backlighter diagnostic. Target capsules had a nominal 120 μm o.d. and a wall thickness ranging from 3 to 8 μm. The cryogenic fuel was formed by the fast re-freeze technique resulting in a solid fuel layer 1 to 2 μm thick. The uniformity and quality of this layer was verified interferometrically prior to each shot.

The primary implosion diagnostic is x-ray radiograph of the imploding shell as recorded by a streak camera. Because the unablated shell is of low atomic number material as well as being relatively cold, there is little x-ray self-emission and high x-ray absorption. An aluminum x-ray source placed adjacent to the target emits line radiation when illuminated by a third laser beam. X-rays from this target are absorbed by the inward moving portion of the shell, providing an image such as is shown in Figure 2.

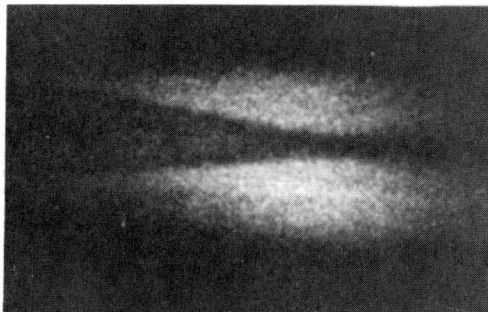

Fig. 2. A bright, laser generated source of aluminum x-ray lines is produced behind the imploding target. Temporally resolved transmission of the x-rays is recorded in one dimension by a pinhole-streak camera combination. Transmission through the core ceases at an interface radius of about 20 μm. The implosion time is about 720 ps for a shaped laser pulse. Time increases to the right, while the radial position is the vertical axis; target center coincides with center of dark band.

Interpretation of the history of the implosion depends upon a careful analysis of these images. Densitometer scans, as a function of position, for several times during the course of the implosion follow the position of the fuel-pusher interface until the x-ray absorptivity causes the core region to become opaque. The position of the outer portion of the compressed material can be followed to the culmination of the implosion. Comparison of these experimental profiles to those generated by the numerical simulation code show excellent agreement. From these data (as well as other experimental quantities) the compressed density can be inferred. Assuming a spherical final state, a density of 35 to 40 g/cm^3 has been achieved.

Not all experiments ended with a uniformly compressed core. Some shots showed significant x-ray self emission from the core region in the time-integrated pinhole picture. No information was obtained as to the mechanism responsible for this "shell breakup;" however, it is tempting to speculate that Rayleigh-Taylor instability may be involved.

During 1987, we will extend these measurements on low preheat implosions using cryogenic fuel and shaped laser pulses. A brighter, higher energy (titanium target) backlighter source will be used in conjunction with higher resolution imaging devices, and possibly including a two-dimensional x-ray framing camera.

ACKNOWLEDGMENT

Prepared for the Department of Energy under Contract No. DE-AC08-82DP40152.

PROSPECTS FOR A GAMMA RAY LASER
BASED UPON UPCONVERSION

C. B. Collins
University of Texas at Dallas

ABSTRACT

Because of a renaissance in concepts, the past few years have seen a substantial increase in the level of research aimed at the determination of the feasibility of a gamma ray laser. The essential concerns are reviewed in this presentation.

INTRODUCTION

At the nuclear level the storage of excitation energies in the Mossbauer range of 1-100 keV can approach tera-Joules (10^{12}J) per liter for thousands of years. Over the past seven years, our research group has described[1-16] several viable means through which this energy might be coupled at will to the radiation fields while maintaining the natural Mossbauer width. In such cases the cross section for stimulated emission around 1A could reach 10^{-17} cm^2, an order of magnitude more favorable than the value for the stimulation of 1.06 μm from Nd^{3+} in YAG. The successful release of such nuclear energies in this way would occur at the rate at which resonant electromagnetic radiation passed through the laser medium and could lead to output powers as great as 3 X 10^{21} Watts/liter. This is an astronomical level of intensity and has not been approached to within five orders of magnitude on earth by any means previously. The peak power from a one liter device would represent 0.03% of the total power output from the sun.

Unfortunately, the quest for a gamma ray laser has been one of the longest unfruitful efforts in the field of laser science. Virtually all of the sustained pioneering work was done by Baldwin and Solem's groups in the US and by Gol'danskii's in the USSR and focused upon the single photon, brute force approach to pumping. Their work dealt extensively with concepts involving the use of a neutron flux for pumping the laser medium, either in situ, in real-time or as a preparatory step to be followed by a rapid separation of isotopes within their natural lifetimes. All proposals were concluded to require infeasibly high levels of particle fluxes to pump the inversions, exceeding even those available from nuclear explosions, and to require neutron moderators having virtually infinite thermal capacities. By 1980 all conceivable variants of the single photon approach had been characterized as hopeless. In 1981 this "traditional" approach to a gamma ray laser was virtually abandoned with Baldwin's publication of the monumental review[17] of all classical efforts.

The involvement of our UTD Center for Quantum Electronics dates back to 1978, arising from previous activity focused upon fundamental interactions of coherent radiation with matter. Concerned at first with the problem of the correct gauge and basis sets to use in describing multiphoton process, we began to consider the impact of the work upon areas other than the usual atomic and molecular. As a result, the modernized concept of coherent pumping with optical radiation was introduced in a sequence of papers[1-7] concerned with nonlinear processes mediated by virtual states of nuclear excitation and included the stimulated anti-Stokes scattering of intense but conventional laser radiation.[6] The theoretical treatment served to estimate matrix elements for a new class of two-photon Mossbauer

© American Institute of Physics 1987

transitions making possible, in principle, the frequency upconversion of optical laser photons to gamma ray energies.

In 1981 the implications of this theoretical renaissance to the prospects for a gamma ray laser based on several variants of upconversion were reviewed in an article[8] appearing the following year. If strengthened by recent infusions of dressed state theory,[18,19] that article still provides the most convenient review of the basic concepts and requirements for a viable gamma ray laser scheme. Subsequently tested in a series of modest experiments, the underlying concepts were confirmed by demonstrating[9-12] that the matrix elements used to obtain the favorable estimates of the threshold for laser output *were correctly estimated* and that extremely large ferromagnetic enhancements of the effective powers applied in the coherent pumping scheme can be obtained. The conclusion from these experiments was *that the gamma-ray laser is definitely feasible if a sufficiently ideal isotope exists in reality.* This is the single most critical issue to the development of a gamma-ray laser--the identity of the most nearly ideal candidate for upconversion.

Despite the many applications of beautiful and involved techniques of nuclear spectroscopy, the current data base is inadequate in both coverage and resolution either to answer the question of whether an acceptable isotope exists or to guide in the selection of a possible candidate medium for a gamma-ray laser. Two new techniques for the measurement of nuclear properties with laser-grade precision have been recently introduced in our laboratory. Reported in the technical sessions will be the construction and evaluation of a flash x-ray device for pumping test materials and a Nuclear Raman Spectrometer to facilitate the search for certain necessary arrangements of nuclear levels.

Such is the enthusiasm accruing from this renaissance in concepts, it can be reasonably estimated that at this time of review about 50 colleagues in the US and Western Europe (33 in our laboratory) are now devoting full time efforts toward the resolution of the interdisciplinary puzzle posed by the gamma ray laser. In the following sections the critical issues are reviewed which guide this unprecedented level of effort.

THEORETICAL CONTEXT

By involving two distinct steps, the schemes we have discussed for pumping a gamma-ray laser avoid the severe relationships between storage times and spontaneous powers wasted at threshold that were imposed on the single-step processes.[17] Replacement power that is required falls within a technically accessible range avoiding damage to the laser medium.

These two-step, upconversion processes can be divided into two basic categories that correspond to the type of pumping employed: coherent and incoherent, as shown in Fig. 1. The critical concept here is that either transfers the stored population to a state at the head of a cascade leading to the upper laser level. To be effective the pumping processes cannot transfer too many quanta of angular momenta from the fields, and the cascade provides a mechanism for further changes that may be necessary to reach the laser levels. Then the ultimate viability of these pump schemes will depend upon: (1) spectroscopic studies locating a suitable configuration of nuclear energy levels, and (2) "kinetic" studies providing an efficient path of cascading from the intermediate or dressed state to the upper laser level.

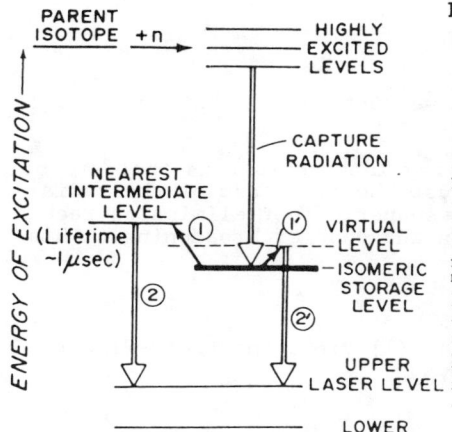

Fig.1:
Schematic diagram showing the energetically excited levels of a typical nucleus of interest to the development of a gamma-ray laser. Lifetimes of the stored energies in the isomeric level produced by the initial capture can range from days to hundreds of years. The first phase of the two-step process for the stimulated release of the stored energy is shown in the figure by the solid arrows. Both correspond to the use of longer wavelength radiation to lift a nucleus from the storage level to a higher level of excitation that has a much shorter lifetime. The arrow marked (1) illustrates the incoherent pumping of the storage level through the absorption of an x-ray that is resonant with the energy separation between the storage level and the next higher level of proper symmetry. The arrow marked (1') represents the alternative process of coherent pumping through the non-resonant absorption of a photon from the radiation field in order to create a virtual or dressed state of excitation shown by the dashed level in the figure. In either case the gamma-ray output ultimately results from the upper laser level populated by a cascade occurring as a second step, as shown in the figure by either of the double arrows, (2) and (2').

A recent variant to the scheme for pumping incoherently with flash x-rays will also be reviewed in a subsequent technical paper. There the transfer from storage to intermediate level is assumed to be excited by the direct coupling of energy from giant collective oscillations of electron shells driven by laser radiation focused to power densities comparable to binding energies. While the magnitudes estimated for the gross rates of energy transfer into the nucleus are encouraging, the likely dominance of the inverse process suggests that little net transfer could be realized. Once correlated motion of a shell is established, unless strong processes dephasing the motions of individual electrons are introduced with attendant losses, free induction decay at the end of the pulse will extract the energy back from the nucleus into the fields. This could only be avoided if some super-allowed transition in the nucleus were assumed to quench the excitation to a lower uncoupled state, but then this is tantamount to the assumption of a type of previously unobserved nuclear transition of very high width. While it is too early for definitive resolution, it is unlikely this recent pumping variant could prove effective and the most likely prospects remain confined to those shown in Fig. 1.

Because of the interdisciplinary nature of the problem, even for an idealized nuclear material, computations of threshold levels of pumping are not without difficulty. It is useful next to review these fundamental concerns within the context of the nucleus.

From very fundamental bases, the cross section for the interaction of polarized radiation with matter leading to stimulated emission can be generally expressed (even for nuclei)

$$\sigma_0 = \frac{\lambda^2}{8\pi} Ag(\nu), \tag{1}$$

where A is the Einstein coefficient for spontaneous emission and $g(\nu)$ is the normalized lineshape function for the transition attributed to the matter,

$$\int g(\nu)d\nu = 1 . \tag{2}$$

Approximating

$$g_{eff}(\nu) = 1/\Delta\nu , \qquad (3)$$

where $\Delta\nu$ is the absorption bandwidth, we must recall that the actual maximum of $g(\nu)$ is only $2g_{eff}/\pi$.

One intrinsic advantage of gamma ray interactions is that at the nuclear level the width $\Delta\nu$ is often just the transform of the radiative lifetime. In those cases the Mossbauer effect eliminates recoil and with it problems of thermal motion and Doppler broadening. This natural width gives

$$\Delta\nu = A/\pi \qquad (4)$$

which substituted with Eq. (3) into Eq. (1) yield the Breit-Wigner cross section for stimulated emission,

$$\sigma_0 = \lambda^2/8 , \qquad (5)$$

a very large value, even at 1A. The strongest homogeneous broadening process will limit access to these benefits but this effect has not yet been observed experimentally. Gamma ray transitions have been found to have natural widths down to instrumental limits as small as 10^5 Hz. Thus, a lifetime down to at least 1 μsec can be safely assumed for a process occurring at σ_0.

To estimate requirements for the incoherent pumping of populations from a storage level to an upper laser level through absorption of incident x-rays, one further complexity must be introduced.

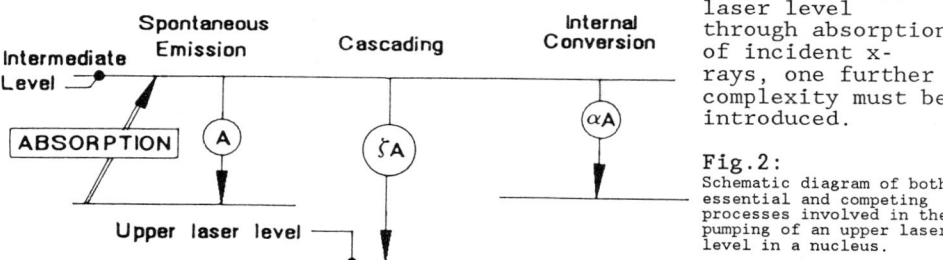

Fig.2: Schematic diagram of both essential and competing processes involved in the pumping of an upper laser level in a nucleus.

As shown in Fig. 2, both cascading and internal conversion are accommodated by transition rate coefficients scaled to the A-coefficient by factors ζ and α, respectively. As a result, the total width for absorption becomes

$$\Delta\nu = (1 + \alpha + \zeta)A/\pi , \qquad (6a)$$

while the cross section for absorption is reduced accordingly,

$$\sigma = \sigma_0/(1 + \alpha + \zeta) . \qquad (6b)$$

Even with broadening, the width for a nuclear transition is so much more narrow than the width for any structure in a non-nuclear source of x-rays, the pump must be considered to be a continuum. As a result the _rate_ of pumping _concentration_ from the initial state is proportional to the product, $\sigma\Delta\nu$ of the terms from Eqs. (6a) and (6b) and is independent of α and ζ (providing broadening is less than

10^3-10^4 so that σ is not reduced below the cross section for photo-electric absorption by the electrons in the material).

The consequent effects of broadening on the production <u>rates</u> of <u>concentration</u> in various levels at the incident surface of the material is summarized in Table I.

Table I

Effect of internal conversion and branching upon production rates of concentrations of levels shown.

Level	Score	Effect	Comment
Intermediate	LOSE	Same production rate	Lifetime over which pump rate can be integrated is *reduced by* $1/(1+\alpha+\zeta)$.
Upper laser level	WIN	Production rate reduced by $\zeta/(1+\alpha+\zeta)$	Usable pump duration is increased up to τ_U, the lifetime of the upper laser level.

Assuming we pump with both polarizations, a total concentration N can be pumped in the upper laser level,

$$N = N_0 F(\nu_0) \frac{\lambda^2 A}{4\pi} \frac{\zeta}{(1+\alpha+\zeta)} \tau_U . \qquad (7)$$

where N_0 is the concentration of absorbers, and $F(\nu_0)$ is the photon flux per unit frequency averaged over τ_U, the lifetime of the upper laser level. As customarily used, A is summed over the degeneracy, Z_L of the final states and averaged over Z_U initial states so that the Breit-Wigner cross section for absorption must be modified from Eq. (5) through multiplication by (Z_U/Z_L). With that modification, Eq. (7) is entirely consistent with the customary expression for the Mossbauer cross section, as usually expressed when it is recognized the latter describes absorption of a narrow line as opposed to continuum.

THRESHOLD ESTIMATES

The critical concept in the design of a gamma ray laser pumped by incoherent x-ray is the realization that in Eq. (7) the width-lifetime product, $A\tau_U/\pi$ can be made much greater than unity. In such a case, population from the storage level is funneled through a broad, but short-lived level to a longer-lived laser level for subsequent stimulation. The threshold requirement for the pump flux can be estimated by simply equating the gain contributed by the population pumped according to Eq. (7) with the loss from nonresonant photoelectric absorption[20] in the matrix into which the nuclei are diluted.

In 1982 we published[8] the details of a basic modeling study incorporating this bandwidth funneling in an idealized nucleus. The resulting threshold requirements were accessible to existing technology and were revised even lower with the incorporation of the Borrmann effect, as described at last year's ILS conference[13]. For the output transition the Borrmann effect allows the development in a crystal of standing waves of such quadrature that coupling to nuclei is enhanced while coupling to electrons is minimized.

Because of recent conflicts in "Private Communications" the actual quantitative level of the enhancement from the Borrmann effect upon threshold requirements is uncertain and the results of last year[13] must be given a larger variance. Current results of the application of Eq. (7) to idealized nuclei diluted to 0.04% concentration in a Be lattice and arranged only for output along a Borrmann mode at 10keV are as follows:
1) Threshold Fluence   100-300 J/cm²/0.1%BW/lifetime
2) Temperature Rise   100-300°C.
Specification of the fluence per unit bandwidth in terms of 0.1% of the transition energy, as shown, is a convenience as it thus roughly corresponds to the fluence within the natural width of an x-ray line.

Estimates of both threshold fluence and temperature rise are extremely sensitive to actual material characteristics. Once the proper nucleus is identified, considerable further improvement is possible. This can be appreciated from the schematic reproduced in Fig. 3 for a typical, but hypothetical case[13].

There it can be seen that the major part of the pump line that is unavailable for nuclear absorption is deposited into the material through the ejection of photoelectrons from a much thicker layer than the one in which the nuclei are pumped. The estimate of temperature rise of 100-300°C assumes full conversion of the photoelectron energy into heat, but the medium could be layered to permit escape of the primary photoelectrons, thus reducing considerably the temperature rise. An optimal configuration can be arranged once the specific characteristics of the best candidate nuclei are known.

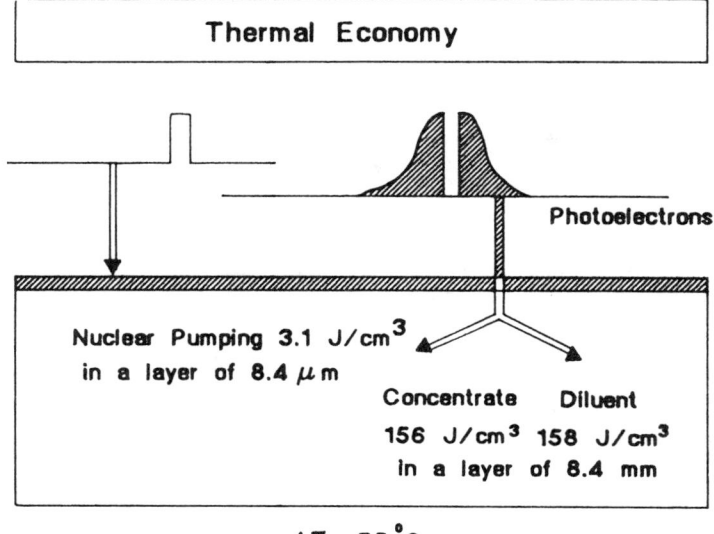

Fig.3: Schematic diagram of the thermal economy of the scheme for pumping a gamma ray laser with broader band x-radiation.

Coherent pumping, the technique depicted in the right of Fig. 1, depends upon the alteration of the properties of the storage level produced by the scattering of large intensities of long wavelength radiation from the nuclei. Again the distinction must be made between this approach and the recent suggestion to use laser radiation to drive collective but localized oscillations of electrons near nuclei. The concept we address uses coherent, long wavelength radiation to drive the non-local correlations of the electrons manifest as well-known phenomena of magnetization or ferroelectric polarization throughout the bulk of the material.

In ferromagnetic and ferroelectric hosts the active nuclei are immersed in extremely large fields capable of developing substantial interaction energies across a nuclear volume when switched by relatively modest applied fields. If transitions to the storage level exist in the nuclei at energies comparable to that of a photon of the driving fields, the properties of the other state of the transition will be mixed into those of the storage level. It is assumed that this other state is better able to radiate gamma radiation. While the driving field need not be precisely resonant with the transition energy, the detuning, ΔE from resonance must be comparable to the interaction energy if properties are to be fully mixed. In such cases the metastability of the storage level against gamma ray emission is switched off by the admixture of properties from the other state of the low energy transition being driven. It is this concept which comprises the foundation of the scheme for coherently pumping a gamma ray laser.

While precise computations of the threshold for coherent pumping are not yet available, estimates from perturbation theory[8,16] suggest that the threshold requirements in idealized cases are comparable to those presented above for the case of incoherent pumping. The principal difficulty in this case is that, again, estimates are extremely sensitive to material specifics. For coherent upconversion to be viable a real nucleus must be found with two accidentally degenerate levels, one being a long-lived isomeric state. Such a combination would be completely invisible to current techniques of nuclear spectroscopy.

CRITICAL EXPERIMENTS

Under the idealized conditions discussed above, the part of the pump energy which must be supplied *in situ* would not be large enough to represent a major impediment to the realization of a gamma ray laser. The real difficulties take peculiar forms. The use of coherent upconversion would require the location of nearly degenerate levels which could not be resolved by conventional techniques of nuclear spectroscopy. The use of incoherent pumping with x-rays would require a level of knowledge about branching ratios and transition probabilities beyond that available from current methodology. In fact, the paucity of laser-grade data describing nuclear properties is so severe that one cannot say which real isotope represents the best approximation to the ideal.

For lifetimes ranging from seconds to infinity there are 1886 real nuclei to consider as candidates for a gamma ray laser. Computer based searches of the existing data base have served to identify 29 first class candidates. Of these, 10 are known to have the necessary (but not necessarily sufficient) arrangement of levels in which there is an isomeric storage level and at lower energies: 1) an upper laser level with lifetime between 1 nsec and 10 μsec and 2) a lower laser level of even less energy. For these materials, the applicability of the favorable threshold estimates will depend upon:
    1)   Spectroscopic studies locating a suitable intermediate or scattering state to which transitions can be made from the isomer,

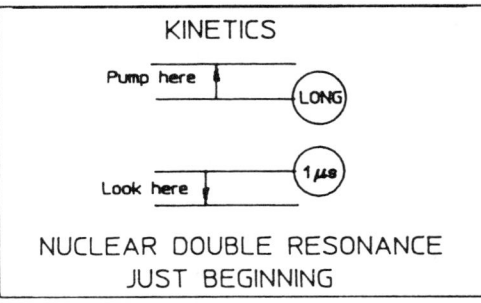

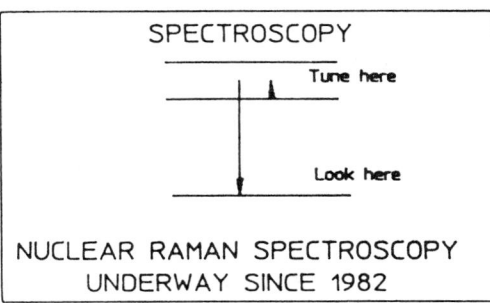

2) "Kinetic" studies providing an efficient coupling from the intermediate or scattering state to the upper laser level.

To meet this need for laser-grade data on nuclear properties we recently introduced[14,16] analogs to the powerful techniques for evaluating spectroscopic and kinetic properties of atoms and molecules at optical wavelengths. These are shown schematically in Fig. 4.

Fig.4:
Schematic representation of nuclear analogs of the optical double resonance and Raman spectroscopic methods that have served so effectively to build a laser related data base for atomic systems at optical energies.

Essential to the success of the double resonance experiments is the accessibility of a source of pulses of x-rays of nanoseconds duration that is very powerful in comparison with conventional devices. A dose of 10^{16} keV/keV of bandwidth is needed from the pump in a reasonably brief working period. Either laser plasmas or large e-beam machines can do this in a single shot, each of which requires about an hour of laboratory time to prepare; but costs are very high. As a result, none of these traditional light sources for the subAngstrom region could be used to complete an evaluation of the 29 most attractive materials before the turn of the century.

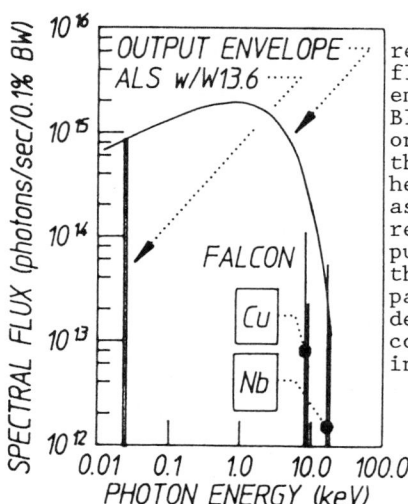

In the following technical section we report recent successes with a prototype flash x-ray device[21], FALCON which can emit 0.3W average power near 0.5A. A Blumlein driven device, it apparently is one of the first which is not choked at the output diode. Reasonable matching of head to line allows pulse durations to be as short as 20 nsec with peak powers reaching 150 kW. About one-third of the pulse energy appears in the K_α lines of the anode material, meaning that at those particular energies, outputs approach the design objective of the ALS synchrotron configured with the 13.6 wiggler, as shown in Fig. 5. Linewidths from FALCON and the

Fig.5
Spectral flux from our Blumlein driven, flash x-ray device, FALCON, in comparison to the design objective of the ALS synchrotron with a 13.6 wiggler. Line output is determined by the composition of an interchangeable anode and two examples are shown.

ALS with wiggler are comparable, but of course the ALS output is collimated and tunable to any energy within the envelope while lines from our device are fixed at the K-line energies of available anode materials. However, costs of our table top device are three to four orders of magnitude less and for the illumination of extended absorbers it offers an interesting alternative to the more conventional sources of x-rays.

Perhaps of greater interest is that 2/3 of the output energy in each pulse from FALCON is distributed in a true continuum in which all wavelengths in a range from about $E(K_\alpha)$ to about $2E(K_\alpha)$ are present. If considered as equivalent to a pulsed radioactive source, when "on," each pulse is the *equivalent of 5-10Ci of strength within the width of an allowed nuclear transition*, simultaneously at every possible transition energy in the range. Since energies of potential pump transitions in most systems are very poorly known, the availability of continua from FALCON actually makes it a more viable source for mounting double resonance experiments, than would be a synchrotron which would have to be tuned over the working range, line at a time.

The other new technique of Fig.4, preserves the high resolution aspects of Mossbauer spectroscopy but replaces the mechanically generated Doppler shifts, conventionally used for tuning, with sum and difference frequencies generated by a mixing process. A computer controlled frequency synthesizer sweeps a radiofrequency used to switch intense magnetization in a host material into which the test nuclei are diluted. Tunable sidebands generated on the intrinsic gamma ray transition then replace the Doppler shifted lines as probes.

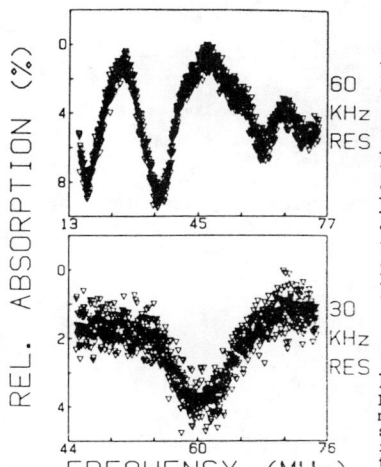

The first spectrum from such a continuously tuned Nuclear Raman Spectrometer (NRS) is shown in Fig.6 to be unexpectedly rich in detail. Not a laser candidate but a simulation, ^{57}Fe was used in this demonstration. Shown in absorption are resonant structures from expected states in the lower panel together with unexpected surface states, yet to be analyzed. Widths of the lines are not instrumental but rather reflect the transform of the lifetime of the final state of the nuclear transition at 14.4 keV.

Fig.6:
First continuously tuned NRS spectrum of a nuclear material, in this case the 14.4 keV transition of the simulation nucleus Fe-57. Instrumental resolution is indicated and observed linewidths correspond to the transform of the lifetime of the nuclear state.

Detailed review of both new experiments is presented in the following technical section and seems to encourage the perception that momentum is building in the development of new technology necessary in the search for the best candidate material for a gamma ray laser.

CONCLUSIONS

The principal conclusion to be reiterated is that *a gamma ray laser is feasible if some real nucleus has properties sufficiently close to the ideals being modeled.* Recent computations have dispelled lingering concerns about material survival at the levels of pump fluence that are sufficiently high to reach the threshold for gamma ray laser output for an ideal material. While many enhancements to configuration of a gamma ray laser might be envisioned ultimately to result from studies of even more exotic effects such as superradiance, giant resonances, and nuclear excimers, the absolutely critical factors determining the feasibility and means of realizing a gamma ray laser remain: 1) the identity of the best real candidate; 2) the threshold level for laser output; and 3) the upconversion driver for that material.

*Research supported in part by the Office of Naval Research and in part by the Strategic Defense Initiative Office (IST) through the Naval Research Laboratory.

REFERENCES

1. C. B. Collins, S. Olariu, M. Petrascu, and I. Popescu, Phys. Rev. Lett. $\underline{42}$, 1397 (1979).
2. C. B. Collins, S. Olariu, M. Petrascu, and I. Popescu, Phys. Rev. C $\underline{20}$, 1942 (1979).
3. S. Olariu, I. Popescu, and C. B. Collins, Phys. Rev. C $\underline{23}$, 50 (1981).
4. S. Olariu, I. Popescu, and C. B. Collins, Phys. Rev. C $\underline{23}$, 1007 (1981).
5. C. B. Collins in *Proceedings of the International Conference on Lasers '80*, edited by C. B. Collins (STS Press, McLean, VA, 1981) p. 524.
6. C. B. Collins in *Laser Techniques for Extreme Ultraviolet Spectroscopy*, edited by T. J. McIlrath and R. R. Freeman (AIP Conference Proceedings No. 90, New York, 1982) p. 454.
7. C. B. Collins in *Proceedings of the International Conference on Lasers '81*, edited by C. B. Collins (STS Press, McLean, VA, 1982) p. 291.
8. C. B. Collins, F. W. Lee, D. M. Shemwell, B. D. DePaola, S. Olariu, and I. I. Popescu, J. Appl. Phys. $\underline{53}$, 4645 (1982).
9. B. D. DePaola and C. B. Collins, J. Opt. Soc. Am. B $\underline{1}$, 812 (1984).
10. C. B. Collins and B. D. DePaola in *Laser Techniques in the Extreme Ultraviolet*, edited by S. E. Harris and T. B. Lucatorto (AIP Conference Proceedings No. 119 New York, 1984) p. 45.
11. C. B. Collins and B. D. DePaola, Optics Lett. $\underline{10}$, 25 (1985).
12. B. D. DePaola, S. S. Wagal, and C. B. Collins, J. Opt. Soc. Am. B $\underline{2}$, 541 (1985).
13. C. B. Collins in *Advances in Laser Science-1*, edited by W. C. Stwalley and M. Lapp (AIP Conference Proceedings No. 146, Dallas, 1985) p. 18.
14. F. Davanloo, T. S. Bowen, and C. B. Collins in *Advances in Laser Science-1*, edited by W. C. Stwalley and M. Lapp (AIP Conference Proceedings No. 146, Dallas, 1985) p. 60.
15. S. S. Wagal and C. B. Collins in *Advances in Laser Science-1*, edited by W. C. Stwalley and M. Lapp (AIP Conference Proceedings No. 146, Dallas, 1985) p. 62.
16. C. B. Collins in *Advances in Laser Science-1*, edited by W. C. Stwalley and M. Lapp (AIP Conference Proceedings No. 146, Dallas, 1985) p. 40.
17. G. C. Baldwin, J. C. Solem, and V. I. Goldanskii, Rev. Mod. Phys. $\underline{53}$, 687 (1981).
18. C. Cohen-Tannoudji and S. Haroche, J. de Physique, $\underline{30}$, 125 (1969).
19. S. Haroche, Ann. Phys. $\underline{6}$, 189 (1971).
20. W. J. Veigele, Atomic Data Tables, $\underline{5}$, 51 (1973).
21. C. B. Collins, F. Davanloo, and T. S. Bowen, Rev. Sci. Instrum. $\underline{57}$, 863 (1986).

THE MULTI-BEAM BORRMANN EFFECT IN CRYSTALLINE γ-RAY LASERS

J. T. Hutton, G. T. Trammell, and J. P. Hannon
Rice University, Houston, Texas 77251

ABSTRACT

The advantages of utilizing the 2-beam Borrmann modes of a crystalline, rather than an amorphous, sample for the lasing modes of a γ-ray laser (reduced lasing threshold and higher gains above threshold, due to reduced photoabsorption and improved coupling between the lasing mode and the nuclei) are well known.[1,2] We discuss here the further improvements which are obtained in multi-beam (3 or more) Borrmann modes. These modes would be fed automatically by emitters of multipole M1 or higher located within the crystal. As an example, the 6-beam mode which couples well to M1 emitters would have a lower threshold than a similar 2-beam mode. More importantly, the slope of the net gain vs. population inversion density curve would be three times greater in the 6-beam mode, allowing practical levels of gain to be reached at much lower inversion densities.

LASING IN A BORRMANN MODE

In a previous paper,[1] we have shown that the critical population inversion requirement at threshold for a Mössbauer crystal, lasing in a 2-beam Borrmann mode, is substantially smaller than that required to induce lasing in an amorphous sample of the same material (or equivalently, off-Bragg in the crystal), so long as the emitters have multipolarity higher than E1.

The 2-beam Borrmann modes,[3] which are composed of the coherent superposition of two waves, $\vec{k}_0$ and $\vec{k}_1 = \vec{k}_0 + \vec{\tau}$, which differ by a reciprocal lattice vector $\vec{\tau}$, have nodes in the E field at the atomic site, but antinodes in the B field and electric field gradient. Because each atom experiences a greatly reduced electric field, the photoabsorption rate is strongly suppressed for these modes. Similarly, for E1 nuclear transitions, there is no coupling to Borrmann modes, for either absorption or emission. However, there is strong coupling to these modes for nuclear transitions of multipolarity M1 or higher. This suppression of the electronic coupling along with enhancement of the coupling to the nuclei is referred to as the "anomalous emission effect" in the case where waves emitted by M1 and higher nuclear transitions emerge from single crystals.[1,4]

Lasing requires that the gain from stimulated emission exceed the losses from nonresonant absorption, and as we have previously shown for 2-beam Borrmann modes,[1] the anomalous emission effect should decrease the required inversion population density for lasing by one or two orders of magnitude.

© American Institute of Physics 1987

Multi-beam Borrmann modes, which consist of the coherent superposition of 3 or more plane waves in the crystal, have properties which are even more advantageous for grasing. These modes can have smaller absorption coefficients, and, more importantly, larger coupling to the nuclear emitters, as well as sharper collimation, and hence require even less population inversion density for lasing, than do corresponding 2-beam modes.

While it is likely that any graser will decay via superradiant, rather than steady-state stimulated, emission,[1,5,6] the steady-state equations are much simpler and serve quite well to demonstrate the advantages of multi-beam modes. Further, the gain-length product which occurs in the steady-state theory is exactly that which is important for superradiance, so that the same advantages to multi-beam modes will be manifest for superradiant decay.

The Schawlow-Townes[7] steady-state lasing condition is $K \geq 0$, where K, the net gain per unit length coefficient, is given by

$$K = K_o - \mu = \sigma_N g_N \Delta n - \sigma_A g_A n_o. \tag{1}$$

In this equation, σ_A is the nonresonant atomic absorption cross section, which for the Mössbauer energies will be almost entirely due to photoelectric absorption, while σ_N is the nuclear absorption-stimulated emission cross section at resonance.[8] Δn is the population inversion density, including spin degeneracy factors. g_N and g_A are the coupling factors of the nuclear resonators and the atomic electrons to the electromagnetic field of the lasing mode, respectively. For the usual plane wave modes (off Bragg or in an amorphous sample) $g_N = g_A = 1$, but this will not be the case for multi-beam modes, which consist of the coherent superposition of two or more plane waves, produced by Bragg reflections in perfect crystals.[2]

Setting $K = 0$ yields the threshold population inversion density, Δn_t, given by

$$\Delta n_t = \frac{\sigma_A}{\sigma_N} \frac{g_A}{g_N} n_o. \tag{2}$$

This threshold may be reduced either by decreasing photoabsorption (reducing g_A) or by improving the coupling to the nuclei (increasing g_N). A practical graser, however, will have to operate with a gain well above threshold, at some population inversion density $\Delta n_p \gg \Delta n_t$, as is shown in Fig. 1a. Reducing g_A will lower the threshold, but does not change the slope of the line K vs. Δn, so that Δn_p is reduced by an amount equal to the reduction in Δn_t, as is shown in Fig. 1b. Increasing g_N, however, both lowers the threshold and increases the slope, as is shown in Fig. 1c, so that there is a correspondingly large reduction of Δn_p. Thus, increases in g_N are particularly significant, since this increased slope will allow the required net gain per unit length to be obtained at a much lower population inversion density.

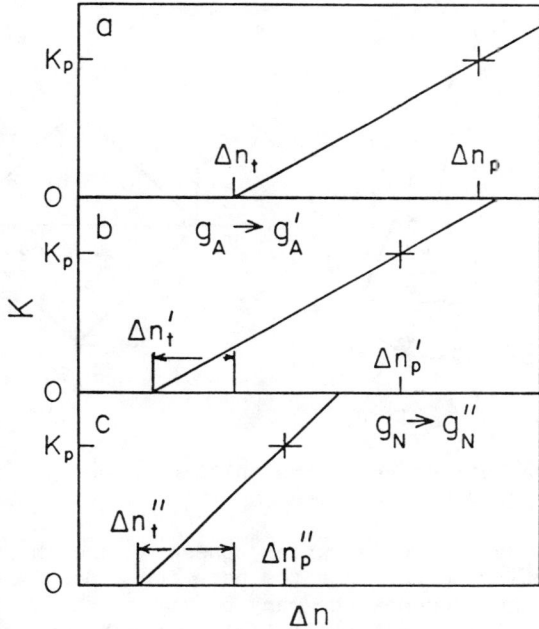

Fig. 1. Diagram showing the effects of decreasing g_A or increasing g_N on the gain of a steady-state lasing system.

MULTI-BEAM BORRMANN MODES

Our recent results show that a substantial increase in the nuclear coupling coefficient, g_N, is obtained in multi-beam Borrmann modes, in which several plane waves simultaneously satisfy the Bragg condition for reflection into one another. Such modes may be formed whenever the reciprocal lattice points involved are coplanar and lie on a circle. The m beams in the mode then lie on the surface of a cone whose base is the polygon of reciprocal lattice vectors so defined, as is shown in Fig. 2. Since there are 2 linearly independent polarization vectors associated with each beam in the mode, there are 2m eigenmodes which correspond to each such geometry of m beams.

The number of available modes depends upon the length of $\lambdabar^{-1}$, which determines the size of the Ewald sphere. Which modes are open for a given choice of mode axis may be determined by taking a cut through the reciprocal lattice of the crystal perpendicular to that axis, and then drawing all possible circles of radius less than $|\vec{k}| = 1/\lambdabar$ on the resulting 2-dimensional plane of reciprocal lattice

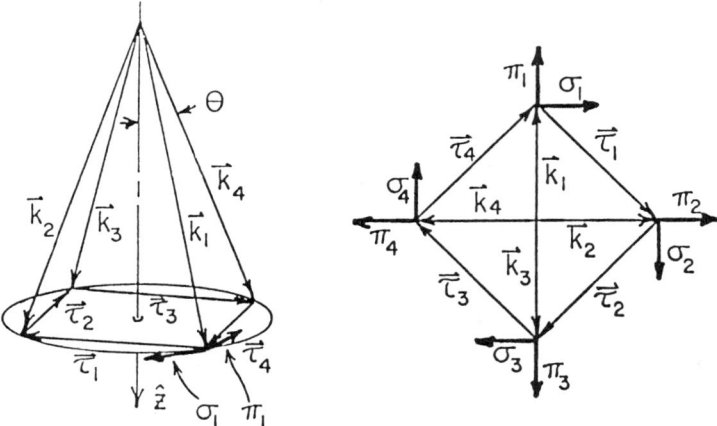

Fig. 2. General structure of the multi-beam modes.

points. Any such circle which passes through more than one reciprocal lattice point forms the base for a possible set of modes.

Even though there can be many beams in some of these modes (12 or more), they will be automatically excited by an internal source. An excited atom in the crystal will emit spherically expanding waves, so that there is always radiation emitted into whatever is the proper set of directions to make up the mode. There are no alignment problems, and the nuclei couple to the modes automatically.

Each multi-beam geometry (i.e. for a given value of m) has several Borrmann modes, and it is possible to find Borrmann modes which couple well to any arbitrary multipole emitter, except E1. (Since Borrmann modes have a node in the electric field at the equilibrium site of the nucleus, they will not couple to E1 Mössbauer transitions.) Of particular interest are modes which couple well to M1 and E2 emitters (a majority of Mössbauer transitions are either M1 or E2). Some Borrmann modes which couple well to M1 nuclear transitions, as well as modes which couple well to E2 transitions, are shown in Fig. 3.

For the M1 multi-beam modes,

$$g_A = \frac{m}{2} \frac{\sin^2\theta}{\cos\theta} \frac{\langle x^2 \rangle}{\lambdabar^2}, \qquad (3)$$

and

$$g_{N,M1} = m \frac{\sin^2\theta}{\cos\theta}, \qquad (4)$$

where m is the number of beams in the mode, θ is the half-apex angle of the cone formed by the wavevectors, and $\langle x^2 \rangle$ is the mean thermal plus zero point displacement of the atom from equilibrium. M1 transitions thus favor modes for lasing which occur at large cone

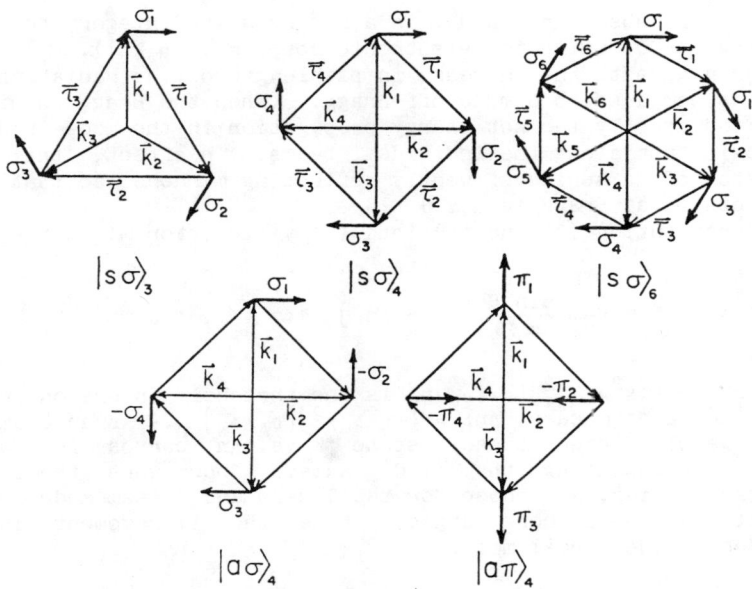

Fig 3. Borrmann modes, $|s\sigma\rangle_m$, which couple well to M1 Mössbauer transitions and 4-beam Borrmann modes, $|a\sigma\rangle_4$ and $|a\pi\rangle_4$, which couple well to E2 Mössbauer transitions.

angles and which have many beams. For the 4-beam E2 modes shown above,

$$g_A = 2 \frac{\sin^2\theta}{\cos\theta} \frac{\langle x^2 \rangle}{\lambdabar^2}, \qquad |a\sigma\rangle_4 \text{ mode} \qquad (5\text{i})$$

$$g_A = 2 \sin^2\theta \cos\theta \frac{\langle x^2 \rangle}{\lambdabar^2}, \qquad |a\pi\rangle_4 \text{ mode} \qquad (5\text{ii})$$

and

$$g_{N,E2} = 4 \frac{\sin^2\theta}{\cos\theta}, \qquad |a\sigma\rangle_4 \text{ mode} \qquad (6\text{i})$$

and

$$g_{N,E2} = 4 \sin^2\theta \cos\theta. \qquad |a\pi\rangle_4 \text{ mode} \qquad (6\text{ii})$$

In general, the coupling factors always have the form

$$g = m \frac{F(\theta)}{\cos\theta} \left(\frac{\langle x^2 \rangle}{\lambdabar^2}\right)^\eta, \qquad (7)$$

where m is the number of beams, $F(\theta)$ is a function which describes the coupling of the mode to the multipole radiation pattern of the emitter or absorber, and η is a non-negative integer (for nuclear coupling η = 0, while for electronic coupling, η ≥ 1). 1/cosθ is the gain due to the increase in pathlength of the radiation in the mode, as compared to a mode off-Bragg. (When the Bragg condition is satisfied exactly the net energy propagation in the mode is down the cone axis, not along the individual beams. In effect, the crystal consists of a series of weakly reflecting mirrors, so that it acts as an etalon at Bragg angles.)

Substituting (3) and (4) into (1) yields, for M1 emitters,

$$K = \sigma_N m \frac{\sin^2\theta}{\cos\theta} [\Delta n - \Delta n_t] \qquad (8)$$

where $\Delta n_t = [<x^2>/2\bar{\lambda}^2][\sigma_A/\sigma_N]n_o$ is the threshold inversion density. (For the 2 beam case, $\Delta n_t = [<x^2>/\bar{\lambda}^2][\sigma_A/\sigma_N]n_o$.) As an example, in Fig. 4 we plot three steady state K vs. Δn curves for an ^{57}Fe needle, grown along the (100) axis. Shown are the lines for off-Bragg lasing, and those for the 2 beam and 8 beam modes at the largest available cone angle. Note the improvement in slope obtained in the 8 beam mode.

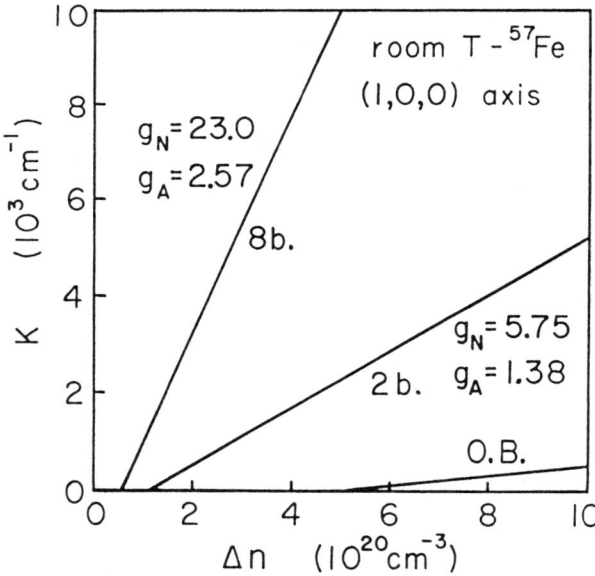

Fig. 4. Plots of K vs. Δn for an ^{57}Fe needle at room temperature, grown along the (100) axis. Lines are shown for off-Bragg, as well as for the M1 symmetry 2 and 8 beam modes with the largest available cone angle (here θ=71.7°).

ANGULAR DIVERGENCE PROPERTIES

These multi-beam Borrmann modes have interesting angular divergence properties as well. The angular divergence of the wave field emerging from an amorphous needle, or a crystalline needle lasing in an off-Bragg mode is determined by the "macro-geometry" of the sample; either by the physical collimation introduced by the size of the needle or by the diffraction limit, which depends on the needle diameter, whichever is greater, as is shown in Fig. 5.

Much better collimation may be obtained by utilizing the natural collimation which takes place during Bragg diffraction. In a 2 beam Borrmann mode, Bragg diffraction limits the angular divergence in the rocking curve direction to $\Delta\theta \approx W_B$, the Borrmann width, typically 10^{-6} to 10^{-5} rad., regardless of the thickness of the crystal in that direction, (so long as the crystal is sufficiently thick to form the Borrmann modes, i.e. a few primary extinction lengths). Thus, in this case the angular divergence is determined by the "micro-geometry" of the sample, rather than the "macro-geometry."

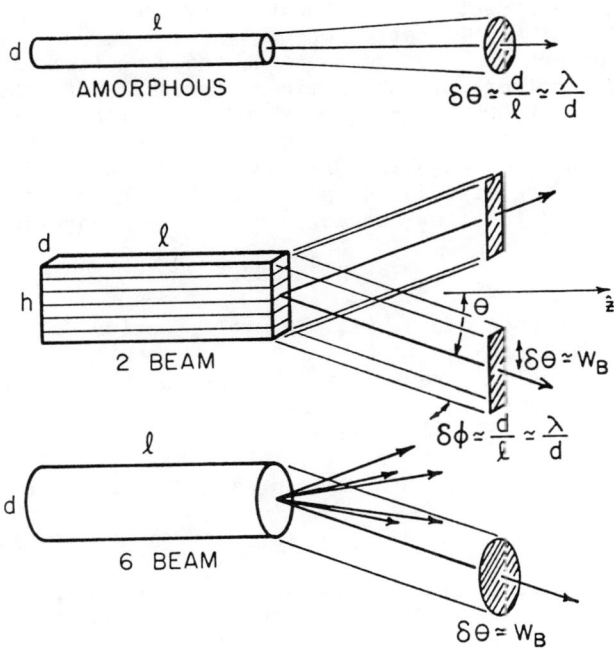

Fig. 5. Angular divergence properties of beams emerging from amorphous and crystalline γ-ray lasers.

In 2-beam modes, the angular divergence in the other direction is still determined by the size of the crystal, either geometry or the diffraction limit, as in the amorphous case. In multi-beam Borrmann modes, however, the radiation emerges in well defined beams, which are collimated in both directions to $\Delta\theta \approx W_B$.

CONCLUSION

In conclusion, the advantages obtained by using a crystalline, rather than an amorphous, sample geometry may be summarized as follows:

i) g_N, the nuclear coupling coefficient to the Borrmann mode, is $\propto m$, where m = number of beams in the mode (for example, for modes which couple well to M1 transitions, $g_N = m\sin^2\theta/\cos\theta$). Thus, for instance, the coupling of the resonant nuclei to a 6-beam mode will be three times larger than the coupling to a 2-beam mode at the same angle.

ii) g_A, the atomic absorption coupling coefficient, $\propto (\langle x^2 \rangle/\bar{\lambda}^2)^\eta$ in a Borrmann mode, where $\langle x^2 \rangle/\bar{\lambda}^2 \ll 1$, and $\eta = 1, 2$.

iii) The crystal acts as an etalon at the Bragg angles, confining the radiation in a manner similar to single mirrors, increasing the effective pathlength in the lasing medium, thus giving a $1/\cos\theta$ amplification of the gain.

iv) The angular divergence $\delta\theta$ of the wave field in the Borrmann mode is determined by the "micro-geometry" of the crystalline planes, not by the "macro-geometry" of the lasing system. The beams emerging in a multi-beam Borrmann mode are collimated by Bragg reflection to an angular divergence of the order of 10^{-5} rad, regardless of the size and shape of the lasing sample (so long as it is large enough to allow formation of the Borrmann modes).

REFERENCES

1. J. P. Hannon and G. T. Trammell, Optics Communications 15, 330 (1975).
2. For an excellent review of all aspects of γ-ray laser research, see G. C. Baldwin, J. C. Solem, and V. I. Gol'danski, Rev. of Mod. Phys. 53, 687 (1981).
3. G. Borrmann, Z. Phys. 42, 157 (1941); 127, 297 (1950).
4. J. P. Hannon, N. J. Carron, and G. T. Trammell, Phys. Rev. B9, 2810 (1974).
5. G. C. Baldwin and M. S. Feld, J. Appl. Phys. 59, 3665 (1986).
6. G. C. Baldwin, M. S. Feld, J. P. Hannon, J. T. Hutton, and G. T. Trammell, Proc. of the Int. Conf. on X-Ray Lasers, Aussois, France, April 1986.
7. A. L. Schawlow and C. H. Townes, Phys. Rev. 112, 1940 (1958).
8. See for example Eq. A4 in Ref. 4.

PROGRESS IN THE GAMMA RAY LASER PROGRAM AT TEXAS 1: FLASH X-RAY TECHNIQUES FOR PUMPING NUCLEAR MATERIALS[*]

F. Davanloo, T.S. Bowen, and C.B. Collins
Center for Quantum Electronics, University of Texas at Dallas
P.O. Box 830688, Richardson, Tx 75083-0688

ABSTRACT

This paper describes the progress in construction and scaling of a flash x-ray device producing intense nanosecond pulses to excite nuclear fluorescence for the evaluation of candidate materials for a gamma ray laser.

INTRODUCTION

Our group has modeled a coherent pumping scheme for a gamma ray laser.[1] The approach requiring flash x-ray techniques was also emphasized in our recent communications.[2,3] As mentioned there, the Modulated Nuclear Radiation (MNR) methodology can be used to produce a data base for nuclear kinetics of suitable candidate materials. Access to an x-ray source with nanoseconds pulse duration and a total of one Joule per keV of linewidth in a reasonable working period is essential to the success of this technique. Here we present a brief review on progress made in our laboratory to scale and characterize a repetitively pulsed flash x-ray device capable of emitting high photon fluxes.

DEVICE DESIGN

As described and shown in our previous report[4], the design of the flash x-ray device centers around three important subassemblies: (1) a low inductance x-ray tube, (2) a Blumlein power source, and (3) a commutation system capable of operation at high repetition rates. Elements (2) and (3) differed little from drivers developed in our laboratory.[5] The design of the x-ray tube has evolved substantially from that used in the first models reported earlier.[4] The first models of the x-ray tube were considered to be expendable, and the cost of construction was low enough to justify that conception. However, a recent refinement of objectives has created a potential need for the use of exotic metals, and it became necessary to develop the capability shown in Fig. 1 for the interchangeability of anodes.

OPERATION AND PERFORMANCE

Precise measurements of time-resolved voltages and currents were rendered difficult by the extremely low impedence of the Blumlein and by the commutation of the thyratron mounted in a grounded-grid configuration, on the time scales of 10-20 nsecs. With resonant pulse charging of the Blumlein it was possible to

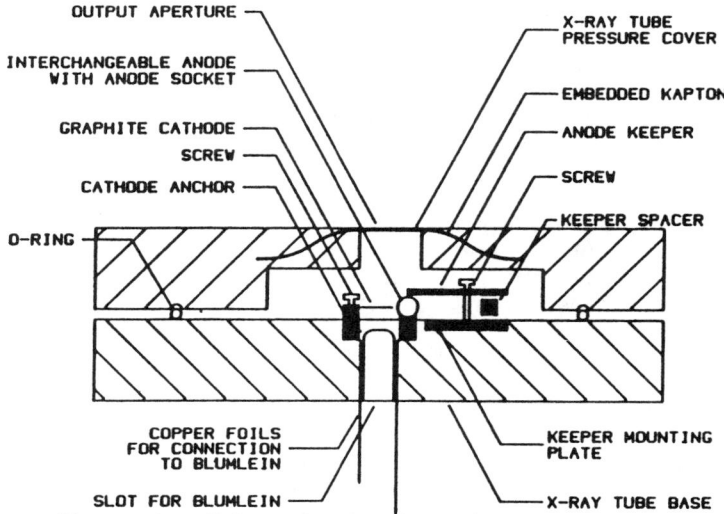

Figure 1. Schematic drawing of a cross section of an x-ray diode used in this work.

shunt the x-ray tube with a voltage divider constructed from a tapped water resistor of sufficiently low impedance, around 200 ohms, so that meaningful measurements of voltage as a function of time could be made. X-ray outputs were detected with a block of fast scintillator plastic equivalent to NE114 with a nominal 7.0 nsec decay time. The resulting light output was measured with a faster photomultiplier with 1.5 nsec resolution and recorded with a Tektronix 7912AD transient digitizer. Calibration was obtained by comparing the time-

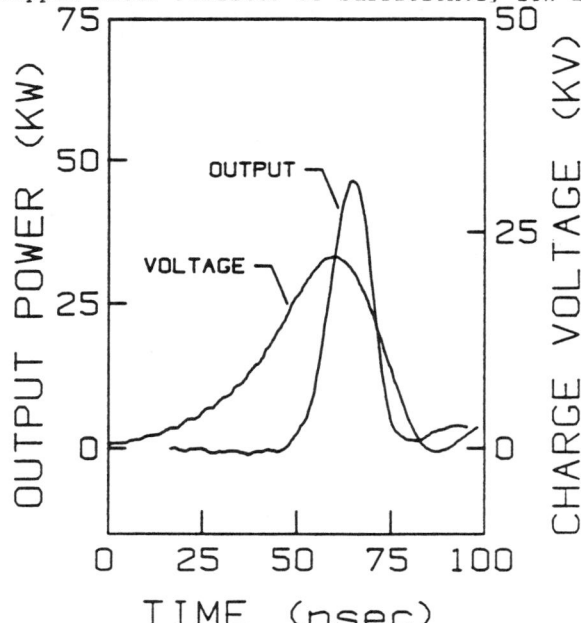

Figure 2. Typical relationship for the voltages and output intensities measured at the x-ray diode.

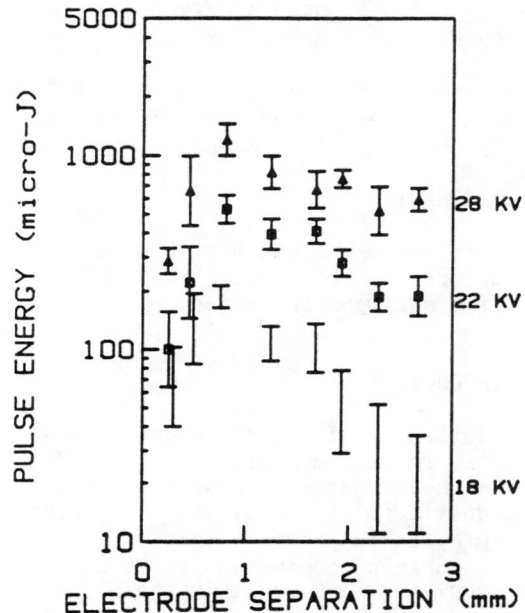

Figure 3. Total x-ray pulse energy emitted under single shot conditions as a function of the separation between anode and cathode at different applied voltages.

integrated fluorescence from the plastic detector when illuminated with geometrically attenuated x-rays from the flash x-ray source with the level of excitation produced by a radioactive source of known characteristics. Relative times at which switching and x-ray emission were found to occur are shown in Fig. 2.

Different scale devices were constructed and performances were characterized. Output energies up to 3.1 mJ per pulse were accessible with the largest of these devices.[6]
An optimum value of electrode separation was found for each of these devices characterized. Fig. 3 shows the typical variation in x-ray pulse energy observed as the separation between electrodes and the applied voltage were varied. The trend displayed in Fig. 3 at smaller values of electrode spacing was entirely reproduceable and indicates a decreasing pulse width resulting from the more narrowed separations.

REFERENCES

*Supported by the Strategic Defense Initiative Office (IST) through NRL and in part by ONR.

1. C.B. Collins, F.W. Lee, D.M. Shemwell, B.D. DePaola, S. Olariu and I.I. Popescu, J. Appl. Phys. $\underline{53}$, 4645 (1982).
2. C.B. Collins, Advances in Laser Science-I, edited by W.C. Stwalley and M. Lapp (AIP Conference Proceedings No.146, New York, 1986) pp.18-21.
3. Our communication TUA1 in the Preceeding Session.
4. F. Davanloo, T.S. Bowen, C.B. Collins, Advances in Laser Science-I, edited by W.C. Stwalley and M. Lapp (AIP Conference Proceedings No.146, New York, 1986) pp.60-61.
5. C.B. Collins, IEEE J. Quantum Elec., $\underline{QE-20}$, 47 (1984).
6. See our poster communication in this conference.

PROGRESS IN THE GAMMA-RAY LASER PROGRAM AT TEXAS 2: COHERENT TECHNIQUES FOR PUMPING A GAMMA-RAY LASER *

S. Wagal, P. Reittinger, E. Juengerman, C. Eberhard and C.B. Collins
Center for Quantum Electronics, University of Texas at Dallas
P.O. Box 830688, Richardson, TX 75083-0688

ABSTRACT

Experiments were performed to determine whether absorption sidebands to the hyperfine structure of ^{57}Fe induced by an Rf magnetic field are the result of a multiphoton process or magnetostriction.

INTRODUCTION

As early as 1968 the observation of Rf field induced sidebands to the hyperfine structure of ^{57}Fe in ferromagnetic thin foils was reported[1] (fig.1). In 1976 it was shown that a similar phenomenon could be produced in nonferromagnetic foils containing ^{57}Fe using ferromagnetic foil drivers[2] (fig. 2). It is possible that magnetostriction could give rise to this phenomenon. It is also possible that a dressing of the nuclear energy levels by the Rf photons could be taking place in the nucleus, in a manner analogous to the dressing of atomic energy levels by a magnetic field.[3] If a nuclear energy level can be dressed, then a gamma-ray LASER can be pumped with currently existing lasers. Three experiments were performed which were designed to determine the cause of these sidebands.

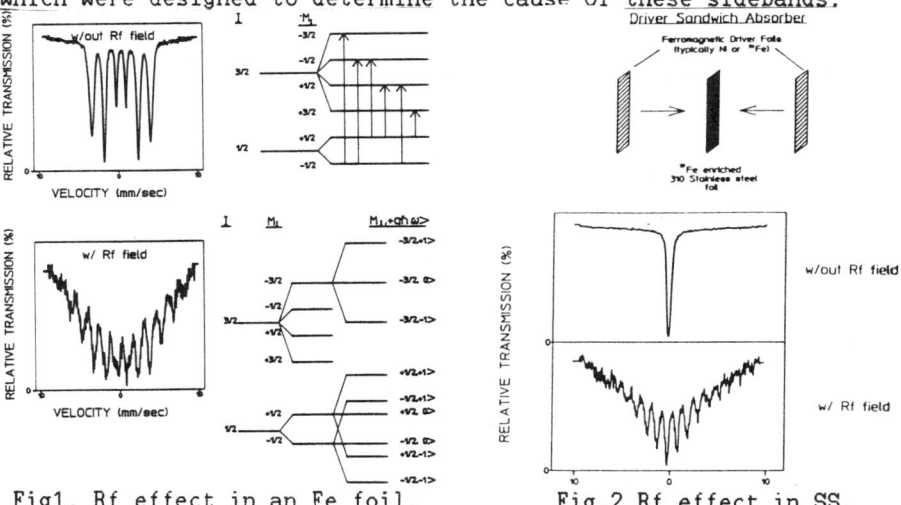

Fig1. Rf effect in an Fe foil.   Fig.2 Rf effect in SS.

PROCEDURE

First it was necessary to determine if sidebands could be produced in a non-magnetostrictive ferromagnetic foil. Two Fe-Ni

permalloy foils were
obtained from ORNL,
one 81% Ni and the other
82% Ni. It can be seen
that an 81% Ni permalloy
is non-magnetostrictive.
(fig.3) Both foils were
subjected to the same field
intensity and it can be
seen that sidebands appear
in the non-magnetostrictive
foil(fig.4). Also, it should
be noted that the relative
intensity of the sidebands is
less in the magnetostrictive
foil.

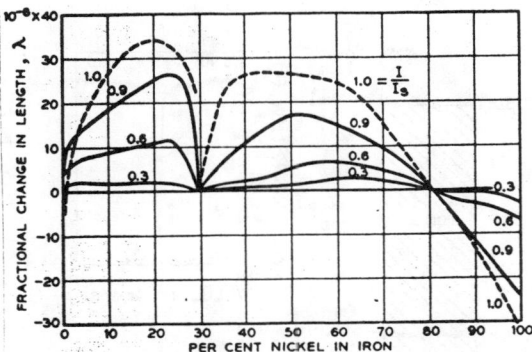

Fig.3 Magnetostriction of Fe-Ni alloy
(R.M. Bozorth, Ferromagnetism,
Van Nostrand Co., New York, 1951)

The second experiment was
intended to establish whether
or not the phenomenon could be
produced in a non-
ferromagnetic foil which was
acoustically but not
magnetically isolated from its
ferromagnetic drivers. The
^{57}Fe enriched stainless steel
absorber foil was separated
from the Ni driver foils by
100 μm of tissue, yet it was
still possible to produce the
sidebands (fig.5).

In the third experiment
the additivity of the effect
in a non-ferromagnetic absorber
foil was studied. Initially,
the linearity of the first order
sidebands with applied power was
used to establish a scale from

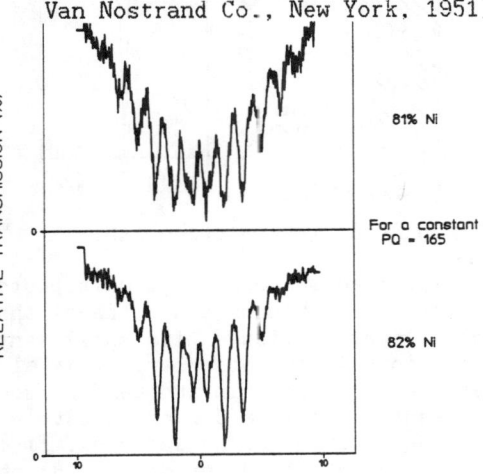

Fig.4 Permalloy Experiment.

which a baseline could be extrapolated.(fig.6a) Then for a constant
field intensity, the reduction in the intensity of the first order
sidebands was measured after one of the ferromagnetic drivers was
removed.(fig.6b) It should be noted that a static B-field has been
added which is orthogonal to both the Rf field and the gamma ray
propagation direction. Without this static field the first order
sideband intensity never decreased by more than a factor of two when
one of the drivers was removed. With the static B-field, however,
more than a factor of three decrease was observed.

CONCLUSION

For a ferromagnetic absorber foil, the Rf sideband phenomenon
appears to be independent of magnetostriction if not hampered by it.
For a non-ferromagnetic absorber foil the results obtained may best

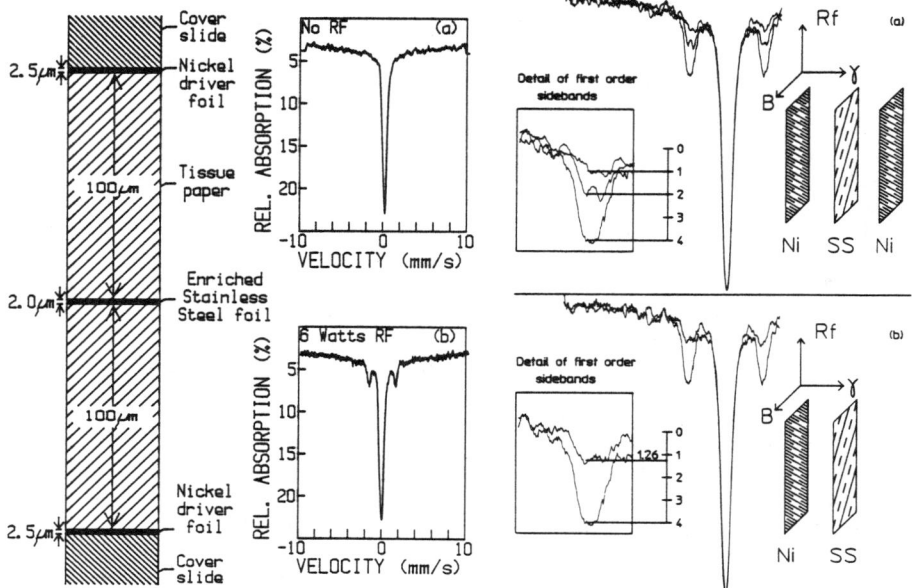

Fig.5 Acoustic Isolation Exp.   Fig.6 Addition of effect.

be explained in terms of a multiphoton effect. It is difficult to believe that phonons could travel through such a disproportionately thick layer of tissue with sufficient intensity to drive lattice vibrations in the absorber. It is also difficult to imagine that lattice vibrations in independent sources could be sufficiently coherent so as to add constructively when propagated into the absorber from opposing surfaces. If one believes that the formation of surface poles on one driver (Rf photons) triggers the development of surface poles on the other driver, then it is reasonable to believe that if the sidebands are the result of a multiphoton process then the effect should add coherently. If the sideband phenomenon is mechanical in nature then it is more difficult to justify coherent addition of the effect or the enhancement by a static B field. Yet, it is reasonable to believe that the static B field magnetically textures the driver foils so that the magnetization vectors of the domains in a single foil flip in the same hemisphere, thereby increasing the magnitude of the surface poles formed on a single driver.

REFERENCES

1. N.D. Heiman, et. al., Phys. Rev. Lett. <u>21</u>, 93 (1968).
2. C.L. Chien and J.C. Walker, Phys. Rev. B <u>13</u>, 1876 (1976).
3. C.B. Collins and B.D. DePaola, Opt. Lett. <u>10</u>, 25 (1985).

* Supported by SDIO (IST) through NRL and in part by ONR.

CLASSICAL AND SEMICLASSICAL CALCULATION OF NUCLEAR-ELECTRON COUPLING

D.W. Noid
Oak Ridge National Laboratory
Oak Ridge, Tennessee

F.X. Hartmann
Institute for Defense Analyses
Alexandria, Virginia

M.L. Koszykowski
Sandia National Laboratories
Livermore, California

ABSTRACT

Energy transfer processes between an excited nuclear proton and an inner core electron are studied semiclassically. The coupled independent particle models exhibit large spectral perturbations.

INTRODUCTION

Interactions between electronic transitions of an atom and nucleonic transitions of its nucleus are of interest in the excitation of nuclei from ground states to low-lying excited states or the excitation of short-lived states from long-lived isomeric states. Such interactions can proceed via nonradiative energy transfer between the excited electronic state and the initial nuclear state.

The excitation of low energy isomeric states in nuclei from their ground states has been observed by Gol' danskii and Namiot [1] in hot plasmas. They reported the laser plasma production of ^{235m}U (73 eV 7/2- to 1/2+) as a result of the capture of continuum electrons to the outer shell in ionized uranium atoms. Their explanation of this excitation differs from that of Izawa and Yamanaka[2] who initially attributed the excitation to a $6p_{3/2}$ (-32.5 eV) to $5d_{5/2}$ (-103.1 eV) atomic transition. Additional evidence for the excitation of nuclear ground states is given by the observation of de-excitation gamma-rays in ^{237}Np by Saito, Shinohara and Otozai[3] and the excitation of ^{189}Os by Otozai, Arakawa and Morita[4]. The possibility of nuclear excitation by laser driven coherent outer electron oscillations is described by Biedenharn, Baldwin, Boyer and Solem[5,6].

In general, nuclear-electron interactions occur in the processes of nuclear orbital electron capture or internal gamma-ray conversion. Ultra-low energy nuclear transitions, in particular, lead to excitations in the outer electron shells; e.g. the 73 eV electromagnetic decay in ^{235m}U is a highly converted transition which leads to ejection of a bound electron near the valence shell; the 2.6 keV weak decay in ^{163}Ho leads to atomic excitations in the M and higher shells (<2 keV regime)[7]. The first case illustrates the predominance of the multipole components of the near Coulombic field of the nucleus over the direct radiation process (which plays the same role in the inverse process); the second example suggests an alternate approach to achieving specific electronic excitations in atoms. The analogous muonic excitations of nuclei have been extensively reviewed by Borie and Rinker[8].

© American Institute of Physics 1987

We have recently devoted our attention to the study of excitation pathways available for the transfer of energy from highly excited long-lived isomeric nuclear states to much shorter-lived states in laser plasmas.[9] In many non-radiative energy transfer processes one treats the nuclear and electron quantum system as states of a separable Hamiltonian and the coupling term is treated as a perturbation. This is illustrated by the model presented by Morita.[10] In this paper we briefly highlight results of our non-perturbative approach used to study coupling between electronic and nucleonic transitions applied to a simple single-particle model for both electrons and nucleons.

SPECTRAL ANALYSIS METHOD

A quantization of systems not permitting separation of variables was first proposed by Einstein:[11] one finds canonical invariants, namely the action variables J_i, and quantizes them:

$$J_i = \oint \tilde{p}\, d\tilde{q} = (n_i + \delta_i) h$$

such that the different J_i's are obtained by integrating over topologically independent paths. Here $\tilde{q}$ ($q_1, q_2, ... q_N$) and $\tilde{p}$ ($p_1, p_2, ... p_N$) denote canonical coordinates and momenta and n_i is a quantum number. Keller[12] introduced the fractional term δ_i, usually 0 or 1/2. Eastes and Marcus[13], and Noid and Marcus[14] showed how to evaluate these action integrals in nonseparable systems having smoothly varying potential energy functions and evaluated eigenvalues semiclassically. Other methods have been developed for systems having degeneracies. These latter results are for cases which are quasiperiodic (systems having action-angle variables).

The differences in eigenvalue frequenices for complicated systems can be obtained by using the coupled trajectory to compute the appropriate autocorrelation function. This is then Fourier-transformed to obtain the spectrum. According to Bohr's correspondence principle the resultant mechanical frequenices are equal to the differences of eigenvalues $(E_i - E_j)/h$. A more complete discussion of the semiclassical method and many numerical tests of these theoretical techniques are discussed in ref. (15).

MODEL HAMILTONIAN

We consider a mathematical model characteristic of nuclear and electronic single particle dynamics. Nuclear transitions are described by single particle transitions of an odd proton in a Woods-Saxon potential well[16]:

$$V(r_n) = V_o [1 + \exp(r_n - R_o)/a_n]^{-1}$$

where r_n is the nucleon position, a_n the well diffusivity and R_o is the nuclear radius. We use $a_n = 0.65$ fm and $R_o = 1.25\, A^{1/3}$ where A is the atomic mass number. Such a picture might best approach reality for a doubly-magic nucleus (such as ^{209}Bi = ^{208}Pb + ^{1}p) with the neglect of spin-orbit effects. Mathematically we can, of course, scale any parameter. The most convenient parameters to scale are the atomic mass A and the proton number Z. The well depth V_o is chosen such that the density of states $\rho = (4\pi/3)(p_n/2\pi h)^3$ [given by a square well approximation to the nuclear volume] leads to a binding energy of the odd nucleon comparable to the experimental value, when filled by the other nucleons. The single particle electronic transitions are modelled by an electron which moves in orbits of a screened nuclear core of charge $\eta(Z - k)e$ and the screened odd proton has charge $\eta k e$. The parameter $k = 1$ models the coupled system; $k = 0$ models the uncoupled system. Here η is a screening parameter which scales the hydrogenic orbits. The complete Hamiltonian, restricted for convenience to a planar geometry is written as:

$$H = \frac{\tilde{p}_n^2}{2M_n} + V_o \{ 1 + \exp[(r_n - R_o)/a_n] \}^{-1} + \frac{\tilde{p}_e^2}{2M_e} - \frac{e^2\eta(Z-k)}{r_e} - \frac{\eta k e^2}{|r_e - r_n|}$$

This Hamiltonian is separable thus all but the last coupling term is easily quantizable. Hamilton's equations of motion are used to numerically generate the classical trajectories for both the coupled and uncoupled system. Initial conditions for the nucleon trajectory are found using a WKB approximation to the Woods-Saxon well to fix the position-momentum at the classical outer turning point. Initial conditions for the Coulombic well are given in ref. (17).

RESULTS

Sample electron and nucleon trajectories are depicted in the left parts of figs. (1) and (2) for both uncoupled and coupled cases. The distance scales are on the order of a hundred fermis. The nucleon trajectories shown there comprise approximately 5000 points; each point is integrated over two "time-unit" time steps. The natural "time-unit" which we use is defined as the time for light to travel one fermi in a vacuum. Trajectories up to a million points are used for the actual computation of autocorrelation functions. In the same figures, the electron orbits are depicted for 10,000 points integrated in 25 time-unit steps. Electron trajectories have drastically visible differences for electron trajectories involving the K shell, the differences in the nucleon trajectories for coupled and uncoupled cases are more subtle. Enlargements of the nuclear centers are shown on the right of figs. (1) and (2) with increases to 10,000 points for both coupled and uncoupled trajectories. The precession rates of the orbits in both cases are not identical.

The time-dependent dipole moment for the total system is used to study the dynamics of the system. Initally, the largest component of the total dipole moment is of lower frequency associated with the electronic motion. Superimposed on the electron contribution is the nucleon contribution which is initially higher in frequency and smaller in magnitude. When the coupling is turned-on, the system dipole moment deviates from its regular evolution in intensity and frequency. Typical Fourier-transforms of such dipole moment autocorrelation functions are depicted in figs.(3) for Z=51 and (4) for Z=83. For Z=29, the coupled spectrum looks identical to the uncoupled spectrum. By scaling Z and A we can examine the dynamics of coupling by examining the eigenvalue spectra. In figs. (3) and (4), we look at Z and A corresponding to near magic nuclei for this mathematical model. With increasing A, the spectra change and can be characterized by a transition from quasiperiodic to chaotic behavior [see ref. (15)].

DISCUSSION

The frequencies in the spectra are understood as follows. In the uncoupled system we see spectral intensities at frequencies characteristic of the nuclear motions and spectral intensities characteristic of the electron motions. In the coupled system we expect to see additional spectral intensity in electron-nucleon sum and difference bands. These spectral lines must be assigned to specific coupled single-particle electron-nucleon transitions to examine a particular transition of interest. All possible transitions are computed for the given nuclear parameters and not all (such as energetically high lying hole states) are of practical interest. The initial starting conditions serve to fix the total energy and total orbital angular momentum of the system.

The relationships of the slight nucleon orbit precession differences and the chaotic electron behavior at higher A has a physical (classical) interpretation. When the electron is traveling fastest, it is nearest the nucleus. In this region the rotating nuclear moment can be in any angular position relative to the electron's near pass. At this distance, the effects of the energy transfer on the spatial alteration of the trajectory are subtle. It is not until the electron travels furthest from the nucleus, that the radial energy dependence of the potential reveals the significant deviations in the electron's orbital trajectory. Since the nucleon-electron energy transfer occurs mostly for the brief time the electron is nearest the nucleus and the nuclear moment is essentially randomly oriented, the electron achieves a chaotic orbit. Ultimately it is the spectral features of the Hamiltonian which correspond to the true observables in the spectral analysis technique, and clearly any classical interpretation of the intermediate steps is only to give physical insight.

Theoretically, in the limit of extreme coupling, the electromagnetic emission associated with the spectrum of such a system is strictly neither a gamma ray nor an x-ray but a coupled emission from a single wavefunction with electron and nucleon components.

CONCLUSION

We have calculated absorption frequenices of coupled and uncoupled electron-nucleon motion in a simple single-particle model. The dynamics of the coupled and uncoupled systems are qualitatively different, and the difference is more pronounced with increasing A and Z. Transition probablities, frequencies and intensities are significant signs of differences between the coupled models and the uncoupled models. We have found some instances of strong coupling and instances of chaotic motion in this simple model. The spectral analysis method is a powerful technique for studying non-linear non-separable systems which model physical systems.

This model is being expanded to treat the deformed rotor and single-particle rotor models to examine behavior of nuclear-electron coupling in more complicated systems. These results should prove useful to the study of upconverted nucleon transitions from isomeric levels. Such transitions, expected to be driven from highly excited electronic states produced in laser plasmas, could lead to enhanced decay rates from nuclear isomers. Experiments are now under consideration which may demonstrate this effect.

The authors acknowlege helpful comments from L. C. Biedenharn. This research was sponsored by the Division of Materials Sciences, Office of Basic Energy Sciences, U. S. Dept. of Energy under contract DE-AC05-84OR21400 with Martin Marietta Energy Systems, Inc. and partially by the Department of Defense (SDIO/IST) and DARPA. The work is approved for public release by the Off. Asst. Sec. Defense (OASD/Public Affairs) Distribution Unlimited through DARPA. One of us (FXH) gratefully acknowledges partial support from the NSF and the particular assistance of the Sandia and Oak Ridge collaborators.

REFERENCES

1. V. I. Goldanskii and V. A. Namiot, Sov. J. Nucl. Phys. 33, 169 (1981).
2. Y. Izawa and C. Yamanaka, Phys. Lett. 88B, 59 (1979).
3. T. Saito, A. Shinohara and K. Otozai, Phys. Lett. 92B, 293 (1980).
4. K. Otozai, R. Arakawa and M. Morita, Prog. Theor. Phys. 50, 1771 (1973).
5. L. C. Biedenharn, K. Boyer and J. C. Solem, Advances in Laser Science I, Eds. W. C. Stwalley and M. Lapp, AIP Conf. Proc. 146, 50 (1986).
6. L. C. Biedenharn, G. A. Baldwin and K. Boyer, Advances in Laser Science I, Eds. W. C. Stwalley and M. Lapp, AIP Conf. Proc. 146, 52 (1986).
7 F. X. Hartmann and R. A. Naumann, Phys. Rev. C 31,1594 (1985).
8. E. Borie and G. A. Rinker, Rev. Mod.Phys. 54, 67 (1982) and refs. therein.
9. F. X. Hartmann, S.Rotman, M. Smith, D.Noid IDA Paper P-1970/M-291(1987).
10. M. Morita, Prog. of Theor. Phys. 49, 1574 (1973).
11. A. Einstein, Verh. Dtsch. Phys. Ges. 19, 82 (1917).
12. J. B. Keller, Ann. Phys. (N. Y.) 4,180 (1984).
13. W. Eastes and R. A. Marcus, J. Chem. Phys. 61,4301 (1974).
14. D. W. Noid and R. A. Marcus, J. Chem. Phys. 62, 2119 (1975), 66,559 (1977), 85, 3305 (1986).
15. D.W. Noid, M. L. Koszykowski, and R. A. Marcus, J. Chem. Phys. 67, 7 (1977) and Ann. Rev. Phys. Chem. 32, 267 (1981).
16. A. Bohr and B. R. Mottelson, Nuclear Structure.(Benjamin, Reading,1969) Vol. 1.
17. R. Langer, Phys. Rev. 51, 669 (1937).

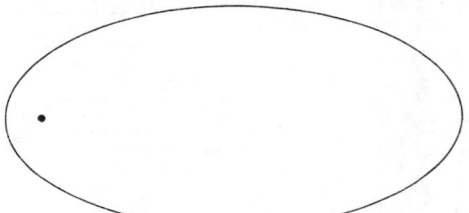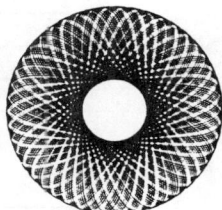

Fig. 1. Uncoupled electron (max r=1188 fm) and nucleon (r=8.66 fm) trajectories.

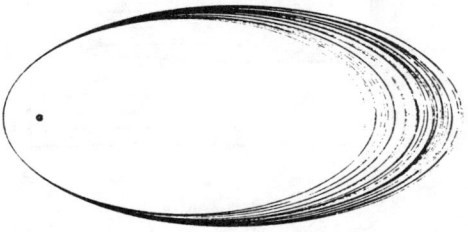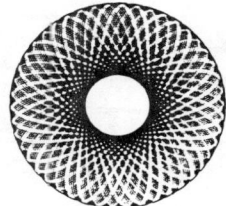

Fig. 2. Coupled electron and nucleon trajectories for comparison to fig. 1.

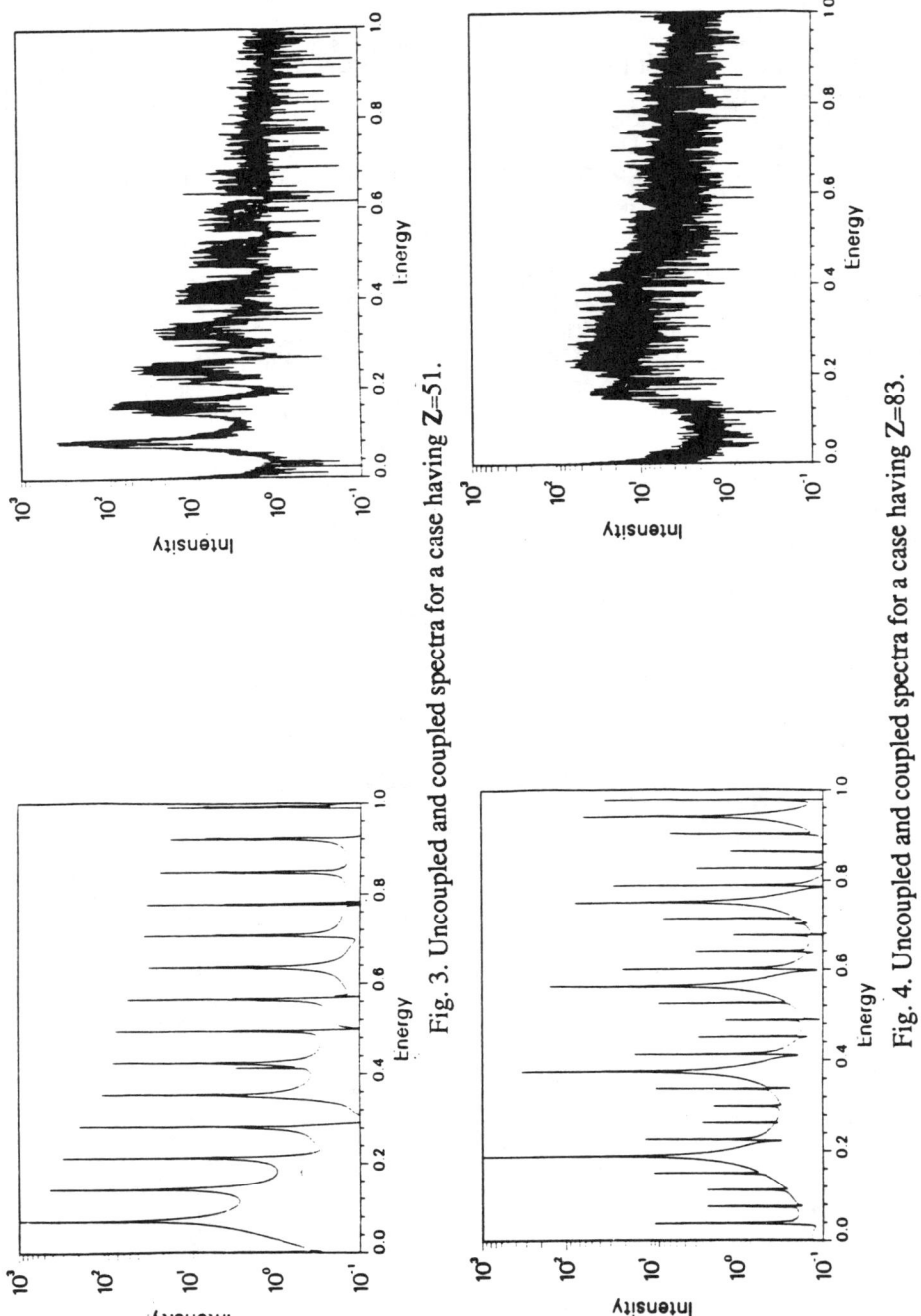

Fig. 3. Uncoupled and coupled spectra for a case having Z=51.

Fig. 4. Uncoupled and coupled spectra for a case having Z=83.

NUCLEAR INTERLEVEL TRANSFER DRIVEN BY COLLECTIVE OUTER SHELL ELECTRON OSCILLATIONS*

G. A. Rinker and J. C. Solem
Theoretical Division, Los Alamos National Laboratory
Los Alamos, New Mexico 87545

and

L. C. Biedenharn
Duke University
Durham, NC 27706

ABSTRACT

We discuss the general problem of dynamic electron-nucleus coupling, and the possibility of using this mechanism to initiate gamma-ray lasing. Single-particle and collective mechanisms are considered. The problems associated with accurate calculation of these processes are discussed, and some numerical results are given. Work in progress is described.

1. The problem of the transfer process

A commonly-proposed gamma-ray laser scheme[1] is shown in Fig.1. One envisions a long-lived storage state, I, which can be populated by some laser[2] or radiochemical[3] means, and which can be pumped with a relatively small amount of energy to a lasing state I'. This state can then decay

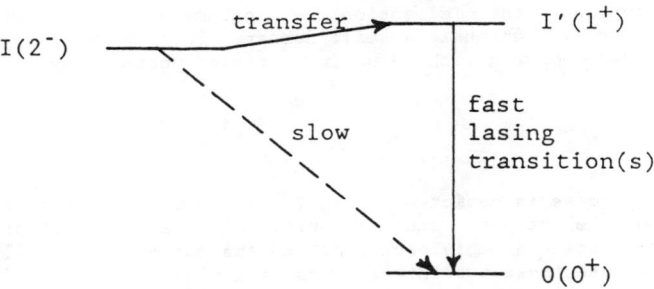

Fig. 1. Proposed gamma-ray laser level scheme.

---

*Work supported in part by the US Department of Energy and the Office of Innovative Science and Technology of the Strategic Defense Initiative.

with one or more radiative transitions, including at least one with the desired lasing characteristics. The graph represents a simple, ideal example. The storage state shown (2⁻) requires a magnetic quadrupole transition to reach the ground state (0⁺) directly. This is strongly inhibited.* The transfer step 2⁻→1⁺ is electric dipole, normally the strongest multipole. The lasing transition 1⁺→0⁺ is magnetic dipole, favored for Borrmann[4] mode propagation in crystals. The physical problem is to transfer the population I→I' efficiently.

1.1 Direct nuclear photoabsorption

One of the simplest transfer procedures would be to use direct nuclear photoabsorption γI→I'. This rate can be investigated on dimensional grounds. In the long wavelength approximation, the photon field is

$$\vec{E} \simeq E_0 \, \hat{\epsilon} \sum_{L=1}^{\infty} (2L-1) \frac{(ikr)^{L-1}}{(2L-1)!!} P_{L-1}(\cos\theta) \tag{1}$$

The interaction matrix element is thus of order

$$M_L \simeq \frac{ZeE_0}{k} (2\pi R_N/\lambda)^L \simeq \frac{ZeE_0}{k} (2\times 10^{-7})^L \tag{2}$$

This expression contains the dimensionless parameter (nuclear radius)/(photon wavelength), which is a small number. Thus the transfer rate is greatly inhibited compared with competing <u>atomic</u> photoabsorption processes.

1.2 Dual process

An alternative process is depicted in Fig.2. Here, photoabsorption occurs via the atomic electrons, and the residual electron-nucleus interaction is used to transfer this excitation to the nucleus. We call this a <u>dual</u> process because these two interactions take place together.[5,6] It has been called a two-step process, but this is not entirely appropriate because there is no well-defined, time-resolved intermediate state.

The photoabsorption matrix element has the same form as Eq.(2), but the nuclear radius is replaced by a length of atomic electron dimensions, and the nuclear charge by Z', the effective number of electrons responding collectively:

* Although, perhaps in practice, M2 transitions are not inhibited enough to be useful here.

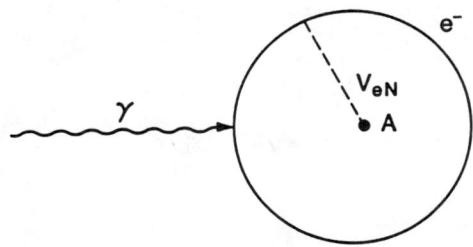

Fig. 2. Dual process for nuclear excitation.

$$M_L \approx \frac{Z'eE_0}{k} (2\pi r_e/\lambda)^L \approx \frac{Z'eE_0}{k} (3\times 10^{-3})^L \quad . \tag{3}$$

This is larger than before by a factor $\approx (Z'/Z)(1.5\times 10^4)^L$. However, we must also include the electron-nucleus interaction

$$V_{eN} \approx ZZ'e^2 \sum_{L'=0}^{\infty} \frac{r_<^{L'}}{r_>^{L'+1}} P_{L'}(\cos\theta) \rightarrow ZZ'e^2 \sum_{L'=0}^{\infty} R_N^{L'} r_e^{-L'-1} P_{L'}(\cos\theta) \quad . \tag{4}$$

The total effective matrix element is thus

$$M_L \rightarrow \frac{Z'eE_0}{k} (2\pi r_e/\lambda)^L \cdot \frac{1}{\Delta E} \cdot ZZ'e^2 R_N^{L'} \langle j|r_e^{-L'-1}|j'\rangle \quad . \tag{5}$$

There are three important factors. The first is just the electron photoabsorption matrix element. The last is the electron-nucleus interaction taken between electron states j and j', which may be single-particle or many-body states. This is not necessarily small, since the interaction tends to diverge for small r and is cut off only by the finite nuclear radius. The fact that this interaction is more nearly singular than the plane wave is the reason that internal conversion dominates radiative decay for a wide class of nuclear transitions. The middle term in Eq.(5) is the energy denominator (propagator) which occurs in any second-order process. Note that if L'=L, a simple dimensional argument gives

$$M_L \approx (Z'eE_0/k)\ (2\pi r_e/\lambda)^L\ (ZZ'e^2/\Delta E)\ (R_N^L/r_e^{L+1})$$

$$= \frac{ZeE_0}{k}\ (2\pi R_N/\lambda)^L \cdot \frac{1}{\Delta E} \cdot \frac{Z'^2 e^2}{r_e} \ . \qquad (6)$$

The last factor is approximately Z'^2 times one atomic energy unit. If this is comparable with ΔE, the expression reduces to precisely what we had before for direct nuclear photoabsorption. Thus we can obtain enhancement only by using collective electron transitions,[6] by making $\Delta E \to 0$ (resonant coupling), or by exploiting the fact that in general,

$$\langle j|r_e^{-L-1}|j'\rangle \neq \langle j|r_e|j'\rangle^{-L-1} \ . \qquad (7)$$

Thus any amplification by the atomic electrons depends upon the exact electron-nucleus matrix elements, which may involve collective electron effects in addition to detailed behavior of the electron wave functions near the nucleus. This is known as the dynamic hyperfine effect in muonic atoms, where it has been studied for many years.[7] It has more recently been applied to electronic atoms by Morita.[8]

2. Quantitative description of electron-nucleus coupling

Figure 3 shows a detailed level diagram for the dual process. The uncoupled nucleus is on the left, with the lasing decay represented by the rate $\Gamma_{I'0}$. Relevant electron states are in the middle, with the photoabsorption rate represented by $\Gamma_{jj'}$. The states of the coupled system are shown on the right. It is assumed that the initial state of the coupled system is $|Ij\rangle$. This state is pumped via the electron components with the rate $\Gamma_{jj'}$ to the mixed states $|a\rangle$ and $|b\rangle$, which then decay via their nuclear components to the ground state $|0j\rangle$. The hamiltonian eigenstates $|a\rangle$ and $|b\rangle$ are described in terms of their mixing angle θ in the nearly-degenerate subspace $|2\rangle$, $|3\rangle$:

$$|a\rangle = \sin\theta |2\rangle + \cos\theta |3\rangle \qquad\qquad |2\rangle = |Ij'\rangle$$
$$|b\rangle = \cos\theta |2\rangle - \sin\theta |3\rangle \qquad\qquad |3\rangle = |I'j\rangle \ . \qquad (8)$$

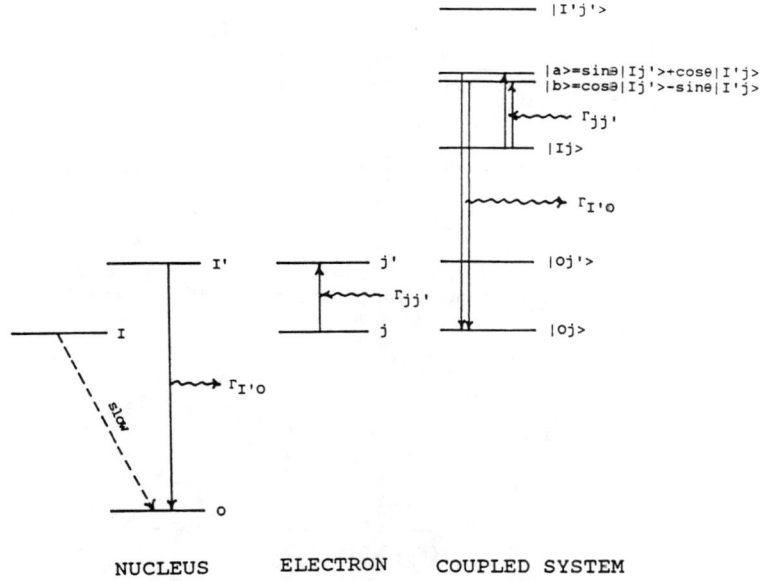

Fig. 3. Level diagram for the dual process.

The hamiltonian in this subspace is

$$H = \begin{pmatrix} E_2 & V_{23} \\ V_{23} & E_3 \end{pmatrix} \qquad \begin{array}{l} E_2 = E_I + E_{j'} \\ E_3 = E_{I'} + E_j \\ V_{23} = \langle 2|V_{eN}|3\rangle \end{array} \qquad (9)$$

A solution for θ is exactly

$$\tan\theta = \frac{E_2 - E_3}{2V_{23}} \left\{ 1 - \left[1 + \frac{4V_{23}^2}{(E_2 - E_3)^2} \right]^{1/2} \right\} . \qquad (10)$$

Note that

$$\tan\theta \to 1 \quad \text{as} \quad E_2 \to E_3, \text{ and}$$

$$\tan\theta \to \theta \simeq -\frac{V_{23}}{E_2-E_3} \quad \text{as} \quad V_{23} \to 0. \tag{11}$$

Finally, the steady-state population N' of the lasing state I' is

$$N' = \frac{2\tan^2\theta}{1+\tan^4\theta} \frac{\Gamma_{jj'}}{\Gamma_{I'0}} \to \frac{\Gamma_{jj'}}{\Gamma_{I'0}}, \quad E_2 \to E_3$$

$$\to \frac{2V_{23}^2}{(E_2-E_3)^2} \frac{\Gamma_{jj'}}{\Gamma_{I'0}}, \quad V_{23} \to 0$$

$$= \text{mixing coefficient} \cdot \frac{\text{electron photoabsorption}}{\text{nuclear decay}}. \tag{12}$$

Note that in the limit of complete mixing ($\tan^2\theta \to 1$), the steady-state population is precisely what one would obtain in balance between photoabsorption and decay, except that the photoabsorption occurs with the electron rate instead of the nuclear rate. For less complete mixing, the population is reduced by a factor $2\theta^2$, where $\theta \propto -V_{23}/(E_2-E_3)$. We define the dimensionless **amplification factor**

$$K = \frac{2\tan^2\theta}{1+\tan^4\theta} \frac{\Gamma_{jj'}}{\Gamma_{II'}}. \tag{13}$$

This is the factor by which the population inversion is increased over what it would be by direct nuclear photoabsorption. This factor varies between 0, for small mixing, and a maximum value of order $(Z'/Z)^2(r_e/R_N)^{2L}$. The naive discussion of Eq.(6) given previously leads to an expected value of order unity.

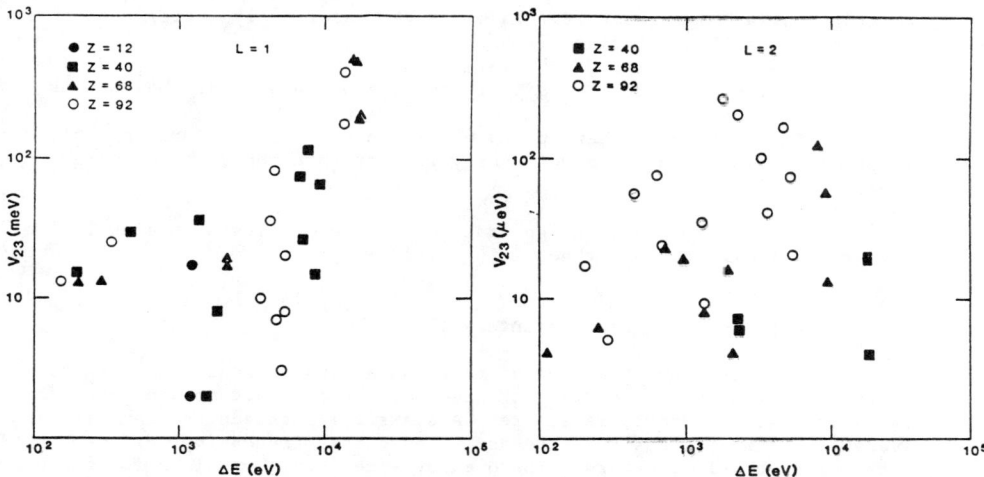

Fig. 4. Dipole and quadrupole matrix elements for selected electron transitions.

2.1 Single-particle electron response

In order to calculate V_{23} for single-particle transitions, we have made use of electron wave functions from atomic Dirac-Hartree-Fock-Slater calculations. We have made a general canvass of L=1, 2, and 3 transitions for atomic numbers Z=12, 40, 68, and 92, using nuclear transition strengths of a few single-particle units. Results[5] for L=1 and 2 are shown in Fig.4, where the matrix element V_{23} is plotted against electron transition energy ΔE. For L=1, the matrix elements all lie between 10^{-3} and 1 electron volt. For L=2, they lie between 10^{-6} and 10^{-3} eV. We found no case for L=3 where the matrix element was greater than 0.5×10^{-6} eV.

3. Collective electron response

3.1 "The Articles of Faith"

The possibility of collective electron motion modifies the above considerations. The formalism remains the same in principle, but the number of electron degrees of freedom increases enormously. In addition, there is a nonlinear interplay between the Coulomb interaction with the nucleus, the self energy of the electron gas, and the interaction with the external photon field. The nonlinear spatial and temporal interactions have led to three conjectures about an atom's response to the fields generated by a high-intensity laser. We facetiously refer to these conjectures as "The Articles of Faith:"

1. Atomic electrons can <u>amplify</u> the field produced at the nucleus;

2. The electrons can produce harmonics of the driving field, inducing nuclear transitions of energy greater than the quantum energy of the driving field; and

3. The electrons can generate electric fields at the nucleus of higher multipolarity than the driving field.

3.2 Calculation of collective interaction

Quantitative investigation of these questions on a quantum-mechanical basis involves many degrees of freedom. In most attempts made so far, basis truncation errors are severe. A classical approach reorganizes and averages the degrees of freedom, lessening truncation effects but introducing physical errors. There exist some atomic problems for which a classical description has proved useful in obtaining a quantitative description.[9] However, we must keep in mind that the electron-nucleus coupling involves discrete states and is inherently quantum-mechanical. For this aspect of the problem, classical approximations can yield only qualitative insight.

Our approach is a stepwise combination of classical and quantum methods. To calculate the interlevel transfer, we

(1) Use a classical model (Vlasov gas) to describe the collective electron response to an applied laser field;[10]

(2) Diagonalize a quantum electron hamiltonian[11] using the applied photon field and the self-consistent electron density response (1); and

(3) Use the mixed quantum electron states to compute electron-nucleus matrix elements as for the single-particle states.

A fourth step, to integrate the time-dependent Dirac equations for the coupled system, is in progress. This problem must be addressed in order to calculate nuclear pumping rates. Other statistical approaches (in particular, the Thomas-Fermi model and a self-interacting Vlasov gas) are being considered for step 1.

Figure 5 shows some typical results of step 1. Plotted are the L=1, 2, and 3 moments of the atomic electron density distribution calculated as a classical Vlasov gas, for an applied dipole laser field in the z direction, of one atomic unit in strength and wavelength λ=198nm. The moments are defined as

$$Q_L = \frac{(2L+3)(2L+1)}{4\pi} \frac{Z}{n} \sum_{i=1}^{n} (r_i/R)^L P_L(\cos\theta) \qquad (14)$$

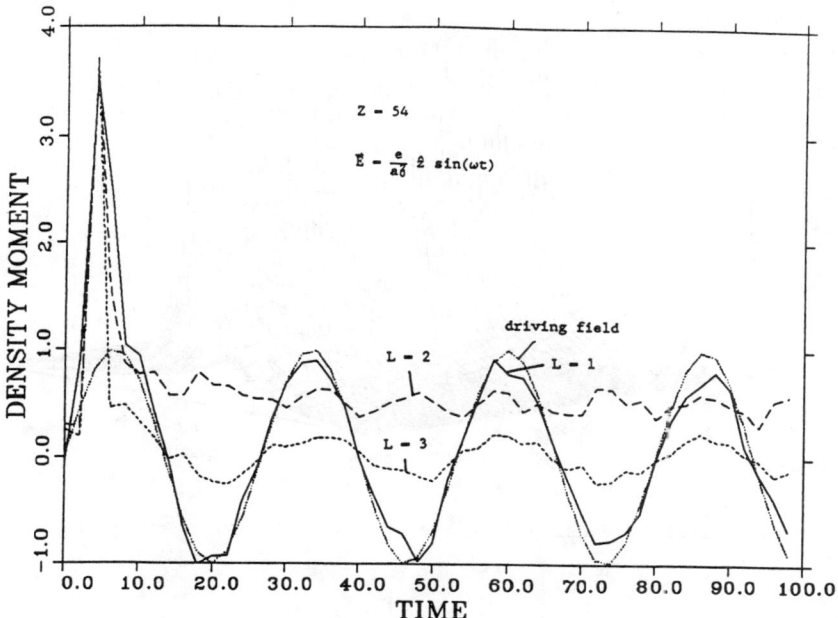

Fig. 5. Density moments for the Vlasov gas as a function of time.

where n is the number of Vlasov test particles (10000), and R is an atomic dimension ($2a_0$) beyond which particles are considered lost. Initial ionization is apparent in the strong peak in the first laser cycle, after which the density settles down to a more regular response. The dipole density closely follows the applied field. The quadrupole and octupole responses, however, show some evidence of higher harmonics and are clearly nonzero, in spite of the fact that the applied field has no components beyond L=1. This supports the second and third articles of faith. The noise is somewhat reduced if the magnetic force due to the photon field $\vec{v}\times\vec{B}=\vec{v}\times(\hat{k}\times\vec{E})$ is included, but no significant change in the oscillation amplitudes or harmonics is produced.

Figures 6 and 7 display the response of the physical atomic electron density to the applied field as a function of time. We have plotted the z-component of the density, integrated over x and y:

$$\rho(z,t) = \int_{-\infty}^{\infty} dx \int_{-\infty}^{\infty} dy \, \rho(x,y,z,t) \quad . \tag{15}$$

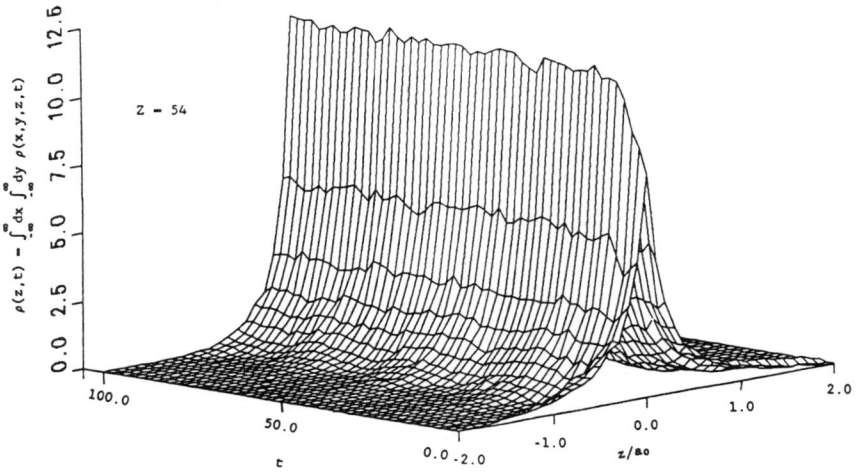

Fig. 6. Electron density response as a function of time.

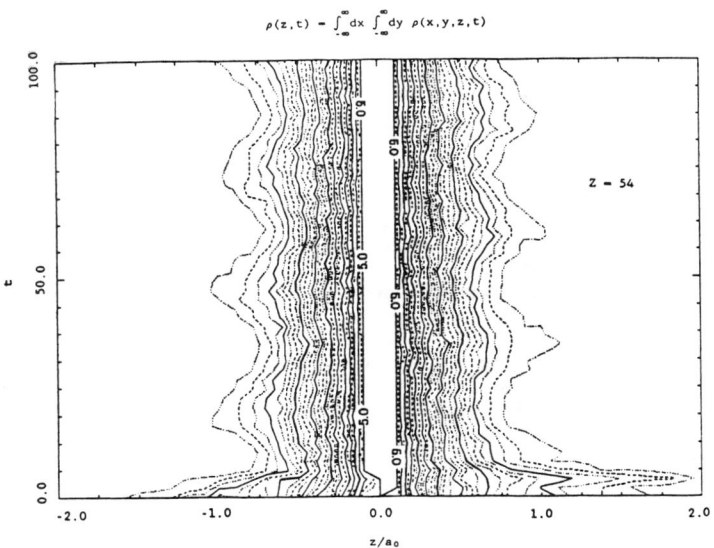

Fig. 7. Electron density response as a function of time.

4. Coupling to the nucleus

The classical electron density distributions in Figs.5-7 could be used in principle to generate time-varying multipole potentials at the nucleus, in order to calculate excitation rates. Such a procedure would be quantitatively inaccurate, however, because the classical approximations fail at distances small compared with the electron Compton wavelength. Even the semiclassical Thomas-Fermi model gives the wrong density behavior near the nucleus. Our step (2) described above is an attempt to avoid this difficulty. These calculations are not complete, but we have discovered several interesting facts: (1) The intense electric field used for the classical results in Figs.5-7 also produces strong mixing among quantum electron states, including effective $\Delta L=2,3...$ couplings arising from higher orders in the interaction potential; (2) The self-consistent dipole interaction from the Vlasov gas is a shielding effect, reducing the effect of the applied field by $\simeq 15\%$; (3) The quadrupole and octupole interactions from the Vlasov gas are smaller, but not zero; (4) Magnetic effects are negligibly small, as in the classical part of the calculation.

5. Conclusions and future work

These heuristic studies of nuclear interlevel transfer driven by the induction-field interaction with atomic electrons have led to the following tentative conclusions:

(1) L=1 pumping transitions are preferred. We find in general that higher multipole rates are down from this by factors of order $(10^3)^{L-1}$.

(2) The dual-process amplification factor is bounded from above by $\Gamma(\gamma j \rightarrow j')/\Gamma(\gamma I \rightarrow I')$ and is strongly dependent upon the number of electrons involved, the specific electron states, and the degree of coupled-state degeneracy.

Our own program for future work includes increasing the basis for the quantum electron treatment, solving the time-dependent wave equations, and investigating the effects of self-interaction in the Vlasov gas.

ACKNOWLEDGMENTS

We would like to thank M. Horbatsch and R. Dreizler for generously sharing their expertise and computer codes for classical and semiclassical atomic calculations, and J. Louck for helpful discussions.

REFERENCES

1. J. W. Eerkens, U. S. Patent 3,430,046 (1969); E. V. Baklanov and V. P. Chebotaev, Zh. Eksp. Teor. Fiz. Pism'a Red. 21, 286 (1975); P. Kamenov and T. Bonchev, C. R. Acad. Bulg. Sci. 28, 1175 (1975); L. A. Rivlin, Sov. J. Quantum Electron. 8, 1412 (1977); L. A. Rivlin, Sov. J. Quantum Electron. 7, 380 (1977); B. S. Arad, S. Eliezer, Y. Paiss, Phys. Lett. A74, 395 (1979).

2. P. Dyer, G. Baldwin, C. Kitrell, D. Imre, E. Abramson, and E. Schweitzer, J. Appl. Phys. 58, 2431 (1985); P. Dyer and G. Baldwin, in *Lasers in Nuclear Physics*, edited by C. E. Bemis and H. K. Carter (Harwood Academic Publications, London, 1983).

3. R. S. Sharpe and R. A. Schmitt, General Atomics Report GA910 (1959); G. Baldwin, J. P. Neissel, and L. Tonks, US Patent 3,324,099 (1966).

4. G. Borrmann, Phys. Z. 42, 157 (1941); H. Campbell, J. Appl. Phys. 22, 1139 (1951); J. P. Hutton, PhD thesis, Rice University, 1986; G. C. Baldwin, M. S. Feld, J. P. Hannon, J. P. Hutton, and G. P. Trammell, Journal de Physique 47-C6, 299 (1986).

5. G. A. Rinker, in *Proceedings of the International Laser Science Conference*, Dallas, Texas, 1985 (AIP Conf. Proc. 146, New York, 1986), p. 48.

6. L. C. Biedenharn, G. C. Baldwin, K. Boyer, and J. C. Solem, in *Proceedings of the International Laser Science Conference*, Dallas, Texas, 1985 (AIP Conf. Proc. 146, New York, 1986), p. 52.

7. L. Wilets, K. Dan. Vidensk. Selsk. Mat.-Fys. Medd. 29, No. 3, 1954; B. A. Jacobsohn, Phys. Rev. 96, 1637 (1954). For a more recent review, see E. Borie and G. A. Rinker, Rev. Mod. Phys. 54, 67 (1982).

8. M. Morita, Prog. Theor. Phys. 49, 1574 (1973).

9. J. E. Bayfield and P. M. Koch, Phys. Rev. Lett. 33, 258 (1974).

10. M. Horbatsch and R. M. Dreizler, Phys. Lett. 113A, 251 (1985).

11. G. A. Rinker, Comput. Phys. Commun. 16, 221 (1979).

NUCLEAR TRANSITIONS INDUCED BY ATOMIC EXCITATIONS

J. A. Bounds, P. Dyer and R. C. Haight
Los Alamos National Laboratory, Los Alamos, NM 87545

ABSTRACT

In the two-step pumping scheme for a gamma-ray laser, an essential step is that of exciting the nucleus from a long-lived storage isomer to a nearby short-lived state that then decays to the upper lasing level. For a crystalline structure host, the radiation must be used efficiently so as not to destroy the crystal. High intensity sources of photons are available only for relatively low quantum energy, but it is difficult to couple a long-wavelength photon directly to the much smaller nucleus. An experiment is proposed to induce this transfer by first exciting the atomic electrons. The nuclear excitation should occur by the exchange of a virtual photon in the near field of the electrons. As a test case, the 73 eV ^{235m}U isomer might be excited by electronic motions induced by a high-brightness UV laser. The conversion electrons from the decay of the isomer would be detected and a 26 minute decay curve would indicate induced nuclear transitions.

It has already been demonstrated that lasers can be used to excite nuclear transitions. In the experiments of Izawa and Yamanaka of Osaka University[1] the isomeric state of ^{235}U was produced by laser bombardment of natural uranium. The laser formed a plasma which, in recombining, excited the nucleus to its 73 eV isomeric state. The experimenters attributed this to the coincidence of electronic energy level differences with the nuclear excitation energy. The coined abbreviation NEET (Nuclear Excitation by Electron Transition) refers to this process. Goldanskii and Namiot, however, attributed the experimental results to inverse internal electron conversion (IIEC) wherein the nucleus is excited when the electrons make a transition from the continuum to a bound state.[2] They argued that the probability of a free-bound induced transition is about three orders of magnitude greater than the bound-bound induced transition probability for the Japanese experiment. In either case the nucleus is believed to be excited by an electron transition in the cooling plasma created by a strong laser pulse.

Yet another mechanism has been proposed to excite a nucleus with a laser pulse. It has been observed that the intense electromagnetic field from a high-brightness laser causes effects that indicate strong coupling of energy to the atoms.[3] It has been postulated[4] that once the laser electric field strength, as seen by the electrons in the focal volume of such a laser, approaches the field strength of the binding energy for these same electrons, chaotic and possibly correlated motion of the electrons may be produced. With such motion high multipolarity fields may be produced at the nucleus. In addition, the field from the electron

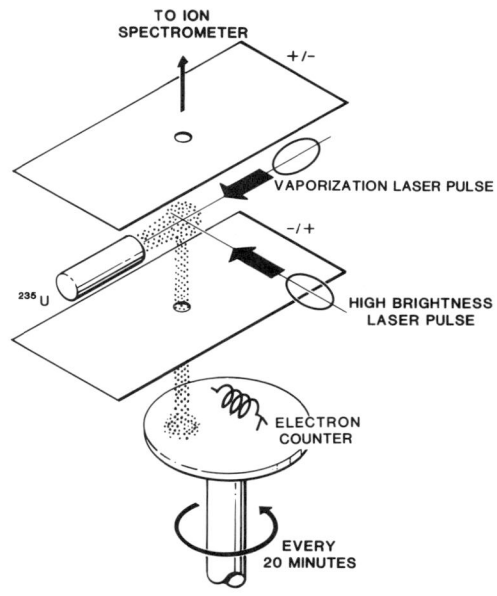

Fig. 1. Ion collection scheme. Gas puff in center is created by a vaporization laser pulse striking a ^{235}U sample. The high-brightness laser pulse strikes the puff. According to plate polarity, the ions are either analyzed in the spectrometer or collected and later counted.

motion would couple to the nucleus much more strongly than an external field alone would, since the electrons and nucleus interact through the near field. This same near-field effect is what causes internal conversion to dominate multipolar electromagnetic transitions at low energies.

The experiment proposed to test these effects is as follows. The nucleus ^{235}U is chosen since its isomeric level at 73 eV is the lowest known and is within the energy transfer range of current high-brightness lasers. The isomer decays by internal conversion, emitting low-energy electrons. In the experiment a low density vapor of uranium is produced by pulsed laser evaporation of a small sample. The high-brightness laser is focused into this vapor. If the laser produces transfers to the excited nuclear state, these excitations can be detected by looking for electrons emitted in the later decay of the isomer. Typically the laser irradiations would occur at a rate of 10 per second for one isomeric half-life (26 minutes) followed by electron counting for a period of one or more half lives. An electron signal falling off in intensity with the half life of the isomer would indicate production of the isomer.

There are two possible collection schemes. In the first, the atoms are collected as the gas puff deposits on a surface. The conversion electrons, whose energies are typically 10 to 20 eV, can penetrate only a few monolayers through the surface. Thus the target gas density is limited since the high-brightness laser intersects only a small portion of the gas puff, and during the deposit the unexcited atoms tend to cover up those that experienced the strongest laser field. After the collection period, the deposit is moved in front of an electron counter for data acquisition. A retarding grid between the deposited atoms and the electron counter can offer rapid evidence whether or not the detected electrons have less than 73 eV energy. In the second collection scheme, the gas puff irradiation occurs in a region of electric field produced by two parallel plates (Figure 1). The high-brightness laser ionizes nearly all of the atoms of interest, i.e. those atoms which experience the strongest field. The ions are extracted by the electric field to a collector where they are accumulated as before until counted. In this ion collection scheme the ratio of atoms collected from the beam interaction region to atoms collected from the rest of the gas puff is much higher, allowing the use of higher gas densities and therefore permitting detection of a smaller effect.

These experiments will be carried out for a wide range of vapor pressures and laser powers to map out the process. At the low vapor pressures proposed in these experiments, inverse internal electron conversion, which depends strongly on the plasma density, should be negligible. A cross-section for the conversion rate will be determined. It is conceivable that this method of exciting nuclear transitions may be of use in gamma-ray laser research,[5] perhaps serving as the required pumping process.

REFERENCES

1. Y. Izawa and C. Yamanaka, Phys. Lett. 88B, 59 (1979).
2. V.I. Goldanskii and V.A. Namiot, Phys. Lett. 62B, 393 (1976); V.I. Goldanskii, R.N. Kuzmin and V.A. Namiot, in Proceedings of the International Conference on Lasers '81, New Orleans, LA, 1981, edited by C.B. Collins (STS Press, McLean, VA, 1982).
3. H. Jara et al., these proceedings; C.K. Rhodes, Science 229, 1345 (1985), and references therein.
4. G. Rinker, J. Solem and L. Biedenharn, these proceedings; J. Solem, in Advances in Laser Science - 1, proceedings of the First International Laser Science Conference, Dallas, TX, 1985, edited by W.C. Stwalley and M. Lapp (AIP Conference Proceedings No. 146, 1986).
5. The NEET process has previously been proposed for gamma-ray lasers by K. Okamato, in Laser Interaction and Related Plasma Phenomena, proceedings of the Fourth Workshop, Troy, NY, 1976, edited by H.J. Schwartz and H. Hora (Plenum, New York, 1977).

WHY GAMMA RAYS ARE DIFFERENT
SOME NOTES FOR PEDESTRIANS

Harry J. Lipkin
High Energy Physics Division
Argonne National Laboratory
Argonne, IL 60439

ABSTRACT

This talk brings the discussion of gamma ray lasers down from the science fiction level to the real world. Flamboyant exciting proposals presented with evangelical zeal obscure the real difficulties to be overcome before a gamma ray laser becomes feasible in the laboratory, and long before any practical applications. Nuclear gamma radiation does not have many of the properties taken for granted in atomic or molecular radiation and necessary for lasers. The basic science and technology underlying these differences and the proposed methods of overcoming difficulties resulting from them are not properly understood. Grandiose proposals for grasers generally presented tend to ignore them. Considerable illumination in this interdisciplinary problem could be provided by some back-of-the-envelope calculations and simple experimental surveys by small groups of students and postdocs with an elementary knowledge of the nuclear and solid state physics which is evidently not familiar these days to laser physicists. However, nobody seems ready to propose or undertake such work. Unfortunately, budgets for unglamorous basic science are being cut across the board, and evangelical glamour seems to be more effective than real physics in obtaining support.

Any realistic discussion of gamma ray lasers must take into account the very different physics present in nuclear gamma ray transitions and not present in the transitions at longer wave lengths used for conventional lasers. Laser action between two energy levels depends upon the fact that the transition from the upper level to the lower level takes place primarily by the emission of a photon, that the photon emitted in the transition has the right properties to cause the inverse transition in another atom or molecule, and that it is relatively easy to produce a population inversion by pumping with photons. This does not happen in normal nuclear transitions.

The 14 KeV nuclear transition in ^{57}Fe does not normally take place by photon emission. The probability of photon emission is less that 10%. This small number of photons normally emitted do not generally have the proper energy to produce the inverse transition in another ^{57}Fe nucleus. Furthermore, there is no simple way to "pump" a population inversion of these two levels by irradiation with photons; there is only a tiny probability that a photon incident on a sample will produce a nuclear transition. The dominant absorption mechanism for photons is electronic, not nuclear, and the probability that a photon incident upon a sample will produce a nuclear transition rather than being absorbed by other means is generally less than one in a million.

© American Institute of Physics 1987

These differences must be understood and taken into account before any gamma ray laser design can be considered. It is necessary to find ways to increase the probability of photon emission, to find ways to get the emitted gamma ray to have the proper energy and to find new efficient ways of achieving a population inversion.

SOME SIMPLE SCALE DIFFERENCES

Some insight into the nature of these differences which indicate where normal intuition from optical transitions breaks down for gamma rays can be seen by examining characteristic scales for the two cases. Let R denote the size of the radiating system; e.g. nucleus, atom or molecule, D denote the distance between neighboring radiators (the lattice constant in a crystal), x denote the amplitude of zero-point and thermal oscillations of the radiator, and k denote the wave number of the radiation. Then for normal optical laser transitions,

$$x << R \sim D << (1/k). \tag{1a}$$

Thus

$$kx << kR \sim kD << 1. \tag{1b}$$

For nuclear gamma ray transitions,

$$R << x \sim (1/k) << D \tag{2a}$$

and

$$kR << kx \sim 1 << kD \tag{2b}$$

In both cases $kR << 1$, indicating that the long wave length approximation and the multipole expansion are valid. But the fact that kx is of order unity for the gamma ray case and kD is large introduces new problems of coherence for lasers. The motion of the radiating nucleus in a lattice introduces a phase modulation that tends to destroy the coherence. The Debye-Waller factor $e^{-k^2 \langle x^2 \rangle}$ expresses the reduction in intensity of the coherent radiation due to zero-point and thermal motion. The fact that kD is large means that there are large phase differences between radiation from nearest neighbors, and that considerable effort is needed to control this phase in order to maintain coherence.

These effects can be seen explicitly by examining the matrix element for the amplitude emitted from two neighboring nuclei

$$A = \{\psi(\vec{x}_1, \vec{x}_2) | e^{i\vec{k}\cdot\vec{x}_1} | \psi(\vec{x}_1, \vec{x}_2)\}, \tag{3}$$

where $\vec{x}_1$ and $\vec{x}_2$ are the coordinates of the nuclei relative to their equilibrium positions. Then for gamma rays

$$\vec{k}_\gamma \cdot \vec{D} >> 1 \tag{4a}$$

and
$$k_\gamma^2 \langle x^2 \rangle \sim 1. \tag{4b}$$

The reduction in the coherent intensity due to thermal and zero-point motion is seen to be given by the Debye-Waller factor $e^{-k^2 \langle x^2 \rangle}$.

We also note that $\vec{k} \cdot \vec{D}$ satisfies Bragg condition for constructive interference between the two terms in Eq. (4a) only at Bragg angle. Coherence is washed out when the intensity is averaged over angles and destroys superradiance.

Additional scale differences are simply seen in the energy domain; namely that the energy of optical radiation is much less than atomic ionization energies, while the energy of nuclear gamma rays is much greater

$$\hbar\omega_{opt} \ll E_I \ll \hbar\omega_\gamma. \tag{5a}$$

This allows energy to be lost by ionization processes absent in conventional optical lasers. The kinetic energy of the recoil of an atom because of the momentum carried by the emitted photon is much less than the natural line width of the radiative transition in the optical case. In the gamma ray case the recoil energy is very much larger than the natural line width and is of the same order as the thermal energy or the Debye temperature of the crystal.

$$\Gamma \ll \frac{(\hbar k_\gamma)^2}{2M} \sim K\theta_D. \tag{5b}$$

For the 14 KeV line in ^{57}Fe

$$\frac{(\hbar k_\gamma)^2}{2M} = \frac{(1.4)^2 \times 10^8}{2 \times 57 \times 10^9} \sim \frac{1}{570} \text{e.v.} \gg \Gamma. \tag{6a}$$

For a 50 KeV line in a nucleus with mass ~ 50

$$\frac{(\hbar k_\gamma)^2}{2M} \simeq KT. \tag{6b}$$

SOME IMPORTANT CONSEQUENCES OF THE SCALE DIFFERENCES

A. Frequency shift effects

1. The electromagnetic radiation emitted in a transition between two energy levels of an isolated atom or molecule has the right frequency to induce the inverse transition between the same two levels in another atom or molecule. The electromagnetic radiation emitted in a transition between two energy levels of an isolated nucleus does not have the right frequency to induce the inverse transition between the same two levels in another nucleus. The recoil kinetic energy, negligible for atomic and molecular transitions, is crucial for nuclear transitions.

2. The ratio of the natural line width to the energy of a nuclear gamma ray can be many orders of magnitude smaller than for the atomic or molecular case. Thus appreciable line shifts which destroy laser action can be produced in nuclear transitions by tiny effects which are completely negligible in atomic or molecular transitions. Some examples are Doppler shifts produced by tiny velocities, or small differences in external fields acting on different atoms. Nuclear transitions have even been affected by the gravitational red shift of a photon passing between two nuclei.

B. Phase Shift Effects

1. The wave length of electromagnetic radiation emitted in atomic and molecular transitions is long in comparison with the distance between nearest neighbors in normal matter. The wave length of electromagnetic radiation emitted in nuclear transitions is short in comparison with the distance between nearest neighbors in normal matter. There is therefore an appreciable phase shift in the propagation of nuclear gamma radiation between neighboring nuclei, which can affect phase coherence.

2. The wave length of electromagnetic radiation emitted in atomic and molecular transitions is long in comparison with the amplitude of thermal or zero point motion in normal matter. The wave length of electromagnetic radiation emitted in nuclear transitions is of the same order as the amplitude of thermal or zero point motion in normal matter. Thus this motion can destroy phase coherence.

C. Loss of Energy by Radiationless Transitions

1. The dominant mechanism for the transition between an upper and a lower energy level of an isolated atom or molecule is the emission of a photon. In most transitions of interest for grasers, the dominant mechanism for the transition between an upper and a lower energy level of an isolated nucleus is the ejection of an atomic electron (internal conversion) not photon emission.

2. The dominant absorption mechanism in matter for photons emitted by electrons in atomic or molecular transitions is absorption by bound electrons which then make other atomic or molecular transitions; hopefully the desired transition for a laser. The dominant absorption mechanism in matter for photons emitted by nucleons in nuclear transitions is still absorption by electrons which recoil into the continuum and cannot induce other nuclear transitions like the desired transition for a graser.

NEW PROBLEMS IN THE NUCLEAR DOMAIN RELEVANT TO GAMMA RAY LASERS

We now discuss the problems arising from these differences. The large differences in wave length, frequency, energy and momentum between gamma radiation and optical radiation have important physical implications both in the energy-momentum domain and in the space-time domain. Gamma rays have large momenta, producing recoil shifts larger than the natural line width, as noted in Eq. (5b) They have energies larger than atomic ionization energies as noted in Eq. (5a). Thus gamma ray emission in nuclear transitions competes with radiationless transitions in which the energy is carried off by an atomic electron. The overwhelmingly dominant absorption mechanism in matter for nuclear gamma rays is electronic, rather than nuclear, and does not produce further nuclear excitation. Gamma ray wave lengths are so much shorter than interatomic distances that phase differences between radiation from nearest neighbors can completely destroy coherence effects as noted in Eqs. (2–4). These effects all deserve serious attention. They lead to difficulties which may prove to be insurmountable.

A. New Problems with the Lasing Transition

Two important effects present in gamma ray transitions are the recoil shift and internal conversion. The recoil shift changes the energy of the gamma ray emitted in a hopefully lasing transition by such a large amount that it makes the gamma ray useless for the excitation of another nucleus. The shifted gamma ray is thus lost to further laser action. Internal conversion takes most of the energy of the lasing transition out of the radiation field and converts it into incoherent electron emission also useless for further laser action. Some ways must be found to eliminate large energy losses due to the emission of recoil-shifted gamma rays and conversion electrons in order to make the construction of a gamma ray laser feasible.

In addition the phenomenon of superradiance is very different for nuclear gamma rays, whose wave lengths are short compared to interatomic spacings, in contrast to the case of optical radiation. Most of the coherence effect is lost in the nuclear case.

1. The Recoil Shift

Any atom or nucleus which emits a photon must recoil with a momentum equal and opposite to the photon momentum. The energy of a photon emitted in the transition between two levels is then less than the energy difference between the two levels by the kinetic energy of the recoil. For optical transitions, the recoil shift is much less than the natural line width and can be ignored. For gamma rays the recoil shift is orders of magnitude larger than the natural width as noted in Eqs.(6).

The only known method for overcoming this difficulty is by using the Mössbauer effect, in which the atom making the transition is bound in a crystal and the recoil momentum is taken up by the whole crystal with negligible energy loss. However, the ability of a crystal to absorb the recoil momentum depends upon the energy of the gamma ray, the ambient temperature and the Debye temperature of the crystal in a function of these variables known as the Debye-Waller factor. The requirement that the Debye-Waller factor must be near unity to avoid wasting energy with the emision of recoil-shifted photons places serious constraints on the values of the above-mentioned variables.

2. Internal Conversion

Unlike atomic and molecular transitions, where the photon energy is smaller than ionization energies, the energy of a nuclear transition is considerably greater than the ionization energy of the atom as noted in Eq. (5a). It is therefore possible for the transition between two nuclear levels to take place with the energy emitted by ejecting an atomic electron, rather than in the form of radiation. In common transitions, like the Mössbauer transition in ^{57}Fe, the probability of this electron ejection process called internal conversion is more than an order of magnitude greater than the probability of photon emission. The only methods proposed for eliminating this loss due to internal conversion are by the use of coherent effects like the Borrmann effect or superradiance which can enhance the relative probability of photon emission. Unlike the Mössbauer effect, which is well understood theoretically and confirmed experimentally, there has been no convincing theoretical prediction confirmed by experimental verification of the possibility of overcoming the loss due to internal conversion. Considerable basic research will be needed to establish whether the Borrmann effect exists and has the desired properties in nearly perfect crystals relevant to gamma ray lasers. There will then remain the open question of how the effect is destroyed in a practical crystal with defects and radiation damage.

3. Superradiance

Dicke has shown how the rate of a spontaneous radiative transition can be enhanced by a collective effect, in which any one of N atoms may be excited and the many-body wave function is a coherent superposition of states in which a different atom is excited. The transition probability is then enhanced by a factor N and the lifetime of the excited level reduced by a factor N from the lifetime of a single atom. However, this depends upon a coherent excitation and a coherent radiation in which all contributions from different atoms are in the same phase. This is possible as long as the wave length of the radiation is large compared to the size of the radiator.

For nuclear gamma transitions this effect no longer occurs, because the interatomic distances are no longer negligible in comparison with the wave length of the radiation as shown in Eqs.(3–4). The relative phases of the contributions of radiation from different nuclei depend upon the wave length of the radiation, the distance between atoms and the angle of the radiation, in a manner similar to the case of X-ray scattering by crystals. Although it is possible to adjust the phases of the radiation to interfere constructively in certain directions in a crystal; e.g. Bragg directions, this enhancement occurs only in a very small solid angle around the Bragg direction, and is mainly at the expense of radiation in other directions. The result of the coherence is primarily to change the angular distribution of the radiation, but not its overall intensity, thus giving little or no change in the lifetime of the excited state.

B. New Problems with the Pumping Transition

In addition to the lasing transition itself, there is also the pumping transition necessary to achieve a high population of the excited state in the lasing transition. Here again there are important differences in the nuclear case. The absorption of gamma rays by matter is very different from the absorption of optical photons, and most of the energy of the pumping transition can be expected to be wasted. Furthermore, because there will be no simple relation between the interatomic distances and the wave lengths of the pumping radiation and the lasing radiation, any collective excitation of nuclei in the sample will not have the desirable coherence properties and phases for coherent emission of laser radiation.

1. Energy considerations.

The pumping transition must have a higher energy than the lasing transition in order to reach a level higher than the lasing level, unless there is an isomeric storage level in the vicinity of the lasing level. There must then be a radiation cascade from the pumping level to the lasing level followed by the lasing transition. Because nuclear cross sections are much lower than photoelectric and Compton cross sections, the probability of absorption of pumping photons by the desired nuclear transition will be very small. Factors of 10^{-8} can be expected in typical cases. This implies an enormous waste of energy in the pumping process, and a necessity to dispose of this energy without excessively heating the sample. The Debye-Waller factor is very sensitive to temperature, and unnecessary heating can destroy the efficiency of the Mössbauer effect. There is also the energy of the cascade radiation from the pumping level to the lasing level, which can heat the sample. In addition to heating there is also the possibility of radiation damage by all the undesired photons which can lead to crystal defects and harm coherence effects.

One might envisage an isomeric storage level very close in energy to the lasing level which would require only a small energy for the transition to the lasing level. However, no isomers having this property are known and it is extremely improbable that they should exist.

2. Phase considerations

Since the wave length of the pumping radiation is not commensurate with that of the lasing transition, Bragg conditions for coherent excitation and radiation of these two transitions are incompatible. Thus there is no possibility of a collective excitation of the lasing level in the sample by means of the pumping radiation to give the phase coherence necessary for constructive interference between the radiation from different nuclei in superradiant or Borrmann directions.

A NANOSECOND FLASH X-RAY DEVICE OF SUBANGSTROM EXCITATION A TABLE TOP ALTERNATIVE TO SYNCHROTRON RADIATION

T. S. Bowen, J. J. Coogan, F. Davanloo and C. B. Collins
Center for Quantum Electronics, University of Texas at Dallas
P. O. Box 830688, Richardson, TX 75083-0688

ABSTRACT

This paper presents the scaling of a repetitively pulsed, flash x-ray source capable of emitting 300-mW average power in 20 nsec pulses of radiation. For some applications this device can offer a laboratory alternative to laser plasma x-rays or synchrotron radiation.

INTRODUCTION

Available x-ray sources fall into the following general categories: (1) x-ray production from laser produced plasmas and E-beam discharges, (2) x-ray production from synchronous electron storage rings, (3) x-ray production from x-ray diode/triode tubes. High power lasers are used to produce plasmas radiating soft x-rays.[1] Conversion efficiencies are substantially reduced in the production of plasmas radiating x-rays in the wavelength regions of less than 10A. E-beam discharges are x-ray sources that offer maximal emitted power per shot in the hard x-ray region. Since these devices have low repetition rates, their application is limited in experiments dependent upon the integration of responses that occur with low probabilities. Synchrotrons have the unique advantage of collimation. However, such devices are few in number and require complex supporting facilities, resulting in experimental time being at a premium. X-ray production from some x-ray diode tubes exhibit CW or quasi-CW operation which is not particularly suited for experiments involving fast phenomena. Other existing pulsed x-ray diodes have either long pulse widths or lower levels of fluence. Here we describe a new pulsed x-ray source with high levels of fluence, short pulse durations and high repetition rate capabilities. Average x-ray powers are great enough to integrate experimental responses above the noise in a reasonable working period.

PERFORMANCE AND OPERATION

Device design is described in our recent reports.[2,3] Three such Blumlein driven x-ray devices were built which differed in length and energy storing capabilities. The scaling of x-ray outputs to larger values with increased system size is illustrated in Fig. 1. The total capacitance of systems A, B, and C were 7, 20, and 60nF, respectively. It can be seen that a larger value of stored energy resulted in greater outputs.

Rather curiously, it was found that the output pulse duration seemed to depend upon the anode materials. Figure 2 shows this

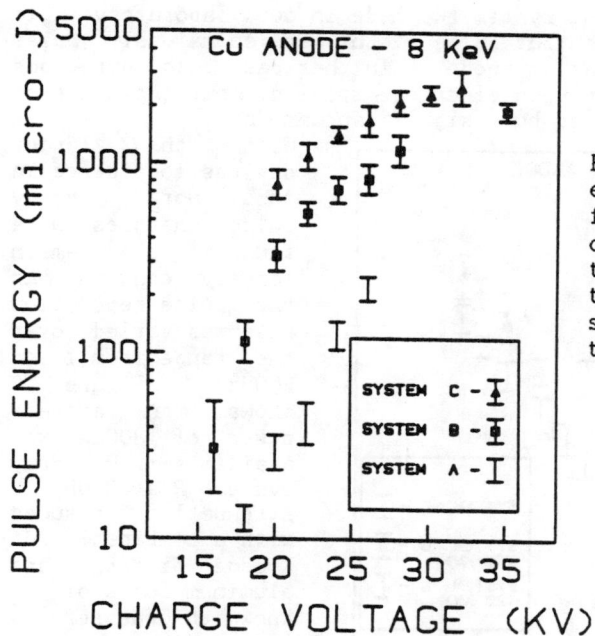

Figure 1: X-ray pulse energies emitted as functions of the charging voltage of the Blumlein for the three different systems described in the text.

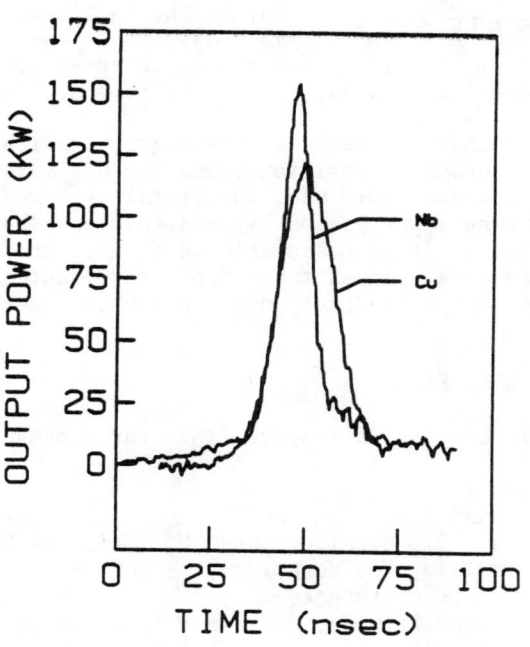

Figure 2: X-ray power emitted as a function of time from two different anodes, Cu and Nb, in system C with electrode spacing of 1.07mm.

behavior in the largest system (C) made in our laboratory. It can be seen that the output pulse widths were anomalously brief when niobium was used as an anode. In that case pulse durations still depended strongly upon electrode spacing, but the scale of the variation was shifted by a significant amount.

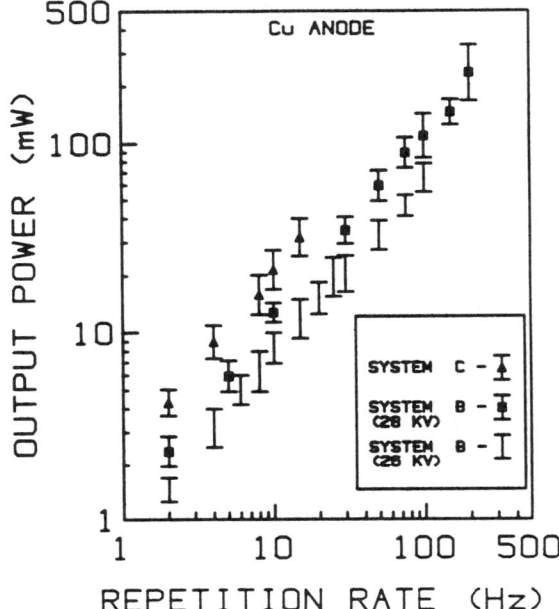

With the larger devices considered in this work, x-ray pulse energies were found to remain largely constant as the pulse repetition rate was varied over the range from 1 to 200Hz. Figure 3 shows an average power of 300mW was available from x-ray system B at 200Hz. Attenuation measured with a combination of K-edge filters and aluminum foils of known thicknesses indicated that at least 25% of the total x-ray energy lay in the K_α lines, in agreement with previous observations and expectations.[2,4]

Figure 3: Average powers emitted as functions of the pulse repetition rates.

Our application concerning the use of the x-ray device described here is to excite nuclear transitions, and the ultimate signal-to-noise ratio will depend only upon the total keV/keV (term customary for reporting laser plasma yields) that can be delivered in a working period. It is reasonable to expect that for some other applications this compact table top x-ray source can offer an alternative to either synchrotron radiation or laser plasma x-rays.

REFERENCES

*Supported by the Strategic Defense Initiative (IST) through NRL and in part by ONR.
1- Communications WG3 and WG4 in this conference.
2- F. Davanloo, T. S. Bowen, and C. B. Collins, Advances in Laser Science-I, edited by W. C. Stwalley and M. Lapp (AIP Conference Proceedings No. 146, New York 1986) pp. 60-61
3- Our communication TUD3 in this conference.
4- L. C. Bradley, A. C. Mitchell, Q. Johnson, and I. D. Smith, Rev. Sci. Instrum. 55, 25 (1984).

NUCLEAR RAMAN SPECTROMETER*

P. Reittinger, S. Wagal and C. B. Collins
Center for Quantum Electronics, University of Texas at Dallas
P.O. Box 830688, Richardson, TX 75083-0688

ABSTRACT

A device has been built, based on a nuclear analog to Raman spectroscopy, which monitors the gamma ray absorption cross section of nuclei in an Fe foil as a function of the frequency of an Rf magnetic field in which the foil is placed.

INTRODUCTION

As early as 1968 it was observed that the Mossbauer absorption spectra of an ^{57}Fe foil subjected to an Rf magnetic field contained additional structure.[1] The Rf field produces absorption peaks (Rf sidebands) at sum and difference frequencies of the Zeeman energies and integral multiples of the frequency of the applied field.(fig.1)

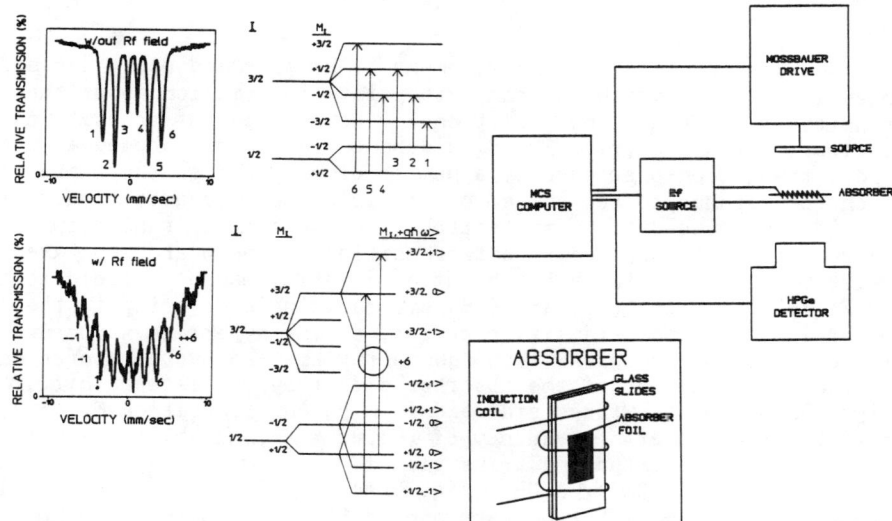

Fig.1 Mossbauer spectra of ^{57}Fe foil.   Fig.2 Automated NRS.

Currently there is uncertainty about the origin of these Rf sidebands. Yet the Rf sideband's dependence on the frequency of the field applied to the Fe foil spawns a new type of nuclear spectroscopy. The term Nuclear Raman Spectroscopy (NRS) was coined in 1985 for a type of Mossbauer absorption spectroscopy designed to monitor changes in the intensity of transmitted single-frequency gamma photons as a function of frequency of the long wavelength photons of the alternating magnetic field in which the absorbing nuclei are immersed.[2] This new tool could lead to an understanding of these Rf sidebands which in turn could prove to be critical in the development of a gamma ray LASER.[3]

APPARATUS

A device has been constructed which automates NRS and the heart of it is a multi-channel scalar (MCS) and IEEE-488 GPIB interface with an Apple II+ computer (fig. 2). The MCS was designed to have a 100% duty cycle, making the data collection time per spectra 1.5 days. The GPIB enables the spectrometer to continuously sweep through frequencies of an Rf magnetic field with a Wavetek frequency synthesizer. The Mossbauer drive allows the frequency of the gamma photon to be biased by a constant Doppler shift. The spectrometer in its present form has an instrumental resolution of 100 Hz and a range of 10^9 Hz with a stability of 0.1 Hz/sec with a stationary source. These characteristics are comparable to having a Mossbauer spectrometer with a means of Doppler shifting the gamma ray source with a resolution of 0.01μm/sec and a range of ± 100 mm/sec with a stability of 0.1 Å/sec/sec.

DATA

NRS provides a direct measurement of Rf sideband positions and intensities, from which one can extrapolate information about the transitions between Zeeman split energy levels (parent transitions), labeled 1 through 6 (fig. 2). Rf sidebands have been labeled as a parent transition preceeded by a number of +'s or -'s, the number of which corresponds to the number of Rf field energy quanta (order of the sideband) added to or subtracted from the parent transition. The symmetrically opposed parent transitions 1 and 6 are separated by 123.7 MHz. Applying a 61.85 MHz alternating magnetic field to the Fe foil produces (+1) and (-6) sidebands which overlap in the symmetric center, or transition center, of the hyperfine structure of the ^{57}Fe. The energies of the gamma ray emitted by the source and the transition center of the absorber differ by the isomer shift, Δ (fig.3a). In NRS the Stokes sideband from parent transition 6, (-6), would be detected at a frequency of $(61.85 - \Delta)$ MHz while the anti-Stokes sideband from parent transition 1, (+1), would be detected at a frequency of $(61.85 + \Delta)$ MHz (fig. 3b and c). Therefore, NRS should produce a spectrum with two peaks around 60 MHz, separated by 2Δ.

If we apply a small Doppler shift, δ, to the source, we should obtain an NRS spectrum with two peaks around 60 MHz, separated by $2(\Delta + \delta)$ (fig. 4). Classically, the frequencies at which sidebands appear, fs, is simply

$$fs(MHz) = [v - (Pj + iso)] * (11.6/ord)$$

where Pj is the position of the j'th parent transition (mm/sec), v is the velocity of the source (mm/sec), iso is the isomer shift (mm/sec), and ord is the order of the sideband of interest. The source used was in a Pd lattice (iso = -0.185).

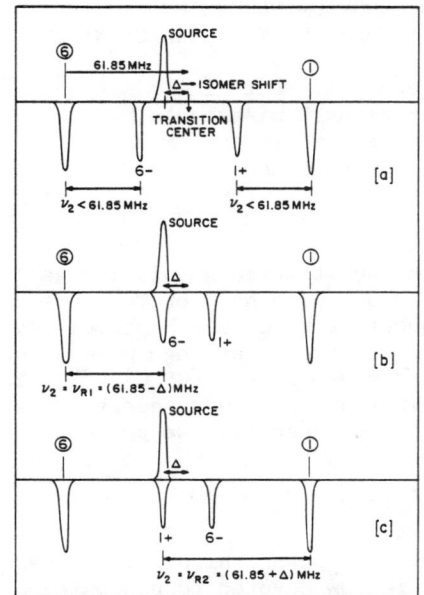

Fig.3 How NRS measures isomer shift.

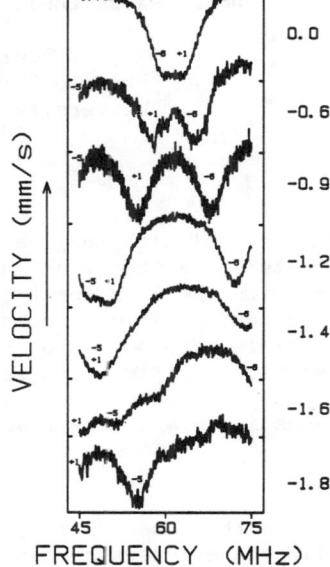

Fig. 4 Typical NRS spectra.

Rf sideband positions are also apparently affected by the intensity of the Rf magnetic field.(fig. 5) It is yet to be determined whether sideband position is a function of intensity as well as frequency of the Rf magnetic field, or whether the temperature shift of the parent transitions is being detected, or both. Since we start with a negative isomer shift, raising the temperature of the absorber should reduce the energy difference between the source transition and the transition center of the absorber. Therefore, increasing the Rf field intensity should raise the temperature of the absorber and in turn decrease the isomer shift.

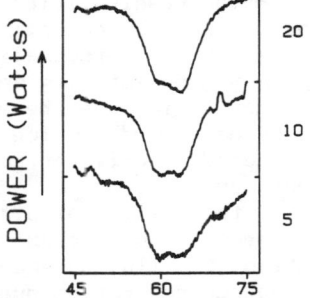

Fig.5 Effect of field intensity on sideband position.

So NRS in addition to providing information about sideband intensity and position could also prove to be a means for direct and accurate measurement of isomer and temperature shifts.

REFERENCES

1 N. Heiman et. al., Phys. Rev. Lett. $\underline{21}$, 93 (1968).
2 S. Olariu, I. Popescu and C.B. Collins, Phys. Rev. C $\underline{23}$, 50 (1981).
3 B.D. DePaola et. al., J. Opt. Soc. Am. B $\underline{2}$, 541 (1985).

* Supported by SDIO (IST) through NRL and in part by ONR.

SPECTROSCOPY OF NEW CHROMIUM/NEODYMIUM DOPED OXIDE LASER MATERIALS : GARNETS AND HEXAALUMINATES

G. Boulon, C. Garapon, A. Monteil
Physico-Chimie des Matériaux Luminescents
Université Lyon I - Unité associée au CNRS
69622 Villeurbanne - France

ABSTRACT

We deal with new laser materials, substituted gadolinium gallium garnets or lanthanum hexaaluminates, doped with Nd^{3+} or Cr^{3+}. We summarize the main spectroscopic properties which may be favorably compared to that of already known efficient high average power laser materials (YAG:Nd^{3+}) or tunable solid-state lasers (GSGG:Cr^{3+}). The possibility of improved laser efficiency by $Cr^{3+} \rightarrow Nd^{3+}$ energy transfer in codoped crystals is also studied. More generally, we present the trends in this active research field.

INTRODUCTION

In the research field of high average power, high efficiency solid state lasers special attention is now devoted to new materials. The principal requirements are :
- the possibilities of growing large crystals of high optical quality. This is determined in part by the crystallographic parameters and more precisely by the segregation coefficient.
- a good matching between pump light and material absorption in order to improve the efficiency. This can be achieved for example by using Cr^{3+}, which has broad absorption bands, as sensitizer for the Nd^{3+} activator, providing that the energy transfer is efficient and fast.
- the ability of high energy storing, which is determined by the spectroscopic parameters such as the stimulated emission crosssection, the fluorescence life time and the maximum doping level without concentration quenching.
- good thermomechanical properties.

The standard material at the present time YAG:Nd^{3+} does not meet completely these criteria.
- the crystal growth remains difficult in particular for large crystals. The segregation coefficient in very low (α=0.2):Y^{3+} (ionic radius 1.015Å) is not easily substituted by Nd^{3+} (1.12Å) in the lattice of small cell parameter (a=12.005Å).
- the efficiency is usually not more than 1-2% in part because of a poor absorption of the pump light. Moreover failure is met in sensitizing the Nd^{3+} fluorescence by Cr^{3+} because the energy transfer is very slow : as Cr^{3+} is located in a strong crystal field site its fluorescence consists in the sharp line $^2E \rightarrow {}^4A_2$ having poor overlap with the Nd^{3+} absorption spectrum and characterized by a long life time (1.5ms).
- high doping levels are prohibited for crystallographic reasons and because of concentration quenching of the Nd^{3+} emission

© American Institute of Physics 1987

In order to remove these limitations but still keeping the good thermomechanical properties of garnets, some research laboratories have tested other compounds of this family. The objective is to get larger crystalline cell by substituting Y^{3+} and Al^{3+} by larger cations. One of the solution is met with the gadolinium gallium garnets (GGG) which can be more or less substituted by other cations. All the expected advantages are present :
- the crystals which are grown by the standard Czochralski technique have a more homogeneous repartition of doping ions.
- the crystal field at Cr^{3+} site is decreased : the emission spectrum consists in the broad band $^4T_2 \to {}^4A_2$ and the line $^2E \to {}^4A_2$ appears only at low temperature showing that the $^4T_2$ level is depressed to a position very close to that of the ^2E level. The fluorescence decay time is short ($\sim 100\mu s$) and the energy transfer to Nd^{3+} fast and efficient.
- the concentration quenching is less than in YAG.

In addition, due to the broad band emission, the Cr^{3+} doped GGG-type garnets are possible materials for solid-state tunable laser in the red and near IR.

Another approach is to look for materials of different structure and such an example is the lanthanum hexaaluminate, which is described in the second part of this paper.

For these two materials we report here the main spectroscopic properties including $Cr^{3+} \to Nd^{3+}$ energy transfer, properties which are of great importance for the design of high average power lasers.

GGG-TYPE GARNETS

The crystallographic structure is cubic with 8 formula molecules per unit cell. The general composition is $C_3A_2D_3O_{12}$: Y^{3+} or Gd^{3+} and Nd^{3+} are located in the dadecahedral site C (coordination number 8), Al^{3+}, Ga^{3+}, Sc^{3+} and Cr^{3+} in the octahedral site A (coordination number 6), Al^{3+}, Ga^{3+} in the tetrahedral site D (coordination number 4). Examples of such materials, derived from YAG($Y_3Al_5O_{12}$) are GGG($Gd_3Ga_5O_{12}$), (a=12.376Å), GSGG($Gd_3Sc_2Ga_3O_{12}$)(a=12.567Å), GGG-Ca,Mg,Zr ($Gd_{3-x}Ca_xGa_{5-x-2y}Mg_yZr_{x+y}O_{12}$) (a=12.497Å), GGG-Ca,Zr without Mg or GGG-Mg,Zr without any Ca. Large scale samples can be grown due to industrial experience in magnetic bubble memory substrates making.

Two approaches have been developped depending on the use of the expansive scandium cation or not.

The best available material for medium power lasers with $\sim$100Hz as repetition rate is GSGG(Nd^{3+}-Cr^{3+}). Russian groups in collaboration with German laboratories have demonstrated a two fold increase in the efficiency of flash-lamp pumped laser rods relative to YAG:Nd^{3+} and studied the important kinetic mechanisms responsible for the observed improvement [1-2-3]. In USA, among others, Laser Propram at Lawrence Livermore National Laboratory have recently determined the spectroscopic, optical and thermomechanical properties of GSGG useful in the design of solid-state lasers [4].

In France, our own approach was to develop first no scandium based garnets as a joint effort by Government (DRET), industry (Compagnie Générale d'Electricité CGE at Marcoussis and Crismatec at Gières) and our laboratory [5] about the Ca, Mg, Zr substituted GGG. Similar research have been undertaken in USA [6] and China [7].

Doped GGG-CaZrMg crystal growth

The first advantage of this material is that the segregation coefficient of Nd^{3+} becomes closer to 1 (table I) so that crystal growth is easier than for YAG. However the Cr^{3+} segregation coefficient is different, so that inhomogeneity problems remain for codoped crystals. We note also the higher value of the lattice constants relative to YAG:Nd^{3+} allowing lower crystal field to be expected and compensation of the lattice distortion in the codoped garnets.

Table I Segregation coefficients α and lattice constants a of Nd^{3+} (or Cr^{3+}) doped GGG - Ca,Mg,Zr crystals

Samples	α	a(Å)
2.8 % Nd^{3+}	0.75	12.5038
0.08% Cr^{3+}	2.96	12.4931
0.33% Cr^{3+}	2.80	12.4942
4.6 % Cr^{3+}	2.48	12.4749
0.33% Cr^{3+} - 2.53% Nd^{3+}	2.8 (Cr^{3+}) 0.63 (Nd^{3+})	12.5003
0.09% Cr^{3+} - 3.20%Nd^{3+}	2.8 (Cr^{3+}) 0.63 (Nd^{3+})	12.5025

Nd^{3+} - GGG(Ca,Mg,Zr)

The spectroscopic properties are shown in table II and the $^4F_{3/2} \rightarrow {}^4I_{11/2}$ fluorescence spectrum at room temperature is represented in figure 1.

The most striking feature lies in the large inhomogeneous broadening of the lines in the absorption and fluorescence spectra. This is due to the presence of sites of various crystal field induced by the various cations Ca^{2+}, Mg^{2+}, Zr^{4+} introduced in the lattice.

Such a disorder has two consequences, interesting for high average power laser : (i) the concentration quenching is rejected to higher concentrations so that the best laser efficiency is obtained for 3×10^{20} Nd^{3+}/cm^3. (ii) The stimulated emission cross-section is decreased as desirable in high average power lasers for which the amplified spontaneous emission must be decreased.

Table II Spectroscopic properties of some Nd^{3+}-doped garnets

	YAG	GSGG	GGG Ca,Mg,Zr
$^4F_{3/2} \to {}^4I_{11/2}$ peak wavelength (nm)	1064	1061.2	1061.4
FWHM $\Delta\nu(cm^{-1})$	6.5	7.2	26
$^4F_{3/2}$ Stark splitting (cm^{-1})	85	63	72
$\tau_{^4F_{3/2}}$ (μs)	250	280	270
$^4F_{3/2} \to {}^4I_{11/2}$ peak cross-section σ (cm^2)	2.8×10^{-19}	1.3×10^{-19}	5×10^{-20}
Nd^{3+} optimum concentration (cm^{-3})	1.4×10^{20}	2×10^{20}	2.5×10^{20}

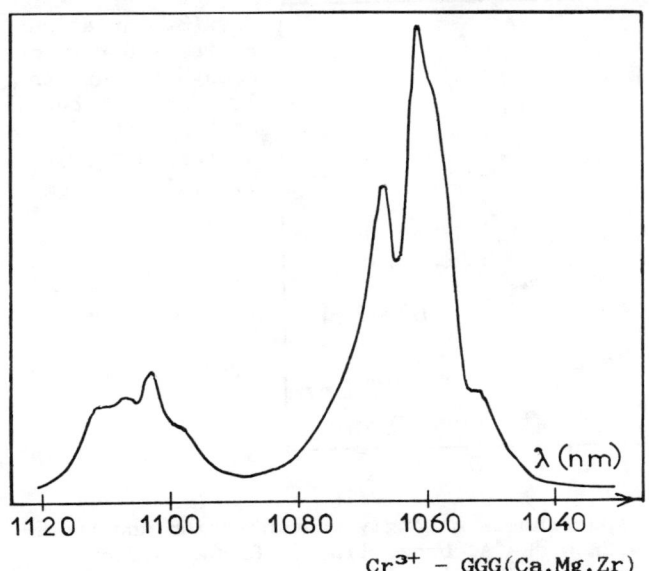

Fig. 1. $^4F_{3/2} \to {}^4I_{11/2}$ fluorescence spectrum of GGG-Ca,Mg,Zr: $3 \times 10^{20} Nd^{3+}/cm^3$

Cr^{3+} - GGG(Ca,Mg,Zr)

The absorption spectrum shows the broad bands $^4A_2 \to {}^4T_1$ and $^4A_2 \to {}^4T_2$ peaking respectively at 470nm and 640nm. At room temperature, the fluorescence spectrum consists in the $^4T_2 \to {}^4A_2$ broad band

Cr^{3+} - GGG(Ca,Mg,Zr)

The absorption spectrum shows the broad bands $^4A_2\to{}^4T_1$ and $^4A_2\to{}^4T_2$ peaking respectively at 470nm and 640nm. At room temperature, the fluorescence spectrum consists in the $^4T_2\to{}^4A_2$ broad band (maximum at 750nm) with a shoulder at about 695nm (figure 2). The Cr^{3+} life time is about 106μs at low concentration. Both these spectra and life time value show that in this compound the Cr^{3+} ions are in weak crystal field sites as in GSGG. For this reason this compound is a good candidate for near IR tunable solid state laser.

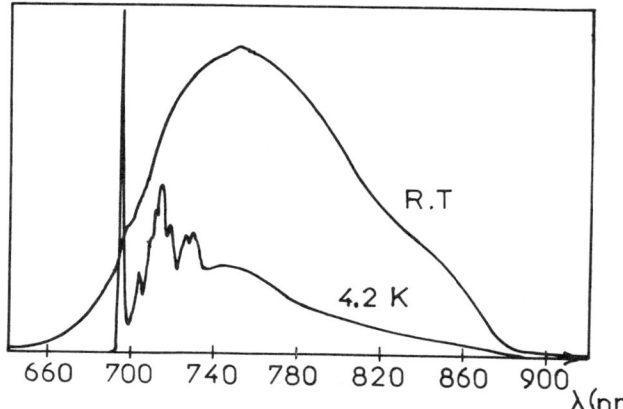

Fig. 2. Fluorescence spectra of GGG(Ca,Mg,Z) : $5.5\times10^9 Cr^{3+}/cm^3$ at room temperature and 4.2K.

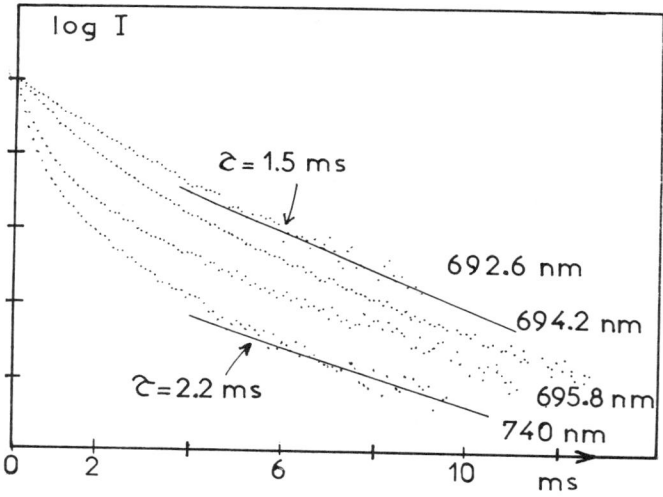

Fig. 3. Variations of the fluorescence decays through the line and the band at 4.2 K for GGG(Ca,Mg,Zr) : $5.5\times10^9\ Cr^{3+}/cm^3$.

The fluorescence spectrum is strongly concentration and temperature dependent. At 4.2K a $^2E\to{}^4A_2$ broad line at 6950Å (width 40 cm^{-1}) associated with its vibronic side-bands is superposed to the $^4T_2\to{}^4A_2$ broad band (figure 2). The excitation spectra and the fluorescence decays are different for the line and for the band (figure 3). This shows that the Cr^{3+} ions are located in two different kinds of sites [8] : some have a rather weak intermediate field and their emission is constituted by the broad band $^4T_2 \to {}^4A_2$ decreasing as temperature decreases and a broad $^2E \to {}^4A_2$ line, weak even at 4.2K, which lies on the weak energy side of the main line. Others have a rather strong intermediate field and their emission is constituted by the main $^2E \to {}^4A_2$ line, increasing strongly as temperature decreases and probably of a weak $^4T_2 \to {}^4A_2$ band hidden by the $^4T_2 \to {}^4A_2$ band of the other kind of sites. Moreover each of these kinds of sites has a broad distribution of crystal field values as is indicated by the large inhomogeneous width of the $^2E \to {}^4A_2$ lines and the non-exponentiality of all decays even at low concentration. Precise determination of the level positions and eventual energy transfer between these sites are underway in our laboratory.

If Cr^{3+} concentration is increased the $^2E \to {}^4A_2$ line and its vibronic structure disappears at 4.2K probably due to the formation both of exchange-coupled Cr^{3+} pairs and Cr^{3+} low crystal field sites.

$$Cr^{3+}-Nd^{3+}-GGG(Ca,Mg,Zr)$$

The first important point to be mentioned is the absence of any residual absorption for wavelength about 1100μm contrarily to what has been observed for GSGG crystals [9]. This should decrease the losses due to reabsorption of the laser emission at 1.06μ.

As expected for a low crystal field material the energy transfer $Cr^{3+} \to Nd^{3+}$ is effective but the observed increase in the fluorescence intensity is not very high : 40% for 3.1×10^{20} Nd^{3+}/cm^3 - $0.55 \times 10^{20} Cr^{3+}/cm^3$, concentrations for which the transfer seems to be maximum.

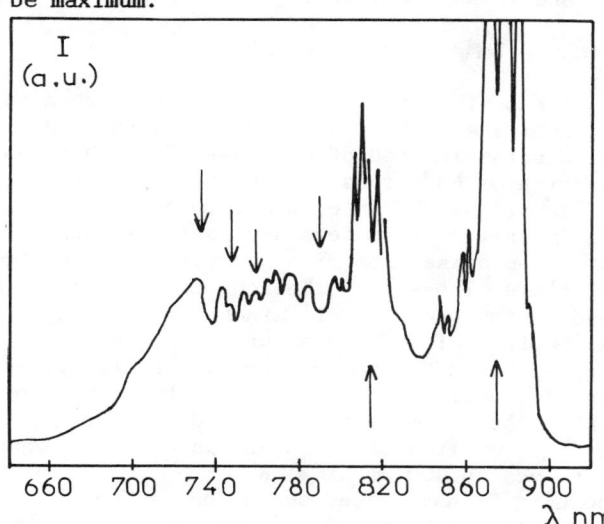

Fig. 5. Nd^{3+} emission lines ↑ and absorption lines ↓ superposed to the Cr^{3+} emission band for GGG-Ca,Mg,Zr: $3.1 \times 10^{20} Nd^{3+}/cm^3$ - $5.5 \times 10^{19} Cr^{3+}/cm^3$.

The energy transfer occurs both by radiative and non-radiative processes. The former is seen by the deeps in the Cr^{3+} broad band emission spectrum coinciding with Nd^{3+} absorption lines whereas the later is observed by the strong quenching of the Cr^{3+} integrated time constant in the presence of Nd^{3+}. The total energy transfer efficiency may be calculated from the intensities of the Cr^{3+} emission in the presence of Nd^{3+} or absence of Nd^{3+}. With $1.09 \times 10^{20} Nd^3/cm^3$ - $1.65 \times 10^{20} Cr^{3+}/cm^3$ we found η = 0.68.

The limitation of the energy transfer efficiency may be attributed to $Nd^{3+} \to Cr^{3+}$ back transfer or to energy transfer from Cr^{3+} (or Nd^{3+}) to some unknown impurity in the host or also to the poor radiative energy transfer relative to the case of $Cr^{3+}-Nd^{3+}$-codoped GSGG where this process seems to play the most important role [10]

LANTHANUM HEXAALUMINATE $LaMgAl_{11}O_{19}$ MATERIAL

A whole serie of compounds $LnMgAl_{11}O_{19}$ (Ln=La,Ce,Pr,Nd,Sm,Eu,Gd) and $LaMAl_{11}O_{19}$ (M=Mg,Mn,Fe,Co,Ni) with magnetoplumbite like structure has been prepared in the form of large single crystals and their characteristics investigated in France [11-12-13-14]. Among these materials lanthanum hexaaluminate $LaMgAl_{11}O_{19}$ appears to be an important crystalline laser host matrix which can be doped with different ions : La^{3+} being substituable by other lanthanide ions [13], Mg^{2+} by divalent transition ions [14] and Al^{3+} for instance by Cr^{3+}. Doped by Nd^{3+}, it forms the compound $La_{1-x}Nd_xMgAl_{11}O_{19}$ ($0<x<1$) a new performant laser material when x=0.1, which gives stimulated emission at 1054nm and 1082nm and even tunable in each of the two lines [15-16-17]. It may be a substitute for the $YAG:Nd^{3+}$. Doped with Ni^{2+} on the Mg^{2+} site it is a potential vibronic laser material which could lead to a tunable laser emission around 1100nm [18-19]. Doped with Cr^{3+} on Al^{3+} site, this compound may be considered as a new potential Cr^{3+} vibronic laser [20] whose present leader is alexandrite $BeAl_2O_4:Cr^{3+}$. The crystals of these compounds have quite high thermal conductivity and high hardness similar to those of $YAG:Nd^{3+}$ and can be grown by using different melting techniques like Verneuil, floating zone modified Bridgman-Stockbarger and Czochralski. At the same time, a very active research is performed both in Soviet Union [21-22] and in United States [23-24]

Nd^{3+}-doped $LaMgAl_{11}O_{19}$ crystal

The lanthanide ion environment consist of a 12-corner polyhedron but the absorption spectra indicates that, in fact, lanthanide ions occupy three closely related sites instead of only one [25] so that the absorption and emission spectra of Nd^{3+} ions are characterized by inhomogeneously broadened lines. The fairly good agreement between calculated and experimental magnetic parameters has confirmed that the real sites of the lanthanide ion arise from minor perturbations of the normal magnetoplumbite structure. These perturbations could arise from $Al^{3+}-Mg^{2+}$ site disorder and/or the presence of oxygen vacancies in the lattice. The fluorescence spectrum of Nd^{3+} is shown in Fig. 5
The concentration quenching of the Nd^{3+} luminescence is relatively weak in lanthanum hexaaluminate. The optimal neodymium concentration found in crystals grown by Czochraski method is ~ 10 at % ($\sim 3.4 \times 10^{20}$ at cm^{-3}). The radiative lifetime of the $^4F_{3/2}$ excited state is 320 µs at room temperature and this lifetime increases to 400 µs at 4.2K.

The laser emission both in CW and pulsed operation has been observed independly by French [16] and Russian [21-22] groups. It should be mentionned that advantages of Nd^{3+}-doped lanthanum hexaaluminate compared to YAG are (i) crystal growth easier with a melting temperature

1830°C instead of 1870°C for YAG, (ii) greater Nd^{3+} content

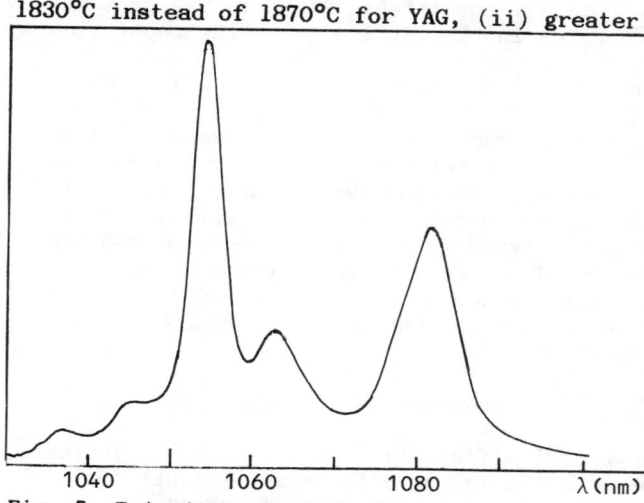

Fig. 5. Emission spectrum of
LaMgAl$_{11}$O$_{19}$: 3.4×10^{20} Nd^{3+}/cm^3.

3.4×10^{20} at. cm^{-3} instead of 1.4×10^{20} at.cm^{-3} in YAG, (iii) cheaper starting materials and (iiii) tunability. This potential specific application over not only 1.054µm but also 1.082µm is very important in a region of the spectrum generally devoid of tunable sources which coincides with the resonance transition in the triplet spectrum of helium for optical pumping in ^{3}He and ^{4}He [16].

Cr^{3+} - doped LaMgAl$_{11}$O$_{19}$ crystal

The position of the maximum of the $^4A_2 \rightarrow {}^4T_2$ absorption band (565nm) close to that of alexandrite (580nm), shows that we may assume an intermediate crystal field for the main Cr^{3+} site yielding the possibility to observe the fluorescent transitions both from ^2E and $^4T_2$ levels.

The emission spectrum is shown in Fig. 6. It consists of a main broad peak at 695nm (bandwidth 64 cm^{-1} at 4.4K) due to the $^2E \rightarrow {}^4A_2$ transition with some satellite lines at 688.6nm and 691.0nm at 4.4K and we observe a broad band extending up to 880nm due to $^4T_2 \rightarrow {}^4A_2$ transition. We have identified Cr^{3+} ions mainly into the 4f antiprism (695nm), into the 12K perturbed octahedron (691nm) and also into the 2a octahedron sites sites (688.6nm).

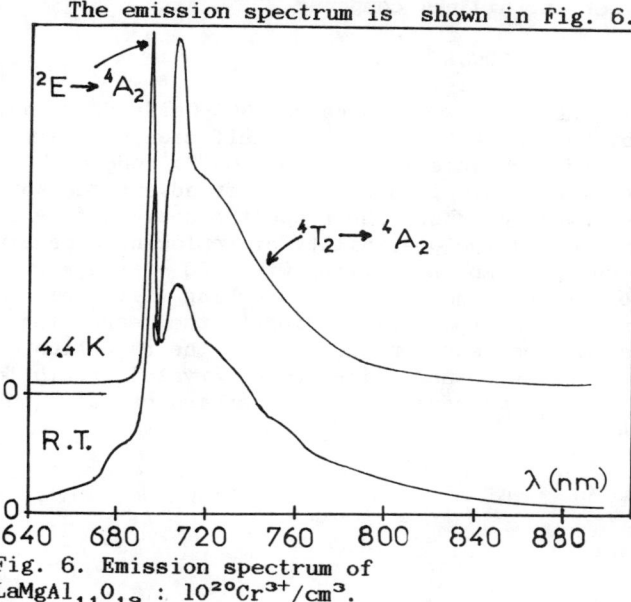

Fig. 6. Emission spectrum of
LaMgAl$_{11}$O$_{19}$: 10^{20}Cr^{3+}/cm^3.

These attributes are essentially consistent with (i) the presence of three octahedral Al^{3+} sites in the lattice, (ii) the population of each kind of site, (iii) and by the usual decreasing dependence of the $^2E \rightarrow {}^4A_2$ transition energy with the increasing of the bond length between Al^{3+} (or Cr^{3+}) and O^{2-} ions.

On the other hand, the fluorescence decays are exponential at 4.4K for low Cr^{3+} concentration but not when either the concentration or the temperature are increasing due to the existence of energy transfer mechanisms between each kind of sites. The average time constant of the ^2E level around 695 nm varies between $\tau=6.13$ms at low concentration and $\tau=5.47$ms for a higher concentration whereas $\tau=31$ms for the regular octahedron site at 688.6nm [20]. These ^2E level longlived govern the energy transfer as it was observed with other Cr^{3+} doped materials [26].

$$Cr^{3+} - Nd^{3+} - \text{codoped } LaMgAl_{11}O_{19}$$

As it has been shown Cr^{3+} efficiently sensitizes Nd^{3+} fluorescence in GSGG [21], enhancing the lasing efficiency, so that the prospect of further improving laser performance of $LaMgAl_{11}O_{19}$ crystal has to be performed. These studies are now in progress in our laboratory in collaboration with French groups [17-20] but first evidence of such $Cr^{3+} - Nd^{3+}$ energy transfer has already been demonstrated by Livermore group [23]. The energy transfer efficiency calculated from the spectral intensities of the Cr^{3+} ions alone and the Cr^{3+} ions in presence of Nd^{3+} yields $\eta=0.87$, a value comparable to GSGG : $Cr^{3+} - Nd^{3+}$. Based on this data it would appear that the Cr^{3+} sensitization of Nd^{3+} in lanthanum hexaaluminate will be about as effective as found for GSGG although the strength of the crystal field appears to be higher in this new compound.

CONCLUSION

The first assay of laser action with garnet GGG-Ca,Zr,Mg-garnet and $LaMgAl_{11}O_{19}$ doped by Nd^{3+} show that the laser efficiency is similar to that of YAG-Nd^{3+} and even superior with GGG-Ca,Zr codoped by Cr^{3+} and Nd^{3+}. These new laser materials present some advantages so that research has to be developed about their spectroscopic and dynamical properties : they are (i) the possibility of replacing expansive Sc cation by some very cheap Ca and Zr cations with GGG - type garnets, (ii) the reduction of the concentration quenching (iii) the increasing of the inhomogeneous broadning leading to the decreasing of the stimulated emission cross-section and then to the reduction of the amplified spontaneous emission and (iiii) easier crystal growth for GGG-type garnets than for YAG, moreover helped by an industrial experience of several years.

REFERENCES

1. V.G. Ostroumov, Y.S. Privis, V.A. Smirnov, I.A. Shcherbakov, J. Opt. Soc. Am. B3 81 (1986).
2. A.I. Denisov, V.G. Ostroumov, Z.S. Saidov, V.A. Smirnov, I.A. Shcherbakov, J. Opt. Soc. Am. B3 95 (1986).
3. D. Pruss, G. Huber, A. Beimowski, V.V. Laptev, I.A. Shcherbakov, Y.V. Zharikov, Appl. Phys. B28, 355 (1982).
4. W.F. Krupke, M.D. Shinn, J.E. Marion, J.A. Caird, S.E. Stokowski, J. Opt. Soc. Am. B3 102 (1986).
5. C. Garapon, A. Monteil, G. Boulon, J. Mareschal, G. Martinez, G. Villela, J.P. Dumas, Report DRET (1985) : CGE - Crismatec - Université Lyon I and CNRS : unpublished results.
6. M.D. Shinn, W.F. Krupke, J.A. Caird, L.K. Smith, International Conference Proceedings CLEO/IQEC (1985).
7. L.H. Zhang, Prog. Crystal growth and charact. 11, 283 (1985).
8. A. Monteil, C. Garapon, G. Boulon (to be published).
9. S. Stokowski, M. Shinn, This conference.
10. G. Armagan, B. Di Bartolo, International Conference proceedings CLEO/IQEC (1986).
11. A. Kahn, A.M. Lejus, J. Thery, D. Vivien, French Patent N801 51 71 (July 8, 1980).
12. A. Kahn, A.M. Lejus, M. Madsac, J. Thery, D. Vivien, J.C. Bernier J. Appl. Phys. 52 (II) 6864 (1981).
13. D. Saber, A.M. Lejus, Mat. Res. Bull. 16 (10) 1325 (1981).
14. F. Laville, A.M. Lejus, J. Cryst. Growth 63, 426 (1983).
15. D. Vivien, A.M. Lejus, J. Thery, R. Collongues, J.J. Aubert, R. Moncorgé, F. Auzel, C.R. Acad. Sc. Paris t. 298, série II, n° 6, 195 (1984).
16. L.D. Schearrer, M. Leduc, D. Vivien, A.M. Lejus, J. Thery, I.E.E.E. J. Quantum Electr. 22 (5) 713 (1986).
17. J.J. Aubert, C. Wyon, D. Vivien, A.M. Lejus, IQEC Conference (1986)
18. F. Laville, M. Perrin, A.M. Lejus, M. Gasperin, R. Moncorgé, D. Vivien, J. Solid State Chem. (in press) (1986).
19. R. Moncorgé, T. Benyattou, D. Vivien, A.M. Lejus, J. of Luminescence 35, 199 (1986).
20. B. Viana, A.M. Lejus, D. Vivien, V. Ponçon and G. Boulon, J. Solid State Chem. (to be published) (1986).
21. K.S. Bagdasarov et al, Sov. J. Quantum Electron, 13, 639 (1983) and 13, 1082 (1983).
22. V.M. Garmash, A.A. Kaminskii, M.I. Polyakov, S.E. Sarkisov, A.A. Filimonov, Phys. Stat. Sol. (a) 75, KIII (1983).
23. M.D. Shinn, W.F. Krupke, J.A. Caird, H.W. Newkirk, Int. Laser Science Conf. Dallas November 21 (1985), Conference Proceedings n° 146 - New-York (1986) "Optical Science and Engineering" - Series 6 Advances in Laser Science - 1 - Ed. W.C. Stalley and M. Lapp. p. 216.
24. M.D. Shinn, J.A. Caird, W.F. Krupke, IQEC Conference (1986).
25. D. Saber, J. Dexpert-Ghys, P. Caro, A.M. Lejus, D. Vivien, J. Chem. Phys. 82, 5648 (1985).
26. G. Boulon, Materials Chemistry and Physics to be published (1986).

AN EXTENDED FOERSTER-DEXTER MODEL FOR CORRELATED DONOR-ACCEPTOR PLACEMENT IN SOLID STATE MATERIALS

S.R. Rotman and F.X. Hartmann
Institute for Defense Analyses
1801 N. Beauregard Street
Alexandria, Virginia 22311

ABSTRACT

The current theory of donor-acceptor interactions in solid-state materials is based on a random distribution of donors and acceptors through the crystal. In this paper, we present a model to calculate the observable transfer rates for the correlated positioning of donors and acceptors in laser materials. Chemical effects leading to such correlations are discussed.

INTRODUCTION

A good phosphor or laser candidate ion, in a particular lattice, must be able to absorb the pumping light efficiently as well as emit at the desired wavelengths. Recently, based on a theory originated by Foerster[1] and Dexter,[2] and further developed by Inokuti, et al.,[3] there has been an increasing desire to separate the absorption and emission processes to different ions, rather than attempt to identify one single dopant which can both absorb and emit. The advantage of such an approach is that a good donor (which absorbs the external radiation flux) can be paired with a good acceptor (the emitting lasing ion). An efficient transfer of energy between the two ions (donor to acceptor) is necessary in this scheme.

In models of non-radiative transfer applied to doped solid-state laser materials, the distance between excited donors and neighboring acceptors critically affects the calculated energy transfer rates. In the generally successful formulation of non-radiative transfer theory, the donors and acceptors are taken to be randomly distributed and are independent (uncorrelated) in position.[4,5] For real crystals this is not necessarily the case, and actual non-radiative transfer rates may deviate from the simple Foerster-Dexter description.

In this paper we present a more general expression to treat donor-acceptor transfer rates. In comparing our results to the Foerster-Dexter theory, we consider two specific cases appropriate to actual laser materials: (1) an "excluded" volume around a donor diminished in acceptor concentration and (2) an "enhanced" volume around a donor in which acceptors preferentially locate. These are simpler cases of our general result and the details are to be published elsewhere.

ENERGY TRANSFER THEORY

For radiationless electromagnetic interactions, the strength of the non-radiative transfer rate is proportional to r^{-s} where r is the radial distance between the donor and the acceptor and s takes on specific integer values. In particular, s = 6 for a dipole-dipole interaction, s = 8 for a dipole-quadrupole interaction, s = 10 for a quadrupole-quadrupole interaction, etc. Energy transfer between uniformly distributed ions was examined by Inokuti et.al.[3] They show that the time-dependent excited donor concentration $N_D(t)$ is

$$N_D(t) = N_D(0) \exp[-t/\tau_0 - \Gamma(1 - 3/s) c/c_0 (t/\tau_0)^{3/s}] \quad (1)$$

where c is the acceptor doping concentration, Γ is the gamma function, c_0 is the critical concentration of acceptors, and τ_0 is the natural decay rate of the donor. The critical concentration c_0 is that concentration at which the energy transfer rate and the natural donor decay rate are equal for the average donor-acceptor distance r_0. For the dipole-dipole interaction (s = 6) the decay rate contains an exponential $t^{1/2}$ factor.

Locally correlated donor decay rates are found by starting with eq. (2):

$$N_D(T) = N_D(0) \exp(-t/\tau_0) \lim_{\substack{N_A \to \infty \\ V \to \infty}} \left\{ \int_V \exp[-tn(r)] u(r) dV \right\}^{N_A} \quad (2)$$

where the number of acceptors N_A and the volume V extend to infinity such that the concentration N_A/V remains finite. Here n(r) is the radially dependent transfer rate. For the "excluded volume" case where there are no acceptors within a volume V_i (of radius r_i), the distribution function u(r) is depicted in Fig. 1a. Solving for $N_D(t)$ we obtain

$$N_D(t) = N_D(0) \exp\left\{-t/\tau_0 - CV_i [1 - \Phi(Z_i)/\exp(Z_i)]\right\} \quad (3)$$

where

$$Z_i = (r_0/r_i)^6 (t/\tau_0) \quad (4)$$

and $\Phi(1, 1 - 3/s; Z)$ is the degenerate hypergeometric function (written as $\Phi(Z_i)$ for the case s = 6).

For the case of dipole-dipole interactions, as r_i goes to zero, Z_i goes to infinity. For large Z_i, the ratio $\Phi(Z_i) \exp(-Z_i)$ approaches $\sqrt{\pi Z_i}$. In this case, eq. (3) reduces to eq. (1) for s = 6 (since $\Gamma(1/2) = \sqrt{\pi}$). As physically required, the Foerster-Dexter solution is obtained in the small r_i limit. Specifically, the correlation effects appear as a deviation from Gaussian behavior of the term-by-term ratio of the series expansion of $\Phi(Z_i)$ to $\exp(Z_i)$.

One possible distribution for an enhanced placement of acceptors near donors is shown in Fig. 1b. The solution for $N_D(t)$ is now:

$$N_D(t) = N_D(0) \exp[-t/\tau_o - A\sqrt{\pi}\,\overline{\frac{C}{C_o}}\left(\frac{t}{\tau_o}\right)^{1/2}$$

$$- (B-A)\,\Phi(Z_i)/\exp(Z_i)$$

$$- (1-B)\,\Phi(Z_D)/\exp(Z_D)] \tag{5}$$

From these results it can be shown that for a generalized distribution u(r) in eq. (2), the donor decay is:

$$N_D(t) = N_D(0) \exp\left[-t/\tau_o - \int \frac{du(r)}{dr} CV(r) \frac{\Phi(Z_r)}{\exp(Z_r)}\,dr\right] \tag{6}$$

In Fig. 2 we show the corresponding donor and acceptor decays obtained by both the excluded-volume and enhanced volume models.

CHEMISTRY OF CORRELATED PLACEMENT

The natural correlation of donors and acceptors may occur due to the chemistry of the codoped ions--both electronic structure and ionic size play a role. Such correlation would be especially useful in crystals in which deliberate macroscopic or microscopic non-uniform doping would help control thermal gradients or facilitate the usage of large crystals.

Size mismatch between dopants and sites leads to local correlations. A dramatic example of this effect has been noted in $Eu:Mn:RbMgF_3$. Shinn, et al.[6] find 95 percent of the Eu^{+2} paired with Mn^{2+} ions.

Aliovalent doping also provides a promising approach to new laser materials. If one codopes both with ions which are effectively negative and ions which are effectively positive relative to the sites they enter, charge neutrality is maintained; moreover, the coulombic attraction will cause the ions to attract each other. For example, codoping Ni^{+2} and Zr^{+4} into trivalent - cation sites leads to Zr-Ni pairs which tend to be correlated and can affect the type of lattice site occupied, i.e. Ni^{+2} goes partially into tetrahedral sites in YAG only when codoped with zirconium.[7]

CONCLUSIONS

A model for analzying Foerster-Dexter non-radiative transfer under the conditions of correlated donor-acceptor placement has been analyzed. Chemical effects leading to such correlation have been discussed.

REFERENCES

1. T. Foerster, Ann. Phys. 2, 55 (1948).
2. D.L. Dexter, J. Chem. Phys. 21 (5), 836 (1953).
3. M. Inokuti and F. Hirayama, J. Chem. Phys. 43 (6), 1978 (1965).

4. L.A. Riseberg and M.J. Weber, "Relaxation Phenomena in Rare-Earth Luminescence" in Progress in Optics XIV, ed. E. Wolf, publ. by North-Holland, pp. 91-158 (1976).
5. W.M.Yen, J. de Phys. C6, 333 (1983).
6. M.D. Shinn and W.A. Sibley, Phys. Rev. B 29 (7), 3834 (1984).
7. S.R. Rotman, "Defect Structure of Luminescent Garnets", Ph.D. thesis in the Department of Electrical Engineering, M.I.T., September 1985.

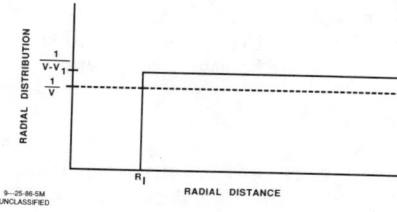

Fig. 1a) Excluded-volume, donor-acceptor radial-distance probability distribution

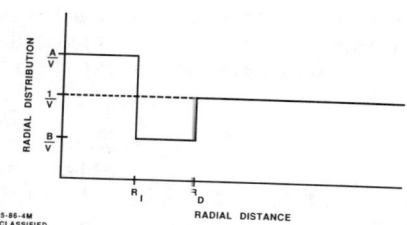

Fig. 1b) Enhanced-volume, donor-acceptor radial-distance probability distribution

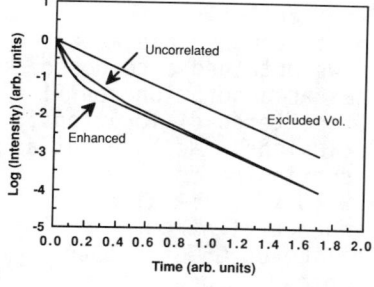

Fig. 2a) Time-resolved excited-donor concentraton

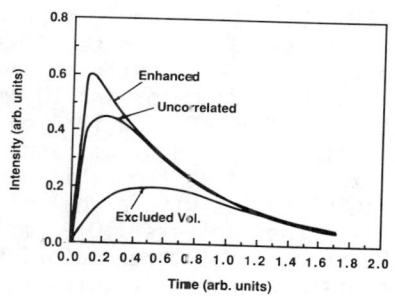

Fig. 2b) Time-resolved excited-acceptor concentration

LARGE Nd,Cr:GSGG BOULE GROWTH AND QUALITY*

S. E. Stokowski and M. D. Shinn
Lawrence Livermore National Laboratory, Livermore, CA 94550

M. Randles and D. Dawes
Allied-Signal Corp., Charlotte, N. C. 28231

ABSTRACT

Gadolinium scandium gallium garnet (GSGG) has been grown in boules of 13-cm diameter by 15 cm long. Three scale-up issues have been identified and progress made toward their solution. Large slabs of Nd,Cr:GSGG are now being fabricated for efficient, medium power lasers.

INTRODUCTION

Nd,Cr:GSGG is an excellent material for use in efficient, medium average power lasers.[1-5] Nd,Cr:GSGG lasers have about twice the efficiency of Nd:YAG lasers in flashlamp-pumped, rod systems because of Cr sensitization. Further, "coreless" Nd,Cr:GSGG can be grown by the Czochralski method in much larger sizes than Nd:YAG, which contains a central "core" with high birefringence.

In 1985 we initiated a program to develop a process for growing large diameter Nd,Cr:GSGG boules for use in slab lasers.

GSGG BOULE GROWTH

We grew GSGG by the Czochralski method using a 23-cm diameter by 23-cm deep iridium crucible to contain the melt, which initially weighed 53 kgf. The Nd and Cr concentrations in the resulting crystal are 2×10^{20} cm^{-3} and 1×10^{20} cm^{-3}, respectively.

The first 13-cm diameter GSGG boule grown was undoped and about 15 cm long (Fig. 1). After a few growth runs we obtained a co-doped boule 13 cm in diameter and 10 cm long. It has an absorption coefficient at 1064 nm of $<5 \times 10^{-3}$ cm^{-1}, about 100 scattered dislocations, a birefringence of 8 nm/cm, and no iridium inclusions. We have cut a $1 \times 7 \times 10$ cm^3 slab from this boule. We are continuing to grow Nd,Cr:GSGG boules with a goal of obtaining $1 \times 10 \times 20$ cm^3 slabs with loss coefficients $<10^{-3}$ cm^{-1} and birefringence <1 nm/cm.

We encountered three major problems in scaling-up GSGG to 13-cm diameters: a spiral or corkscrew growth morphology, a one-micron absorption band, and boule cracking during cool-down.

Spiral growth morphology is a common problem in Czochralski growth. Previously, GGG crystal growers determined that adding calcium to the melt prevents spiraling. Thus, we use Ca (about 100 pm by weight) in our GSGG growth runs. We have also found that the tetravalent ions Si and Zr induce spiraling; therefore, we are

---

*Work performed by Lawrence Livermore National Laboratory under the auspices of the Defense Advanced Research Projects Agency, ARPA Proj. No. 5358 and DOE/UC Contract No. W-7405-ENG-48.

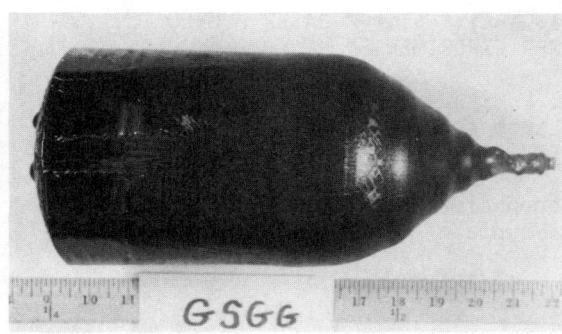

Fig. 1. First 13-cm diameter GSGG boule grown. Length at full diameter is 15 cm. (Scale is in inches.)

using the purest raw material available, which contains <5 ppm Si and <1 ppm Zr by weight.

A one-micron absorption band sometimes appears in garnet crystals doped with Cr. We have correlated its intensity with Ca content in the crystal.[4] Also the one-micron absorption is reduced by adding tetravalent ions (Si or Zr) to the melt, or by heating Nd,Cr:GSGG in a reducing atmosphere. We suggest that the absorption band is that of Cr^{4+} in the tetrahedral site, which forms to charge-compensate for the presence of divalent ions.

Because we use Ca to avoid boule spiraling, we need to have an ion other than Cr^{4+} to provide charge compensation. We found that adding the redox couple (Ce^{3+}, Ce^{4+}) to the melt is effective in compensating for Ca^{2+} and reducing the one-micron absorption band, without adversely affecting boule morphology or laser efficiency.

Boule cracking during cool-down is apparently due to thermal stresses and not to internal stresses generated by dislocations. Our plan is to decrease the cool-down rate and to make the furnace more isothermal during cool-down.

CONCLUSIONS

We have demonstrated that large, good-quality Nd,Cr:GSGG boules can be grown by the Czochralski technique. Adding Ca and Ce to the GSGG melt has proven effective in preventing boule spiraling and one-micron absorption. Nd,Cr:GSGG slab lasers with outputs approaching 1 kW average power are now a real possibility.

REFERENCES

1. E. V. Zharikov, V. V. Laptev, V. G. Ostroumov, Yu. S. Privis, V. A. Smirnov, and I. A. Shcherbakov, Sov. J. Quantum Electron. 14, 1056-1062 (1984).
2. A. Beimowski, G. Huber, D. Pruss, V. V. Laptev, I. A. Shcherbakov, and Y. V. Zharikov, Appl. Phys. B28, 234-235 (1982).
3. D. Pruss, G. Huber, A. Beimowski, V. V. Laptev, I. A. Shcherbakov, and Y. V. Zharikov, Appl. Phys. B28, 355-358 (1982).
4. J. A. Caird, M. D. Shinn, T. A. Kirchoff, L. K. Smith, and R. E. Wilder, Appl. Opt. 25, 4294 (1986).
5. W. F. Krupke, M. D. Shinn, J. E. Marion, J. A. Caird, and S. E. Stokowski, J. Opt. Soc. Am. B3, 102 (1986).

SINGLE CRYSTAL $Na_3Ga_2Li_3F_{12}:Cr^{3+}$ LASER PUMPED LASER EXPERIMENTS[*]

J. A. Caird, P. R. Staver, and M. D. Shinn
University of California, Lawrence Livermore National Laboratory,
P. O. Box 5508, Livermore, CA 94550

H. J. Guggenheim and D. Bahnck
AT&T Bell Laboratories, 600 Mountain Avenue, Murray Hill, NJ 07974

ABSTRACT

We report for the first time tunable quasi-CW lasing from single crystal $Na_3Ga_2Li_3F_{12}:Cr^{3+}$. A free lasing wavelength of 791 nm and a tuning range from 741 nm to 841 nm was obtained. Measurements of the slope efficiency, and gain and loss coefficients are reported.

LASER EXPERIMENTS

A crystal was grown with 4 at.% Cr^{3+} substituted for Ga^{3+} by horizontal zone melting. A section 4.5 mm thick was pumped by a krypton ion laser at the center of a 5 cm radius near spherical resonator similar to that developed for previous experiments.[1] A single birefringent tuner plate was used to provide variable output coupling by adjustment of the angle of incidence, θ, away from Brewster's angle. The round trip output oupling, C_{out}, is given by

$$C_{out} = 4R(\theta) + T_{M1} + T_{M2} \qquad (1)$$

where T_{M1} and T_{M2} are the transmissions of the resonator mirrors M_1 and M_2 respectively. $R(\theta)$ is the single surface reflectivity of the tuner plate.

Output power levels at 786 nm were determined through measurements with calibrated thermopiles, including all sources of output coupling indicated by (1). Typical slope efficiencies of 15% were obtained with a C_{out} of 2%. The observed slope efficiency is considerably less than the quantum defect limited value of 82%, which led to speculation on whether passive losses were responsible for the degradation. Passive losses such as scattering from imperfections and impurity absorptions can potentially be eliminated since they are not intrinsic material properties. The levels of passive loss were measured by a modified Findlay-Clay technique[2] as described below.

The tuning range was measured at different input power levels, parametric in output coupling, and the end points are plotted versus the corresponding absorbed power in Fig. 1. The curves drawn through these points represent the threshold power as a function of wavelength. The five data sets shown in Fig. 1 correspond to different settings of the tuner plate reflectivity.

---

[*]Work performed under the auspices of the U.S. Department of Energy by Lawrence Livermore National Laboratory under Contract No. W-7405-ENG-48.

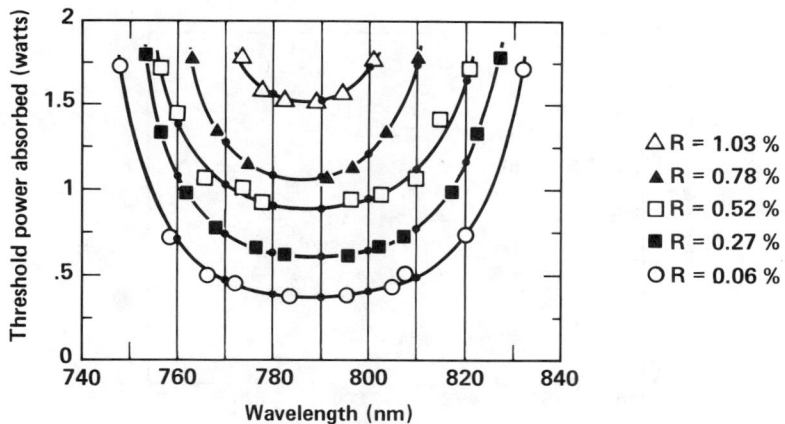

Figure 1. Threshold power absorbed vs. output wavelength parametric in tuner plate reflectivity.

ANALYSIS AND DISCUSSION

In the limit of small gains and losses, the threshold condition for laser oscillations can be written as

$$G_p \, P_{th} = C_{out} + L \qquad (2)$$

where G_p is the gain per unit pump power absorbed and P_{th} is the absorbed power at threshold. The dependence of P_{th} on C_{out} can be determined for any output wavelength from the data in Fig. 1. The intersections of the vertical lines with the curves give the threshold power at different values of $R(\theta)$. Values of C_{out} are obtained from (1), using measured transmission data for the mirrors. G_p is then calculated as the derivative of C_{out} w.r.t. P_{th} at any output wavelength. The passive loss, L, is obtained from a linearized plot as the extrapolated value of $-C_{out}$ at $P_{th} = 0$.

Values of the passive insertion loss obtained by this method varied from 0.2 to 1.0 % cm^{-1}, and were generally higher at the short wavelength end of the tuning range. Spectrophotometer traces indicated losses several times greater than the laser measurements. This result is apparently due to the fact that the spectrophotometer averages the loss over a relatively large area of the crystal (>10 mm^2), while the laser mode probes a very small area (<0.1 mm^2). Experiments performed over a large number of locations in the sample showed that the average threshold for lasing was more than 3 times higher than the minimum threshold. All laser pumped laser measurements of gain, loss, and efficiency were performed at a location where the threshold and therefore passive losses were minimized. The loss levels determined by the laser experiments are thought to be more representative of intrinsic material properties.

The values of gain per unit pump power absorbed derived from the data in Fig. 1 are plotted vs. wavelength in Fig. 2. Theoretically, G_p is related to the cross section for stimulted emission, $\sigma_e(\lambda_\ell)$ by

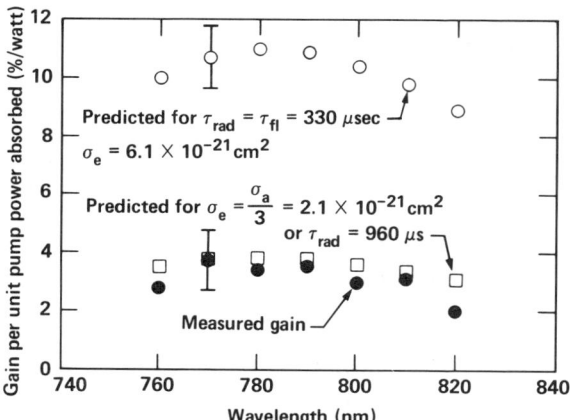

Figure 2. Predicted and measured round trip gain per unit pump power absorbed.

$$G_p = \frac{4\lambda_p \tau_{fl} \sigma_e(\lambda_\ell)}{\pi \omega_0^2 hc} \qquad (3)$$

where τ_{fl} is the room temperature fluorescence lifetime (330 μs), ω_0 is the gaussian mode radius of the pump beam (87 μm), and λ_p and λ_ℓ are the pump and output laser wavelengths, respectively. The emission cross section was estimated in two different ways. First, the room temperature fluorescence lifetime was assumed to be purely radiative, and the fluorescence lineshape was used to calculate the emission cross section as a function of wavelength. This model, however, leads to a predicted gain 3 or 4 times greater than the measured gain as shown in Fig. 2. Since the degeneracy of the $^4T_2$ upper laser level is 3 times that of the $^4A_2$ ground state, it could be postulated that the peak emission cross section is actually only one third of the peak absorption cross section. This value for σ_e leads to gain predictions very comparable to the measured gain but would indicate a high non-radiative relaxation rate at room temperature. Recent spectroscopic measurements of the lifetime and absorption linestrength as a function of temperature provide no evidence to support the presumption of strong nonradiative relaxation.

As a check on the experimental methodology, an identical series of measurements was performed on a sample of $GSGG:Cr^{3+}$ which is known to have near unity quantum efficiency. In these experiments the measured and predicted gains were nearly equal for short wavelengths. The measured gain fell to a maximum deviation of 30% less than the predicted gain at longer wavelengths.

CONCLUSION

Passive insertion loss levels were found to be on the order of 1% cm^{-1} in our laser experiments using a modified Findlay-Clay analysis. With 2% output coupling this level of passive loss would reduce the achievable slope efficiency, n_s, by no more than one-third, according to the relation[3]

$$n_s = \frac{\lambda_p}{\lambda_\ell} \cdot \frac{C_{out}}{C_{out} + L} \qquad (4)$$

Thus, the passive losses are inadequate to explain the low observed slope efficiencies. Possible explanations for the low slope efficiency include the presence of additional losses due to excited state absorption, or very rapid nonradiative relaxation of ions in defect sites.

The levels of gain produced in the material were low compared to a prediction based on the assumption of unity quantum efficiency at room temperature. The reduced gain could also be indicative of a high non-radiative relaxation rate at room temperature and/or the presence of a significant amount of excited state absorption.

REFERENCES

1. J. A. Caird, P. R. Staver, M. D. Shinn, H. J. Guggenheim, and D. Bahnck, to be published in the Proceedings of the Third International Conference on Tunable Solid State Lasers, Zigzag, Oregon, June 4-6, 1986.
2. D. Findlay and R. A. Clay, Phys. Lett., 20, 277 (1968).
3. H. G. Danielmeier, in Lasers Vol. 4, A. K. Levine and A. J. DeMaria eds., Marcel Dekker, New York, 1976, p. 28.

Nonlinear Refractive Index Measurements of Glasses and Crystals

Robert Adair, L. L. Chase, and Stephen A. Payne*

Lawrence Livermore National Laboratory
University of California
Livermore, California 94550

We have measured the nonlinear index for numerous glasses and crystals with the goal of determining the material properties that contribute to the magnitude and dispersion of n_2. We have found that a nearly-degenerate three-wave-mixing (TWM) process, similar to that first used by Maker and Terhune[1], is the most useful method since it can provide both rapid and accurate n_2 measurements.

The TWM process combines the two input fields at frequencies ω_1 and ω_2 such that new frequencies at $2\omega_1-\omega_2$ and $2\omega_2-\omega_1$ are generated. For phase-matched mixing of identically polarized beams of intensities I_1 and I_2 in an isotropic medium of length ℓ, the TWM intensity at $\omega_s=2\omega_1-\omega_2$ is given by

$$I(\omega_s) = \frac{k}{n_0} \mid \chi^{(3)}_{1111}(\omega_s,\omega_1,\omega_1,\omega_2) \mid^2 I_1^2 I_2 \ell^2$$

where k is a constant and $\chi^{(3)}_{1111}(\omega_s,\omega_1,\omega_1,\omega_2)$ is a component of the tensor describing the third order nonlinear susceptibility. The nonlinear refractive index is related to the third-order susceptibility by the relation

$$n_2(LP) = \frac{12\pi}{n_0} \chi^{(3)}_{1111}(-\omega;\omega,\omega,-\omega)$$

where LP indicates that we are considering the case of a linearly polarized laser beam. The magnitude of n_2 can

---

*Work performed under the auspices of the U.S. Department of Energy by Lawrence Livermore National Laboratory under Contract No. W-7405-ENG-48.

© American Institute of Physics 1987

therefore be determined from the intensity of the signal at
ω_s. Ideally, one would want to have $\omega_1 = \omega_2 = \omega_s$ to
obtain n_2. TWM uses frequency discrimination to separate the
signal from the pump beams, however, so ω_s must be shifted
from ω_1 by a small amount. We have found that a shift of
$\Delta = \omega_1 - \omega_2 = 60$ cm^{-1} is sufficient to separate out the
signal beam, while not so large as to cause the value of n_2
obtained to differ from that at $\Delta = 0$ cm^{-1}.

In the past, other methods have been used to measure n_2.
The most extensive measurements have involved the technique of
time-resolved interferometry (TRI).[2-3] This method provides
absolute measurements of n_2 at $\Delta = 0$ cm^{-1} with good
accuracy, but requires large samples (several centimeters) of
high optical quality, and precise control of the laser beam
both temporally and spatially. In contrast, TWM measures the
magnitude of n_2 relative to some known standard and requires
special techniques to determine the sign. Offsetting this is
the speed of the measurement (about two hours per data point),
and the more modest optical requirements on the sample and
laser beams.

Our measurements were calibrated by matching our relative
results to eight TRI n_2 measurements.[2,3] We can compare
these calibrated measurements with those calculated by an
empirical formula developed by Boling, et.al.[4] Their equation
is based on a simple model for the nonlinear refractive index
of isotropic materials and is known to work reasonably well for
a large number of glasses. The model assumes that the solid
contains a single polarizable constituent and, in analogy with
"Millers Rule" for $\chi^{(2)}$, that the hyperpolarizability is
proportional to the square of the linear polarizability. Within
this approximation, n_2 can be determined solely from
information regarding the linear index. Fig. 1 gives a graphic
comparison of the values calculated by this model and our

measured values. The agreement is quite good except for the high index glasses, for which the measured values fall below the calculated ones. We have also measured n_2 for a number of garnet crystals that are of interest for laser applications. The n_2 values for the garnets, as well as the calculated values obtained from the formula of Boling, et.al., are given in Table 1. For these materials, the measured values are also somewhat smaller than the calculated ones.

Our work using 1.06 μm light has shown that nearly-degenerate three-wave mixing provides a rapid and reliable method for measuring the nonlinear refractive index.[5] Results obtained by this method agree well with those obtained by other methods and with values calculated for low index glasses using the empirical formula of Boling, et.al.[4] We now plan to perform these measurements at the Nd laser harmonics, and possibly as a continuous function of wavelength near one- and two-photon absorption thresholds in various media.

References

1. P. D. Maker and R. W. Terhune, , Phys. Rev. 137, A801 (1965).
2. M. J. Moran, C.-Y. She, and R. L. Carman, , IEEE J. Quantum Electron. QE-11, 259-263 (1975).
3. M. J. Weber, D. Milam and W. L. Smith, , Opt. Eng. 17, 463-469 (1978).
4. N. L. Boling, A. J. Glass, and A. Owyoung, , IEEE J. Quant. Elec. QE-14, 601-608 (1978).
5. Robert Adair, L. L. Chase, and Stephen A. Payne, , to be published in the J. Opt. Soc. Am. B.

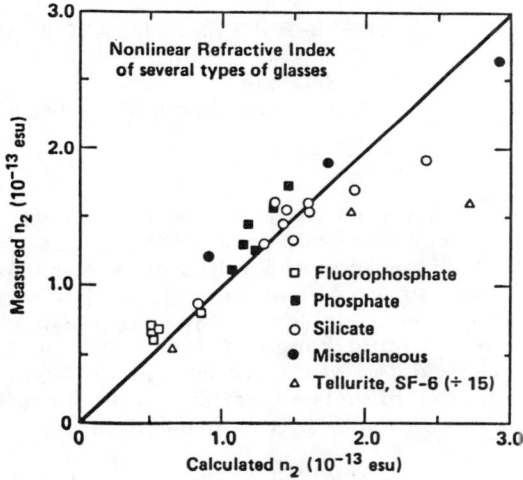

Fig. 1. Comparison of the calculated and measured nonlinear refractive indices of the glasses listed in Table 1. Note that the n_2 values of the SF-6 and tellurite glasses, represented by triangles, have been divided by 15.

Table 1
Calculated and measured nonlinear refractive indices of Garnets (10^{-13} esu)

Sample	$n_2^{meas.}$ (this work)	$n_2^{calc.}$
$Gd_3Sc_2Al_3O_{12}$	4.0	5.7
$Gd_3Sc_2Ga_3O_{12}$	5.2	8
$Gd_3Ga_5O_{12}$	5.8	8
$Y_3Al_5O_{12}$	2.7	5.0
$Y_3Ga_5O_{12}$	5.2	5.8
$La_3Lu_2Ga_3O_{12}$	5.8	8.2

HIGH EFFICIENCY LASER-PUMPED EMERALD LASERS

S.T. Lai
Allied-Signal, Inc., Morristown, NJ 07960

ABSTRACT

Highly efficient laser operation has been achieved in emerald. In a quasi-cw laser-pumped emerald laser, 64% output slope efficiency has been measured at 768nm, corresponding to a laser quantum yield of 76%. An output power of 1.6 W was reached at 3.6 W of pump power at 647.1nm from a krypton laser, and was pump power limited. The emerald laser has a tuning range of 720 to 842nm. The round trip loss excluding the excited state absorption (ESA) is 0.4%/cm. These results indicate the high laser efficiency and the high optical quality of the emerald attainable in the present laser.

INTRODUCTION

We demonstrated in alexandrite nearly conversion-limited laser operation in a finely tuned laser-pumped laser. A 51% slope efficiency and a projected 85% quantum yield was measured[1]. Such a high efficiency laser cavity has been used in the critical evaluation of other new laser materials such as $SrAlF_5$:Cr and $ScBO_3$:Cr. A laser-pumped emerald laser has been studied earlier with lesser quality emerald crystals[2]. The optical quality has since been significantly improved under a crystal development program. The results from the present study show that emerald has a laser efficiency close to that of alexandrite, and is significantly higher than all other Cr doped room temperature lasers including GSGG, GSAG, $KZnF_3$, $ScBO_3$, and $SrAlF_5$.

LASER EFFICIENCY

The laser efficiency of emerald can be evaluated through its output power slope and the threshold pump power. An expression for the slope efficiency has been derived for vibronic lasers with provision for ESA losses [3]:

$$\eta = \left(\frac{h\omega_L}{h\omega_p}\right) \frac{T_o}{T_o + L + \Delta} \cdot \xi, \qquad (1)$$

All parameters were defined in Ref. 3. Our interest is to measure L, an extrinsic loss term, and Δ, the ESA loss at the laser wavelength. The extrinsic losses include the scattering due to the less than ideal optical quality of the crystal, the absorption loss due to impurities, and thermally induced lensing and birefringence. As indicated by equation (1), the higher the the output coupler transmission T_o, the higher the output slope η, providing that the sum of the ESA loss and the extrinsic loss L is comparable or higher than T_o. However, if T_o is significantly greater than L, the slope η is then limited by the ESA, and it approaches a limiting value determined by $\sigma_e/(\sigma_e + \sigma_{2a})$, where σ_{2a} and σ_e are the ESA cross section and the emission cross section at the laser wavelength respectively. A good laser performance would therefore comprise of: (1) high attainable η (low σ_{2a}/σ_e), and (2) a reasonably small T_0 with which the limiting slope can be reached (low extrinsic losses).

EMERALD LASER MEASUREMENTS

Our focus in the laser measurements was on exploring the operating limits in emerald lasers. The laser oscillator was formed by two concave mirrors in a nearly concentric configuration[1]. The minimum beam waist was about 60um and was mode matched with the pump beam in a 3mm long emerald. With a single element birefringent filter, the laser was tuned from 720 to 842nm. A 64% output slope was measured at 768nm with an output coupler of 1.16% transmission (Fig. 1). The maximum output was 1.65 W at 3.6 W of raw pump power from a krypton laser. These results are among the highest in laser efficiency reported in laser-pumped Cr lasers. The effect of ξ is very small (.99) in our case[4]. Output slopes at other free running wavelengths and their respective output coupler transmission are listed in Table I. In general, emerald lasers operate with high efficiency and high gain in the 760 to 790nm range. The round trip loss including the ESA loss is 0.36%. The extrinsic loss L can be isolated if the ESA cross section at the laser wavelength is known. We use the results reported earlier[5] from a single-pass gain measurement in emerald. The L is calculated to be 0.08% in a 2mm thick emerald, corresponding to a laser round trip loss of 0.4%/cm. This value is comparable to the optical loss in the well developed laser crystals such as YAG:Nd and alexandrite.

Acknowledgement: This work was supported by the U.S. Army Research Office under grant DAAG 29-83-C-0015.

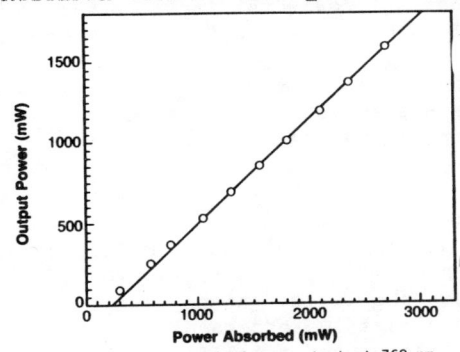

Fig. 1 Emerald laser output at 768 nm.

Table I. Output Slope Efficiency in Emerald.

Free running Wavelength	Output coupler Transmission (%)	Slope Efficiency (%)
768	1.16	64
771	0.85	56
776	1.76	56.7
790	0.25	38.7*
822	0.89	38

*Measurement was made in a 2 mm thick sample.

REFERENCES

1. S.T. Lai, and M.L. Shand, J. Appl. Phys. 18, 5642, 1983.
2. M.L. Shand, and S.T. Lai, IEEE J. Quantum Electr. QE-20, 105, 1984.
3. S.T. Lai, and M.L. Shand, Proc. Int'al Conf. Laser '83, STS Press, Mclean, VA, 1983, p.165.
4. S.T. Lai, Proc. Int'al Optical Engineering SPIE, 622, 146, 1986.
5. M.L. Shand, and S.T. Lai, Springer Series in Optical Sciences, 47, 76, Springer-Verlag, NY, 1985.

New generation of high power laser systems based on multiple-pass amplifiers

S. Jackel, R. Lalluz, E. Yarkoni, M. Givon, B. Arad, S. Eliezer, and A. Zigler
Plasma Physics Dept, Soreq NRC, 70600 Yavne, Israel

ABSTRACT

Low, medium, and high power Nd:glass, Nd:YAG lasers were built at substantially reduced cost using large fill-factor, non-resonant, polarization and angularly multiplexed multiple-pass (& multiple-beam) amplifiers. Incorporation of phase conjugate mirrors (stimulated Brillouin backscatter in organic liquids) was successful and proved the potential for building ultra-high performance units.

INTRODUCTION

Small, medium, and large glass rod amplifiers have typical diameters d and small signal gains g of I: d=0.5-3 cm, g=50-15, II: d=3-7 cm, g=15-5, III: d=7-12 cm, g=5-2. Many applications require overall system gain of 10^3-10^6 terminating in class II or III sized amplifiers. Conventional laser systems achieve this overall gain using chains of increasingly larger (I-II-III) single-pass amplifiers. These systems are inefficient because to keep the gain high (small signal gain region), the percentage of stored energy transferred to the beam is kept low (i.e., far from saturation).[1]

One solution to the inefficiency of single-pass designs is a multiple-pass amplifier where the beam passes through the same volume of lasent many times. During initial passes, gain is high but energy extraction is low. On the last pass, the amplifier is in the saturated region where gain is low but the extraction of stored energy is high.

We here report on very successful designs of multi-pass amplifiers that proved themselves in laser systems of varying size[2], and on recent advances obtained by incorporating phase conjugate mirrors into multi-pass amplifiers.

MULTIPLE-PASS AMPLIFIERS

Performance of a typical class II sized triple-pass amplifier is shown in figure 1 and the optical layout is depicted in figure 2. Figure 1 shows small signal gain after the first, second, and third passes. The maximum gain of 600 was almost two orders of magnitude greater than that achieved from any single pass amplifier of comparable size. In daily operation, 10 GW output was obtained from class II amplifiers for 1-4 ns trapezoidal laser pulses. Beam quality was measured at maximum gain conditions using a far-field lens and an infra-red TV camera. With a diffraction limited input beam, a near diffraction limited output beam was obtained. Thermal lensing shifted the far-field focus and the triple-pass unit had a (-)200m effective focal length.

Figure 2 shows a layout of the triple-pass amplifier. The class II sized head contained a 1% doped silicate glass rod, pumped by eight 46cm long flashlamps. Optics provided passive polarization and angular control of the beam path. The polarizer and quarter-wave plate acted as a one-way mirror. Mirrors M1 & M2 were at 0° & 1° angles of incidence respectively. M3 extracted the beam after it propagated 4m from the laser head.

Significant design features of the triple-pass amplifier were: 1- The input beam was of large diameter so that non-uniform pump depletion and intensity dependent nonlinear beam breakup were insignificant on the first two passes. 2- Beam extraction was through a bare glass surface with circular polarized light so that the damage threshold was maximized. 3- Rod fill-factor could be high. 4- A resonator geometry wasn't used so prepulses were minimal. Self-oscillations between M1 at 0° and the oscillator were controlled with Pockels cells near the oscillator, but will be eliminated by replacing the dielectric mirror at M1 with a phase-conjugate mirror. 5- The triple-pass layout can be retrofitted onto any rod or slab amplifier with input/output faces nearly normal to the beam. 6- The triple-pass optics cost much less than the single-pass amplifiers it replaced so the design was very cost effective.

A single/double-beam double-pass amplifier was built using a class III size, 0.5% Nd doped, silicate glass rod pumped by sixteen 46cm long flashlamps with 88 kJ of electrical energy(figure 3). Maximum gain was 30. Double-beam damage threshold was maximized by placing sensitive optics prior to the second pass, time delaying one beam, and exiting through bare glass with circular polarized light. The triple & double pass amplifiers form the large amplifier section in our ALADIN (Advanced Laser Amplifier Design INtegration) laser, nominally rated at 50-100 GW for 1-4 ns duration flattop pulses.

PHASE CONJUGATION

Phase conjugate mirrors based on stimulated Brillouin backscatter find a natural place in multi-pass amplifiers because they have a threshold for operation and so can eliminate problems with amplified spontaneous emission and self-oscillations, they correct non-intensity dependent phase distortions in the beam, they have high damage thresholds, and they are inexpensive and easy to use. For instance, the triple-pass amplifier (figure 2) could be enhanced by replacing the conventional mirror M1 with a PCM. New designs are now possible.

We did a series of experiments using a small, high gain (g=40) class I amplifier and a phase conjugate mirror based on stimulated Brillouin backscatter of light focused with a simple lens into a cell containing either carbon tetrachloride or benzene. Reflectivity measurements (figure 4) showed threshold behavior and a saturated, internal reflectivity of up to 90%. For Gaussian duration input pulses, figure 5 shows that at threshold the conjugated pulse was compressed, but for saturated input energies the pulse was undistorted except for a sharper leading edge. The entire double-pass configuration is shown in figure 6. A conventional 0° double-passing mirror was precluded by self-oscillations. With the phase conjugate mirror, the high gain double-pass amplifier produced prepulse free, diffraction limited pulses. This work is now being extended.

1. W. Koechner,"Solid-State Laser Engineering"(Springer,NY) 1976
2. S. Jackel,et al, Laser & Particle Beams, accept for publ 1987

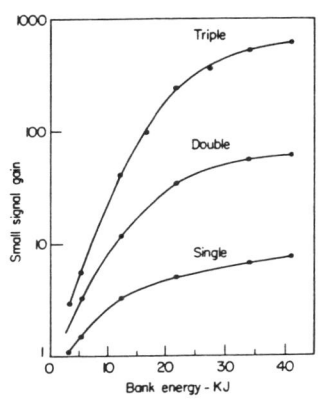

Fig 1 Gain of triple-pass amp

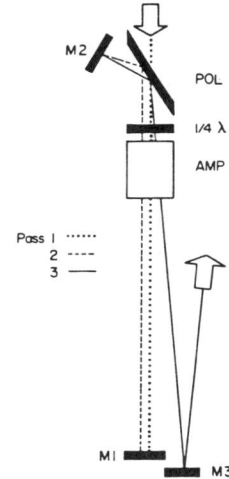

Fig 2 Triple-pass amp

(a) Single beam double pass amplifier

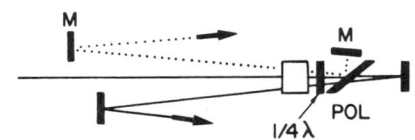

(b) Double beam double pass amplifier

Fig 3

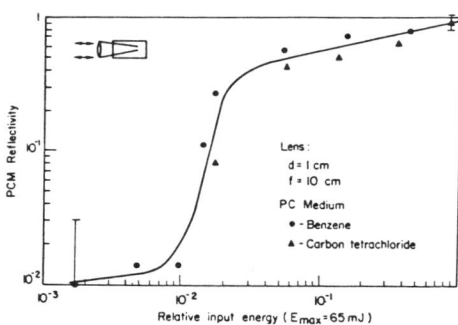

Fig 4 Phase conjugate mirror reflectivity's dependence on input energy

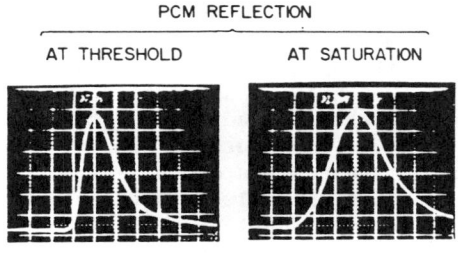

Fig 5 PCM reflection's temporal behavior

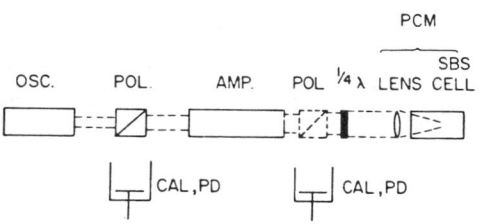

Fig 6 Close-coupled OSC & double-pass amp (Using PCM)

Intracavity Frequency-Doubling of Quasi-CW Pumped YAG Laser

Yao Jianquan, Li Yu and Liu Yan-ming
Dept. of Precision Instrument Engineering
Tianjin University, China

Sammry

We designed a quasi-CW pumped intracavity frequency-doubling YAG laser, which has a higher average power and peak-power output, better beam quality and less thermal effect than that of the CW pumped laser.

I. The Gaussian-like beam and its frequency-doubling

For mixing modes $E(r, \varphi, z) = \sum_{mn} E_{mn}(r, \varphi, z)$ the spot radius is[1]

$$W_M^2(z) = \frac{\sum_{mn} C_{mn}^2 \left[(m+2n+1)\cdot(m+n)!/n!\right]\cdot \cos^2(m\varphi)}{\sum_{mn} C_{mn}^2 \left[(m+n)!/n!\right] \cos^2(m\varphi)} W^2(z),$$

where $w(z)$ is the spot radius of TEM_{00} mode.

The power of second harmonic wave with high conversion efficient.

a) $\Gamma = 0$, $\mathcal{E}_2(r,z) = e^{-\frac{r^2}{W_0^2}} \mathcal{E}_{10} \, th\left(\frac{4d\pi \omega_1^2 L}{k_1 \cdot c^2} \mathcal{E}_{10} e^{-\frac{r^2}{W_0^2}}\right)$,

$P_2 = P_1 - \frac{\sqrt{cn} k_1 c^2 W_0^2}{8 d \pi \omega_1^2 L} \sqrt{P_1} \cdot th\left(\frac{16 d \pi \omega_1^2 L}{k_1 c^2 W_0} \sqrt{\frac{P_1}{cn}}\right) +$

$\frac{cn k_1^2 c^4 W_0^2}{128 d^2 \pi^2 \omega_1^4 L^2} \ln\left[ch\left(\frac{16 \pi \omega_1^2 L}{k_1 c^2 W_0} \sqrt{\frac{P_1}{cn}}\right)\right]$

b) $\Gamma \neq 0$, $\mathcal{E}_2(z) = sn(Az, p\cdot g)$,

$P_2(z) = \frac{cn}{4} \int_0^{\infty} sn^2(Az, p\cdot g) \cdot r \, dr.$

Frequency doubling of Gaussian-like beam at high conversion coefficiency: $P_2 \propto P_1$.

II. Intracavity frequency doubling with quasi-CW pump

Fig.1 Currents of Kr lamp for CW ① and quasi-CW pumping ② ③.

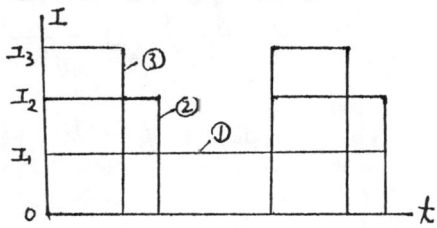

Based on the conclusion of $P_2 \propto P_1^2$, we prove a new intracavity frequency doubling scheme of mixing modes beam, i.e. quasi-CW pumping. The output power of 0.532μm can be increased by a factor n.

Experimental results (for ϕ 5 × 98mm YAG rod, with acusto-optical Q-switch)

pump	0.532μm output power (watts)		
CW	2.8	3.4	4.2
quasi-CW	5.8	7.1	8.9
(n)	(2.07)	(2.08)	(2.1)

III. Analysis and experiment of thermal effect of YAG laser with quasi-CW pump

From thermal conduction equation:

$$\frac{\partial^2 u(r,\tau)}{\partial r^2} + \frac{1}{r}\frac{\partial u(r,\tau)}{\partial r} - \frac{1}{a}\frac{\partial u(r,\tau)}{\partial \tau} + \frac{A(r,\tau)}{K} = 0,$$

$$\left.\frac{\partial u}{\partial r}\right|_{r_0,\tau} = -h(u-u_F)\Big|_{r_0,\tau}$$

The temperature distributions at pumping period and cooling period are $u_1(r,\tau)$ and $u_2(r,\tau)$:

$$\frac{\partial^2 u_1}{\partial r^2} + \frac{1}{r}\frac{\partial u_1}{\partial r} - \frac{1}{a}\frac{\partial u_1}{\partial \tau} + \frac{A_1}{K} = 0,$$

$$\frac{\partial^2 u_2}{\partial r^2} + \frac{1}{r}\frac{\partial u_1}{\partial r} - \frac{1}{a}\frac{\partial u_2}{\partial \tau} = 0,$$

$$u_1(r,\tau) = u_F + (aA_0/K)\tau + \sum_{m=1}^{\infty} C_m J_0(\alpha_m, r)\exp(-a\alpha_m^2 \tau),$$

$$u_2(r,\tau) = u_F + \sum_{m=1}^{\infty} D_m J_0(\alpha_m, r)\exp(-a\alpha_m^2 \tau).$$

Thermal effect in YAG rod can be expressed by $\phi(\tau)$:

$$\phi(\tau) = 2\pi\beta \int_0^a r[2R(r,\tau) - u(r,\tau)]dr.$$

Assuming same evarage input power i.e. $A(r,\tau) = \frac{A_0}{2}$, the everage thermal effect

$$\overline{\phi(\tau)} = \frac{1}{2}[\overline{\phi_1(\tau)} + \overline{\phi_2(\tau)}]$$

$$= 2\pi\beta \frac{A_0}{K} \frac{h}{r_0} \sum_{m=1}^{\infty} \frac{1}{(h^2 + \alpha_m^2)\alpha_m^2 J_0(\alpha_m, r_0)} \cdot \frac{1}{1} \left[\frac{2 - 2J_0(\alpha_m, r_0)}{\alpha_m^2} - \frac{r_0 J_1(\alpha_m, r_0)}{\alpha_m}\right]$$

For CW pump $\phi = 2\pi\beta \cdot (A_0 r_0^2/64K)$, taking $r_0 = 0.3$ cm, $K_0 = 0.13$ W/cm²·K°C, $H = 0.8$ W/cm²·K°C, $a = 0.046$ cm²/S,

$T_0 = 100$ ms, then $\overline{\phi(\tau)}\big|_{q\text{-}CW} = 1.04 \times 10^{-4} (A_0/k) 2\pi\beta$,

$\overline{\phi(\tau)}\big|_{CW} = 1.265 \times 10^{-4} (A_0/k) 2\pi\beta$.

So the beam propagation parameter through a YAG rod under quasi-CW pumping is 22% lower than that under CW pumping.

Conclusion of quasi-CW pump: 1). to get high average output power of second harmonic wave (8.9 watts). 2). Increase the gain and get more narrow pulse width (70-80ns) and high peak power of 0.532μm, 3). Decrease the thermal effect inside YAG rod.

[1] Yao Jianquan etal., CLEO'85 ThC6

THERMO-OPTICAL MODELING OF FLASHLAMP-PUMPED ZIG-ZAG SLABS*

Robert J. Gelinas
Said K. Doss
Susarla S. Murty
Lawrence Livermore National Laboratory

ABSTRACT

This article describes the thermo-optical mechanics of wave front phase distortions in flashlamp-pumped, convectively cooled zig-zag slabs. We identify, from both computations and laboratory measurements, the causes of beam phase distortions that are frequently observed in output beams from zig-zag slabs.

INTRODUCTION

The zig-zag slab concept was invented in order to minimize the effects of thermal and stress lensing upon beam quality in both slab and rod laser outputs. Although this concept has provided effective phase compensation in slab interior regions, other effects that are associated with surface deformations at slab entrance, exit, and transition regions have remained as significant sources of beam degradation. We have applied optical footprint analyses in this work in order to identify those remaining sources of beam degradation and to develop improved design criteria for zig-zag slabs.

TEXT

Slab geometry determines zig-zag beam quality to a very great extent. Referring to Fig. 1, the following relationships are found to be essential in zig-zag slab designs:

(i) <u>Beam/slab symmetry</u>. Total number of beam bounces on total internal reflection (TIR) surfaces must be an even integer M. M must be the same for all rays. The tip-to-tip slab length $W = (M+1)HFP$, where the half-footprint distance between beam bounces on TIR surfaces is HFP = slab height (HT) • tan ψ.

(ii) <u>Unit beam fill</u>. $\psi = \pi/2 - \theta$.

(iii) <u>Unit beam size</u>. Incident rays are parallel to slab TIR surfaces; $\cos\theta = [1 \pm (1+8n^2)^{1/2}]/4n$, where n is the ratio of slab to ambient indices of refraction.

---

*Work performed by the Lawrence Livermore National Laboratory under the joint auspices of the U.S. Department of Energy under Contract No. W-7405-Eng-48 and the U.S. Department of Defense Advanced Research Projects Agency under ARPA Order No. 5358.

(iv) <u>Reflector symmetry</u>. The reflector aperture LREFL should be equal to an integral multiple of HFP. For unit fill, LREFL = W - (2m+3)HFP, where the offset distances of reflector apertures from slab end surfaces are (m+1/2) HFP, m = 1 or 2.

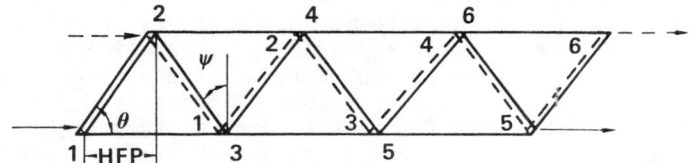

Fig. 1. Schematic view of ray trajectories in a 6-bounce unperturbed zig-zag slab.

It can be shown that, when both unit fill and unit size conditions are satisfied, zig-zag optical path lengths are equal to within $O(\delta\phi^2)$ for all rays, where $\delta\phi$ represents small phase aberrations that may occur as rays refract and/or reflect from locally deformed slab surfaces. A primary objective in our zig-zag slab designs is to ensure that optical path differences (OPDs) due to thermal and stress deformations are maintained at sufficiently small values that diffraction-limited beam quality is attained.

Figure 2 presents computed results for a benchmark design of a 70 x 25 x 2.12 cm Nd-doped glass zig-zag slab (M=16). This slab is pumped uniformly on each TIR surface by flashlamps with radiative fluences of 1.2 W/cm^2 at the slab surfaces. The TIR surfaces are cooled by a turbulent water film at 20°C, and slab end faces are exposed to ambient air at 20°C. The slab wedge angle θ = 27.25°, and the reflector aperture is 49.40 cm with 12 beam traversals under the reflector. Diffraction-limited beam quality is attained over approximately 80% of the beam aperture. The OPDs that appear near the slab edges are caused by a thermal boundary flux mismatch between the pumped slab and the unpumped elastomer and ASE cladding. Such a flux mismatch can be eliminated in practice by either modifying pump source distributions near slab edges or modifying thermal flux conditions near slab edges (to be discussed in future articles).

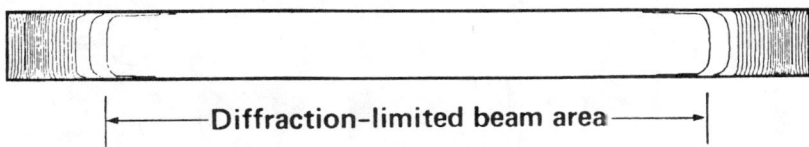

Fig. 2. Fringes computed in benchmark design for a double-pass He-Ne Fizeau interferometer.

The fringes in Fig. 2 were computed with the 3-D ray trace code
BREW. Thermal, stress, and surface deformation data are provided to
BREW from numerical solutions at designated slab operating
conditions by the TOPAZ, NIKE, and TECATE thermal stress codes. Ray
trajectories are solved from the eikonal equation. Essential
physical processes that are implemented in BREW include: refraction
at deformed slab entrance and exit surfaces, ray bending due to
thermal (dn/dT) and stress ($\pi \cdot \sigma$) gradients, geometrical path
decrements to/from deformed slab surfaces, and propagation in
multi-pass cavity applications. Figure 3 presents detailed
bounce-by-bounce OPD data for optical path differences that are
incurred during each sample ray traversal through the zig-zag slab.
The accumulation of OPDs is shown immediately below the individual
footprint data. The OPDs are dominated by surface deformation
effects; dn/dT and $\pi \cdot \sigma$ effects are truly imperceptible in these
results. The major source of optical phase deviations occurs during
propagation in the pumped-to-unpumped transition regions denoted by
traverses 3 and 15 in Fig. 3. Slab/reflector symmetry
considerations are extremely important for achieving compensation of
OPDs in these pump transition regions, as well as at slab entrance
and exit regions. Final compensation of accumulated OPDs occurs
during the propagation of the redirected rays (relative to
unperturbed trajectories) from their final TIR points within the
slab to the far-field reference plane, denoted by 16-R in Fig. 3.
This zig-zag beam compensation process yields output OPDs on the
order of $(\delta\phi)^2$ <0.25 µm over the central 80% of the slab
aperture.

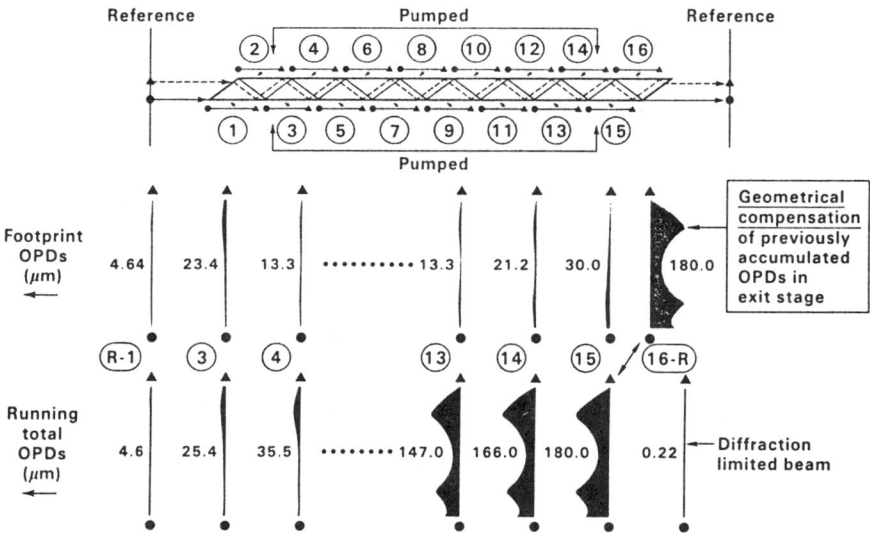

Fig. 3. Optical footprint data in 70-cm zig-zag slab.

NEAR-FIELD PHASE MEASUREMENTS OF DIODE LASER ARRAYS

Gregory C. Dente
G.C.D. Associates, 2100 Alvarado N.E., Albuquerque, NM 87110

Kimberley A. Wilson, David Depatie, John Querns
Kirtland Air Force Base, NM 87117-6008

ABSTRACT

We will describe a simple self-referencing interferometer for near-field phase measurements at the output facet of diode laser arrays.

INTRODUCTION

We have developed a self-referencing interferometer that will allow phase measurements of the near-field of diode arrays. The method uses a double-slit aperture that is scanned across a magnified image of the diode near-field. The far-field double-slit fringe locations then give a direct measurement of the phase difference between the two slits. We can then reconstruct the phase using the sampled difference data. The method has several advantages over other interferometric techniques, including: (1) The wavefront is interfered with itself, so that a separate reference wavefront is not needed. (2) The method is very insensitive to array near-field intensity variations so that fringe maps need not be deconvolved from intensity variations. (3) The method works well in the presence of broadband radiation. The soure needs very little coherence length which is quite an important consideration for diode lasers. Data from several commercially available devices will be presented.

METHOD

As shown in figure 1, a double-slit mask that only transmits a small portion of the wavefront is scanned across a magnified image of the array near-field. (The present set-up uses a magnification ~ 250.) When illuminated by a plane wave, a lens creates the far-field of the scan aperture. If a local phase difference or a local slope exists between the center-lines of the two slits, the Young's fringes will shift. We then track the position of the central fringe as the double-slit aperture is scanned across the diode near-field. We define the diode array output phase in waves as $W(x)$. With the slit centerline separation taken as $\underline{a}$ (typically .015" to .025" in our measurements), the raw data at near-field location x_i is

$$\Delta W(x_i) = W(x_i + a) - W(x_i) = \xi_i,$$

in which ξ_i is the shift of the fringe position divided by the fringe spacing. We reconstruct the function W(x) by assuming that the phases can be expanded as a Fourier series with an additional tilt term as

$$W(x) = Cx + \Sigma[a_n \cos(2\pi nx/P) + b_n \sin(2\pi nx/P)];$$

P is a periodic window that is placed on the diode near-field. We then determine the coefficients so that $\Delta W(x_i)$ approximates ξ_i (data) in a least-squares sense.

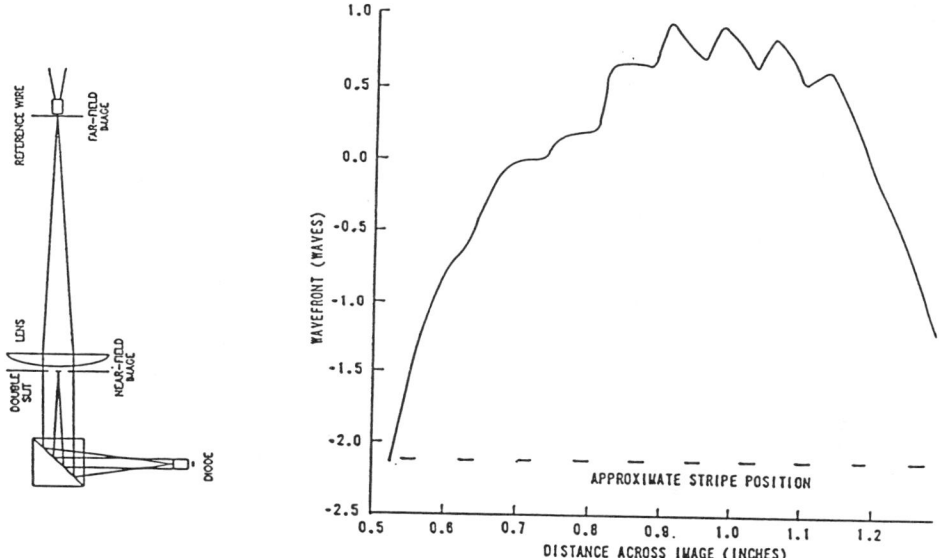

Fig. 1. Experimental LayoutFig. 2. Near-field Phase

RESULTS

We have made phase measurements on several commercially available ten-stripe gain-guided arrays (SDL model # 2410-C). In figure 2 we show results that are representative of these devices. The near-field intensity follows the stripe geometry while the far-field shows the expected twin-lobes; both of these results have been commented on many times and are taken as evidence of operation in the highest-order supermode (N=10)[3]. However, the figure 2 near-field phase measurements show an unexpected result that we will discuss in the remainder of this letter.

The device output phase appears to be that of two tilted waves; one propagates to the right and is concentrated in the right half of the exit aperture, while the other propagates to the left and is concentrated in the left half of the exit aperture. The right travelling wave will dominate the total phase on the right edge of the aperture, while phase on the left edge will be determined by the left-travelling wave. The central region will have phase contributed by both wavefronts that interfere to produce a more oscillatory phase structure. The coherent sum of the distorted amplitude plane waves would yield the measured phase of figure 2.

The far-field data is also better explained by the distorted amplitude plane waves. In this case, each lobe of the twin-lobed far-field will have an angular width determined by the fraction of the aperture filled by each distorted amplitude tilted wave. These lobe widths are in agreement with measurements, while the supermode picture predicts lobes that are narrower than measured[1].

CONCLUSION

We have demonstrated a simple method for obtaining the output phase of semiconductor lasers and laser arrays. Our measurements on commercial arrays suggest that the supermode description of gain-guided laser arrays is not correct. Furthermore, the more recent uniform amplitude Fabry-Perot mode description is also inadequate[4]. A distorted amplitude tilted wave picture of the device operation appears to explain the measured near and far-field results.

REFERENCES

1. D. Botez and D.E. Ackley, "Phase-Locked Arrays of Semiconductor Diode Lasers", IEEE Circuits and Devices Magazine, pp.8-17, Jan. 1986. (This provides a good introduction to the literature of diode laser arrays.)

2. J. Yaeli, "Phase Measurements of Laser Diode Array Radiation", Appl. Phys. Lett. 49, 427(1986).

3. T.L. Paoli, W. Striefer, and R.D. Burnham, "Observation of Supermodes in Phase-Locked Diode Laser Arrays", Appl. Phys. Lett. 45, 217(1984).

4. J.P. Hohimer, G.R. Hadley, and A. Owyoung, "Interelement Coupling in Gain-Guided Diode Laser Arrays", Appl. Phys. Lett. 48, 1504(1986).

PREVENTION OF SIDEBAND-INDUCED DETRAPPING IN TAPERED-UNDULATOR FREE-ELECTRON LASERS

David C. Quimby
Spectra Technology, Inc., Bellevue, WA 98004

ABSTRACT

Sideband generation results in electron detrapping in free-electron laser oscillators with long, highly-tapered undulators. This result contrasts with the enhanced extraction observed in untapered systems. It is shown that multilayer dielectric mirror coatings can provide the intracavity wavelength selectivity necessary for sideband suppression with full extraction recovery in tapered systems.

INTRODUCTION

The sideband instability of free-electron laser (FEL) oscillators was originally predicted in 1979 by Kroll and Rosenbluth.[1] The instability was subsequently observed in the untapered Los Alamos experiments of 1984.[2] This result emphasizes the expected need for sideband suppression in tapered-undulator oscillators. While the instability is predicted to actually enhance the power output of untapered systems, it is a possible threat to the potential high extraction of tapered systems. Simulations predict[3] that the instability will lead to detrapping in FELs with relatively long, highly-tapered undulators in which trapped electrons undergo multiple synchrotron oscillations while traversing the undulator. Sideband growth is associated with loss of extraction by one-half or more. Means for sideband suppression using multilayer dielectric mirror coatings are investigated. It is found that the introduction of wavelength selectivity can suppress the instability, with recovery of full extraction.

SIDEBAND MODELING

Gain only occurs in an FEL when the electrons are bunched at a particular range of phases in the ponderomotive well. Under such conditions, the photons also experience a phase shift which locally retards the wavefront. The sideband instability results because trapped electrons oscillate back and forth within the ponderomotive well. Thus the rate of gain and phase shift also varies along the length of the undulator. The oscillations of the trapped electrons are known as synchrotron motion. As the electrons slip along the optical wave, the gain and phase shift are applied nonuniformly, causing the optical wave amplitude and phase to become modulated. This modulation is equivalent to formation of an optical sideband in the frequency domain, hence the terminology, "sideband" instability. The number of optical wave modulation periods within the slippage length of an electron is given by the number of synchrotron oscillations along the undulator length. Thus the wavelength offset, $\Delta\lambda_s$, of the resonant sideband is proportional to the ratio of the number of synchrotron oscillations, N_{sy}, to the number of undulator periods, N,

$$\frac{\Delta\lambda_s}{\lambda_s} = \frac{N_{sy}}{N} . \qquad [1]$$

The fractional wavelength offset of the sideband is typically in the range of 1 or 2 percent.

Our approach to modeling the sideband evolution has been to numerically integrate the coupled equations describing the self-consistent electron and photon beam dynamics, directly in the space-time domain. This provides a one-dimensional simulation of the longitudinal pulse structure. Effects due to the transverse size of the beams are included in an approximate manner based on appropriate filling factors. Rather than simulating the entire pulse, Colson and Freedman[4] have suggested the use of a periodic boundary condition model, and this is the approach which we have used at Spectra Technology. A detailed description of our simulation model has been given elsewhere.[3] The longitudinal structure of the optical pulse phase and amplitude is examined within a short section of the pulse which is a few slippage distances in length. This greatly reduces the computer time requirements at the expense of some loss of frequency resolution, and short pulse effects are neglected. In the usual case where the slippage length is small compared to the pulse length, short pulse effects are expected to be small and the periodic boundary condition model is appropriate. At visible wavelengths, for example, 20 ps micropulses are about 200 modulation periods in length and short pulse effects are relatively unimportant.

The emittance of the electron beam is accounted for in an approximate way by driving the electrons with the radially-averaged optical field, rather than the on-axis value. This accounts for loss of overlap due to the comparable transverse dimensions of the 2 beams. E-beam energy spread is included directly by dispersing the initial electron energies. The spatial variation of amplitude and phase due to diffraction is included using a complex fill factor.[3]

UNTAPERED UNDULATORS

Results of periodic boundary condition modeling for parameters representative of the Los Alamos untapered oscillator experiments[5] are shown in Figure 1. The amplitude and phase modulation of the optical pulse is examined within a window which is four electron slip lengths in width. Parameters of the simulation are listed in Table I. It is found that the sideband instability develops when the circulating power in the resonator reaches a certain threshold value.[6] This instability threshold occurs when the number of

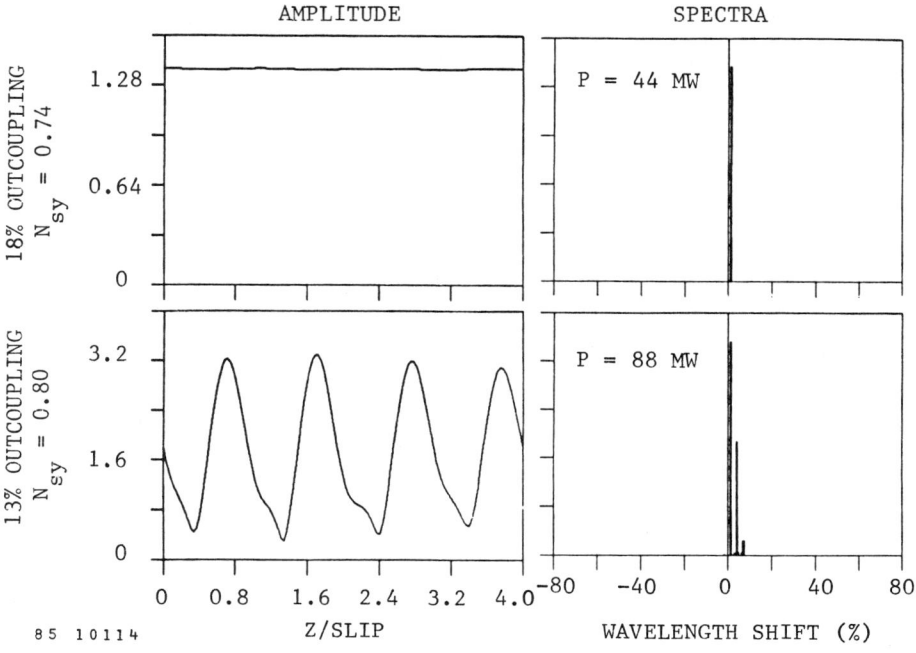

Fig. 1. Untapered FEL is unstable to sideband growth when ≥ 1 synchrotron oscillation along undulator length.

Table I Parameters for Simulation of Los Alamos Untapered Experiment.

Undulator Length	1 m
Undulator Period	2.73 cm
RMS Undulator Parameter, a_w	0.546
Optical Wavelength	10.6 μm
Optical Beam Radius, w	1.145 cm
Electron Beam Radius	0.090 cm
Electron Beam Energy	20 MeV
Electron Beam Current	40 A
Electron Beam Energy Spread	0

synchrotron oscillations along the undulator length is approximately equal to unity. The steady-state circulating power can be increased by lowering the outcoupling, and the onset of the instability is nnumerically found to occur when there are about 0.8 synchrotron periods along the undulator length. The modulation has a periodicity equal to the electron slip length, which is consistent with resonance with about one synchrotron period along the undulator length.

In the untapered undulator case, the extraction efficiency, or the fraction of electron beam power which is converted to light, is actually improved by the instability.[6] Figure 2 shows the extraction efficiency as a function of time for the two simulations shown in Figure 1. With 18 percent round trip loss, the untapered FEL saturates with about 1 percent extraction, which is near the 1/2N theoretical value. As the outcoupling is reduced, the circulating power is higher but the extraction is nearly unchanged. However, when the sideband begins to grow, the extraction actually improves by a factor of about 1.5 in this case. These untapered results are in sharp contrast to the loss of extraction which will be shown for tapered systems with multiple synchrotron periods.

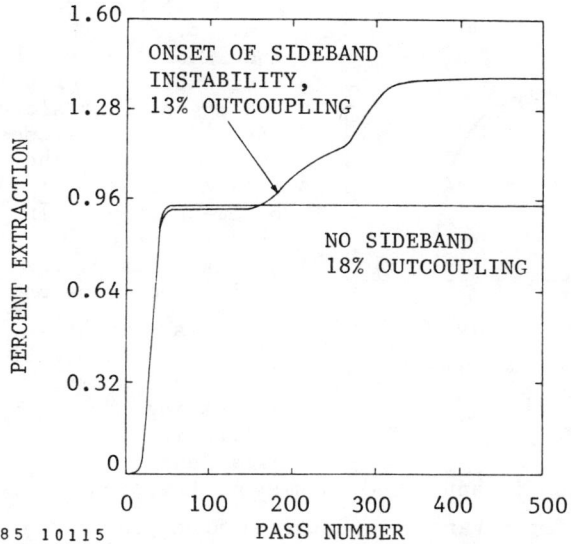

Fig. 2. Onset of the sideband instability leads to enhanced extraction in untapered FEL.

The results shown in Figures 1 and 2 precisely reproduce the previously published results of Colson[6] in all respects including the instability threshold, optical and electron spectra, extraction efficiency, and power levels. These parameters serve as an excellent benchmark case for comparison of various periodic boundary condition models.

The uniform modulation seen in Figure 1 is only observed at power levels close to the instability threshold. If the output coupling is further reduced, the sideband structure becomes more complex. At higher circulating power levels the optical pulse modulation becomes irregular and the spectrum becomes more broadband. The optical spectrum may become chaotic[3] with many random spikes resulting from further sideband mixing.

Figure 3 summarizes the results for untapered undulators. The sideband instability is only active at relatively low cavity loss values. For low outcoupling, the circulating power builds to the point where the number of synchrotron periods along the undulator length becomes comparable to unity. When the synchrotron period approaches equality with the undulator length, the oscillation of gain and phase shift along the undulator becomes substantial and the sideband instability becomes active. In the untapered undulator case, the addition of sideband power is cumulative,[7] with the presence of a strong sideband increasing the extraction efficiency.

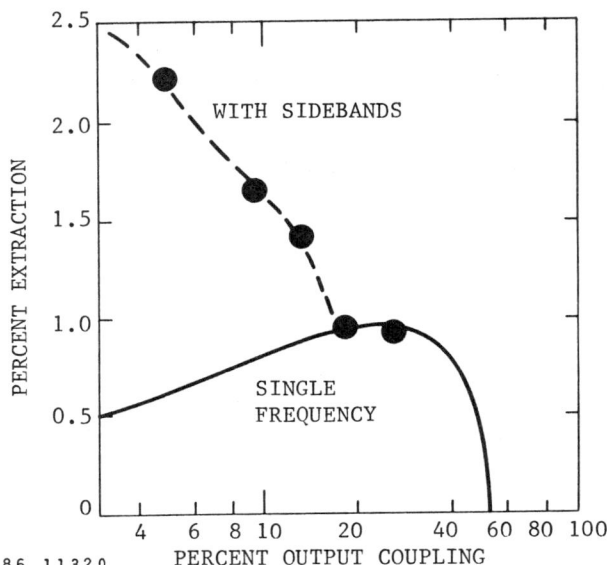

Fig. 3. Sideband generation improves the extraction of untapered undulators.

The increased extraction in untapered undulators has been explained qualitatively by Roger Warren.[8] Saturation occurs in an untapered undulator when enough energy is removed from the electrons so that they become detuned from the FEL resonance condition. A tapered undulator maintains resonance by tailoring the undulator parameters as the electrons lose energy. Warren has pointed out that the frequency modulation resulting from the sideband instability tracks the energy changes with corresponding changes in the wavelength of the light.

To illustrate this process, Figure 4(a) shows a frequency modulated optical wave and its spectrum. The wave is moving to the right, overtaking electrons as it goes. An electron which is initially exposed to short wavelength light experiences an increasing wavelength. This tends to maintain resonance over its entire path, so that this electron loses a maximum amount of energy without saturation. On the other hand, an electron which starts at long wavelength converts less of its energy to light than if the wave was not modulated.

When averaged over all electrons, it is found that the total energy extracted is increased by the modulation in second order. Furthermore, the extracted energy is added nonuniformly to the optical wave, causing an amplitude modulation like that shown in

Figure 4(b). This correlated amplitude and frequency modulation modifies the original spectrum, emphasizing the long-wavelength sideband over the short-wavelength one, and this is consistent with the spectra observed in simulations and experiments.

TAPERED UNDULATORS

In untapered undulators it is found that the sideband instability is not particularly harmful to laser performance, other than causing considerable broadening of the laser linewidth. Much different behavior is observed for systems with several synchrotron periods along the undulator length. In long, highly tapered undulators the number of synchrotron periods can be considerably larger than one. This is because the spatial period of the synchrotron motion depends inversely on the square root of the electric field. There is a threshold E-field to achieve trapping in a tapered undulator, and the threshold E-field is proportional to the fractional energy taper, $\Delta\gamma_r/\gamma_r$. When these two relationships are folded together, one finds that the number of synchrotron periods along the undulator length at saturation is proportional to the square root of the product of the number of undulator periods and the taper,

$$N_{sy} = \left[\frac{N}{\pi \tan \psi_r} \frac{\Delta\gamma_r}{\gamma_r} \right]^{1/2} , \qquad [2]$$

so that long, highly-tapered undulators experience multiple synchrotron periods.

The number of synchrotron periods at saturation in two example experiments is compared in Table II for a particular resonant phase

Table II Comparison of Tapered-Undulator Experiments

	N	$\Delta\gamma_r/\gamma_r$	N_{sy} at $\psi_r = 30°$
Los Alamos	40	7%	1.24
Boeing/Spectra Technology	220	10%	3.5

angle, ψ_r. The 10 μm tapered experiments at Los Alamos[9] have used a relatively short, mildly tapered undulator and the number of synchrotron periods remains close to unity. However, in a system such as the Boeing/Spectra Technology visible oscillator experiment[10] which has a relatively long, highly tapered undulator, there are roughly three times as many synchrotron periods for any given ψ_r

value. Such a system is susceptible to sideband-induced detrapping of the electrons. The more synchrotron periods are present, the more likely it is for the sideband modulation to cause the trapped electrons to be resonantly shaken out of the ponderomotive well of the FEL interaction.

The effect of the sideband instability in the multiple synchrotron period regime will be illustrated using a simulation for the parameters of the visible oscillator experiment. The objective of this experiment is demonstration of the potential high extraction of tapered undulators at visible wavelengths. This is a collaborative program with Boeing having responsibility for the RF linac and Spectra Technology developing and operating the undulator and optical cavity. This experiment is forced into the multiple synchrotron period regime since it requires a large taper for high extraction and a long undulator for adequate gain at visible wavelengths and high extraction.

Figure 5 depicts the results of a simulation of sideband evolution for conditions of the visible oscillator experiment. The parameters of the simulation are listed in Table III.

Table III Parameters for Simulation of Boeing/Spectra Technology Tapered Experiment.

Undulator Length	5 m
Constant Section Length	0.88 m
Linearly Tapered Section Length	4.12 m
Energy Taper, $\Delta\gamma_r/\gamma_r$	10%
Undulator Period	2.18 cm
RMS Undulator Parameter, a_w	1.33
Optical Wavelength	0.5145 μm
Rayleigh Range	2.4 m
Output coupling	8.5%
Electron Beam Radius	0.034 cm
Electron Beam Energy	123 MeV
Electron Beam Current	100 A
Electron Beam Energy Spread	1%

Several snapshots of the optical spectrum at various times during the simulation are shown. The simulation is started at a low power level representative of the approximate spontaneous emission power level. Even though the simulation is initiated with all possible optical frequencies present, the laser quickly selects a narrow line at the wavelength of peak small-signal gain. After about 200 passes, the circulating optical power reaches several GW, the level at which electrons begin to be trapped and undergo synchrotron oscillations.

149

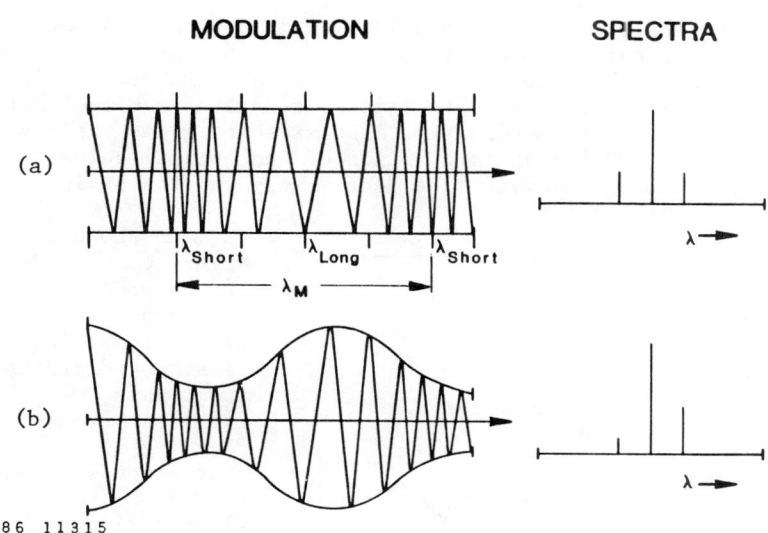

Fig. 4. Enhanced extraction of (a) frequency modulated wave and (b) amplitude and frequency modulated wave is explained by simple model of Warren.[8]

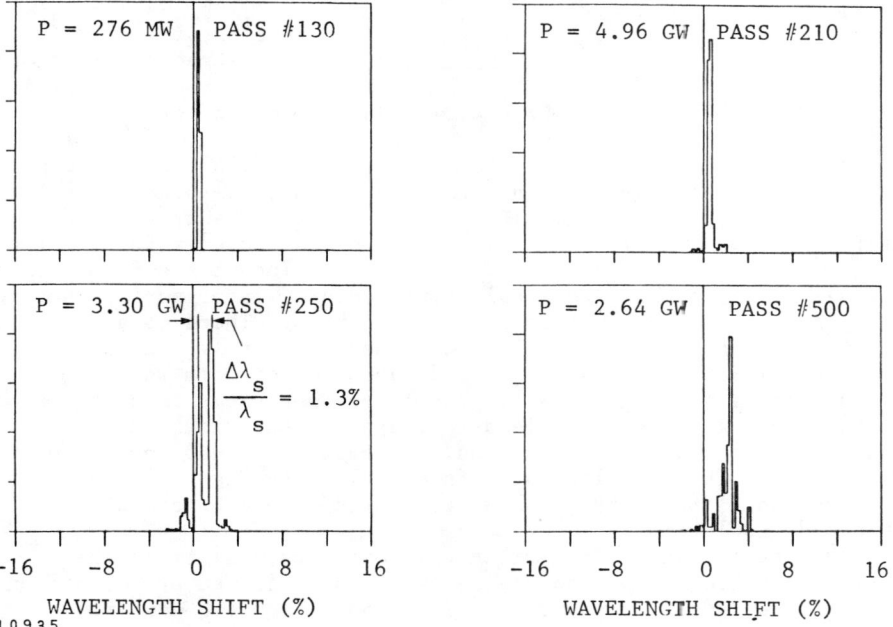

Fig. 5. Sideband forms when electron trapping conditions reached after 200 passes in long, highly-tapered FEL.

Sidebands begin to grow immediately. The initial modulation of the optical pulse occurs with about three modulation periods per electron slippage length, which corresponds to three synchrotron oscillations along the undulator length. The initial sideband is offset about 1.3 percent toward longer wavelengths. The sideband grows rapidly until there is nearly 100 percent modulation of the optical pulse. Eventually the original main line decays away but the optical spectrum remains complex.

As shown in Figure 6, the addition of the sideband severely degrades the net extraction. The onset of sideband growth at about pass 200 prevents the laser from ever reaching its ideal extraction of about 5 percent. In fact, the sideband-induced detrapping causes the extraction to decay to less than one-half of the ideal value.

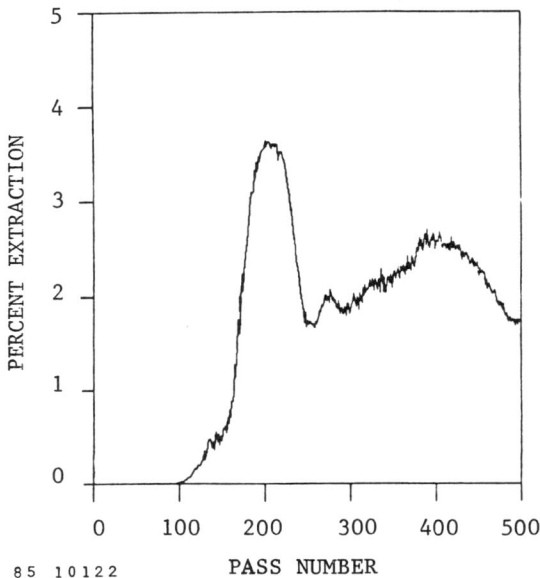

Fig. 6. Sideband growth leads to loss of extraction in long, highly tapered FEL.

Figure 7 shows results of a study[3] of the extraction loss due to the sidebands as a function of output coupling. The solid line in the figure shows the calculated steady-state extraction efficiency for a narrow line at the frequency of peak gain. This shows the ideal saturated oscillator performance for a single narrow line which is allowed to chirp. There is a range of output coupling values for which the laser will evolve to power levels sufficient for trapping. Sidebands do not form when the outcoupling is too high to allow the laser to achieve trapping. For all outcoupling values that lead to trapping, the sideband instability reduces the extraction to approximately 40 percent of the ideal value. This result contrasts sharply with Figure 3, in which the sideband instability is shown to enhance the extraction of untapered undulators.

The deleterious effect of the sideband instability in the multiple synchrotron period regime has been confirmed by comparison with an independent pulse propagation code. Figure 8 compares results from a Los Alamos simulation[11] of the full 20 ps pulse in the visible oscillator experiment with results from the Spectra Technology periodic boundary model.[3] Provided that the periodic

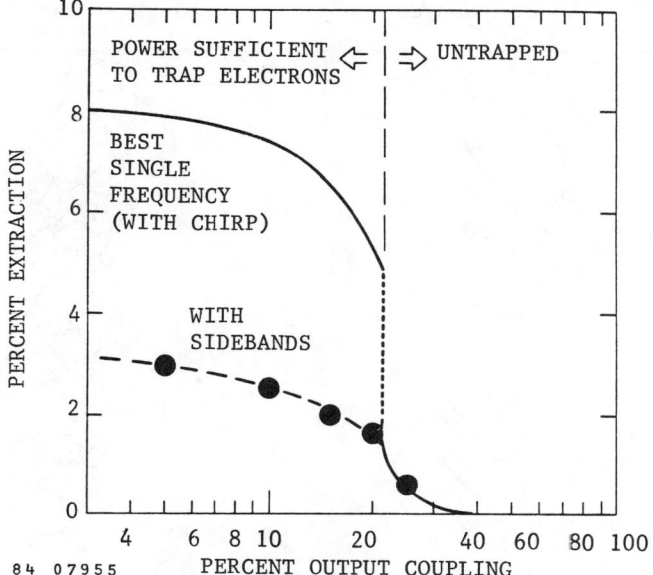

Fig. 7. Loss of extraction exceeds one-half in long, highly tapered FEL.

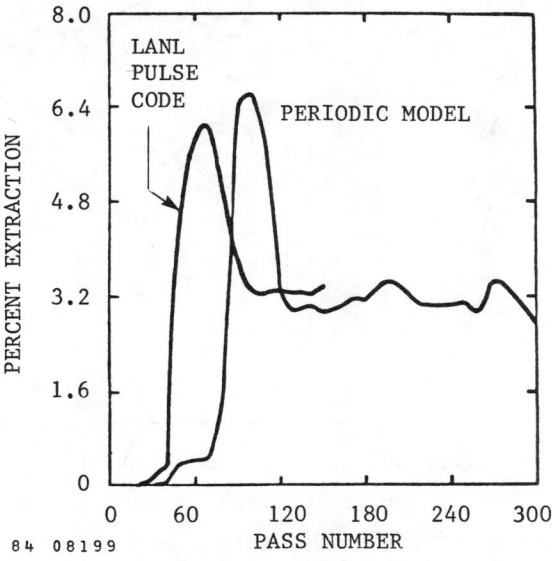

Fig. 8. Periodic Model is in excellent agreement with full pulse simulation.

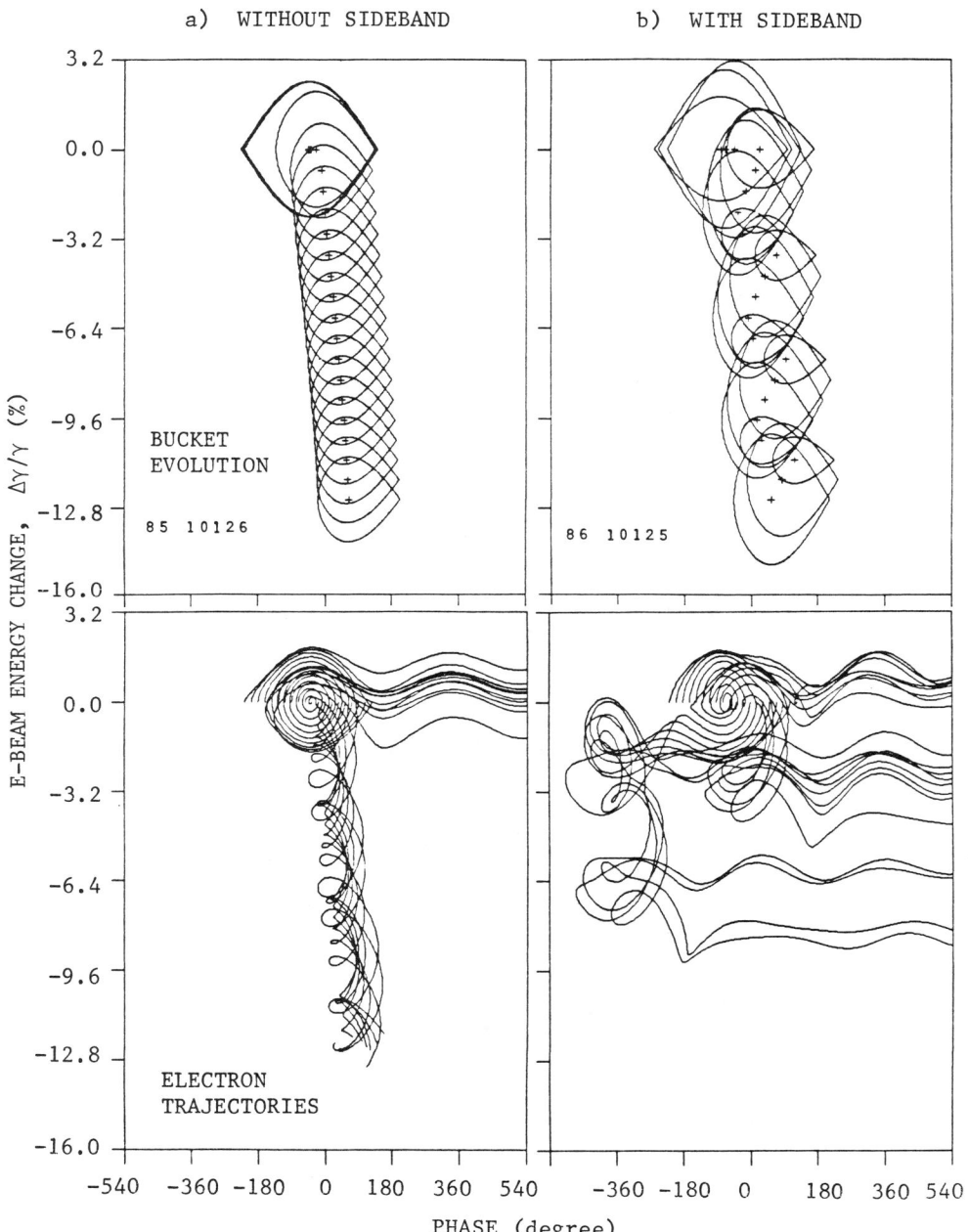

Fig. 9. Comparison of FEL ponderomotive well dynamics (a) without and (b) with sideband.

model is applied at an e-beam current level which is typical of the average current in the pulse, the qualitative agreement is excellent. The only deviation between the two models is a slight difference in startup time. Since startup is very sensitive to the lasing wavelength, this discrepency can probably be explained by a slight variation in the choice of the initial wavelength in the pulse simulation.

The loss of extraction observed for the case of multiple synchrotron oscillations can be understood by examination of the electron dynamics. The ideal evolution of the ponderomotive potential well at full saturated power is shown in Figure 9(a). The initial symmetric bucket corresponds to the short untapered section at the undulator entrance, which is provided to enhance the small-signal gain. The bucket then decelerates about 12 percent corresponding to the maximum taper of the undulator.

The electron trajectories corresponding to the ponderomotive well dynamics are also shown in Figure 9(a). Roughly 50 percent of the electrons are trapped and decelerated the full amount, giving a net extraction exceeding 5 percent. About one-half of the electrons are trapped and the four synchrotron oscillations undergone by the trapped electrons are clearly seen.

Figure 10 shows the optical E-field amplitude and phase seen by a typical electron as it traverses the undulator. The dashed line is for a single optical frequency. The electron sees slightly higher amplitude at the center of the undulator since that is where the photon beam has its minimum waist. The solid line shows the modulation which results if a sideband is added with 50 percent additional power. The offset of the sideband places it in resonance with the four synchrotron oscillations of trapped electrons.

The impact of the modulation on the ponderomotive well dynamics is shown in Figure 9(b). The amplitude and phase modulation results in rather severe jitter in the bucket size and position.

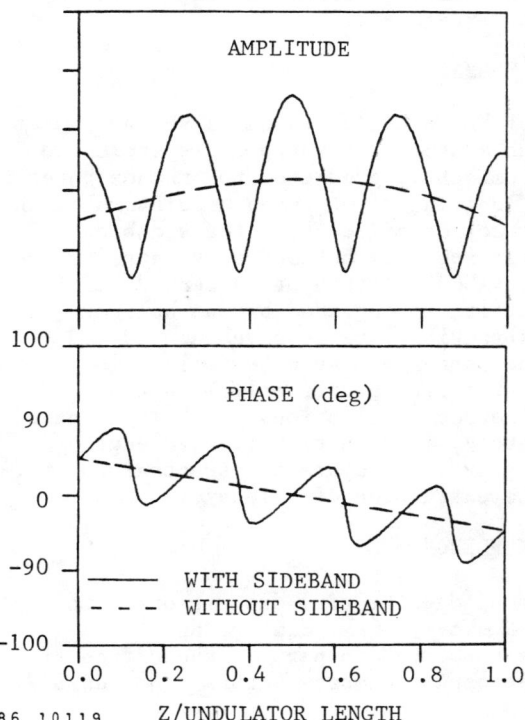

Fig. 10. Addition of sideband modulates E-field seen by electrons.

The electrons are resonantly driven out of the bucket. Some actually find themselves momentarily retrapped in an adjoining bucket, but they still never undergo full deceleration.

SIDEBAND PREVENTION

Introduction of wavelength selectivity into the optical cavity is a means for suppressing the sideband instability. Figure 11 shows the reflectivity curve for a multilayer dielectric coating which could be used on one of the cavity mirrors. The filter function is configured to give large loss at the initial sideband frequency. Such a coating could serve the dual purpose of sideband suppression and transmissive outcoupling at the main line. While the coating design shown here provides a narrow band-pass filter, an edge filter was shown to successfully suppress sidebands in the Los Alamos experiments.[2] Edge filters are of interest because the initial sideband occurs only at longer wavelengths.

As shown in Figure 12, the use of this wavelength selectivity function successfully suppresses the sideband instability. A narrow optical spectrum is maintained and full extraction efficiency is restored. It is anticipated that similar mirror coatings will be tested in the visible oscillator experiment.

SUMMARY

The sideband instability of FEL oscillators has a varied impact depending on the taper of the undulator. In untapered systems the instability broadens the spectrum but is predicted to enhance power output. On the other hand, in long, highly-tapered undulators which have the potential for high extraction efficiency, the sideband instability not only broadens the spectrum but severely degrades the power output. For a system with about 4 synchrotron periods along the undulator length, the extraction is degraded by one-half or more. The mechanism for the extraction loss is sideband-induced detrapping of electrons from the ponderomotive potential wells of the FEL interaction. Deleterious detrapping occurs only when the electrons undergo several synchrotron oscillations in a single pass through the undulator. Fortunately, introduction of wavelength selectivity into the optical cavity can restore the spectral purity and lead to full recovery of the extraction efficiency.

ACKNOWLEDGMENTS

The author gratefully acknowledges helpful discussions with W.B. Colson, J.C. Goldstein, B.D. McVey, J.M. Slater, and L.C. Steinhauer. This work was supported in part by the Office of Naval Research through Boeing Aerospace Company Contract No. GD2527.

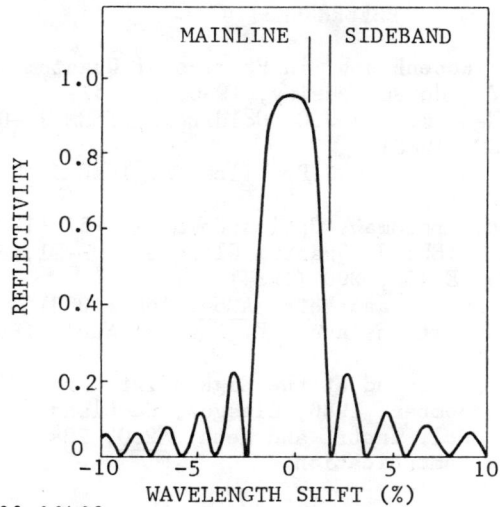

Fig. 11. Multilayer dielectric mirror coating preferentially outcouples unwanted sideband.

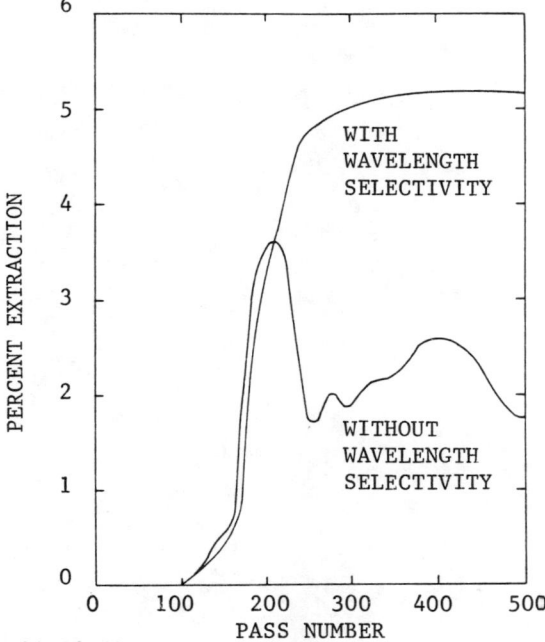

Fig. 12. Introduction of wavelength selectivity restores full extraction.

REFERENCES

1. N.M. Kroll and M.N. Rosenbluth, in Physics of Quantum Electronics, Vol. 7 (Addison-Wesley, 1980) p. 147.
2. R.W. Warren, B.E. Newnam, and J.C. Goldstein, IEEE J. Quantum Electron. QE-21, 882 (1985).
3. D.C. Quimby, J.M. Slater, and J.P. Wilcoxon, IEEE J. Quantum Electron. QE-21, 979 (1985).
4. W.B. Colson and R.A. Freedman, Optics Comm. 46, 37 (1983).
5. B.E. Newnam, et al., IEEE J. Quantum Electron. QE-21, 867 (1985).
6. W.B. Colson, Proc. SPIE 453, 290 (1984).
7. W.B. Colson, Nucl. Instr. and Meth. A250, 168 (1986).
8. R.W. Warren, J.C. Goldstein, and B.E. Newnam, Nucl. Instr. and Meth. A250, 19 (1986).
9. R. Warren, et al., presented at the Eighth Int. Free-Electron Laser Conf., 1-5 September, 1986, Glasgow, Scotland.
10. J. Slater, et al., Nucl. Instr. and Meth. A250, 228 (1986).
11. B.D. McVey, private communication.

X-RAY CHARACTERIZATION OF PICOSECOND LASER PLASMAS

O. L. Landen, E. M. Campbell and M. D. Perry
University of California, Lawrence Livermore National Laboratory
Livermore, California 94550

ABSTRACT

Short-lived plasmas, produced by focussing a 0.58 μm, 1 ps, 1 mJ dye laser beam to $10^{13} - 10^{14}$ Wcm^{-2} onto solid planar targets, are diagnosed by a photoconductive detector, pinhole camera, x-ray streak camera and by shadowgraphy.

INTRODUCTION

The interaction of > 20 ps high intensity (> 10^{13} Wcm^{-2}) laser pulses with solid targets has been extensively investigated in the context of laser fusion and XUV lasers over the past 15 years.[1-8] It is well known that within 10 ps, a high density:

$$(N_e < N_{critical} = 10^{21} \lambda^{-2}, N_e \text{ in cm}^{-3}, \lambda \text{ in μm})$$

rapidly expanding (V > 10^7 cms^{-1}) plasma is formed above the target.
The laser energy is then absorbed through a variety of processes such as inverse bremsstrahlung, resonance absorption and other parametric processes in the plasma corona ($N_e < N_{critical}$). Energy is then transported into the dense solid both by electron conduction and radiation transport. However, the initial (t < 10 ps) interaction of a strong EM field with solid matter is not well understood, despite a few early experiments,[9,10] and is the focus of the investigation reported here utilizing a picosecond laser. In addition, the behavior of the dense plasma created by such ultra-short laser pulses is examined. These plasmas give rise to short (< 20 ps)[11] intense bursts of XUV and x-ray radiation which can be used to determine temporal resolutions of photoconductive and pyroelectric detectors, diodes, framing and streak cameras. Short x-ray pulses could also be used in solid state physics to probe avalanche phenomena.

EXPERIMENTAL DETAILS

The laser system consists of a well characterized synchronously pumped mode-locked dye laser followed by four dye amplifier cells pumped at 10 Hz by a frequency-doubled Q-switched Nd:Yag laser. Laser energies and pulse lengths are continuously monitored for both the dye oscillator and amplified beams by a power meter, energy probe and high and low repetition rate autocorrelators. A typical autocorrelator trace of a 1.5 mJ dye laser pulse (resolution of 50fs) yields a 0.87 ps FWHM for a sech^2 fit. Single shot pulse shapes recorded over a longer time-scale by a 2 ps resolution optical streak camera show a fast rising leading edge with ~ 20% of the laser energy distributed after the initial pulse. Laser focal spot distributions for the 7.5 cm and 25 cm spherical focal length lenses used yield 25 μm and 50 μm FWHM respectively, 2.5 times the diffraction limit. Typical operating parameters are 2 mJ in 1.5 ps, 1 mJ in 0.8 ps and 0.5 mJ in 0.5 ps with a maximum irradiance on target of 2×10^{14} W cm^{-2}. Forty percent conversion efficiency to $\lambda = 0.29$ μm could be achieved in a 1 cm KDP crystal, focusable to a 20 μm FWHM diameter spot by the 7.5 cm lens.

The laser beam was focused onto planar targets equipped with variable tilt and height adjustment mounted inside a small vacuum chamber evacuated to 10^{-5} Torr.

X-ray diagnostics included x-ray streak cameras, a pinhole camera equipped with a 7 μm diameter pinhole, an Indium Phosphide photoconductive detector[12] and a 2 cm diameter filter/film pack for shadowgraphy. Al, C and V filters in conjunction with 2497 film were used to record 100-500 eV radiation. Be and thicker Al foils were used in conjunction with DEF film for > 500 eV radiation.

EXPERIMENTS AND RESULTS

The soft x-ray yield (250 ± 25 eV radiation) was investigated as a function of target material for a constant laser irradiance of 1.5×10^{14} W cm^{-2}, corresponding to the tightest focus using the 7.5 cm lens and 1 mJ in 1.5 ps of laser energy normally incident on target. The photon yields were recorded with a 300 μg cm^{-2} C filter and 2497 film. The absolute yields over 2π were estimated using published film calibration curves and deconvolving filter and film responses.[13-15] Al, Mo, Ta and Au produced 7×10^{11} - 1.3×10^{12} keV keV^{-1}, whereas C gave only 3.5×10^{10} keV keV^{-1}.

Twenty-five μm-thick Tantalum (z=73) targets were then chosen for all further experiments unless stated otherwise. Figure 1 plots the conversion efficiencies to 250 ± 25 eV radiation as a function of laser irradiance for constant laser spot size (25 μm FWHM at $\lambda = 0.58$ μm). The soft x-ray yield is proportional to $I^{2.7}$ above an apparent threshold at $\sim 10^{13}$ Wcm^{-2}. By contrast, a maximum frequency doubled laser output of 350 μJ on target focused to 7×10^{13} Wcm^{-2} yielded < 1% of the 250 eV output measured using the fundamental frequency, $\lambda = 0.58$ μm, at the same laser irradiance. For UV laser outputs of 200 μJ and less, no soft x-ray signals could be detected on film, even for 10 shot accumulation. For the same value of $I \lambda^2$ (6×10^{12} Wcm^{-2} μm^{-2}) which is proportional to the oscillatory velocity of electrons in the laser field, the 250 eV yield for $\lambda = 0.29$ μm is still a factor of ~ 30 lower than for $\lambda = 0.58$ μm.

A coarse x-ray spectrum between 200 and 2000 eV (Fig. 2) was recorded for $\lambda = 0.58$ μm focussed to $1.3 \pm 3 \times 10^{14}$ W cm^{-2} at normal incidence using a variety of filters and films placed 4-10 mm away from the plasma. The integrated area under the curve represents a 1 - 5% conversion efficiency to XUV and x-ray radiation.

A single-shot streak record was taken using an x-ray streak camera equipped with a CsI photocathode placed 40 cm away from the plasma and masked by a 96 μg cm^{-2} filter. 60 ± 10% of 100 - 270 eV x-rays imaged were within the initial spike whose FWHM (40 ps) is the instrument function; the longer-lived, ~ 200 ps long x-rays are probably created by the tail in the dye laser emission. A streak camera

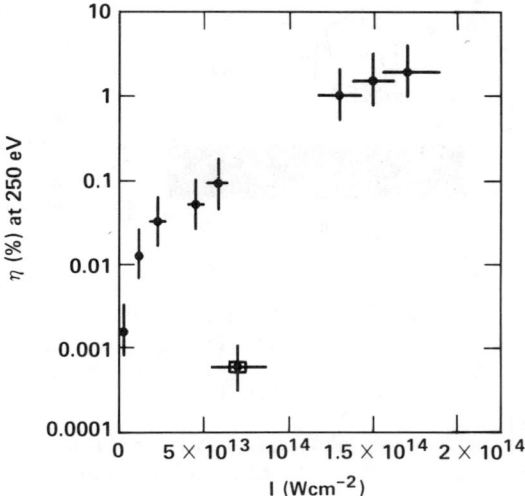 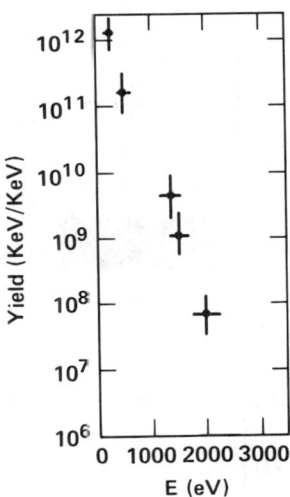

Fig. 1 250 eV yield versus laser irradiance for constant laser spot size at $\lambda = 0.58$ μm (o) and $\lambda = 0.29$ μm (■).

Fig. 2 Radiation spectrum for $I = 1.3 \pm .3 \times 10^{14}$ W cm^{-2} over 2π steradians.

with better temporal resolution (5 ps FWHM in theory) was also employed with a 1 μm Be filter to set an upper limit of 20 ps on the pulse length of the initial 100 eV radiation burst. Such short signals suggested that the x-ray emitting region would expand less than 1 μm in 10 ps for typical ion velocities of 10^7 cms^{-1}.

A series of x-ray imaging experiments were then undertaken, initially using a 7 μm diameter pinhole with an image magnification varying between five and seven on different shots. Single shot images for $I = 1.8 \times 10^{14}$ Wcm^{-2} at 250 eV viewed at ~ 10° to the target surface, (Fig. 3a), show a 27 μm wide emitting region with < 5 μm spatial extent above the target surface. For a given laser spot size on target, the width of the emitting region decreases with decreasing laser irradiance, (Fig. 3b), as expected from the strong dependence of 250 eV radiation yield on irradiance shown in Fig. 1. A side view at 1 keV (Fig. 3c) and 1.5 keV photon energies accumulated over 40 shots, however, is unresolved implying the higher energy x-rays come from a region < 5 μm in all directions.

To increase spatial resolution and keep diffraction effects negligible, a 20 μm diameter W wire placed parallel to and 5 - 10 μm above the target surface was imaged with a magnification of 100 -200 (Fig. 4a), Figs. 4b and 4c show unsaturated shadowgraphs of the wire at 250 eV and 1.5 keV photon energies taken with 0.2 mJ and 1 mJ of laser energy on target respectively and recorded with 2497 and DEF film respectively. The resolution at the wire edge furthest from the target surface is poorer due to the higher angles subtended to the target surface. The apparent curvature at this wire edge is quantitatively explained by calculations including the inverse square law

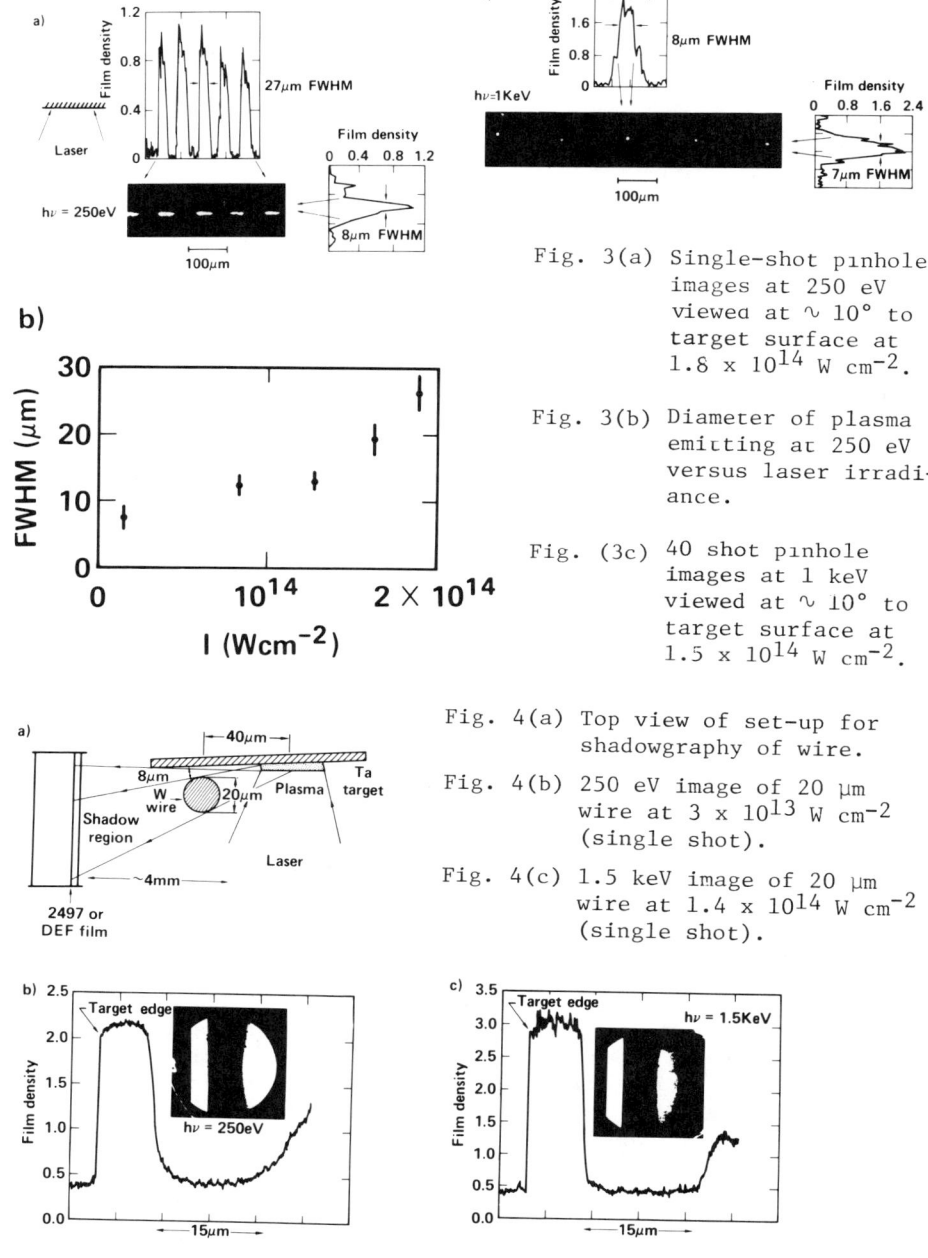

Fig. 3(a) Single-shot pinhole images at 250 eV viewed at ∿ 10° to target surface at 1.8×10^{14} W cm^{-2}.

Fig. 3(b) Diameter of plasma emitting at 250 eV versus laser irradiance.

Fig. (3c) 40 shot pinhole images at 1 keV viewed at ∿ 10° to target surface at 1.5×10^{14} W cm^{-2}.

Fig. 4(a) Top view of set-up for shadowgraphy of wire.

Fig. 4(b) 250 eV image of 20 μm wire at 3×10^{13} W cm^{-2} (single shot).

Fig. 4(c) 1.5 keV image of 20 μm wire at 1.4×10^{14} W cm^{-2} (single shot).

for distance between plasma and film, and increasing angles subtended through filter and at film. From these images, the diameters of the emitting regions are calculated to be 7 and 2.5 μm at 250 eV and 1.5

keV respectively. The upper limits on the thickness of the emitting regions are 2 μm and 3000 Å respectively.

Heated areal densities for 1 ps pulses at $I = 3 \times 10^{13}$ Wcm^{-2} were then inferred from the XUV yield of a high Z, strongly emitting target, Ta, overcoated by evaporation with a low Z weakly emitting element, C. Figure 5 plots the ratio of the XUV yield at 50 - 70 eV, measured by an Indium Phosphide detector filtered with 0.39 μm Al, with and without carbon overcoat versus carbon thickness. Each data point is averaged over at least 10 shots. The best fit curve yields 200 Å as the heated depth, equivalent to a measured optical depth of 1 for C at $\lambda = 0.58$ μm.

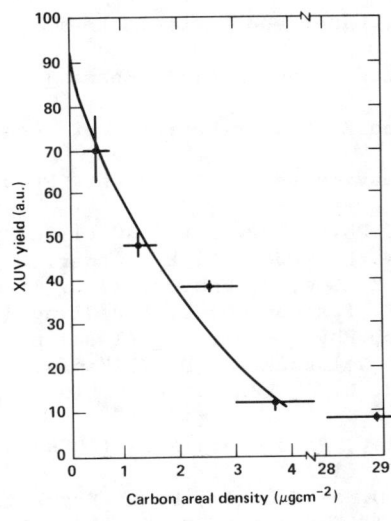

Fig. 5 XUV yield versus carbon overcoat areal density. Curve is an experimental fit.

SUMMARY

1 - 5% coupling of 0.5 μm of laser energy into plasma radiation has been observed on a picosecond timescale. This should be compared with the 40 - 50% conversion efficiencies observed for longer pulse laser plasmas at $\lambda = 0.53$ μm, $I = 3 \times 10^{13}$ W/cm^{-2}, $\tau = 3$ ns[16] and $I - 10^{14}$ W cm^{-2}, $\tau = 600$ ps.[17] The 10^9 photons produced at 1 keV should be sufficient to test the spatial and temporal resolution of x-ray framing cameras being developed. The 3000 Å resolution in one dimension offered by the 1 - 1.5 keV x-ray burst produced is an order to magnitude better than for conventional long pulse (100 ps) point backlighters[18] and comparable in resolution to Kirkpatrick-Baez x-ray microscopes.[19,20] The high energy, up to 2000 eV, x-rays measured suggests efficient production of hot electrons at $\lambda = 0.58$ μm for 1 ps pulses at 10^{14} Wcm^{-2}. Poor (< 0.1%) x-ray conversion efficiencies obtained with the frequency doubled ($\lambda = 0.29$ μm) laser beam at 7×10^{13} Wcm^{-2} suggests the predominance of an $I\lambda^2$ absorption mechanism.

Future planned experiments to better understand ultrashort pulse laser plasma production include soft x-ray spectroscopy using a transmission grating, optical reflectivity and absorption measurements as a function of laser irradiance and target material, and time resolved ion accelerations in laser plasmas.

ACKNOWLEDGMENT

We thank G. Glendinning, J. Trebes and C. Wang for the loan of photoconductive detectors and streak cameras. Work performed under the auspices of the U.S. Department of Energy by Lawrence Livermore National Laboratory under contract #W-7405-Eng-48.

REFERENCES

1. P.H.Y. Lee and H.G. Ahlstrom, Laser and Particle Beams 2 (1984) p. 303.
2. P. Ladrach and J.E. Balmer, Laser and Particle Beams 1 (1983) p. 67.
3. M.D.J. Burgess, R. Dragila and B. Luther-Davies, Opt. Comm. 52 (1984) p. 189.
4. M. D. J. Burgess, B. Luther-Davies and K. A. Nugent, Phys Fluids 28 (1985) p. 2286.
5. K. Estabrook and W. L. Kruer, Phys. Rev. Lett. 40 (1978) p. 42.
6. W. C. Mead, E. M. Campbell, W. L. Kruer, R. E. Turner, C. W. Hatcher, P. S. Bailey, P. H. Y. Lee, J. Foster, K. G. Tirsell, B. Pruett, N. C. Holmes, J. T. Trainor, G. L. Stradling, B. F. Lasinski, C. E. Max and F. Ze, Phys. Fluids 27 (1984) p. 1301.
7. M. H. Key, W. T. Toner, T. J. Goldsack, J. D. Kilkenny, S. A. Veats, P. F. Cunningham and C. L. S. Lewis, Phys. Fluids 26 (1983) p. 2011.
8. D. W. Phillion and C. J. Hailey, Phys. Rev. A 34 (1986).
9. H. Salzmann, J. Appl. Phys. 44 (1973) p. 113.
10. N. G. Basov, S. D. Zakharov, O. N. Krokhin, P. G. Kryukov, Yu V Senat-skii, E. L. Tyurin, A. I. Fedosimov, S. V. Chekalin and M. Ya. Shchelev, Sov. J. of Quant. Elec. 1 (1971) p. 2.
11. D. J. Bradley, A. G. Roddie, W. Sibbett, M. H. Key, M. J. Lamb, C. L. S. Lewis and P. Sachsenmaier, Opt. Comm. 15 (1975) p. 231.
12. D. R. Kania, A. E. Iverson, D. L. Smith, R. S. Wagner, R. B. Hammond and K. A. Stetler, to be published.
13. B. L. Henke, F. J. Fujiwara, M. A. Tester, C. H. Dittmore and M. A. Palmer, J. Opt. Soc. Am. B. 1 (1984) p. 829.
14. B. L. Henke, S. L. Kwok, J. Y. Uejio, H. T. Yamada and G. C. Young, J. Opt. Soc. Am. B. 1 (1984) p. 818.
15. B. L. Henke, J. Y. Uejio, G. F. Stone, C. H. Dittmore and F. G. Fujiwara, J. Opt. Soc. Am. B. 3 (1986) p. 1540.
16. K. Eidmann and T. Kishimoto, Appl. Phys. Lett. 49 (1986) p. 377.
17. W. C. Mead, E. M. Campbell, K. G. Estabrook, R. E. Turner, W. L. Kruer, P. H. Y. Lee, B. Pruett, V. C. Rupert, K. G. Tirsell, G. L. Stradling, F. Ze, C. E. Max and M. D. Rosen, Phys. Rev. Lett. 47 (1981) p. 1289.
18. J. D. Kilkenny, (private communication).
19. J. F. McGee in X-ray Microscopy and Microradiography, V. E. Cosslett, A. Engstrom and H. H. Pattee, Jr., Eds. (Academic, New York, 1957) p. 164.
20. J. H. Underwood, T. W. Barbee and C. Frieber, Appl. Opt. 25 (1986) p. 1730.

SOFT X-RAY POPULATION INVERSION OF Na XI LEVELS BY INTERCOMBINATION LINE RESONANT PHOTOEXCITATION

Zhengquan Zhang, Renxiang Lu and Guangyu Yin

Shanghai Institute of Optics
and Fine Mechanics
P. O. Box 8211
Shanghai, People's Republic of China

The possibility of constructing a soft x-ray laser by utilizing coincidence between resonance lines of multiply charged ions has been amply studied. Here we report a new scheme utilizing near coincidence between resonance line and intercombination line, by which a population inversion of the n = 5-4 transition of hydrogen-like NaXI ions has been achieved using intercombination line of helium-like AlXII ions as pumping source.

Experimental facility was a standard neodymium glass laser, giving maximum energy of 10 J, pulse duration 600 ps, and peak intensity 10^{14} W/cm^2 at the target surface. X-ray emission from the laser-produced plasma was recorded photographically using a spatially resolving TlAP crystal spectrograph (4.5-15 A range) and a lead stearate crystal spectrograph (55-91 A range). Sensitometric calibration of 5F medical x-ray film used in the spectrograph was carried out before hand. The target was designed as a flat NaCl crystal coated with aluminum, optimum thickness of Al being about 1000 A.

From the obtained spectra the population inversion was deduced; the population densities were n_4 = 6.6 x 10^{12}/cm^3, n_5 = 5.0 x 10^{13}/cm^3, and the inversion ratio $n_5 g_4/n_4 g_5$ = 7.5. Electron temperature of about 500 eV and electron densities of 10^{19}-10^{20}/cm^3 were measured. The intensity ratio of the AlXII $1s^2$-$1s2p$ transition between intercombination line and resonance line reached 0.6. Therefore it is generally feasible considering intercombination line radiation as resonant photoexcitation source too.

© American Institute of Physics 1987

SHORT WAVELENGTH LIMITATIONS
OF FOUR WAVE MIXING IN GASES

H. SCHEINGRABER AND C. R. VIDAL

*Max-Planck-Institut für Extraterrestrische Physik,
D-8046 Garching, Federal Republic of Germany*

ABSTRACT

After a short review of the different intensity regimes of four wave mixing in gases the different competing nonlinear processes and their wavelength dependence are discussed.

INTRODUCTION

In recent years four wave parametric processes have been successfully used to extend the tuning range of lasers into the vacuum ultraviolet (VUV) spectral region. Most of these VUV systems have so far been based on two-photon resonant sum frequency mixing in gases [1,2,3]. Mostly dye lasers have been used which are pumped by excimer laser, nitrogen laser or frequency doubled Nd laser. The dye lasers generally consist of an oscillator amplifier arrangement in order to achieve the small linewidth and the beam quality as required for high resolution laser spectroscopy. In this case one of the dye lasers is tuned to a suitable two-photon resonance of the nonlinear medium, whereas the other dye laser is continuously tunable. For high resolution work the dye lasers may be equiped with air-spaced etalons which allow an accurate pressure scan covering the spectral region of interest.

For the nonlinear medium a gaseous two-component system is generally employed. It is generated either inside a heat pipe oven [4,5] containing a metal vapor inert gas mixture or inside a cell containing a mixture of two different inert gases. Sources of this kind have reached energies of $10^{11} - 10^{13}$ photons per shot in the VUV [6], whereas in the more distant XUV ($\lambda < 100$ nm) only about $10^6 - 10^8$ photons per shot [7] were typically achieved. In the XUV region which is, for example, of particular interest for the spectroscopy of the astrophysically important H_2 molecule, highly efficient sources in the XUV spectral region are of interest. For this reason we have investigated the origin of the significantly lower conversion efficiency in the XUV region compared to that of the VUV region.

THIRD ORDER NONLINEAR SUSCEPTIBILITIES

In describing four wave mixing in gases it was shown [3] that it is useful to distinguish three different intensity regimes: 1. the small signal limit, 2. the onset of saturation and 3. the high intensity saturation regime. With growing electric field amplitudes an increasing number of nonlinear processes are required for a quantitative description and the limits between the three different regimes are therefore determined to lowest order by the value of the corresponding third order nonlinear susceptibilities.

In general the third order nonlinear susceptibility is given by the following expression [8, 3]

$$\chi^{(3)}(-\omega_s;\omega_1,\omega_2,\omega_3) = \frac{S_T}{6\hbar^3} \sum_{gabc} \rho(g) \frac{\langle g|\vec{e}_s\mu|a\rangle\langle a|\vec{e}_1\mu|b\rangle\langle b|\vec{e}_2\mu|c\rangle\langle c|\vec{e}_3\mu|g\rangle}{(\Omega_{ag}-\omega_1-\omega_2-\omega_3)(\Omega_{bg}-\omega_2-\omega_3)(\Omega_{cg}-\omega_3)} \quad (1)$$

The summation has to be carried out over all atomic (or molecular) states a, b, c and g and the permutation operator S_T requires that the sum has to be taken over all possible permutations of the interacting waves specified by the individual $\vec{e}_k$ and ω_k where $\vec{e}_k$ defines the polarization of wave k and where the index k runs from 1, 2, 3 to s. The factor $\rho(g)$ takes care of the weighting over the unperturbed equilibrium distribution of the initial states. For a single, thermally populated atomic groundstate, $\rho(g) = 1$ and can be dropped. However, for molecular systems this is in general not true. The wavelength dependence is completely described by the frequency denominators with Ω_{pq} being the complex atomic or molecular transition frequency containing the homogeneous linewidth Γ_{pq}.

Depending on the sign of the ω_k, very different nonlinear processes are contained in the third order nonlinear susceptibility of Eq. (1). A number of competing processes have therefore to be considered besides the sum frequency mixing of interest such as: 1. the difference frequency mixing, 2. the two photon absorption and 3. the field dependent changes of the refractive index. In addition, other processes have to be considered which are a natural consequence of the population densities generated by the two-photon absorption such as 4. stimulated emission, 5. parametric processes and 6. a redistribution of the population densities [9].

For the optimum conditions of the sum frequency mixing it was shown [3] that one should always try to stay in the regime of the onset of saturation where field dependent changes of the refractive index start to destroy the phase matching condition. In going to shorter wavelengths this is still correct. For a Gaussian beam all of these processes, in addition, depend on space and time giving rise to an intricate convolution which is associated with a self-(de)focusing of the incident beams [9].

Since in the small signal limit and for optically thin ystems the intensity Φ_s at the sum frequency $\omega_s = \omega_1 + \omega_2 + \omega_3$ is given by

$$\frac{\Phi_s}{n_s} \sim \left(NL\chi_T^{(3)}(-\omega_s;\omega_1\omega_2\omega_3)\right)^2 \frac{\Phi_{10}\Phi_{20}\Phi_{30}}{n_1 n_2 n_3} F(\Delta k\ L) \tag{2}$$

where $F(\Delta k\ L) \leq 1$ is the phase matching factor, it is clear that a decrease of the optimum incident intensities by one to two orders of magnitude will lead to a decrease of the sum frequency intensity by three to six orders of magnitude as observed in the experiments. This may be due to a wavelength dependent increase of the critical nonlinear susceptibility which is responsible for the onset of saturation.

RESULTS

In recent experiments [9] several low-lying two-photon resonances in strontium have been investigated at input intensities which are well above the onset of the saturation. Significant population densities of the two-photon resonant states caused a population inversion with respect to the lower lying excited states. A major redistribution of the population densities and a subsequent self-defocusing was observed. In the high intensity saturation regime the two-photon transition could even be bleached and strong stimulated emission was observed on several transitions starting from the two-photon resonant state and from subsequent states. Since the stimulated emission is colinear with the incident wave a large number of parametric processes can occur in which one or two laser photons combine with the photons from the stimulated emission. Some of these parametric processes turned out to be so efficient that their intensities were comparable to the intensity of the harmonic wave. In a theoretical model the ac Stark effect and power broadening on the two-photon resonant transition had to be considered.

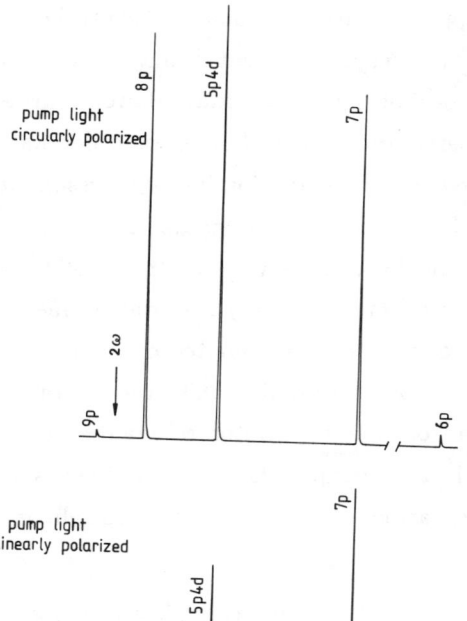

Fig. 1: Two-photon resonant excitation of the 5s 8d ^{1}D level in strontium showing the subsequent stimulated emission. The transition from the 9p ^{1}P level is due to an accidental pumping of the 6d ^{1}D - 9p ^{1}P transition by the stimulated emission from the 8d ^{1}D to the 7p $^3P_1$ level. The two scans are obtained with linearly polarized and circularly polarized radiation, respectively.

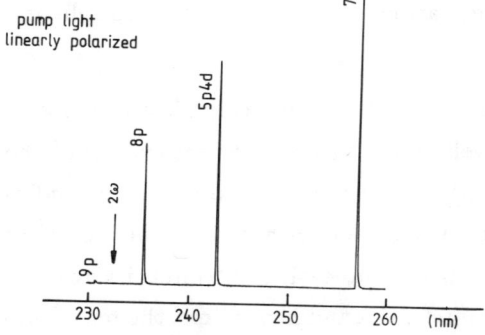

In going to shorter wavelengths we now investigated higher lying two-photon resonances in strontium, magnesium, cadmium and mercury. All of these experiments showed very similar results to what was observed in strontium on the lowest two-photon resonances. A striking example is shown in Fig. 1, which shows the emission after pumping the two-photon resonant 5s 8d ^{1}D level. In particular, one sees a series of stimulated emissions from the p-levels which are populated due to stimulated emission from the laser excited two-photon resonant state, and no second harmonic generation is observed as indicated by the arrow in the figure. None of the other experiments showed any trace of a second harmonic signal. As a peculiarity one also notices a transition from the 9p ^{1}P level to the ground state which lies above the 8d ^{1}D state and is due to an accidental coincidence between the 6d ^{1}D - 9p ^{1}P transition and the stimulated emission from the 8d ^{1}D to the 7p $^3P_1$ level.

The second harmonic generation [10, 11] has been the subject of several controversies. Miyazaki and coworkers [12] first suggested that it may be due to charged particles as generated by the two-photon resonant multiphoton ionization which destroy the spherical symmetry and which allow even order harmonics. Bethune et al. [13] even resorted to an explanation implying resonant quadrupole transitions. Freeman et al. [14] finally favored spatial density gradients removing the spherical symmetry. In this context we saw strong selfdefocusing due to density gradients [9], but never detected in any of our experiments a trace of a second harmonic. Instead we saw strong stimulated emission very close to, but never at the position of the second harmonic. This process is particularly efficient for an initial transition from the two-photon resonant state to the closest n p ^{1}P level in the infrared. In this manner the stimulated emission may be erroneously mistaken as a second harmonic if the fine structure splitting is very small.

These processes apparently become particularly important if according to Eq. (1) the two-photon excitation as well as the field dependent change of the refractive index are resonantly enhanced by the resonance line of the nonlinear medium. A large transition moment together with a favorable resonant denominator shift the transitions between the three different intensity regimes mentioned above to significantly maller intensities and hence limit the maximum conversion efficiency.

In summary we note that the wavelength dependence of the nonlinear susceptibilities favors a number of competing processes to an extent where the onset of saturation determining the optimum operation of the nonlinear medium, occurs at significantly lower intensities and where as a result of this the conversion efficiency in the XUV becomes significantly smaller.

REFERENCES

[1] C.R.Vidal, *Appl. Opt.* **19**, 3897 (1980)

[2] W.Jamroz and B.P.Stoicheff, *Generation of tunable coherent vacuum ultraviolet radiation*, in *Progress in Optics* ed. by E.Wolf (North Holland, Amsterdam 1983), vol. **20**, p. 326-380.

[3] C.R.Vidal, *Four wave frequency mixing in gases*, in *Tunable Lasers* edited by L.F.Mollenauer and J.C.White, Topics in Applied Physics, **59**, pp.19-75 (Springer, Heidelberg 1986).

[4] C.R.Vidal and F.B.Haller, *Rev. Scient. Instr.* **42**, 1779 (1971)

[5] H.Scheingraber and C.R.Vidal, *J. Opt. Soc. Am. B* **2**, 343 (1985)

[6] H.Scheingraber and C.R.Vidal, *Rev. Scient. Instr.* **52**, 1010 (1982)

[7] P.R.Herman and B.P.Stoicheff, *Opt. Lett.* **10**, 502 (1985)

[8] D.C.Hanna, M.A.Yuratich and D.Cotter: *Nonlinear optics of free atoms and molecules*, Springer Series in *Optical Sciences*, **17**, (Springer, Berlin, Heidelberg, New York 1979)

[9] H. Scheingraber and C. R. Vidal, *IEEE J. Quant. Electron.* **Q E-19**, 1747 (1983)

[10] K.Miyazaki, T.Sato and H.Kashiwagi: *Phys. Rev. Lett.* **43**, 1154 (1979)

[11] W.Jamroz, P.E.LaRocque and B.P.Stoicheff: *Opt. Lett.* **7**, 148 (1982)

[12] K.Miyazaki, T.Sato and H.Kashiwagi: *Phys. Rev. A* **23**, 1358 (1981)

[13] D.S.Bethune, R.W.Smith and Y.R.Shen: *Phys. Rev. A* **17**, 277 (1978)

[14] R.R.Freeman, J.E.Bjorkholm, R.Panock and W.E.Cooke: *Optical second harmonic generation by a single laser beam in an isotropic medium*, in *Laser Spectroscopy V* edited by A.R.W.McKellar, T.Oka and B.P. Stoicheff, Optical Sciences vol.30 (Springer, Heidelberg 1981) pp.453 - 457

PHOTOFRAGMENT SPECTROSCOPY WITH COHERENT VUV: PRODUCT CORRELATIONS AND ALIGNMENT

G.E. Hall, N. Sivakumar, G. Chawla and P.L. Houston,
Department of Chemistry, Cornell University,
Ithaca, N.Y. 14853 U.S.A.

I. Burak,
School of Chemistry, Tel-Aviv University, Israel

I.M. Waller, H.F. Davis and J.W. Hepburn,
Centre for Molecular Beams and Laser Chemistry,
University of Waterloo, Waterloo, Ontario N2L 3G1 Canada

ABSTRACT

The use of vacuum ultraviolet laser-induced fluorescence for photofragment spectroscopy is illustrated by reference to several recent experiments. In these studies, molecules are photolyzed under molecular beam conditions, and the photofragments are detected by vacuum ultraviolet laser-induced fluorescence. The detailed information about photofragmentation dynamics that can be obtained from such studies is discussed.

INTRODUCTION

The experimental study of photofragmentation dynamics is one of the many results of the application of lasers to experimental chemical physics. The use of lasers to dissociate molecules has allowed for the detailed study of very low density gases in a molecular beam.[1] More recently lasers have found another important role, that of detecting the photofragments resulting from laser photolysis[2]. The use of laser spectroscopic detection enables one to learn a great deal about the dynamics of the photodissociation event. Detailed analysis of the spectra of photofragments reveals not only the product internal energy distribution, but can also provide information on alignment of electronic orbitals in the fragments,[3] anisotropy of the product recoil velocity,[4] and alignment of the product rotation vector.[5,6] As well as being able to gather all of these separate facts about a photodissociation event, proper use of laser spectroscopy can provide information on correlations between different product properties, an example being the resolution of one product's internal energy distribution into two different distributions correlating with two different internal energy states of the other product.[6]

Of course there is no such thing as the perfect detector and in the case of laser spectroscopy the major drawback is the lack of general applicability. As direct absorption spectroscopy is not sufficiently sensitive to detect the low densities of products under molecular beam conditions, excitation spectroscopy must be used for product detection. The most sensitive, and therefore most widely used, laser spectroscopic techniques are laser-induced fluorescence and multiphotonionization. Both are methods for recording the electronic spectrum of an atom or molecule and the problem arises that for most small atoms or diatomic molecules, the lowest excited electronic state accessible from the ground state by an allowed transition is at high energy. As examples, the longest wavelength allowed absorptions in H, C, N, O and Cl atoms are at 121.6, 156.1, 113.5, 135.6 and 135.2 nm respectively, while for H_2 and CO the lowest energy excited singlet states are at 11.2 eV and 8 eV. While any of these states can be excited by multiphoton absorption, this approach has several drawbacks. The most serious is the loss in sensitivity that results from relying on multiphoton excitation, as the absense of intermediate excited states make the two and three photon absorption cross-sections quite small. Use of tight focussing results in a small detection volume, and high laser powers can lead to a swamping of the desired signal by probe laser photolysis, as the molecule under study frequently dissociates as a result of ultraviolet excitation. Coupled with these problems is the necessary complication of the data analysis resulting from driving two and three photon transitions rather than one photon transitions. These factors and others make it better to excite the resonant transition in a one photon step, thus using single photon LIF or one photon resonant, two photon ionization for product detection. This requires a reliable, high powered, broadly tunable source of coherent vacuum ultraviolet (VUV) light ($\lambda < 2000 Å$) to detect a large number of possible photofragments. While generation of this VUV is not a trivial matter, the techniques are well established and the methodology has been proven in several experiments.[7] In this paper, we shall discuss some of our results from recent photofragmentation experiments which use coherent VUV for product detection.

The intention of this paper is not to discuss the details of the photofragmentation dynamics in these systems, but rather to bring together a variety of results to illustrate the power of the VUV detection method and to show the detailed information that can be obtained on these systems. We apologize at the outset for the eclectic nature of this paper. All of the data discussed here is being published in more complete fashion elsewhere, and we refer those who are interested to the appropriate articles for more detailed discussions of these results.

In these experiments coherent VUV is used to detect the CO or S products from a variety of photofragmentations:

$$CS_2 + 193 \text{ nm} \rightarrow CS + S(^3P_J, ^1D_2) \tag{1}$$

$$OCS + 222 \text{ nm} \rightarrow CO + S(^1D_2) \tag{2}$$

$$(CHO)_2 + 430 \text{ nm} \rightarrow H_2CO + CO \tag{3a}$$

$$\rightarrow H_2 + 2CO \tag{3b}$$

$$\rightarrow HCOH + CO \tag{3c}$$

$$Fe(CO)_5 + 193 \text{ nm} \rightarrow Fe(CO)_x + y\, CO \tag{4}$$

In the case of the first three fragmentations, we shall discuss what can be learned from a detailed examination of the doppler profile of the photofragment spectral lines, while in the $Fe(CO)_5$ experiment, we shall focus on what can be learned from the details of the product energy distribution.

An interesting aspect of these different experiments that will be highlighted in this paper is a measurement of correlation between photofragment properties. The correlation between product channel and recoil velocity will be discussed for reaction (1). In the case of reactions (2) and (3) correlation between the recoil velocity vector, **v**, and the CO rotation vector, **J**, will be described. Results on reaction (4) will be discussed in terms of the correlation between the CO internal energy and dissociation step in a complex mechanism.

EXPERIMENTAL

The techniques used for these experiments have been described in detail elsewhere[9]. Briefly, a pulsed supersonic jet of the molecule of interest seeded in He (typically with concentrations of 1% to 5%) was intersected by the pulsed photolysis laser. A short time delay later, a second laser pulse was sent through the volume defined by the crossing of the molecular beam and the photolysis laser to detect the products. The detection laser was tunable over the 142-170 nm range with linewidths between 0.15 cm^{-1} and 0.20 cm^{-1} and the S and CO products were detected by laser-induced fluorescence. The coherent VUV was generated by resonant four-wave sum mixing in Mg vapour[8]. The S atoms were detected by exciting the $^1P_1 \leftarrow ^1D_2$ resonance line at 144.82 nm and $^3D_{J'} \leftarrow ^3P_{J''}$ lines between 147.4 nm and 148.7 nm. The CO(v,J) products were detected by exciting several different bands of the $A^1\Pi \leftarrow X^1\Sigma^+$ spectrum in the 144 nm to 160 nm region. Doppler profiles were recorded for selected absorption lines using a polarized photolysis laser, with the plane of polarization either parallel or perpendicular to the VUV probe laser axis. In order to determine the spectral resolution of the LIF detection system under our experimental conditions, the LIF spectrum of a CO/He beam was recorded.

RESULTS AND DISCUSSION

CS_2: The photofragmentation of CS_2 at 193 nm has been the topic of several recent studies.[10] These have led to contradictory conclusions about the photochemistry of CS_2, particularly with regard to the relative importance of the two possible product channels:

$$CS_2 + 193 \text{ nm} \rightarrow CS(^1\Sigma^+) + S(^1D) + 45 \text{ kcal/mole} \qquad (1a)$$

$$CS_2 + 193 \text{ nm} \rightarrow CS(^1\Sigma^+) + S(^3P) + 19 \text{ kcal/mole} \qquad (1b)$$

Although this very important concern is addressed by our experiments, we would like to focus for now on resolving the dynamics of the two different product channels from one another.

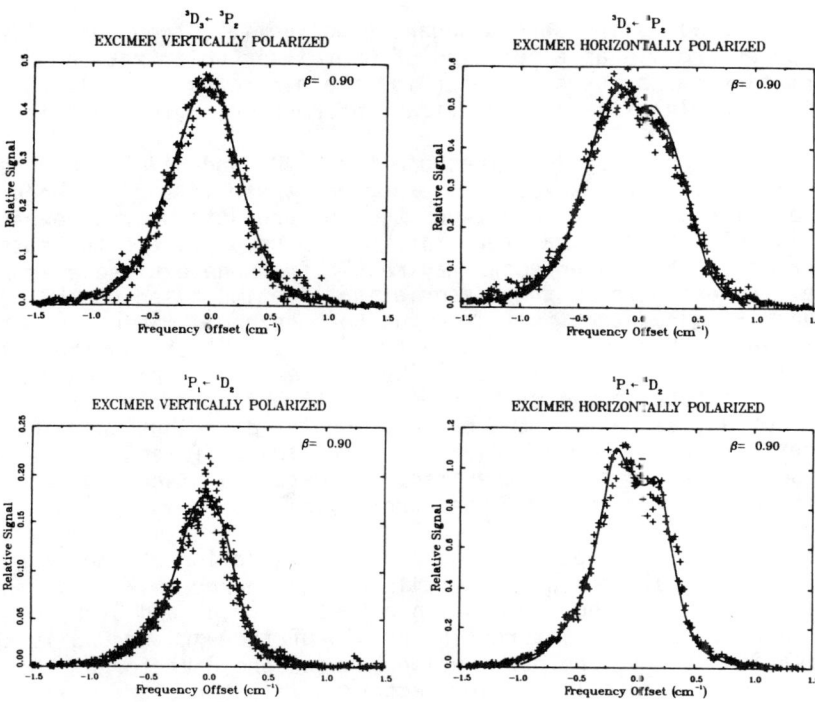

Figure 1 - Doppler profiles for S(^{1}D) and S(^{3}P)

Data points are given with several scans superimposed. Solid line is the result of convoluting the known laser lineshape with a product velocity distribution, and an angular distribution given by equation (5), with β=0.90

This can be done by doing doppler spectroscopy on the atomic S products. Since different spectral lines are used to detect S(^{3}P) and S(^{1}D), this approach automatically separates the two channels. In fact, the doppler lineshape probes two things simultaneously. The first is the velocity distribution of the products, which provides information on how the energy release in the dissociation is correlated with product channel. Since the doppler shift is determined by the projection of the recoil velocity onto the probe laser axis, the doppler lineshape can also provide information about the angular distribution of photofragments with respect to the photolysis laser polarization. This is done by measuring the doppler line profile for different orientations of the photolysis laser polarization with respect to the probe laser axis. If one assumes that the probability, P(θ), of the recoil velocity vector having an angle θ with respect to the photolysis laser polarization is given by:[11]

$$P(\theta) = 1 + \beta P_2(\cos\theta) \tag{5}$$

where $P_2(\cos\theta)$ is the second Legendre polynomial, then the variation in doppler profile as a function of photolysis polarization orientation can be used to determine the parameter β, while the detailed doppler lineshape provides information on the velocity distribution.

The observed doppler lineshape for $S(^1D)$ and $S(^3P)$ products are shown in figure 1. As would be expected, there is a difference in the velocity distribution between the two products but, somewhat surprisingly, the β parameters for the two channels are the same, indicating that both products may result from one excited electronic state. An estimate of the dissociation lifetime based on the measured β parameters is 1 psec for both channels, which is consistent with previous lifetime estimates.[12] Although there is insufficient space for discussion here, the solid lines drawn through the data in figure 1 require the translational energy distribution, as well as the product angular distribution. Thus, the details of the energy release for the two different channels is also determined by these experiments. A detailed discussion of our results on CS_2 will be published shortly.[13]

OCS: In recent papers on the photofragmentation of the $^1\Delta$ state of OCS we have discussed the details of the energy release in the dissociation[14] and the correlation between $CO(v,J)$ and $S(^1D)$ or $S(^3P)$ product[6]. One interesting result of the work on OCS is the observation of a strong correlation between the **J** vector of the CO product and the recoil velocity vector[6]. This correlation is expected on the basis of conservation of angular momentum. To conserve angular momentum in the photodissociation the following must be true:

$$\mathbf{J}_{OCS} + \mathbf{J}_{h\nu} = \mathbf{J}_S + \mathbf{J}_{CO} + \mathbf{L} \qquad (6)$$

where **L** is the orbital angular momentum of the S + CO products. Since the OCS is cooled by supersonic expansion, $\mathbf{J}_{OCS} \approx 0$, and $\mathbf{J}_{h\nu} = 1$, $\mathbf{J}_S = 2$ can be ignored with respect to $\mathbf{J}_{CO}$ ($\mathbf{J}_{CO} \approx 60$). Thus eqn. (6) simplifies to $\mathbf{J}_{CO} + \mathbf{L} = 0$, meaning that $\mathbf{J}_{CO} = -\mathbf{L}$. Since **v** is perpendicular to **L**, **J** must be perpendicular to **v**.

The result of any correlation between **v** and **J** is a different doppler line profile for Q and R branch lines (P and R lines are identical).[5b] For $\mathbf{v} \perp \mathbf{J}$, Q branch lines will have a dip at line centre, and P and R branch lines will not. In figure (2) the doppler profiles for the P and Q branch lines for the CO(v=0, J=59) product of OCS + 222 nm $\rightarrow$ CO + $S(^1D)$ are given, with the photolysis laser polarization perpendicular ($\theta' = \pi/2$) and parallel ($\theta' = 0$) to the VUV probe axis. The doppler profile expected in the absense of **v, J** correlation is given in the top two figures, along with the probe laser line profile (dotted line). These profiles are calculated from the known CO recoil velocity, and using the angular distribution given in eqn (5) with $\beta = 0.6$. In the absence of **v, J** correlation the line profiles for P, Q and R branch lines should be

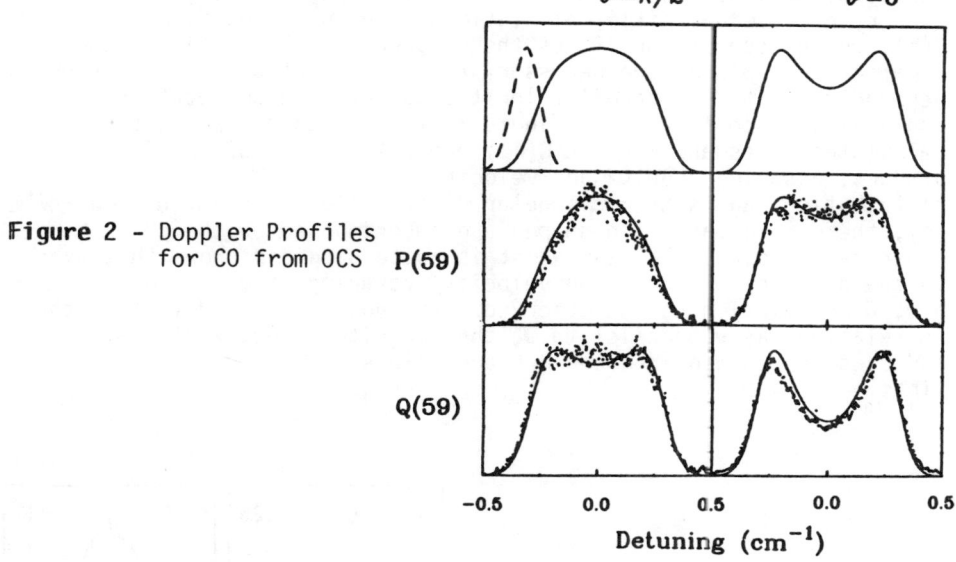

Figure 2 - Doppler Profiles for CO from OCS

identical. In the next two rows, the experimentally determined doppler profiles are shown and the P and Q line profiles are quite different, even though they result from excitation of identical CO products (CO(v=0, J=59)). The difference in profile is caused by the **v** $\perp$ **J** correlation, and the solid lines drawn through the data are calculated using the known recoil velocity, the angular distribution with $\beta = 0.6$, and with **v** $\perp$ **J**.

GLYOXAL: The photodissociation of glyoxal is far more complex than OCS or CS_2. There are 6 atoms in glyoxal, and photodissociation at 430 nm results in three chemically distinct sets of products (eqn.(3)). Furthermore, the dissociating state is quite long-lived, with a lifetime of 800 nsec. We have recently completed an extensive study of the photofragmentation of glyoxal[15] and will only discuss one aspect of it here. In the case of OCS → CO + S(^{1}D), when the (v,J) state of CO was specified, conservation of energy determined the recoil velocity. This is clearly not the case with glyoxal, where one cannot even unambiguously correlate the CO(v,J) state with product channel. The very long dissociation lifetime, and the fact that K,J ≠ 0 are excited ensures that the recoil velocity angular distribution will be isotropic. However, when the doppler profiles for product CO lines are measured, there is a consistent difference in shape between Q branch lines and P branch lines (see figure 3) with the Q lines always showing a dip at line centre, reminiscent of the CO from OCS. In fact, the cause is the same: a correlation between **v** and **J**, with **v** $\perp$ **J**. This correlation is not caused by simple conservation of angular momentum considerations, as all products are structured and can have rotational angular momentum. The cause of the **v**, **J** correlation is that the dissociation occurs in one plane, with little out of plane torsional motion. This planar dissociation is, in fact, predicted by very precise ab initio calculations[16] for both reactions 3(a) and 3(b), which account for 95% of the CO product. By confining the dynamics to a plane, one necessary result is that **J** will always be perpendicular to the recoil velocity, as both the CO rotation plane and **v** will be coplanar. In figure 3, the bottom row shows the calculated lineshape for the Q(41) and P(41) line of the CO(v=0) product, assuming a Boltzmann velocity distribution, an isotropic angular distribution, and **v** $\perp$ **J.** In row (c), these calculated lineshapes are superimposed on the experimental data. The agreement is quite good and shows that even in the presence of angle and velocity averaging, the presence of a **v, J** correlation can be observed. It should be noted that if the correlation was **v** parallel to **J**, the opposite effect would be expected i.e. a dip in P and R branch lines, and no dip in the Q lines.

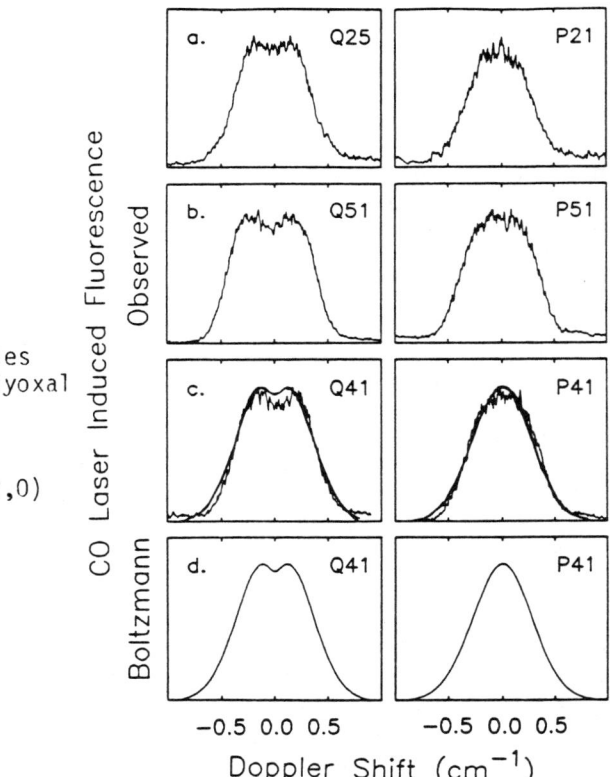

Figure 3 - Doppler profiles for CO from Glyoxal

Data for Q(25), P(21); Q(51), P(51) and Q(41), P(41) lines of the A←X (2,0) band are shown

Fe(CO)$_5$: Even in the case of a large molecule like Fe(CO)$_5$, it is still possible to observe product correlations. The photolysis of Fe(CO)$_5$ at 193 nm leads to several different unsaturated Fe(CO)$_x$ products, and several recent studies have attempted to determine the detailed mechanism for this process[17]. The consensus about the mechanism is that the CO elimination occurs sequentially, in a series of fragmentation steps each one of which involves loss of a single CO from Fe(CO)$_x$. This suggests that it may be possible to correlate CO products with the various steps in the dissociation process. For 193 nm photolysis, an average of about 3 CO products are formed for each Fe(CO)$_5$ fragmented under collision-free conditions. When we investigated the CO product energy distribution[18] by the methods already described, we determined the rotational and vibrational distributions shown in figure 4. The solid lines drawn through the experimental data are the result of a statistical calculation based on the idea that there are three different CO photofragments, each resulting from an RRKM-like dissociation of vibrationally excited Fe(CO)$_x$. There were no adjustable parameters used in the calculation, and the agreement with the data is striking.

This agreement provides strong support for our model for the dissociation, and indicates that even in this very complex, statistical dissociation process, one can still see the effect of having three different CO products, and can in some cases resolve a product energy distribution into three separate distributions, each corresponding to a different fragmentation step. This data is described in more detail in reference 18, and a more extensive and detailed discussion of Fe(CO)$_5$ photofragmentation is in preparation[19].

Figure 4 - CO from Fe(CO)$_5$

Product rotational and vibrational distributions, plotted as Boltzmann plots are given.

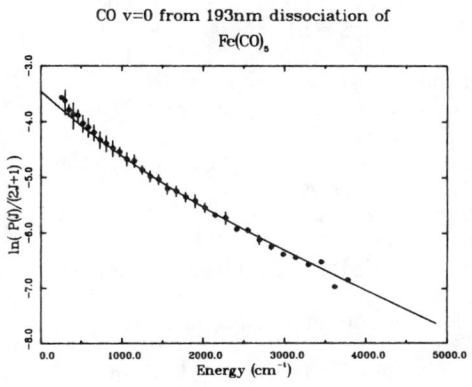
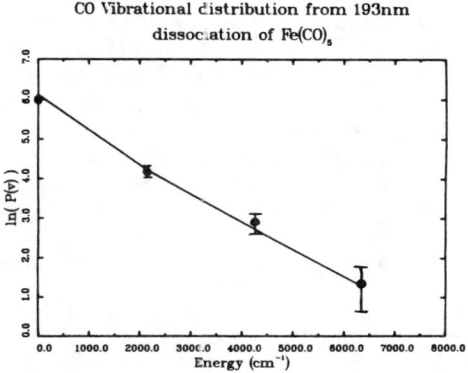

ACKNOWLEDGEMENTS

The work on CS_2 and $Fe(CO)_5$ was carried out at Waterloo by IMW, HFD and JWH and was supported by NSERC (Canada), the Petroleum Research Fund (17811-AC6) and a grant from Imperial Oil. IMW thanks NSERC for a graduate scholarship.

The work on OCS and glyoxal was carried out at Cornell by NS, GC, GH, PLH, IB and JWH, and was the result of a NATO collaboration between PLH and JWH (637/83). The work at Cornell was supported by the NSF (CHE-8314146) and AFOSR (F49620-83-K-0012). The research used lasers funded through the Department of Defense Instrumentation program (DAAG29-84G-0076) and the Dow Chemical Foundation.

REFERENCES

1. A.E. de Vries, Comments At. Mol. Phys. 11, 157 (1982).
2. S.R. Leone, Adv. Chem. Phys., 50, 255 (1982).
3. P. Andresen, G.S. Ondrey, B. Titze and E.W. Rothe, J. Chem. Phys. 80, 2548 (1984).
4. S. Klee, K.-H. Gericke and F.J. Comes, J. Chem. Phys. 85, 40 (1986).
5. a) R.N. Dixon, J. Chem. Phys. 85, 1866 (1986).
   b) G.E. Hall, N. Sivakumar, P.L. Houston, and I. Burak, Phys. Rev. Lett. 56, 1671 (1986).
6. G.E. Hall, N. Sivakumar, R. Ogorzalek, G. Chawla, H.-P. Haerri, P.L. Houston, I. Burak and J.W. Hepburn, Faraday Disc. 82, xxxx (1986).
7. J.W. Hepburn, Israel J. Chem. 24, 273 (1985).
8. S.C. Wallace and G. Zdasiuk, Appl. Phys. Lett. 28, 449 (1976).
9. I.M. Waller, H.F. Davis and J.W. Hepburn, "Short Wavelength Coherent Radiation: Generation and Applications", D.T. Atwood and J. Bokor eds. (AIP, 1986), p. 430.
10. V.R. McCrary, R. Lu, D. Zakheim, J.A. Russell, J.B. Halpern and W.M. Jackson, J. Chem. Phys. 83, 3481 (1985), and references quoted in the introduction.
11. a) R.N. Zare and D.R. Herschbach, Proc. IEEE 51, 173 (1963).
    b) S. Yang and R. Bersohn, J. Chem. Phys. 61, 4400 (1974).
12. K. Hara and D. Phillips, Trans. Faraday Soc. 74, 1441 (1978).
13. I.M. Waller and J.W. Hepburn, manuscript in preparation.
14. N. Sivakumar, I. Burak, W.-Y. Cheung, P.L. Houston and J.W. Hepburn, J. Phys. Chem. 89, 3609 (1985).
15. I. Burak, J.W. Hepburn, N. Sivakumar, G.E. Hall, G. Chawla and P.L. Houston, to be published J. Chem. Phys.
16. Y. Osamara, H.F. Schaefer, M. Dupuis and W.A. Lester, J. Chem. Phys. 75, 5828 (1981).
17. a) T.A. Seder, A.J. Ouderkirk and E. Weitz, J. Chem. Phys. 85, 1977 (1986).
    b) J.T. Yardley, B. Gitlin, G. Nathanson and A.M. Rosan, J. Chem. Phys. 74, 370 (1981).
18. I.M. Waller, H.F. Davis and J.W. Hepburn, to be published, J. Phys. Chem.
19. I.M. Waller and J.W. Hepburn, manuscript in preparation.

Measurements of Subpicosecond Laser-produced Plasmas

H. Milchberg, R. R. Freeman, S. C. Davey

AT&T Bell Laboratories
Murray Hill, NJ 07094

ABSTRACT

A high power subpicosecond laser ($10^{16} W/cm^2$, 7 mJ, 400 fs FWHM) has been used to create solid density plasmas in a number of target materials, including carbon and copper. Plasma soft x-ray spectra indicate temperatures of 100-200 eV. Resulting soft x-ray bursts from a copper target are measured at the limit of a streak camera's resolution to be at most 20 psec in duration, the shortest burst of x-rays measured to date. Preliminary evidence is presented of novel "inverted" reflectivities of these solid density plasmas at the highest laser power densities.

INTRODUCTION

Laser-produced plasmas are a promising source of high flux, short pulse soft x-rays. Among the many uses of such a source would be picosecond time resolved studies of solid state structure[1] and pumping of possible x-ray laser transitions.[2] The present work deals with characterization of the plasma source itself.

EXPERIMENT AND DISCUSSION

The front end of the high power subpicosecond laser used in these experiments is a synchronously pumped mode-locked oscillator operating at 616 nm, which is followed by a fibre/grating compressor. Pulses from this laser are cavity dumped and amplified by $\sim 10^6$ to 0.5 mJ in a 4-stage pulsed dye amplifier (PDA) pumped by a Q-switched, frequency doubled YAG laser operating at 25 Hz. These PDA pulses, approximately 600 fs FWHM, are frequency doubled in a KDP crystal to 308 nm and are amplified by two XeCl excimer modules. These amplifiers have a spatial filter between them to remove amplified spontaneous emission (ASE). The resulting pulses at 308 nm are 7 mJ, 400 fs FWHM and focusable to $10^{-6} cm^2$, giving peak power densities of $10^{16} W/cm^2$.

These laser pulses were focused by a 200 mm lens onto rotating and translating cylindrical targets inside a vacuum vessel with background pressure of 5×10^{-7} torr. The target rotated at a sufficient rate so that new surface area was exposed to the focused beam for each pulse as the laser strobed at 25 Hz. Figure 1 shows an XUV spectrum from a carbon target taken with a 1 m grazing incidence spectrograph, viewing the plasma spot at 45 ° to the target normal. The distribution of helium-like and hydrogen-like carbon states, and the $CVI-CV$ recombination continuum indicate a temperature of 100-150 eV.

X-ray streak camera photos, not shown here, have been taken of a copper target plasma. A 1000Å Al filter had been placed in front of the camera's aperture to cut off light above 200Å. The observed x-ray pulse FWHM ($\sim$20 ps), is close to the camera's temporal resolution. The actual x-ray pulse duration is expected to be closer to the laser pulse duration, considering that radiation and thermal conduction cooling at these solid densities will greatly accelerate plasma recombination. Note, for instance, that at $n_e = 10^{23} cm^{-3}$ and $T_e = 150$ eV, a still rather high temperature, the calculated three body recombination time constant[3] is 0.5 ps.

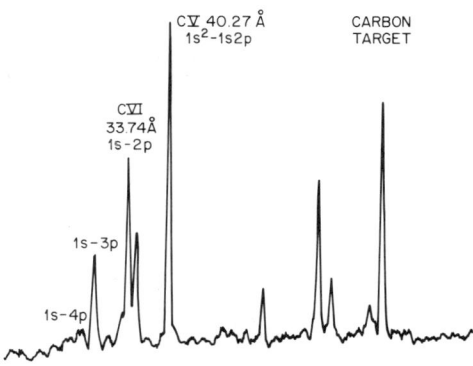

Fig. 1. carbon spectrum.

An indication of the different nature of these solid density plasmas was revealed in a preliminary specular reflectivity experiment using a rotating cylindrical copper target, which was highly polished. Part of the incident ultrashort 308 nm pulse was split off and measured with a joulemeter as a reference while the remainder was focused onto the target and collected specularly by another joulemeter at 45°. As an interesting comparison, a 60 ps frequency-tripled YAG pulse at 355 nm was used in the same geometry. Figures 2(a) and 2(b) show the results using the 60ps and 400fs pulses respectively, where the highest reflectivities have been normalized to unity. In the long (60 ps) pulse case, the reflectivity decreased with increasing power, as expected, due to both enhanced absorption (resulting from elongated plasma density gradients due to more energetic flow out of the target) and non-specular reflection from the ripples in the critical density surface that certainly have time to develop in 60 ps. By contrast, for the 400 fs pulses [Fig. 2(b)], the reflectivity starts out low, then increases and saturates for power densities above $\sim 5 \times 10^{12} W/cm^2$. Note that for power densities below $10^{11} W/cm^2$, the instruments were unable to record the incident and specular energies accurately, although the trend was for the reflectivity to increase back to its cold metal value.

An interpretation of this result considers that reflection remains specular at all times during the laser pulse (target material moves at most $\sim$40Å, or $\sim \lambda/10$ during 400 fs) so that the data reveals conditions in the solid density plasma. At low power densities, there is evidence of much stronger absorption due to a colder solid density plasma, in which the electron mean free path is very short. However, if Spitzer plasma resistivity[4], η, is valid at higher power densities (and therefore, temperatures), then since $\eta \sim T_e^{-3/2}$, greater reflectivity would be expected in this regime. In effect, the early part of the laser pulse establishes the hot plasma conditions which cause the latter part of the pulse to strongly reflect. Especially in the lower power (temperature) region, a much more complex theory is needed to describe these strongly coupled, highly collisional plasmas.

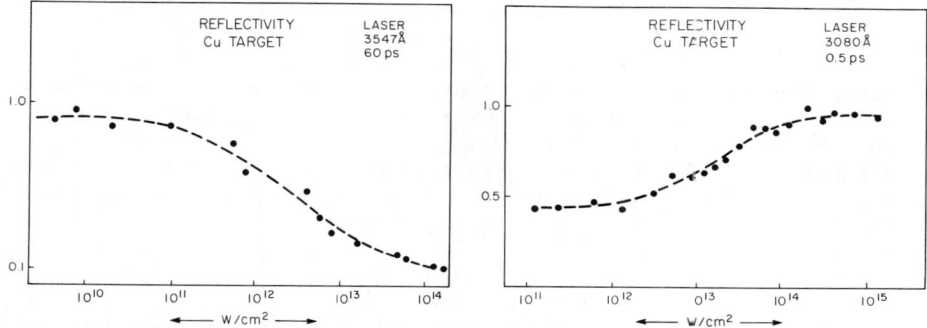

Fig. 2(a), (b): normalized reflectivities for 60 ps, 0.5 ps pulses.

CONCLUSIONS

Experimental results thus far indicate the production of hot, solid density plasmas with short pulse x-ray emission. Future work will include an improved measurement of x-ray pulse duration, more detailed reflectivity experiments with assorted materials, and spectroscopic studies of solid density plasmas.

ACKNOWLEDGEMENTS

The authors would like to thank S. Darack and P. Bucksbaum for useful help and discussions, and E.M. Campbell, of Lawrence Livermore Labs, for the use of a streak camera.

REFERENCES

1. P. A. Lee, P. H Citrin, B. M.Kincaid, Rev. Mod. Phys. 53, 769 (1981).
2. M. Duguay and P. Rentzepis, Appl. Phys. Lett. 10 , 350 (1967).
3. F. V. Bunkin, V. I. Derzhiev, and S. I. Yakovlenko, Sov. J. Quant. Electron 11 , 981 (1981).
4. L. Spitzer, Jr., Physics of Fully Ionized Gases , (John Wiley & Sons, N.Y., 1962).

A LASER PRODUCED PLASMA LIGHT SOURCE FOR HIGH RESOLUTION SPECTROSCOPY AND SOFT X-RAY LITHOGRAPHY

M. L. Ginter and T. J. McIlrath
Institute for Physical Science and Technology
University of Maryland, College Park, MD 20742

ABSTRACT

Laser produced plasmas have been demonstrated[1] to be convenient laboratory light sources for absorption spectroscopy below 50nm. Under proper conditions no line structure appears in the intense continua from laser sources when observed at the highest possible spectral resolutions[2] and we have used this source for high spectroscopy. The source also was used with modified conditions for soft x-ray lithography[3].

INTRODUCTION

Laser produced plasmas using heavy metal targets and high repetition rate lasers are high average power source of XUV (<100nm) and soft x-ray radiation. We have made intensity measurements and studied outputs under very high resolution in order to characterize these sources for several applications. The plasmas were produced by focusing the output of a Nd:YAG laser (1064nm, 600mj/pulse, 10Hz) or an excimer laser (248nm, 300mj/pulse, 150Hz) onto a metal target using an f=30cm or an f=10cm lens. The maximum power density on the target with both sources was $10^{11} - 10^{12}$ W cm^{-2}. The cylindrical metal targets[4] were rotated to produce fresh target material for every pulse. The plasmas were viewed at right angle to the incident laser beam with the normal to the target surface at the focal spot being approximately 45° to both the incident laser beam and the viewing direction.

SOURCES FOR HIGH RESOLUTION XUV SPECTROSCOPY

The outputs of the laser plasmas sources have been studied in the normal incidence spectral region (~30-100nm) using a 6.65m spectrograph equipped with a 4800g/mm grating[5]. Targets included Cu(Z=29), Yb(Z=70), Hf(Z=72), W(Z=74 and Pb(Z=82). The continuum intensities from Yb, Hf, W and Pb were similar yet of somewhat different spectral distributions. The four higher Z element plasmas can be made to produce clean continua with very few emission lines. However, in order to obtain intense line-free continua it is necessary to carefully image the hottest portion of the plasma onto the entrance slit of the spectrograph. Optimization of the continuum output and emission line suppression required careful placement (~ 100 microns for the source described above) onto the 50 micron entrance slit of the spectrograph. The continuum intensity from the Cu target was found to be weaker than those obtained from the high Z element targets by approximately an order of magnitude with the Cu plasma emission dominated by strong line spectra.

The emission spectra from W, Yb and Cu targets were also observed on a 10m grazing incidence spectrograph[6] equipped with a 1200g/mm grating. The spectral region from 4.5nm to 60nm was studied. Both the W and Yb sources produced almost line free continua with comparable intensities. The intensity appeared to be strongest in the 10-20nm region. The falloff of intensity at longer wavelengths was such that absorption spectra of He in the 17-21nm region could be recorded in third order with no order separator and no significant contamination by first or second order light.

Intensity studies have been made in the 8 to 40nm region with a 1.5m grazing incidence spectrometer and a channel electron multiplier detector[7]. Nine different elements were compared in intensity and spectral purity. These studies confirmed the desirability of using high Z targets and provide a basis for choosing the target most suited to the particular spectral region being studied.

SOFT X-RAY LITHOGRAPHY

In soft X-ray lithography the continuum purity of the source is no longer of concern. We have studied the exposure of photoresists using Cu, W or steel targets driven by the Nd:YAG laser[8]. The photoresist used was a copolymer of polyglycidyl methacrylate and ethyl acrylante (COP), which was spun to a thickness of 4700Å on the surface of 75-mm diameter silicon wafers. COP is a negative photoresist with a sensitivity of $15mj/cm^2$. Masks consisted of polymide membranes overlayed with circuit patterns in gold. The shortest exposure times were obtained with steel (Fe) with Cu targets being only slightly less effective. Exposure times with W targets were approximately twice as long. The lithographs showed greater uniformity of exposure than that obtained using synchrotron radiation exposure. Exposure times were ~10 times longer with the laser plasmas than with the synchrotron. Preliminary studies of the output using the 150Hz excimer laser suggest that exposure times would be reduced by a factor of ~10 over the time using the 10Hz Nd:YAG laser. There were no obvious differences in the quality of the lithographs made using either light source.

SOURCE DEBRIS

A major problem in using laser plasma sources is the production of large amounts of debris. We have begun studies of the effect of buffer gases in the source chamber on the resultant debris. A glass sample plate was placed 10cm from the source which was irradiated by the 150Hz excimer laser driver. A steel target was used and the Fe deposited on the sample plate was quantitatively analyzed by chemical means. It was found that the introduction of 20 microns of He gas reduced debris by a factor of 2 and above 100 microns pressure the debris was reduced by more than a factor of 10 to near our detection limits. It was also found that raising the laser pulse energy from 100mj to 300mj increased the debris level by more than a factor of 6. Since the XUV output is nearly linear with the driving energy, this indicates that the debris problem can be reduced by using a higher repetition rate with a lower pulse energy. Work is continuing on optimizing the output with minimum debris.

This work was supported by the Air Force Office of Scientific Research under contract F49620-83-C0130.

REFERENCES

1. P.K. Carroll, E.T. Kennedy and G. O'Sullivan, Appl. Opt. 19, 1454 (1980).
2. M.L. Ginter and T.J. McIlrath, Nucl. Inst. and Meth., A246, 779 (1986).
3. P. Gohil, H. Kapoor, D. Ma, M.C. Peckerar, T.J. McIlrath and M.L. Ginter, Appl. Opt. 24, 2024 (1985).
4. G. O'Sullivan, P.K. Carroll, T.J. McIlrath and M.L. Ginter, Appl. Opt. 20, 3043 (1981).
5. F.B. Orth, K. Ueda, T.J. McIlrath and M.L. Ginter, Appl. Opt. 25, 2215 (1986).
6. P. Gohil, V. Kaufman and T.J. McIlrath, Appl. Opt. 25, 2039 (1986).
7. J.M. Bridges, C.L. Cromer and T.J. McIlrath, Appl. Opt. 25, 2208 (1986).

XUV LASER STARK SPECTROSCOPY OF Xe AUTOIONIZING RYDBERG STATES

W. E. Ernst[*], T. P. Softley[+], L. Tashiro, and R. N. Zare

Department of Chemistry, Stanford University, Stanford, CA 94305

ABSTRACT

Tunable coherent radiation between 90 and 110 nm is generated by frequency tripling the output of a pulsed UV laser. The 3rd harmonic XUV beam is separated from the high power fundamental UV laser radiation and focussed into an atomic Xe beam. Single-photon transitions from the ground state of Xe to the autoionizing Rydberg series are observed by collecting and detecting Xe ions in a time-of-flight mass spectrometer. An electric field is applied in the interaction region and the line intensities, shifts, and splittings have been investigated for $0 < E < 2$ kV/cm.

INTRODUCTION

For most atoms and molecules the investigation of one-photon transitions from the ground state to Rydberg states or of single-photon-ionization requires working in the vacuum ultraviolet (VUV) or extreme ultraviolet (XUV) spectral region. In recent years frequency upconversion of high power dye lasers has extended the application of laser spectroscopy into this short wavelength range. In this work we generated XUV radiation by tripling the frequency of a frequency doubled dye laser in a free jet of the rare gases Ar and Xe[1]. As the third harmonic XUV beam is collinear with the high power fundamental UV laser radiation, the spectroscopy of many species is obscured by UV multiphoton absorption processes. We have developed a setup in which the XUV is separated from the UV by using two dichroic beam splitters, as demonstrated by Falcone and Bokor[2], and is then focussed into an atomic or molecular beam. In a first application of our XUV laser spectrometer we have studied the Stark effect of Xe autoionizing Rydberg states in the vicinity of the second ionization limit around 92.5 nm.

XUV LASER SPECTROMETER

A pulsed dye laser operating in Rhodamine 590 is pumped by a Nd-YAG laser of 6 µs pulse length at 10 Hz repetition rate. Without the use of an intracavity etalon the linewidth of the dye laser is about 0.35 cm^{-1}. The dye laser output is frequency doubled in a KDP crystal. The generated UV radiation of about 12 mJ/pulse is focussed into a pulsed free jet of rare gas atoms in a vacuum

---

[*]Present address: Institut für Molekülphysik, Freie Universität Berlin, Arnimallee 14, D-1000 Berlin 33, West Germany.

[+]Present address: University Chemical Laboratory, Lensfield Road, Cambridge, CB2 1EW, England.

chamber which serve as a gaseous nonlinear medium.[1] Frequency tripling in a Xe jet yields about 10^8 photons per pulse in the 93 μm region, while the conversion efficiency in Ar is about one order of magnitude lower. Behind the jet the UV fundamental and the XUV are incident on a beam splitter consisting of a coated quartz plate. The light is s-polarized with the angle of incidence being 70°. With a quarter wavelength layer of MgF_2 the reflection is about 45% in the XUV and 1% in the UV.[2] The light passes a second beam splitter with the result that the fundamental UV intensity is reduced by four orders of magnitude and about 20% of the generated XUV can be used for spectroscopy. The first beam splitter has a toroidal surface focusing the XUV on to the sample under investigation which is in our case a pulsed atomic or molecular beam. The XUV is detected behind the sample with an electron multiplier. Potential fluorescence from the sample is collected and focussed on to a photomultiplier. Generated ions are collected and detected in a 50 cm time-of-flight mass spectrometer.

STARK SPECTROSCOPY OF Xe RYDBERG STATES

Xenon atoms in a pulsed atomic beam were excited by XUV radiation between 92.3 and 92.8 nm, into Rydberg states close to the second ionization limit, i.e., $Xe^{+2}P_{1/2}$. These states autoionize and Xe ions are detected in the time-of-flight mass spectrometer. Due to the different electric potentials of the repeller and extractor plates mounted around the laser-atom interaction region, a finite electric field is always present. Single-photon excitation of Xe under zero-field conditions would lead only into ns' and nd' Rydberg series whereas in the presence of the field the states mix and np' series are also observed. We have varied the electric field strength between 30 V/cm and 2000 V/cm and recorded the series from n = 14 to the ionization limit. Figure 1 shows a portion of the spectrum at different field strengths. At about 150 V/cm np' lines appear in addition to the ns' and nd' series (Fig. 1a). Above 1kV/cm states with $\ell > 2$ start to mix in and ℓ is no longer a good quantum number in this region (Fig. 1b). Instead hydrogen-like Stark multiplets start to be observed. In Table I the positions of np' lines, extrapolated to zero field, and effective principal quantum numbers are given. The ns' and nd' series have been measured by Yoshino and Freeman.[3] Furthermore, we have determined the splittings and shifts of Rydberg lines which should be compared with the results of Stark multichannel quantum defect theory calculations.

This work was funded in part by ONR under N00014-78C-0403 and NSF PHY 85-06668.

REFERENCES

1. C. T. Rettner, E. E. Marinero, R. N. Zare, and A. H. Kung, J. Phys. Chem. 88, 4459 (1984).
2. R. W. Falcone and J. Bokor, Opt. Lett. 8, 1 (1983).
3. K. Yoshino and D. E. Freeman, J. Opt. Soc. Am. B2, 1268 (1985).

Table I: Extrapolated zero field positions of the np′ Rydberg series (experimental accuracy 1 cm^{-1}) and effective principal quantum numbers.

n	wavenumber	n*
18p	107849.7	14.51
19p	107914.0	15.50
20p	107967.7	16.50
21p	108012.7	17.51
22p	108050.1	18.50
23p	108082.4	19.51
24p	108109.6	20.50
25p	108109.6	20.50
26p	108171.6	23.47
27p	108171.6	23.47
28p	108188.0	24.50
29p	108201.2	25.44
30p	108214.9	26.53
31p	108225.8	27.51

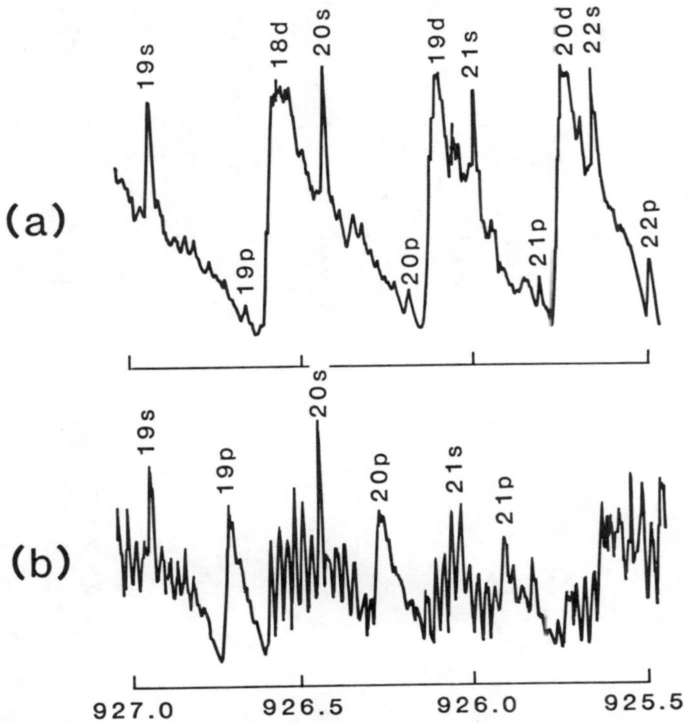

Fig. 1. Part of the Xe Rydberg spectrum at (a) 157 V/cm and (b) 1312 V/cm wavelength given in Å.

III. Nonlinear Optical Phenomena and Applications
- A. Fluctuations, Noise, and Chaos in Laser Systems 190
- B. Stochastic Fluctuation Effects in Nonlinear Optics 202
- C. Nonlinear Optical Techniques for Gas-Phase Measurements 208
- D. Lasers and Nonlinear Optics 223

OPTICAL BISTABILITY SWITCHING WITH EXTERNAL NOISE

E. Arimondo, D. Dangoisse *, L. Fronzoni
U. Pisa, Physics Department, Pisa 56100, Italy
O. Incani, N. K. Rahman
U. Pisa, Chemistry Department, Pisa 56100, Italy

ABSTRACT

In a bistable system the switching time between the off and the on states is a measure of noise and fluctuations. For a CO_2 laser containing a molecular absorber, the distribution of the switching times has been measured at different noise levels in the control parameter. The first passage time from the Fokker-Planck equation has been compared to the experiment.

INTRODUCTION

In an optically bistable device the switching between the two bistable states is realized by modifying a control parameter to a value outside the bistability region. In the case of Fig. 1 starting from the lower bistability branch A_o control parameter, a transition to the upper branch occurs if the control parameter is switched to the A_f value larger than the switch-up value $A_\uparrow$. The switching time, defined as the first-passage time required to develop from the initial I=0 state to an output intensity I_f, depends on I_f and on the $A_f - A_\uparrow$ difference. With A_f close to $A_\uparrow$ a critical slowing down occurs. In this regime the switching response presents large fluctuations, as typical of a phase transition behaviour.

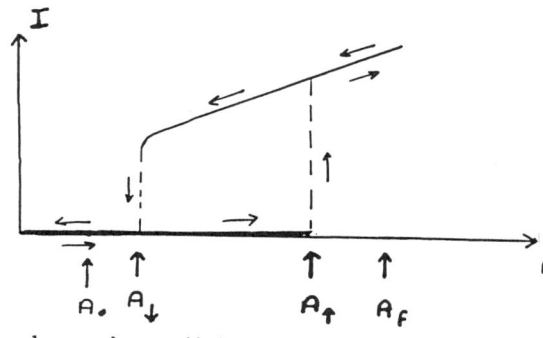

Fig. 1. LSA output power I versus control parameter A.

* on leave from Universite de Lille I, Villeneuve d'Ascq, France

We have investigated the switching time distribution in the bistable operation of a CO_2 laser containing a saturable absorber (LSA)[1,2]. To determine the influence of noise, an external noise, controlled in amplitude and frequency distribution was applied to the laser system.

Measurements of the first-passage time distributions have been performed monitoring the statistical fluctuations in superfluorescence[3], Rydberg maser switching[4] and build-up of a dye-laser[5].

EXPERIMENT

The CO_2 laser containing SF_6 as saturable absorber has been described previously[1]. The laser operation point is determined by the control parameter A, defined as the unsaturated small gain of the CO_2 amplifier rapported to the cavity losses. The A parameter is controlled by modifying the current in the discharge amplifier, and in the low current regime the A control parameter is linearly proportional to the discharge current. Thus a linear sweep, a step function and random noise are electronically applied to the A parameter.

Fig. 1 represents schematically the laser output power I versus the linearly swept A parameter. The optical bistability region depends on the sweep rate of the control parameter[2].

The A parameter was modified by a noise with r.m.s. amplitude $\sqrt{<\delta A^2>}$ and a frequency distribution determined by the input electronic noise and the frequency response of the amplifier system. To derive the frequency distribution of the noise on the control parameter A, the output power correlation function of the laser operating without saturable absorber was investigated. There the laser output power is given by:

$$I = A - 1$$

Because the cavity damping rate is much higher than the amplifier population damping rate, the laser power follows the amplifier noise. Thus the laser power correlation function provides the frequency distribution of the A parameter noise. In the experiments to be presented, the amplifier noise is coloured with a correlation time $\tau_c = 0.15$ μs.

RESULTS

The following features of the LSA bistable system have been investigated in presence of externally applied noise: i) bistability cycles; ii) time-averaged output power; iii) correlation function of the output power; iv) switching times; v) transient bimodality; vi) switching-time distributions.

Experimental results for the switching time in absence or presence of noise are reported in top and bottom records of Fig. 2 respectively. The laser starts to operate with a laser output power I different from zero with a time delay τ_D in respect to the step function in the current, i.e. the step function in the A control parameter. By comparing the single shot record of Fig. 2a to the superimposed shots of the record in Fig. 2b with noise level $\sqrt{<\delta A^2>}=0.4$, it appears that the average delay time $<\tau_D>$ is shortened by the noise presence. Furthermore the noise increases the fluctuations in the switching time delay.

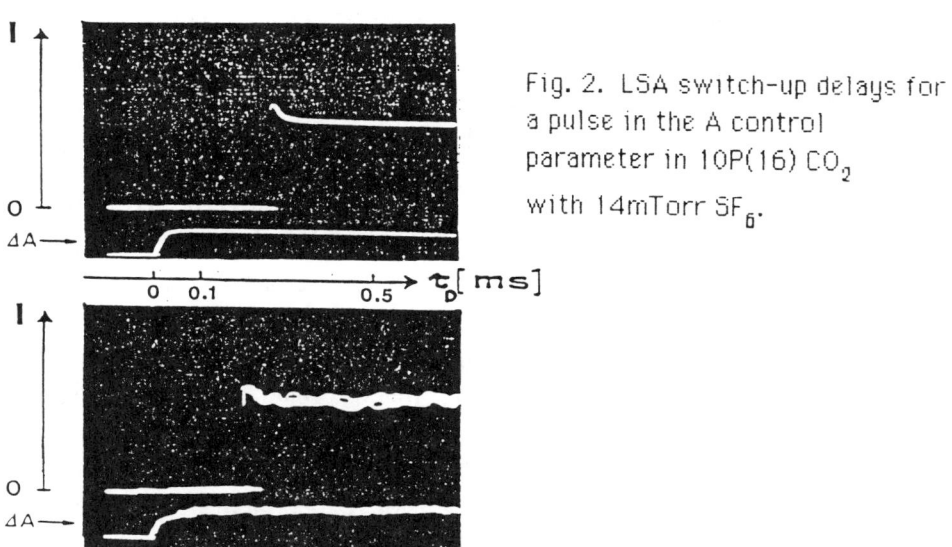

Fig. 2. LSA switch-up delays for a pulse in the A control parameter in 10P(16) CO_2 with 14mTorr SF_6.

The switching time required to reach the seventy percent value of the final laser power was measured in 2000 switching processes.

A histogram of the switching times, representing the probability distribution $P(I,\tau_D)$, was accumulated in a microcomputer. It should be noticed that owing to the fast laser transition between the off and the on states, the probability distribution does not depend significantly on the I value where the measurement is performed. Probability distributions for a fixed $A_f - A_t = 0.02$ value and different noise levels are reported in Fig. 3.

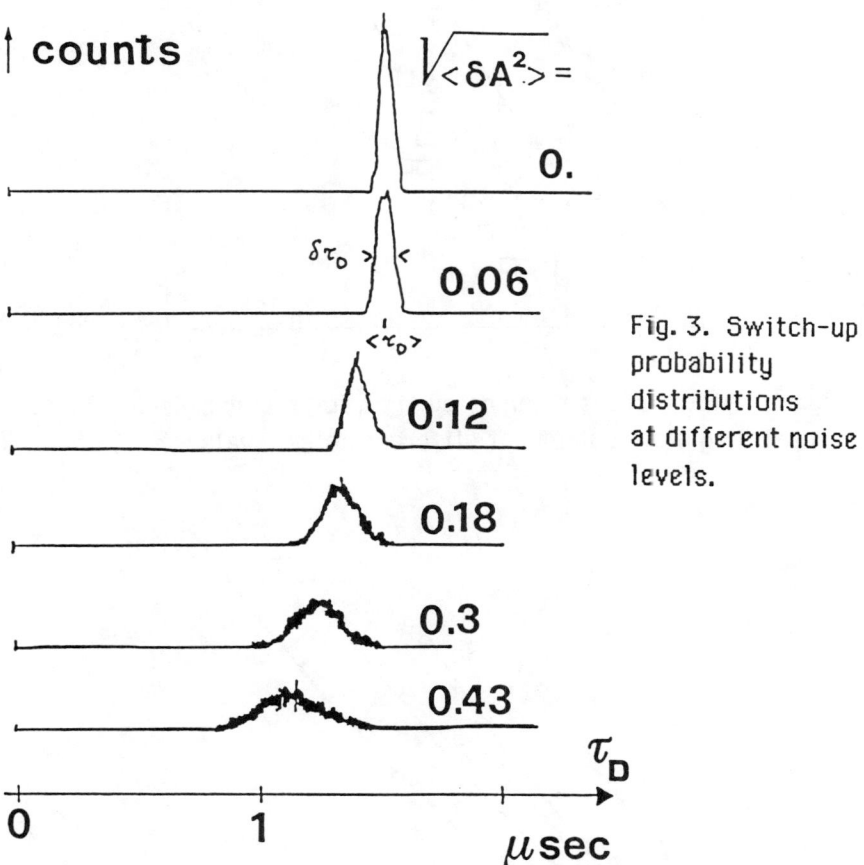

Fig. 3. Switch-up probability distributions at different noise levels.

To test the modifications of the switching-time distributions we have plotted the width $\delta\tau_D$ of the distribution as a function of the average time delay $\langle\tau_D\rangle$, i.e. the center of the distribution. The relation between width and the average delay time is modified by the noise level in the LSA as plotted in Fig. 4. In order to make a comparison with theory the width and the average delay time have been measured in dimensionless quantities, corresponding to the population decay rate $\gamma = 10^4$ sec^{-1}.

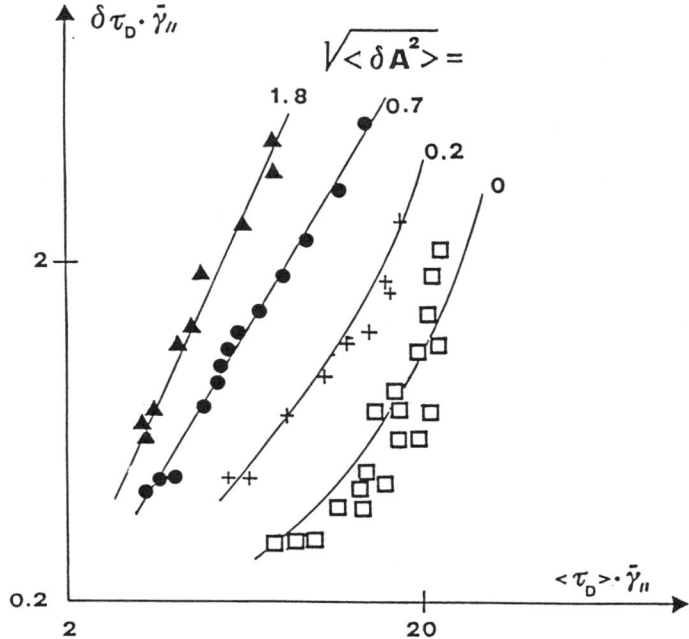

Fig. 4. Width versus mean delay time for the measured probability distributions at different noise levels.

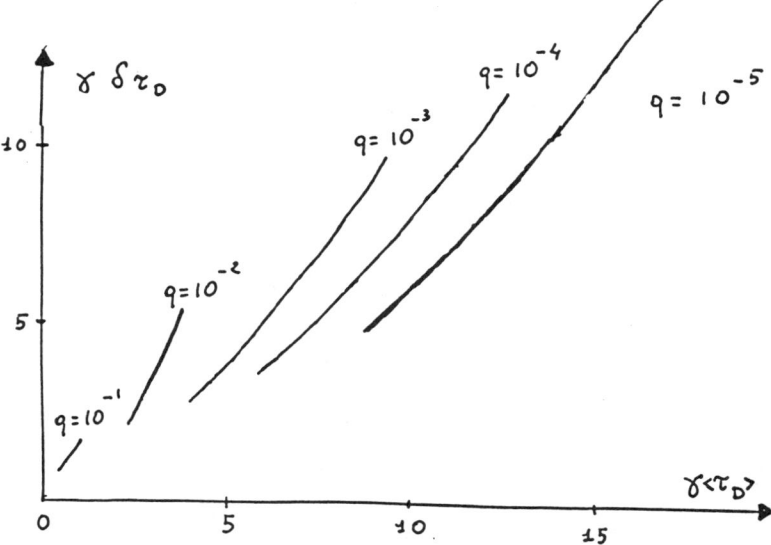

Fig. 5. Theoretical dispersion versus first passage time as obtained from the Fokker-Planck equation.

An effective Fokker-Planck equation may be written for the probability distribution $P(I,\tau_D)$ of the laser with a potential $V(I)$ describing the LSA operation. The bistable potential we are considering has two minima, and is obtained through an adiabatic elimination of the amplifier and the absorber population in the LSA rate equations[2]. Modifying the A control parameter by a step function the laser is placed into a state of unstable equilibrium in the potential V. The first passage time in which the system initially located at the unstable equilibrium position arrives at the stable position is calculated through the Fokker-Planck equation. Furthermore the dispersion, or variance of the mean arrival time, is calculated through the Fokker-Planck equation.

The numerical results for the dispersion versus the first passage time are plotted in Fig. 5 with the quantities measured in dimensionless units. It may be noticed that increasing the noise level the relation between the dispersion and the first passage time becomes more steep and shifted towards the origin. A comparison with the experimental results of Fig. 4 shows that theory predicts the trend observed in the experiment for what concerns the width and the center of the probability distribution. On the contrary the theory predicts that these times should be determined by the cavity relaxation rate which is much higher than the absorber population rate, the latter being the one that has been introduced in the plot of Fig. 4. This result comes from the adiabatic elimination of the absorber and amplifier populations to derive the potential V(I). A more exact analysis of the experimental results for the LSA should be performed without introducing in the theory an adiabatic elimination.

REFERENCES

1. E.Arimondo and B.M.Dinelli, Opt. Commun. <u>44</u>, 277 (1982)
2. E.Arimondo, D.Dangoisse, C.Gabbanini, E.Menchi and F.Papoff, submitted to J. Opt. Soc. Amer. B
3. D.Polder, M.Schuurmans and Q.Vrehen, Phys. Rev. <u>A19</u>, 1192 (1979)
4. P.Goy, L.Moi, M.Gross, J.M.Raymond, C.Fabre and S.Haroche, Phys. Rev. <u>A27</u>, 2065 (1983)
5. S.Zhu, A.W.Yu and R.Roy, Phys. Rev. <u>A34</u>, November 1986

BIDIRECTIONAL OSCILLATION AND BICHROMATIC EMISSION IN A RING DYE LASER: EVIDENCE FOR A NEW MODE STRUCTURE HIERARCHY

N. M. Lawandy

Division of Engineering and Department of Physics, Brown University,
Providence, Rhode Island 02912

ABSTRACT

We have performed a careful experimental study of a high-Q Rhodamine 6G ring dye laser and have measured bichromatic emission with wavelength spacings as large as 80 angstroms when the laser operated bidirectionally. The bichromatic emission vanished at all excitations when the laser was forced into unidirectional operation. The experiments show that no discontinuities occur in the intracavity power as a function of excitation and that the intracavity fields which are completely consistent with the measured thresholds are not capable of producing Rabi splittings larger than a few angstroms. We have been able to quantitatively fit the threshold and wavelengths of the bichromatic emission using a distributed feedback analysis with counter-propagating ring modes driving the susceptibility grating.

INTRODUCTION

In this paper, we present experimental measurements on a high-Q Rhodamine 6G ring dye laser which was found to exhibit bichromatic emission with splittings on the order of 15 - 80 Å while operating in the range of 2-10 times above threshold in the bidirectional mode.

The intracavity laser power in an experiments was found to be of the order of a watt or less. This value is consistent with the saturation intensity of the laser transition ($0.25\,MW/cm^2$) and the laser being about five times above threshold.[1]

It is worth noting that in other reports of bichromatic emission, it is claimed that the intracavity power is of the order of ~ 100 watts and that the saturation intensity is calculated based on $T_1 = 10\,T_2$.[2,3,4,5,6] This is in disagreement with the known spontaneous lifetime ($T_1 = 5\times10^{-9}$ seconds) and the dephasing rate of Rh6G ($T_2 \sim 10^{-12}$ seconds) .[1] With the proper values of T_1 and T_2 ($T_1 = 5\times10^3 T_2$) the theory given in Reference (4) leads to no significant modification of the gain lineshape.

In the next sections we will describe the experiments, and the results which we obtained. In addition, we offer a theory based on a new mode structure hierarchy which may explain our observations and those of others.

© American Institute of Physics 1987

DESCRIPTION OF THE EXPERIMENTS:

The experiments were carried out on a modified Spectra Physics 380 ring dye laser. The laser is a figure eight ring with a 200μm thick free flowing dye jet of 2×10^{-3}M solution of rhodamine 6G in ethylene glycol. The jet is at the center of two curved mirrors which produce an average focused Gaussian beam spot radius of approximately 10–15μm at the laser wavelength. In addition there is a pump focusing mirror which concentrates the argon-ion laser pump beam to a spot radius of about the same size as the laser mode beam waist. The multi-line argon-ion pump laser is an actively power stabilized Spectra Physics Model 2020-05 with a range of 0 - 7 watts. Each of the ring dye laser mirrors was controlled by two Starrett Inc. micrometers with ±2.0μm positioning resolution.

The laser was operated in a totally sealed off cavity containing only the dye jet, an astigmatism compensator (optional), and mirrors with the highest reflectivity coatings commercially available. The mirrors had a measured power transmitivity of $T = 10^{-3.8}$ in the wavelength range 5700 Å - 6100 Å. The transmitivity of the mirror out of which the power measurement was made was measured at an angle of 8° (position in the cavity) against a National Bureau of Standards blank in a dry nitrogen pumped spectrophotometer. These values were obtained from absorbance measurements and therefore result in a maximum reflectivity. This is acceptable as we are interested in estimating the highest possible values for the intracavity fields. Using the measured output power in a given direction, the maximum possible intracavity power associated with that beam was directly found by dividing by the transmitivity factor $T = 10^{-3.8}$. In addition to this technique, intracavity powers were measured using calibrated intracavity scattering. The measurements were performed in the presence of the average dusty environment using two overlapping copropagating beams (the intracavity laser beam and an external beam of known power).

Some of the experiments were performed with the insertion of a Faraday isolator to insure unidirectional operation. The Faraday rotator had an insertion loss of about 1.25% which doubled the oscillation threshold pump power of the sealed-off cavity. The dye laser was modified with external reflectors to allow for the simultaneous monitoring of the wavelength output and the power coupled out in each direction of the ring.

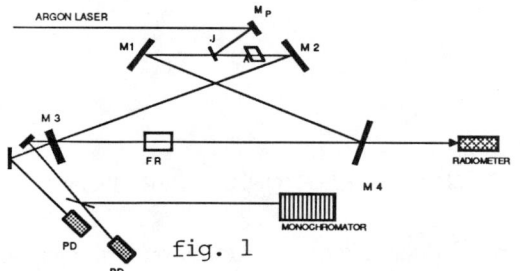

fig. 1

The wavelength measurements were performed using a Jarrel-Ash monochrometer with a resolution of 1.5 Å and a S - 20 photomultiplier tube. The experimental apparatus is shown in Fig. 1.

BICHROMATIC EMISSION EXPERIMENTS WITH BIDIRECTIONAL OPERATION

The Rhodamine 6G dye laser output was examined spectrally with a resolution of 1.5 Å. The laser output was sent through a diffuser before entering the monochromoter in order to eliminate any spatial ghost effects which could appear as spectral structure. The results of a typical experiment to measure the spectral output versus intracavity power are shown in Figure (2). The measurements in Figure 2 were undertaken in order to measure the bichromatic separation as well as to search for chromatic hysteresis. Within the resolution of the experiments we were able to measure large splittings but found no evidence of hysteresis. The behavior of the splitting as a function of pump power could be altered by adjusting the astigmatic compensator and its dispersion losses.

In order to insure that the laser was operating in a single transverse mode at all times (including the regime of bichromatic behavior) qualitative and quantitative measurements of the transverse mode profile were performed. Qualitatively the intensity pattern was studied using the Newton's rings which formed when the output beam traversed a thermal lensing plastic. In addition, the far-field transverse mode profile of the laser was scanned with the laser operating monochromatically and with the laser operating bichromatically. The superimposed transverse mode profiles are shown in Figure 3.

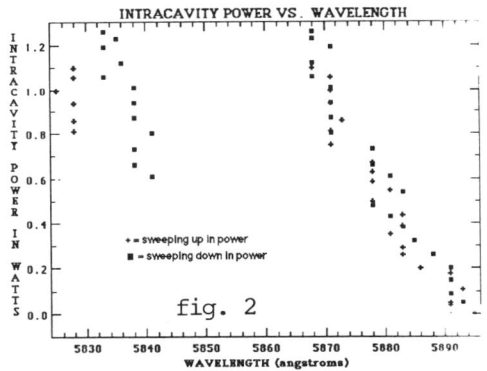

fig. 2

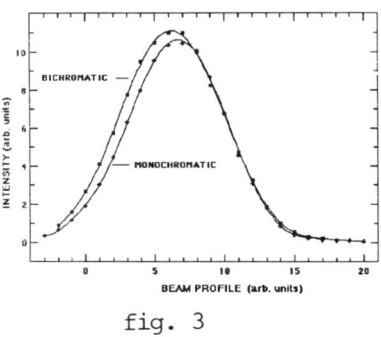

fig. 3

EXPERIMENTS WITH FORCED UNIDIRECTIONAL OPERATION

In order to unambiguously test for the effects of saturation gratings on the bichromatic emission we operated the high-Q ring laser with an intracavity antireflection coated Faraday isolator to force unidirectional operation. Since the circulating power of the laser goes up with the isolator this should enhance any bichromatic splitting due to Rabi effects.

When the laser was operated with the Faraday isolator and the astigmatism compensating Brewster angle rhomb, the threshold pump power increased about 0.7 watts. These experiments were operated with as much as nine watts of pump power bringing the laser to an excitation which is twelve times above threshold. In every experiment that we performed with forced unidirectional operation, there was no bichromatic emission.

When an intracavity glass surface was used to reflect the strong unidirectional beam in the counterpropagating direction for a single pass through the active dye region, bidirectional emission occurred at high pump powers (≥ 5 watts). When the weak reflected beam was slightly misaligned, the bichromatic emission vanished. Measurements of the power of the counterpropagating beam verified that it was only seeing the saturated single pass gain of the jet. These experiments clearly demonstrate that saturation gratings are necessary and are likely to be at the heart of the bichromatic state observed in our experiments.

TOWARDS A THEORY OF BICHROMATIC EMISSION

In this section we will describe the essentials of a theory of steady state bichromatic emission which is consistent with our experimental observations. The theory is only valid at the threshold point of any given bichromatic state and therefore cannot explain energy extraction by the actual bichromatic mode solution. Therefore, this treatment should be viewed as a theory for the origin of a type of new solution which we call a hybrid mode.

We have formulated a theory for the longitudinal mode profiles and frequencies of a periodically modulated, resonant medium in a ring cavity.[5] The above mentioned theory is applied to the problem of bichromatic emission of a ring dye laser by using the standing wave cavity mode saturation grating to effect the periodic modulation of the medium. Since two counterpropagating modes with a wavevector k_0 result in an intensity grating with a fundamental spatial modulation at $2k_0$, a near Bragg condition is present in the medium for the possible evolution of a reflection assisted bulk mode.

The theory we derived is for the configuration shown in Figure 4. The clockwise and counterclockwise field envelopes are assumed to have the following forms respectively:

$$R(z) = r_1 e^{\gamma z} + r_2 e^{-\gamma z} \qquad (1)$$

$$S(z) = s_1 e^{\gamma z} + s_2 e^{-\gamma z} \qquad (2)$$

In addition we impose the following boundary conditions:

$$(r_1 e^{-\gamma L/2} + r_2 e^{\gamma L/2}) = \rho'(r_1 e^{\gamma L/2} + r_2 e^{-\gamma L/2}) \qquad (3)$$

$$(s_1 e^{\gamma L/2} + s_2 e^{-\gamma L/2}) = \rho'(s_1 e^{-\gamma L/2} + s_2 e^{\gamma L/2}) \qquad (4)$$

where

$$\rho' = \rho e^{-\iota(\beta_0 + \delta)(L_c - L) - \iota\beta_0 L} \tag{5}$$

L_c is the total cavity length, L is the active dye jet thickness, ρ is the lumped mirror reflectivity, and $2\beta_0$ is the modulation wavevector in the medium. The boundary conditions yield two dispersion relations from which the threshold gain and frequency of the bulk mode solution can be found:

$$\gamma^2 = \kappa^2 + (\alpha - \iota\delta)^2 \tag{6}$$

$$\gamma = \frac{\pm i\kappa \sinh\gamma L(1 - \rho'^2)}{(1 - \rho' e^{\gamma L})(1 - \rho' e^{-\gamma L})} \tag{7}$$

The dispersion relations were solved using a complex root finder (ZANLYT) to find the frequencies and threshold for the first few solutions. The model calculations utilized a flat loss with a Lorentzian shaped dispersive loss superimposed upon it. The minimum loss of 4% was centered at 66Å above gain linecenter. The laser mode then oscillated at 60Å above line center. The dispersive loss Lorentzian had a full width at half maximum of 300Å and a height of 6%. The gain was modelled by a Lorentzian with FWHM equal to 250Å. The peak gain-length product as a function of pump power was found from our measurements to be 0.14P. This along with medium parameters quoted earlier resulted in a ±10% fit to the experimental dye laser output power curve.

The model produced results which were quite plausible in terms of threshold excitations of the laser modes and bulk solutions with wavelengths which were 2-69Å away from the conventional ring laser modes. Figure 5 shows the detuning (bichromatic separation) between the new bulk mode at threshold and the ring laser modes as a function of pump power. This simple model has resulted in splittings which are consistent with our observations as well as those of Ref (2,3,4) with the measured intracavity intensity values present in the laser mode. In addition, since a symmetric gain curve was utilized, a shift in the sign of the detuning of the initial cavity mode results in a corresponding shift their sign of the detuning of the distributed feedback mode relation to the cavity mode.[6]

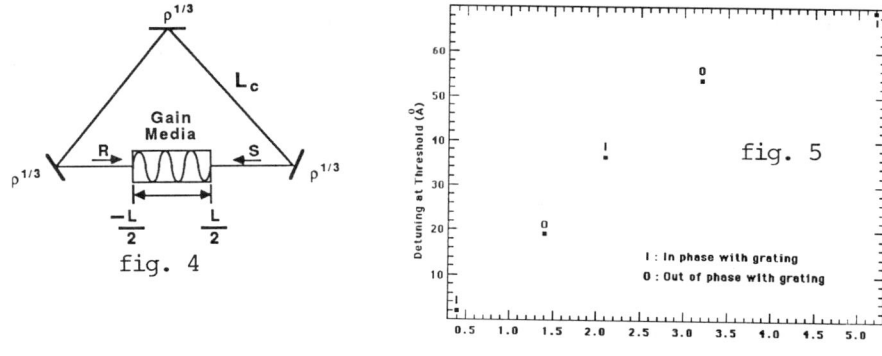

fig. 4

fig. 5

I : In phase with grating
0 : Out of phase with grating

Pump Power (watts)

The stability of these modes was examined by a rate equation model which utilizes a DFB mode lifetime which is a function of the population grating produced by the cavity mode.[6] The model utilizes five equations to model the system, two field equations and three equations to account for the time evolution of the various Fourier components of the population. The dynamic simulations indicate that regimes exist where modes of different symmetry can coexist and that the bichromatic state is a hybrid mode which can be viewed as two DFB ring cavity modes which scatter into each other in a stable manner. In addition this theory predicts that for low excitations the first two modes can be less than an angstrom apart and that the so called hole filling mode is a nearly degenerate bichromatic solution. The success of this model we feel supports the conjecture that there exists a yet unexplored longitudinal mode hierarchy in the high gain ring laser. These hybrid states are stable solutions of the Maxwell equations and the matter equations and should be present in a full Maxwell-Bloch formalism which allows for stable bidirectional solutions. This may require the inclusion of transverse effects which could alter the cross-saturation problem.

CONCLUSIONS

We have observed bichromatic emission from a ring dye laser only when it was operated bidirectionally. The wavelength splittings as a function of pump power was found to be sensitive to the dispersion and tuning of the initial cavity modes. We have quantitatively explained the bichromatic state as the coexistence of two modes of a ring cavity dressed by a modulated medium.

REFERENCES

1. O. Teschke, Andrew Dienes and John R. Whinnery, IEEE JQE, QE-12, No. 7, p. 383 (1976).
2. Lloyd W. Hillman, J. Krasinski, R. W. Boyd and C. R. Stroud, Jr., Phys. Rev. Lett., Vol. 52, No. 18, p. 1605 (1984).
3. Lloyd W. Hillman, J. Krasinski, K. Koch, and C. R. Stroud, Jr. JOSA B, Vol. 2, No. 1, p. 211, 1985.
4. Karl Koch, Stephen Chakmakjian, Lloyd W. Hillman and C. R. Stroud, Jr., Optical Soceity of America 1986 Annual Meeting, Tech. Digest, (FT3), p 138, (1986).
5. W. S. Rabinovich and N. M. Lawandy, IEEE JQE, QE-23, February 1987.
6. N. M. Lawandy, W. S. Rabinovich, Optical Instabilities Cambridge Studies in Modern Optics 4, eds. R.W. Boyd, M. G. Raymer, L. M. Narducci, Cambridge University Press, p. 280-282, 1986.

ROLE OF FLUCTUATIONS IN NONLINEAR OPTICAL ABSORPTION PROCESSES

Stephen J. Smith
Joint Institute for Laboratory Astrophysics, University of
Colorado and National Bureau of Standards, Boulder, CO 80309-0440

ABSTRACT

Fully quantitative experimental investigations of nonlinear optical absorption processes can only be carried out with optical fields for which higher order correlation functions are known. A technical capability for synthesizing such laser power spectra by introducing Gaussian frequency fluctuations to the external field of a well-stabilized single-mode laser has been developed and applied to double-optical-resonance measurements in atomic sodium. Experimental results agree well with predictions of a full theoretical treatment.

INTRODUCTION

The first observations of asymmetry reversal in double optical resonance (DOR) measurements of ac Stark splitting were obtained in this laboratory[1] in the mid-1970s. This work, carried out using an atomic sodium beam illuminated at right angles with multi-mode radiation from flashlamp pumped dye lasers, was necessarily rather qualitative, but the results were an unexpected and unequivocal violation of the predictions available at that time, all based on a model which assumed a monochromatic radiation field. The optical processes involved are illustrated in Fig. 1 for the relevant levels of atomic sodium. For a given irradiance and detuning Δ of the intense laser field in near resonance with the $3S_{1/2} \rightarrow 3P_{3/2}$ transition, an ac Stark splitting Ω is obtained which may be studied by tuning the weak probe-laser field coupling the two components of the Stark split doublet to a higher state (nD). The resulting population of the nD state was sampled through the photoionization current produced by the same laser fields. Exactly on the $3S_{1/2} \rightarrow 3P_{3/2}$ resonance the observed Stark split doublet is symmetrical. For the monochromatic

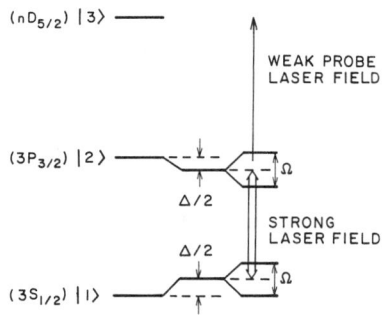

Fig. 1. Energy levels of atomic sodium relevant to the DOR measurements described in this paper. Δ is the detuning of the saturating field from exact resonance. $\Omega = (\Delta^2 + \Omega_o^2)^{1/2}$ is the ac Stark splitting and Ω_o is the on-resonance Rabi frequency.

field model the peak amplitude asymmetry develops monotonically with detuning, Δ; but for the experimental results the asymmetry, A, defined as $(I_1-I_2)/(I_1+I_2)$, where I_1 and I_2 are the respective amplitudes of the individual components of the doublet, was essentially a third order curve (similar to that shown in Fig. 3d), i.e., included three points for which A = 0, one on exact resonance and two for finite detunings.[2] Thus, for small detunings the peak asymmetry was "reversed" from that expected from the monochromatic model.

These observations stimulated considerable interest by theoreticians. It was shown that a pure Lorentzian power spectrum (with full width at half maximum (FWHM) greater than the width of the absorbing spectral line) would result in a reversed asymmetry at all detunings Δ, however large.[2] While a pure Lorentzian is unphysical, this identified the reversed asymmetry effect as arising from field fluctuations, and stimulated further theoretical work including a comprehensive treatment[3] which took account of the inevitable departure of laser power spectra from the Lorentzian ideal, using the phase-diffusion model (constant amplitude, fluctuating phase and frequency). The phase fluctuations may be described as a two-dimensional Markov process: a random Gaussian process with constant spectral density 2b on which a finite correlation time $1/\beta$ has been imposed. The lowest order correlation function then may be written:

$$\langle \dot{\phi}(t) \dot{\phi}(t+\tau) \rangle = b\beta \, e^{-\beta|\tau|} \,. \qquad (1)$$

Because a Gaussian process is specified, all higher order correlation functions are prescribed as products of the lowest order correlation function.[4] In this paper we discuss the realization of a laser field in the laboratory for which the fluctuations conform to this model and this correlation function [Eq. (1)]; and we describe its application to double optical resonance measurements.

EXPERIMENTAL METHODS

Extrapolating from the preceding description of a two-dimensional Markov process, what is necessary for the generation of an optical field with the desired properties is: 1) a noise (voltage) generator, for example a shot-noise diode, which satisfies the conditions of the central limit theorem[5] and generates a Gaussian distribution with spectral density independent of frequency over a wide range; 2) a means for imposing a short term memory on the output of the noise generator, such as an RC filter where RC = $1/\beta$, the required finite correlation time; 3) a means for converting these (voltage) fluctuations to phase or frequency fluctuations of specified rms amplitude; and 4) an extra-cavity device (or devices) for superposing these fluctuations on the beam from a very well-stabilized (nearly monochromatic) continuous-wave laser. Such devices may include acousto-optic and electro-optic modulators.

The technical details of the experimental configuration which incorporates these principles, are beyond the scope of this paper and have been described elsewhere.[6] The major point to be emphasized here is the essential requirement that the statistical distribution of the fluctuations must be Gaussian through all steps of processing of the noise, a requirement necessary to ensure that higher order correlation functions are known. This means that all elements used in processing the noise signal must be quite linear and must have conservatively rated dynamic ranges.

Frequency correlation functions of all orders are required to construct the field correlation function $\langle E^*(t)E(t+\tau)\rangle \propto \langle e^{i[\phi(t)-\phi(t+\tau)]}\rangle$. Since, by the Wiener-Khintchine Theorem, the power spectrum of the laser field is just the Fourier transform of this field correlation function, and vice versa, it is apparent that a measurement of the power spectrum, which can be accomplished by mixing the modulated and unmodulated laser beams on an optical diode provides a useful experimental test. By means of such measurements we can confirm that power spectra are realized which are specified within ±1 db of a form corresponding to some selected pair of values b and β, over a frequency range to ±1 GHz from line center.

With this system, nonlinear absorption processes can be studied as a function of irradiance (or, equivalently, of Rabi frequency), of detuning Δ, of laser linewidth, and of the quality of the laser power spectrum (Lorentzian, Gaussian, or in between). For many purposes the 3s-3p transition in atomic sodium has proved to be convenient because all of the parameters listed above can be controlled over a range from zero (or a few MHz) to several times the natural linewidth of 10 MHz. Laser bandwidths are variable up to about 20 MHz, Rabi splitting to about 100 MHz, detuning to well over 100 MHz, and the inverse correlation time 1/β to about 100 MHz.

Of particular interest are the values of b and β, or more specifically, their relative values. There are two limiting cases: 1) β >> b. Relatively rapid phase fluctuations are allowed. The Fourier transform of Eq. (1), the power spectrum, according to the Wiener-Khintchine theorem, is a Lorentzian. 2) β << b. Only relatively slow phase fluctuations are allowed. The power spectrum of the laser field approaches a Gaussian form. Therefore, by choice of b and β we control the power spectrum, its shape and, as well, its bandwidth.

The DOR results reported here pertain to the 3s-3p transition in atomic sodium, as did the old flashlamp-pumped dye laser measurements, but differ in many respects. The present measurements were carried out with the continuous-wave noise-modulated system discussed above, using a single-mode ring laser with a residual linewidth ~150 kHz. Furthermore, the atomic sodium beam was optically pumped into the F=2, M_F=2 ground state, and excited with circularly polarized light, thus defining a true two-level system ($3^2S_{1/2}$, F=2, M_F=2 → $3^2P_{3/2}$, F=3, M_F=3). The probe laser then coupled to the $4^2D_{5/2}$ (F=4, M_F=4) state, again using circular polarization. In the old experiment of Moody,[1] photoionization current was detected. In the new work 330 nm radiation due to radiative decay of the $4^2D_{5/2}$ state through the $4^2P_{3/2}$ state was observed.

The power spectrum of the saturating optical field was monitored continuously and compared repeatedly with the form calculated for the designated values of b and β. Figure 2 shows the measured power spectrum for β/2π = 80 MHz and FWHM = 14 MHz, and indicates the Lorentzian and Gaussian limits.

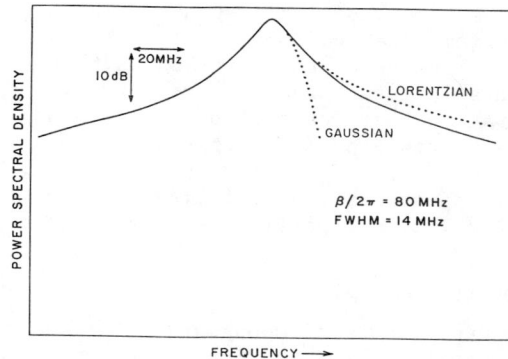

Fig. 2. The trace of a measured power spectrum with parameters $\beta/2\pi$ = 80 MHz and FWHM = 14 MHz is compared with the Lorentzian ($\beta \gg b$) and Gaussian ($\beta \ll b$) limits. The ordinate is logarithmic.

EXPERIMENTAL RESULTS

The DOR technique was applied, varying all the various parameters (b,β,Δ,Ω) over the widest possible ranges. The results provide fully quantitative tests[7,8] of theoretical predictions for the first time. One of the most interesting asymmetry curve sequences is shown in Fig. 3. Here, for a fixed power density (Ω_0 = 67 MHz) and a fixed laser field bandwidth (FWHM = 14 MHz), $\beta/2\pi$ is varied over a range from 10 MHz to 80 MHz, which corresponds to an evolution from a nearly Gaussian power spectrum to a nearly Lorentzian power spectrum. The development of the reversed asymmetry is displayed

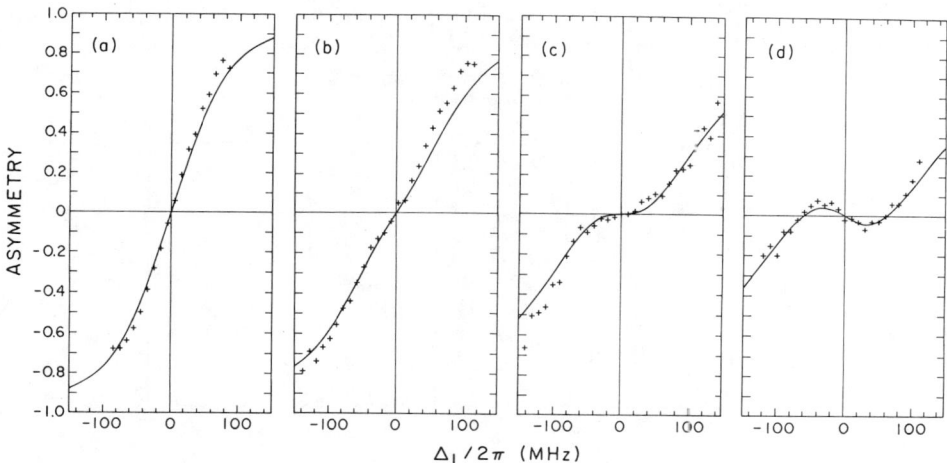

Fig. 3. A sequence of experimental results (+) compared with theoretical predictions of Refs. 3,8 (solid lines) for fixed values Ω_0 = 67 MHz and FWHM = 14 MHz, with values of $\beta/2\pi$ of (a) 10 MHz, (b) 30 MHz, (c) 60 MHz, and (d) 80 MHz.

and compared to predictions of the theory of Dixit et al.[3] The calculations were performed by Dziemballa and Zoller.[8]

Other sequences show the development of the reversed asymmetry with increasing bandwidth of a nearly Lorentzian power spectrum.[8] These results confirm the expectation that reversal occurs only as the bandwidth <u>exceeds</u> the natural linewidth. A third sequence, in which the parameter varied was the irradiance of the optical field, illustrates that the reversal disappears as the Rabi frequency Ω_0 exceeds β, the inverse of the correlation time imposed on the noise spectrum.[8]

CONCLUSION

From the early qualitative experiments of Moody and M. Lambropoulos[1] have evolved an appreciation of the importance of higher order correlation functions in the description of nonlinear optical absorption processes, and the development of experimental techniques for synthesizing fields for which the statistical properties are fully characterized. Figure 4 shows a sequence of DOR spectra measured through the application of this technique. These

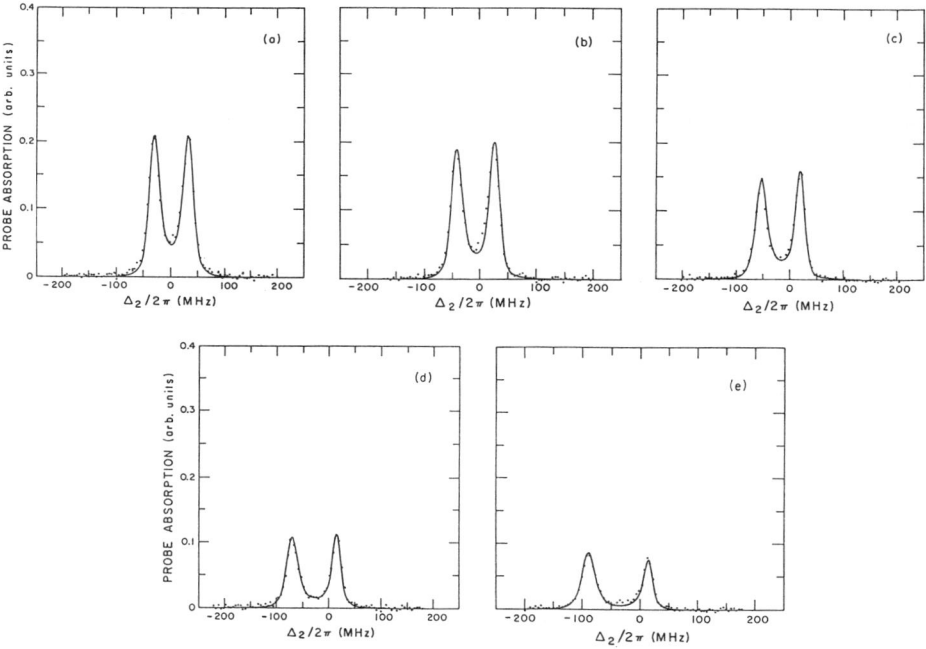

Fig. 4. Measured DOR spectra (dots) compared with theoretical predictions (solid lines) for the power spectrum shown in Fig. 2. Here $\Delta_2/2\pi$ is the probe laser detuning. The detuning $\Delta_1/2\pi$ of the saturating laser field takes values (a) 0, (b) −20 MHz, (c) −40 MHz, (d) −60 MHz, (e) −80 MHz.

spectra were measured with the power spectrum shown in Fig. 2, nearly Lorentzian and with a bandwidth exceeding the 10 MHz natural linewidth. The parameter varied is the detuning of the saturating laser field. The evolution of the reversal is evident. In this figure the dots represent experimental results. The solid lines are calculated from the theory of Dixit et al.[3] Small departures near the centers of some of the spectra result from some radial variation in intensity of the saturating laser field not taken into account in the calculation. Otherwise, the agreement is excellent.

These results mark the convergence of a fully quantitative theory with the first DOR measurements in which the field fluctuations were fully characterized.

ACKNOWLEDGMENTS

The author acknowledges the essential contributions of M. W. Hamilton, D. S. Elliott, and K. Arnett to the accomplishment of these experimental measurements. This work was supported by the U.S. Department of Energy Office of Basic Energy Sciences.

References

1. S. E. Moody, Ph.D. Thesis, University of Colorado, 1975; S. E. Moody and M. Lambropoulos, Phys. Rev. A $\underline{15}$, 1497 (1977).
2. P. B. Hogan, S. J. Smith, A. T. Georges, and P. Lambropoulos, Phys. Rev. Lett. $\underline{41}$, 229 (1978).
3. S. N. Dixit, P. Zoller, and P. Lambropoulos, Phys. Rev. A $\underline{21}$, 1289 (1980).
4. M. C. Wang and G. E. Uhlenbeck, Rev. Mod. Phys. $\underline{17}$, 323 (1945).
5. S. O. Rice, Bell Tel. J. $\underline{23}$, 282 (1944).
6. D. S. Elliott, M. W. Hamilton, K. Arnett, and S. J. Smith, Phys. Rev. A $\underline{32}$, 887 (1985).
7. M. W. Hamilton, D. S. Elliott, K. Arnett, and S. J. Smith, Phys. Rev. A $\underline{33}$, 778 (1986).
8. M. W. Hamilton, D. S. Elliott, K. Arnett, S. J. Smith, M. Dziemballa, and P. Zoller, Phys. Rev. A (in press).

PARAMETRIC PROCESSES AND GAIN SATURATION IN RESONANTLY ENHANCED OPTICAL PHASE CONJUGATION IN NA VAPOR NEAR A TWO-PHOTON RESONANCE*

Reiner K. Wunderlich,** W. R. Garrett, and M. G. Payne
Chemical Physics Section, Oak Ridge National Laboratory
P.O. Box X, Oak Ridge, Tennessee 37831-6125

ABSTRACT

Conjugate wave generation has been studied in Na vapor near two-photon resonance with the 4D state. Measurements of reflection gain at the laser frequency, stimulated Raman scattering, amplified spontaneous emission, and parametric four-wave mixing intensities are used to delineate the role of parasitic processes in determining conjugate ʺreflectionʺ gain profiles.

INTRODUCTION

In studies aimed toward maximizing reflection coefficients in conjugate wave generation, several investigators have used gas phase atomic systems, particularly metal vapors, as nonlinear media where use can conveniently be made of resonant enhancements in degenerate four-wave mixing by tuning near one photon or two photon atomic transitions. We report the results of a study of phase conjugation by degenerate four-wave mixing (DFWM) in Na vapor, enhanced through the 3s-4d two photon resonance at ~578.89 nm (see Fig. 1).

We utilized a Na heat pipe with a 20 cm vapor column with variable Na pressures from 10^{-4} to 2 Torr. The laser system consisted of an excimer pumped dye laser with 4 ns pulses of 5 mJ maximum pulse energy at 0.2 cm^{-1} bandwidth in a 0.3 cm unfocused beam diameter.

Though two photon resonant DFWM has been said to lead to an unbleachable reflection coefficient,[1] measurements on the two-photon resonantly enhanced process involving the Na $4D_{3/2,5/2}$ states show saturation and bleaching of conjugate wave intensity profiles with increasing pump intensity at constant

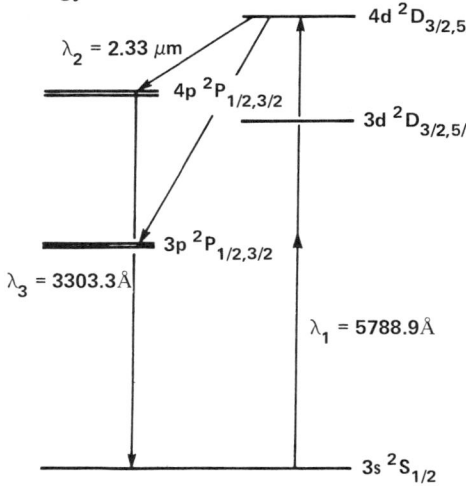

Fig. 1. Na energy levels (Some transitions are not shown.)

*Research sponsored by the Office of Health and Environmental Research, U.S. Department of Energy under contract DE-AC05-84OR21400 with Martin Marietta Energy Systems, Inc.

**Present address: Max-Planck Institut fur Kernphysic, D6900 Heidelberg 1, FRG.

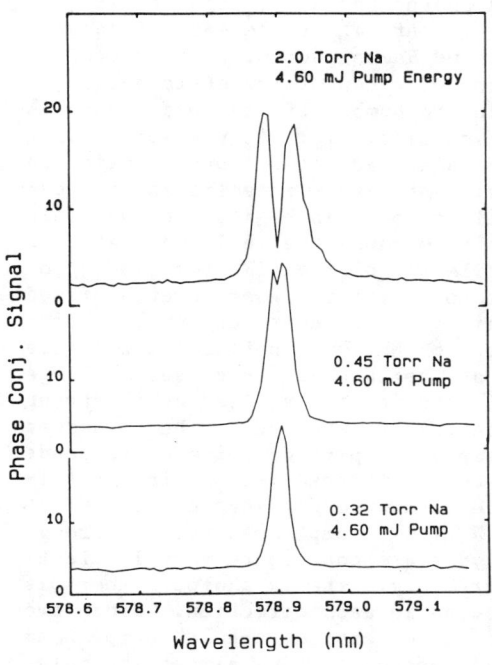

Fig. 2. Conjugate wave reflection profile at constant laser intensity for three Na pressures.

pressure and probe intensity; or alternatively at constant pump and probe intensities but increasing number density, as in the example shown in Fig. 2. In the present study we make detailed measurements of conjugate beam gain profiles, pump beam absorption, amplified spontaneous emission (ASE), and parametric four-wave mixing (FWM) intensities (see Fig. 1) to reveal the mechanisms that lead to gain saturation and loss in the experiments which we describe.

RESULTS AND CONCLUSIONS

First we note that pump absorption, resonant multi-photon ionization, and self focusing effects have been considered, among others, as sources for degradation of gain coefficients in resonant DFWM in metal vapor studies. However, little attention has been given to the possible role in gain reduction of parametric processes involving frequencies other than the laser frequency, which are generated within the nonlinear medium. If stimulated Raman, ASE, and parametric FWM are ignored, the three photon (two photon resonant) ionization yield of Na 4d can easily be calculated. Using the known 4d photoionization cross section ($\sim 1.5 \times 10^{-17}$ cm^2) and a calculated two photon Rabi frequency, we obtain an ionization probability of 70% per atom with our laser at 10 MW/cm^2 per pulse. Thus if the number of atoms in the beam, N_T, is greater than the number of photons, N_γ, the medium should be rather opaque. Instead, we find $\sim$18% absorption on resonance at 0.2 Torr Na (where $N_T > N_\gamma$). Our measurements indicate pump absorption through ionization cannot account for the observed loss of gain in DFWM. Neither is there any evidence of self focusing. However, we have measured the internally generated IR and UV photon beams in forward and backward directions, along with DFWM gain and pump depletion to characterize the contributions of parametric process to resonant conjugate wave generation.

Under conditions where gain begins to saturate and ultimately decrease at exact two-photon resonance, we observe strong production of infrared radiation from ASE and stimulated Raman (SR) emission to the 4p levels of Na. We also see UV emission in the forward direction originating from FWM involving a SR photon and a UV photon at the difference frequency $\omega_{UV} = 2\omega_L - \omega_{SR}$ where ω_L is the laser frequency and ω_{SR} the frequency of the stimulated Raman photon. With 0.2 Torr Na, e.g., at 13 MW/cm^2 we measure power conversion efficiencies in the forward direction of 1.5% (5% in the number of photons) and 3% for the IR and UV generation, respectively. At pressure and/or intensities where IR production from ASE saturates, probe gain in DWFM measurements also saturates. At higher intensities or pressures, IR emission remains flat while probe gain begins to decrease at line center (see Fig. 2). We also measure the full width at half maximum of the pump absorption profile and of the IR emission profile. We find that the width is proportional to laser power at fixed pressure, but with a coefficient which is dependent on number density. At pressures below 10^{-2} Torr each Na atom is found to generate at least one IR photon. We find that the IR emission causes a large a.c. Stark splitting of the 4D and 4p levels, leading to strong suppression of the two photon excitation of the 4d. The observed width of the IR emission with respect to pump detuning corresponds exactly to the IR induced width of the 4d-4p transition. The resultant splitting of the upper state also strongly suppresses DFWM at exact resonance. In regions where $N_T < N_\gamma$ pump absorption through excitation and ionization of the 4d state contribute very little to gain suppression. Indeed at 0.2 Torr, e.g., the IR photon number is 5% of pump beam in number of photons in each direction. Since two laser photons are required for each IR conversion, the 18% pump beam depletion (single pass) mentioned above is due almost entirely, within experimental error, to conversion of laser photons to new frequencies rather than to ionization. Thus atomic excitation of the 4d, and three photon ionization out of that state as well as DWFM at the 4d resonance are all suppressed by the a.c. Stark splitting which develops in atoms downstream from sources of strong IR emission occuring early in the laser pulse propagation. Space does not allow comparison which show that results are in quantitative agreement with theoretical estimates of the observed effects.

REFERENCES

1. T.-Y. Fu and M. Sargent III, Optics Lett. 5, 433 (1980).

By acceptance of this article, the publisher or recipient acknowledges the U.S. Government's right to retain a nonexclusive, royalty-free license in and to any copyright covering the article.

MULTIWAVE MIXING AND MULTIPHOTON IONIZATION IN STRONTIUM VAPOR

K. Böhmer, J. Reif, and E. Matthias
Freie Universität Berlin, D-1000 Berlin 33, F.R. Germany

ABSTRACT

The coherent interaction of a two-photon-resonant, pulsed dye laser with strontium vapor leads to both multiphoton ionization and the emission of coherent light simultaneously at several other frequencies. For a laser intensity of about 4×10^8 W/cm^2 and a vapor pressure around 10^3 Pa, these frequencies are all of the same order of intensity, about three orders weaker than the laser. Interferences of different channels are observed, which render the usual characterization of non-linear optical processes rather ambiguous.

INTRODUCTION

Resonant multiphoton interaction in atomic vapors has been the subject of intensive studies for many years. On the one hand, the electronic structure of the atoms was investigated by multiphoton ionization[1], on the other hand, it was used to convert coherent light from the visible to other spectral regions by means of non-linear optical processes[2]. Only recently, it was realized that the interrelation of the two different kinds of processes is important to the understanding of saturation and cancellation effects[3-5].

In this contribution we report on experiments which show that the polarization generated by a two-photon-resonant laser in strontium vapor is much more complicated than usually considered, also enabling many new channels of mutual interaction for different processes.

EXPERIMENTAL CONDITIONS

Our experiments were performed in atomic strontium vapor contained in a thermionic diode/heat-pipe oven at a pressure of about 10^3 Pa. Light from a tunable, Nd:YAG pumped rhodamine dye laser in resonance with the $6s^2 \rightarrow 5s5d$ $^1D_2$ two-photon transition (cf. Fig.1) was slightly focused into the vapor to yield an intensity of about 4×10^8 W/cm^2. Outcoming light passed a 4 mm BG24 filter in order to block the residual laser light by about three orders of magnitude. After a distance of about 1.50 m which permitted us to get rid of isotropically emitted fluorescence light, the directed radiation was focused into a 60 cm monochromator and measured by a photomultiplier. The signals from the photomultiplier, a reference Fabry-Perot wavelength marker and the unbiased thermionic diode were processed by gated integrators and then recorded.

RESULTS

The most striking result is shown in Fig. 2, where the laser is tuned to exact two-photon resonance. While in previous experiments the third harmonic of the laser[6] (cf. "b" in Fig. 1) and the stimulated 16 μm hyper-Raman transition to the 6p state[7] (cf. "a" in Fig. 1) had been observed, here we see a great number of brilliant

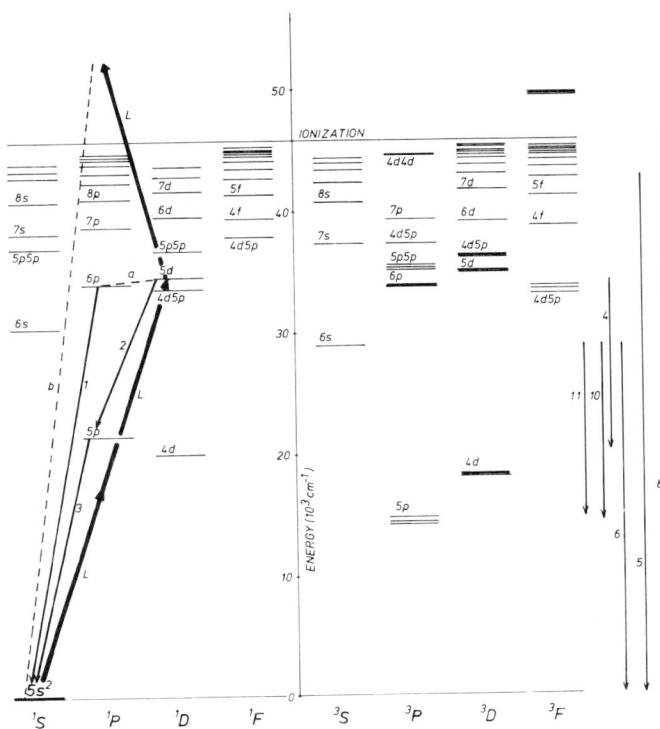

Fig. 1. Energy level scheme for atomic strontium (after Ref. 8). Arrows indicate photon energies for laser (L) and observed light (numbers as in Fig. 2).

(some μJ/pulse, as can be deduced from the attenuation of the laser line by three orders by the BG24 filter) lines throughout the investigated range between 200 and 800 nm. Besides the frequencies from the two expected 4-wave/stimulated hyper-Raman schemes via the 6p (a+1) and 5p (2+3) states a number of lines appear which cannot be easily ascribed to such simple processes, reaching to frequencies even above the one-photon ionization level.

When the laser is scanned, the emitted light reaches its maximum intensity at exact two-photon resonance, while the simultaneously observed ionization signal drops to a minimum (Fig. 3). Furthermore, the red line to the 5p state ("2" in Figs. 1,2,3) is excitable in a wider range of laser frequencies than the expected matching blue line ("3").

DISCUSSION

In view of the many generated light frequencies, each of considerable intensity, there exist many possible channels which may lead to ionization of the atoms. They range from one-photon over two-, three-photon interaction to many-photon processes and may start from the ground state as well as from occupied higher states like the metastable 4d $^1D_2$. It is obvious that all these channels can interfere which in turn may lead to cancellation of the actual ionization[5]. Thus, the dip in the ionization signal may be explained in terms of such destructive interference. However, the situation of the light generation seems much more complicated. The additional arrows on the right hand side of Fig. 1 indicate possible resonances which might enhance some of the observed lines. In fact many combinations of some generated frequencies may lead to the generation of others which are observed. Beyond that, more generated frequen-

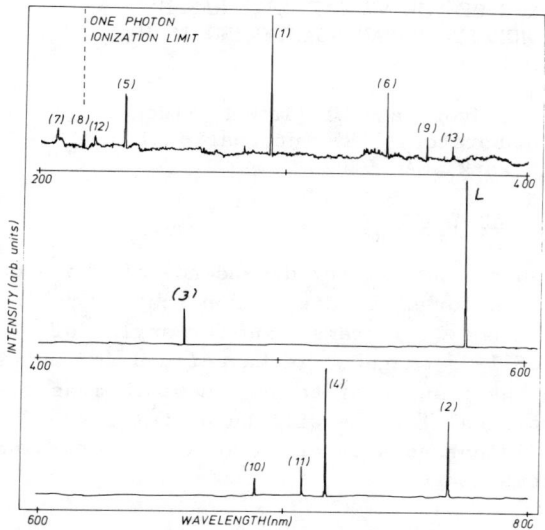

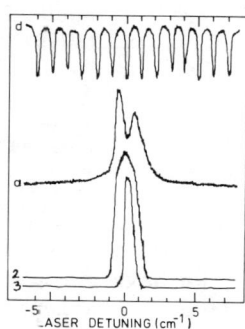

Fig. 3. Ionization signal (a) and emitted lines (2) and (3) (cf. Figs. 1,2) as a function of laser detuning. Trace d) comes from reference Fabry-Perot.

Fig. 2. Spectrum of observed coherent emission when laser is tuned to exact two-photon resonance. "L" denotes residual laser line (BG 24 filter).

cies in the VUV and IR spectral ranges must be expected which have not been measured yet because of experimental limitations. All these interrelations make it very doubtful to explain restricted aspects only in terms of classical few-wave interaction like e.g. four-wave mixing (cf. lines "2" and "3"). The situation becomes even harder to understand in view of the different excitation regions for the correlated lines (cf. Fig. 3). All this demands theoretical reconsideration of resonant multiphoton interaction in dense vapors.

ACKNOWLEDGEMENTS

We gratefully acknowledge support from the Deutsche Forschungsgemeinschaft (Sfb 161).

REFERENCES

1. D. Normand, J. Reif, and J. Morellec, pp. 471-485 in J. Eichler, I.V. Hertel, and N. Stolterfoth (eds). Electronic and Atomic Collisions, (Elsevier, 1984)
2. D.C. Hanna, M.A. Yuratich, and D. Cotter, Nonlinear Optics of Free Atoms and Molecules, Springer Series in Optical Sciences, Vol. 17 (Springer, 1979)
3. J.C. Miller and R.N. Compton, Phys. Rev. A25, 2056 (1982)
4. D. Normand, J. Morellec, and J. Reif, J. Phys. B16, L227 (1983)
5. D.J. Jackson and J.J. Wynne, Phys.Rev. Lett. 49, 543 (1982)
6. H. Scheingraber, H. Puell, and C.R. Vidal, Phys.Rev.A18, 2585(1978)
7. J. Reif and H. Walther, Appl.Phys. 15, 361 (1978)
8. C.E. Moore, Atomic Energy Levels, NSRDS-NBS 35 (1971)

ANGULAR DEPENDENCE OF THE VIBRATIONAL RAMAN LINEWIDTHS FOR STIMULATED RAMAN SCATTERING IN H_2

G. C. Herring, Mark J. Dyer, and William K. Bischel
Chemical Physics Laboratory, SRI International
Menlo Park, CA 94025

ABSTRACT

We have measured the angular and density dependence of the Dicke narrowed vibrational line shapes for stimulated Raman scattering in H_2. Measurements were made on the Q(1) transition for angles of 0-165 deg and for densities of 1-25 amagat. We have found an empirical formula that agrees with the results of these linewidth measurements to within 5%. This empirical formula will be useful for modeling the Raman gain coefficient as a function of angle for Raman beam clean-up and aperture combining.

INTRODUCTION

Raman beam clean-up and aperture combining is currently a topic of intense research within the laser community. This technique improves the spatial mode of coherent light sources by using stimulated Raman scattering in H_2 to amplify a high quality stokes beam. Energy is simultaneously transferred from several high intensity, poor quality pump beams to the high quality Stokes seed beam by crossing the pump beams at small angles to the stokes beam. Thus the angular dependence of the Raman gain needs to be well characterized before this type of beam clean-up process can be accurately modeled. For pump and stokes beams that are linearly polarized in the same directions, the angular dependence of the gain comes only from the angular dependence of the linewidth (the peak value of the steady-state gain coefficient is inversely proportional to linewidth). Previous experimental work reports linewidth measurements for only the forward and backward directions.[1] This work presents experimental results for the angular dependence of the vibrational Raman linewidth in gas phase H_2.

EXPERIMENTAL DETAILS

Using a standard quasi-cw stimulated Raman spectrometer,[2] we have measured the linewidth of the Q(1) line for angles of 0-165 degrees and densities of 1-25 amagat. Both the cw probe (683 nm) and the pulsed pump (532 nm) are single mode lasers, with the 10 ns pulsed laser limiting the spectral resolution to about 100 MHz. This laser linewidth was determined from the excess linewidth not accounted for by pressure or Doppler broadening in a copropagating geometry, where the lineshape is well known, and from assuming that the laser line shape is Gaussian. For small angles (less than 30 deg) and small densities (less than 10 amagat), this laser linewidth was accounted for by fitting the data to Voigt line shapes. For large angles or large densities the 100 MHz width of the laser is negligible. These data were fit to pure Lorentzian line shapes.

EMPIRICAL MODEL

The density dependence of these measured linewidths was compared to an empirical model given by

$$\Delta\nu = \frac{A}{\left[\rho^n + \left(\frac{A}{\Delta\nu_D}\right)^n\right]^{\frac{1}{n}}} + B\rho \qquad (1)$$

and to the diffusion model[3] given by

$$\Delta\nu = \frac{A}{\rho} + B\rho \qquad (2)$$

where $\delta\nu_D$ is the Doppler broadened linewidth, B is the density broadening coefficient, ρ is the density. The quantity A is given by

$$A = \frac{D}{\pi}\left[2k_p k_s(1 - \cos\theta) + k_R^2\right] \qquad (3)$$

where θ is the crossing angle, D is the self-diffusion coefficient, and k_p, k_s, and k_R are the pump, stokes (probe), and Raman shift wave numbers respectively.

We find that Eq. 1, with n = 1.5, gives the best fit to our data and predicts the linewidths to within 5% of our measurements for all angles and densities. At densities below 1 amagat, the differences between the data and the model become larger. Because Eq. 1 is empirical, no physical significance should be attached to the value of 1.5 for n. The experimental data and the empirical fit of Eq. 1 for four different crossing angles is illustrated in Fig. 1.

Although useful for modeling purposes, the empirical Eq. 1 does little to help the understanding of the physics of the collisional contribution to the line shape. However, line shape measurements with uncertainties of less than 1% are necessary to compare with more physically meaningful models such as the Galatry and related line shapes.[4] We plan to make improved measurements (modifying our pulsed pump laser with a single mode seed laser) to make this comparison.

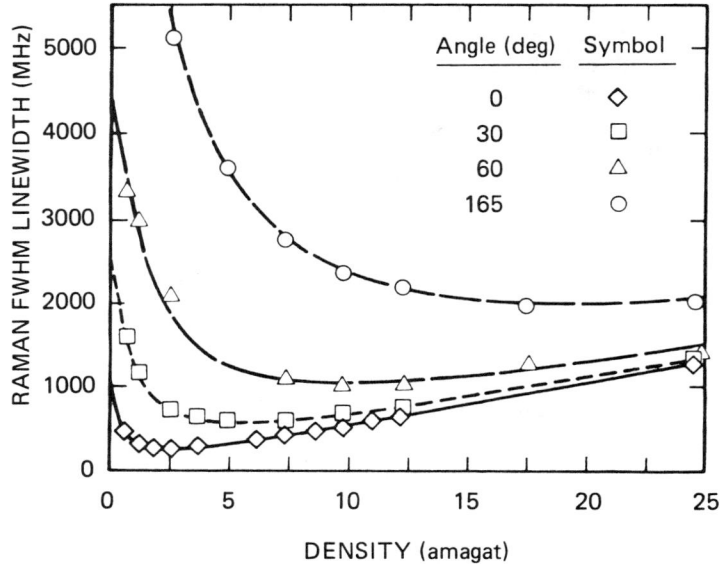

Figure 1. Comparison of the experimental data (symbols) and the empirical model (solid and dashed lines) of Eq. 1, with n = 1.5.

ACKNOWLEDGEMENT

This work was supported by the Defense Advanced Research Agency under Contract N0014-84-C-0256, through the Office of Naval Research.

REFERENCES

1. J. R. Murray and A. Javan, J. Mol. Spectrosc. $\underline{42}$, 1 (1972).
2. P. Esherick and A. Owyoung, "High-Resolution Stimulated Raman Spectroscopy," in <u>Advances in Infared and Raman Spectroscopy</u>, R.J.H. Clark and R. E. Hestor, eds. (Heydon, London, 1983), pp. 130-187.
3. W. K. Bischel and M. J. Dyer, Phys. Rev. A $\underline{33}$, 3113 (1986).
4. P. L. Varghese and R. K. Hanson, Appl. Opt. $\underline{23}$, 2376 (1984).

OBSERVATION OF LONG-LIVED COLLISION-INDUCED COHERENCES AND GROUND-STATE-SPIN GRATINGS IN A FLAME

Rick Trebino and Larry A. Rahn
Combustion Research Facility, Sandia National Laboratories, Livermore, CA 94550

ABSTRACT

We observe collision-induced four-wave-mixing resonances—Zeeman and hyperfine coherences and ground-state-spin gratings—in a sodium-seeded hydrogen-air flame. We see several new characteristics of collision-induced phenomena at high intensity.

I. COLLISION-INDUCED RESONANCES IN FOUR-WAVE-MIXING

In four-wave-mixing and induced-grating processes, collisions can dephase quantum-mechanical amplitudes that would otherwise destructively interfere.[1] As a result, nonlinear-optical spectra of species undergoing collisions can exhibit normally forbidden resonances with line-strengths that increase with pressure. Previous experimental work[1-5] has demonstrated these "extra resonances" in sodium in the presence of various buffer gases. In this paper, we extend these measurements to a flame environment, where collision-sensitive processes are of great interest. In addition, we observe several new characteristics of collision-induced phenomena.

II. EXPERIMENTAL APPARATUS

Two argon-laser-pumped Coherent 699 cw single-mode ring dye lasers provide the 590-nm light for these studies. This light is pulse-amplified by separate three-stage, Nd:YAG-pumped pulsed amplifiers, yielding about 2-mJ, 20-nsec, temporally Gaussian pulses in each beam with linewidths of < 60 MHz. We split one of these beams into two pulses of equal energy and allow all three beams to propagate, unfocused, into an approximately stoichiometric hydrogen-air flame seeded with sodium. The beam geometry is three-dimensional, with the four beams of the interaction appearing as corners of a square if observed end-on. Nearly Doppler-free spectra result from this geometry.

III. COLLISION-INDUCED POPULATION GRATINGS

For this set of experiments, parallel-polarized excitation beams at frequencies 2 to 100 cm^{-1} below the Na $3\,^2S_{1/2} - 3\,^2P_{1/2}$ transition illuminate the flame from slightly different angles, producing a spatially sinusoidal excitation intensity. Due to the detuning, the beams do not excite isolated sodium atoms, but they do excite sodium atoms that are undergoing collisions. Thus, a spatially sinusoidal distribution of collision-induced excited states—a collision-induced excited-state grating—occurs. After excitation, collisional quenching then returns excited atoms to the ground electronic state. Because the excitation process involves some spin-selectivity, gratings also occur in the various ground-electronic-state $| F, m_F \rangle$ population densities.[2,4]

© American Institute of Physics 1987

Diffraction off such gratings yields four-wave-mixing spectra in the flame typified by that shown in Fig. 1, having both a broad and a very narrow component. The very narrow component represents a ground-state-spin grating;[2,4] its width is due mostly to the laser linewidths and residual Doppler broadening, indicating an extremely long lifetime—which is reasonable in view of the spin flip required for relaxation between the F=1 and F=2 hyperfine levels. The broad component is thought to be due to an excited-electronic-state (3 $^2P_{1/2}$) grating, the width of which is the inverse of the collisional quenching time of this state. The observed linewidth of this latter component indicates a collision-quenching timescale of about 500 psec. Extending the detuning to as much as 100 cm^{-1} below the 3 $^2P_{1/2}$ level, where intracollisional effects should become important, yields no significant changes in the lineshape. We obtain qualitatively similar results for detunings 2 cm^{-1} above the 3 $^2P_{1/2}$ state.

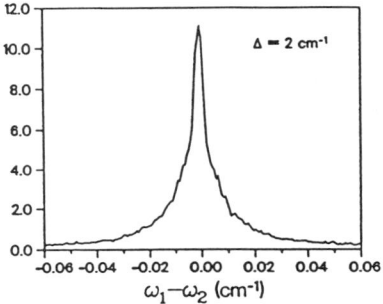
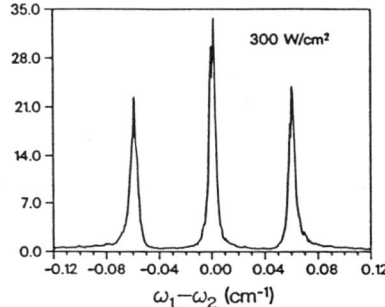

Fig. 1. Four-wave-mixing spectrum of collision-induced population gratings in sodium in a flame. Two components—a broad pedestal and a narrow spike—are present, corresponding to a rapidly decaying collision-quenched excited-state grating and a slowly decaying ground-state-spin grating, respectively.

Fig. 2. Four-wave-mixing spectrum of collision-induced Zeeman and hyperfine coherences at low intensity. All linewidths are limited by residual Doppler broadening and the laser linewidths.

IV. ZEEMAN AND HYPERFINE COHERENCES

In another set of experiments, orthogonally polarized excitation beams, at frequencies a few wavenumbers above or below the Na 3 $^2S_{1/2}$ - 3 $^2P_{1/2}$ transition, illuminate the flame. The uniform intensity prohibits intensity-dependent gratings from forming in the medium, but now the two Raman-type effects, Zeeman and hyperfine coherences, can be seen. For zero magnetic field, the resonance condition for the Zeeman coherence is $\omega_1-\omega_2=0$, and for the ground-state hyperfine coherence in sodium, resonance occurs at $\omega_1-\omega_2=\pm 0.059$ cm^{-1}.

Figure 2 shows a typical spectrum of these coherences for a detuning of 2 cm^{-1} at a relatively low intensity (about 300 W/cm^2 peak per beam). The observed linewidths are due partly to residual Doppler broadening and partly to the finite laser linewidths. For higher laser intensities (about 4.5 kW/cm^2 peak per beam), the intensity and lineshape of the Zeeman coherence show no saturation effects (See Fig. 3). The hyperfine coherences, however, exhibit significant broadening. Finally, Fig. 4 shows the spectrum at a peak intensity of 20 kW/cm² per beam, where nonlinear scaling of the line-strengths with the three beam intensities begins to occur. Observe that additional lines have appeared at $\omega_1-\omega_2 \approx \pm \omega_{hyperfine}/2$. These additional spectral lines are most likely due to fifth-order-susceptibility ($\omega_{signal}=2\omega_1-\omega_2+\omega_2-\omega_2$) effects. Specifically, this process has the resonance condition: $2(\omega_1-\omega_2)=\pm\omega_{hyperfine}$. In addition, optical pumping of the ground-state-spin populations appears to play an important role in these high-intensity spectra. Finally, because of the long-lived nature of the gratings and coherences in these experiments, it is clear that we are operating in a highly transient regime, where the predictions of the well-known steady-state theory may not be accurate.

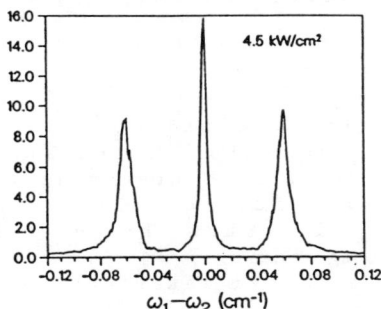

Fig. 3. Four-wave-mixing spectrum of collision-induced Zeeman and hyperfine coherences at intermediate intensity. Observe that the hyperfine coherences have broadened somewhat compared with their widths at lower intensity.

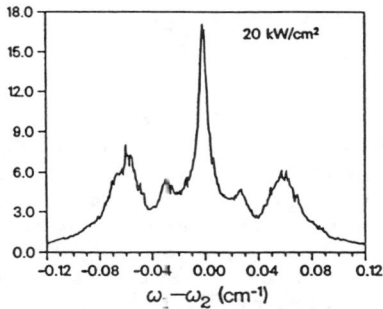

Fig. 4. Four-wave-mixing spectrum of collision-induced Zeeman and hyperfine coherences at high intensity. Observe the increased broadening in all three lines. Also, observe the additional spectral components—subharmonics of the hyperfine resonances—at ±.03 cm⁻¹.

V. CONCLUSIONS

Our studies show that collision-induced extra resonances in four-wave mixing—long-lived ground-state-spin gratings and Zeeman and hyperfine coherences—are easily observed in flames. In addition, our observations of the high-intensity behavior of the Zeeman and hyperfine coherences reveal preferential broadening of the hyperfine coherences and additional resonances in the collision-induced spectrum. We hope that this work will help to clear the way for diagnostic applications of these effects.

REFERENCES

*This work was supported by the U.S. Department of Energy, Office of Basic Energy Sciences, Division of Chemical Sciences.
1. Y. Prior, A.R. Bogdan, M. Dagenais, and N. Bloembergen, Phys. Rev. Lett. 46, 111 (1981).
2. N. Bloembergen, A.R. Bogdan, and M.W. Downer, in *Laser Spectroscopy V*, edited by A.R.W. McKellar, T. Oka, and B.P. Stoicheff (Springer-Verlag, Berlin, 1981), p. 157.
3. L.J. Rothberg and N. Bloembergen, Phys. Rev. A 30, 820 (1984).
4. L.J. Rothberg and N. Bloembergen, Phys. Rev. A 30, 2327 (1984).
5. Y.H. Zou and N. Bloembergen, Phys. Rev. A 33, 1730 (1986).

GENERATION OF SOLITONS IN TRANSIENT STIMULATED RAMAN SCATTERING BY OPTICAL PHASE SHIFTS

D.C. MacPherson and J.L. Carlsten
Physics Dept., Montana State University
Bozeman, Montana 59717

ABSTRACT

Initiation of solitons in a stimulated Raman scattering amplifier using a π phase shift produced by modulating the envelope of the Stokes seed through zero is discussed. We find that by modulating the envelope slowly compared to T_2 a soliton pulse is generated which is much shorter than T_2.

DISCUSSION

Several groups[1-2] are studying the use of optical phase shifts to initiate soliton formation in transient stimulated Raman scattering. The equations governing SRS admit soliton solutions if damping due to collisional dephasing is neglected. This is valid when the pulse is short compared to the dephasing time $T_2(1/\Gamma)$. Our experiments use a generator-amplifer Raman laser pumped by a frequency doubled Nd:YAG laser at 532nm. The Raman medium is hydrogen at 1-100 atmospheres. We use an electrooptic crystal to place a π phase shift near the center of a gaussian temporal Stokes pulse between the generator and the amplifier. This shift momentarily reverses the direction of gain. The phase shift is produced by modulating the amplitude of the Stokes envelope through zero using polarizers and a Pockels cell. In this configuration the π phase shift in the Stokes field occurs very rapidly as the envelope goes through zero.

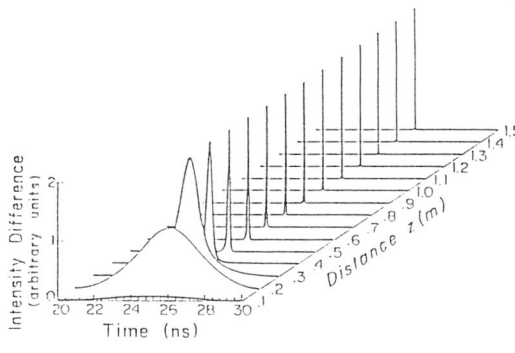

Fig. (1) Evolving Soliton

We have used numerical calculations to investigate the evolution of the pump and Stokes pulses in the amplifier when an amplitude modulated phase shift (AMPS) is placed in the Stokes field. Our calculations indicate that solitons can be initiated by switching the Pockels cell voltage over a time of several nano seconds when T_2 is considerably less than a nano second. The development of the temporal pump pulse is shown in Fig. (1). Here we have plotted the difference between pump pulses which evolve with and without a π phase shift in the Stokes seed. The initial pulse formation can be understood from steady state theory which is valid because the field envelope resulting from the AMPS changes slowly compared to T_2. The AMPS leaves a hole in the Stokes seed intensity. The input temporal gaussian pump pulse first depletes where the product of the Stokes seed and the input pump is the largest leaving a pump pulse whose initial width depends on the duration of the AMPS. The newly formed pulse will narrow as its wings are depleted.

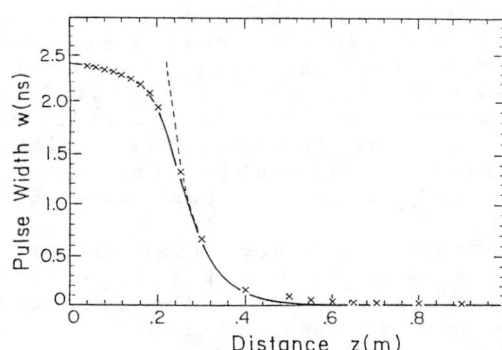

Fig. (2) Pulse Width

Since the seed is zero at the center of the AMPS the peak amplitude will be unchanged. The width W (HWHM) of the evolving pulse is shown in Fig. (2) as the X's. The solid curve is the pulse width predicted by steady state theory which works well until the pulse width becomes comparable to T_2. The broken curve is an approximation to the steady state predictions which has the form of a decaying exponential. This approximation is valid after the pump first depletes on both sides of the AMPS. Surprisingly the decay is faster for smaller damping.

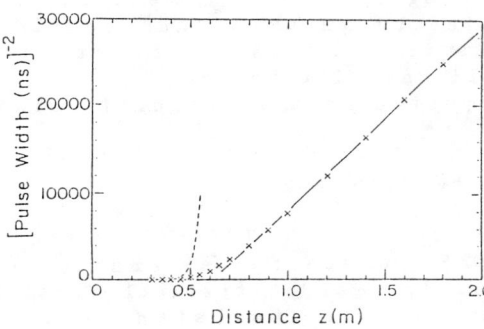

Fig. (3) $1/W^2$

Numerical results for $1/W^2$ are plotted in Fig. (3) as the X's. Also shown is an analytical hypertransient approximation (solid curve). The analytical result is obtained by applying energy conservation. For the exact soliton solution ($\Gamma=0$) energy is supplied by the fields during the leading half of the pulse by converting pump photons to Stokes photons. This energy is stored in the polarization of the medium. During the trailing half of the pulse energy is recovered from the polarization due to the π phase shift and used to convert Stokes photons to pump photons. When damping is present ($\Gamma>0$) some of the molecules in the medium are dephased before their energy is returned to the fields. Once a molecule's phase has been randomized it's energy cannot on the average be retrieved. This effective energy loss to dephasing is supplied by a net conversion of pump photons to Stokes photons. By applying photon conservation the peak of the pump pulse remains unchanged; therefore its width must decrease. The broken curve in Fig. (3) is the steady state result which narrows too rapidly because it does not take account of gain reversal.

Numerical calculations also indicate that as the pulse narrows it approaches the hyperbolic secant form. When the pulse becomes much narrower than T_2 its width and shape change slowly due to its soliton character. These results indicate that pulses considerably shorter than those produced using standard electrooptical switching techniques can be generated.

Ackerhalt and Milonni[2] have predicted that placing a π phase shift in the Stokes seed will initiate soliton formation in the second Stokes wave. If observed this would be the first example of four wave solitons.

England and Bowden[3] have predicted that solitons will be initiated by quantum fluctuations in the spontaneous emission in the Raman generator. They predict this will occur about once in fifty shots.

We are presently setting up an experiment to study soliton formation initiated by an amplitude modulated phase shift. Experimental studies of fluctuation initiated and four wave solitons will follow.

This research is supported by the National Science Foundation (PHY-8516110).

REFERENCES

1. D.J. Kaup, Physica D <u>19</u>, 125 (1986); H. Steudel, Optics Comm. <u>57</u>, 285 (1986): K.J. Druhl, J.L. Carlsten, and R.G. Wenzel, J of Statistical Phys. <u>39</u>, 615 (1985); R.G. Wenzel, J.L. Carlsten, and K.J. Druhl, J of Statistical Phys. <u>39</u>, 621 (1985); K. Druhl, G. Alsing, Physica <u>20D</u>, 429 (1986).

2. J.R. Ackerhalt and P.W. Milonni, Phys. Rev. <u>A33</u>, 3185 (1986).

3. John C. England, Charles M. Bowden, to be published.

HIGH EFFICIENCY DISTRIBUTED FEEDBACK LASERS OPTIMALLY DESIGNED FOR STABLE SINGLE MODE OPERATION

H. Ishikawa, H. Soda, Y. Kotaki, K. Kihara, and H. Imai
Fujitsu Laboratories Ltd., Atsugi, 10-1 Morinosato-Wakamiya, Atsugi, Japan

ABSTRACT

Optimum design for stable single mode operation has been done for an asymmetric facet reflectivity type InGaAsP/InP DFB laser emitting at 1.3 μm. The design takes into account phase corrugation at both facets and the effect of spatial hole burning in the longitudinal direction realized in single mode operation with high efficiency

INTRODUCTION

In the DFB laser, the main problem has been the effect of corrugation phase at both facets which gives only a finite single mode output. In addition, we have recently found out that the spatial hole burning in the longitudinal direction has a serious effect on the stability of the mode above the threshold current! Here, a design of a DFB laser with asymmetric facet reflectivity emitting at a wavelength of 1.3 μm has been used to achieve stable single mode operation up to high optical output powers. The design takes into account the effect of the spatial hole burning and the corrugation phase realized for high power single mode operation above a 10 mW optical output power.

STRUCTURE AND THE DESIGN

Figure 1 shows schematically the side view of the DFB laser and the distribution of the optical field and carrier in the DFB laser. One end of the laser has a low reflectivity facet coating and the other end is cleaved. The design concerns the reflectivity of the low reflectivity end and the coupling coeffcient between the optical field and the corrugation. Figure 2 shows the calculated single mode operation yield in the absence of the spatial hole burning. The yield is defined as a percentage of cases of various phase configurations which gives a normalized gain difference between the main mode and the next mode larger than 0.05. The yield is higher for the larger KL and smaller facet reflectivity.

© American Institute of Physics 1987

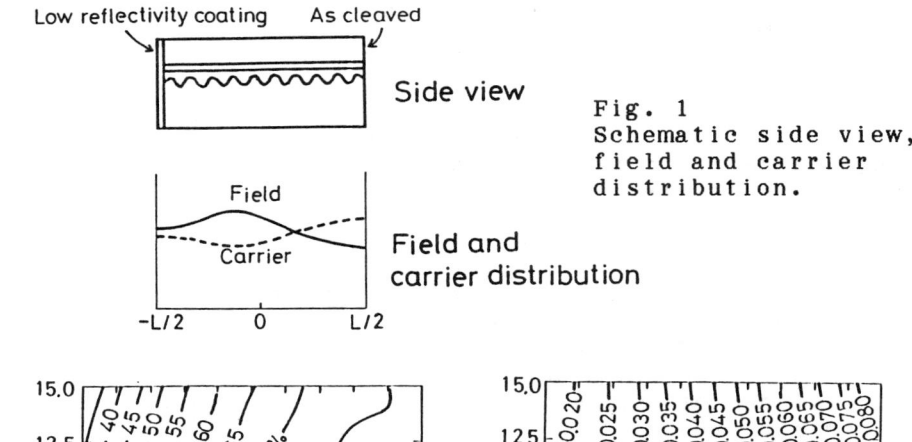

Fig. 1 Schematic side view, field and carrier distribution.

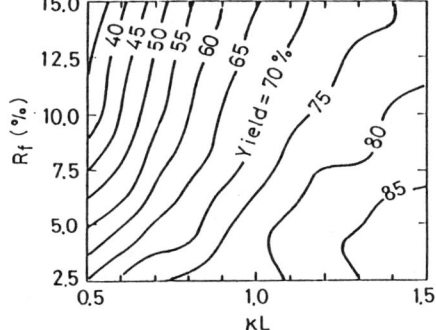

Fig.2 Calculated single mode operation yield without consideration of spatial hole burning.

Fig.3 Calculated field flatness in the laser cavity.

However, experimentally the yield is very low for large κL. Then we considered the effect of the spatial hole burning as illustrated in Fig.1. As a measure of the spatial hole burning we used the following integral.

$$F = \int_{-L/2}^{L/2} (I(z)-1)^2 \, dz/L \qquad (1)$$

$$\int_{-L/2}^{L/2} I(z) \, dz/L = 1 \qquad (2)$$

where $I(z)$ is the field intensity averaged over all possible phase configurations at both facets. The smaller the value of F, the flatter the field distribution in the cavity. Figure 3 shows the calculated F values. The value of F increases steeply for κL larger than 1 and for Rf smaller than 5 %. For a large Rf, although we can get a small F value, Fabry-Perot modes tend to lase. Then we choose κL between 0.5 and 1, and a Rf around 5% for which a small reproducible F value can be obtained.

EXPERIMENTAL

We have fabricated FBH-DFB lasers with different values of κL and compared the single mode operation yield. The reflectivity of one end was ajusted to be 5 % by the Al_2O_3 coating. For a κL of 1.2 large percentage of samples showed multiple mode operation at low optical output power. The single mode yield above a 10 mW optical output was only 27 %. While for κL between 0.5 to 1, the yield was 61 %. This result clearly shows the effect of the spatial hole burning. The distributed feedback lasers should be designed so that the optical field in the longitudinal direction is as flat as possible. Figure 4 shows a spectrum of optimally designed DFB laser. Sufficient side mode suppression has been attained. The optimized DFB laser showed a threshold current of 15 mA, and high differential efficiency of 0.25 mW/mA. High efficiency DFB lasers with a high single mode operation output have been realized by the optimization of κL and the facet reflectivity.

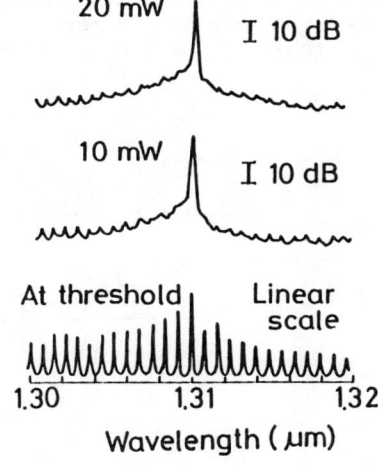

Fig.4 A typical lasing spectra of an optimized DFB laser.

REFERENCE

1. H. Soda, H. Ishikawa, and H. Imai, Electron. Lett., 22, 1047 (1986)

LONGITUDINAL MODE WIDTH IN EXCIMER LASERS

Gabriel G. Lombardi and William H. Long, Jr.
Northrop Research and Technology Center
One Research Park, Palos Verdes Peninsula, CA 90274

ABSTRACT

The frequency drift of the modes of a XeCl laser oscillator was deduced by measuring the contrast of fringes produced by the laser in a Michelson interferometer. A calculation of the drift produced by the time-dependent refractive index of the gain medium is in reasonable agreement with the measured value.

INTRODUCTION

The time-averaged width of individual longitudinal modes of a multi-mode, XeCl, discharge laser oscillator was measured using time-delayed interferometry, a technique related to Fourier-transform spectroscopy. The duration of the optical pulse was approximately 120 ns (FWHM). Usually, the oscillator was injection locked to reduce the linewidth to approximately 1 GHz, or ten longitudinal modes. However, similar results were obtained with the free-running oscillator, which has a linewidth of several hundred gigahertz. This is because the technique is being used to measure the width of individual modes, rather than the total bandwidth of the oscillator. The beam from the laser was divided by the beamsplitter of a Michelson interferometer. The arms of the interferometer were adjusted so that their lengths differed by an integral multiple of the laser cavity length. Interferograms were recorded using a charge injection device (CID) camera and a video cassette recorder. Figure 1 is a schematic diagram of the optical system.

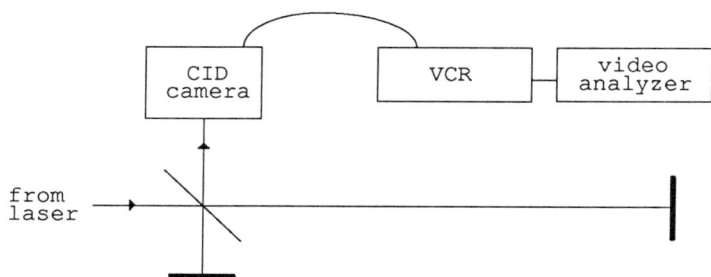

Fig. 1. Schematic diagram of optical system.

© American Institute of Physics 1987

The visibility, V, of interference fringes produced by a source as a function of optical delay is related to its frequency spectrum.[1] Fringe visibility in time-delayed interferometry is analogous to the fringe intensity as a function of path length in Fourier-transform spectroscopy. Measurements of the fringe visibility, made at path-length differences of up to three times the cavity length, were used to calculate the width of the longitudinal modes, averaged over the laser pulse. The visibility was measured using a video analyzer.

RESULTS

Figure 2 is a plot of the fringe visibility, normalized to the visibility at zero path length difference, as a function of interferometer's optical delay, measured in units of the laser oscillator cavity roundtrips, N. The filled circles are the measured data; the light curve is a fit to the measurements of a visibility curve obtained by assuming that the spectral broadening of the modes is caused by the refractive index of the gain medium having a quadratic time dependence:

$$n(t) = n_0 + \alpha_0 t + \beta_0 t^2. \tag{1}$$

The fit was obtained by minimizing the sum of the square of the deviations of the curve from the measurements. The spectral width (FWHM) corresponding to this visibility curve is 25 MHz, compared to a transform-limited width of 4 MHz.

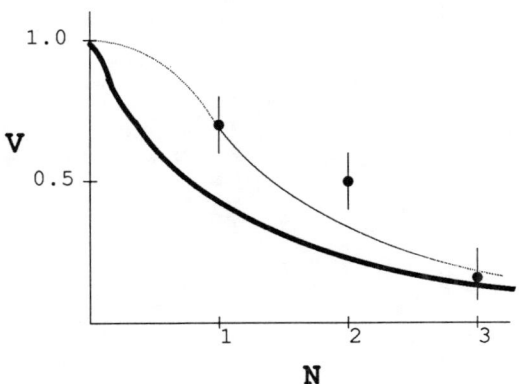

Fig. 2. Fringe visibility as a function of optical delay.

A computer model of the laser gas kinetics was used in conjunction with values of the refractive index of atoms, molecules, and ions obtained from the literature to calculate the time dependence of the refractive index. The model predicts the time-dependent values of the densities of HCl, H, Cl, Ne, Ne*, Xe, Xe*, and electrons. A quadratic fit to n(t) predicted by the model resulted in coefficients $\alpha=2\alpha_0$ and $\beta=4\beta_0$. The visibility curve for these values of α and β are plotted as the heavy curve in Fig. 2. The lack of close agreement between the measured data with the heavy curve is not surprising, considering the uncertainties in the refractive index data. The principal cause of refractive index variation was found to be the dissociation of HCl.

REFERENCE

1. M. Born and E. Wolf, *Principles of Optics* (Pergamon, N.Y., 1959), p. 266. The visibility is defined by $V = (I_{max} - I_{min})/(I_{max} + I_{min})$, where I_{max} and I_{min} are the maximum and minimum fringe intensities, respectively.

High Power He-Cd⁺ White Light Laser

A.Fuke, K.Masuda and Y.Tokita
Koito Manufacturing Co.,Ltd. Shimizu, Japan

ABSTRACT

The new structure of hollow cathode He-Cd⁺ white light laser was investigated. It was improved from a flute type and experiments on the laser tube of 61 cm in length and 4 mm in bore diameter were made. As a result, three colors, red, green, and blue, were oscillated simultaneously. The total output power was 194.2 mW at an anode current of 2A and the oscillation efficiency was 0.032%. The noise levels in rms between 10 Hz and 10 MHz were 0.086%, 0.096%, and 0.066% for red, green, and blue, respectively.

INTRODUCTION

The hollow cathode He-Cd⁺ laser has 12 spectral oscillation lines.[1] Visible lights with wavelengths of 441.6 nm (blue), 533.7 nm and 537.8 nm (green), and 635.5 nm and 636.0 nm (red) were selected from the oscillation lines. Using broad-band reflection mirrors, these lines are oscillated simultaneously to allow white light laser to be obtained. It is expected that the laser will be applied to image equipment, photograph, and printing because it oscillates at the three primary colors of light, red, green, and blue, simultaneously in a single laser tube and their wavelengths are close to the ideal wavelengths of the three primary colors of light, 450 nm, 540 nm, and 610 nm.[2] Though various types of hollow cathode structures have been proposed, no hollow cathode type lasers have been put to practical use. The authors used a new hollow cathode structure introduced by improvement of the flute type that makes it possible to obtain a continuous white light laser with high power and high output stability, which is reported below.

STRUCTURE OF LASER TUBE

The hollow cathode type laser uses high energy electrons and a negative glow area with high electron density that cannot be provided by positive column. The laser is divided into flute,[3] coaxial,[4] and modular types,[1] etc. according to the structure, which have been studied so far.

The authors designed the new structure shown in Fig.1. In the figure, the cathode is 4 mm in bore diameter, 14 mm in outer diameter, and 61 cm in active cathode length and 20 anodes are attached at intervals of 3 cm. On the opposite side of the anodes, the Cd reservoirs are located midway between anodes. The circumference part is made of pyrex glass, which is connected closely to the cathode at the part other than anode holes and Cd vapor introduction holes. Thus, Cd vapor is introduced into the bore rapidly and is distributed uniformly. On the other hand, the structure facilitates the exhaust of spatter products from the bore. Sub anodes are arranged at the both ends of the bore to prevent Cd vapor from being dispersed to Brewster windows. Pure iron was used as a cathode material. The comparison of 7 types of metals, titanium, stainless steel, aluminum, copper, molybdenum, tungsten, and iron, made it clear that iron can provide the highest output power and stability of laser. The cathode was assembled after sufficient gas discharge in a vacuum and high temperature furnace. The resonator is constructed by dielectric multi-layer film mirrors with reflectivity of 100% and 99% in the spectral range between 440 nm and 650 nm.

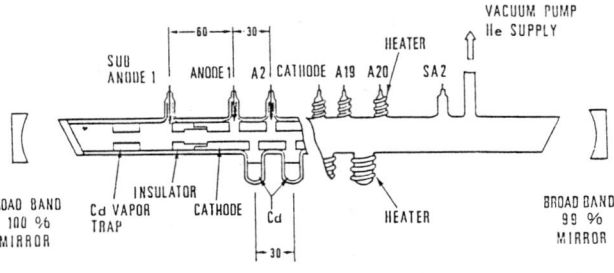

Fig.1. Laser tube structure.

EXPERIMENTAL RESULTS AND DISCUSSIONS

Fig.2 shows the output of each color light during the simultaneous oscillation of three colors on the basis of the peak values when the He pressure is changed. The peak values of red, green, and blue light appear at He pressures of 10.0, 11.5, and 16.0 torr, respectively and the peak value of white light laser appears at a He pressure of 11.5 torr.

Fig.3 shows the output power for the change of anode current at the He pressure of 10 torr. The He pressure is set at 10 torr because the value is optimum for the red light that has the minimum output among the three colors. At the anode current of 2A, the output powers of red, green, and blue are 41.3 mW, 60.4 mW, and 92.5 mW, respectively.

Fig.4 shows the laser output efficiency for the power entered directly into the anodes and sub

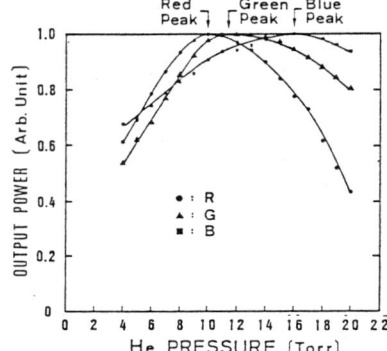

Fig.2. He pressure dependency of laser output power.

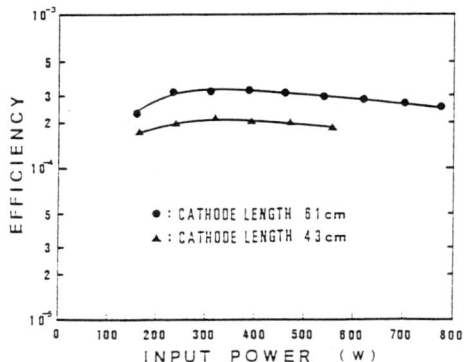

Fig.4. Output power efficiency for input power.

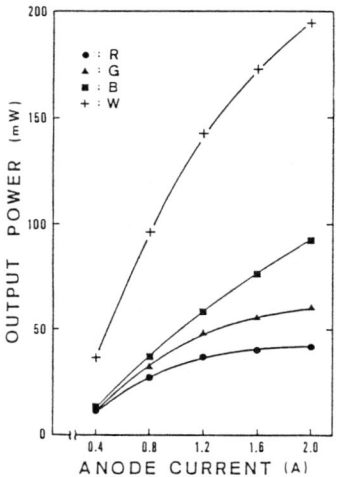

Fig.3. Laser output power as a function of anode current.

anodes. The input power does not include the power that enters the Cd reservoir heaters. The figure also shows the data for the cathode length of 43 cm. It indicates the efficiency of 0.032% for the cathode length of 61 cm.

Fig.5 shows the output stability when the total output power is about 62 mW. The output fluctuation is within 0.6%. For the same conditions, the noise levels in rms between 10 Hz and 10 MHz are 0.086%, 0.096%, and 0.066% for red, green, and blue lights, respectively.

Fig.6 shows the frequency characteristic of noise for blue output. The thin line in the figure indicates the background noise. There is much noise with frequency components of up to about 300 kHz. The characteristic is almost the same as that of red and green lights. The hollow cathode type He-Cd$^+$ laser does not produce noise by moving striations that appear in the positive column type. The noise levels in rms is within 0.1% for any color of red, green, and blue, which is much less than that in the positive column type.

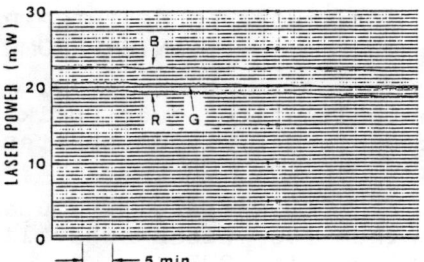

Fig.5. Output stability (R:Red, G:Green, B:Blue)

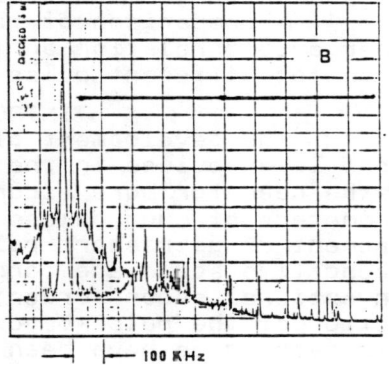

Fig.6. Noise characteristic of frequency for blue output.

CONCLUSION

The use of a new hollow cathode structure enabled continuous oscillation of He-Cd$^+$ white light laser with high output of 194 mW to be obtained. The noise levels in rms were as low as 0.1% for three colors and the laser has high output stability. Moreover, the simple structure is suitable for mass production.

To put white light lasers to practical use, continuous oscillation, high output power, high output stability, and long life are required. The authors are now researching ways to enhance the length of the laser life.

REFERENCES

(1) S.C.Wang:SPIE 232, 42(1980)
(2) W.A.Thornton:J.Opt.Soc.Am. 61, 1155(1971)
(3) K.Fujii, T.Takahashi, Y.Asami:IEEE J.QE 11, 3(1975)
(4) K.Fujii:Jpn.J.Appl.Phys. 14, 1339(1975)

ULTRASHORT PULSE CHIRP PARAMETER DETERMINATION BY INTERFEROMETRIC METHODS

R. Fischer
Zentralinstitut fuer Optik und Spektroskopie,
Akademie der Wissenschaften der DDR,
1199 Berlin, GDR

C. Rempel, J. Gauger, J. Tilgner
Zentrum fuer Wissenschaftlichen Gerätebau,
1199 Berlin, GDR

ABSTRACT

A comparison of the fringe resolved second-order autocorrelation and the linear autocorrelation is given for different pulse shapes and chirp functions.

INTRODUCTION

The development of techniques to generate ultrashort laser pulses in the femtosecond region[1] has led to the need for more accurate methods of determining the parameters of such pulses. The only method for the measurement of ultrashort pulse parameters on the subpicosecond time scale are correlation techniques. Besides conventional second-order autocorrelations and the detection of the pulse-bandwidth-product two interferometric techniques have been proposed: The fringe resolved second-order autocorrelation (FRACF)[2] and the linear autocorrelation (LA)[3,4] which is the Fourier spectrum of the pulse. These methods allow the measurement of the phase dependence on time within the pulse duration (chirp). Of special interest is the knowlegde of the (time dependent) phase of the (chirped) pulses for two reasons, one is the better understanding of the generation process within the active medium and the other is information about the variation in time under the influence of nonlinear optical effects of self-action, especially during the propagation in fibers.

RESULTS

The calculations of the FRACF and the LA have been performed with a fast Fourier transform algorithm for several pulse shapes as Gaussian, rectangular, and secans-hyperbolic. Different functions for the time dependent phase have been used. The computations show for both the FRACF and the LA the typical narrowing of the mean peak when the chirp increases. It is known that the FRACF shows side maxima in the presence of chirp but we have shown that such side maxima occur in the LA too when the chirp is nonlinear. Because of the simpler mathematical expression the LA is easier to interprete.
In Figs. 1 and 2 the results for the second-order

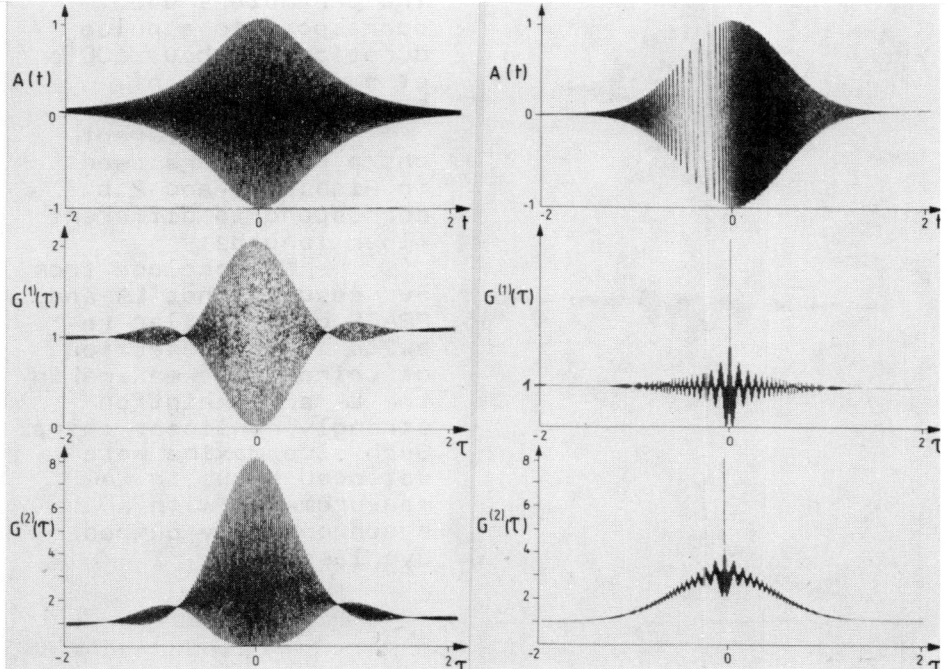

Fig. 1   Fig. 2.a

Fig. 1. $G^{(2)}(\tau)$ and $G^{(1)}(\tau)$ for the pulse function A(t) of Eq. (1). The chirp parameter is B=1.

Fig. 2. $G^{(2)}(\tau)$ and $G^{(1)}(\tau)$ for the pulse function A(t) of Eq. (2). The chirp parameter is B=10 (b) or B=150 (a).

autocorrelation function $G^{(2)}(\tau)$ and the first-order autocorrelation function $G^{(1)}(\tau)$ in dependence on the delay-time τ are shown for two different pulse functions A(t). In Fig. 1 the electric field strength is chosen as

$$A(t) = \operatorname{sech}(at) \cos\{D\pi t + 2B \ln[\cosh(at)]\} \quad (1)$$

(a=2.17); this function is typical for the case of colliding-pulse mode locking of dye lasers[5]. In Fig. 2. the results are given for

$$A(t) = e^{-2t^2} \cos\{D\pi t - B e^{-4t^2}\} \quad ,(2)$$

i.e. a Gaussian pulse where the time dependence of the phase is caused by self-phase modulation, e.g. in fibers[6].

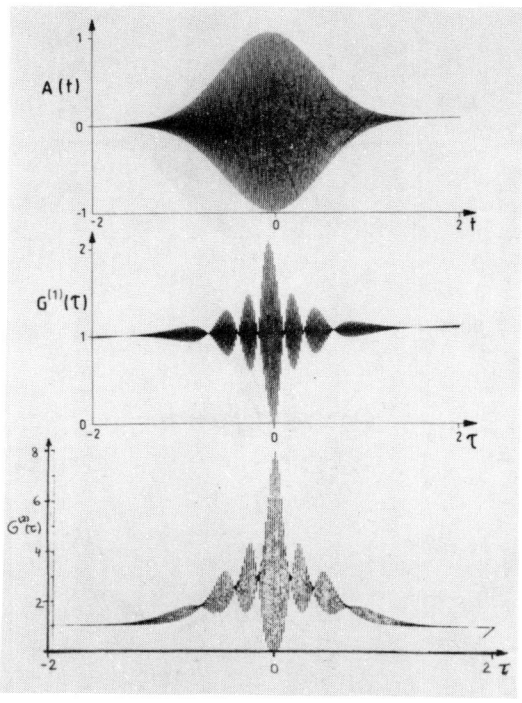

Fig. 2.b

The parameters used correspond to a pulse duration of about 100fs at a wavelength of 600 nm, i.e. D=100.

The different chirp parameters used in Figs. 2.a and 2.b correspond to different fiber lengths.

We conclude from our results that LA and FRACF give similar results in the detection of chirp. Side maxima in the LA are a hint on strongly nonlinear chirp. Such side maxima were detected by us in LA-measurements with a synchroneously pumped dye laser[4].

REFERENCES

1. D.H.Auston and K.B.Eisenthal,Eds.,Ultrafast Phenomena IV(Springer Series in Chemical Physics, Vol. 38, Springer-Verlag, New York, 1984)
2. J.C.Diels,J.J.Fontaine,I.C.Mc Micheal and F.Simoni, Appl.Optics 24, 1270 (1985)
3. J.A.Blodgett and R.A.Patten,Appl.Optics 12,2147(1973)
4. C.Rempel,J.Gauger,J.Tilgner and R.Fischer,IV.International Symposium"Ultrafast Phenomena in Spectroscopy" 1985(Conference Proceedings,Teubner Texte zur Physik, Teubner-Verlag, Leipzig, 1986)
5. D.Kühlke and W.Rudolph,Optical and Quantum Electr. 16, 57 (1984)
6. B.H.Kolner,D.M.Bloom,J.D.Kafka and T.M.Baer, in ref. 1., p.19

PHASE CONJUGATION OF 2.91 μm HF LASER RADIATION VIA STIMULATED BRILLOUIN SCATTERING

Michael T. Duignan,[†] B.J. Feldman, and W.T. Whitney,
Naval Research Laboratory, Code 6540, Washington, DC 20375

ABSTRACT

A flashlamp-pumped pulsed hydrogen fluoride laser, operating single-line at 2.91 μm, was used to generate backward stimulated Brillouin scattering in high pressure xenon gas with power reflectivities of ~50%. Far field spatial profiles indicate correction of random phase front aberrations through phase conjugation via the SBS process.

INTRODUCTION

Numerous experimental studies[1] in the UV, visible and near-IR have confirmed that stimulated Brillouin scattering (SBS) can result in wavefront reversal or phase conjugation of the input beam. The potential for correction of phasefront aberration arising from a variety of sources is also of considerable interest for use in laser systems in the mid-IR. However, despite the availibility of powerful sources (HF, DF, CO, and CO_2), very few SBS studies in this region of the spectrum have been reported. Considerations such as a limited choice of non-absorbing SBS media, longer phonon build-up time, optical breakdown, and competition from other nonlinear processes make SBS in the mid-infrared somewhat more difficult than at shorter wavelengths. We have recently demonstrated[2] high reflectivity SBS of HF laser radiation, operating single-line at 2.91μm, using ~45 atm of xenon as the nonlinear media. We have now confirmed preliminary experimental evidence indicating that the SBS return wave is the conjugate of the input.

EXPERIMENTAL

An atmospheric pressure, flashlamp initiated, pulsed hydrogen fluoride laser used in the experiments was tuned to the 2P8 line at 2.91 μm, chosen because it is one of the stronger HF lines and because of its good atmospheric transmission. Laser output was ~3 J in a ~1-μs pulse. The laser cavity was aligned to maximize focused intensity. The laser was optically isolated from the SBS return by means of a ZnSe Brewster stack polarizer and a zero-order magnesium fluoride quarter-wave plate. The measured

---

[†]Potomac Photonics, Inc., College Park, MD, 20742

extinction of the Brillouin return was approximately 15:1. With this isolation there was no detectable interference between the SBS and laser output. The far field intensity profiles of the input and SBS reflected beams were measured by means of a 256-element linear pyroelectric array consisting of 80 μm x 1000 μm elements on 100 μm centers. In order to maximize spatial resolution, the image of the focal plane of a 3-m focal length mirror in the sampling arm was magnified 5 times and projected onto the array. Both the input and SBS far-field spots were projected side-by-side simultaneously on the same array with each shot, and thereby could be easily compared.

RESULTS AND DISCUSSION

The measured width of the far-field of the laser output showed the beam to be nearly diffraction-limited as seen in Fig. 1 (a). Due to the pointing instability of this single-shot laser, a precise quantitative measure of beam quality was difficult. However, an aperture corresponding to ~2 times the calculated Airy disk diameter at the magnified image plane was found to pass up to 70% of the beam energy and roughly half of the beam energy could be found within the calculated $1/e^2$ diameter.

A sodium chloride flat that had been wetted and re-polished to a shiny but textured finish was inserted in the beam path to act as a phase aberrator. The aberrator transmitted ~88% of the laser pulse on axis. Separate experiments showed that a double-pass of a good quality laser beam through it had the effect of decreasing the far-field peak intensity by a factor of ~4 and increasing the half-width of the central intensity lobe by ~50% (see the dashed line of Fig. 1). The aberrated beam was then focused into a cell containing xenon at 61 amagat. The resulting SBS reflection passed back through the NaCl aberrator and the near diffraction-limited far-field intensity profile of this reflected beam was found to be restored, as indicated by the top trace, Fig. 1 (b). The dashed line in Fig. 1 corresponds to the intensity profile observed using the sodium chloride aberrator and a plane mirror instead of the stimulated Brillouin conjugate.

REFERENCES

1. B. Ya. Zel'dovich, V.I. Popovichev, V.V. Ragul'skii, and F.F. Faizullov, Zh. Eksp. Teor. Fiz. Pis'ma Red. 15, 160 (1972) [Sov. Phys. JETP 15, 109 (1972)]; *Optical Phase Conjugation*, Robert A. Fisher, ed., (Academic Press, New York, 1983).

2. M.T. Duignan, B.J. Feldman, and W.T. Whitney, Opt. Lett. 12, (1987), in press.

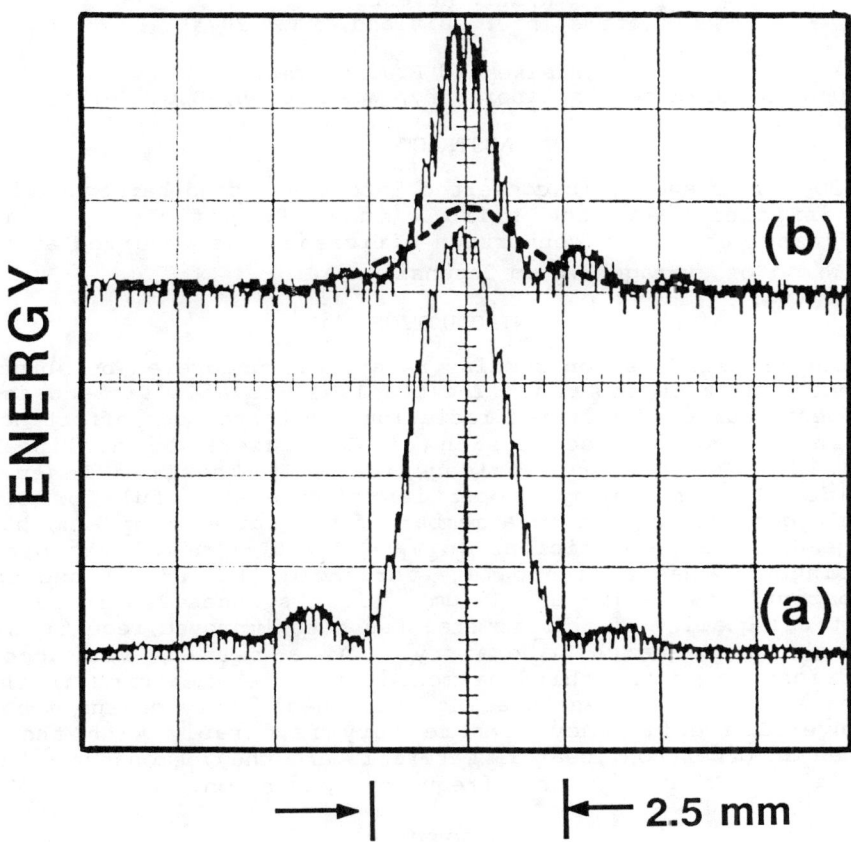

Figure 1. The display of the linear pyroelectric array indicating the far-field spatial profile of the output of the laser before aberration (a), and the profile of the SBS return. The 2.5 mm distance indicated corresponds roughly to the calculated diffraction-limited $1/e^2$ diameter. The intensity of (b) has not been normalized to (a). Note that the optical layout caused a left-right reversal between laser output and SBS in the target plane. The dashed line indicates the far-field profile observed using the same aberrator and a plane mirror instead of the SBS, with the intensity normalized to the relative energy in each.

EFFICIENT HARMONIC GENERATION OF CO_2 LASER RADIATION IN THALLIUM ARSENIC SELENIDE

R.C.Y. Auyeung
Geo-centers Inc., Suitland, Md. 20746

D.M. Zielke and B.J. Feldman
U.S. Naval Research Laboratory, Washington, D.C. 20375

ABSTRACT

CO_2 laser energy is converted into the second harmonic with ~28% efficiency and the third with ~5.5% efficiency. No significant change in conversion efficiency was observed at up to ~3 W/cm^2 of average 9.6 μm intensity.

INTRODUCTION

The infrared region remains relatively uncovered by strong coherent radiation sources. The CO_2 laser is an efficient and convenient source of infrared radiation and hence much effort has gone into frequency conversion of CO_2 lasers by nonlinear materials. The ternary chalcogenide salt Thallium Arsenic Selenide (TAS) has many properties that are useful for CO_2 harmonic generation. It is a member of the space group R3m, has a large nonlinear coefficient (d_{eff} = 40x10^{-12} m/V), exhibits birefringence, has a high damage threshold (1.7 J/cm^2) and is transparent between 1.3 and 17 μm. First synthesized in 1972, the optical quality of this crystal has been improved recently by Westinghouse Research Laboratory. As a result, very good conversions up to the third harmonic have been measured in the present study. It is encouraging that these early measurements of conversion efficiency compare very favourably with those attained in other nonlinear IR materials and they demonstrate the potential of TAS in CO_2 laser frequency conversion.

RESULTS

The TAS crystals were grown, polished and coated at Westinghouse Research Laboratory. The dimensions of the crystals used in these experiments were 2.5 cm in diameter and ~6 cm in length. The crystals were cut at an angle of ~19° with respect to the c-axis. The end surfaces have hygroscopic AR coatings at 9.6 and 4.8 μm and therefore were sealed with ZnSe windows which had non-hygroscopic AR coatings at these two wavelengths.

A CO_2 laser (Lumonics 822HP), operating on the 9.6 μm P(20) transition in the TEM$_{00}$ mode, provided ~300 mJ of linearly polarized output in a 50 ns FWHM pulse. Under Type I doubling conditions phase-matched at ~19°, 28% of the total fundamental energy was converted into the second harmonic at an incident fluence of 680 mJ/cm^2. This result is the highest second harmonic conversion of CO_2 laser radiation reported to date. A log-log plot of the second harmonic energy with that of the fundamental yields a least squares fit slope of 2.0, in excellent agreement with theory.

Third harmonic conversion was achieved by Type I sum frequency generation at a phase-matched angle of 20.6°. A special CdS waveplate ($7\lambda/2$ at $\lambda = 9.6$ μm and 8λ at $\lambda = 4.8$ μm) rotated the 9.6 μm polarization into coincidence with the 4.8 μm polarization before the two collinear beams entered a second crystal. At an incident 9.6 μm fluence of 370 mJ/cm^2, ~5.5% of the fundamental energy was converted into the third harmonic. If corrected for 3.2 μm losses at the crystal surface, this conversion efficiency would approach 10%. A log-log plot of the third harmonic energy with the fundamental energy yields a slope of ~2.7 in reasonable agreement with the cubic dependence predicted by theory.

The effect of increasing the average fundamental power incident on the crystal was examined by operating the CO_2 laser near 9 Hz. At an average intensity of a few watts/cm^2, no decrease in either the second or third harmonic conversion efficiency was observed. (The 9.6 μm pulse energy remained constant with repetition rate over this range).

CONCLUSIONS

This work has shown that high second and third harmonic conversion efficiencies are attainable in TAS. These conversion efficiencies do not decrease at a few watts/cm^2 of average 9.6 μm intensity and indicate the potential of TAS in high repetition rate conversion. The results in this study demonstrate TAS to be a promising new source of infrared radiation. Further improvement in the optical quality of these crystals is expected and should lead to higher conversion efficiencies and the development of a TAS-based optical parametric oscillator.

Radiative Trapping Effects in Ruby: 77°K to 300°K

Milton Birnbaum and Curtis L. Fincher
The Aerospace Corporation

Jason Machan and Michael Bass
Center for Laser Studies
University of Southern California

ABSTRACT

Observations of complete trapping of the 0-phonon fluorescence at 77°K and 300°K (a first) in ruby have resulted in determination of the fluorescent lifetime of the 0-phonon fluorescence (694 nm) and the vibronic fluorescence.

SUMMARY

Radiative trapping effects are important in the operation of a number of lasers. Ruby ($Cr^{3+}:Al_2O_3$) was utilized in this study because of the immediate availability of laser rods of varying size. The benchmark paper of Nelson and Sturge[1] was used in interpreting our results.

An estimate of the magnitude of the radiative trapping effect was obtained from the analysis of Holstein.[2] For a Gaussian line with a peak absorption coefficient of α_m, Holstein finds that in a long cylinder of radius, R, the lifetime is lengthened by the factor

$$\tau_{tr}/\tau_{utr} = 0.63\alpha_m R \left[\ln(\pi\alpha_m R/2)\right]^{1/2} \qquad (1)$$

where τ_{tr} = the trapped or lengthened lifetime and τ_{utr} = the untrapped lifetime. Eq. (1) predicts that an opaquely silvered ruby rod 5/8" diameter and 7.5" will show a large lifetime lengthening factor resulting in practically complete trapping of the 0-phonon lines of ruby.

The experimental arrangement was similar to that of Ref. 3. A cold finger arrangement was utilized for measurements of the smaller rubies (see Table 1) over the temperature range of 77-300°K. The larger rubies (see Table 1) were studied at 300°K only. All samples in this work had Cr^{3+} concentrations of ~ 0.05%. At 77°K, data was obtained with bare ruby rods and heavily silvered rubies. At 300°K, the rubies were packed tightly in a $BaSO_4$ powder which provided a diffuse reflectivity greater than 99.5%. The effects of amplified spontaneous emission were readily observed at 77°K and procedures were developed to circumvent this effect which results in reducing the observed lifetime. We utilized a computer controlled data processor which permitted signal averaging for improved accuracy in the lifetime determinations. The relevant measurements obtained and the derived lifetimes are summarized in Table 1.

Table 1. Lifetimes of Fluorescent Emissions in Ruby

Ruby No.	T °K	Dimensions (in.) diam. × length	τ_{obs} trapped (ms)	τ_m (ms)	τ_1 (ms)	τ_2 (ms)	τ_3
1	77	3/8 × 9/8	15.4	4.3	6.0	15.4	--
2	77	1/4 × 3/2	16.6	4.3	5.8	16.6	--
3	77	1/4 × 2	15.5	4.3	6.0	15.5	--
4	77	1/2 × 2	16.8	4.3	5.8	16.8	--
5	300	3/4 × 10	8.4	3.3	5.5	15.0	20
6	300	1 × 13	8.7	3.3	5.4	15.4	20

T = Temperature,
τ_{obs} = experimentally observed lifetime,
τ_m = measured lifetime in the thin sample limit[1] (trapping ≈ 0)
τ_1 = R-line lifetime, 0-phonon line,
τ_2 = R-vibronic lines lifetime,
τ_3 = non-radiative lifetime of the ^2E level, at 77° >> 50 ms[1].

Measurement of the trapped lifetimes (column 4) of rods 3 and 4 with and without silver coated sides (barrels) yielded similar values confirming that trapping factors (Eq. 1) were sufficient to provide the required trapping of the 0-phonon line fluorescence. The measured lifetime, τ_{obs}, is

$$1/\tau_{obs} = 1/\tau_1 + 1/\tau_2 + 1/\tau_3 \qquad (2)$$

At 77°K, $\tau_3 \gg \tau_2$ and for strong trapping of the 0-phonon line, $\tau_1 \gg \tau_2, \tau_3$; thus, $1/\tau_{obs} = 1/\tau_2$. At 300°K, with strong trapping of the 0-phonon line, $1/\tau_{obs} = 1/\tau_2 + 1/\tau_3$; i.e., $\tau_1 \gg \tau_2, \tau_3$. The spontaneous radiative lifetimes, τ_1, τ_2 are thus shown to be independent of temperature (Table 1) as expected. Explicit values of τ_1 and τ_2 have been obtained by utilizing an experimental arrangement which provided practically complete trapping of the 0-phonon radiation.

The superior performance of our $BaSO_4$ reflector,[4] in conjunction with the large ruby rods (5, 6 of Table 1),[5] enabled us to achieve complete trapping of the 0-phonon radiation at 300°K.[6]

References

1. D. F. Nelson and M. D. Sturge, "Relation Between Absorption and Emission in the Region of the R-Lines of Ruby," Phys. Rev., <u>137</u>, No. 4A, 15 Feb. 1965, A1117-A1130.

2. T. Holstein, "Imprisonment of Resonant Radiation in Gases," Phys. Rev., Vol. <u>172</u>, 1212 (1947).

3. M. Birnbaum, A. W. Tucker and C. L. Fincher, "CW Ruby Laser Pumped by an Argon Ion Laser," IEEE T. Quant. Electr., Vol. <u>QE-13</u>, No. 10, Oct. 1977, pp 808-9.

4. We acknowledge participation in these experiments by Jackson Jung, a Summer High School Fellow at USC.

5. Loaned to us by Dr. John McMahan, Optical Physics Division, Naval Research Lab., Wash., D.C.

6. We acknowledge the support of the Air Force Office of Scientific Research under AFOSR Grant No. AFOSR-84-0378.

A Study of Multiphoton Resonances in Kr and Ar under Intense Laser Field

J. J. Tiee, M. J. Ferris, and G. K. Anderson
Los Alamos National Laboratory, Los Alamos, NM. 87545

Abstract

Multiphoton ionization and vacuum ultraviolet (vuv) light generation through 3- and 4-photon resonances in Kr and Ar are investigated as a function of gas pressure and laser fluence. The observed spectral broadening and shifting in the MPI spectra near 3- and 4-photon resonances are interpreted as a result of the ac Stark effect and the competing harmonic generation processes. 4-photon resonances was observed to enhance and deplete third harmonic generation under certain pressure conditions. In addition, vuv light at wavelength shorter than the third harmonic wavelength was produced when several 4-photon resonances in Ar were excited.

Introduction

A competition between resonant multiphoton ionization (MPI) and third harmonic generation (THG) involving three-photon resonances was first reported in Xe by Miller et. al.[1] The process was later studied more extensively[2,3] and was reasonably well understood. Recently, a similar competing process involving 4-photon resonances was observed in the multiphoton ionization of Hg,[4] and it was concluded that the 4-photon resonances could enhance THG. In this paper, we report a similar observation in the 3- and 4-photon excitation spectra of Kr and Ar, and the detection of vacuum ultraviolet (vuv) light at wavelength shorter than the third harmonic wavelength when some 4-photon resonances of Ar are excited.

Experimental

The experimental apparatus is shown schematically in Fig 1. A Nd:Yag laser pumped dye laser was frequency mixed or doubled in order

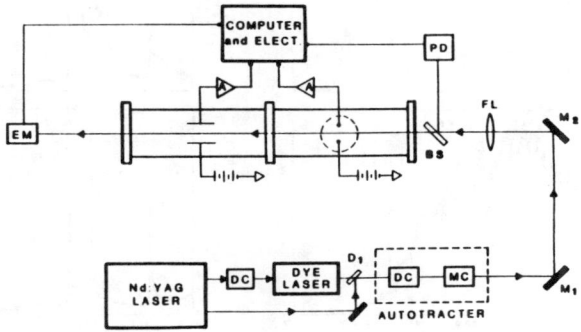

Fig.1 MPI and vuv photoionization apparatus

to generate the desired tunable light excitation in the 300 - 370 nm region. A typical laser output energy of 2 to 4 mj was focused into the experimental cell with a 10 cm focal length lens. The experimental cell was composed of two chambers separated by a LiF window: the first chamber with biased pin electrodes for detecting multiphoton ionization (MPI), and the second chamber was fitted with plate electrodes, and typically, it was filled with 0.5 torr of NO used for monitoring vuv light.

Results and Discussion

Both the multiphoton ionization and the vuv radiation generated in the focal volume were measured over a broad range of gas pressures (0.01 to 100 torr) and laser fluences (0.5 to 50 J/cm^2). The resonant ionization processes are five-photon ionization through either three-

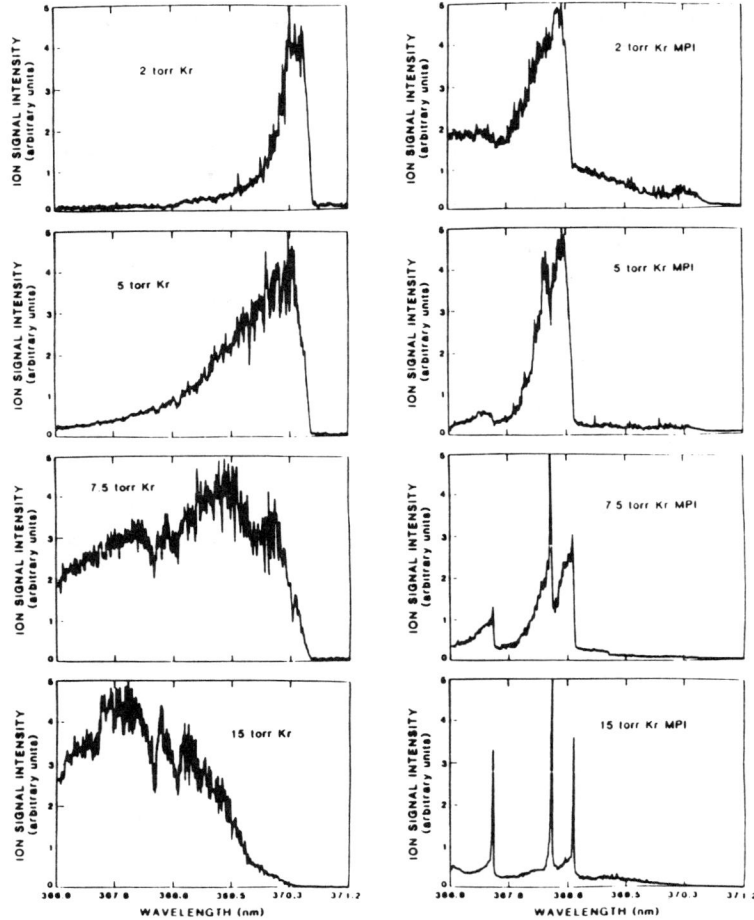

Fig. 2 (a) vuv light output and (b) MPI excitation spectra

or four-photon resonances. The 5s, 5f, and 6p' levels of Kr and the 4s, 4p', 6d', 7p, 7f, and 9p levels of Ar are investigated. Fig 2. shows excitation spectra of Kr at a few selected pressures by monitoring both the MPI and vuv light intensity. The spectral lines were observed to broaden and to shift from line center as a function of gas pressure and laser intensity. Under relatively higher pressure conditions (>5 torr), it appears that the observed effect was more dominated by the gas pressure change than that of the laser intensity. Only under low pressure conditions could the observed linewidth broadening be attributed predominantly to ac Stark Effect in non-homogeneous field. At higher pressures, the linewidth of three-photon resonant lines all became broader and more shifted toward the blue part of the spectrum. This is consistent with the conclusion that this is a result of the competition between MPI and THG, MPI dominates at low pressures, whereas THG becomes more important at high pressures and is dictated by the phase matching condition.

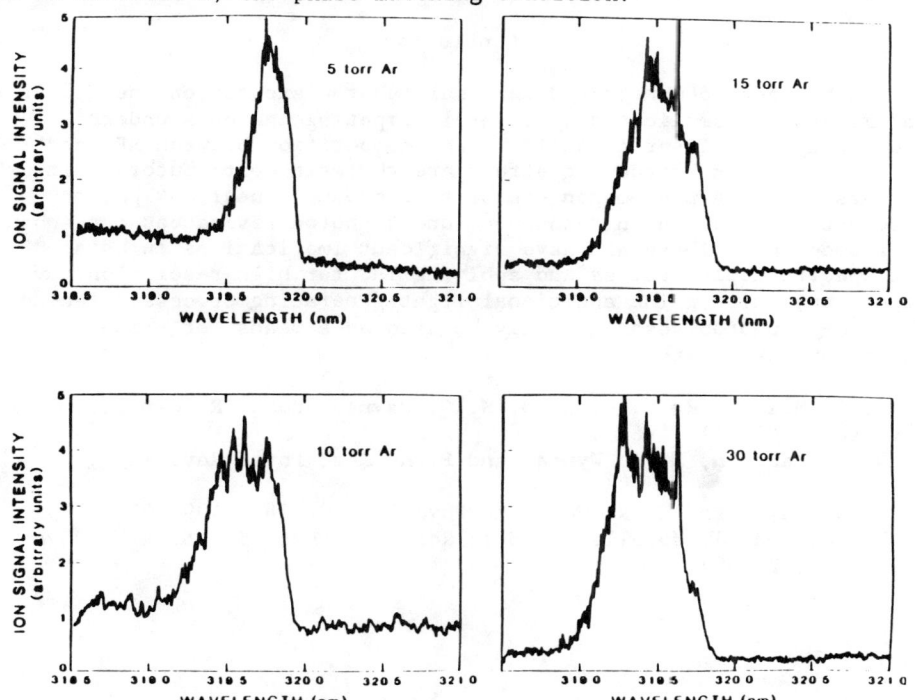

Fig. 3 The excitation spectra of Ar by monitoring vuv light output

It was observed that the four-photon resonant lines behave like the three-photon resonant lines at low pressures, however, differently at higher gas pressures. It was determined that at the intermediate pressures the broadening in MPI spectra was due partly to the occurrence of THG as it was enhanced by the four-photon resonances - a six-photon process, and perhaps more strongly to ac Stark effect.

However, it is not clear why the linewidth of the MPI spectra become narrower when the pressure is further increased. It is believed that space charge generated in the focal volume plays a role and could be a contributor to the observed result.

VUV light at wavelength shorter than the third harmonic wavelength was also detected at some of the 4-photon lines in Ar as shown in Fig 3 (the sharp lines). These spectral lines were observed only when the LiF window between the two chambers was removed, indicating that the vuv light wavelength is shorter than the LiF window cut-off wavelength. It is assumed tentatively that the vuv light originates from levels in the proximity of the excited 4-photon states, producing light at wavelength shorter than 100 nm. By examining the conditions under which the shorter wavelength light was generated, it was deduced that the process could occur only when the intense laser field was present. Detail investigations of this observation are currently underway in our laboratory.

Conclusion

It has been demonstrated that multiphoton excitation spectra of Kr and Ar can be complicated by several competing channels under moderately high laser intensity. The competition between MPI and THG, and the ac Stark broadening effect are the main contributors. It appears that the phenomenon can be as pronounced near a 4-photon resonance as a 3-photon resonance, and 4-photon resonances can enhance or reduce THG. These all have significant implications in using MPI for spectroscopic studies and applying THG for high-resolution vuv light generation. The additional light generating process detected for some 4-photon resonances may be used as a means for producing short vuv wavelength.

1. J. C. Miller, R. N. Compton, M. G. Payne, and W. R. Garrett, Phys. Rev. Lett. 45, 114(1980).
2. D. J. Jackson, J. J. Wynne, and P. H. Kes, Phys. Rev. A 28, 781(1983).
3. M. G. Payne and W. R. Garrett, Phys. Rev. A 28, 3409(1983).
4. D. Normand, J. Morellec, and J. Reif, J. Phys. B: At. Mol. Phys. 16, L227-232(1983).

SIMULTONS VERSUS RAMAN SOLITONS: DIFFRACTION AND PULSE SHAPE EFFECTS FROM A NUMERICAL EXPERIMENT POINT OF VIEW

F.P. Mattar#,† and J. DeLettrez#
Department of Physics,# New York University, New York, NY 10003
and George R. Harrison Spectroscopy Laboratory,†
Massachusetts Institute of Technology, Cambridge, MA 02139
and
J.P. Babuel-Peyrissac*, J.P. Marinier*, and C. Bardin**
Departement de Physico-Chimie* and Informatique Internationale**
Centre d'Etudes nucleaires de Saclay, 91191 Gif/Yvette, France

ABSTRACT

The purpose of this communication is to focus on *shape* as a determinant in nonlinear propagation of two beams in a three-level atomic system. As an example, we examined the distortionless propagation (as a function of pulse shape) of a coherent pulse in a two-level absorber known as a self-induced-transparency SIT [1] phenomenon. The Lorentzian pulse was found to propagate with a group velocity much smaller than that of the supergaussian. When transverse effects are included, one finds that the pulse shape effects the SIT self-focusing [2] as well.

We are concerned with the study of two pulses of comparable optical area propagating in a three-level system. It is useful to interpret (i) the observation of temporal coincidence of two-color superfluorescence [3]; and (ii) to understand the coherent exchange of energy between the Raman pump and a buildup Stokes of comparable Rabi frequencies [4].

This paper should be viewed as an extension of Konopnicki et al's simulton calculation. Before proceeding one should note that we and Konopnicki et al feel that the statement that numerical experiments have a narrower impact on understanding the underlying physics than would an analytical approach is only partially true. Any analytical approach requires that approximations be made. When the approximations ignore physical effects which are truly unimportant a valuable insight results. However, when the physical problem is such that analytical progress can be made only by ignoring important physical effects, the analytical result can be of doubtful value, or even mislead. In much of the two-field three-level work which has been studied numerically, analytic results already existed. Such results cannot be compared with experiment, because the approximations necessary to carry out the analysis also mutilated the physics. Our approach should be viewed as a numerical experiment.

The concomitant propagation of two coherent pulses in a three-level medium was first analyzed by Konopnicki and Eberly [5]. They discovered that under particular conditions, (which restrict the initial level populations by the ratio of oscillator strengths and the applicable Rabi frequencies), simultaneous solitons with equal group velocity may result. The results they reported were both analytical and numerical and reflected a lossless evolution in a uniform plane wave regime. This result is consistent with our physical intuition, since according to Brewer and Hahn's symmetrization [6], the two-laser three-level interaction reduces to an effective single field interacting with a two-level atom.

The two pulses will propagate with equal group velocity and without any distortion. The pulses will preserve their shape while being transmitted through the nonlinear medium which is prepared in a specific population scheme. One can qualify the *simul*taneous propagation of two different wavelength soli*tons* of comparative strength (the *simultons*), as the first example of Raman solitons. However, one ought to stress the intrinsic differences between the two processes. Although they are different, the simultons do appear ephemerally in the nonlinear evolution of the Stokes in the Raman amplifiers. More explicitly, the two pump and Stokes pulses will have, at some propagation distances, comparable Rabi frequencies and time-integrated areas of the same order of magnitude. The energy transfer from one beam to the other is significantly larger for the Raman amplifier (from the initially-strong depleting pump to the initially-weak building up Stokes) than it is for the simultons case (where the two comparable-time-integrated-area solitons preserve their lossless and distortion free character).

On one hand, the simultons' pulses must have their Beer's length equal, and they interact with a three-level medium whose atomic levels are partially populated (which makes *both* active transitions absorptive). On the other hand, the Raman pump and the Stokes seed are at the input plane of distinctively different optical areas and experience an initially grounded three-level atom with quite different oscillator strengths. The pump experiences an absorption while the Stokes sees an amplification.

Moreoever, since Newstein et al [7] in their study of transverse effects in SIT predicted an on-resonant single SIT self-focusing in a two-level absorber that was verified [8] by independent observations in sodium, in neon and iodine, it is expected that in this two-field three-level coherent interaction the interplay of diffraction and the medium inertial response will inevitably redistribute the beam energy both spatially and temporally. This interplay is characterized by F_g, the Beer's length Fresnel number that is associated with Beer's length α^{-1}. One would thus expect that even for identical transitions and identical beams the distortionless character is lost. However, the two beam energy temporal- and spatial-coincidence remains unaltered.

Furthermore, as reported by Mattar et al [9], when diffraction is accounted for in the simulton propagation, their shape-preserving characteristic is at least destroyed (under specific constraints such as unequal Fresnel number and equal Beer's length) while the temporal coincidence is maintained. The two beams' profiles will get distorted simultaneously. In particular, they have studied the role of transverse effect for the simulton evolution for two restricted physical situations: (i) the Beer's length nonlinearities are equal while the transitions are not identical, i.e., $\mu_a \neq \mu_b$ and $\lambda_a \neq \lambda_b$, $\alpha_a = \alpha_b$, (this leads to $\lambda_a^{-1} \mu_a^2 = \lambda_b^{-1} \mu_b^2$ and unequal Beer's length

Fresnel number); (ii) the Beer's length are unequal: $\alpha_a \neq \alpha_b$, but with equal Beer's length Fresnel numbers: $F_{ga} = F_{gb}$ (which leads to $\mu_a/\lambda_a = \mu_b/\lambda_b$). The resultant interplay of free space diffraction with the medium nonlinearity for beams larger than unity Fresnel number is identical for the two beams. Co-propagation and reshaping lead to self-action phenomena that occurs simultaneously for each of the moderately diffracting beams (where the Fresnel numbers are larger than unity). The synchronization prevails even though the beam transverse structure gets distorted. Both beams experience self-focusing or self-defocusing simultaneously. This result is a generalization of KE simulton condition since synchronization is maintained.

Subsequently, calculations with either (a) the equal Beer's length Fresnel for strong diffraction (i.e., both Fresnel numbers F_a and F_b are smaller than one-half and unequal) or (b) the most general transverse variations have been carried out. Both the shape preservations and the synchronization characteristics are eliminated. That is, for more general initial conditions, the beams not only reshape transversely, but they do it independently, i.e., without maintaining any temporal- or spatial-synchronization. One of the beams goes through a self-defocusing while the second one is still in the process of focusing. Both characteristics of the simulton disappear. Moreover, calculations for non-gaussian beam profiles have shown that diffraction can be enhanced or reduced. Thus, both pulse and beam shapes are effective in changing the departure from the soliton features.

On the other hand, for the Raman soliton [10] process, the diffraction is an absolute necessity. Without it the nonlinear gain cannot be compensated for. The formation of the soliton is conditioned to the occurence of a linear balancing loss.

To complement our understanding, we have calculated the influence of the longitudinal pulse shape on the propagation characteristics. The response of the atomic medium is shown to depend on whether or not the input pulse shape is conserved when transformed into the frequency domain. The group velocity of a Lorentzian or sub-Lorentzian pulse is significantly different from that of a hyperbolic secant or a hyper-gaussian pulse. Moreover, the differential velocity across the beam varies for one shape to create a different relative motion between the two lasers.

The additional contribution of the pulse shape is unambiguously illustrated through the spatial dependence of the power spectrum in the planar geometry (see Figure 1) and through the radial profile of the time-integrated energy, i.e., the fluence for two different pulse shapes (see Figure 2). The light-matter interaction must be carried out in the frequency domain as well. A family of universal parametric curves in both the physical (r,z,t) and momentum (k_t,z,w) spaces have been compiled and will be published elsewhere

The numerical algorithm can be found in reference [11].

This work was sponsored by NSF, ARO/AFOSR and ONR.

REFERENCES

[1] S.L. McCall and E.L. Hahn, Phys. Rev. 183, 457 (1969)
[2] N. Wright and M.C. Newstein, Opt. Commun. 1, 8 (1973), F.P. Mattar and M.C. Newstein, Opt. Commun. 18, 70 (1976).
[3] R. Florian, L.O. Schwan, and D. Schmid, Phys. Rev. A23, 2709 (1984); F. Haake and R. Reibold, Phys. Rev. A29, 3208 (1984); and F. Haake, R. Reibold, and F.P. Mattar, J. Opt. Soc. Am. 81, 547 (1984).
[4] F.P. Mattar, SPIE, 540, 588 (1985).
[5] M.J. Konopnicki and J.H. Eberly, Phys. Rev. A24, 2567 (1981).
[6] R Brewer and E.L. Hahn, Phys. Rev. A11, 1641 (1975).
[7] F.P. Mattar and M.C. Newstein, IEEE, J. Quantum Electron. QE-13, 505 (1977).
[8] (a) F.P. Mattar, M.C. Newstein, P.E. Serafim, H.M. Gibbs, B. Bölger, G. Forster and P.E. Toschek in *Coherence and Quantum Optics IV*, ed. L. Mandel and E. Wolf (Plenum, 1978) p. 143-164, (b) H.M. Gibbs, B. Bölger, F.P. Mattar, M.C. Newstein, G. Forster and P.E. Toschek, Phys. Rev. Lett. 37, 1743 (1976); and J.J. Bannister, H.J. Baker, T.A. King and W.G. McNaught, Phys. Rev. Lett. 44, 1062 (1980).
[9] F.P. Mattar et al, in *Advances in Laser Science II*, ed. W.C. Stwalley and M. Lapp (Am. Inst. Phys., to be published).
[10] E.J. Robinson and F.P. Mattar, J. Opt. Soc. Am. B3, 170 (1986).
[11] F.P. Mattar, Appl. Phys. 17, 53 (1978); F.P. Mattar and M.C. Newstein, Comp. Phys. Commun. 20, 139 (1980).

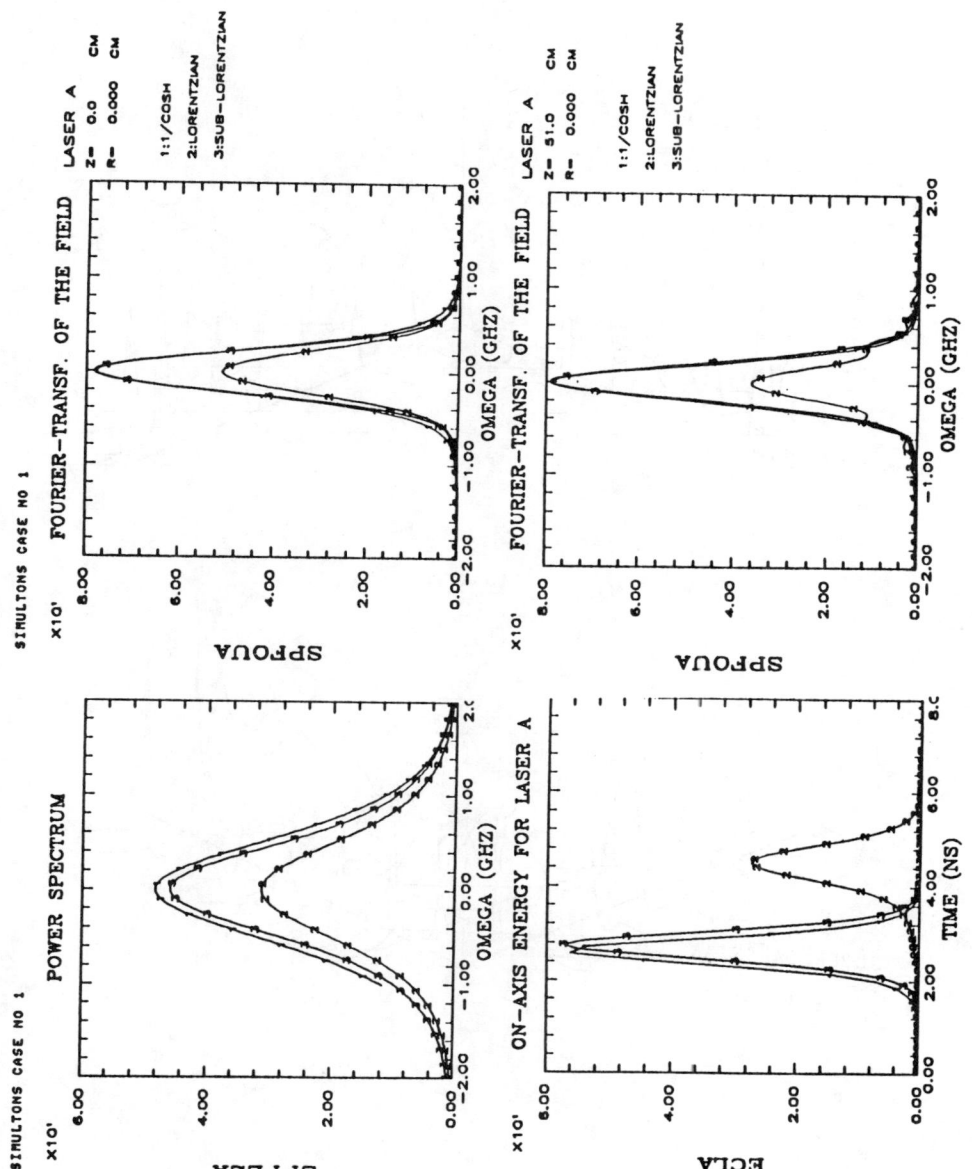

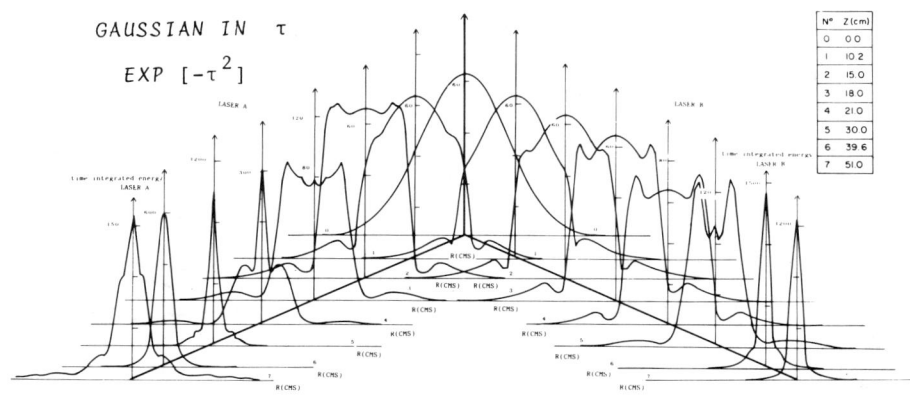
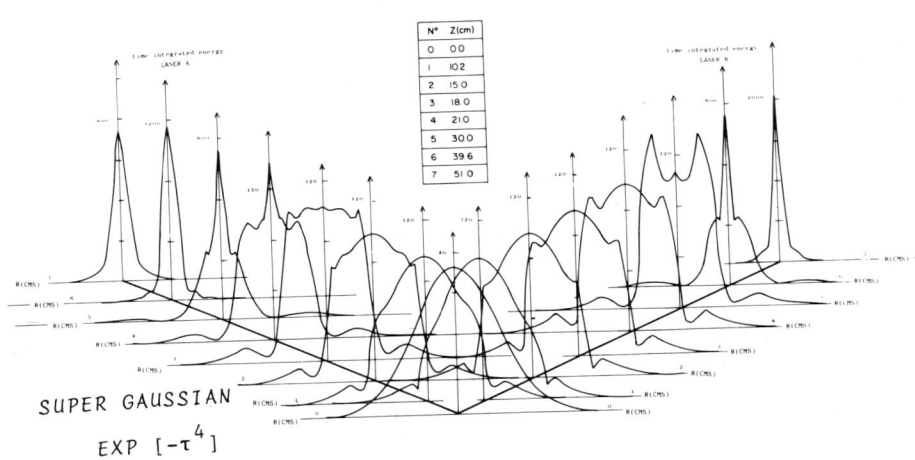

SIMULTON SUMMARY

Since

2 fields, 3 level medium ⟶ | Brewer and Hahn's symmetrization | ⟶ effective one field, 2 level atom that can propagate à la McCall and Hahn's SIT

Thus

plausibility of simultons:
shape-preservation and temporal-coincidence resulting from equal group velocity

subjected to the availability of the following initial conditions:
    i - pulses of comparable area and Rabi frequencies
   ii - ratio of Beer's length comparable to unity
  iii - partial population which make the active transitions absorptive

However

SIT ⟶ | Transverse effects ∇_T^2 | ⟶ Coherent SIT self-focusing a la Newstein et al

Therefore

Simultons distortionless character disappear while they can remain locked to each other as they propagate. In general, both distortionless and temporal- or spatial-coincidence disappears.

DIRECT MEASUREMENT OF NONLINEAR ENERGY DEPOSITION FROM AN INTENSE 532 nm PHOTON FIELD INTO ALKALI HALIDES

Scott C. Jones, Xiao-An Shen, and Peter Braunlich
Department of Physics, Washington State University, Pullman, WA 99164

Paul Kelly
Division of Physics, National Research Council, Ottawa, Ontario K1A OR6

ABSTRACT

Prebreakdown temperature increases exceeding 300 K in NaCl exposed to 80 psec pulses at 532 nm are reported along with four-photon absorption cross sections in NaCl and KBr.

INTRODUCTION

The primary interaction between a wide-gap insulator and high intensity laser pulse has been debated for a number of years. The argument has centered around the question of at what order N (the number of photons required to bridge the forbidden gap at a given wavelength) does the generation of conduction band carriers cease to be dominated by the simultaneous absorption of N laser photons by valence electrons, and become largely due to electron impact ionization. In this paper we present evidence of four-photon absorption in NaCl (8.6 eV bandgap) for nondamaging pulses derived from a frequency-doubled Nd:YAG laser (2.33 eV photon energy). Our findings show that dramatic heating occurs in the focal volume of the sample without indicating the formation of impact ionization avalanche, thereby extending to N = 4 the regime where multiphoton absorption is the primary process of energy deposition.

PHOTOACOUSTIC MEASUREMENTS

We use the photoacoustic method to obtain the measure of the absolute energy deposition in the sample. It is known that an acoustic pulse has an amplitude that is proportional to the absorbed energy,[1] so that calibration of the acoustic detector is accomplished by measuring the signal generated by a known energy absorption in a known interaction volume. In our case this is done by determining the photoacoustic signal due to two-photon absorption in NaCl at 266 nm, where the deposited energy is obtained by measuring pulse attenuation. Success of the photoacoustic method depends critically on the quality of the NaCl crystals. Consistent results were obtained only from crystals whose OH⁻ contamination had been significantly reduced by reactive-atmosphere processing. We have repeatedly measured prebreakdown signals in five different crystals from different specially purified boules. Calibrated data are shown in Fig. 1 for two crystals from the boule exhibiting the highest damage threshold. All data points represent signals from nondamaging laser pulses. To avoid the inclusion of data measured at or beyond breakdown, all data points were rechecked for reproducibility at fluxes lower than the highest indicated.

The temperatures indicated in Fig. 1 are those calculated for the geometrical focal point of the laser spot in the bulk of the crystal. The temperature is obtained from the known total absorbed energy and the spatial distribution calculated from

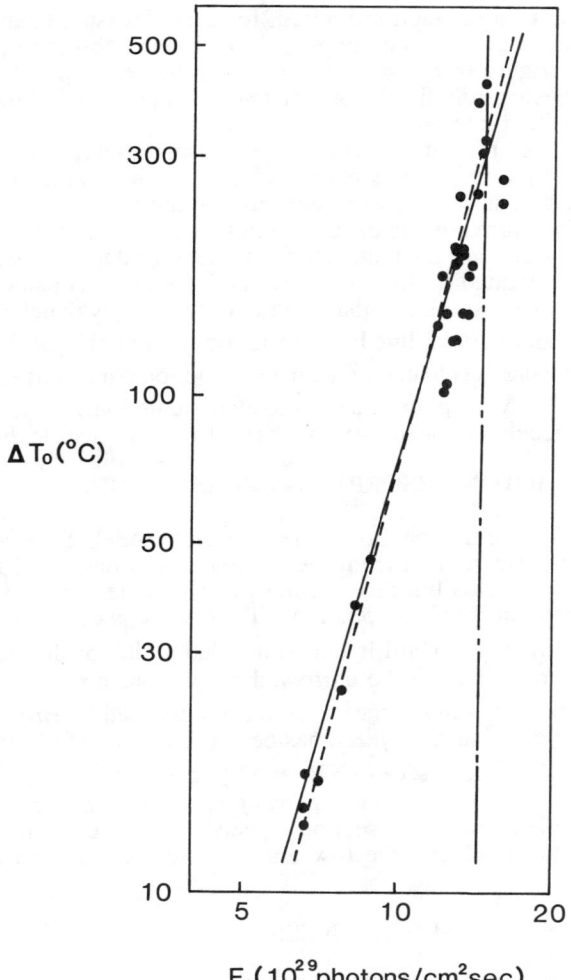

Figure 1. Temperature rise at focal point (ΔT_o) vs peak laser flux (F_p) in NaCl exposed to 85 psec 532 nm laser pulses. The solid line is a fit to the data for the four-photon-polaron-defect model with $\sigma^{(4)} \approx 2 \times 10^{-113}$ cm^{8}sec^3. The dashed line is the fit for "free electron" heating in the four-photon model with $\sigma^{(4)} \approx 1.5 \times 10^{-114}$ cm^{8}sec^3. The slope of both fits on the double-logarithmic plot is approximately four, indicating four-photon absorption is the primary mechanism for energy deposition. The nearly vertical line (——— — ———) is a calculation of the avalanche model using a starting conduction band density of $\approx 10^{11}$ cm^{-3}, with parameters chosen to fit the highest prebreakdown temperature obtained.

fourth-order absorption of a diffraction-limited, focused Gaussian beam. In any case, the reported temperatures are simply proportional to the absorbed energy, so the vertical axis in the figure is a rescaling of the absorbed energy. The slope of nearly four on the double logarithmic plot of temperature vs peak flux of the incident laser pulse indicates the four-photon nature of the process.

Model calculations of the total interaction are also presented in the figure. Secondary, single-photon absorption by excited electrons, either in the conduction band or in transient defect states, are also registered in the photoacoustic signals. The two models closely approximating the data represent an overall process initiated by four-photon valence electron excitation. One of these is a calculation wherein the conduction carriers are treated as free polarons, and the other considers energy absorption by free electrons, similar to that used to derive the avalanche theory of laser damage.[2] The nearly vertical line is the result of a calculation of the electron impact ionization avalanche mechanism,[2] with a conduction band starting electron density of 10^{11} cm^{-3}. Multiphoton-assisted avalanche ionization was also considered. All such models are inconsistent with the data, as is clear in the figure.

FOUR-PHOTON ABSORPTION CROSS SECTIONS

The validity of either the free polaron or free electron model of the conduction band carriers cannot be ascertained from the photoacoustic data, and therefore, model interpretations of the data limit the accuracy of the value of the four-photon absorption cross section in NaCl at 532 nm. Thus, we can conclude only that $1 \leq \sigma^{(4)} \leq 20 \times 10^{-114}$ cm^{8}sec^3. Until it can be decided which of the free carrier models is correct, this range cannot be narrowed by photoacoustic experiments. However, the self-trapped exciton luminescence technique, used by Brost et al.[3] to obtain the three-photon cross section in KI, has been used successfully to obtain a value of $\sigma^{(4)} = (2\pm 1) \times 10^{-112}$ cm^{8}sec^3 in KBr at 532 nm. Due to the peculiarities of self-trapped exciton luminescence among the various alkali halides, however, this approach has not yet yielded a cross section in NaCl, as our current cryogenic apparatus is inadequate to achieve the low temperatures required to make the measurements.

REFERENCES

1. A. C. Tam, Rev. Mod. Phys. **58**, 381 (1986).
2. A. S. Epifanov, Sov. Phys. JETP **40**, 897 (1975).
3. G. Brost, P. Braunlich, and P. Kelly, Phys. Rev. **B30**, 4675 (1984).

Degenerate Stimulated Parametric Scattering in $LiNbO_3$:Fe

G. Y. Zhang, S. M. Liu, Z. K. Wu
Department of Physics, Nankai University, Tianjin, P. R. China

Q. X. Li, P. P. Ho, and R. R. Alfano
Institute for Ultrafast Spectroscopy and Lasers, Departments of Physics
and Electrical Engineering, The City College of New York, N. Y. 10031

Abstract

Polarization-anisotropic photoinduced light scattering with a large scattered angle $\sim 47°$ caused by degenerate stimulated parametric forward four-wave mixing of waves with orthogonal polarization was first observed in $LiNbO_3$:Fe by illuminating the crystal with an ordinary polarized krypton ion laser beam at 414 nm wavelength.

Several forms of photoinduced light scattering were observed from $LiNbO_3$:Fe crystal, when it was illuminated with a coherent light beam[1-8]. Most studies have been performed for polarization-isotropic photoinduced light scattering of extraordinary wave[1,2,4,7]. When an ordinary wave is incident, a small-angle dispersing light scattering in the direction perpendicular to the polar axis of the crystal with the form of extraordinary wave have been observed[3,5,6]. It was suggested[5] that the photoinduced light scattering with orthogonal polarization change arises from recording the photorefractive noise phase grating (PNPG) by the excitation of specific spatially oscillating photovaltaic current in the crystal. Recently, Odoulov et al[9] claimed that degenerate stimulated parametric scattering (DSPS) in $LiTaO_3$:Cu has been observed. The scattered light was located in a cone pattern where apex angle is defined by the phase matching condition for forward four-wave mixing of ordinary and extraordinary polarized light. In this paper, we report on the observation of the strong degenerate stimulated forward parametric four-wave scattering in a $LiNbO_3$:Fe crystal at a well-defined angle using ordinary polarized light and its relative competition process with the scattering caused by PNPG.

In our experiments, an ordinary polarized krypton ion laser beam at 414 nm normally illuminated X- or Y-cut samples of $LiNbO_3$:Fe crystal (0.08 wt%) with thickness 1 mm. Two types of extraordinary wave photoinduced scattering patterns were observed under various conditions. One type of the scattered light was the small-angle light scattering (PNPG) with angular spread from 2° to 20°. This type of light scattering has been previously observed[3,5,6]. The second is a new type which occurs at higher illuminaing power level and is located at a larger scattering angle spread from 44° to 49°.

In Fig. 1, the intensity dependence of both types of the photoinduced scattering patterns are displayed. These photographs of scattering patterns were taken by illuminating the crystal with krypton ion laser beam at different power levels. In Fig. 1a, the incident laser power was set at 100 mw and only the small-angle scattering pattern (PNPG) occurred. As the laser power was increased to 200 mw, the large-angle scattering (DSPS) appeared, as shown in Fig. 1b. Increasing the laser power to 300 mw, the large-angle scattering became stronger and the small-angle scattering disappeared. It is clear that the larger power of the illumination beam is more favorable in DSPS. Decreasing the laser power, the competition process reversed. The competition process also depends on the illuminating beam size. A larger illumination beam size results in a favorable competition of the DSPS. In addition, when the illuminating beam size in crystal was very small, for example, 12 μm, neither DSPS nor PNPG was induced.

Fig. 1. Photographs of the incident laser intensity dependence of the competition process between the PNPG and DSPS: (a) P_{in}= 100 mw; (b) P_{in}= 200 mw; (c) P_{in}= 300 mw. θ_s- scattering angle. Beam spot size at sample is 0.4 mm. The middle up-down scattering in fig. is ordinary PNPG. The 2°-20° side scattering is extraordinary PNPG; the side scattering at $\theta_s \sim 47^o$ DSPS.

The large-angle scattering in a $LiNbO_3$:Fe crystal is similar to the DSPS in a $LiTaO_3$:Cu crystal caused by forward four- wave mixing of waves with ordinary and extraordinary polarizations[9]. The uniqueness of the narrow scattering angular window of DSPS indicated the necessity of the phase-matching condition.

In Fig. 2, a proposed vector diagram for the formation of DSPS process is presented. Fig. 2a displays the formation of the initial weak noise phase grating in the crystal due to the interference of the incident beam $\vec{K}_{oi}$ with weak scattered beam: $\vec{K}_{os}^{(1)}$, $\vec{K}_{os}^{(2)}$, $\vec{K}_{os}^{(3)}$, ... originating from imperfections in crystal where $\vec{K}_{oi}$, $\vec{K}_{os}^{(1)}$, $\vec{K}_{os}^{(2)}$, $\vec{K}_{os}^{(3)}$, ... represent the ordinary wave vectors of the incident beam and the scattered beams, respectively. Vectors $\vec{K}^{(i)}$ are defined to be the recorded noise phase grating vectors as $\vec{K}_{os}^{(i)} - \vec{K}_{oi}$, where i = 1,2,...,N. They are all terminated at the sphere surface S with radius equal to $\vec{K}_{oi} = 2\pi n_o/\lambda$, where n_o is the ordinary refractive index of the crystal. In Fig. 2b, when one of these noise phase gratings in Fig. 2a has met the phase-matching condition, the orthogonal polarization scattering occurs through the four-wave mixing, where $\vec{K}_{es}$ is the wave vector of the extraordinary scattered wave and $\vec{K}_g = \vec{K}_{oi} - \vec{K}_{os}$ is the phase-matching grating vector. This four-wave mixing process of orthogonal polarization waves, $2\vec{K}_{oi}$ and $2\vec{K}_{es}$, in turn increases the amplitude of the phase-matching grating due to the standard photorefractive origin (δx_{13}, δx_{31}, δx_{23} and δx_{32} are nonzero)[8]. This process will continue to amplify the phase-matching grating. Therefore, the extraordinary wave scattering at a well-defined scattering angle gradually increases until it is saturated. Due to the termination of $\vec{K}_g$ at the sphere surface S the phase matching condition of the vectors $\vec{K}_g$, $2\vec{K}_{oi}$ and $2\vec{K}_{es}$ constitutes a right triangle. From this relationship, the scattering angle is derived as following:

$$\theta_s = \sin^{-1}[\, n_e/n_o(n_o^2 - n_e^2)^{1/2}\,]. \qquad (1)$$

Using $n_o = 2.4223$ and $n_e = 2.3103$ at $\lambda = 414$ nm in equation (1), the calculated scattering angle $\theta_s = 44°$, which agrees well with the experimental result.

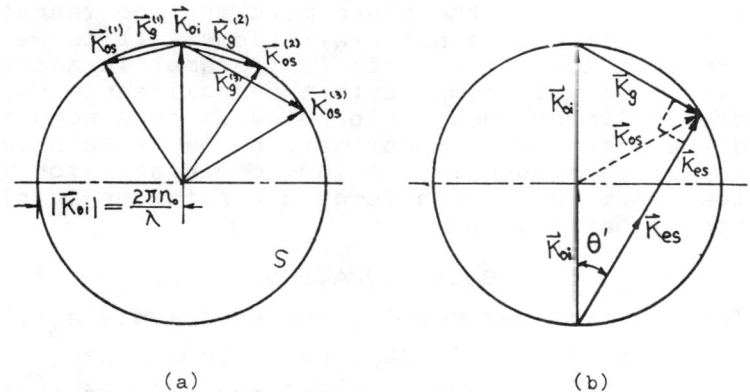

(a)　　　　　　　　　　(b)

Fig. 2. Hypothetical vector diagrams of DSPS. (a) Formation of noise phase gratings $\vec{K}_g^{(i)}$; (b) Phase-matching condition of DSPS.

References

1. W.Phillips, J.J.Amodei and D.L.Staebler, RCA Rev. 33, 94 (1972).
2. R.Magnussen and T.Gaylord, Appl. Opt. 13, 1545 (1974).
3. E.M.Avakyan, S.A.Alaverdyan, D.G.Belabaev, V.K.Sarkisov and D.M.Tumanyan, Sov. Phys. Solid State 20, 1401 (1978).
4. L.F.Kanoev, V.K.Malinovsky and B.I.Sturman, Opt. Commun. 34, 95 (1983).
5. E.M.Avakyan, K.G.Belabaev and S.G.Odoulov, Sov. Phys. Solid State 25, 1980 (1983).
6. S.Liu, G.Zhang, Z.Wu, G.Li, S.Feng, J.Zhang, J.Zhao and L.Xu, Acta Phys. Sin. 33, 105 (1984) (Chinese Physics 4, 593 (1984)).
7. G.Zhang, Q.Li, P.Ho, S.Liu, Z.Wu, and R.Alfano, Appl. Opt. 25, 2955 (1986).
8. D.A.Temple and C.Warde, J.Opt. Soc. Am. B3, 337 (1986).
9. S.Odoulov, K.Belabaev and Z.Kiseleva, Opt. Lett. 10, 31 (1985).

THIRD HARMONIC GENERATION WITH HIGH EFFICIENCY

R. Fischer
Zentralinstitut fuer Optik und Spektroskopie,
Akademie der Wissenschaften der DDR,
1199 Berlin, GDR

ABSTRACT

It is shown by numerical analysis that in plane-wave approximation, taking into account the depletion of the pump wave, it is possible to nearly compensate a nonlinear mismatch if one neglects the dispersion of the third-order susceptibility. The dispersion of the nonlinear susceptibility can lower the maximum possible efficiency.

INTRODUCTION

For direct third harmonic generation (THG) in cubic-nonlinear media the influence of intensity-dependent parts of the refractive index must be taken into account because the corresponding susceptibilities for the parametric process and the process of self-action are of the same order. The intensity dependence of the index of refraction may destroy the phase matching and therefore lower the maximum attainable efficiency. It is well known that at low efficiencies (in parametric approximation) it is easy to compensate the nonlinear mismatch by a proper linear one. The problem is more complicated beyond the parametric approximation.[1,2,3] We have calculated for this case the maximum efficiency for THG in loss-less media under quasistatic conditions in plane-wave approximation.

BASIC EQUATIONS

The relation determining the efficiency $a_3(z)/a_{10}$ (a_j - real amplitudes) in dependence on the length z of the nonlinear medium and the amplitude a_{10} of the incident fundamental wave is given by

$$\frac{z}{L_{NL}} = \int_0^{a_3(z)/a_{10}} \frac{dx}{\left[(1-x^2)^3 - x^2\left(\frac{\Delta k \cdot L_{NL}}{2} + \frac{\alpha}{2\chi_3} + \frac{\beta-\alpha}{4\chi_3}x^2\right)^2\right]^{1/2}} \quad ,(1)$$

where $L_{NL} = \dfrac{cn}{6\pi\omega\chi_3 a_{10}^2}$ is the nonlinear interaction length,

$\Delta k = k_3 - k_1$ the linear mismatching and $\alpha = 6\chi_{31} - 3\chi_{11}$, $\beta = 3\chi_{33} - 6\chi_{31}$. The χ_{ij} are the (effective) Fourier components for harmonic generation (χ_3) and self-action:

$$\chi_3 = \vec{e}_3 \cdot \chi^{(3)}(-3\omega; \omega, \omega, \omega) : \vec{e}_1 \vec{e}_1 \vec{e}_1 ,$$

$$\chi_{11} = \vec{e}_1 \cdot \chi^{(3)}(-\omega; \omega, \omega, -\omega) : \vec{e}_1 \vec{e}_1 \vec{e}_1 , \quad \chi_{33} = \vec{e}_3 \cdot \chi^{(3)}(-3\omega; 3\omega, 3\omega, -3\omega) : \vec{e}_3 \vec{e}_3 \vec{e}_3 ,$$

$$\chi_{13} = \vec{e}_1 \cdot \chi^{(3)}(-\omega; \omega, 3\omega, -3\omega) : \vec{e}_1 \vec{e}_3 \vec{e}_3 \equiv \chi_{31} = \vec{e}_3 \cdot \chi^{(3)}(-3\omega; 3\omega, \omega, -\omega) : \vec{e}_3 \vec{e}_1 \vec{e}_1.$$

The maximum possible efficiency is found from Eq. (1) as the smallest real root of the cubic equation in the dominator of Eq. (1) (or one can make use of the corresponding analytical solution given in ref.[3]). If there is no dispersion of the third-order susceptibility, one has $\alpha = -\beta = 3\chi_3$. To discuss the influence of the dispersion of the third-order susceptibility, it is convenient to define $\alpha = 3\chi_3(1+a)$, $\beta = -3\chi_3(1+b)$; i.e. a = b = 0 means no dispersion.

RESULTS

By choosing a proper linear mismatch it is possible to almost compensate the nonlinear mismatch and so to attain a higher efficiency in comparison with the case $\Delta k = 0$.

No dispersion of the third-order susceptibility: In this case the efficiency is determined only by one parameter $\Delta k \cdot L_{NL/2}$. Whereas for $\Delta k = 0$ the maximum

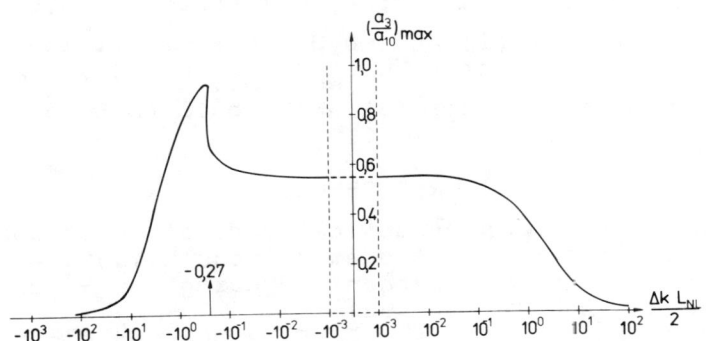

Fig. 1. Dependence of the maximum efficiency on $\Delta k \cdot L_{NL}$ for a=b=0.

efficiency (defined here as the ratio of amplitudes) is 0.55, for an optimum linear mismatch $(\Delta k)_{opt} = -0.27 \cdot 2/L_{NL}$ one gets 0.92 as maximum attainable efficiency (Fig.1).

Influence of the dispersion of the third-order susceptibility: In Fig. 2 the real roots of the cubic equation in the dominator of Eq. (1) are shown in dependence on $\Delta k \cdot L_{NL}/2$ for different a and b.

The maximum attainable efficiency is given by the smallest root. This Fig. shows that the dispersion lowers the maximum attainable efficiency.

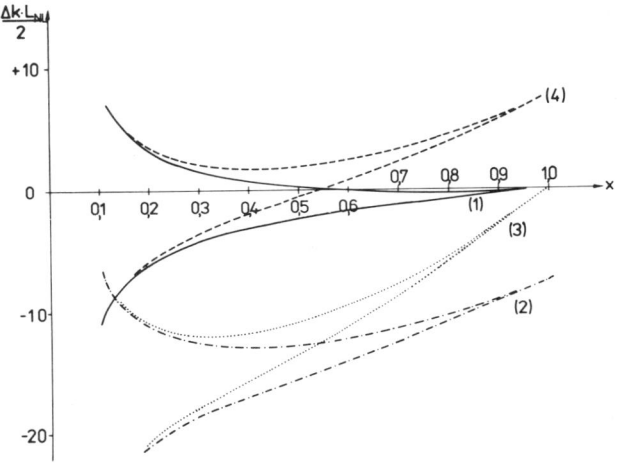

Fig. 2. Dependence of the real roots on $\Delta k \cdot L_{NL}$ for different a,b: (1) a=0, b=0 (maximum attainable efficiency 0.92 at $(\Delta k \cdot L_{NL}/2)_{opt} = -0.27$); (2) a=10, b=0 (0.67 at -13.2); (3) a=10, b=10 (0.56 at -12.2); (4) a=0, b=10 (0.67 at 1.8)

REFERENCES

1. J.F.Reintjes,Nonlinear optical parametric processes in liquids and gases (Academic Press,N.Y.,1984),ch.4
2. J.A.Armstrong,N.Bloembergen,J.Ducuing and P.Pershan, Phys. Rev. 122, 1918 (1962)
3. H.P.Puell and C.R.Vildal, IEEE J. Quantum Electron. QE-14, 364 (1978)

Intensity Dependence of the Polarization Tensor in a Quantum Treatment

N. Chencinski* and A. N. Weiszmann
College of Staten Island of CUNY, N.Y. 10301

ABSTRACT

An exact quantum mechanical expression for the polarization tensor is obtained, the photon number (intensity) dependence of the component oscillating at ω_p frequency is presented and the line shapes analyzed.

The third order nonlinear susceptibility $\chi^{(3)}$ has been used in the treatment of four wave mixing problems[1]. Recently an exact semiclassical treatment has been added to the purtubative density matrix operator approach [2]. We consider a quantum mechanical solution for $\chi^{(3)}$ using a Hamiltonian formulation [3].

A five level system interacting with four coherent signals is considered. The states $|g\rangle$, $|t\rangle$, $|k\rangle$, $|j\rangle$ and $|g'\rangle$ are dressed atomic states with fields in photon number representation[3]. All one photon exchanges between the levels are allowed in any order. The t-g and k-j one photon transitions are forbidden. Normal resonances are the material resonance of the Na doublet achieved only by one photon transitions, while extra resonances are two photon transitions between the split levels. States $|1\rangle$ and $|5\rangle$ have the same atomic state, g, but differ in photon number. The Hamiltonian is fully quantized with the interaction term given by

$$H' = \hbar \sum_{K=a,b,c,p} [K(J_- a_K^+ + J_- a_K) + \chi^*(J_+ a_K^+ + J_+ a_K)]$$

The form of the Hamiltonian matrix is different for Bloembergen's diagram 31, 32, and 33 following the correct time sequence, cabp, acbp, and cbap respectively. The matrix elements for diagram 31 are:

$\langle 1|2\rangle = \hbar i \mu_{jg} \sqrt{n_c+1}\, e^{i\phi_c}; \quad \langle 2|3\rangle = \quad g k \sqrt{n_a}\, e^{i\phi_a};$

$\langle 3|4\rangle = \hbar i \mu_{tj} \sqrt{n_b}\, e^{i\phi_b}; \quad \langle 4|5\rangle = i\, kt\sqrt{n_p+1}\, e^{i\phi_p};$

$\langle 1|5\rangle = \hbar \mu_{gg'}; \quad \langle 2|2\rangle = \omega_{gj} + \omega_c; \quad \langle 3|3\rangle = \omega_{kj} - \omega_a + \omega_c;$

$\langle 4|4\rangle = \omega_{kt} - \omega_p$ with zeros and complex conjugates completing the matrix.

* Supported by PSC-CUNY Research Award Program

An exact solution of the Schroedinger equation for the eigenvalues λ_i and eigenvectors a_{ij} has been obtained. The expectation value for the time-dependent polarization vector has been calculated, and the exact component for the term oscillating at frequency ω_p was extracted:

$$<P(t)> = \sum_{k,j=1}^{5} A^*_{kj}(t) A_{ij}(t) \mu_{ki}(0)$$

$$P(\omega_p) = \mu_{32} \sum_{\ell,m=1}^{5} a^*_{3\ell} a^{*'}_{\ell 5} a_{2m} a'_{m5}$$

where $A_{ij} = <\psi|\psi(t)> = \sum a_{ik} a_{kj} e^{-i\lambda_k t/\hbar}$ $i,j=1,\ldots,5+\text{C.C.}$

The lowest order of the $P(\omega_p)$ expression again exactly recovers the 31, 32, 33, Bloembergen terms.

Computer simulated plots for the polarization oscillating at ω_p, are presented using the 3S-3P transition in Sodium. The transition frequencies are : $\omega_{tg} = 0.059$ cm^{-1}; $\omega_{kg} = 16956$ cm^{-1}; $\omega_{jg} = 16973$ cm^{-1} with $\omega_a = \omega_b = 16961$ cm^{-1}; $\Delta\omega_a = -5$ cm^{-1} while ω_c is varied. The plot shown in fig. 1 illustrates some of the features of the quantum mechanical solution for low photon numbers $n_a = 15$, $n_c = 3$, $n_p = 0$. From this and plots with varying values of n_a, n_c, and $\Delta\omega_a$ we conclude:
1) All resonances are present and remain finite for all values of ω_c.
2) The resonance line is shifted and split for both weak and strong fields similar to the AC start effect.
3) Broadening and merging of the peaks, 'stirring', occurs as n_c/n_a increase to 1.
4) New extra resonances appear for small n_a's due to the fields taken to all orders.
5) By changing the ratio of n_c/n_a we find that the relative amplitudes of the split peaks as well as the ratio of the extra resonance to normal resonance peak changes. By example for $n_a = 15$, $n_c = 1$ and $\Delta\omega_a = -5$ cm^{-1} the split peak at 16977 cm^{-1} is greater than the peak at 16979 cm^{-1}.
6) The ratio of the peak heights is also a function of the detuning.
7) Higher resolution calculation show the resonance peaks as the envelope of higher frequencies similar to infinite Fourier series behavior.

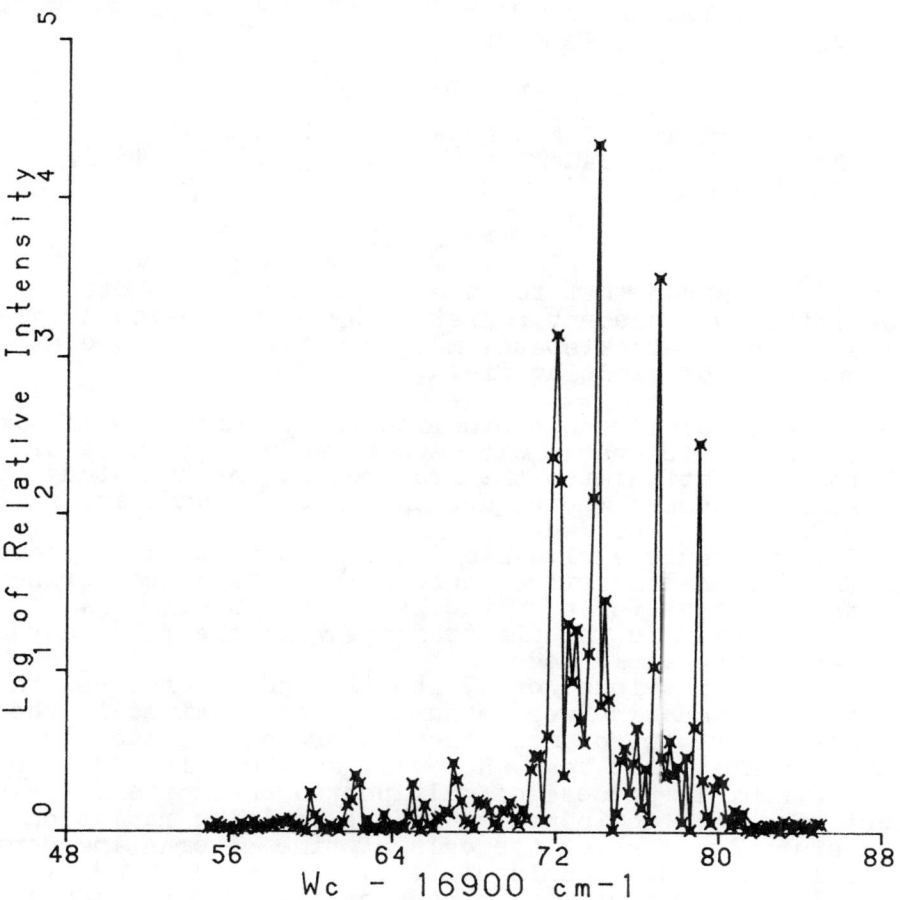

Fig. 1 Plot of Log (1+P (ω_p)) vs ω_c

REFERENCES

1. N. Bloembergen and Y. R. Shen, Phys. Rev. A37,133,(1964).
   N. Bloembergen, H. Lotem and R. T. Linch Jr., Ind. Journ. Pure and Appl. Phys. 16,1151,(1978).
   T. U. Yee and T.V. Gustafson, Phys. Rev. A18,1597,(1978).
   Y. Prior, IEEE J. Quant. Elec. QE20,37,(1984).
2. W.M.Schreiber, N.Chencinski, A.M.Levine, and A.N.Weiszman Methods of Laser Spectroscopy ed. Y. Prior et al (Plenum Press, New York, 289,(1986).
3. Y.Prior and A.N. Weiszmann, Phys. Rev. A29 2700,(1984).

TRANSFORMATION OF COHERENT STATES OF ELECTROMAGNETIC FIELD TO QUASI-FOCK STATES

V. I. Zakharov

The Institute of Atmospheric Optics, Siberian Branch, USSR Academy of Sciences, Tomsk, 634055, U.S.S.R.

ABSTRACT

It is shown that for the process of two-photon absorption of coherent radiation by open two-level systems the quasi-Fock states may be obtained for the strong attenuation of exciting field.

The recently shown experimental possibility of obtaining the field states with sub-Poisson statistics of photons [1] stimulates the researchers for the study of optical processes where quantum noise of radiation could be effectively reduced.

In the states with sub-Poisson statistics of photons their antibunching effect occurs which is characterized by the negative value of the parameter $Q = <\Delta n^2>/<n> - 1$, where $<\Delta n^2>$ is the dispersion, $<n>$ is the mean photon number in the field mode.

In the previous works, see [2] and references, the states of such kind were assumed to be obtained in the process where two or more photons are simultaneously removed from a light beam. However, as shown in [3] for the reversible process of multiphoton absorption, the effect of photon antibunching in the exciting radiation appears only in some intervals of time alternating periodically with the bunching effect.

In the present work a possibility of reduction of quantum photon fluctuations of cw laser radiation is discussed. This effect can appear as a result of two-photon absorption using a resonance transition in atom excited from the ground state $|0\rangle$ to the upper level $|2\rangle$ relaxing mainly to the other levels. It may be $3S \rightarrow 5S$ and $3S \rightarrow 4D$ two-photon transitions in Na atom, which are pumped by one mode resonance field.

In order to understand the physical nature of occurrence of the effect in the case considered, we represent the incident field as a mixture of Fock states with various number of photons $|n = 0\rangle$, $|n = 1\rangle$, $|n = 2\rangle$, ... It is evident that for the strong attenuation of exciting field only the states $|n = 0\rangle$ and $|n = 1\rangle$ remain in the field, since the resonance absorption of these states by the atoms considered is forbidden by the law of energy conservation. On the contrary, the states with the photon number $n \geq 2$ are continuously reduced because the photons are emitted to the other field modes after relaxing the upper level of the system. Moreover, the Fock-states $|2n\rangle$ of exciting field are transformed to $|n = 0\rangle$ and $|2n +1\rangle$ to $|n = 1\rangle$. Therefore, the probabilities of appearing of the Fock-states $|n = 0\rangle$ and $|n = 1\rangle$ in the mode of exciting field for the case of coherent initial pumping are :

$$P^K(|n=0\rangle) = \sum_{K=0} e^{-\langle n \rangle} \frac{\langle n \rangle^{2K}}{(2K)!} = e^{-\langle n \rangle} ch\langle n \rangle;$$

$$P^K(|n=1\rangle) = \sum_{K=0} e^{-\langle n \rangle} \frac{\langle n \rangle^{2K+1}}{(2K+1)!} = e^{-\langle n \rangle} sh\langle n \rangle;$$ (1)

and for the case of thermal initial field are :

$$P^T(|n=0\rangle) = \sum_{K=0} \frac{\langle n \rangle^{2K}}{(1+\langle n \rangle)^{2K+1}} = \frac{1+\langle n \rangle}{1+2\langle n \rangle};$$

$$P^T(|n=1\rangle) = \sum_{K=0} \frac{\langle n \rangle^{2K+1}}{(1+\langle n \rangle)^{2K+2}} = \frac{\langle n \rangle}{1+2\langle n \rangle};$$ (2)

The parameter Q is negative both for coherent and thermal sources.

$$Q^K = -\frac{1}{2} + \frac{1}{2} e^{-2\langle n \rangle}$$

$$Q^T = -\frac{\langle n \rangle}{1+2\langle n \rangle}$$ (3)

We have $Q^K \simeq -1/2$ and $Q^T \simeq -1/2$ for the initial mean number of photons $\langle n \rangle \gg 1$. Differences for coherent and thermal sources are only in the frequency intervals where quantum noise is effectively reduced.

The presence of chaotic orientation of spins of the photons in the exciting field may give the additional noise for transformed radiation. Really, the transition 3S→5S is excited only by pairs of photons ($\lambda \simeq 6022$ Å) with antiparallel spins orientation. Therefore, in the transformed radiation all Fock-states of photons with parallel spins orientation are conserved. On the contrary, at the excitation of transition 3S→4D the pairs of photons ($\lambda \simeq 5787$ Å) with parallel spins orientation are removed from the exciting field. As a result, in this case the Fock state $|n = 2\rangle$ with antiparallel spins orientation appears in the transformed field. Therefore, the second case is more perspective for the production of states of the field where the quantum noise is strongly reduced at some frequency intervals.

CONCLUSION

In conclusion attention should be paid to the fact that if the S→S transition frequency of one type atoms is equal to S→D transition of other type atoms then the gas mixture of these atoms can serve as an effective reducer of photon fluctuations of passing through the gas mixture radiation. Only it is necessary to create the conditions of resonance two-photon field interaction with gas when the strong attenuation of passing radiation takes place. In this case, only the Fock states of photon $|n = 0\rangle$ and $|n = 1\rangle$ in transformed field are conserved independently from orientation of spins of the photons in the initial field. Therefore, for the initial mean photon number $\langle n \rangle \gg 1$ (intense field) the quantum noise is maximally reduced and the parameter Q, in this case, may be equal to $Q \simeq -1/2$.

Acknowledgements

The author thanks Drs. V.M.Mitchenkov, V.G.Sokovikov, A.S.Shelekhov and especially V.M.Klimkin and A.Z.Fazliev for their helpful suggestions and fruitful discussions.

REFERENCES

1. R. Short and L. Mandel, Phys. Rev. Lett., 51, 384 (1983).
2. H. Paul, Rev. Mod. Phys. 35, 710 (1982).
3. V. I. Zakharov, Vl. G. Tyuterev, Dep. VINITI, N5619-84 (1984); Journ. Opt. Soc. Amer., 2B, 387 (1985); Laser and Particle Beams (1987), to be published.

Application of the Split Operator Fourier Transform method to the Solution of the Nonlinear Schrödinger Equation

Paul L. DeVries
Department of Physics, Miami University, Oxford, Ohio 45056

The nonlinear Schrödinger (NLS) equation arises in the description of a variety of physical processes, including the evolution of optical pulses in single-mode fibers.[1] In this note, we discuss the application of the Split Operator Fourier Transform (SOFT) method to the nonlinear equation

$$i\frac{\partial \psi}{\partial t} = \frac{\partial^2 \psi}{\partial x^2} + 2\,\psi^*\psi\,\psi \tag{1}$$

which has been particularly well studied as an example of a soliton-producing equation that is exactly solvable by the inverse scattering technique, if the initial condition $\psi(x,0)$ vanishes sufficiently fast at large $|x|$. The numerical solution of (1) has also received considerable attention, both with finite difference methods and with methods employing Fourier transforms. An excellent review of these methods was recently presented by Taha and Ablowitz,[2] who tested them on various soliton propagation problems. They conclude that the split step Fourier method of Hardin and Tappert[3] is the best method for the NLS equation, in the sense of obtaining a given level of accuracy with the least amount of computing time. We will demonstrate that the SOFT method, originally developed by Fleck, Morris, and Feit,[4] can be extended to the nonlinear Schrödinger equation and that it is superior to the split step method of Hardin and Tappert. (A similar method has recently been used by Agrawal and Potasek[5] to investigate nonlinear pulse distortion in optical fibers.)

THE NUMERICAL METHOD

Many physically significant problems are described by the linear Schrödinger equation

$$i\frac{\partial \psi}{\partial t} = -\frac{\partial^2 \psi}{\partial x^2} + V(x)\psi \tag{2}$$

with a potential $V(x)$ that is independent of time. The solution to this equation can be written as

$$\psi(x, t+\delta_t) = e^{-i\left(-\frac{\partial^2}{\partial x^2}+V(x)\right)\delta_t}\psi(x,t). \tag{3}$$

Although attractive to write the exponential factor as

$$e^{i\frac{\partial^2}{\partial x^2}\delta_t - iV\delta_t} = e^{i\frac{\partial^2}{\partial x^2}\delta_t}e^{-iV\delta_t}, \tag{4}$$

this is not an equality unless the derivative operator and the potential commute. The correct expression relating exponential operators, known as the Baker-Cambell-Hausdorf (BCH) theorem, is that

$$e^{\mathbf{A}}e^{\mathbf{B}} = e^{\mathbf{C}} \tag{5}$$

if and only if

$$\mathbf{C} = \mathbf{A} + \mathbf{B} + [\mathbf{A},\mathbf{B}] + \cdots. \tag{6}$$

© American Institute of Physics 1987

Using this theorem, the error in (4) is found to be of $O(\delta_t^2)$. A better approximation can be obtained by writing the exponential operator of (3) as

$$e^{i\frac{\partial^2}{\partial x^2}\delta_t - iV\delta_t} \approx e^{i\frac{\partial^2}{\partial x^2}\delta_t/2} e^{-iV\delta_t} e^{i\frac{\partial^2}{\partial x^2}\delta_t/2}. \tag{7}$$

Splitting the operator in this symmetric way leads to cancellation of the first commutator term, so that the leading error-contributing term is eliminated — the error in (7) is only of $O(\delta_t^3)$. This critical step was first demonstrated by Fleck et al.

The exponential of the derivative operator is a formidable object when expressed in coordinate space, but is easily evaluated in the Fourier transform space. Beginning with the expression for the transform of a derivative, one quickly finds that

$$e^{d^2/dx^2}\psi = \mathcal{F}^{-1}\big[e^{-k^2}\mathcal{F}[\psi(x,t)]\big] \tag{8}$$

where $\mathcal{F}$ denotes the Fourier transform and k is the variable conjugate to x. The time evolution of the wavefunction is thus accomplished as

$$\psi(x, t + \delta_t) = e^{i\frac{\partial^2}{\partial x^2}\delta_t/2} e^{-iV\delta_t} e^{i\frac{\partial^2}{\partial x^2}\delta_t/2} \psi(x,t). \tag{9}$$

This can be interpreted as a half time-step propagation with the derivative operator, a full time-step propagation with the potential, and a final half time-step propagation with the derivative operator.

Returning to the original problem, we observe that equation (2) is nearly identical to (1), with the potential replaced by $2|\psi(x,t)|^2$. Without further discussion, we propose to use the SOFT method for the nonlinear Schrödinger equation after making the obvious substitutions. Thus the solution is first propagated a half time-step via the derivative operator:

$$\hat{\psi}(x,t) = \mathcal{F}^{-1}\big[e^{-ik^2\delta_t/2}\mathcal{F}[\psi(x,t)]\big]. \tag{10}$$

The next step is to propagate with the "potential" term; note that $\hat{\psi}(x,t)$, which has just been evaluated, and not $\psi(x,t)$ is used in this step:

$$\tilde{\psi}(x,t) = e^{-2i|\hat{\psi}(x,t)|^2\delta_t}\hat{\psi}(x,t). \tag{11}$$

Finally, the wavefunction at $t + \delta_t$ is obtained by a second half time-step with the derivative operator:

$$\psi(x, t+\delta_t) = \mathcal{F}^{-1}\big[e^{-ik^2\delta_t/2}\mathcal{F}[\tilde{\psi}(x,t)]\big]. \tag{12}$$

RESULTS

Taha and Ablowitz investigated both one and two soliton problems with various initial conditions, comparing such methods as the classical explicit method, the hopscotch method, the Crank-Nicolson method, the Ablowitz-Ladik method, and the pseudospectral method of Fornberg and Whitham, before concluding that the split step Fourier method was the best. Since the primary purpose of this poster is to demonstrate that the SOFT method is applicable to these problems, our numerical investigations are not extensive. We compare only to the split step Fourier method, and limit our investigation to the one soliton problem specified by the wavefunction

$$\psi(x,t) = e^{-i[2x - 3t + \pi/2]}\,\text{sech}(x - 4t)$$

on the infinite interval, which is an exact solution to (1). We take this as the initial condition at time $t = 0$, and at subsequent times compare the numerical solutions obtained by the SOFT method and by the split step Fourier method to this solution by evaluating the average root mean square (RMS) error. In all calculations, the wavefunction was propagated from $t = 0$ to $t = 1.0$, and the Fourier transforms required by both methods were evaluated with the FFT algorithm on a 128 point grid on the interval $-20 \leq x \leq 20$.

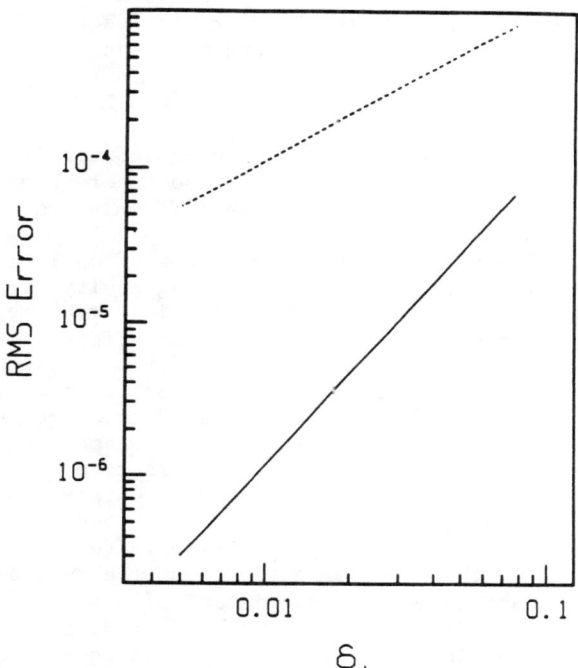

FIGURE 1. The average RMS error in the calculated wavefunction is plotted versus the step size for both the SOFT method (solid line) and the split step method (dashed line). For a time step of 0.02, used in Table I of Taha and Ablowitz, the error in the SOFT method is more than an order of magnitude less than for the split step method.

The essential results of these investigations are presented in Figure 1, in which the RMS error is plotted as a function of the time step δ_t. The figure leaves little doubt as to which method is superior in terms of accuracy. Noting the slope of the error versus time step curves on this logarithmic plot, we find that the SOFT method is more rapidly convergent as well. We conclude that the SOFT method is superior to the split step method and, by induction, to all the methods investigated by Taha and Ablowitz. With these results in hand, we anticipate that the SOFT method will rapidly become an important tool in the investigation of nonlinear problems.

This research is supported by the donors of the Petroleum Research Fund, administered by the American Chemical Society, and by the National Science Foundation under grant PHY-85-06679.

REFERENCES

1. W.J. Tomlinson, R.H. Stolen, and C.V. Shank, *J. Opt. Soc. Am. B* **1**, 139 (1984).
2. T.R. Taha and M.J. Ablowitz, *J. Comput. Phys.* **55**, 203 (1984).
3. R.H. Hardin and F.D. Tappert, *SIAM Rev. Chronicle* **15**, 423 (1973).
4. J. Fleck, J.K. Morris, and M.J. Feit, *Appl. Phys.* **10**, 129 (1976).
5. G.P. Agrawal and M.J. Potasek, *Phys. Rev. A* **33**, 1765 (1986).

THE EFFECTS OF DIFFRACTION, DISPERSION, AND TRANSIENT PROPAGATION ON OPTICAL BISTABILITY IN A ONE-DIRECTIONAL RING CAVITY

F.P. Mattar*,#, J. Teichmann*, Y. Claude* and C. Goutier*
Department of Physics*, New York University, New York, NY 10003
and George R. Harrison Spectroscopy Laboratory#,
Massachusetts Institute of Technology, Cambridge, MA 02139

ABSTRACT

We extend a model of optical bistability in a unidirectional nonlinear ring cavity to include coherent transients, diffraction, self-phase modulation, feedback mirrors, and atomic dynamics.

One of the prevailing models of optical bistability [1] is the uni-directional nonlinear ring cavity first suggested and analyzed by Bonifacio and Lugiato [2] in a treatment which is restricted to one-dimensional propagation effects. The role of diffraction has been examined by Maloney et al [3] in a CW simulation and in a more elaborate analysis by LeBerre et al [4] although still under a transient condition restricted to an adiabatic approximation for the medium. Subsequently, it became apparent that the simultaneous presence of temporal and diffractive effects is important, as pointed out by Gibbs and co-workers [5]. Accordingly, we have analyzed the model again with the goal of meeting these reservations and extending the analysis to the general case.

In this re-examination, the theoretical model is semi-classical in approximation, consisting of a paraxial wave equation coupled to the two-level density matrix which, for simplicity only, assumes a sharp-line system. The material equations allow for detuning and thus admit mixed absorptive and dispersive bistability. The cavity itself is included by satisfying boundary conditions at the feedback mirrors.

The methodology of solution, which represents the primary contribution of this study, is based on recognizing and exploiting the power of a computational program [6] formulated to treat the problem of self-induced-transparency (SIT) in an earlier study [7a]. In the adaptation to the present problem, in which we admit either CW or transient propagation, the SIT code has the virtue of already including the complete atomic dynamics for an absorbing two-level medium and *diffraction coupling* which automatically includes self-phase modulation. *Spatial averaging* is also expected to influence the propagation in a pre-excited medium. As in superfluorescence study [7b], the initial inversion density is radially dependent since the pump has a gaussian-like profile. This feature causes radially nonuniform Beer's length and leads to a differential refraction from shell to shell across the beam. The input signal may be either a single pulse or train of pulses whose separation may be any factor of the round trip time. Thus the interplay of diffraction and the inertial response of the medium in optical bistability under quite general conditions can be studied directly for the first time. For completeness we note that the program also possesses both an implicit

Bloch solver [8] for large detuning and short relaxation times, and a provision to include frequency broadening and quantum fluctuations if desired.

The free space propagation of the field between the mirrors is efficiently calculated for small Fresnel numbers with a modification of the quasi-Fast Hankel Transform [9a-e] previously reported as a finite Hankel Fourier-Bessel series [9f]), while for large Fresnel numbers a Hermite-Gaussian à la Kogelnik and Li [10] is used. We find that in the intermediate range, only a higher order finite difference method with non-uniform meshes yields the desired accuracy.

We have calculated a range of effects which comprise the dynamic properties of pulse propagation, pulse generation, and pulsed optical bistability for finite beams in the uni-directional configuration. Specific examples in which one can see how feedback and multiple passes affect the evolution of self-induced-transparency in a resonant absorber and vary the growth of super-radiance emission in an amplifying medium have also been calculated.

The problem has been examined for a variety of ring cavities with different lengths ℓ such that the round-trip time τ_{rt} can be less than, comparable to, or much greater than the input pulse length τ_p or the medium relaxation times T_1 and T_2 (with $T_1 = 2T_2 = 20\tau_p$). In all cases, the dependence is studied as a function of beam widths r_p and absorption lengths (α_{eff}) of the signal.

We should like to add as a final note that the analytical methodology is richer in capability than may be apparent from our examples and that we are also able to examine the influence of cavity properties on experimental results [11] and to probe the effects of quantum statistics [12].

This work was supported by NSF, ARO/AFOSR and ONR.

REFERENCES

[1] (a) H.M. Gibbs, *Optical Bistability: Controlling Light with Light*, Academic Press (1985); (b) L.A. Lugiato, in *Progress in Optics*, Vol. XXXI, ed. E. Wolf, North-Holland Pub. (1984), p. 69; (c) J.A. Goldstone, in *Laser Handbook IV*, ed. M. Stitch and M. Bass, North-Holland Pub. (1984); and (d) H.J. Carmichael, in *Optical Instabilities*, ed. R.W. Boyd, M.G. Raymer, L.M. Narducci (Cambridge Univ. Press, 1986) p. 111.
[2] L.A. Lugiato, Contemp. Phys. 24, 333 (1983).
[3] (a) J.V. Maloneu, M.R. Belic and H.M. Gibbs, Opt. Commun. 41, 379 (1982); (b) J.V. Maloney and H.M. Gibbs (i) Phys. Rev. Lett. 48, 1607 (1982) and (ii) Appl. Phys. B28, 100 (1982); and (c) J.V. Maloney, M. Sargent III and H.M. Gibbs, Opt. Commun. 44, 289 (1982).
[4] (a) M. LeBerre, E. Ressayre, A. Tallet, F.A. Hopf, H.M. Gibbs and J.V. Maloney, post-deadline paper, DD-AS IQEC '84; and (b) M. LeBerre, E. Ressayre, A. Tallet, K. Tai and H.M. Gibbs, IEEE J. Quantum Electron. QE-21, 1404 (1985).
[5] (a) J.F. Valley, H.M. Gibbs, M.W. Derstine, R. Pon, K. Tai, M. LeBerre, E. Ressayre and A. Tallet, in *Optical Bistability III*, ed. H.M. Gibbs, P. Mandel, N. Peyghambarian and S.D.

Smith (Springer-Verlag, 1986) p. 327; (b) H.M. Gibbs, M.W. Derstine, K. Tai, J.F. Valley, J.V. Maloney, F.A. Hopf, M. LeBerre, E. Ressayre and A. Tallet, in *Optical Instabilities III*, p. 340 (1986); (c) M. LeBerre, E. Ressayre, A. Tallet and H.M. Gibbs, in *Optical Instabilities III*, p. 312 (1986) and (d) J.V. Maloney, in *Optical Instabilities III*, p. 315 (1986).

[6] (a) F.P. Mattar, Appl. Phys. $\underline{17}$, 53 (1968); (b) F.P. Mattar and M.C. Newstein, Comp. Phys. Commun. $\underline{20}$, 139 (1980), and ibid $\underline{32}$, 225 (1984); and (c) F.P. Mattar, in *Optical Bistability*, ed. C.M. Bowden, M. Ciftan and H.R. Robl, Plenum Press (1981), p. 503-555.

[7] (a) F.P. Mattar and M.C. Newstein, IEEE J. Quantum Electron. $\underline{QE-13}$, 507 (1977) and F.P. Mattar, M.C. Newstein, P.E. Serafim, H.M. Gibbs, B. Bölger, G. Forster and P.E. Toschek, in *Coherence and Quantum Optics IV*, ed. L. Mandel and E. Wolf, (Plenum Press 1978) p. 143-164; (b) F.P. Mattar, H.M. Gibbs, S.L. McCall and M.S. Feld, Phys. R-v. Lett. $\underline{46}$, 1121 (1981); and (c) E.A. Watson, H.M. Gibbs, F.P. Mattar, M. Cormier, Y. Claude, S.L. McCall and M.S. Feld, Phys. Rev. $\underline{A27}$, 1427 (1983).

[8] (a) F.P. Mattar and J.H. Eberly, in *Laser Induced Process in Molecules: Physics and Chemistry*, ed. by K.L. Kompa and S.D. Smith, (Springer-Verlag 1979), p. 61; and (b) B.R. Suydam and F.P. Mattar, *Refereed Proceedings of the Los Alamos Conference on Optics '83*, publ. SPIE, vol. 380, p. 433 (1983).

[9] (a) A.E. Siegman, Opt. Lett. $\underline{1}$, 13 (1977); (b) M. Lax, J.H. Batteh and G.P. Agrawal, J. of Appl. Phys., $\underline{52}$, 109 (1981); (c) E.M. Wright, W.J. Firth and I. Galbraith, J. Opt. Soc. Am. $\underline{B2}$, 383 (1985); (d) W.J. Firth and E.M. Wright, Opt. Commun. $\underline{40}$, 233 (1982) and Phys. Lett. $\underline{92}$, 211 (1982); (e) D. Weaire, J.P. Kermode and V.M. Dwyer, Opt. Commun. $\underline{55}$, 223 (1985); and (f) C. Bardin, J.-P. Babuel-Peyrissac, J.P. Marinier and F.P. Mattar, SPIE, $\underline{540}$ 581 (1985).

[10] (a) H. Kogelnik and T. Li, Appl. Optics $\underline{5}$, 1550 (1966); (b) A. Yariv, *Quantum Electronics*, (John Wiley, 1975) Chapt. 6.9 p. 118 and Chapt. 6.5 p. 109; (c) A.E. Siegman and E. Szieklas Appl. Opt. $\underline{13}$, 2775 (1974) and ibid $\underline{14}$, 1874 (1974) and Proceeding of IEEE $\underline{62}$, 410 (1974); (d) A.E. Siegman, Opt. Commun. $\underline{31}$, 369 (1979); (e) B.J. Coffey, M. Lax and C.J. Elliot, IEEE J. of Quantum Electron. $\underline{QE-19}$, 297 (1982); and (f) B. Perry, P. Rabinowitz and M.C. Newstein, Phys. Rev. A___, (1984).

[11] (a) A.T. Rosenberger, L.A. Orozco, R.J. Brecha and H.J. Kimble, in *Optical Bistability III*, p. 306 (1986); and (b) A.T. Rosenberger, L.A. Orozco and H.J. Kimble, in *Optical Instabilities III*, p. 325 (1986).

[12] H.J. Carmichael, in *Optical Bistability III*, p. 217 (1986).

COMPARISON OF LASER PHASE AND SUPPRESSION OF NOISE

Xuanmin Yang, Ming Yi, Zonghua Huan
Department of Physics, Nanjing University, Nanjing, China

ABSTRACT

By slightly shifting one of the laser beams and recombining them, we have developed an electronic phase modulating method; we can then convert the high frequency phase change of the laser beams to the lower frequency of the optical beating waves. By analyzing the sources of the phase modulating noise in the optical path, we have suppressed much phase modulating noise by a double optical path differential method.

INTRODUCTION

For arbitrary mode ultra low frequency vibration or mechanical displacement, using laser heterodyne phase modulating technique, we have analysed the properties of different noise which are always neglected by other authors and we constructed a setup in which much noise can be suppressed by a sophisticated technique. The results show the potential applications for modern applied physics fields.

PRINCIPLES

Let E_1 and E_2 be the electric field of two plane wave emitted by two laser beams from A-O frequency shifter and propagating in the same direction x, We may write:

$$E_1 = E_{o1} \exp[j(\omega_o t + k_1 x_1 + \varphi_1)]$$

$$E_2 = E_{o2} \exp\{j[(\omega_o + \omega_A)t + k_2 x_2 + \varphi_2]\}$$

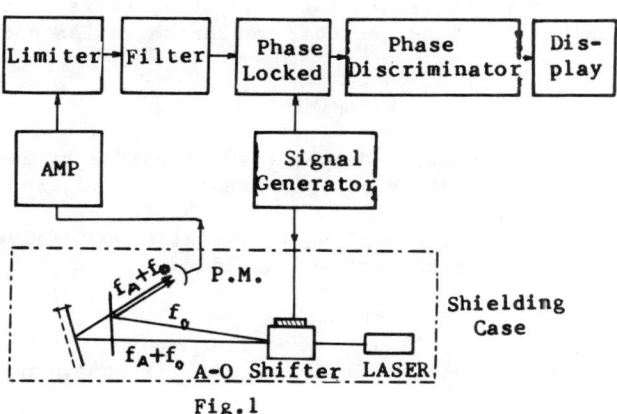

Fig.1

where $k_1 = \omega_0/c$, $k_2 = (\omega_0 + \omega_A)/c$ and ω_0 is the angular frequency of unshifted laser beam, $\omega_0 + \omega_A$ is the angular frequency of shifted laser beam.

The two laser beams are mixed by means of a photomultiplier. The output photoelectric current i_0 will be:

$$i_0 = \alpha \left\{ E_{01}^2 + E_{02}^2 + 2E_{01} E_{02} \cos\left[\omega_A(t + x_2/c) + \omega_0/c(x_2 - x_1) + (\varphi_2 - \varphi_1)\right] \right\} \quad (1)$$

In fact, from eq. 1, E_{01} and E_{02}, the amplitude of two beams, are fluctuated. As we know, the noise power spectrum [1] of laser is

$$W(\Omega) \simeq \frac{r^3 E_0^2 h}{\pi [\Omega^2 + (\frac{1}{4}\gamma E_0^2)^2]} \left(\frac{N_{2t}}{N_{2t} - N_{1t}} \frac{g_2}{g_1} + n_{th} \right) \quad (2)$$

Therefore, the noise power spectrum of our stable frequency He-Ne laser are shown in Fig. 3. The half width is about 5 MHz.

To avoid the high level noise, we have selected the heterodyne frequency of 30 MHz, and its noise can be filtered by a narrow band filter except 30 MHz + 0.1 MHz.

The amplitude modulated noise of the amplitude modulated factor $2E_{01} E_{02}$ of ω_A in third term of eq. (1) would be cut out by means of an amplitude limiter.

Thus a considerable improvement of signal to noise ratio is one of the main advantages of this method.

The modulated noise in the term of $\omega_0(t)(x_2 - x_1)/c$ may be minimized by decreasing the optical path difference $\Delta = x_2 - x_1$ and using a more stable laser.

Ultrasonic frequency is unstable caused by signal generator and the noise of thermal effect of shifter. We have reduced the term of $\omega_A(t)(t + x_2/c)$ as low as we can.

The noise of vibration of optical components and the variation of refractive index in optical path caused by air flow in term ($\varphi_1 - \varphi_2$) which involves the measured physical quantitied could be minimized by isolating the setup from environment and cancelled by another similar optical setup.

Therefore we have increased the S/N ratio, it is better than the theoretical analysis by other authors which the varios noise is unavoidable.

EXPERIMENT

Fig. 2 is a slow variation diaplacement recorded by X-Y recorder. It may be seen, the resolution of this method is 1Å, and may promote better result.

Photo 1,2,3 are three examples of crystal $LiNbO_3$ driven by rectangular wave in different voltage amplitude.

REFERENCE

(1) Yariv, A., Quantum Electronics (John Wiley & Sons, In C.,)

RESULTS OF EXPERIMENTS

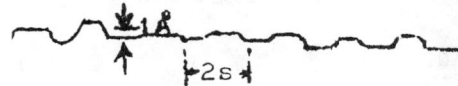

Fig. 2

Slow Variation Displacement Recorded by X-Y Recorder

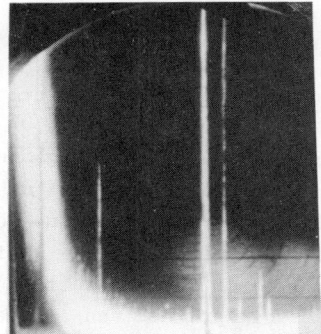

Fig. 3

Noise Spectrum Distribution of Heterodyne Method (Heterodyne frequency is 30 Mc, Central peak is 10 Mc.)

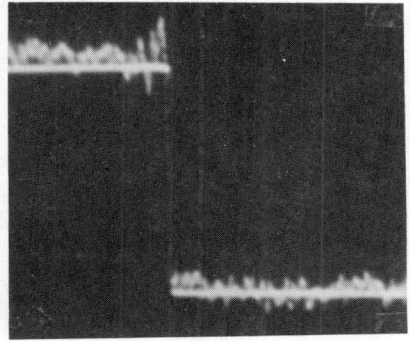

Photo 2.

Waveform of 30 Å, Pulse Width: 22 ms.

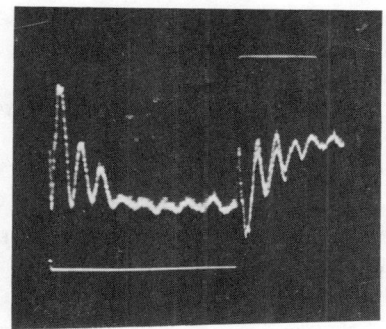

Photo 3.

Damping Wave of Micro-Displacement 100 Å driven by a Rectangular Voltage.

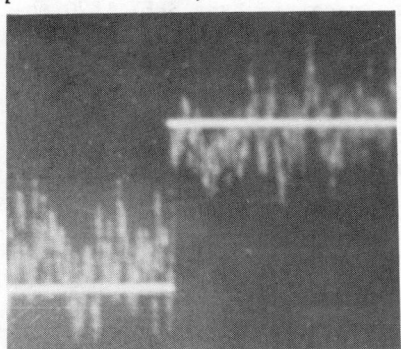

Photo. 1

Micro-Displacement Waveform of 3 Å (Rectangular wave for comparison, Pulse Width: 5 ms.)

OBSERVATION OF A CONE RADIATION IN SODIUM VAPOR

Han Xiaofeng Lü Zhenguo Ma Zuguang
Laser Division, Harbin Institute of Technology, China

ABSTRACT

This paper reports the observation of a yellow cone radiation in sodium vapor pumped by an UV dye laser. It is proposed that a Raman resonant spontaneous four wave parametric process is generated and a doublet emission on resonant transition of 3p-3s is induced.

INTRODUCTION

We have observed a yellow cone radiation in sodium vapor pumped by an excimer laser-dye laser(334.0-338.8 nm). Simultaneously, we have also observed the 3s-4s stimulated electronic Raman scattering(SERS) [1] tunable from 2.38 μm to 2.65 μm and a strong radiation at 1.14 μm in sodium vapor. On the hollow ring of the forward cone radiation, there spectral lines appear at 5890Å, 5896Å, and 5894Å, respectively. It is proposed that the radiations at 5894Å and 1.14 μm are a result of a Raman resonant spontaneous four wave parametric process and induce a doublet emission at 5890 Å and 5896 Å as a transient light fields.

EXPERIMENTAL SET UP

The experimental set up is similar to that in Ref.1. The yellow cone radiation and infrared radiation are received by a SPEX 1870 monochromator with a RCA 8852 photomultiplier and a WDG30 monochromator with a PbS detector, respectively. The electrical signals from RCA8852 or PbS are processed by a Boxcar integrator and recorded by a chart recorder(K200). In the forward direction the cone radiation in the form of a hollow light ring can be observed on a screen.

RESULTS AND DISCUSSION

In the case of temperature of 500-600 °C and buffer gas pressure of 5-20torr, we simultaneously observed the 3s-4s SERS tunable from 2.38 μm to 2.65 μm, a strong radiation at 1.14 μm and a yellow cone radiation as the dye laser was tuned from 334.0 to 338.8 nm. Only one spectral line was observed at 1.14 μm. The yellow cone radiation includs three spectral lines 5890Å, 5896Å and 5894Å. The generation of the yellow cone radiation had a sharp energy threshold(259 μJ). The measured cone angle was about 20 mard. As the sodium vapor pressure rises, the cone angle and ring width of cone radiation increased. We propose that a Raman resonant four wave mixing process based on a spontaneous parametric amplification coupling among 1.14 μm (ν_3), 5894 Å(ν_4), Stokes wave(ν_s) and the pumping wave(ν_1) was generated in sodium vapor, as shown in Fig.1.

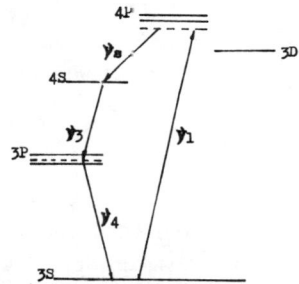

Fig.5, Energy level diagram of sodium atom showing the Raman-resonant spontaneous four wave mixing process.

Considering the third order nonlinear effect, the nonlinear polarization at ω_4 ($\omega_i = 2\pi c \nu_i$) in the wave equation

$$P_x^{(3)}(\omega_4) = 6\chi_{xxxx}^{(3)}(\omega_4)E_x(\omega_1)E_x(-\omega_s)E_x(-\omega_3) \quad \cdots \cdots \cdots (1)$$

where the nonlinear susceptibility

$$\chi_{xxxx}^{(3)}(\omega_4) = i3.145\times10^7 \frac{N}{\sqrt{4s3s}} \langle 3s0\tfrac{1}{2}|x|4p0\tfrac{1}{2}\rangle \langle 4p0\tfrac{1}{2}|x|4s0\tfrac{1}{2}\rangle \times$$

$$\langle 4s0\tfrac{1}{2}|x|3p0\tfrac{1}{2}\rangle \langle 3p0\tfrac{1}{2}|x|3s0\tfrac{1}{2}\rangle \left[2(v_{4p\tfrac{3}{2}}-v_1)^{-1}+(v_{4p\tfrac{1}{2}}-v_1)^{-1}\right] \times$$

$$\left[2(v_{3p\tfrac{3}{2}}-v_3)^{-1}+(v_{3p\tfrac{1}{2}}-v_3)^{-1}+2(v_{3p\tfrac{3}{2}}-v_4)^{-1}+(v_{3p\tfrac{1}{2}}-v_4)^{-1}\right] \cdots (2)$$

In the absence of initial light wave field at ν_3 and ν_4, they might be generated from spontaneous noise or quantum fluctuations [3] by resonance enhancement to form the parametric radiation if the frequency and phase matching conditions

$$\nu_4 = \nu_1 - \nu_s - \nu_3 \quad \cdots \cdots \cdots (3)$$

$$k = k_1 - k_s - k_4 \qquad k = 2\pi n(\nu)\nu \quad \cdots \cdots \cdots (4)$$

could be satisfied.

In formula (2), $\chi_{xxxx}^{(3)}(\omega_4)$ is enhanced not only by resonance of ν_1 to $4p_{3/2, 1/2}$-3s transitions but also by that of ν_4 (or ν_3) to $3p_{3/2, 1/2}$-3s transitions. From formulas (1) and (2), it is clear that in case of very weak light wave field $E_x(\omega_3)$, $P_x^{(3)}(\omega_4)$ would be caused and then generate light wave field $E_x(\omega_4)$ if the pumping wave $E_x(\omega_1)$ and the Stokes wave $E_x(\omega_s)$ are strong enough.

Once $E_x(\omega_4)$ is generated, $E_x(\omega_3)$ is enhanced through the nonlinear interaction and so enhanced each other with the $E_x(\omega_4)$. The light wave field at ν_4 may be expressed by [2]

$$E_4(\vec{r},t) = E_x(\omega_4)\exp(-i\omega_4 t)$$

$$=E_{xo}(\omega_4)\exp(i(\vec{k}_1-\vec{k}_s-\vec{k}_3)\cdot\vec{r}-i\omega_4 t)\exp(|\Delta\eta|z) \quad \ldots \ldots (5)$$

By using the wave equation and longitudinal dematching relationship ^{2} a solution was obtained as follows

$$E_{xo}(\omega_4) = -(2\pi\omega_4/c)\, 6\chi^{(3)}_{xxxx}(\omega_4)E_{xo}(\omega_1)E_{xo}(-\omega_s)E_{xo}(-\omega_3)/(\Delta k+i\Delta\eta) \quad \ldots \ldots (6)$$

From equation (6), the light wave at γ_4 can be seen as a cone radiation and the maximum of γ_4 intensity should appear at angle of $\Delta k=0$ and the intensity rapidly decreases as Δk increases.

Using the longitudinal dematching relationship at $\Delta k=0$ and assuming that light waves at γ_4 and γ_3 approximately have the same small cone angle θ, we obtain tentatively

$$\theta^2 \sim -2(n(\gamma_1)\gamma_1 - \Delta n(\gamma_3)\gamma_3 - \Delta n(\gamma_4)\gamma_4)/(\Omega_{4s3s} + \Delta n(\gamma_3)\gamma_3 + \Delta n(\gamma_4)\gamma_4) \quad \ldots \ldots (7)$$

From (6) and (7), the angle θ is related to the atom density N and increases as N increases. This is consistent with our experiment observation, in which the cone angle increases as the sodium vapor pressure rises.

As discussed above, the radiation at 1.14 μm (γ_3) may be also thought of as a cone originated in transition from 4s to a virtual level between doublet levels $3p_{3/2}$ and $3p_{1/2}$ and would create some gain for inducing the light wave fields at 5890 Å and 5896 Å. The wave vectors of the acting wave 1.14 μm and induced waves 5890 Å, 5896 Å combined on the conical shell are collinear. A detail interpretation about the radiation at 1.14 μm and polarization features of those observed radiations is going to be carried out.

REFERENCES

1. Han Xiaofeng, Lü Zhenguo, Ma Zuguang, Cheng Yongkang, "SERS in sodium vapor", ILSC, Seattle, 1986.
2. Y-R Shen, Nonlinear **Infrared Generation**, Berlin, Springer, 1977.
3. Y,P.Malakyan, Sov.J. Quantum Electron. 15(7), 905(1985)

IV. Atomic, Molecular, and Ionic Spectroscopy
A. Chaos in Molecular Systems 282
B. Orientational and Nonadiabatic Effects in Collision
   Dynamics... 295
C. Laser Cooling and Trapping 307
D. Molecular Ion Spectroscopy and Applications 347
E. Molecular Spectroscopy 359
F. Multiple Excitation of Atoms—Summary Session
   (not published in this volume)
G. Atomic Spectroscopy 396

QUASIPERIODIC AND CHAOTIC MOTIONS IN INTENSE FIELD MULTIPHOTON PROCESSES*

Shih-I Chu
Department of Chemistry, University of Kansas
Lawrence, Kansas 66045

ABSTRACT

The question of the behavior of quantum systems in time-dependent fields whose classical counterparts exhibit chaotic behavior is addressed. For any nondissipative bounded quantum system under the influence of polychromatic (i.e. quasiperiodic) fields, it is proved by means of the many-mode Floquet theory that the auto-correlation function will recur infinitely often in the course of time, indicating no strict quantum stochasticity is possible. In particular, for an N-level quantum system undergoing multiphoton transitions, its dynamic behavior is described by the quasiperiodic motion of an (N^2-1)-dimensional coherence vector $\vec{S}$ in accord with the SU(N) dynamic symmetries. On the other hand, for any dissipative quantum system, SU(N) symmetries are broken, and non-quasiperiodic behavior is observed as the coherence vector $\vec{S}$ evolves from an initially (N^2-1)-dimensional space to a lower-dimensional space. The dynamical behaviors are illustrated for two- and three- level quantum systems driven by intense bichromatic laser fields.

I. INTRODUCTION

The dynamical evolution of quantum systems under the influence of time-dependent perturbation is a subject of fundamental interest in many areas.[1,2] In this paper, the problem of the (quasiperiodic or chaotic) behavior of bounded quantum systems perturbed by intense polychromatic laser fields is discussed. The theoretical analysis of this problem involves the use of the recently developed many-mode Floquet theory[3] which allows <u>exact</u> reformulation of the time-dependent problem of the interaction of finite-level systems with <u>polychromatic</u> fields as an equivalent <u>time-independent</u> infinite-dimensional eigenvalue problem. The main results are presented in Sec. II A and II B. In Sec. II A, the dynamical behavior of a bounded, non-dissipative N-level system is analyzed. It is found that the system exhibits quasi-periodic behavior under the SU(N) dynamical symmetry constraint. For dissipative systems (Sec. II B), on the other hand, SU(N) symmetries are broken, and the systems evolve from an initially (N^2-1) dimensional space to a low dimensional space in a quasi-random manner.

---

* Work supported by Department of Energy, Division of Chemical Sciences.

II. DYNAMICAL BEHAVIOR OF DISCRETE QUANTUM SYSTEMS UNDER PERTURBATIONS BY INTENSE POLYCHROMATIC FIELDS

Under any time-periodic Hamiltonian (i.e. perturbation by a monochromatic field), it is known that a non-resonant bounded quantum system exhibits quasiperiodic behavior and reassembles itself infinitely often in the course of time.[4] The behavior of the corresponding quantum systems perturbed by bichromatic or polychromatic fields (where the Hamiltonian is non-periodic in time) is less clear and has been investigated only recently. In this section, we discuss the behavior of two- and three-level quantum systems driven by intense bichromatic fields. In Sec. II A, we discuss the behavior in non-dissipative systems, and in Sec. II B, we discuss the behavior in dissipative systems.

A. Quasiperiodic Behavior in Nondissipative Systems

It is known that the dynamical evolution of any bounded N-level quantum system is governed by the SU(N) dynamical symmetries, in which one can describe the time evolution in terms of a coherence vector $\vec{S}(t)$ of constant length rotating in an (N^2-1)-dimensional space.[5] Using the many-mode Floquet theory (MMFT), Ho and Chu[3] have recently exploited both analytically and pictorially the vivid dynamical evolution of the eight-dimensional coherence vector corresponding to a three-level system (N=3) perturbed by bichromatic fields. The results show that the coherence vector $\vec{S}(t)$ displays quasiperiodic behavior when the external fields are not very strong. The behavior of quantum systems in very intense bichromatic fields is somewhat controversial and is discussed below.

Consider a two-level system ($|a\rangle$, $|b\rangle$, $E_a < E_b$) driven by two intense linearly polarized monochromatic fields with amplitudes ($\vec{\varepsilon}_1$, $\vec{\varepsilon}_2$) and frequencies (ω_1, ω_2) respectively. The time-dependent autocorrelation function for the wave function $\Psi(t)$ of the two-level system is defined by

$$C(\tau) \equiv \lim_{T \to \infty} 1/T \int_0^T \Psi^*(t) \Psi(t + \tau) \, dt. \quad (1)$$

If the system exhibits quasi-periodic motion, $C(\tau)$ will be an almost periodic function of the correlation time τ. On the other hand, a rapid decaying auto-correlation function is a signature of quantum chaos.[6] In Fig. 1, we present the modulus of $C(\tau)$ corresponding to three different field intensities. The physical parameters used are $E_b - E_a = 1.0$ (arbitrary units), laser frequencies $\omega_1 = 0.85$, $\omega_2 = \omega_1/\sqrt{15}$, and electric-dipole coupling strengths ($b_i = |\vec{\mu}_{ab} \cdot \vec{\varepsilon}_i/2|$, $i = 1,2$) $b_1 = b_2 = 0.1$ (top figure), 0.5 (middle figure), and 2.0 (bottom figure), arbitrary units. As shown here, the autocorrelation function $|C(\tau)|$ displays a quasi-periodic pattern in the weaker field case (top figure). For very strong fields (bottom figure), $|C(\tau)|$ shows a rapid decrease at small τ and then exhibits noise-like structure at larger τ. Pomeau et. al.[7] have studied numerically the similar problem and they described this phenomenon as "chaotic rabi flopping." However, a close mathematical study of

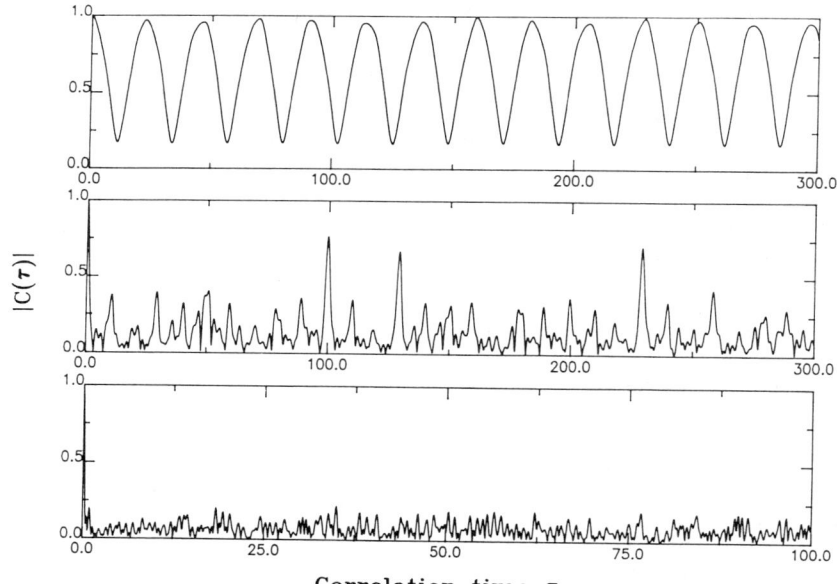

Fig. 1. Modulus of the auto-correlation function $C(\tau)$ for a two-level system in the bichromatic fields. From the top to bottom: weak, medium strong, intense field cases.

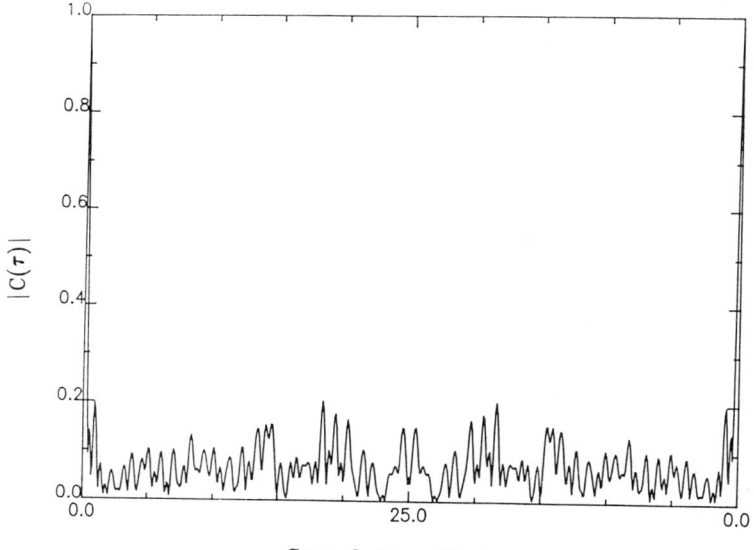

Fig. 2. Time-reversibility of the auto-correlation function (intense field case in Fig. 1).

this noise-like pattern via MMFT reveals that $C(\tau)$ never decays to zero at large correlation times (τ) (as seen in Fig. 1), and in fact it is an almost periodic function of τ regardless of the field strengths.[8] Figure 2 shows that the auto-correlation function in strong fields (bottom figure in Fig. 1) is reversible in time. This augments the conjective that for nondissipative bounded quantum systems, the dynamical evolution in intense polychromatic fields is intrinsically quasi-periodic in nature and no strict quantum stochasticity is possible. More detailed analysis can be found in ref. 8.

B. Non-Quasiperiodic Behavior in Dissipative Systems

Without loss of generality, let us consider now the dynamical behavior of a dissipative three-level system driven by intense bichromatic fields. The dynamical evolution is governed by the Liouville equation with dissipation,[9]

$$id\hat{\rho}(t)/dt = [\hat{H}(t), \hat{\rho}]_{-} - i\{\hat{G}, \hat{\rho}(t)\}_{+}, \quad (2)$$

where $\hat{H}(t)$ is the total Hamiltonian, $\hat{\rho}(t)$ the density matrix of the system, $\hat{G}$ the damping operator, $[\hat{A}, \hat{B}]_{-} \equiv \hat{A}\hat{B} - \hat{B}\hat{A}$ is the commutator, and $\{\hat{A}, \hat{B}\}_{+} \equiv \hat{A}\hat{B} + \hat{B}\hat{A}$ is the anti-commutator. The eight-dimensional coherence vector $\vec{S}(t)$ can be determined through the relationship

$$S_j(t) = \text{Tr}[\hat{\rho}(t)\hat{s}_j], \quad j = 1, 2, \ldots, 8 \quad (3)$$

where $\vec{s}_j$'s are SU(3) generators.[9] Eq. (2) can be readily solved by appealing to non-Hermitian many-mode Floquet theory.[9]

Consider now the example of a dissipative Λ-type three-level system whose level 2 (intermediate level) is subject to irreversible decay with damping rate g_2 (Fig. 3). It is interesting to see how the dissipative coherence vector $\vec{S}(t)$, now a shrinking rotating vector in eight-dimensional space, evolves in time. Under two-photon resonant conditions (Fig. 3) and rotating wave approximations (RWA), the coherence vector $S_j(t)$ ($j = 1,2,\ldots,8$) can be solved exactly analytically. In the limit $t \to \infty$, we find that

$$S_i(t_\infty) \to 0, \quad i = 1,2,\ldots,7 \quad (4)$$

but

$$S_8(t_\infty) \to 2\beta^2/\sqrt{3}\Omega^2, \quad (5)$$

where $\Omega = (\alpha^2 + \beta^2)^{\frac{1}{2}}$, and α and β are Rabi frequencies given by $\alpha = 1/2 <1|\vec{\mu} \cdot \vec{\epsilon}_1|2>$ and $\beta = 1/2 <2|\vec{\mu} \cdot \vec{\epsilon}_2|3>$. Note that the population scalar $S_0(t) (\equiv \rho_{11} + \rho_{22} + \rho_{33})$ in this case reduces to

$$S_0(t_\infty) \to (\sqrt{3}/2) S_8(t_\infty) = \beta^2/\Omega^2. \quad (6)$$

Thus at exact two-photon resonance condition, in the RWA limit, the eight-dimensional coherent vector eventually evolves to a one-dimensional scalar and the population is trapped in the S_8 component only.

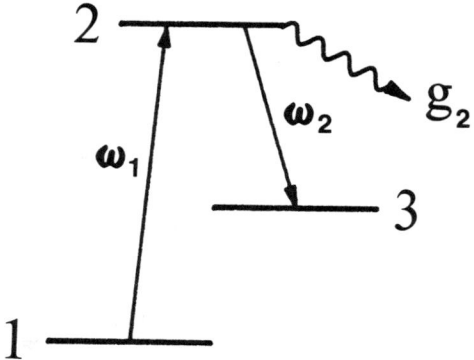

Fig. 3. Λ-type dissipative three-level system.

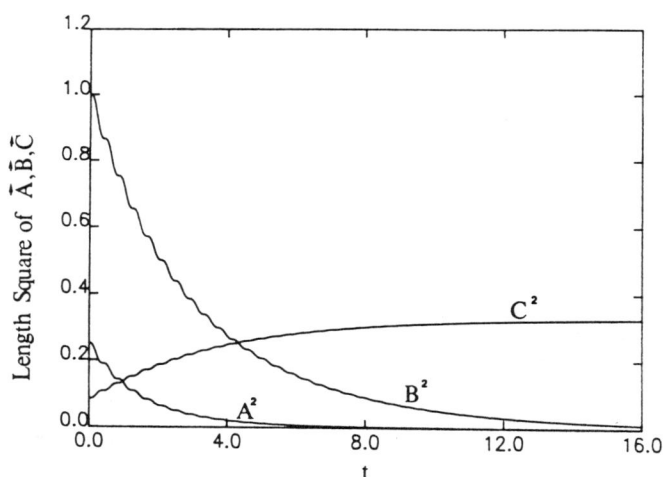

Fig. 4. The time-evolution of the length square of the subvectors $\vec{A} \equiv (S_1(t), S_2(t), S_3(t))$, $\vec{B}(t) \equiv (S_4(t), S_5(t), S_6(t), S_7(t))$ and $\vec{C} \equiv (S_8(t))$, where $\vec{S} = \vec{A} + \vec{B} + \vec{C}$.

If we now relax the RWA limit and perform the "exact" MMFT analysis, we find the coherence vector $\vec{S}(t)$ first evolves qualitatively similar to the RWA case, namely, the population will be temporarily trapped (quasi-trapped) mainly in the S_8 component (Fig. 4). However, at still larger times ($t_\infty \ggg 1/g_2$), the population in the system will eventually completely decay away to the surroundings.

The qualitative difference of the dynamical evolution of dissipative ($g_2 \neq 0$) and non-dissipative ($g_2 = 0$) systems is illustrated in Fig. 5. Shown here are the projections of the trajectories of the coherence vector $\vec{S}(t)$ onto the $S_5 - S_6$ plane. One sees that the coherence vector exhibits quasi-periodic behavior in the nondissipative case (Fig. 5b) in accord with the SU(3) dynamical symmetries. On the other hand, for the dissipative case (Fig. 5a), SU(3) symmetries are broken, and $\vec{S}(t)$ evolves in a non-quasiperiodic manner from an initially (N^2-1)-dimensional space (N = 3 in this case) to a lower-dimensional space.

REFERENCES

1. See, for example, the special issue on laser chemistry, Physics Today, 33, No. 11, 25-59 (1980).
2. For a recent review on Floquet approaches to intense field multiphoton processes, see S.-I. Chu, Adv. At. Mol. Phys. 21, 197 (1985).
3. T.-S. Ho and S.-I. Chu, Phys. Rev. A 31, 659 (1985), and references therein.
4. See, for example, T. Hogg and B.A. Huberman, Phys. Rev. Lett. 48, 711 (1982).
5. J. Elgin, Phys. Lett. 80A, 140 (1980); F.T. Hioe and J.H. Eberly, Phys. Rev. Lett. 47, 838 (1981).
6. M. Shapiro and G. Goelman, Phys. Rev. Lett. 53, 1714 (1984).
7. Y. Pomeau, B. Dorizzi, and B. Grammaticos, Phys. Rev. Lett. 56, 681 (1986).
8. K. Wang and S.-I. Chu, Phys. Rev. A (submitted).
9. T.-S. Ho and S.-I. Chu, Phys. Rev. A 32, 377 (1985).

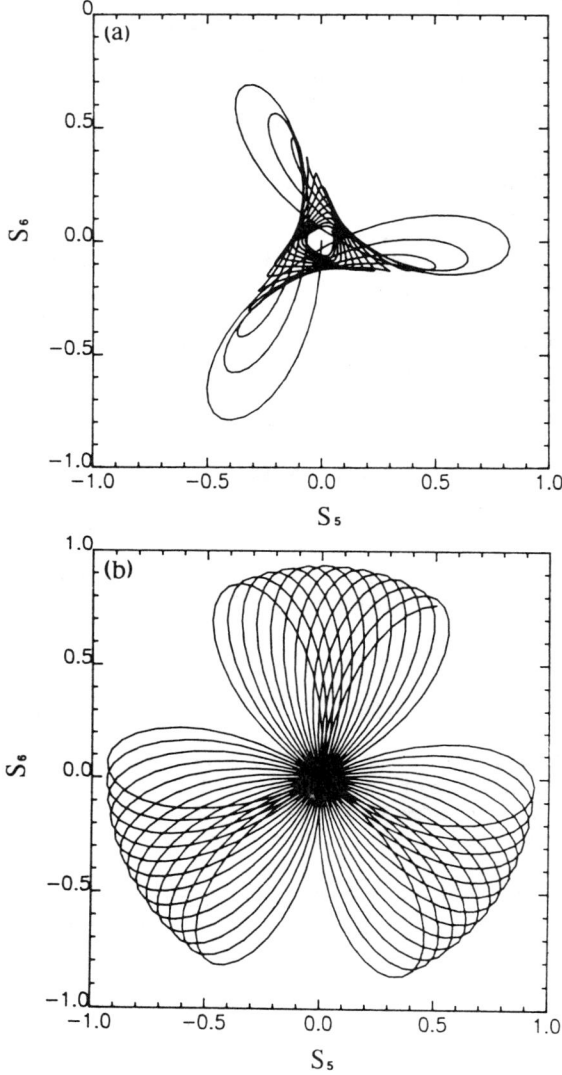

Fig. 5.
Projection of the trajectory of the coherence vector $\vec{S}(t)$ on the $S_5 - S_6$ plane for (a) the dissipative three-level system ($g_2 \neq 0$) and (b) the non-dissipative system ($g_2 = 0$).

DISSIPATIVE MOLECULAR DYNAMICS: QUANTAL VERSUS CLASSICAL TREATMENT

Mingwhei Tung
Swarthmore College, Swarthmore, PA 19081

Jian-Min Yuan and James. F. Heagy
Drexel University, Philadelphia, PA 19104

ABSTRACT

We have studied by using quantum mechanics the dynamics of a driven damped Morse oscillator, whose classical solutions exhibit bistable and chaotic behavior. In this treatment the Morse Hamiltonian, expressed in terms of the generators of an su(2) algebra, is coupled to a bath of harmonic oscillators and driven by a classical field, simulating an infrared laser. We have derived the generalized master equation from the Liouville equation within the Markoffian and weak-coupling approximation. The time dependence of the average energy of the Morse oscillator is obtained by solving the time evolution equations for the matrix elements numerically as a function of the relaxation times, laser amplitude and frequency. Bistability does not show up as a hysteresis loop in the quantum solution, but as a bimodal population distribution in vibrational energy levels. As the field amplitude increases the time series of the average Morse energy becomes increasingly irregular.

INTRODUCTION

A straightforward numerical integration of the classical equation of the driven damped Morse oscillator,

$$\ddot{x} + \gamma\dot{x} + (1 - e^{-x})e^{-x} = A\cos\omega t, \tag{1}$$

as a function of the driving frequency or amplitude shows hysteretic behavior. The bistable domain forms two cusp-shape regions in the $A\omega$-parameter space.[1] The bigger region corresponds to a 1:1 resonance between a vibrational quantum and a photon, the smaller one a 1:2 resonance. Similarly, regions for higher-order subharmonic resonances may exist. For $\gamma=0.4$ and $A=0.45$[1,2] we have found chaotic domains on both branches, the scenario to chaos is that via period-doubling bifurcations. Some essential features of the classical model such as, the response curves of the oscillator as functions of the driving frequency or amplitude and the cusp-shape region of bistability, also manifest themselves in a phenomenological quantum model,[3] which we shall call the semiclassical model. In this model the Hamiltonian of the anharmonic oscillator is constructed by replacing the quantum number in the well-known Morse eigenvalue expression by the number operator $a^\dagger a$ of a harmonic oscillator. Quantum simulation of the classical damping term, $\gamma\dot{x}$ or a more general form, is not a trivial problem.[5] A safe way of treating this problem is to couple the oscillator with a bath of harmonic oscillators. A more rigorous procedure will be followed for the Morse oscillator in the next section, but in the semiclassical model Narducci et al.[4] include in the generalized master equation relaxation contributions derived for a harmonic oscillator coupled to a harmonic bath under the assumption that the anharmonic oscillator interacts with each bath mode only weakly. Based on this generalized master equation Heisenberg equations of motion for the expectation values of operators like a, $a^\dagger$, $a^\dagger a$ .... can be derived, which yields a hierarchy of an infinite number of equations. They truncate the hierarchy by introducing certain factorization ansatz. The result is a set of three nonlinear equations, of which the steady state energy value takes the shape of cusp catastrophe when plotted against laser frequency and amplitude. The theme of this talk is to report to you results of a more rigorous quantum mechanical treatment[6] of the driven damped Morse oscillator and to compare them with the classical and semiclassical ones.

QUANTUM EQUATIONS

According to Levine,[7] the Morse Hamiltonian can be expressed as

$$H_s = \hbar\omega (A^+A^- + I_0/2) \qquad (2)$$

where A^+, A^- and I_0, generators of an su(2) algebra, satisfy the following commutation relations

$$[A^-, A^+] = I_0$$

$$[A^\pm, I_0] = \pm 2x_0 A^\pm. \qquad (3)$$

As mentioned in the Intrduction we couple the Morse oscillator with a reservoir of an infinite number of harmonic oscillators. We assume that each bath mode is weakly coupled to the anharmonic oscillator, but this weakness of coupling does not necessarily mean that the interaction between the anharmonic oscillator and the bath is weak, because the number of bath modes is very large.[5] With the interaction with a classical field added in the Hamiltonian of the entire system becomes

$$H = H_s + H_r + H_{sr} + H_{sf} \qquad (4)$$

where H_r, H_{sr} and H_{sf} are given by

$$H_r = \Sigma \hbar\omega_j a_j^\dagger a_j$$

$$H_{sr} = \hbar(\Lambda_1 A^+ + \Lambda_1^\dagger A^-) + \hbar\Lambda_2 A^+A^- \qquad (5)$$

$$H_{sf} = \mu\varepsilon(A^+ + A^-)\cos\omega t.$$

Λ_1 and Λ_2 are operators in the space of bath modes. Invoking Markoffian and weak-coupling approximation, we arrive at the following generalized master equation[8]

$$d\rho(t)/dt = -iL^S\rho(t) - iL^{sf}\rho(t)$$

$$+\int d\tau \exp(-i\omega x_0\tau) \{<\Lambda_1(\tau)\Lambda_1^\dagger>_0 [\exp(i\omega l_0\tau) A^-\rho(\tau), A^+]$$

$$+<\Lambda_1^\dagger \Lambda_1(\tau)>_0 [A^+, \rho(\tau)\exp(i\omega l_0\tau) A^-]$$

$$+<\Lambda_1\Lambda_1^\dagger(\tau)>_0 [A^-, \rho(\tau)\exp(-i\omega l_0\tau) A^+]$$

$$+<\Lambda_1^\dagger(\tau)\Lambda_1>_0 [\exp(-i\omega l_0\tau) A^+\rho(\tau), A^-]\}$$

$$+\eta\{[A^+A^-\rho(t), A^+A^-] + [A^+A^-, \rho(t)A^+A^-]\}, \quad (6)$$

where $L^S\rho$ denotes $[H^S, \rho]/\hbar$, and similarly for L^{sf} and η s the scaled phase relaxation rate related to Λ_2. Based on Eq. (6), the time evolution equation for the matrix element between the mth and nth Morse eigenstate becomes[6]

$$d\rho_{k,m}/dt = -i\omega_{km}\rho_{k,m} + ((\overline{k+1})(\overline{m+1}))^{1/2}[\gamma_{\downarrow k+1} + \gamma_{\downarrow m+1}]\rho_{k+1,m+1}$$

$$-[\overline{k}\gamma_{\downarrow k} + \overline{m}\gamma_{\downarrow m} + (\overline{k+1})\gamma_{\uparrow k} + (\overline{m+1})\gamma_{\uparrow m} + \eta(\overline{k-m})^2]\rho_{k,m}$$

$$+(\overline{km})^{1/2}[\gamma_{\uparrow k-1} + \gamma_{\uparrow m-1}]\rho_{k-1,m-1} - i\Omega\cos\omega t [\overline{k}^{1/2}\rho_{k-1,m}$$

$$+\overline{k+1}^{1/2}\rho_{k+1,m} - \overline{m}^{1/2}\rho_{k,m-1} - \overline{m+1}^{1/2}\rho_{k,m+1}], \quad (7)$$

where $\gamma_\uparrow$ and $\gamma_\downarrow$ are the transition rates up and down the ladder. $\overline{m}$ is defined as $m[1-x_0(m-1)]$ and Ω is the scaled Rabi rate for the 0th to 1st level transition.

RESULTS AND DISCUSSIONS

We have solved Eq. (7) numerically for parameter values simulating an HF molecule. In Fig. 1 we show the time evolution of the average energy, $<E(t)>$, of the Morse oscillator for several phase relaxation rates (η). As η increases, transient oscillations damp out more quickly as expected. Furthermore Fig. 1 shows that the asymptotic $<E(t)>$ oscillates around a mean value which is not a monotonic function of η. This asymptotic mean (averaged over several periods), E_{av}, first increases with η, but eventually decreases as η becomes too large. This observation is consistent with the interpretation that η is related to the level width. As level width increases it compensates for the detuning caused by the anharmonic defect and thus facilitates the excitation. But as the level width gets too large, the oscillator strength spreads out so thin that excitation becomes less effective.

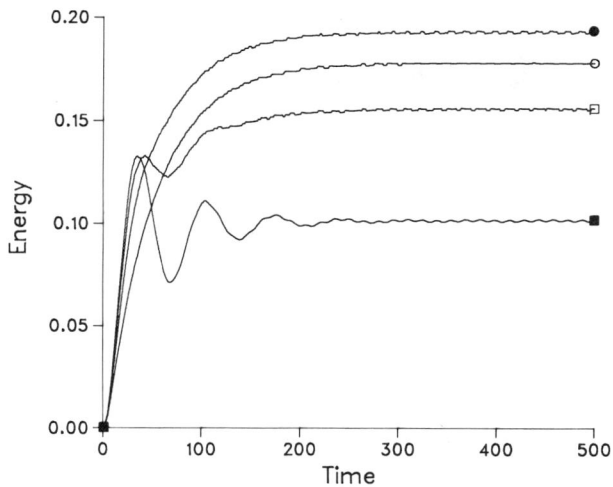

Fig. 1 Time evolution of the average Morse oscillator energy, $\langle E(t) \rangle$, for several η values. • denotes the curve with η fixed at 0.05, ¤ for $\eta = 0.15$, • for $\eta = 0.35$, and o for $\eta = 0.80$.

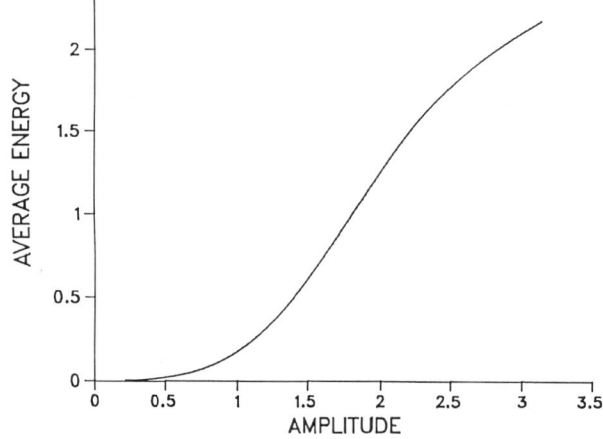

Fig. 2 Asymptotic mean oscillator energy, E_{av}, plotted against the field amplitude

In Fig. 2 we show that E_{av} grows with the driving amplitude with a relatively sudden threshold. Threshold behavior is associated with hysteretic bistability in the classical and semiclassical results. Hysteresis does not exist in the quantum case, because the equation is linear in ρ. But the sudden increase here seems to suggest that the quantum results represent an average between two branches. This interpretation is in fact supported by the asymptotic level population distributions, P_n, plotted in Fig. 3 for several field strength values as measured by Ω. For a very weak field, not shown in the figure, P_n peaks at the ground state and drops quickly to zero as n increases. As Ω increases P_n still peaks at the ground state, but a shoulder seems to grow at some finite quantum numbers. At a high Ω value (3.1) we see that a second peak appears at n=3 to form a bimodal distribution. Further increase of Ω yields a distribution with a single maximum at n=3. Thus the combination of quantum and classical results seems to indicate that the lower branch can be represented by a level population peaking at the ground state, the upper branch, on the other hand, corresponds to a population peaking at a finite quantum number. In quantum mechanics bistability does show up, not as a hysteresis loop but as a bimodal population distribution or a distribution representing a mixture of the above two kinds.

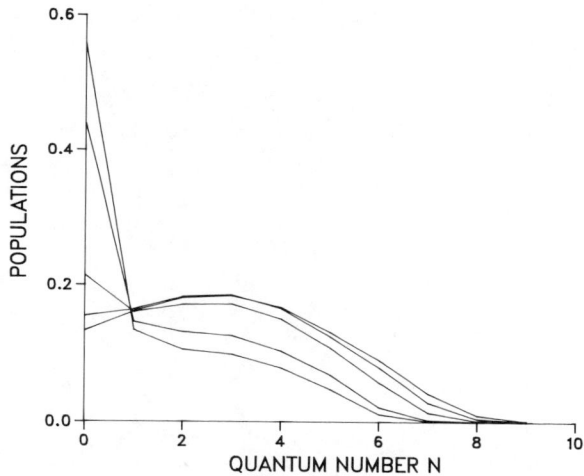

Fig. 3 Average asymptotic level population P_n for several Rabi rates. From the top curve down on the left (or form the bottom curve up from the right) Ω is given by 1.9, 2.2, 3.1, 3.8, and 4.4.

Thus we have found evidences of molecular bistability in classical, semiclassical and quantum mechanical studies. From a practical point of view molecular bistability may be a useful property based on which mirrorless memory devices can be made. Experimental verification of this property, however, is yet to be done. Possible such experiments include measurements of the shape change of a single IR laser pulse[9] and the threshold behavior of stimulated Raman scattering.

At increasing field amplitude the time series of <E(t)> first show periodic, then quasiperiodic, and eventually irregular behavior. It is possible that "quantum chaos" exists for the present dissipative system, but more work needs to be done to confirm it.

REFERENCES

1. G. C. Lie and J. M. Yuan, J. Chem. Phys. 84, 5486 (1986).
2. R. Kapral, M Schell, and S. Fraser, J. Chem. Phys. 86, 2205 (1982).

3. J. M. Yuan, E. Liu, and M. Tung, J. Chem. Phys. $\underline{79}$, 5034 (1983).
4. L. M. Narducci, S. S. Mitra, R. A. Shatas, and C. A. Coulter, Phys. Rev. A$\underline{16}$, 287 (1977).
5. A. O. Caldeira and A. J. Legget, Ann. Phys. (N. Y.) $\underline{149}$, 374 (1983).
6. M. Tung and J. M. Yuan, to be published.
7. R. D. Levine, Chem. Phys. Lett. $\underline{95}$, 87 (1983).
8. M. Tung, E. Eschenazi, and J. M. Yuan, Chem. Phys. Lett. $\underline{115}$, 405 (1985).
9. J. F. Heagy, J. M. Yuan, and S. L. Chin, unpublished.

INELASTIC COLLISIONS IN LASER EXCITED ALKALI ATOMS

Maria Allegrini, Silvia Gozzini and Luigi Moi
Istituto di Fisica Atomica e Molecolare del C.N.R.
Via del Giardino 7, 56100 Pisa, Italy

ABSTRACT

This paper will briefly outline inelastic collisions between two laser excited alkali atoms. Special emphasis will be on recent progress made in our laboratory for heteronuclear systems.

INTRODUCTION

Resonant laser excitation of alkali atomic vapors generally produces several different phenomena. Among these we have studied the collisions between two alkali atoms, both laser-excited to the first p-resonance level. Products of these collisions are highly excited atoms[1] (sometimes highly excited molecules as well[2]) and molecular ions[3] plus electrons:

$A^*(nP) + A^*(nP) \longrightarrow A^{**}(n'L) + A(nS)$ energy transfer

$\longrightarrow A_2^+ + e$ associative ionization

Energy balance between entrance and exit channnels is provided by kinetic energy and is of the order of few kT. Electrons produced in the associative ionization are therefore "slow" electrons and do not cause any secondary effect, unless they gain energy through superelastic collisions with the excited atoms.

Excited atoms A^{**} are detected by their fluorescence emission, while ions can be identified and counted by well established and efficient techniques such as mass spectrometry. More recently, electron spectrometry suitable for these very low energy electrons has been developed[4]. Final goal of these investigations is to provide rate coefficients or cross sections for energy transfer and associative ionization, which, when compared to calculated values, give information on long range interactions.

In our laboratory we have extensively investigated the energy transfer collisions in alkali vapors. To discriminate this particular phenomenon from all others processes, there are some constraints on the main parameters of the experiment: i) atom density has to be high enough to make the collision products observable, but not too high in order to avoid superelastic collisions; a reasonable value for experiments is 10^{12}-10^{13} cm^{-3}. ii) the laser merely serves as a tool to prepare the reactants for the collision and therefore its power can not be too high in order to avoid direct multiphoton ionization of the atoms and any laser induced or laser assisted collisional process. In our experiments we have been working with 10^1-10^3 W/cm^2.

© American Institute of Physics 1987

These collisions are thermal; this fact has several consequences. Rate coefficients drop rapidly with increasing energy defect/excess or, in others words, only excited levels n'L with energy within few kT of the nP+nP energy are populated by the collision. Rate coefficients K are related to cross sections σ, (K=vσ), simply through the mean interatomic velocity v=(8kT/πμ)$^{1/2}$, which in our experiments is about 10^{5}cm/sec. Collision time duration, which very roughly is given by the ratio atom dimension/thermal velocity, is such that a rate equation approach is appropriate to evaluate σ from data measured in the experiments. From a theoretical point of view, the production of highly excited states n'L by thermal collisions between two excited atoms nP/nP may be considered as due to transitions among different electronic molecular states corresponding to the atomic dissociation limits nP+nP and nS+n'L. The nonadiabatic energy transfer usually occurs at the internuclear distances of avoided crossings. Calculations of electronic excited molecular potentials for alkali dimers have been flourishing in the last few years and some of them extend also to large internuclear distances of interest for energy transfer collisions. However specific calculations of the coupling terms and consequent evaluation of cross sections for nP+nP transfer to nS+n'L have been rather limited.

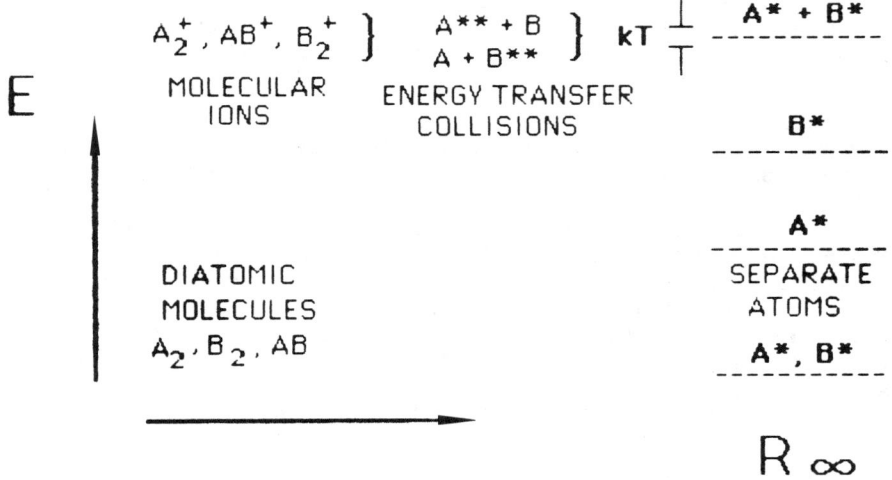

Fig. 1. Schematic representation of energy transfer collisions between two excited atoms.

Recently we have tried to complement our experimental results for energy transfer collisions 3p/3p to 4d and 4f levels in sodium with a calculated value of σ^5. We have made some approximations because almost every pair of molecular states having the right symmetry are coupled by some terms of the total Hamiltonian. Of course approximations which are reasonable for sodium cannot immediately be extended to other systems, however the theoretical approach should be valid in the general case.

Our experiments are based on the detection of the fluorescence emitted by the highly excited states populated by np/np energy transfer collisions. The intensity of these lines may be several orders of magnitude weaker than the resonance fluorescence, i.e. the fluorescence emitted by the nP state directly excited by the laser. Therefore, even if the spectrum is usually simple because atomic alkali lines are very well known, the apparatus for spectroscopy must be rather sensitive. We have detected fluorescence lines in the visible region with intensity up to 10^{-8} compared to resonance fluorescence; however detection in the infrared is not so efficient and sometimes we have not been able to observe some expected lines. The fluorescence intensity of collision-produced high-lying states varies quadratically with both the atomic density and the laser power because the process depends upon two laser excited atoms; this characteristics may be sometimes useful to discriminate levels which are produced by other mechanisms.

Energy transfer in collisions between two excited atoms have been observed with almost all alkali vapors and measurements of cross sections have been made in several laboratories. Problems related to these measurements have been solved in various laboratories with different approaches, which can be found in the original papers. A table with all measurements up to 1984 has been reported[6] and more recent results in sodium, taking into account also the dependence on J value of the colliding 3P-atoms, have been published[7]. Here we only wish to make a brief comment about the size of the cross sections. For example, cross sections for sodium 3p/3p energy transfer collisions are of the order of $10^{-15} cm^2$, which is a significant fraction of ordinary geometrical cross section for 3p atoms ($\approx 10^{-14} cm^2$). Collisions between two excited atoms are therefore important and cannot be disregarded whenever there is a resonant laser excitation of the vapor. In comparison associative ionization cross section in sodium from 3p/3p atoms[8,9] is of the order of $10^{-17} cm^2$. Measurements of these cross sections and calculations of potential energy curves at long interatomic distances have improved understanding of long range interactions, at least in sodium. Since for heteronuclear alkali systems there are less spectroscopic data and theoretical calculations, we have started experimental studies of collisions between two excited alkali atoms of different species, which is briefly described in what follows.

COLLISIONAL EXCITATION TRANSFER IN ALKALI MIXTURES

The process we consider now is

$$A^* + B^* \longrightarrow A^{**} + B$$
$$\longrightarrow A + B^{**}$$

where A and B are alkali atoms of different species, A^* and B^* indicate excitation to the first P-resonance level, A^{**} and B^{**} are high-lying states produced by the collisions. Exit channels are obviously more in number than for pure alkali collisions. Altogether there are about 60 levels for the 10 possible pairs of alkali atoms. The general method to investigate these collisions in a mixture is the same as for a pure alkali vapor: two dye lasers (cw in our experiments) provides the excited reactants A^* and B^* and transfer to A^{**} and B^{**} is monitored through fluorescence emissions. Rate equations involving fluorescence intensity ratios (which are measured in the experiment) are easily written down and steady-state solutions give rate coefficients and total averaged cross sections, exactly in the same way as in pure vapor experiments.

There are however specific problems to be solved for these new experiments. Atomic densities of two different alkali atoms can be very different at the same temperature; moreover the relative densities in a mixture are not simply related to the temperature of the sealed cell containing the saturated vapors. In the presence of A^*/B^* collisions there are also A^*/A^* and B^*/B^* pure collisions which produce excited atoms; in the experiment we have to select fluorescences originated from A^*/B^* collisions from all other fluorescence, in particular the very strong resonance fluorescence of both laser excited atoms.

We have started our investigation in a sodium-potassium mixture for which the process to be investigated is

$$Na^*(3p) + K^*(4p) \longrightarrow Na^{**}(4p,3d) + K(4s)$$
$$\longrightarrow Na(3s) + K^{**}(7s,6p,5d,5f) \quad (1)$$

An amalgam made of 90% pure Na and 10% pure potassium gives a comparable density for the two alkalis at the temperatures of our interest, according to the empirical Raoult law[10]. The amalgam is sealed in a capillary pyrex cell[11] which is heated in a temperature controlled oven. A Rhodamine cw dye laser (yellow laser) excites sodium atoms to $3P_{1/2}$ or $3P_{3/2}$ level and an Oxazine cw dye laser (red laser) excites potassium atoms to $4P_{1/2}$ or $4P_{3/2}$ level. The yellow laser is modulated at frequency v_Y, the red laser is modulated at frequency v_R and detection is made at the sum frequency $v_Y + v_R$. This intermodulated technique assures that only fluorescence from levels populated by Na(3P)+K(4P) collisions is observed. A schematic diagram of the apparatus is shown in figure 2.

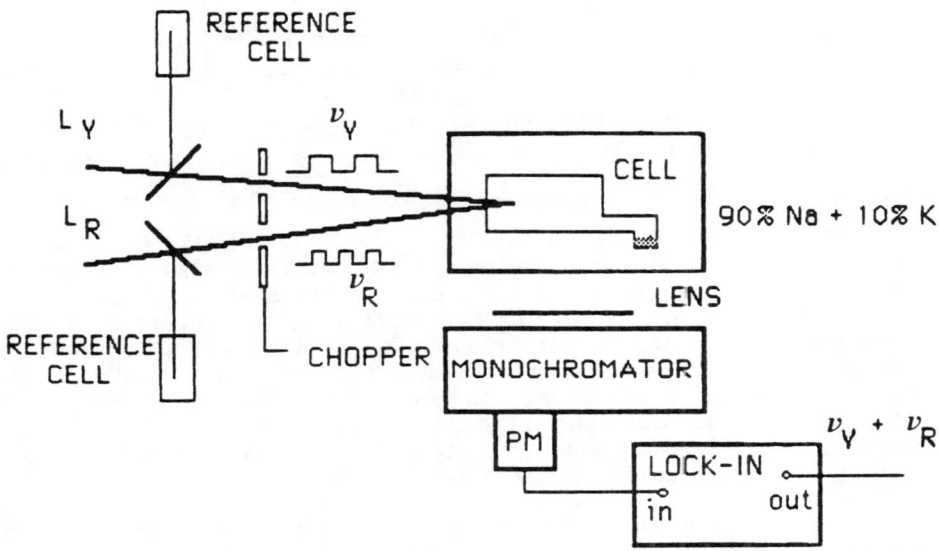

Fig. 2. Sketch of the experimental apparatus.

Fluorescence has been observed from all the expected levels of process (1), except the 5F state of potassium because it emits mainly in the infrared, where our apparatus is not very sensitive. Fluorescence from Na(4P) and K(6P) is in the u.v. and it has been observed without any problem. However, since we can not have precise corrections for glass absorption of these u.v. lines, cross sections have not been evaluated. Densities of Na(3P) and K(4P), (necessary for cross section determination), have been deduced from measurements of the effective lifetimes from time-resolved resonance fluorescence observation under pulsed laser excitation. We have found it convenient to measure the cross sections for process (1) relative to the cross section for production of Na(5S) from sodium 3P/3P collisions for which absolute measurements exist[11,12]. Typical rate equation approach gives, for example, the following relation for the 3D state production of sodium by process (1)

$$\frac{I_{3D-3P}}{I_{5S-3P}} = \frac{K_{3D}}{K_{5S}} \frac{\alpha_{3D-3P}}{\alpha_{5S-3P}} \frac{\gamma_{3D-3P}}{\gamma_{5S-3P}} \frac{\omega_{3D-3P}}{\omega_{5S-3P}} \frac{[K_{4P}]_{\nu_R}}{[Na3P]_{\nu_Y}} \qquad (2)$$

Fluorescence intensities I_{i-j} are measured; α_{i-j} are correction parameters for the detection system, γ_{i-j} is the transition branching ratio, ω_{i-j} is the transition frequency and $[K_{4P}]_{\nu_R}$, $[Na_{3P}]_{\nu_Y}$ are the densities of atoms excited by the laser beams modulated at frequency ν_R and ν_Y respectively. Excited atom densities are independently measured in the pulsed experiment from intensity of resonance fluorescence lines and effective radiative lifetimes. Relation (2) gives $K_{3D} = 0.6 K_{5S}$, which means $\sigma_{3D} = (1.4 \pm 0.6) \times 10^{-15} cm^2$. In a similar way we have measured $\sigma_{5D} = (1.9 \pm 0.9) \times 10^{-15} cm^2$ and $\sigma_{7S} = (0.9 \pm 0.4) \times 10^{-15} cm^2$ for potassium. The size of these cross sections indicates that the energy transfer in collisions between two excited alkali atoms of different species is essentially due to the same mechanism as in pure alkalis, in spite of the different symmetries of the corresponding molecular potentials. The next alkali mixture under investigation in our laboratory is potassium plus rubidium. In this case there is one level, the 5D of rubidium, whose energy is in the midst of the four fine structure levels $K^*(4P) + Rb^*(5P)$. Cross section measurements for production of this state may turn out to be particularly interesting.

We are grateful to Prof. N.K. Rahman for reading the manuscript and to Consiglio Nazionale delle Ricerche for support, under Progetto Bilaterale n. 19931.

REFERENCES

1. M. Allegrini, G. Alzetta, A. Kopystynska, L. Moi and G. Orriols, Opt. Commun. 19, 96 (1976)
2. M. Allegrini, G. Alzetta, A. Kopystynska, L Moi and G. Orriols, Opt. Commun. 22, 329 (1977)
3. A. Hellfeld, J. Caddick and J. Weiner, Phys. Rev. Lett. 40, 1369 (1978); G.H. Bearman and J.J. Leventhal, Phys. Rev. Lett. 41, 1227 (1978)
4. J.L. LeGouet, J.L. Picque, F. Wuilleumier, J.M. Bizeau, P. Dhez, P. Koch and D.L. Ederer, Phys. Rev. Lett. 48, 600 (1982)
5. M. Allegrini, C. Gabbanini, L. Moi and R. Colle, Phys. Rev. A32, 2068 (1985)
6. M. Allegrini, C. Gabbanini and L. Moi, J. Phys. (Paris) Colloq. 46, C1-61 (1985)
7. S.A. Davidson, J.F. Kelly and A. Gallagher, Phys. Rev. A33, 3756 (1986)
8. R. Bonanno, J. Boulmer and J. Weiner, Phys. Rev. A28, 604 (1983)
9. J. Huennekens and A. Gallagher, Phys. Rev. A28, 1276 (1983)
10. R. Seiwert, Ann. Physik 18, 54 (1956); J. Ciurylo, Acta Phys. Pol. A50, 105 (1976)
11. M. Allegrini, P. Bicchi and L. Moi, Phys. Rev. A28, 1338 (1983)
12. J. Huennekens and A. Gallagher, Phys. Rev. A27, 1851 (1983)

ISOTOPE DEPENDENCE OF FINE-STRUCTURE BRANCHING IN He-Na OPTICAL COLLISIONS

Linda L. Vahala
Old Dominion University, Norfolk, Va. 23508

Paul S. Julienne
U. S. National Bureau of Standards, Gaithersburg, Md. 20899

ABSTRACT

Isotopic effects on the branching ratio dependence on detuning is considered for He-Na system. The close-coupled calculations compare favorably to experimental results for detuning at energy $E=200$ cm^{-1}.

SUMMARY

We consider collisional redistribution that occurs in an optical collision of two atoms in scattering state i (by absorption of a photon through an isolated molecular state a to an isolated state b) followed by separation to excited product scattering state f. If the separation velocity of the two atoms is sufficiently large, the final state can be approximated by the same recoil limit which has been developed for photofragmentation [Singer et al., J. Chem. Phys. **79**, 6060 (1983)]. Our close coupling calculations on Na + He systems show how the recoil limit for fine-structure branching ratios is approached for both red and blue detuning as product separation velocity increases. Isotopic effects, at thermal energies, on the branching ratio D1/D2 are calculated for the lighter He^3-Na system and the heavier He^4-Na system.

For blue detuning at energy $E=300$ cm^{-1}, we see that the branching ratio for He^3-Na is always higher than that for He^4-Na, see Fig. 1. However, the isotopic effect is more complicated for red tuning. As the red detuning increases, D1/D2 initially becomes larger for He^3-Na (see inset of Fig. 1) and then larger for He^4-Na, and then back to He^3-Na as $\Delta \to -200$ cm^{-1}cm. At $E=200$ cm^{-1}, Fig. 2, we find good agreement between the experimental results of Havey & Delahanty and theory for blue detuning.

© American Institute of Physics 1987

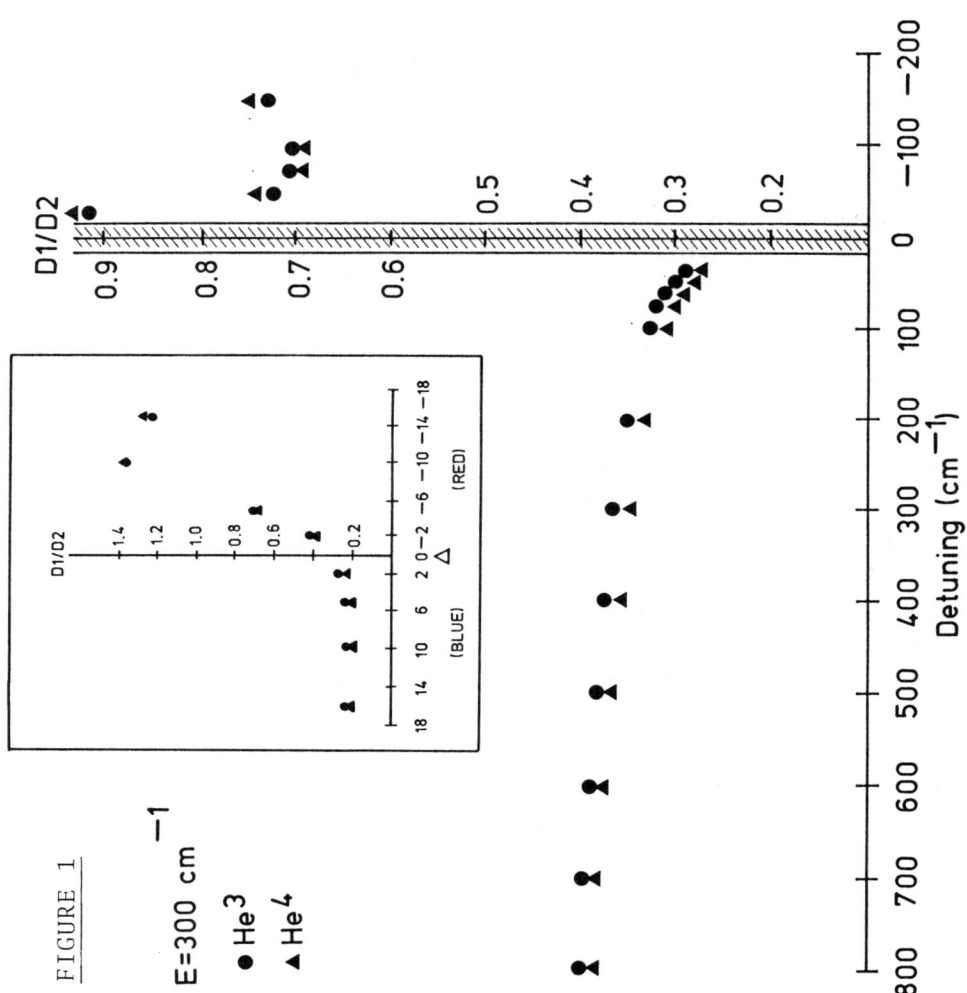

FIGURE 1

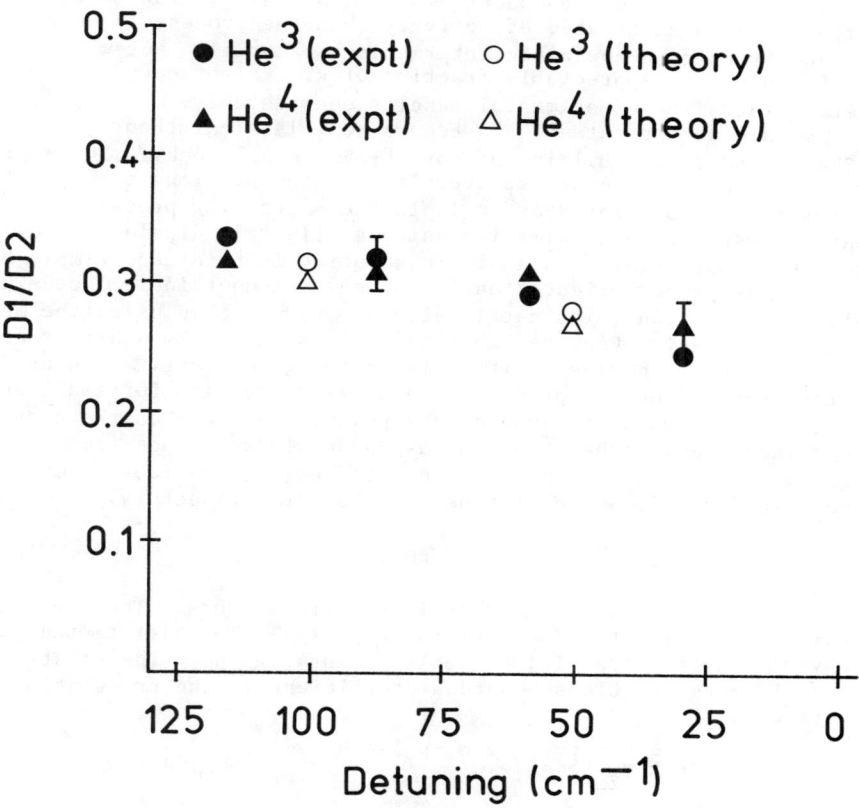

FIGURE 2

NEAR RESONANT ENERGY TRANSFER THEORY APPLIED TO NON-RESONANT PROCESSES

D. E. Godar, K. L. McNesby, and R. D. Bates, Jr.
Department of Chemistry, Georgetown University, Washington, DC 20057

ABSTRACT

The coulomb interaction gives nearly Lorentzian behavior for rates of intermolecular vibrational energy transfer versus energy mismatch, even when this exceeds kT. Cross sections agree with experiment to within a factor of two for CO_2 and a series of linear molecules with defects from 0 to 500 cm^{-1}.

INTRODUCTION

To date collision-dependent vibrational energy flow between molecules has been treated effectively when the process is near resonant by using a coulombic interaction potential. But when energy defects become an appreciable fraction of kT, this theory underestimates the experimental results considerably.
The accepted treatment has been a semiclassical theory, developed by Sharma and Brau[1] building on work by Van Kranendonk.[2] The approach here offers a significant improvement over the previous theory. The problem is now formulated as an inelastic scattering process in the quantum first Born approximation using an electric dipole-dipole interaction potential. An analytic solution is obtained, simply by choosing the proper orientation of the spatial coordinate system when evaluating the transition matrix element, rather than using the method of partial waves.[3] Pack has obtained this solution by a different approach, but did not apply it to large energy defects.[4] The cross section provided here reproduces the previous results for small energy defects. It shows an inverse square power dependence as the energy defect increases, rather than the exponential dependence found for the large impact parameter contribution,[1] or the inverse fourth power dependence for head-on collisions (hard sphere trajectory.)[5]

THEORY

The focus here is on dipole-allowed transitions. The transition matrix element is the product of three parts.[6] The vibrational part is given[1,3] as the transition dipole moments, $\mu_a \mu_b$. The rotational matrix elements are Clebsch-Gordon coefficients. The translational part is given by:

$$\langle r^{-3} Y_{2m} \rangle = \int e^{i(p_i - p_f) \cdot r/\hbar} r^{-3} Y_{2m}(\Omega) \, d^3r. \quad (1)$$

Transforming the coordinate system so that the z axis of the space fixed frame lies along the momentum transfer vector, $q = (p_i - p_f)$, gives

$$\langle r^{-3} Y_{20} \rangle = 2\pi (5/4\pi)^{1/2} \int_d^\infty r^{-3} \left[\int_0^\pi e^{iqr\cos\theta/\hbar} P_2(\cos\theta) \sin\theta \, d\theta \right] r^2 dr, \quad (2)$$

© American Institute of Physics 1987

where m is required to be zero. The radial integral is cut off at the distance of closest approach, <u>d</u>, which is taken as the collision diameter. Equation 2 may be evaluated and used to give the differential scattering cross section:

$$\sigma(p_i, p_f) = (8/3\hbar^4)(\mu_a^2 \mu_b^2) \cdot C_a C_b \cdot \mu^2 (p_f/p_i)[Q^{-2} \cos Q - Q^{-3} \sin Q]^2 \quad (3)$$

with $Q = |p_i - p_f| d/\hbar$ and $C_a = C^2(J_{ai} 1 J_{af}; 000)$ and $C_b = C^2(J_{bi} 1 J_{bf}; 000)$.

From this the total cross section can be obtained as an integral over scattering angles, actually performed over Q. This result is:

$$\sigma = (32\pi/3)(\mu_a^2 \mu_b^2)(\mu/p_i d\hbar)^2 \cdot C_a C_b \cdot [G(u) - G(\ell)] \quad (4)$$

where $u = 2[p_i + (p_i^2 + 2\mu\omega\hbar)^{1/2}]d/\hbar$, $\ell = 2[p_i - (p_i^2 + 2\mu\omega\hbar)^{1/2}]d/\hbar$, ω is the energy defect, and $G(x) = -(1/2)x^{-2} + x^{-3}\sin x - x^{-4}(1 - \cos x)$. For experiments performed in bulk, a thermal average must be performed with the Maxwell-Boltzmann velocity distribution to obtain a cross section related to the experimental rate constant, as $k = \rho \langle v\sigma \rangle$.

This averaging can be done numerically for the whole range of molecular parameters: μ, d, ω, and T. However, the cross section is very nearly a function of a single reduced variable $x = 2\omega d(\mu/2kT)^{1/2}$, just as was found to be exactly true (rather than nearly true) for the semiclassical theory. Thus the velocity-averaged cross section is:

$$\langle v\sigma \rangle / \langle v \rangle = (2\pi/3)\mu_a^2 \mu_b^2 (\mu/d^2\hbar^2 kT) \cdot C_a C_b \cdot H(x,y) \quad (5)$$

where $y = \mu d^2 kT/\hbar^2$, and where the resonance function $H(x,y)$ is scaled so that $H(0,y) = 1$. The function may be approximated by a very useful form which is accurate to within a factor of two for all molecules:

$$H(x,y) \simeq H(x) \simeq 1/(1 + 0.2606 x^{1.83}). \quad (6)$$

RESULTS AND DISCUSSION

When this function is averaged over thermal rotational distributions, rate constants for specific pairs of molecules can be obtained. In Table I cross sections calculated by this treatment are given along with experimentally obtained values and values calculated by using the semiclassical version of the theory. The striking result is the significant improvement in the agreement between theory and experiment even at fairly substantial energy gaps. Agreement is generally to within a factor of two, though not for CO_2-OCS.

Several features of these results merit amplification. First, the functional form of the resonance function is now best approximated as a Lorentzian, which is much more satisfying than the previous exponential form found in Sharma-Brau[1] and Tam[9] because the energy transfer may be thought of as an emission-absorption process, which would be characterized by a natural lineshape. Second, vibrational energy transfer is seen theoretically to remain an efficient process for much larger energy defects than was predicted previously. Cross sections remain comparable to those measured experimentally. This extends the regime in which coulombic interactions are considered to

TABLE I Rates at 300K of V-V processes from $CO_2(001)$

Collision Partner	ΔE (cm^{-1})	Transition Dipole (Debye)	Cross Sections Experimental (Å^2)	Quantum (Å^2)	Semiclassical (Å^2)
CO_2	0.0	0.3321	31[7]	41.3	41.7
$C^{16}O^{18}O$	17.0	0.3321	24[8]	35.4	35.8
$^{13}CO_2$	65.6	0.3178	8.5[8]	4.8	3.9
N_2O	125.4	0.2485	0.60[8]	0.32	0.017
$^{15}N^{15}NO$	194.4	0.2485	0.072[8]	0.13	0.0015
CO	205.9	0.1055	0.030[8]	0.024	0.00047
^{13}CO	253.1	0.1055	0.008[9]	0.015	0.00015
OCS	287.1	0.3740	0.015[8]	0.12	0.00043
NO	473.0	0.0794	0.0036[8]	0.0032	0.0000064
^{15}NO	506.2	0.0794	0.0034[8]	0.0029	0.0000048

be a physically important mechanism. Third, these results are obtained analytically using a simple, first order theory as contrasted with the much more complicated second order theories,[10,11] or with numerical trajectory calculations.[12,13] Finally, the ability to calculate cross sections or rate constants now is comparable to the ability to determine these values experimentally, particularly for multilevel systems for which the observed rate constants are often complicated composites of several specific rate constants.

The authors gratefully acknowledge the valuable insight provided by Dr. Ralph Weston who shared his notes on the existing theory.

REFERENCES

1. R.D. Sharma and C.A. Brau, J. Chem. Phys. 50, 924 (1969).
2. C.G. Gray and J. Van Kranendonk, Can. J. Phys. 44, 2411 (1966).
3. R.J. Cross and R.G. Gordon, J. Chem. Phys. 45, 3571 (1966).
4. R.T. Pack, J. Chem. Phys. 72, 6140 (1980).
5. J.T. Yardley, J. Chem. Phys. 50, 2464 (1969).
6. C.G. Gray, Can. J. Phys. 46, 135 (1968).
7. I. Burak, Y. Noter, and A. Szöke, J. Quant. Elect. QE9, 541 (1973).
8. C.B. Moore in Adv. in Chem. Phys., Vol. 23, I. Prigogine and S.A. Rice, Eds., (Wiley, New York, 1973), p. 41 and references therein.
9. W.G. Tam, Can. J. Phys. 50, 2691 (1972).
10. H.A. Rabitz and R.G. Gordon, J. Chem. Phys. 53, 1815 (1970).
11. R.D. Sharma, Phys. Rev. A2, 173 (1970).
12. T.A. Dillon and J.C. Stephenson, Phys. Rev. A6, 1460 (1972); T.A. Dillon and J.C. Stephenson, 50, 2056 (1973).
13. B. Seoudi, L. Doyennette, and M. Margottin-Maclou, J. Chem. Phys. 81, 5049 (1984).

SINGLE ATOMIC PARTICLE AT REST IN FREE SPACE: NEW VALUE FOR ELECTRON RADIUS

Hans Dehmelt
Department of Physics, University of Washington
Seattle, Washington 98195, USA

ABSTRACT

Zero-point level confinement in a trap is the quantum-mechanical equivalent of the classical single particle at rest in free space. Such confinement has been demonstrated - by the Continuous Stern-Gerlach Effect - for only the 150 GHz cyclotron motion in geonium, a single electron permanently confined in a cold Penning trap. This confinement made possible a new determination of the radius of the electron, some 10 000 times smaller than the accepted upper limit of $\approx 10^{-16}$ cm. Localizing the electron to ≈ 60 μm in the node of a standing wave in the trap cavity approximated free space, however decreasing spontaneous emission ten-fold. Driving the 60 MHz axial motion on a side band higher by the ≈ 12 kHz magnetron freqency, forced the magnetron motion to absorb the excess photon energy, and shrank its radius to ≈ 15 μm. Analogous laser cooling has reduced the oscillation amplitude of a Ba^+ ion in a Paul rf trap to < 120 nm. As in Nuclear Magnetic Resonance, confinement here is much smaller than the wave length and side bands disappear. Quantum jumps to and from an electronic metastable level of Ba^+ have been demonstrated. Using In^+ or similar, these techniques may make a mono-ion oscillator optical frequency standard with a 1000-day reproducibility of 10^{-18} possible. Further experiments are proposed.

ZERO-POINT CONFINEMENT

Here, apologizing to others, I will review only work of three stored-ion spectroscopy groups at the University of Washington, and propose some new experiments.

I begin by discussing zero-point confinement[1] in a suitable trap as a quantum-mechanical equivalent of the classical single particle at rest in free space. Even when servos keep a satellite lab centered on an antihydrogen atom floating freely in it, the atom will be enclosed by some form of non-material box with perfectly reflecting walls. Thus, it can, at best, be in

© American Institute of Physics 1987

the zero-point level of this textbook system. For a box 0.5 mm wide, the zero-point kinetic energy corresponds to a velocity[1] of ≈ 0.5 mm/sec. So far, such zero-point confinement has been demonstrated - by means of the Continuous Stern Gerlach Effect[2] - for only the ν_c=150 GHz cyclotron motion in geonium[3,4], a single electron permanently confined in a Penning trap. The apparatus, which continuously reads out the quantum number sum n + m, is calibrated by inducing spin flips or quantum jumps m = -½ ↔ +½. The data confirm the expectation, that at an ambient temperature 4 K predominantly the n = 0 cyclotron level, of rms radius 100 Angstrom, is populated. Amplitudes of axial and magnetron motions have been reduced also, and on 15 Sept 1984 VAN DYCK, SCHWINBERG & DEHMELT were able to announce "Here, in a cylinder, about 60 μm long and 30 μm in diameter, in the center of our trap, right now, resides positron PRISCILLA and has been executing simple spontaneous ballet routines for the last three months."

NEW VALUE FOR ELECTRON RADIUS

The most important result of stored ion spectroscopy is a new radius for the electron[5], some 10 000 times smaller than the accepted value. This result was obtained by analyzing our g-factor data[3]

$$g/2 = 1.\ 001\ 159\ 652\ 193(4)$$

on the basis of a near-Dirac particle model[4]. A mathematically convenient ideal, unphysical Dirac particle is pointlike, R_{Dirac} = 0. This point particle of mass M carries out a spontaneous quasi-circular Zitter-Bewegung (trembling motion) at the speed of light, as discussed by SCHROEDINGER, by DIRAC, and by HUANG. Its radius is the Compton wavelength, $\bar{\lambda}_C$ = ℏ / Mc. For a particle charge e this motion produces the correct spin magnetism of one particle magneton, eℏ / 2Mc. The less a near-Dirac particle of finite radius R looks like a point in relation to its Zitter-Bewegung radius $\bar{\lambda}_C$, the more should its g deviate from the Dirac value 2. For the proton R approaches $4\bar{\lambda}_C$, as found by HOFSTADTER, and g - 2 reaches ≈ 3.6, as first measured by STERN. In the absence of better guiding principles we conjecture, that, without radiative corrections, the dimensionless ratio

$$(R - R_{Dirac})\ \bar{\lambda}_C^{-1}\ /\ (g - g_{Dirac}) \equiv \rho_D$$

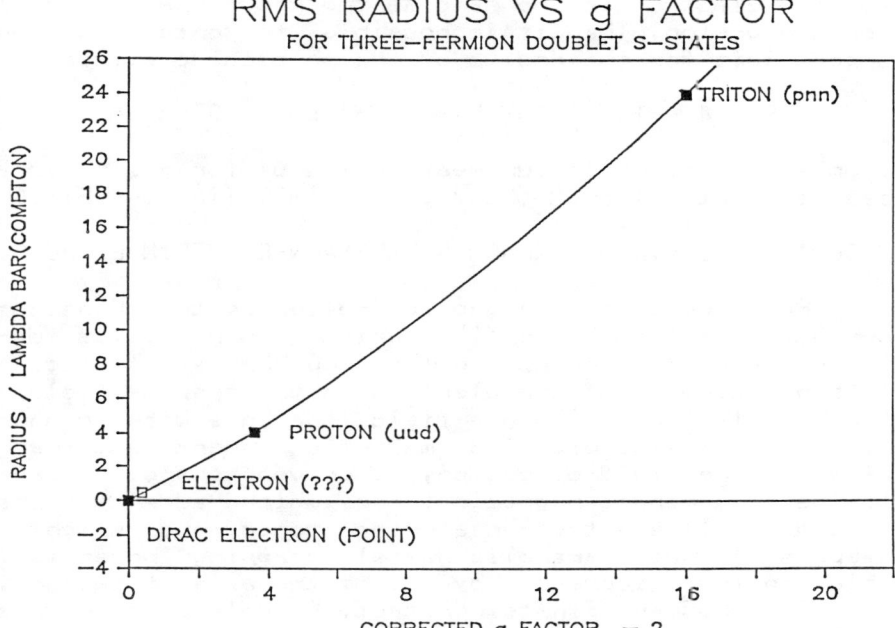

Fig. 1. RMS radius R vs g-2 for near-Dirac particles. In this graph g denotes a corrected empirical g-factor, g = $2\omega_s/\omega_c$ - (QED radiative shift). The simplest way to build up a small composite Fermi-particle is by tightly binding in a doublet S-state three, more primitive, fermions. The proton (3 quarks) and the the triton (3 nucleons) are suitable examples. Stimulated by reference 5, I have plotted radii and g-factors measured for these particles, together with a data point for the theoretical Dirac point electron. A parabola may be fitted to these data. According to the current orthodox conjecture, the physical electron is a true point particle. However, as reviewed by HARARI in 1983, theories have been advanced that, similar to proton and triton, the electron is composed of three smaller fermions, namely ad-hoc prequarks in this case. As a conjecture less idealized than the orthodoxy, I propose, that the electron is not a point particle, but merely too small to measure with today's techniques, and falls close to the curve shown here. Thus, precise measurement of g can yield an (approximate) value for R. For clarity, size and g-2 value for the physical electron have been grossly exaggerated in the plot.

for the electron is the same as the known one $\rho_D \approx 1$ for the proton. Then it is possible to obtain a new upper limit for the radius of the physical electron

$$R \approx 4 \times 10^{-11} |g^{exp} - g^{qed}| \text{ cm} < 10^{-20} \text{ cm}$$

from a comparison of our measured value for g with the best theoretical one[6] obtained by KINOSHITA, see Fig.1.

CONTINUOUS STERN-GERLACH & KAUFMANN-EINSTEIN EFFECTS

RF spectroscopy in geonium relies on the Continuous Stern Gerlach Effect[2], in which a spin flip is detected as a small change in the $\approx$ 60 MHz axial oscillation frequency of the electron in the trap. An electron slowly moving along a field line in a weak magnetic bottle field, with its magnetic moment anti-parallel to the field direction, is driven towards weaker fields by a minute magnetic force, which adds to the strong axial electric restoring force and slightly deepens the net parabolic axial trapping potential. This, in turn, increases by $\approx$ ½ Hz the axial frequency.

The Kaufmann-Einstein Effect, or relativistic mass shift of the electron[3], may become a superior alternative: operating a geonium apparatus as a frequency selective mini-synchro-cyclotron has produced a shift[7] in the axial frequency of 50 Hz. The response of the electron resembles that of a driven anharmonic oscillator. Resolution[4] of the relativistic fine structure, with spacing s_n from 1st to n+1st line, but washed out by the strong cyclotron drive here, may yet yield an accurate non-QED value of the finestructure constant α from 3 frequencies, ν_c and s_n measured by us, and the Rydberg (cR_∞) measured to 1 part in 10^{-9} by others: $\alpha^2 = 2(cR_\infty)s_n/n\nu_c^2$. Potential cavity-reduced line widths are << 1mHz, narrow enough to see individual flux quanta.

ELECTRON-RESONANT CIRCUIT INTERACTION

Like in HABANN's 1926 split-anode magnetron, energy is quickly transferred from the oscillating electron to a resonant LCR-circuit. We use this for the purpose of detection, damping and cooling of the oscillatory motion. For atoms, such increased radiation damping was first discussed by PURCELL in 1946. Conversely, localizing the electron to $\approx$ 60 μm in the node of a standing wave excited by the cyclotron motion[8] in the trap cavity, where the cavity looks like a short circuit, may make it possible to decrease the natural line width hundred-fold. In fact increases in the cyclotron decay time ten-fold and more have been demonstrated[3,9]. Such line-narrowing was suggested for atoms by DICKE in 1960. Both phenomena have been discussed for a microwave antenna in a cavity by SLATER and by SCHELKUNOFF as early as 1942. The electron-cavity system may be modeled by an effective circuit diagram[4] when cyclotron motion and a cavity series resonance are near-resonant. The effective circuit is formed by connecting - in parallel - an lc series circuit resonant at ω_c for the electron and an LC series circuit resonant at Ω_n' for the cavity. For the electron in vacuum the resistance R replaces the series-resonant LC circuit. By tuning[8] ω_c to Ω_n', the resonant frequencies of the lcLC and lcR effective circuits for electron in cavity and in vacuum will become very nearly the same, and no cavity shift will falsify the measured g-value, $g = 2\omega_s/\omega_c$.

SIDE BAND COOLING / EXTENSIONS OF GEONIUM WORK

To an atom slightly oscillating at a frequency ω in a trapping well in the direction of a monochromatic laser beam of frequency Ω_d illuminating it, the laser spectrum will have side bands spaced at $\Omega_d \pm \omega$, due to the Doppler effect. Therefore, the atom, on a sharp internal transition frequency Ω, will be able to absorb a light quantum emitted by the laser even when tuned off-resonance, provided the laser is tuned to $\Omega - \omega$. In this process energy $\hbar\omega$ extracted from the motion of the atom in the trap helps with the internal excitation of the atom and is dissipated in the subsequent re-emission, cooling the ω-motion[4,10,15]. Conversely, by driving the geonium axial motion - here taking the place of the "internal excitation" - on a side band higher than resonance by the ≈ 12 kHz magnetron freqency, it has been possible to force the magnetron motion to absorb the excess in the photon energy and thereby shrink the magnetron radius to ≈ 15 μm[4,11].

Extension of geonium techniques to positron and antiproton were proposed early[12,13]. The first anti-matter experiments in a Penning trap have yielded the positron g-factor[3] since then. Also, catching of antiprotons in a Penning trap by an international group led by GABRIELSE was reported at this conference.

Earlier, an individual proton was isolated in a Penning trap and continuously detected by VAN DYCK et al[14].

OPTICAL SPECTRA / ELECTRON SHELVING / QUANTUM JUMPS

Using laser side band cooling[15], the oscillation amplitude of a Ba^+ ion in a different Paul rf trap has been reduced to $\approx$ 120 nm or less. As in Nuclear Magnetic Resonance, confinement is here much smaller than the wave length and no Doppler side bands spaced at the $\approx$ 6 MHz vibrational frequency in the trap appeared in the well resolved 2070 nm spectrum of Ba^+ recorded[16] by JANIK et al.

Actually two lasers are needed to obtain a strong fluorescence spectrum of Ba^+: one to excite the $6\ S_{1/2} \rightarrow 6\ P_{1/2}$ resonance transition and one to clean out the metastable $5\ D_{3/2}$ level via $5\ D_{3/2} \rightarrow 6P_{1/2}$ transitions. The strong $P_{1/2} \rightarrow S_{1/2}$ fluorescence at 493 nm may be shut off by shelving the optical electron in the other metastable level, $5\ D_{5/2}$, of 30 sec lifetime. This has recently been done by NAGOURNEY et al.[17] by exciting the $6\ P_{3/2}$ level with a hollow cathode lamp, and then letting the optical electron drop down into the $5\ D_{5/2}$ shelf level. When the electron jumps finally into the groundstate, the strong fluorescence turns on again.

FUTURE EXPERIMENTS

The shelving just described is obviously a very effective way to detect the highly forbidden, 5.3 milliHz wide, $S_{1/2} \rightarrow D_{3/2}$ transition potentially useful in optical frequency standards. It may even be possible to operate an ultra-narrow stored ion laser on the forbidden $5\ D_{5/2} \rightarrow 6\ S_{1/2}$ transition just mentioned. To this end, a linear string of Ba^+ ions would be cooled and confined in a 1-meter rf race track trap[1,15]. As in the shelved electron experiment, but by more intense 455 nm $S_{1/2} \rightarrow P_{3/2}$ excitation, the ions would rapidly be pumped into the $D_{5/2}$ level in a manner quite analogous to MAIMAN's original three-level Ruby laser. Such mono-ion oscillator techniques, by using In^+ or similar ion, may make an optical frequency standard with a 1000-day reproducibility of 10^{-18} possible[15]. Various systems of interest may be synthesized[13] and retained in a Paul rf trap soon. I mention hydrogenic U^{91+} stabilized in a circular Rydberg orbit by a superconducting spherical cavity. Further on, some physicists may do spectroscopy and free fall experiments on an antihydrogen atom $\bar{H}$, synthesized from a positron-antiproton plasma in a rf trap and then boxed in by a huge ≈ 100 μm diameter field, also with central $\langle E^2 \rangle_{space}$ __minimum__, but frequency far __above__ Lyman α. Currently most promising for atomic antimatter physics is the antiparticle of the molecular hydrogen ion H_2^+, with rich rf[18] and infrared spectra,

$$\bar{H}_2^-.$$

In a Paul rf trap, loaded with a plasma of many LCR-cooled positrons and a few antiprotons, synthesis and confinement of it and even a singly charged

polymolecular antihydrogen cluster $\bar{H}_n^-$

seem feasible today. Future trapping of now available antideuterons, and antihelium 3, the latter both as nucleus and singly charged ion with optical spectra, appears worthwhile also.

I enjoyed discussions with my colleagues J. BARDEEN, G. GABRIELSE, D. LICHTENBERG, G. MILLER, W. NAGOURNEY, R. VAN DYCK. D. DUNDORE read the manuscript. THE NATIONAL SCIENCE FOUNDATION supports the SEPARIS/SAPARIS projects.

REFERENCES

1. H. Dehmelt, Adv. At. Mol. Phys. __3__, 55 (1967).
2. H. Dehmelt, Proc. Natl. Acad. Sci. USA __83__, 2291 (1986).
3. R. S. Van Dyck, Jr., P. B. Schwinberg & H. G. Dehmelt, Phys. Rev. D __34__, 722 (1986) & in ATOMIC PHYSICS 9, Eds. R. S. Van Dyck and E. N. Fortson (World Scientific Book Publishers, New York, 1984).
4. H. Dehmelt, Ann. Phys. Fr. __10__, 777 (1985).
5. S. J. Brodsky & S. D. Drell, Phys. Rev. D __22__ (1980) 2236.
6. T. Kinoshita, in Atomic Physics 9, reference 3.
7. G. Gabrielse, H. Dehmelt & W. Kells, Phys. Rev. Lett. __54__ (1985) 537.
8. H. Dehmelt, Proc. Natl. Acad. Sci. U.S.A. __81__, 8037 (1984) & __82__, 6366 (1985).
9. G. Gabrielse & H. Dehmelt, Phys. Rev. Lett. __55__, 67 (1985).
10. D. Wineland & H. Dehmelt, Bull. Am. Phys. Soc. __20__ (1975) 637.
11. R. S. Van Dyck, Jr., P. B. Schwinberg & H. G. Dehmelt, in New Frontiers in High Energy Physics, Eds. B. Kursunoglu, A. Perlmutter & L. Scott (Plenum, New York, 1978).
12. D. Wineland, P. Ekstrom & H. Dehmelt, Phys. Rev. Lett. __21__, 1279 (1973).

13. H. G. Dehmelt, R. S. Van Dyck, Jr., P. Schwinberg & G. Gabrielse, Bull. Am. Phys. Soc. 24, 757 (1979).
14. R. S. Van Dyck, Jr., F. L. Moore, D. L. Farnham & P. B. Schwinberg, Bull. Am. Phys. Soc. 31, 974 (1986).
15. H. Dehmelt, in Advances in Laser Spectroscopy, Eds. F. T. Arecchi, F. Strumia & H. Walther (Plenum, New York, 1983).
16. G. Janik, W. Nagourney & H. Dehmelt, J.Opt.Soc.Am. B 2, 1251 (1985).
17. W. Nagourney, J. Sandberg & H. Dehmelt, Phys. Rev. Lett. 56, 2797 (1986).
18. K. B. Jefferts, Phys. Rev. Lett. 23, 1476 (1969).

Diffraction of Atomic Waves by Non-Orthogonal Standing Waves

Peter J. Martin, Phillip L. Gould,[*] Bruce Oldaker, and David E. Pritchard

Department of Physics and
Research Laboratory of Electronics
Massachusetts Institute of Technology
Cambridge, Massachusetts 02139

ABSTRACT

We present data exploring the diffraction of sodium atoms which have a velocity component along the $\vec{k}$-vector of a standing wave of light. The decrease in the net momentum transfer to the atoms is attributed to the dephasing between the two Doppler-shifted counterpropagating traveling waves.

INTRODUCTION

We present experiments to investigate momentum transfer between atoms and a standing wave radiation field in the case where the atoms initially have a velocity component up to 2 m/s along the $\vec{k}$-vector of the standing wave. We also present a theory which yields an analytic expression for the final momentum distributions as a function of this velocity component.

THEORY

The calculation for the momentum transferred to a two level atom by a standing wave electric field has been given in previous papers.[2,3] In the experiment there was ample detuning from resonance so that the dipole force dominated and that the number of spontaneous decays was less than unity.[1] The electric field that the atom experiences as it travels through the Gaussian waist of a standing wave field profile at an angle $\theta = \frac{v_x}{v}$ with respect to the nodes (see Figure 1) is given by:

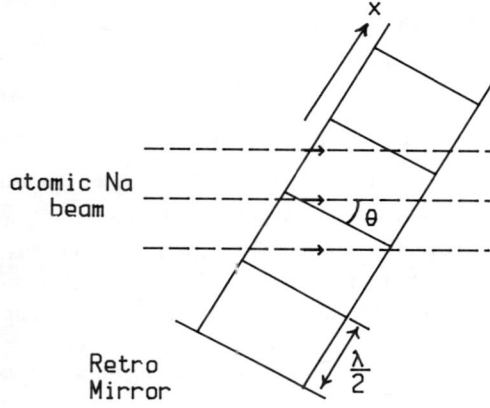

Fig. 1. Diagram of the interaction region.

$$E(x,t) = 2E_o e^{-(t/\tau)^2} \cos(kx - k v_x t)\cos\omega t$$

where x is the value of the coordinate where the atom enters the waist and τ is the time to traverse the waist in orthogonal alignment. Solving the Schrodinger equation for a two level atom in this time dependent potential in the limit where $\Delta >> \Gamma$ yields a relation concerning the amplitude for the ground state atom, $a_1^n(t)$, to have momentum $p_n = n\hbar k$:

$$i\dot{a}_1^n = \frac{\Omega_o^2 e^{-2(t/\tau)^2}}{4\Delta} [a_1^{n-2} \cos 2k v_x t + 2a_1^n + a_1^{n+2} \cos 2k v_x t] \tag{1}$$

The solution to this equation yields the final probability, $P_n(t = +\infty)$, for the atom initially ($t = -\infty$) in the ground state to gain momentum $p_n = n\hbar k$ as $P_n = |a_1^n(\infty)|^2 = J_{n/2}^2(z)$ where

$$z = \frac{\Omega_o^2}{2\Delta} \int_{-\infty}^{\infty} e^{-2(t/\tau)^2} \cos 2k v_x t \, dt = \sqrt{\frac{\pi}{8}} \frac{\Omega_o^2 \tau}{\Delta} \exp\left\{\frac{-(k v_x \tau)^2}{2}\right\}. \tag{2}$$

The *rms* momentum transfer for this distribution is given by

$$p_{rms}(v_x) = \sqrt{2} \cdot \hbar k \cdot z = \sqrt{\frac{\pi}{4}} \frac{\Omega_o^2 \tau}{\Delta} \exp\left\{\frac{-(k v_x \tau)^2}{2}\right\}. \tag{3}$$

The dependence of Eqs. 2 and 3 on tilt angle can be interpreted in this way. As an eigenstate of transverse momentum, $a_1^n(t)$, traverses the standing wave field at a given tilt angle the rate for this momentum state, a_1^n, to transfer to a neighboring state, a_1^{n+2} or a_1^{n-2}, via absorption/stimulated emission of photon pairs from the two counterpropagating traveling waves is modulated at the frequency that the atom perceives as the Doppler shift between these two waves, i.e., $\Delta\omega_{Doppler} = 2k v_x$. This leads to a reduction of the rms momentum transfer as the tilt angle is increased.

EXPERIMENT

The high resolution experimental apparatus, described in a previous paper[1], consists of a supersonically cooled sodium source ($\frac{\Delta v}{v}$ = 11% FWHM), optically pumped to a two level system and collimated with two 10 μ slits spaced 0.9 m apart. The final momentum distributions are measured 1.2 m downstream from the interaction region by a 25 μ scanning hot wire detector. The overall momentum resolution of the machine is approximately 0.7 $\hbar k$ (FWHM).

Figure 2 displays typical data for various tilt angles, $\theta = \frac{v_x}{v}$. As predicted, the amplitudes of the even order peaks fit a Bessel function distribution of an argument, z, that falls off as a Gaussian with respect to tilt angle.

Various data points of p_{rms} vs. tilt angle are shown in Figure 3 for a constant laser power and detuning.

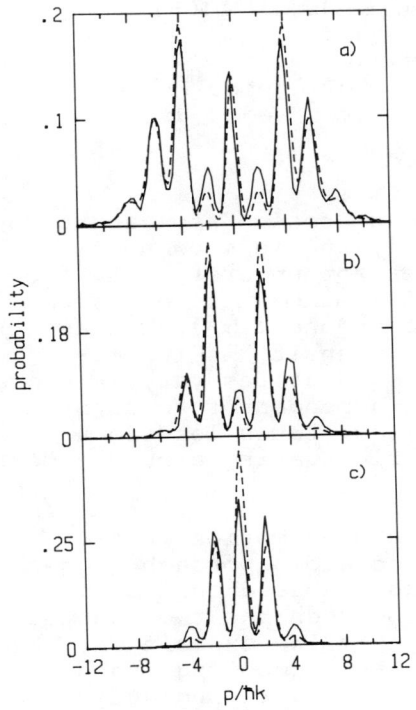

The data points were fit to a Gaussian function, $p_{rms}(v_z) = p_{rms}(0)\exp\left\{-\dfrac{v_x^2}{2v_0^2}\right\}$ which yielded the values $p_{rms}(0) = 4.81 \pm .24\ \hbar k$, and $v_0 = 1.30 \pm .04$ m/s. The independently measured values of the Gaussian waist, laser power, and detuning yield

$$p_{rms}(0) = \sqrt{\dfrac{\pi}{4}\dfrac{\Omega_o^2 \tau}{\Delta}}\ \hbar k = 4.81 \pm .1\ \hbar k$$

and $v_0 = \dfrac{1}{k\tau} = 1.17 \pm .05$ m/s. The predicted and observed values for v_0 agree to within 10%.

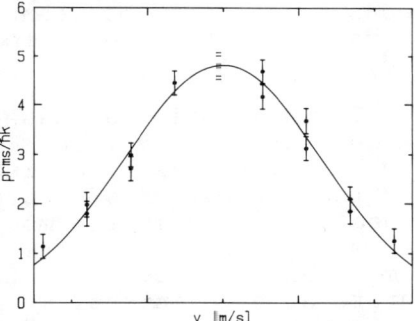

Fig. 2. Diffraction patterns for a) $v_x = -0.06$ m/s b) $v_x = -1.22$ m/s c) $v_x = 1.68$ m/s.

Fig. 3. p_{rms} vs. velocity, v_x, for $\Omega_o = 5.5\ \Gamma$, and $\Delta = 28\ \Gamma$.

CONCLUSION

In conclusion, we have investigated the behavior of the standing wave dipole force on an atomic beam as a function of the velocity component of the atoms along the $\vec{k}$-vector of the standing wave.

REFERENCES

* Current Address: National Bureau of Standards, Gaithersburg, Maryland 20899.

1. P.L. Gould, G.A. Ruff and D.E. Pritchard, Phys. Rev. Lett. 56, 827 (1986).

2. P.J. Martin, P.L. Gould, and D.E. Pritchard, Phys. Rev. Lett. (to be published).

3. A.F. Bernhardt and B.W. Shore, Phys. Rev. A 23, 1290 (1981).

THEORY OF ATOMIC MOTION IN LASER LIGHT

C. Cohen-Tannoudji
Ecole Normale Supérieure et Collège de France
24 rue Lhomond F-75231 Paris Cedex 05

ABSTRACT

We present in this lecture a dressed atom approach to atomic motion in laser light [1], and we show how it provides simple physical pictures for understanding the main features of dipole or intensity gradient forces (mean value, fluctuations, velocity dependence...). The essential idea is that, if the laser intensity is inhomogeneous, the energies of the dressed states vary in space. This gives rise to dressed state dependent forces. Spontaneous transitions between dressed states lead to a multivalued instantaneous force fluctuating around a mean value.

In the particular case of atomic motion in an intense standing wave, the dressed atom approach suggests a new efficient laser cooling scheme based on a stimulated redistribution of photons between the two counterpropagating waves by the moving atom. We explain why, by contrast with usual radiation pressure cooling, these "stimulated molasses" work for a blue detuning and do not saturate at high intensity. Finally, we report a recent experimental observation of this effect on a Cesium atomic beam [2].

REFERENCES

1. J. Dalibard and C. Cohen-Tannoudji, J.O.S.A. **B2**, 1707 (1985).
2. A. Aspect, J. Dalibard, A. Heidmann, C. Salomon and C. Cohen-Tannoudji, Phys. Rev. Lett. **57**, 1688 (1986).

© American Institute of Physics 1987

LASER COOLING AND TRAPPING OF ATOMS

J. E. Bjorkholm, S. Chu, A. Ashkin, and A. Cable
AT&T Bell Laboratories
Holmdel, N.J. 07733

ABSTRACT

We recently demonstrated the first optical trapping of atoms. An important prelude to that work was our earlier demonstration of "optical molasses," a technique for creating a long-lived collection of ultra-cold atoms. This paper presents a pedagogical discussion of those experiments and of the background material needed to understand them.

INTRODUCTION

We recently demonstrated the optical trapping of atoms for the first time [1]. Crucial to that experiment was our prior demonstration of a technique now commonly referred to as "optical molasses" [2]. Optical molasses is a means for providing strong viscous damping of atomic motion. Using it we were able to create the long-lived, reasonably dense collection of ultra-cold atoms needed to load the optical trap. In this paper we present a pedagogical discussion of our recent work and the background needed to understand it.

It has been well understood for many years that a beam of light can exert forces on objects that it illuminates. Before the invention of the laser, however, the available conventional incoherent light sources were so weak and spectrally broad that applications of radiation pressure were limited. The advent of the laser, with its intense and highly directional light beams, dramatically changed the situation and gave the experimentalist many new options. This was clearly demonstrated for the first time in 1970 when Arthur Ashkin used a laser beam to accelerate, trap, and manipulate micron-sized, transparent dielectric spheres suspended in water [3]. In the same paper Ashkin also pointed out that similar radiation-pressure forces could be exerted on an atom when the frequency of the light is tuned near the frequency of one of the atomic resonance transitions. Ashkin's paper initiated the revival of interest in radiation pressure. Since then a great deal of effort has been expended in studying and demonstrating the ways in which laser radiation pressure can be used to advantage with both macroscopic- and microscopic-sized particles [4]. It is not the purpose of this short paper to review that large body of work. Instead, we will discuss the basic ideas which led to our recent experiments while attempting to adequately reference the preceding work which most strongly impacted ours.

Before proceeding, we point out that there is a strong similarity between the light forces exerted on transparent, dielectric spheres and those exerted on atoms. In both cases the forces are due to an exchange of momentum between the light and the illuminated object. They arise out of the generalized scattering of the light by the object; that is, reflection and refraction for a dielectric sphere and spontaneous and stimulated scattering for an atom. There is, however, a major fundamental difference between the two situations, which can be described qualitatively as follows. For macroscopic-sized particles and intense light beams, the rate at which photons are scattered is so large (many photons are simultaneously scattered) that statistical fluctuations of the forces are negligible and a deterministic description of the interaction is possible. For an atom, however, a complete description of the forces is more complicated since the quantized nature of light is of crucial importance. That is, an atom scatters only one photon at a time and, even though the scattering rate may be high, this leads to statistical fluctuations of the radiation-pressure forces in direction and in time. These "quantum fluctuations" of the forces tend to "heat" the atoms and are a limiting factor in some applications of resonance-radiation pressure. Learning how to minimize, or at least counteract, quantum heating was crucial to the success of our trapping experiments.

THE FORCES OF RESONANCE-RADIATION PRESSURE

There are two basic types of resonance-radiation-pressure forces; they arise from the spontaneous and stimulated scattering of light by the atom. Consider an atom illuminated by a Gaussian (TEM$_{00}$) laser beam of frequency ν and wavelength λ which propagates along the z-axis (the longitudinal direction). The transverse intensity distribution of the light is $I(r) = I_0 \exp(-2r^2/w^2)$, where w is the spot size of the laser beam.

The spontaneous force arises from the spontaneous scattering of light by the atom. In a single scattering event the atom absorbs an incident photon and sometime later reemits another photon in some random direction. Averaged over many scattering events, the outgoing photons carry away no momentum since the reemission is symmetrical. Thus the average spontaneous force, F_s, is in the direction of the light propagation. Its magnitude is simply the rate at which photon momentum is absorbed from the incident beam by the atom. For an idealized two-level atom,

$$F_s = (h/\lambda\tau) \left[\frac{1}{2} \left(\frac{p}{1+p} \right) \right], \qquad (1)$$

where τ is the natural lifetime of the atomic excited state and p is the saturation parameter for the transition. The quantity in brackets is the fraction of time that the atom spends in its excited state. The dependence of the force on the optical frequency is obtained using

$$p = \frac{I}{I_s} \left(\frac{1}{1+q^2} \right)$$

where I is the optical intensity; $I_s = \pi h\nu/\lambda^2\tau$ is the saturation intensity for the transition; $q = 2\Delta\nu/\Delta\nu_N$ is the normalized detuning; $\Delta\nu = \nu' - \nu_0$, where ν' is the light frequency in the rest frame of the atom and ν_0 is the atomic resonance frequency; and $\Delta\nu_N = 1/2\pi\tau$ is the natural linewidth (FWHM) of the transition. For p>>1 the average spontaneous force saturates to a maximum value of $h/2\lambda\tau$. In the low intensity limit, p<<1, the force is unsaturated and its dependence on $\Delta\nu$ is Lorentzian. While the spontaneous force is small, it can have significant effects. Consider the sodium atom and its first resonance transition at 0.589 nm, for which $\tau = 16$ nsec. The fully saturated spontaneous force is 3.5×10^{-15} erg/cm, which is about 10^5 times the force of gravity. This force will accelerate a sodium atom initially at rest to a speed of 1.4×10^5 cm/sec (about 0.25 eV of energy) in a distance of 1 meter.

Now consider the fluctuations of the instantaneous spontaneous force about the average. These fluctuations occur because, in each scattering event, an incident photon is scattered into a random direction and the total momentum transferred to the atom is a random variable. Thus an atom illuminated by z-propagating light can acquire transverse velocity components even though the average force is strictly longitudinal. This is the origin of the "quantum heating" due to the spontaneous force that was referred to earlier. Detailed theoretical studies of the subject have been described in several papers [5,6], and we will only refer to the results for simple situations. For an atom illuminated by a traveling plane wave and $p \ll 1$, the rate of quantum heating is given by [5]

$$\frac{dW}{dt} = \frac{m}{2} \left(\frac{h}{m\lambda}\right)^2 \frac{p}{\tau} , \qquad (2)$$

where W is the energy associated with the random motion of the atom. The quantity $h/m\lambda$ is the atomic recoil velocity due to the absorption of a single photon and for the first resonance line of sodium it is about 3 cm/sec.

The so-called dipole force of resonance-radiation pressure is the same as the force exerted on an induced dipole situated in an electric field gradient. It can also be thought of as arising from the stimulated scattering of light by the atom. For the situation depicted in Fig.1 it is primarily transverse to the direction of light propagation. The average dipole force can be written as

$$\vec{F}_d = - \frac{4\pi}{c} \alpha \vec{\nabla} I \qquad (3)$$

where α is the atomic polarizability. For the idealized two-level atom,

$$\alpha = - \frac{1}{2}\left(\frac{\lambda}{2\pi}\right)^3 \frac{q}{1+q^2} \left(\frac{1}{1+p}\right) . \qquad (4)$$

The frequency dependence of the dipole force is dispersive in character. For $\nu < \nu_0$ the direction of the force is such as to pull the atom into the high-intensity regions of the light beam; conversely, for $\nu > \nu_0$ the atom tends to be expelled from the light. Importantly, the dipole force has no upper bound and it can easily be made much larger than the spontaneous force, a fact which is crucial to the trap which we demonstrated.

The dipole force can be derived from a conservative potential given by

$$U(r) = \frac{h\Delta\nu_N}{4} q \log_e [1 + p(r)] . \qquad (5)$$

This potential energy is the same as the shift in energy of the atomic ground state caused by the optical Stark effect. Whenever there are dipole forces exerted on an atom there will be optical Stark shifts of the atomic energy levels. As will be discussed later, it is important to recognize and account for this fact in many applications. A nice discussion of the dipole force and its relation to the optical Stark effect and dressed atom states has been given by Dalibard and Cohen-Tannoudji [7].

The fluctuations of the dipole force are more difficult to understand than are those of the spontaneous force. They are best viewed as arising because of spontaneous decay of the atom between its various dressed states. The quantum heating in our experiments was dominated by the fluctuations of the spontaneous force. Thus, further discussion of the fluctuations of the dipole force will not be given and the interested reader is referred elsewhere for the details [5-7].

THE SINGLE-BEAM, DIPOLE-FORCE TRAP

We are now in a position to describe the single-beam, dipole-force trap proposed by Ashkin in 1978 [8] and demonstrated by us in 1986 [1]. It is simply formed by a sharply focused Gaussian laser beam tuned far below the atomic resonance ($\nu < \nu_0$ and $q \ll -1$). Transverse confinement of the atom is provided by the transverse dipole forces. Longitudinal confinement is brought about by balancing the longitudinal spontaneous force with the longitudinal dipole force which exists because of the strong longitudinal gradients of the intensity (strong focusing). The equilibrium is stable in all directions. It is useful to consider the shapes of the potential wells seen by the atom. One must be careful in doing this, however, since the spontaneous force <u>cannot</u> be derived from a conservative potential. This means that a true three-dimensional, conservative potential cannot be given for the trap. Instead, we must treat the transverse and longitudinal "potentials" separately. They are given by

$$U_t(r) = \frac{h\Delta\nu_N}{4} \frac{I_o}{I_s} \frac{1}{q} \exp[-2r^2/w^2]$$

and

$$U_\ell(z') = \frac{h\Delta\nu_N}{4} \frac{I_o}{I_s} \frac{1}{q} \left\{ \frac{1}{1+z'^2} + \frac{k^2 w_0^2}{q} \left(\frac{\pi}{2} + \tan^{-1} z' \right) \right\}$$

where $I_o = 2P/\pi w_0^2$ is the peak laser intensity, P is the power in the trap beam, w_0 is the focal-spot size, $k = 2\pi/\lambda$, and $z' = 2z/kw_0^2$. Figure 1 shows the shape of the normalized longitudinal potential for several values of laser detuning and parameters typical of our experiment; they are P=200 mW, w_0=10 μm, $\Delta\nu = -150$ GHz ($q=-3\times 10^4$), and I_s = 20 mW/cm. The well depths for those same parameters are $\Delta U_r = 3.5\times 10^{-18}$ ergs = 25 mK and $\Delta U_\ell = 2.2\times 10^{-18}$ ergs = 15 mK. In reality, the above estimates for the potential well depths are overestimates for our experiments. First, the sodium atom is not a good approximation to a two level atom; the trap depths are reduced when one accounts for the actual level structure. Secondly, for reasons to be explained shortly, the trap beam was chopped with a duty cycle of 1/2. Thus for the actual trap that we demonstrated the average transverse well depth was the equivalent of only 10 mK and the average longitudinal well depth was only 5 mK.

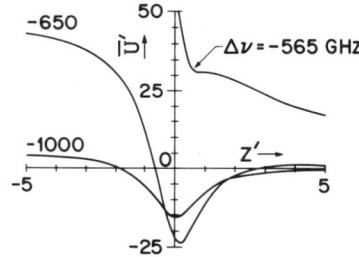

Fig. 1

It can now be readily understood why it took a period of years to learn how to demonstrate this optical trap. The reasons are as follows: a) since the trap is very shallow, ultra-cold atoms (T≈1mK) are required to load them. No such source of cold atoms existed prior to 1985. b) Since the volume of the trap is very small, on the order of 10^{-7} cm^3, the source must provide a dense collection of ultra-cold atoms in order to facilitate loading of the trap. c) Finally, even if a cold atom were loaded into the trap the quantum fluctuations of the forces would cause it to rapidly heat up. Without an effective means of damping the atomic motion the atom would "boil" out of the trap about 10 msec after being loaded. The solution of these three problems is "optical molasses", which we demonstrated in 1985 [2].

OPTICAL MOLASSES

The basic idea behind the technique now widely referred to as optical molasses was suggested by Hansch and Schawlow in 1975 [9]. Consider an atom constrained to move along the x axis. Let it be illuminated by oppositely directed laser beams of equal intensity propagating along the x-axis and tuned slightly below the atomic resonance. If the saturation parameter for each beam is small ($p \ll 1$), then the force exerted on the atom is just the sum of the forces of each beam acting individually. When the atom is at rest the average spontaneous forces exerted by the two light beams are equal and opposite. However, when the atom is moving the laser beam propagating against the atomic motion is Doppler shifted closer to resonance and the co-propagating light beam is shifted further from resonance. As a result, there is a net average force which opposes the atomic motion and which, for small velocities, is given by

$$f = \frac{8\pi h}{\lambda^2} p_o \frac{q}{(1+q^2)^2} v_x = -\beta v_x$$

where p_o is the on-resonance saturation parameter for each beam individually. The above approximation is valid as long as the Doppler shifts are small compared with the natural linewidth; for sodium, the expression is valid for velocities less than 150 cm/sec. The viscous damping force is maximized for $q = -1$, which corresponds to a detuning of one-half the natural linewidth below resonance. The average atomic velocity damps to zero with a decay time m/β (16 μsec for sodium, $p_o = 0.1$, and $q = -1$). While its average velocity is being damped to zero, the atom travels over the distance mv_o/β, where v_o is its initial velocity; for optimum detuning and an initial velocity equal to the atomic recoil velocity due to the absorption or emission of a photon this distance is $\lambda/2\pi p_o$ (about 1 μm for sodium). While $<v_x>$ damps to zero, the value of $<v_x^2>$ is nonzero in equilibrium because the effects of quantum heating cannot be ignored. The equilibrium is established when the rate of heating due to quantum fluctuations, twice Eq. 2, equals the average cooling rate $\beta<v_x^2>$. This yields an equilibrium atomic temperature given by

$$kT_{eq} = m<v_x^2> = h\Delta\nu_N/2 \ .$$

This result is rigorously valid for a 2-level atom situated in a one-dimensional standing wave for which $p_o \ll 1$. On less firm grounds the above reasoning can be extended to three dimensions, yielding the same equilibrium temperature. Treating sodium as a 2-level atom, we find $T_{eq} = 240$ μK.

Optical molasses is simply the extension of these ideas to three dimensions, using three sets of counter-propagating laser beams along the three axes. A slowly moving atom situated in the mutual intersection of the six beams experiences three-dimensional damping and its average velocity is rapidly reduced to zero. As a result, an atom finds it difficult to escape from the optical molasses and it is confined there for a long time; it eventually escapes because of the quantum fluctuations. In spite of the long confinement times that we will describe, optical molasses is not an optical trap because there are no restoring forces. Invoking a simplified picture of the quantum fluctuations once again, an atom in optical molasses experiences velocity changes of $h/m\lambda$ each time a photon is absorbed or emitted. Each velocity kick is viscously damped out and the net result is that the atom executes a random walk in three dimensions with a step size of $\lambda/2\pi p_o$. This motion is similar to the Brownian motion of a dust particle in a very viscous liquid, such as molasses! After a time t, the mean-square deviation of the atom from its starting point is

$$<r^2> = \frac{t}{T}(\lambda/2\pi p_o)^2.$$

where $T \approx \tau/3p_o$ is the average time between scattering events (six beams, p_o each). Thus, if the radius of the roughly spherical optical molasses volume is R, the typical time required for an atom to escape is

$$t_{esc} \approx 4\pi^2 p_o \tau R^2/3\lambda^2$$

Using $p_o = 0.1$ and $R = 0.4$ cm (as in our original experiment) we obtain $t_{esc} \approx 1$ sec for sodium. This is an overestimate of the actual confinement time since it assumes that the random-walk motion takes place throughout all of space. In reality, an atom which reaches the boundary of optical molasses escapes forever since it flies off unimpeded (the boundary is a sink). Including this consideration reduces the storage times by roughly a factor of 3 for the conditions of our first experiments [2].

All of our experiments have been carried out using a pulsed atomic beam. This has several benefits. First, measurements of atomic velocities and storage times are greatly simplified. Secondly, the fraction of slow atoms ($v \approx 2 \times 10^4$ cm/sec) in our pulsed beam seems to be much larger than the fraction in a cw atomic beam. This is an important consideration since we inject slow atoms into optical molasses. Our optical molasses can only capture atoms that are injected with speeds of less than about 3×10^3 cm/sec; the pulsed atomic beam does not contain a significant number of atoms traveling that slowly. To produce the required slow atoms, we reduce the speed of some of the atoms in the beam using the spontaneous force exerted by a laser beam propagating against the atomic beam. In order to achieve significant slowing of the atoms it is necessary to compensate for the changes in the Doppler shift of the light as the atomic velocity changes. Two methods for accomplishing this compensation and dramatic slowing of atomic beams have been demonstrated. Phillips and his co-workers [10] have used a tapered magnetic field to produce a spatially dependent Zeeman shift of atomic energy levels to bring about compensation. We have used the technique demonstrated by Ertmer, et al. [11] in which the frequency of the laser is electro-optically swept at a rate chosen to compensate for the changing Doppler shift as the atoms slow down. In our experiments, atoms with initial velocities of about 2×10^4 cm/sec were slowed down to final speeds of about 2×10^3 cm/sec over a distance of only 6 cm using a laser intensity of about 120 mW/cm². In order to avoid optical pumping of the sodium ground state, which would severely limit our ability to interact repetitively with the atom, we use electro-optic frequency modulation to impose sidebands separated by 1712 MHz on the light beams; it is these sidebands which interact with the atom. The frequency sweeping of the slowing beam is accomplished using a travelling-wave electro-optic modulator. The three incident optical molasses beams are spatially filtered and accurately collimated with a spot size of about 4mm (radius). Counter-propagating beams are obtained by accurately retro-reflecting the incident beams back on themselves; the power in each incident beam typically is about 30 mW.

The pulsed atomic beam is intense enough that atoms confined within the optical molasses were easily visible by eye. We also monitored the resonance fluorescence of the atoms in optical molasses using a collection lens and a photomultiplier. An example of the averaged fluorescence signal as a function of time is shown in Fig. 2 (on following page). It is seen that more than 1/2 the atoms are confined within the optical molasses for times greater than 0.1 sec., in rough agreement with expectations. The baseline in the figure is due to scattered light within the vacuum chamber. As a check, we find that the atomic confinement disappears if any one of the six optical molasses beams is blocked or significantly misaligned. Tuning of the laser with respect to the resonance frequency is also critical and a detuning of about -5 MHz gives best results, also as expected.

Finally, longer confinement times were obtained when larger spot sizes were used for the molasses beams.

A direct measurement of the temperature of atoms in optical molasses was made using a time-of-flight technique. In this measurement we determined the fraction of atoms which escape when the optical molasses beams are all turned off for a period of several msec. The fraction remaining as a function of the off-time was then compared with a calculation which assumed a spherical shaped, well-defined optical molasses initially filled uniformly with atoms having a Maxwell-Boltzmann distribution of velocities. By fitting the results of the calculation to the experimental data using the atomic temperature as the only adjustable parameter we found $T_{eq}=240\mu K$, in excellent agreement with the quantum heating predictions.

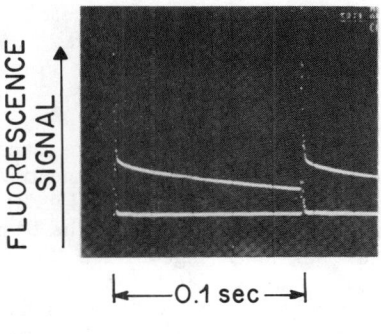

Fig. 2

Further results about optical molasses are given in our original report [2]. Suffice it to say by way of summary that we are able to routinely cool approximately 10^6 atoms to a temperature of about 240 μK while confining them in a volume of about 0.5 cm^3 for times of several tenths of a second.

OPTICAL TRAPPING

The atoms confined in optical molasses constitute the dense ($\sim 10^6$cm^{-3}) source of ultra-cold ($\sim 240\mu K$) atoms needed to demonstrate an optical trap. It should be noted that optical molasses has an added advantage for small, hard to fill traps. That is, since the atoms in optical molasses undergo random-walk motion, each atom has many chances to fall into the optical trap. Atoms escape from optical molasses by reaching the outer boundary of the molasses or by falling into the trap. Thus the fraction of atoms in optical molasses that eventually get trapped is roughly given by the ratio of the surface area of the trap to the area of the optical molasses boundary.

Optical molasses was also used as the required cooling mechanism for trapped atoms. However the trap and optical molasses beams were not simultaneously applied to the atoms. This is because of the optical Stark shifts associated with the trap beam, which were as large as $20\Delta v_N$ in our experiments. Optimum damping occurs for $\Delta v = \Delta v_N/2$; to ensure that the Stark shifts did not prevent efficient damping, we used square-wave modulation of the beams [12] to periodically turn the trap beam off and simultaneously turn on the optical molasses beams. If the chopping cycle is short compared with the transverse oscillation time for a trapped atom, then the atomic motion approximates damped harmonic motion in a potential well having one-half the cw well depth. If the chopping rate is too slow, the atomic motion becomes unstable; if the chopping rate is too fast, the sidebands generated by the square-wave modulation are too far from the atomic resonance and the efficiency of the damping is reduced.

To demonstrate optical trapping, we first loaded optical molasses with atoms as previously described. After a delay of several msec, to allow the atoms to come to equilibrium, we introduced the trap beam while simultaneously turning off the molasses beams. The beams were subsequently turned on and off at a rate of about 1 MHz, as will be detailed later. The trap laser beam was obtained from a second dye laser and it entered the vacuum chamber in a direction nearly parallel to one of the optical molasses beams. The trap beam was linearly-polarized, its power was about 220 mW, and it was focused to a spot size $w_o = 10\mu m$, situated within the optical molasses. The optically trapped atoms were detected visually, by video camera, and photographically as a small, intense spot of fluorescence situated within the much larger and much weaker cloud of fluorescence from atoms in optical molasses (See Fig. 3). The brightness of the small spot indicated that the density of the trapped atoms was much higher than the density of atoms in optical molasses. Since the trap laser beam was tuned very far below the atomic resonance, virtually all the fluorescence was caused by the optical molasses beams.

Fig. 3

For the same reason, the atomic heating caused by the trap beam was so small that the temperature of the atoms in the trap was expected to be close to the cw molasses temperature.

Several tests were made to confirm that we were in fact observing three-dimensional trapping; based on previous observations [13], there was no question that two-dimensional, transverse trapping could be observed. First, the bright, oval shaped spot coming from the trapped atoms occured only within the range of detunings expected for axial trapping. For example, trapping was observed between -570 to at least -1300 GHz below the D_2 resonance at 0.5890 nm; best trapping occurred around -650 GHz. These observations are in excellent agreement with the calculated dependence of the normalized axial well depth on detuning, as shown in Fig. 4. Secondly, the shape of the trap fluorescence varied with detuning and laser intensity as expected from calculations of the axial potentials. Finally, we observed that the lifetime of the trapped atoms was longer than for the atoms confined in optical molasses. Quantitative measurements were made by using a video waveform monitor to analyze the decay of fluorescence with time. Lifetimes for the trapped atoms were around 1 sec, while those for atoms confined in optical molasses were about 0.5 sec. While much longer lifetimes are predicted for the trapped atoms, we surmise that the actual lifetimes are determined by the imperfect vacuum in our chamber. Trapped atoms are ejected from the trap by even very weak collisions with the residual gas atoms which are at a temperature of about 300 K.

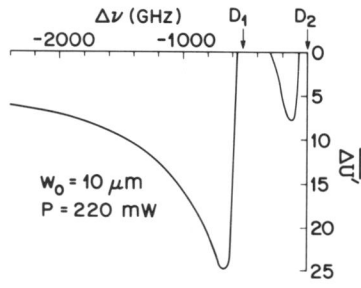

Fig. 4

Measurements were also made on the effect of varying the chopping period for the trap and optical molasses beams. We found excellent agreement with our computer simulations. At a detuning of -650 GHz and P = 220 mW, we were able to obtain good trapping for chopping periods between 0.4 and 10 μsec.

We were also able to infer the temperature of the trapped atoms. This was done by comparing the observed power threshold for trapping with calculations [1]. The inferred temperature was approximately 360 μK, only about 50% greater than the temperature of atoms in optical molasses. Based on this temperature, we can calculate the fraction of the trap volume in which the atoms reside. For a detuning of -650 GHz and a trap beam power of 220 mW, the atoms are confined within a cylinder 210 μm long and diameter of 2.5 μm in diameter, giving a total volume of $1\times 10^{-9} cm^3$. We deduced the number of atoms in the trap and their density in several ways [1], all of which yielded consistent results. We concluded that we had trapped about 500 atoms and that their density in the trap was about $1\times 10^{12} cm^{-3}$. Finally, the collection of trapped atoms could be moved at speeds of about 1 cm/sec by manually scanning the location of the trap beam focal spot.

CONCLUSION

The use of resonance-radiation pressure to cool, trap, and manipulate atoms represents a new and potentially very useful tool for experimental atomic physics. A host of possibilities immediately come to mind; we mention a few of them here. Using atomic beam slowing and cooling techniques it should be possible to produce reasonably intense beams of very slow atoms. Such atoms would be of interest for high resolution spectroscopy and time standards work, for studying ultra-low energy atom-atom collisions and possibly molecule formation (using two slow atomic beams), for studying the collision of cold atoms with surfaces, and for atomic "fountain" experiments[14]. The combination of atomic density and low temperature we have already achieved in the optical trap represents a new regime for gas physics. Pushing to higher densities and even lower temperatures may allow new phenomena to be observed. For instance, when the interatomic spacing in a gas sample is comparable to or smaller than λ_d, the deBroglie wavelength for the atom, interesting quantum phenomena will become important. For the sodium atom, $n\lambda_d^3 \approx 1.45\times 10^{-22} n/T^{3/2}$. In the trap which we demonstrated $n\lambda_d^3 \approx 2\times 10^{-5}$ and the atomic sample is quantum-mechanically dilute. However, a density of 10^{16} cm^{-3} and a temperature of 120 μK would yield $n\lambda_d^3 \approx 1$; achieving this combination does not seem an unreasonable extension of present techniques. Thus we look forward to exciting new possibilities and applications as progress continues in the field of resonance-radiation pressure on atoms.

REFERENCES

[1] S. Chu, J. E. Bjorkholm, A. Ashkin, and A. Cable, Phys. Rev. Lett. **57**, 314 (1986).
[2] S. Chu, L. W. Hollberg, J. E. Bjorkholm, A. Cable, and A. Ashkin, Phys. Rev Lett. **55**, 48 (1985).
[3] A. Ashkin, Phys. Rev. Lett. **24**, 156 (1970) and **25**, 1321 (1970).
[4] For a review of applications through 1980, see A. Ashkin, Science **210**, 1081 (1980); for more recent work see J. Opt. Soc. Am. B2, special issue on "The Mechanical Effects of Light", P. Meystre and S. Stenholm, eds. (1985).
[5] J. P. Gordon and A. Ashkin, Phys. Rev. **A21**, 1606 (1980).
[6] R. J. Cook, Phys. Rev. **A20**, 224 (1979) and Phys. Rev. Lett. **44**, 976 (1980).
[7] J. Dalibard and C. Cohen-Tannoudji, J. Opt. Soc. Am. B2, 1707 (1985).
[8] A. Ashkin, Phys. Rev. Lett. **40**, 729 (1978).
[9] T. W. Hansch and A. L. Schawlow, Opt. Comm. **13**, 68 (1975).
[10] W. D. Phillips and H. J Metcalf, Phys Rev. Lett. **48**, 596 (1982); J. Prodan, A. Migdall, W. D. Phillips, I. So, H. Metcalf, and J. Dalibard, Phys. Rev. Lett. **54**, 992 (1985).
[11] W. Ertmer, R. Blatt, J. L. Hall, and M. Zhu, Phys. Rev. Lett. **54**, 996 (1985).
[12] J. Dalibard, S. Reynaud, and C. Cohen-Tannoudji, Opt. Commun. **47**, 395 (1983).

[13] J. E. Bjorkholm, R. R. Freeman, A. Ashkin, and D. B. Pearson, Optics Lett. **5**, 111 (1980).
[14] R. G. Beausoleil and T. W. Hansch, Opt. Lett. **10**, 547 (1985).

TRAPPING ATOMS WITH RADIATION PRESSURE

D. E. Pritchard and E. L. Raab
Physics Department and Research Laboratory of Electronics,
Massachusetts Institute of Technology, Cambridge, Massachusetts 02139

ABSTRACT

We show that the optical Earnshaw theorem does not always apply to atoms and that it is possible to confine atoms using spontaneous light forces produced by static laser beams. A necessary condition for such traps is that the atomic transition rate is not a function of the light intensity only. We give general approaches which satisfy this condition and present a new trap design utilizing these principles. This trap has a depth on the order of a kelvin and a volume of several cm^3.

A cold, dense cloud of atoms is an ideal subject for experiments in high-resolution spectroscopy, ultra-low energy collisions, collective phenomena, and so forth. Various types of neutral atom traps have been constructed in recent years to produce and isolate such a cloud utilizing either magnetic or optical confinement forces[1-3]. Magnetic traps, especially those with a well-defined region of constant field, are better suited for spectroscopy. Optical traps, however, using commercial dye lasers, can produce a trap as deep as a magnetic "bottle" of several tesla, in a much smaller region. They can therefore more readily achieve the much higher densities favorable to atomic interaction type experiments.

The radiation forces used to optically trap neutral particles can be can be grouped into two types[4]. The first is the "dipole" force arising from the interaction of the particle's induced dipole moment with the field intensity gradient. The second is the "scattering" force associated with the transfer of momentum from photons to particles by the scattering of light. This latter force was used to cool beams of thermal atoms[5-7] and to viscously damp a collection of already cold atoms[8].

Ashkin and Gordon[9] showed that, in analogy with the electrostatic Earnshaw theorem, a trap relying solely on the scattering force would not be stable for a simple polarizable object, such as a dielectric sphere. The reason is that, in the absence of sources and sinks of radiation, the divergence of the Poynting vector of a static laser field must be zero. Hence, if the force is proportional to the laser intensity, it must also be divergenceless, ruling out the possibility that it points inward everywhere on a closed surface. Traps built to date have utilized the relatively weak gradient-force or an ac spontaneous-force for confinement. The latter type was first proposed by Ashkin[10] and uses time-varying light intensities and/or frequencies to circumvent the "optical Earnshaw theorem" (OET) in much the same way rf traps overcome the traditional Earnshaw's theorem.

The (OET) does not apply to atoms, however. Atoms, unlike simple particles, have a multiplicity of states whose populations and energies can be manipulated to circumvent the OET. In order for an object to be trapped optically, the lasers must provide a restoring force towards the trap equilibrium point whenever the object is displaced from that point. The goal, therefore, is to contrive a scheme which coerces the atom to interact most strongly with a laser beam opposing its motion.

There are several schemes which would accomplish this; they can be grouped into two classes. The first class relies on an external influence, such as a magnetic field, to manipulate the atom's preference for a certain type of light. The field can be used, for example, to Zeeman tune the atom's resonant frequency, or to shift its quantization axis. The second

© American Institute of Physics 1987

class depends upon unusual optical pumping effects which occur in certain atoms, such as cesium. These effects conspire to cause the atom to absorb more photons from a preferred beam, even though it may be less intense than a competing beam of opposite polarization.

Several specific applications of these ideas were described in a recent publication[11]. We would also like to propose a new scheme which seems particularly attractive. Consider first one-dimension: a σ+ beam propagates along the +z-axis and a σ- beam moves along -z, superposed on a magnetic field of the form B(z)=constant×z. The dependence of the various sodium D2 transition frequencies on this field is plotted in figure 1.

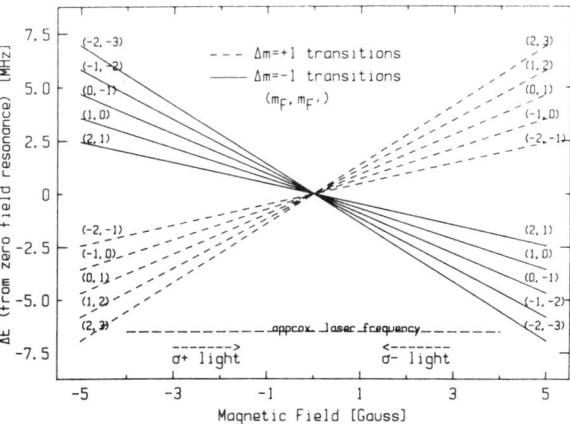

Figure 1: Field dependence of the F=2 to F'=3 transition frequencies in the sodium D2 line.

Note that each transition frequency depends linearly on the field, changing sign as the field changes direction. It is clear that for the most common transitions between hyperfine levels (F=2 to F'=3), tuning the lasers slightly to the red of the zero field resonance will coerce an atom to interact more strongly with the beam opposing its position and velocity, since it is closer to resonance. A Monte Carlo simulation was used to determine the force on the atom versus its position for various velocities; it accounts for all possible transitions in the $3S_{1/2}$ - $3P_{3/2}$ manifold (figure 2).

The trap is extended to three dimensions by placing a pair of coils of opposing current along the z-axis, separated by a radius (figure 3). By directing counter-propagating beams along each of the three Cartesian axes, then, the one-dimensional conditions for trapping are

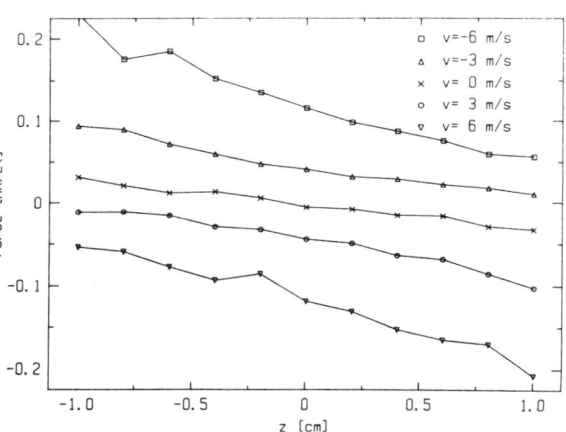

Figure 2: Monte Carlo simulation of the forces felt by a sodium atom in a one-dimensional trap with a magnetic field gradient of 5 G/cm. The trap provides both a restoring force to the origin and velocity damping.

satisfied independently along x,y, and z. Monte Carlo simulations were used to show that this configuration provides a restoring force off-axis as well. The trap size and depth is governed by the diameter of the laser beams and the magnitude of the magnetic field at the edge of the beams. Our current configuration is about a kelvin deep and several cubic centimeters in volume.

It may also be possible to avoid the restrictions of the optical Earnshaw theorem by using saturation[9] or absorption. The latter is particularly attractive: an optically dense cloud of Na will experience a maximum spontaneous radiative pressure of $\sim 10^{-7}$N/m^2 (corresponding to 5 mW/cm^2), enough to contain an atom density of 5×10^{18}/cm^3 at T = 0.25 mK according to the perfect gas law. Indeed, gas pressure limits the confinement density of a spontaneous force trap, but this limit may be increased by using a weaker transition with a correspondingly lower ultimate temperature.

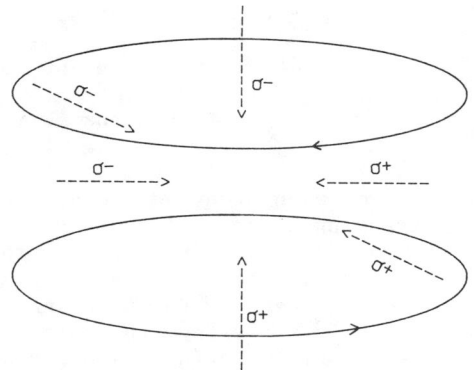

Figure 3: Trap configuration in three dimensions. The field lines at the origin point inward along the axis of symmetry and radially outward (with half the gradient) in the x-y plane. For a 5 cm coil diameter, about 50 amp-turns are needed per coil.

We gratefully acknowledge discussions with Jean Dalibard which motivated the basic principles of this trap.

References:

[1] A. Migdall, J. Prodan, W. Phillips, T. Bergeman, and H. Metcalf, Phys. Rev. Lett. **54**, 2596 (1985).

[2] S. Chu, J. Bjorkholm, A. Ashkin, and A. Cable, Phys. Rev. Lett. **57**, 314 (1986).

[3] V. Bagnato, G. Lafyatis, A. Martin, E. Raab, and D. Pritchard, to be published.

[4] See, for example, the feature issue on the mechanical effects of light, J. Opt. Soc. Am. B **2**, 1707-1860 (1985).

[5] J. Prodan, A. Migdall, W. Phillips, I. So, H. Metcalf, and J. Dalibard, Phys. Rev. Lett. **54**, 992 (1985).

[6] W. Ertmer, R. Blatt, J. Hall, and M. Zhu, Phys. Rev. Lett. **54**, 996 (1985).

[7] R. Watts and C. Wieman, Opt. Lett. (to be published).

[8] S. Chu, L. Hollberg, J. Bjorkholm, A. Cable, and A. Ashkin, Phys. Rev. Lett. **55**, 48 (1985).

[9] A. Ashkin and J. P. Gordon, Opt. Lett. **8**, 511 (1983)

[10] A. Ashkin, Opt. Lett. **9**, 454 (1984).

[11] D. Pritchard, E. Raab, V. Bagnato, C. Wieman, and R. Watts, Phys. Rev. Lett. **57**, 310 (1986).

MAGNETIC TRAPPING OF NEUTRAL ATOMS

T. Bergeman and H. Metcalf
Physics Dept., S.U.N.Y.
Stony Brook, N.Y. 11790

We have been studying the use of inhomogeneous magnetic fields for the trapping of neutral atoms via the interaction with their magnetic moments. Apart from gravitational forces (which may turn out to be of great importance in atom traps), only electromagnetic forces are of real interest for trapping neutral atoms, and these forces must necessarily arise from the interaction of multipole moments and inhomogeneous fields. In this paper we restrict ourselves to static traps and magnetic dipoles, even though optical, quasi-static, and Stark effect traps have been discussed. We present some discussion of the field configurations that might be used, of the motion of "warm" trapped atoms (about 1 K), and of the states of "cold" trapped atoms (about 0.1 μK). The basic scale of the interaction is set by the ratio of the Bohr magneton to the Boltzmann constant, about 1.5 Tesla/Kelvin.

The dominant reason for the study of magnetic traps is that they have been demonstrated to work, and therefore provide the opportunity for experiments. In addition, atoms in such a trap can be kept in the dark and in a cold environment, where the only perturbation to their internal energy levels is from their motion in the trap field. Furthermore the application of very weak, well controlled light beams for additional cooling may be possible in some cases, resulting in an opportunity for the spectroscopy of a totally new, macroscopic quantum system. There are very many new problems to study and new applications to develop.

Since the interaction between a dipole and a field depends on the field gradient, the center of a trap must be at a local extremum in the field magnitude, and for free space this can only be a minimum of the field. Therefore traps can only contain atoms in positive-going Zeeman states and for practical magnets, the atoms must be cooled to about 1 K. These criteria can be satisfied by laser cooling of an atomic beam, which optically pumps atoms into the correct sublevels and cools them to about 0.1 K. Further cooling of an atomic gas may be accomplished by optical molasses.

Calculation of the magnetic field from a particular configuration of current-carrying coils is a straightforward exercise using Maxwell's equations. In order to facilitate numerical computations, especially when considering a field that increases in all directions from a local trap center, we have developed a multipole expansion of such fields in terms of the various orders of Legendre polynomials that is particularly convenient for the cylindrical symmetry associated with magnet coils. This not only provides a convenient method for calculating the field quickly and easily, but also gives insight into certain properties of various trap designs. For example, it is impossible to build a fully isotropic trap; realizable traps therefore do not conserve orbital angular momentum.

This method has been applied to several trap designs and compared with exact calculations. We have studied the two coil (quadrupole) trap, the three coil (octupole) trap, the Ioffe trap and its variations, the "baseball" trap, and others. Various parameters for optimizing the depth, the symmetry, or the ease of construction have been considered.

The study of trap designs is important for several reasons. First, the trap must be deep enough to contain the available atoms. Second, the atomic orbits must be such as to minimize the liklihood of transitions to other Zeeman sublevels that are not confined. Thus energy considerations alone are not sufficient for trap stability: orbits must also be considered. Third, various methods for further cooling may depend strongly on the atomic orbits. Fourth, various experiments may depend upon knowledge of the field seen by the trapped atoms.

We have studied the classical motion of a classical dipole as it moves in the field of the two-coil quadrupole trap. This conical potential does not yield an analytic solution, so the study of these orbits demands numerical computation. Ours have been greatly simplified by the field expansion technique described above. As the dipole moves in the field, it experiences a torque that rotates it with respect to the field, as well as a force that directs its orbit. Away from the origin where the field is large enough so that the magnetic precession frequency is large compared with the orbital frequency, the angle between the moment and the field remains fixed (adiabatic region). But each encounter (i.e., close fly-by) with the origin of the trap results in a nutation of the dipole moment, a corresponding change in the magnitude of its interaction with the trap fields, and therefore dramatic changes in the orbits in the region away from the origin where nutation does not occur.

We have also studied the classical motion of a quantized dipole (i.e., having a fixed orientation with respect to the field) in the two coil, quadrupole trap. Direct integration of the Newtonian equation of motion has produced a variety of motions. We have found open orbits, both simple and complicated closed orbits, and the coexistence of chaotic and regular orbits at energies near and above the trap threshold. We have studied the Poincare plots of these orbits in order to determine the criteria for the onset of chaos.

The choice of an appropriate geometry for further laser cooling of atoms in these orbits remains to be done. However, it may be better to choose a totally different trap geometry in order to minimize optical pumping to untrapped hyperfine sublevels or Majorana transitions near the origin to other Zeeman sublevels. Perhaps the baseball trap, with its non-zero field at the origin and unidirectional field along one axis will turn out to be best. We plan to extend our orbit calculations to this geometry.

If trapped atoms can be cooled on a very narrow optical transition to achieve a very low temperature, their deBroglie wavelengths approach their orbit diameters and quantized motion in the trap occurs. In those traps that have zero field at the origin such as the quadrupole trap, adiabatic following of the field may not be possible for the atomic moment. Atoms in those quantum states with an antinode in their wavefunction at the origin may therefore be so strongly coupled to the continuum that they escape in a very short time, thereby broadening the state considerably.

Solution of the Schroedinger equation for this "new hydrogen atom" is complicated by the inhomogeneous magnetic field of the trap. We have begun the solution by applying a spatially dependent rotation operator that rotates in the field to the z direction everywhere in order to diagonalize the potential energy term. Of course, its spatial dependence means the operator does not commute with the kinetic energy operator so the transformation leaves residual dynamic effects. These both shift the energy levels and broaden them by coupling to continuum states.

We have computed eigenfunctions and eigenvalues for that part of the Hamiltonian that is diagonal in these dynamic terms. Evaluation of the off-diagonal terms requires dealing with the continuum wave functions, and this formidable task is in the future. Nevertheless, we have calculated and classified very many states in the spectrum of this quantized conical potential problem.

Magnetic trapping of neutral atoms is a new area that promises to provide very many new questions for basic research in both theory and experiment, as well as new applications to spectroscopy, collision studies, and other areas of atomic physics.

Supported by the O.N.R. and the N.S.F.

Progress Toward an Alexandrite Laser Trap for Potassium Atoms

Kuo-Ho Yang*, Xizhi Zeng and William C. Stwalley+

Iowa Laser Facility and Department of Chemistry
University of Iowa
Iowa City, Iowa 52242-1294

A detailed study on the feasibility and limitations of a corner-cube laser trap concept using a cw alexandrite laser for trapping neutral potassium atoms is given in a recent paper.[1] In particular, the confinement of the atoms in the two dimensions perpendicular to the laser will be provided in the cavity of an alexandrite laser operating in the CW TEM^*_{01} mode ("doughnut mode") tuned slightly to the blue side of the resonance line of the K atom. By reflecting the TEM^*_{01} laser back on itself with two mirrors, one "caps" the ends of the cylindrical trap, resulting in a slightly weaker end plug. This trap concept employs not laser cooling, but rather counterstreaming ^{4}He atoms, which are cooled to ≤ 1.5 K, to drastically cool the K atoms to thermal energies well below the trap depth (expected to be ~8.6 K). We have also examined various loss mechanisms for the trapped atoms. In particular, K atoms can be lost to the trap if they are multiphoton ionized, if they are heated by absorption and emission of many photons ("recoil" or "diffusional" heating), if they simply have much higher energy than the vast majority of other atoms at 1.5 K or if they form KHe (or KHe_2, etc.). Results from these investigations suggest crude lifetimes for trapped atoms of the order of 1 second.

The alexandrite ($BeAl_2O_4:Cr^{3+}$) laser is a broadly tunable solid-state vibronic laser operating at wavelengths between 700 and 830 nm.[2] Our alexandrite rods measure 3 mm in diameter and 12 cm in length, with a chromium concentration of 0.20 at. %. The rod is pumped by a capillary mercury-arc lamp in the longitudinal pumping geometry inside a single silver-plated elliptical cavity, with the rod and the lamp located at the foci of the elliptical reflectors. Power is supplied to the mercury-arc lamp by a 4.5 kW power supply, and the pump cavity is cooled by deionized water. The output laser light is polarized along the b axis of the alexandrite crystal.

Toward the implementation of the trap concept, we have successfully operated the alexandrite laser in the TEM^*_{01} mode. The maximum output power we have obtained in this mode, when an output coupler with a reflectance of 98.4% is used, is between 2.0 and 2.5 W, which is below the original expectation. To compensate for this

---

* Permanent address: Department of Physics, St. Ambrose College, Davenport, Iowa 52805.
+ Also Department of Physics.

decrease in power, we have designed a new laser cavity with a more tightly focused beam at the trap region. It is estimated that, with a new beam waist of 13.6 μm and the replacement of the output coupler by a high reflector, the original estimated trap depth of ~8.6 K can be attained. The new cavity design also incorporates an optimization requirement to minimize the effects on the beam waist due to fluctuations and uncertainties in other parameters, such as radii of curvature of mirrors, focal lengths of lenses, and positions of optical elements. The effects of thermal lensing of the alexandrite rod can be compensated for by positioning a concave lens in the cavity.[3]

1. K.-H. Yang, W. C. Stwalley, S. P. Heneghan, J. T. Bahns, K.-K. Wang and T. R. Hess, Phys. Rev. A **34**, 2962 (1986).

2. J. C. Walling, D. F. Heller, H. Samelson, D. J. Harter, J. A. Pete and R. C. Morris, IEEE J. Quantum Electronics **QE21**, 1568 (1985).

3. L. M. Osternik and J. D. Foster, Appl. Phys. Lett. **12**, 128 (1968).

"QUANTUM JUMPS" OBSERVED IN THE FLUORESCENCE OF A SINGLE ION

Th. Sauter, R. Blatt, W. Neuhauser,
and P. E. Toschek*
Universität Hamburg, D-2000 Hamburg, Fed.Rep.Germany

ABSTRACT

We demonstrated interruptions of macroscopic duration in a single trapped and cooled Ba^+ ions's 493-nm fluorescence. They are caused by transitions of the ion into the "dark" $^2D_{5/2}$ state. — Multiple simultaneous jumps of three ions indicate cooperative interaction with the light.

INTRODUCTION

One of the major consequences of BOHR's model of the hydrogen atom |1| is the necessity to accept intantaneous transitions from one stationary state of internal energy to another one. These presumptive transitions, said to go along with the emission or absorption of a light quantum, have been dubbed "quantum jumps". At first glance, their postulation seems to be at variance with the observed continuous time variation of atomic variables, and also with their quantum-mechanical description which admits the existence of superposition states. However, those observations have dealt with atomic ensembles so far. Recently, the preparation of single, cold atomic particles - ions - in specified internal and external states has been demonstrated |2-4|. Ensemble averaging, unavoidable in conventional measurements, is absent in experiments with single atomic particles: these experiments allow, for the first time, direct proof of one of the most basic concepts in quantum mechanics by repeated identical preparation and observation of an individual atomic system. Intuitive arguments predict that transitions on a very weak line are detectable via the excitation of resonance fluorescence on a strong transition coupled by a common level |5|. When such a rare transition, say, an absorption event, occurs on the weak signal line, the fluorescence on the neighbouring line is supposed to be quenched since the atom - or ion - in the upper, metastable level of the weak line is no longer available for the fast excitation and fluorescence cycles.

These conclusions have been confirmed by rigorous quantum-statistical calculations |6-11|. They show that

---
*Also: JILA, University of Colorado and NBS, Boulder, Colo., 80309, USA

interruptions in the fluorescence are expected which are on the order of the lifetime of the metastable level. These "dark" or "off" periods are the signature of quantum jumps; they have been observed recently |12-15|.

In our experiments |12,15|, a single Ba+ ion is localized in an electrodynamic ion trap and optically cooled to less than 10 mK |16,17|. The relevant energy levels of Ba+ are shown in Fig. 1. Resonance fluorescence is excited on the 493-nm line with a cw laser, and a second laser beam at 650 nm couples the $^2D_{3/2}$ level to the continuously excited $^2P_{1/2}$ level.

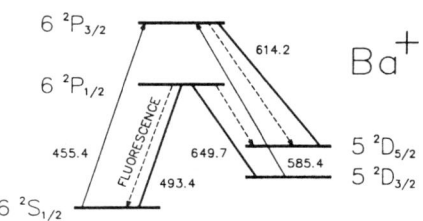

Fig. 1: Simplified energy level scheme of Ba+. Wavelength values in nm.

When the ion, once in a while, drops into the 5 $^2D_{5/2}$ level - upon laser-pumped electronic Raman-Stokes transitions -, the fluorescence becomes suppressed. Thus, the transitions into and out of the "dark" 5 $^2D_{5/2}$ state are observed with 100% detection efficiency, and with time resolution which corresponds to the mean time separation of the photoelectron counts of the fluorescence signal.

EXPERIMENTAL

A thermal beam of barium atoms is ionized by impact with 1-s pulses of a very weak electron beam. After several unsuccessful attempts to generate an ion, eventually green fluorescence signals the apprearance of a Ba+ ion in the center of the 1-mm trap, where the foci of the two coaxial laser beams are located. One beam is generated by a cw Coumarine-102 laser, the other one by a cw DCM laser. The green laser is down-tuned from resonance by 150 MHz for optimum cooling the ion, whereas the red laser is kept at the center of the 6 $^2P_{1/2}$ - 5 $^2D_{3/2}$.

The fluorescence signal is detected by a microscope, cooled photo-multiplier selected for low dark current, and a photon counter.

An additional Ba hollow cathode lamp permits us to weakly excite the 6 $^2P_{3/2}$ level from the ground or 5 $^2D_{3/2}$ metastable states |13|, and a third laser at 614 nm serves for the selective release of the ion from the "dark" level.

INTERRUPTED FLUORESCENCE

Fig. 2 shows a recording of the ion's fluorescence at 493 nm.

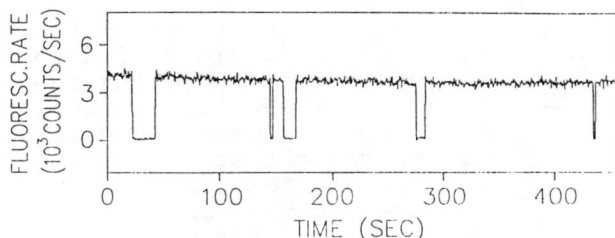

Fig. 2: Recording of the laser-excited fluorescence, at 493 nm, of a single Ba$^+$ ion.

The mean "on" time τ_+ is 136 s ± 13 s, determined by the probability for off-resonant Raman-Stokes transitions (via 6 $^2P_{3/2}$) excited by the green and red laser light at 60 and 10^3 fold saturation of the respective transitions $^2S_{1/2} - {}^2P_{1/2}$ and $^2D_{3/2} - {}^2P_{1/2}$ (Fig. 3). The mean "off" time, $\tau_- = 8$ s, is dominated by Raman-anti-Stokes transitions. There is also a small contribution from collisionally quenching the "dark" state 5 $^2D_{5/2}$.

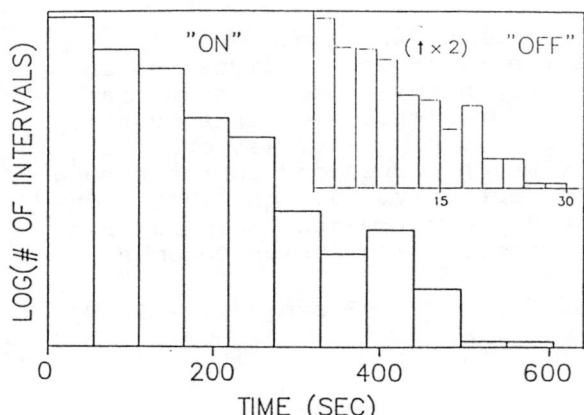

Fig. 3: Distributions of the lengths of "on" and "off" times.

Irradiation with the light of the Ba hollow cathode lamp excites the ion to the real $^2P_{3/2}$ level with a chance to decay into the "dark" state that is higher than for the far off-resonant Raman excitation. Thus, this irradiation reduces the mean "on" time to 24 ± 4 s.

For unambiguous identification of the "dark" state, we have irradiated the ion by additional cw laser light at 614 nm corresponding to the 5 $^2D_{5/2} - 6\ {}^2P_{3/2}$ transition. When a 0.4-s pulse is applied after the observed fluorescence went off, the jump is immediately undone by re-excitation, and fluorescence reappears (Fig. 4). With continuous irradiation, no jumps appear at all, since the $^2D_{5/2}$ level is coupled to the superposition of levels which forms the "on" state. This kind of manipulation or "shelving prevention" represents the active control of

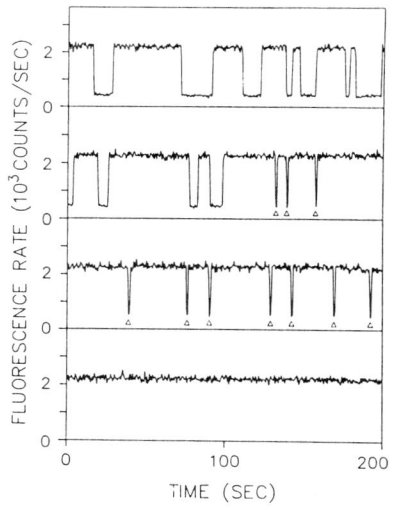

Fig. 4: Single-ion "on" and "off" intervals of fluorescence (top). Removing the ion from the "off" state ($^2D_{5/2}$) by manually pulsed laser light at 614 nm ($\triangle$, center). Coupling the $^2D_{5/2}$ level to the "on" state by continuous 614 nm laser light, which results in "shelving prevention" (bottom). Full length of uninterrupted fluorescence recording: 20 min.

an internal degree of freedom of the ion.

From recordings of the fluorescence signal as in Figs. 2 and 4, the two-time intensity correlation has been calculated. There are predictions of this quantity |9,10|: a superposition of two exponentials, one fast and one slowly decaying, which correspond to the strong and weak transitions, respectively. Only the latter one is observable with the 0.4-s experimental sampling time, and the correspondingly modified expression becomes

$$\langle I(t)I(t+\tau)\rangle / \langle I(t)\rangle^2 = 1 + \frac{\tau_=}{\tau_+} \exp\left[-(\tau_-^{-1}+\tau_+^{-1})T\right]$$

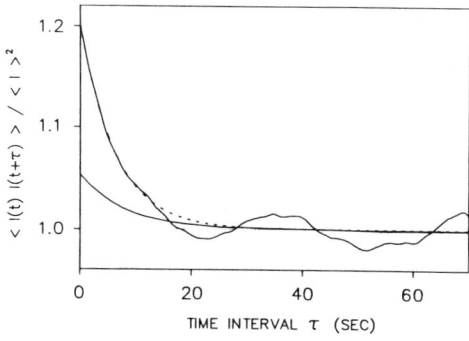

Fig. 5: Two-time intensity correlation function calculated from fluorescence recordings with laser excitation only (lower trace), and with additional lamp excitation of the $^2P_{3/2}$ level (upper trace). The oscillatory feature indicates the existence of population pulsations.

Fig. 5 shows two-time intensity correlations with and without excitation of $^2P_{3/2}$ by coherent light. They are in agreement with the values of τ_+ and τ_- derived from the distributions of "on" and "off" times. The total rate $\tau_0^{-1} = \tau_+^{-1} + \tau_-^{-1}$ is dominated by τ_-^{-1}. The combination of $^2P_{3/2}$ excitation and subsequent decay into the $^2D_{5/2}$ level with anti-Stokes back pumping forms a cyclic process which shows up as modulation of the intensity correlation function (s. Fig. 5), the signature of population pulsations. Note that without excitation of the $^2P_{3/2}$ state by incoherent light the generalized Rabi frequencies at the pump and Stokes transitions, and also the effective two-photon Rabi frequency, are very large due to the lasers being detuned far off the transitions involving the $^2P_{3/2}$ relay level |18|. Thus, the population cannot adiabatically follow, and it does not pulsate.

MULTIPLE JUMPS

The fluorescence of a small cloud of three ions shows four discrete intensity levels corresponding to three, two, one, or no ions in the "on" state (Fig. 6). Upon inspection of the recorded traces it is obvious that simultaneous jumps of two or even three ions happen much more often than expected as random coincidences. This phenomenon has been substantiated by quantitative evaluation of the rates of random multiple jumps |15|. It turns out that the observed rates exceed the random rates by more than two orders of magnitude. This observation indicates that the ions interact collectively with the light fields. Moreover, this collective action does not require ensemble averaging in order to become detectable - as in concentional experiments on super-radiance |19|. It involves real coupling of individual particles, as is proved by the huge excess of simultaneous jumps. Macroscopic collective phenomena, on the other hand, are described in terms of enhanced probability for a microscopic process as caused by the presence of the entire ensemble.

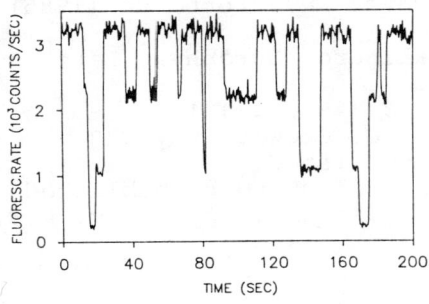

Fig. 6: Multiple jumps documented in the laser-excited 493-nm fluorescence of *three* Ba$^+$ ions.

CONCLUSIONS

We have observed interruptions of random time duration in the fluorescence of a single Ba$^+$ ion stored and optically cooled in an electrodynamic trap. The mean duration of the bright and dark intervals has been evaluated for the ion interacting with two resonant laser beams, and also for additional excitation to the $^2P_{3/2}$ state by incoherent light. The $^2D_{5/2}$ level has been unambiguously identified as the "off" state by re-exciting the dark ion to the $^2P_{3/2}$ state in order to make it join the "on" state - a superposition of $^2S_{1/2}$, $^2P_{1/2}$, and $^2D_{3/2}$ - again. This procedure establishes "quantum-manipulation" of an internal degree of freedom of a single atomic particle. The observed two-time intensity correlation function agrees with the time distributions, and indicates the existence of population pulsations. Three trapped ions show simultaneous multiple jumps at a rate vastly exceeding random coincidence. The cloud interacts collectively with the light by coupling the individual particles. The novel type of measurements exercised in these experiments does not rely on ensemble averaging. Instead, one particle is prepared under specified conditions over and over again, and a very large number of individual measurements is carried out - essentially one measurement for each photoelectron counted.

This work was supported by the Deutsche Forschungsgemeinschaft. - One of us (P.E.T.) thanks the JILA Visiting Fellows Program for support.

REFERENCES

1. N. Bohr, Phil. Mag. 26 (1913) 1, 476.
2. W. Neuhauser, M. Hohenstatt, P.E. Toschek and H. Dehmelt, Phys. Rev. A 22 (1980) 1137.
3. D.J. Wineland and W.M. Itano, Phys. Lett. 82A (1981) 75.
4. G. Janik, W. Nagourney and H. Dehmelt, J. Opt. Soc. Am. B2 (1985) 1251.
5. H. Dehmelt, Bull. Am. Phys. Soc. 20 (1975) 60.
6. R.J. Cook and H.J. Kimble, Phys. Rev. Lett. 54 (1985) 1023.
7. C. Cohen-Tannoudji and J. Dalibard, Europhys. Lett. 1 (1986) 441.
8. J. Javanainen, Phys. Rev. A 33 (1986) 2121.
9. A. Schenzle, R.G. DeVoe and R.G. Brewer, Phys. Rev. A 33 (1986) 2127, and to be published.
10. P. Zoller, M. Marthe and D.F. Walls, to be published.
11. D.T. Pegg, R. Loudon, and P.L. Knight, Phys. Rev. A33, 4085 (1986)

12. Th. Sauter, R. Blatt, W. Neuhauser and P.E. Toschek, IQEC '86 post-deadline paper, San Francisco, June 1986, and Phys. Rev. Lett. 57, 1696 (1986).
13. W. Nagourney, J.Sandberg and H.G. Dehmelt, loc. cit., and Phys. Rev. Lett. 56 (1986) 2797.
14. J.C. Bergquist, R.G. Hulet, W.M. Itano and D.J. Wineland, IQEC '86, loc. cit., and Phys. Rev. Lett. 57, 1699 (1986)
15. Th. Sauter, R. Blatt, W. Neuhauser and P.E. Toschek, Opt. Communic. 60, 287 (1986)
16. W. Neuhauser, M. Hohenstatt, P.E. Toschek and H. Dehmelt, Phys. Rev. Lett. 41 (1978) 233.
17. D.J. Wineland, R.E. Drullinger and F.L. Walls, Phys. Rev. Lett. 40 (1978) 1639.
18. R.G. Brewer and E. L. Hahn, Phys. Rev. A 11, 1641 (1975).
19. See, e.g.,"Cooperative Effects in Matter and Radiation", C.M. Bowden, D.W. Howgate, and H.R. Robl, eds., Plenum, New York 1977 - Q.H.F. Vrehen and H.M. Gibbs, in "Topics in Current Physics 27, Dissipative Systems in Quantum Optics", R. Bonifacio, ed., Springer, New York 1981, p. 111.

The "Containerless" Condensation of "Mirror" Matter

J. T. Bahns
University of Dayton Research Institute, Dayton, OH 45469

ABSTRACT

The problem of concentrating antiprotons and positrons into a high energy density form is analyzed from the viewpoint of the "containerless" condensation of cluster ions of hydrogen in ion traps using ion-neutral association. The constraints that lead to the proposed method and the available ion-neutral association channels are discussed. It is found that the condensation method may require the use of cold neutral hydrogen dimers. It is concluded that the condensation to produce the first seed (critically sized) cluster ions need only be done once in order to surmount the problems associated with "containerless" condensation.

INTRODUCTION

The central problem is the "containerless" assembly of subatomic, atomic, and molecular fragments into bulk antimatter (in this case solid antihydrogen) (1). It is necessary to *condense* antimatter into this form to achieve favorable mass payload fractions (when it is used as a fuel for propulsion) as well as have a new means of high-energy density storage (when it is used for energy storage). Since this condensation has not been done before and must be done under "containerless" conditions, severe restrictions are placed on the methods to be used. Fortunately, normal matter can be used in all aspects of the investigation since the only relevant difference in the two forms of matter (in this case) is the sign of the charge. Hence, this discussion will refer to normal hydrogen. Because the process by which antiprotons are produced involves relativistic velocities (kinetic energy much greater than interatomic and intermolecular potential energy) the antiprotons must have virtually all of this translational energy removed before condensation. In addition, the process of condensation also requires the removal of significant amounts of chemical energy. The primary emphasis is with the condensation of precooled (this includes kinetic, electronic, vibrational, and rotational) hydrogen fragments in devices such as ion traps to ultimately obtain large cluster ions of hydrogen.

DISCUSSION

The problem of laser cooling hydrogen atoms is reasonably well understood; the primary task here is the construction of an appropriate vacuum ultraviolet laser of high average (and peak) power. In recent years several methods have been used to construct Lyman-Alpha lasers that have the right wavelength (121.57 nM) to cool hydrogen atoms. Laser manipulation (cooling, deflecting etc.) of neutral molecules of hydrogen involves the development of multiline cooling laser resembling hundreds of Lyman-Alpha sources. This is unavoidable for these systems due to redistribution among the numerous available rovibration states. However, none of these laser systems come close to meeting the power (and duty cycle) requirements. At present no such cooling laser exists. However, the development of a cooling laser for hydrogen atoms will not solve fundamental problems associated with <u>condensation</u> and <u>storage</u>. It seems clear that research is needed to investigate the character and manipulation of cluster ions (positively or negatively charged) leading to a definite antimatter synthetic (or condensation) method. This will be necessary to answer the question of what to do with cold hydrogen atoms after they have been produced.

There are five primary constraints that the method (or process) must employ in order to be feasible with present and near future technology. The resulting product must be in high-energy density (or mass density) form (implying clusters). The product should be chemically stable (implying cluster ions). The process must at all times confine the antimatter in a tight "containerless" environment that is stable for very long times (months or years) (implying ion traps). It must allow for storage in a near perfect vacuum such that annihilation with normal matter is minimized. To achieve such a near-perfect vacuum ($<10^{-5}$ cm^{-3}) environment, it will be necessary to maintain the system, as well as the product cluster ions, at very low temperatures (<2K). It is thus concluded that ion clusters of hydrogen used in conjunction with ion traps are capable of satisfying these constraints. The resulting "*cluster ion approach*" (2), (formulated in collaboration with Professor W. C. Stwalley of the University of Iowa) has become the first comprehensive plan for the containerless condensation (and confinement) of antimatter into a high-energy density form.

The primary consequences of satisfying the given constraints are that: First, the method will be limited to low densities (10^7-10^{10} cm^{-3}). Second, the association rates will also be low (<10^5 cm^{-3} s^{-1}). These two difficulties can only be overcome by using large volumes and/or times respectively.

Basic research is needed that will lead to an understanding of the interactions between hydrogen cluster ions and laser fields such that upon collision chemical energy may be released through stimulated or spontaneous emission of a photon (or the ejection of some small fragment). Because wall collisions and expansions cannot be used, special condensation techniques utilizing available third-bodies are necessary. The primary goal is to obtain a detailed understanding of the available association channels. The simultaneous requirements of the association reactions are: The release of energy, conservation of momentum, and the generation of a larger cluster ion. Generally, the association pathways may be categorized as radiative, for example,

$$H_2^+ + H_2^* = H_4^+ + h\nu,$$

or dissociative,

$$H_2^+ + H_2 = H_3^+ + H$$
$$\text{or } H_4^+ + H_2 = H_5^+ + H \text{ etc.},$$

where the (*) denotes an electronically excited state. In the first case a larger cluster ion is obtained because the channel allows the excess energy (kinetic, electronic etc.) to be radiated as a photon. In the second case the excess energy is released in the form of kinetic energy of the ejected atom (certain channels may exist that involve <u>both</u> processes simultaneously).

CONCLUSION

The current analysis of the problem of "containerless" condensation indicates that the "cluster ion" approach is a potential solution that could lead to the containment of high densities of cluster ions. However, it is clear that there still remain many problems to overcome before the proposed method can be adopted with reasonable confidence.

1. R. Forward,"Antimatter Annihilation Propulsion", AFRPL TR-85-034, Final Report on F04611-83-C-0046, Subcontract RI-32901 (Sept 1985). Obtain from AFRPL/LKC, Edwards AFB, CA 93523-5000.

2. J. T. Bahns, K. M. Sando, D. C. Tardy, and W. C. Stwalley, "Proceedings of the Hydrogen Cluster Ion Study Group", University of Dayton Special Publication, in Press (Oct 1986).

AUTODETACHMENT SPECTROSCOPY OF NEGATIVE IONS

D. M. Neumark
Department of Chemistry, University of California, Berkeley, CA 94720

K. R. Lykke, T. Andersen,[a] and W. C. Lineberger
Department of Chemistry and J.I.L.A., University of Colorado,
Boulder, CO 80309

INTRODUCTION

One of the most exciting developments in spectroscopy over the last few years has been the application of high resolution spectroscopic techniques to the study of molecular negative ions. These highly reactive species play an important role in both gas phase and solution chemistry, and spectroscopy provides an excellent probe of the link between their structure and reactivity. In addition, one can study the interaction between the weakly bound outermost electron and the vibrational and rotational degrees of freedom of the negative ion.

Prior to 1983, rotationally resolved spectra had been obtained for only two negative ions, C_2^- [1] and OH^-,[2] using photodetachment techniques. Since then, spectra showing rotational structure at varying degrees of resolution have been obtained for about a dozen negative ions. The techniques used have been velocity-modulated infrared absorption,[3] photodetachment in an ion cyclotron resonance trap,[4,5] and, the subject of this article, autodetachment spectroscopy in a coaxial laser-ion beam spectrometer. This last technique provides extremely high resolution ($\Delta \nu < 20$ MHz) and has been used to obtain the first infrared vibration-rotation spectrum of a negative ion, NH^-,[6] as well as rotationally-resolved electronic spectra for $C_2H_3O^-$,[7] $C_2H_2OF^-$,[8] FeO^-,[9] CH_2CN^-,[10] and PtN^-.[11] Only the CH_2CN^- results are discussed in detail here.

In autodetachment spectroscopy, a tunable laser excites transitions between rotational levels of the ground electronic state and an excited vibrational or electronic state of the ion. The upper state rotational level must lie above the detachment threshold so that it can autodetach to the (neutral + e^-) continuum, and the autodetachment products are detected with high efficiency. Thus, a spectrum obtained by tuning the laser while monitoring the neutral or electron signal shows sharp structure wherever an autodetaching state is assessed. The structure in such a spectrum is similar to the absorption spectrum of the ion in the same frequency range, although only states lying above the detachment threshold can be observed. In addition, in favorable case the linewidths of the transitions can be related to the autodetachment lifetime of the excited state, thereby probing the dynamics of the autodetachment process.

---

[a] 1984-85 JILA Visiting Fellow, Permanent Address: Institute of Physics, University of Aarhus, Denmark.

In CH_2CN^-, as well as several other ions for which electronic spectra have been obtained, the excited electronic state is a dipole-bound state lying just below the detachment threshold. The properties of these states have been discussed in detail elsewhere.[7,12,13] These states result from the long range interaction between an electron and a neutral, dipolar molecular core with a dipole moment $\geq 2D$. In some ways, they are the negative ion analog of Rydberg states, but there are important rotation-electronic coupling effects in dipole-bound states due to the anisotropic electron-dipole interaction which is absent in Rydberg states. These effects show up most strongly in the variation of the autodetachment rates with rotational level in the dipole-bound state and will be discussed in more detail below. A preliminary account of the CH_2CN^- results has appeared previously,[10] and a more complete discussion is forthcoming.

EXPERIMENTAL

The coaxial laser-ion beam spectrometer used for the present work has been previously described in detail.[7] A 1 nA beam of $^-CH_2CN$ (cyanomethyl anion) is formed by extraction from a hot cathode discharge source containing CH_3CN and NH_3, mass selection with a 90° sector magnet, and acceleration to 2650 eV. The ions are then bent 90° by a transverse electric field and merged with the output of a home-built tunable ring dye laser pumped by all lines of an Ar II laser. The neutrals that are formed by detachment are separated from the remaining ions by another transverse electric field and strike a KDP plate. This produces secondary electrons which are then detected by an electron multiplier and counted to yield the total cross-section. The electrons that are detached in the interaction region are collected by a weak solenoidal field (~5 Gauss) and counted with another electron multiplier.

The data were taken by scanning the laser in frequency (measured with a λ-meter) while monitoring the neutrals and electrons formed as a function of photon energy and normalized to the ion current and laser power. The laser using styryl 9 dye operates from about 790 to 870 nm with ~400 mW broadband and ~150 mW single-mode output when pumped with ~5 W all lines from the Ar II laser. The dye laser is easily configured in either a standing wave, broadband mode (birefringent tuner only, $\Delta \nu \sim 1$ cm^{-1}) or in single mode ($\Delta \nu < 1$ MHz). The resolution of this spectrometer is <20 MHz, limited by the Doppler spread in the kinematically compressed ion beam.[7,14]

RESULTS AND DISCUSSION

The total photodetachment cross section for CH_2CN^- with the laser in its broadband ($\Delta \nu \cong 1$ cm^{-1}) configuration is shown in figure 1. This is similar to the spectrum obtained by Marks et al.[4] The spectrum shows sharp structure due to autodetachment,

$$CH_2CH^- + h\nu \rightarrow (CH_2CN^-)^* \rightarrow CH_2CN + e^-,$$

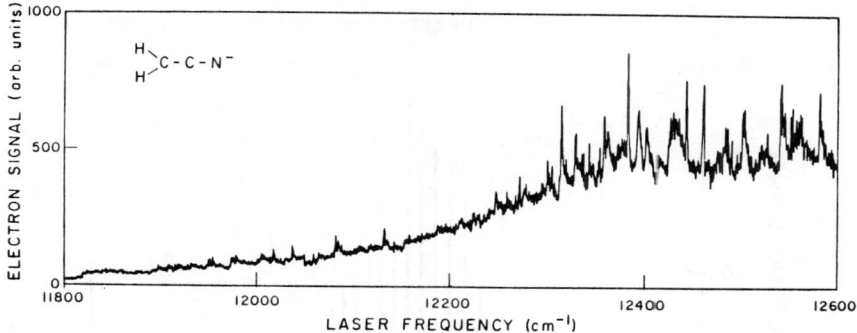

Figure 1. Broadband scan of CH_2CN^- at ~1 cm^{-1} resolution.

where $(CH_2CN^-)^*$ is the dipole bound state. This structure is superimposed on a smoothly rising background due to direct photodetachment.

$$CH_2CN^- + h\nu \rightarrow CH_2CN + e^-,$$

consisting of unresolved transitions to states in the (neutral + e^-) continuum. The spectrum shows two vibronic bands. The autodetachment structure in the vicinity of 12,400 cm^{-1} is due to transitions between the vibrationless levels of the two electronic states of CH_2CN^-, and the structure near 12,000 cm^{-1} consists of hot-band transitions in which the ground electronic state ions are in the $v = 2$ level of the hydrogen out-of-plane bending mode. The structure in this low resolution spectrum represents unresolved Q-branches and is characteristic of a perpendicular electronic transition for a slightly asymmetric prolate rotor.[15]

A low resolution spectrum was obtained for CD_2CN^- as well, and spectra covering the 0-0 and hot band regions of CH_2CN^- and CD_2CN^- were taken with the laser in its single mode configuration. Approximately 10,000 transitions were recorded. A single mode scan of a Q-branch is shown in figure 2. The individual rotational transitions are clearly resolved, and the high J transitions (at higher frequency) are considerably broader than the low J transitions. This indicates that autodetachment is faster for the higher rotational levels of the dipole-bound state.

The observed transitions were fit to an asymmetric top Hamiltonian, including centrifugal distortion, and the complete set of constants thereby obtained for both isotopes will be listed in a future publication. Table I gives A, B, and C for the $v = 0$ levels of the ground and dipole-bound states of CH_2CN^-. From these constants we see the ion is nearly a symmetric prolate rotor with A >> B ≅ C, so each rotational level can be labeled by J, the total rotational angular momentum, and the nearly good quantum number K, the projection of J along the $\underline{a}$ axis of the ion (the C-C≡N axis). The inertial defect Δ obtained from the rotational constants indicates the dipole-bound state is planar and the ground state is either planar or has a small barrier to inversion. A small inversion barrier has been inferred from a recent photoelectron spectroscopy study of CH_2CN^-.[16]

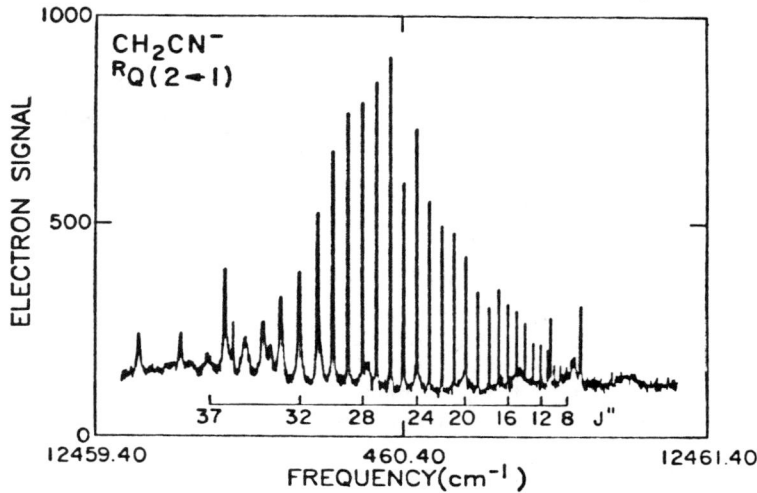

Figure 2. High resolution (<20 MHz) scan of the $^RQ(2 \leftarrow 1)$ branch in CH_2CN^-. Notice the broadening of the lines toward high J.

Table I. Partial list of rotational constants for v = 0 levels of ground and dipole-bound electronic states of CH_2CN^-

	ground state	dipole-bound state
$A(cm^{-1})$	9.29431(14)	9.51035(17)
B	0.338427(20)	0.341049(21)
C	0.327061(21)	0.328764(21)
$\Delta(amu\text{-}Å^2)$	-0.0827	0.0744

Finally, an analysis of the asymmetry doubling for the low K subbands show that this is a $^1B_1 \leftarrow {}^1A_1$ electronic transition, and that the symmetry of the dipole-bound orbital in the upper state is a_1. Bearing in mind that this orbital is localized on the electropositive end of the neutral CH_2CN core, figure 3 qualitatively depicts the shape of the orbital for the dipole-bound electron.

The linewidths of the transitions are plotted as a function of J' for different values of K' in figure 4 for CH_2CN^- and figure 5 for CD_2CN^-. Here J' and K' are the rotational quantum numbers for the dipole-bound state. The striking feature of these plots is the abrupt rise in the autodetachment rates near J' = 33 for CH_2CN^- and J' = 38 for CD_2CN^- which is independent of K'. This is not due to total rotational energy since, for example, K' = 8 rotational levels in CH_2CN^- lie about 600 cm^{-1} above K' = 1 levels with the same J'.

Instead, it appears to depend on what autodetaching transitions to the (neutral + e⁻) continue are energetically allowed.

Figure 3. Qualitative picture of the location and shape of the dipole-bound orbital.

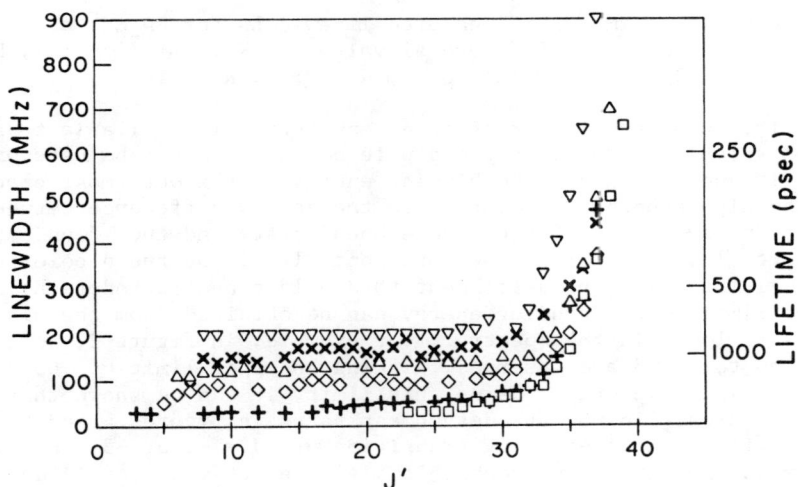

Figure 4. Experimentally measured linewidths for CH_2CN^- as a function of J' for several values of K'. □, K' = 1; +, K' = 3; ◊, K' = 5; △, K' = 6; x, K' = 7; ▽, K' = 8.

The various mechanisms responsible for autodetachment in negative ions[1,6-11,17] and autoionization in neutrals[18] have been discussed previously. In the present case, autodetachment is occurring from the v = 0 rotational levels of the dipole-bound state, and rotational autodetachment should be the dominant mechanism. In this process, the rotating ion undergoes a rotational de-excitation, and the energy thereby released ejects the dipole-bound electron. In general, autodetaching transitions involving small values of ΔJ and ℓ are favored, where ℓ is the orbital angular momentum of the ejected electron.

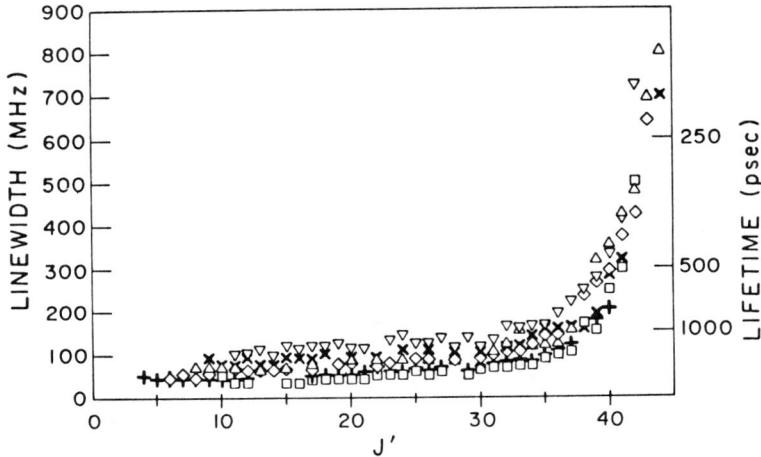

Figure 5. Experimentally measured linewidths for CD_2CN^- as a function of J' for several values of K'. □ K' = 3; +, K' = 4; ◊, K' = 6' △, K' = 8; x, K' = 9; ▽, K' = 11.

The autodetaching transitions energetically available to a given rotational level of the dipole-bound state can be determined readily once one knows the binding energy of the outermost electron in the dipole-bound state; this is the energy difference between the rotationless levels of the dipole bound state and the ground state of neutral CH_2CN. Since the rotational levels of the dipole-bound state lying below the detachment threshold cannot autodetach, an upper limit to the binding energy can be obtained from the missing lines at low J in the spectra. For example, in figure 2, no transitions to J' < 8 are observed, placing an upper limit of ~60 cm^{-1} on the binding energy. Using this limit, it can be shown that the rapid increase in the autodetachment rates in figures 4 and 5 coincides with an autodetaching transition of ΔJ = -2 or -3, ΔK = 0, becoming energetically accessible. (The exact value of ΔJ depends on the binding energy, which may be lower than 60 cm^{-1}.)

For lower values of J' where the autodetachment rate is slow, autodetachment can only occur by a larger ΔJ transition, which is expected to be slow, or by a transition in which K as well as J changes. From figure 3, one can qualitatively see why an autodetaching transition requiring a change in K might be very slow. Rotational autodetachment requires coupling between rotational and electronic motion, and rotational angular momentum about the a axis is going to be poorly coupled to the diffuse orbital of the dipole-bound electron which is nearly cylindrically symmetric about the a axis.

The strong dependence of autodetachment rate upon rotation has been observed in other cases involving dipole-bound states. In each system, abrupt rate increases can be correlated with the availability of low ΔJ, ℓ autodetaching transitions. This type of autodetachment rate behavior appears to be a signature of a sipole-bound state.

REFERENCES

1. P. L. Jones, R. D. Mead, B. E. Kohler, S. E. Rosner, and W. C. Lineberger, J. Chem. Phys. $\underline{73}$, 4419 (1980); W. C. Lineberger and T. A. Patterson, Chem. Phys. Lett. $\underline{13}$, 40 (1972); G. Herzberg and A. Lagerqvist, Can. J. Phys. $\underline{46}$, 2363 (1968).
2. P. A. Schulz, R. D. Mead, P. L. Jones, and W. C. Lineberger, J. Chem. Phys. $\underline{77}$, 1153 (1984).
3. L. M. Tack, N. H. Rosenbaum, J. C. Owrutsky, and R. J. Saykally, J. Chem. Phys. $\underline{84}$, 7056 (1986); K. Kawaguchi and E. Hirota, J. Chem. Phys. $\underline{84}$, 2953 (1986); B. D. Rehfuss, M. W. Crofton, and T. Oka, J. Chem. Phys. $\underline{85}$, 1785 (1986).
4. J. Marks, D. M. Wetzel, P. B. Comita, and J. I. Brauman, J. Chem. Phys. $\underline{84}$, 284 (1986).
5. R. C. Stoneman and D. J. Larson, J. Phys. B $\underline{19}$, L405 (1986).
6. D. M. Neumark, K. R. Lykke, T. Andersen, and W. C. Lineberger, J. Chem. Phys. $\underline{83}$, 4364 (1985); M. Al-Za'al, H. C. Miller, and J. W. Farley, Chem. Phys. Lett. $\underline{131}$, 56 (1986).
7. R. D. Mead, K. R. Lykke, W. C. Lineberger, J. Marks, and J. I. Brauman, J. Chem. Phys. $\underline{81}$, 4883 (1984).
8. J. Marks, J. I. Brauman, R. D. Mead, K. R. Lykke, and W. C. Lineberger, J. Chem. Phys. (submitted).
9. T. Andersen, K. R. Lykke, D. M. Neumark, and W. C. Lineberger, J. Chem. Phys. (accepted).
10. K. R. Lykke, D. M. Neumark, T. Andersen, V. J. Trapa, and W. C. Lineberger, in Laser Spectroscopy VII, edited by Y. R. Shen and T. W. Hänsch (Springer Verlag, Berlin 1985), pp. 130-133.
11. K. R. Lykke, K. K. Murray, and W. C. Lineberger (to be published).
12. R. L. Jackson, P. C. Hiberty, and J. I. Brauman, J. Chem. Phys. $\underline{74}$, 3705 (1981).
13. W. R. Garrett, Phys. Rev. A $\underline{3}$, 961 (1971).
14. B. A. Huber, T. M. Miller, P. C. Cosby, H. D. Zeman, R. L. Leon, J. T. Moseley, and J. R. Petersen, Rev. Sci. Instrum. $\underline{48}$, 1306 (1977).
15. G. Herzberg, Electronic Spectra of Polyatomic Molecules (Van Nostrand, New York, 1967) pp. 247-261.
16. S. Moran, H. B. Ellis, Jr., D. J. DeFrees, and G. B. Ellison, J. Am. Chem. Soc. (submitted).
17. J. Simons, J. Am. Chem. Soc., $\underline{103}$, 3971 (1981).
18. R. S. Berry, J. Chem. Phys. $\underline{45}$, 1228 (1966).

HIGH-RESOLUTION MEASUREMENT OF THE INFRARED ROTATION-VIBRATION SPECTRUM OF NH⁻

H. C. Miller, M. Al-Za'al[*] and J. W. Farley
University of Oregon, Eugene, OR 97403

ABSTRACT

An extensive series of measurements has been performed of the infrared spectrum of the $X^2\Pi$ electronic state of the molecular anion NH^-. A coaxial ion beam/laser beam spectrometer is employed, in which fundamental rotation-vibration transitions are induced by a color-center laser operating near 3000 cm^{-1}. The vibrationally excited ions autodetach, and the fast neutral NH is detected. Hyperfine structure (hfs) has been observed and analyzed in the P, Q, and R branches of $^{14}NH^-$ and $^{15}NH^-$. The hfs parameters are compared with atomic N and the isoelectronic neutral OH. This is the first analyzed case of hfs in any negative molecular ion.

INTRODUCTION

Negative molecular ion are important in terrestrial and laboratory plasmas, but relatively little is known about their structure, compared to cations or neutral molecules. Hfs depends on the behavior of the electronic wavefunction near the nuclei, and is complementary to the other information commonly available, the electron affinity and the threshold photodetachment behavior, which depend on the wavefunction at large values of r. Mead et al.[1] made the only previous observation of hfs in a molecular anion, in $^{13}C^{12}C^-$, but the complexity of spectrum defied analysis.

EXPERIMENT

A coaxial beams apparatus has been used, in which a 3-keV mass-selected ion beam coaxially overlaps a single-mode infrared laser beam, and the ions are Doppler-tuned into resonance. Autodetachment of laser-excited ionic states produces fast neutral NH particles, which are detected by collisional ejection of secondary electrons. The coaxial beams technique has sub-Doppler resolution (5 MHz), because of a well-known velocity-bunching effect.
There have been several previous experimental studies of $^{14}NH^-$. Laser photoelectron spectrometric studies were performed by Celotta, Bennett and Hall and by Engelking and Lineberger. Neumark et al.[2] used a

*Permanent address: Dept. of Natural Sciences, Mu'tah University, P.O. Box 7, Mu'tah, Karak, Jordan.

coaxial beams apparatus to make the first direct observation of a rotation-vibration spectrum in a negative molecular ion. None of these studies resolved hyperfine structure. In other work on $^{14}NH^-$, the present authors observed rotationally excited metastable states in the R-branch[3], and extensive measurements were made[4] in the P-, Q-, and R-branches. We measured[5] the spectrum of $^{15}NH^-$, analyzed the isotopic shifts, and extracted the equilibrium vibrational frequency ω_e and the anharmonicity $\omega_e x_e$.

RESULTS AND DISCUSSION

Hfs was observed in the Q-branch transitions and in the lowest-J members of transitions in the the P- and R-branches. We have reproduced the hyperfine lineshapes in $^{14}NH^-$ and $^{15}NH^-$. Table I shows the hyperfine structure (Frosch-Foley) parameters for both isotopes. Table I also shows the hfs parameters for the hydrogen in neutral OH, and also the hfs constants from an atomic 2p orbital localized on the nitrogen.

The ionic hyperfine parameters are physically reasonable: the hydrogen parameters are the same within 1% in the two isotopes, and the nitrogen parameters are proportional to the nuclear magnetic moments within 0.5%. There are no theoretical calculations with which to make comparison. We hope that our experimental work stimulates theoretical interest in this problem.

ACKNOWLEDGMENTS

Financial support by the National Science Foundation and the Chemical Physics Institute of the University of Oregon is gratefully acknowledged.

Table I. Hyperfine parameters (in MHz) for both isotopes of NH⁻ and other relevant species. The nitrogen parameters in the last column agree with the ratio of atomic parameters $g_I(^{15}N)/g_I(^{14}H) = 2(-0.2831)/0.40375 = -1.402$

param.		^{14}NH⁻	other	^{15}NH⁻	other	$\dfrac{^{15}\text{NH}^-}{^{14}\text{NH}^-}$
N	a−(b+c)/2	79.89	167.54(a)	−111.72	−234.9(b)	−1.3984
	d	107.61	167.54(a)	−150.525	−234.9(b)	−1.3988
H	a−(b+c)/2	50.79	93.03(c)	50.87	93.03(c)	1.00158
	d	44.47	56.62(c)	44.86	56.62(c)	1.00877

(a) atomic ^{14}N; (b) atomic ^{15}N; (c) hydrogen parameters for neutral OH.

REFERENCES

1. R. D. Mead, doctoral thesis, Chemistry Depart., Univ. of Colorado, 1984 (unpublished); R. D. Mead, A. E. Stevens, and W. C. Lineberger, in Gas Phase Ion Chemistry, Vol. 3, M. T. Bowers, ed. (Academic Press, NY, 1984), pp. 213-248.
2. D. M. Neumark, K. R. Lykke, T. Andersen, and W. C. Lineberger, J. Chem. Phys. 83, 4364 (1985) and references therein.
3. M. Al-Za'al, H. C. Miller, and J. W. Farley, Chem. Phys. Lett. 131, 56 (1986)
4. M. Al-Za'al, H. C. Miller, and J. W. Farley, accepted for publication in Phys. Rev. A.
5. H. C. Miller and J. W. Farley, accepted for publication in J. Chem. Phys.

SO^+ Emission in a Supersonic Jet

I.W. Milkman, J.C. Choi, J.L. Hardwick, and J.T. Moseley
University of Oregon, Eugene, OR 97403

ABSTRACT

Rotationally cold emission has been recorded in the $A^2\Pi_i$-$X^2\Pi_r$ band system of SO^+ at a resolution of 0.3 cm^{-1} using a source consisting of a supersonic expansion excited by a corona discharge. The observed spectra, including over 50 bands with v'=0-11 and v"=0-10, have been rotationally analyzed and the derived constants have been merged with those obtained from previously published spectra.

A study has been made of rotationally cold A-X emission in SO^+ using a corona-excited supersonic expansion (CESE) source. The CESE has been developed and described as a tool for the observation of radicals by Droege and Engelking.[1] We wish to explore the CESE source as a tool for studying molecular ions.

The source consists of a glass capillary narrowed to a 100μm nozzle, into which a thin tungsten wire is fitted. A potential of 15kV is applied to the wire, and a corona discharge forms at the tip as the gas expands through the nozzle into the chamber. By constraining the discharge to pass through the throat of the nozzle as the gas expands, ions and radicals are produced which have considerable rotational cooling, yet are vibrationally and electronically excited. The resulting emission spectrum is simple and uncongested when compared with spectra from other types of discharges and this simplicity can be exploited to characterize previously intractable molecular ion spectra.

The SO^+ molecular ion is an important component of plasmas containing sulphur and oxygen. Despite the interest in this ion, no spectroscopic information was available until recently. The $A^2\Pi_i$-$X^2\Pi_r$ transition in the visible and near ultraviolet was identified by Tsuji et al. in 1980.[2] Cossart et al.[3] and Coxon and Foster[4] performed the first rotational analysis of a number of bands in the A-X system. The only high resolution data are due to Hardwick et al.[5] who measured several A-X bands at Doppler-limited resolution, and recently Woods[6] has made measurements of some pure rotational transitions in the X state of SO^+. We have been able to observe the SO^+ A-X transition in the CESE source by seeding a trace of SO_2 in a helium expansion and recording the emission photographically at medium resolution through a 0.75m spectrograph.

© American Institute of Physics 1987

The resulting spectra displayed an intensity distribution maximized at the lowest rotational quantum number in each branch,[6] implying a rotational temperature which is less than 10^6K. However, the high-J lines are more intense than would be predicted by such a temperature, indicating that the rotational population distribution is non-Boltzmann in nature. In contrast to the low rotational temperature, the degree of vibrational excitation is extremely high, allowing us to significantly extend the range of observations for this electronic transition. Bands with v'=0-11 and v"=0-10 have been observed and analyzed. Another interesting feature of the observed spectra is that the $\Omega=1/2$ branch was much less intense than the $\Omega=3/2$ branch in all bands; this is presumably due to expansion cooling of the spin-orbit distribution in the upper electronic state, which depopulates $\Omega'=1/2$. Our observation of strong, but not complete, spin-orbit cooling is consistent with a study performed by Carrick and Engelking,[7] which indicates that spin-orbit cooling will be present in the CESE source but will be less important for Hund's case (a) molecules than for molecules with intermediate coupling.

The reduced congestion of the spectrum produced by the CESE source allowed us to rotationally analyze for the first time over 50 bands of SO^+. In addition, our new information enabled us to assign portions of several bands which were previously recorded at high resolution[5] but not assigned because they are weak and overlapped by stronger bands. Molecular constants were derived for all the assigned bands and then were merged with previously reported values to obtain the best set of molecular constants. Plans for future work include improving the quality of the data by using a 7m spectrograph with Doppler-limited resolution.

REFERENCES

1. A.T. Droege and P.C. Engelking, Chem. Phys. Lett. 96, 316 (1983).
2. M. Tsuji, C. Yamagiwa, M. Endoh, and Y. Nishimura, Chem. Phys. Lett. 73, 407 (1980).
3. D. Cossart, H. Lavendy, and J.M. Robbe, J. Mol. Spectrosc. 99, 369 (1983).
4. J.A. Coxon and S.C. Foster, J. Mol. Spectrosc. 103, 281 (1984).
5. J.L. Hardwick, Y. Luo, D.H. Winicur, and J.A. Coxon, Can. J. Phys., 62(12) 1792 (1984).
6. R.C. Woods, H.E. Warner, and N. Carballo, private communication.
7. P.G. Carrick and P.C. Engelking, Chem. Phys. Lett. 108, 505 (1984).

DOUBLE RESONANCE TECHNIQUES FOR THE HIGH RESOLUTION SPECTROSCOPY OF UNSTABLE MOLECULES

Wolfgang E. Ernst[*]

Department of Chemistry, Stanford University, Stanford, CA 94305

ABSTRACT

New microwave-optical double resonance techniques have recently been developed and offer the high resolution and the sensitivity needed to study chemically unstable species at low concentrations. The methods have been demonstrated in applications to alkaline earth monohalide radicals. Dense optical spectra were simplified by labeling individual lines in a double resonance scheme, and hyperfine structure splittings as well as Stark effect were investigated at high accuracy.

INTRODUCTION

Chemically unstable molecules have attracted much interest among both spectroscopists and groups studying reaction dynamics. Free radicals have usually one or more unpaired electrons and interactions of the nonzero spin within the molecule can lead to complex spectra, which, once they have been analyzed, provide many details of the molecular structure. The alkaline earth monohalides represent a group of radicals which has been the subject of many reaction dynamic studies.[1] To good approximation they consist of two closed shell ions and a single nonbonding electron, which is mainly metal centered in the ground and first excited electronic states. As the potential curves and internuclear distances are very similar for these states, in the electronic transitions vibrational bands with $\Delta v=0$ are favored by the Franck-Condon principle and form extremely congested optical spectra. For the heavy species like strontium and barium monohalides the rotational structure can only be resolved by Doppler-free laser spectroscopy techniques. For our investigations we chose Doppler-free polarization spectroscopy[2] which provides the high sensitivity needed for the study of species at low number densities.[3]

In all alkaline earth monohalides the unpaired electron spin interacts with the nuclear spin of the halogen nucleus. Since the electron is mainly centered at the metal atom, the hyperfine structure (hfs) is very weak and can barely be resolved with sub-Doppler laser spectroscopy. Independent measurements of the ground state hfs with microwave techniques are necessary. Microwave-optical polarization spectroscopy (MOPS)[4] was used for the hfs studies in reaction cells. Furthermore, with the knowledge of

---

[*]Heisenberg Fellow, on leave from Institut für Molekülphysik, Freie Universität Berlin, Arnimallee 14, D-1000 Berlin 33, West Germany

ground state rotational constants, microwave radiation tuned to a particular rotational transition was used to label individual lines in optical spectra recorded with Doppler-free polarization spectroscopy.

When a molecular beam of the unstable species is generated, the molecular beam laser-microwave double resonance method[5] can be used to yield much narrower linewidths than a cell experiment (where the resolution is limited by pressure broadening). We applied this technique to alkaline earth monohalides and, in an external electric field, performed highly precise Stark effect measurements.[6]

POLARIZATION SPECTROSCOPY METHODS

In a free space reaction cell alkaline earth monohalides were produced by reacting alkaline earth metal vapor with a halogen donor in a stream of argon buffer gas (Fig. 1). A total number density of 10^{11} radicals per cm^3 is produced in a reaction zone of 25 cm diameter at a background pressure of 0.1 Torr. The experimental arrangement shown in Fig. 1 allows the investigation of optical and microwave transition. Microwave radiation of vertical polarization is introduced via a horn antenna. Two counterpropagating laser beams of the same wavelength pass through a slit in the top of the horn radiator. The best overlap of the three electromagnetic waves is given in the reaction zone. The probe laser beam has a linear polarization inclined by 45° with respect to the polarization plane of the counterpropagating pump laser beam and the microwaves. In this way the probe laser beam which is passed through a nearly crossed polarizer behind the cell can be used to detect an optical anisotropy induced in the gas by the absorption of pump laser or microwave radiation or both. If the pump laser and the microwaves are amplitude modulated at frequencies f_p and f_m, respectively, and phase sensitive detection of the probe signal is employed, the following experiments can be performed:

1) detection at f_p, laser scanning: optical spectra are recorded at sub-Doppler resolution using Doppler-free polarization spectroscopy;[2,3]

2) detection at f_m, laser fixed to an optical transition, microwaves scanning: microwave transitions are recorded which are connected to a level of the induced optical transition in a double resonance scheme = microwave optical polarization spectroscopy (MOPS);[4]

3) detection at $f_p + f_m$ or $|f_p - f_m|$, microwaves fixed to a rotational transition, laser scanning: optical transitions which are connected to a level pumped by microwaves are detected at sub-Doppler resolution = microwave modulated polarization spectroscopy (MMPS).[7]

Polarization spectroscopy techniques require less intense laser and microwave radiation than the corresponding nonlinear methods based on fluorescence detection. Power broadening is largely avoided. MMPS (mode 3) has been used to assign lines in the dense optical spectra of SrCl and SrBr. Doppler-free polarization spectroscopy (mode 1) has been our tool for recording optical

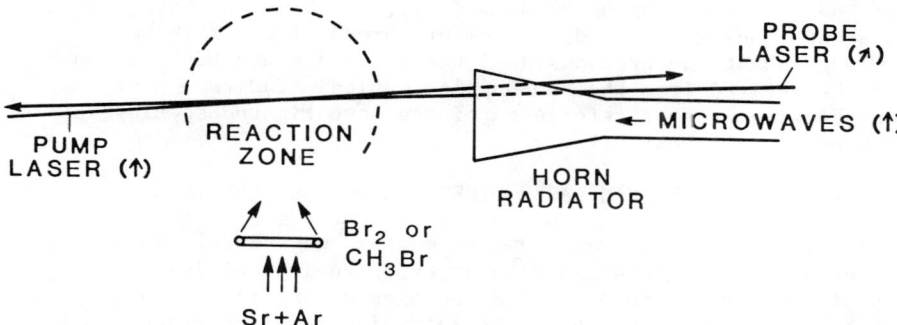

Fig. 1. Experimental arrangement for polarization spectroscopy in a reaction cell: interaction zone between gas sample and laser and microwave radiation.

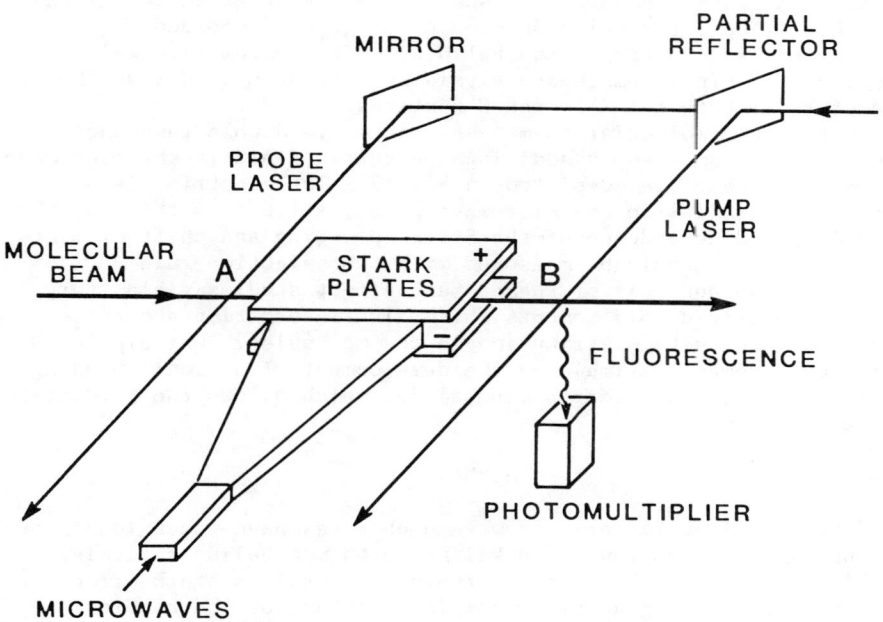

Fig. 2. Set-up for molecular-beam laser-microwave double resonance.

spectra of many alkaline earth monohalides. Systematic hyperfine structure studies in the ground states of alkaline earth monohalides have been performed using MOPS (mode 2), in which case the linewidth was about 1 MHz determined by pressure broadening. With the knowledge about the ground states the excited state hfs constants could be derived from the sub-Doppler optical studies. A more comprehensive list of references of the investigated species is contained in a recent article.[8]

MOLECULAR BEAM LASER-MICROWAVE DOUBLE RESONANCE

At low background pressure molecules can travel in a beam over considerable distances without experiencing collisions. Depletion of the population by optical pumping at one place can be probed at another place by observing the laser-induced fluorescence. If the level is repopulated via radiofrequency or microwave transitions in the region between the two places the laser-induced fluorescence increases. Childs and Goodman applied this scheme to measure radiofrequency transitions in alkaline earth monohalides.[9] As shown in Fig. 2, we adopted this pump and probe experiment for the study of microwave transitions.[5] In this case the linewidth is mainly determined by the time-of-flight of the molecules through the microwave interaction region and is on the order of 10 kHz under our experimental conditions. Among other species we investigated the ground state hfs of BaI, for which Johnson et al.[10] had recorded sub-Doppler optical spectra of the BaI $C^2\Pi - X^2\Sigma^+$ system with well resolved hfs. In a combined analysis we were able to derive the hfs parameter for both the $X^2\Sigma^+$ and $C^2\Pi$ states.[11]

In the same molecular beam laser-microwave double resonance experiment, we applied an additional electric field in the microwave interaction region, as depicted in Fig. 2. The electric field vector was parallel to the microwave polarization. In this way the electric field dependence of the Stark splitting and shift of microwave transitions could be measured with the selection rule $\Delta M_F = 0$.[6] The analysis of these Stark effect studies yields very precise electric dipole moments which give information about the total change-density distribution in the molecules. Our dipole moment measurements stimulated the development of an ionic bonding model for the alkaline earth monohalides which allows the prediction of the dipole moment.[12]

SUMMARY

A number of new laser-microwave double resonance techniques have been applied to the group of alkaline earth monohalide radicals. Rotational, fine and hyperfine structure as well as Stark effect were studied yielding a fairly complete picture of the bonding nature of these molecules. The described spectroscopic techniques are very versatile and can be used for the investigation of many other free radicals which can be produced at similar concentrations in reaction cells or molecular beams.

ACKNOWLEDGEMENTS

The author wishes to thank Dr. J. O. Schröder and Dipl. Phys. J. Kändler for their extraordinarily strong experimental support, and Profs. T. Törring and R. N. Zare for many stimulating discussions. Financial support from the Deutsche Forschungsgemeinschaft is gratefully acknowledged.

REFERENCES

1. e.g. C. Noda, J. S. McKillop, M. A. Johnson, J. R. Waldeck, and R. N. Zare, J. Chem. Phys. 85, 856 (1986) and references therein.
2. C. Wieman and T. W. Hänsch, Phys. Rev. Lett. 36, 1170 (1976).
3. W. E. Ernst, Opt. Comm. 44, 159 (1983).
4. W. E. Ernst and T. Törring, Phys. Rev. A 25, 1236 (1982).
5. W. E. Ernst and S. Kindt, Appl. Phys. B 30, 79 (1983).
6. W. E. Ernst, S. Kindt, and T. Törring, Phys. Rev. Lett. 51, 979 (1983).
7. W. E. Ernst, Opt. Comm. 46, 18 (1983).
8. W. E. Ernst, J. O. Schröder, and J. Kändler, in: Methods of Laser Spectroscopy, eds. Y. Prior, A. Ben-Reuven, and M. Rosenblum (Plenum, New York, 1986) p. 191.
9. W. J. Childs and L. S. Goodman, Phys. Rev. Lett. 44, 316 (1980).
10. M. A. Johnson, C. Noda, J. S. McKillop, and R. N. Zare, Can. J. Phys. 62, 1467 (1984).
11. W. E. Ernst, J. Kändler, C. Noda, J. S. McKillcp, and R. N. Zare, J. Chem. Phys. 85, 3735 (1986).
12. T. Törring, W. E. Ernst, and S. Kindt, J. Chem. Phys. 81, 4614 (1984).

REDUCTION OF 1+1 REMPI SPECTRA TO POPULATION DISTRIBUTIONS:
SATURATION AND INTERMEDIATE STATE ALIGNMENT EFFECTS

D. C. Jacobs, R. J. Madix, and R. N. Zare
Department of Chemistry, Stanford University, Stanford, CA 94305

ABSTRACT

A two-step methodology is presented for extracting ground state population distributions from 1+1 resonance enhanced multiphoton ionization (REMPI) spectra. In the first step the ion signal is corrected for variation with laser intensity as it is collected, generating an iso-power spectrum. In the second step populations and alignments are derived from the iso-power spectrum by correcting for the interdependent effects of saturation and intermediate state alignment. This procedure is applied to a room temperature thermal distribution of nitric oxide using the 1+1 REMPI process in which lines of the NO $A^2\Sigma^+$-$X^2\Pi$ (0,0) band constitute the resonant transition. The present treatment is able to recover the known rovibrational population distribution, independent of branch choice, over a wide range of practical operating conditions.

INTRODUCTION

Resonance enhanced multiphoton ionization (REMPI) is generally regarded to offer more sensitivity than laser induced fluorescence (LIF) for detecting low concentrations of small gas-phase molecules. In addition, REMPI may be more widely applicable. However, before the benefits of REMPI can be routinely realized in the quantum state analysis of molecular samples, it is necessary to be able to relate unambiguously ion yields to ground state populations. This task is by no means a trivial one, because the REMPI technique is inherently a nonlinear process requiring high laser powers, and hence is more susceptible to saturation effects, power broadening, AC Stark broadening, laser intensity variations, etc.[1]

This paper describes a methodology for reducing 1+1 REMPI spectra to accurate population distributions and alignment factors and applies this procedure to the 1+1 REMPI spectra of the (0,0) band of the NO $A^2\Sigma^+$ - $X^2\Pi$ transition as a test case. Proper spectral reduction is achieved in two steps: first, ion yields are recorded as a function of laser wavelength in a manner such that all ion intensities correspond to the same effective integrated laser intensity; second this so-called iso-power spectrum is then corrected for the combined effects of saturation and intermediate state alignment.

EXPERIMENTAL

The experimental details are more fully described elsewhere.[2] A time-of-flight tube is used for MPI ion collection. The tunable UV radiation (224-227 nm) is generated through Raman shifting the doubled output of a Nd:YAG pumped dye laser.

THE NO $A^2\Sigma^+ - X^2\Pi$ TEST CASE

A 1+1 REMPI system of choice is the (0,0) band of the NO $A^2\Sigma^+ - X^2\Pi$ transition. The intermediate state lifetime has been measured to be 216±4 ns,[3] and the NO A - NO^+ X photoionization cross section as (7.0±0.9) x 10^{-19} cm^2.[4] Total pressures as low as 10^{-11} torr can be detected under tightly focused conditions.[5] Figure 1 shows a typical 1+1 REMPI spectrum of NO recorded at a temperature of 299 K and a pressure of 10^{-7} torr. The subbands arising from each spin-orbit ground state are apparent and each rotational line is readily assigned using the known molecular constants of the NO A and X states.[6]

Figure 2 presents an expanded portion of the 1+1 REMPI spectrum (see dashed box in Fig. 1) taken at three different laser powers. Note the marked change in intensity of the same lines under the three different laser power conditions. It is seen that the power dependence exhibits a functional form that is neither linear nor quadratic in the laser intensity. Moreover, the power variation appears to be different for each line in the spectrum and to vary even across a line profile.

Figure 3 shows the Boltzmann plot (logarithm of the intensity/ line strength vs. the internal energy) of a room temperature spectra recorded using a high laser fluence without adequate power normalization. The resulting branch-dependent temperatures vary over a wide range, and the individual branch contributions do not coalesce. This indicates the need for a more comprehensive data collection/ analysis routine.

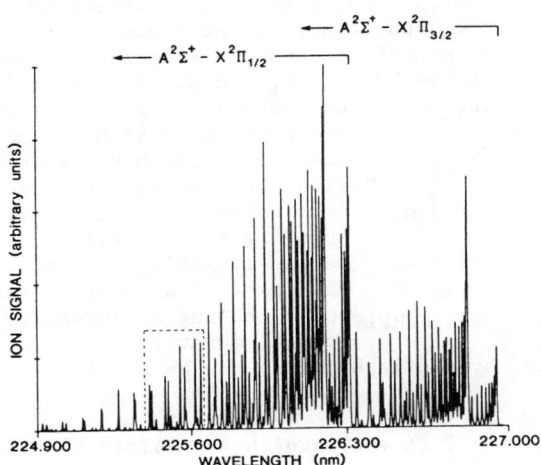

Fig. 1. Thermal(299 K) 1+1 REMPI spectrum of the NO $A^2\Sigma^+ - X^2\Pi$ (0,0) band.

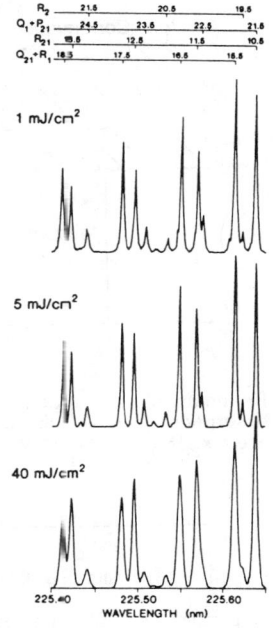

Fig. 2. An expanded portion of the NO $A^2\Sigma^+ - X^2\Pi$ (0,0) band [see dashed box in Fig. (1)], recorded under different laser fluences.

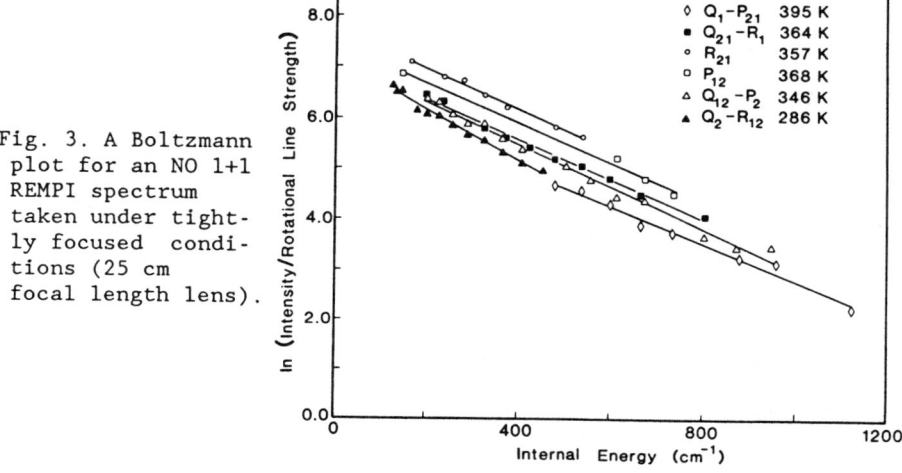

Fig. 3. A Boltzmann plot for an NO 1+1 REMPI spectrum taken under tightly focused conditions (25 cm focal length lens).

METHODOLOGY

The following draws from the theory presented in detail elsewhere.[7] A simple rate equation model (Fig. 4) can be employed to describe the 1+1 REMPI system. We examine the common case where the two transitions (resonant and ionization) are excited by photons of the same color from the same source, characterized by intensity I. The populations of the ground, resonant intermediate, and ion states are represented by N_0, N_1 and N_2, respectively. The rate constants k_{01}, k_{10} and k_{12} are associated with the processes of resonant absorption, stimulated emission, and ionization. The corresponding rate equations can be analytically solved for the case of a square laser pulse having an intensity I for a duration Δt. The total number of ions produced by the laser pulse can be written as:

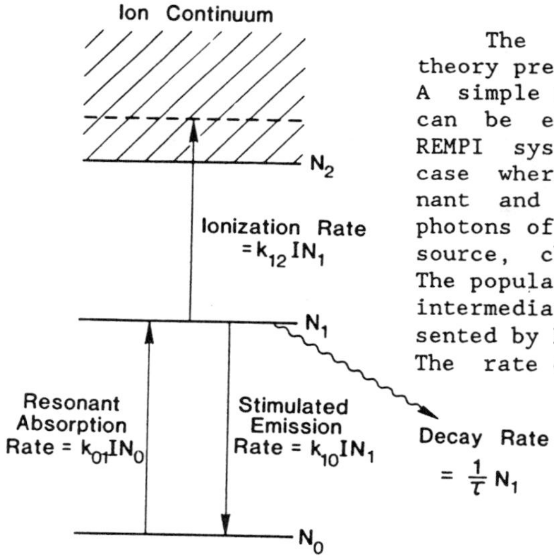

Fig. 4. Schematic Diagram of 1+1 REMPI.

$$N_2(\Delta t) = N_0 \left\{ 1 - \frac{1}{2B} \left[(A + B) \exp[-\tfrac{1}{2}(A - B)I\Delta t] \right. \right.$$
$$\left. \left. - (A - B) \exp[-\tfrac{1}{2}(A + B)I\Delta t] \right] \right\}, \quad (1)$$

where

$$A = 2k_{01} + k_{12}, \quad (2)$$

and

$$B = (4k_{01}^2 + k_{12}^2)^{1/2}. \quad (3)$$

It can be shown that the assumption of a square laser pulse matters little to the final analysis.[7]

Figure 5 illustrates the 1+1 REMPI power dependence for the $R_{21}(10.5)$ line of the NO A-X (0,0) band. Each data point represents the average ion signal for individual laser shots having the specified integrated laser intensity. The saturation function (Eq. 1) fits the data well. It can be seen that linear or quadratic fits are not appropriate for extrapolating the data to an arbitrary integrated laser intensity.

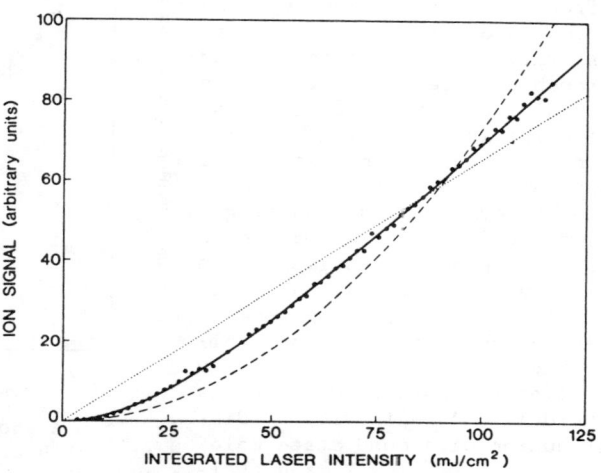

Fig.5. Laser power dependence of the 1+1 REMPI ion signal for the $R_{21}(10.5)$ resonant transition of the NO A-X (0,0) band. The dashed, dotted and solid lines represent the best quadratic, linear and saturation function (Eq. 1) fits to the data, respectively.

During an experimental run, the laser intensity can vary dramatically because of shot-to-shot fluctuations and because of variations in laser gain when scanning across a dye curve. Thus a spectrum must be normalized for laser power changes. The parameters used in the saturation function are expected to change across a wavelength scan. For different spectral transitions, the variations of N and k_{01} correspond to a change in the ground state population and the rotational line strength factor, respectively. At each laser wavelength the ion signal and integrated laser intensity are recorded as data pairs. A group of these data pairs are then fitted in a least squares routine to the functional form of Eq (1) letting N and k_{01} vary. The resulting two parameters (N and k_{01}) may then be used to extrapolate the ion yield to any predetermined constant laser power. This functional fit can be performed at every wavelength point so as to record the spectrum at constant integrated laser intensity. We refer to spectra recorded in this manner as "iso-power spectra".

Figure 6 shows a Boltzmann plot of an iso-power spectra recorded at 20 mJ/cm^2. Here, the laser power normalized ion yields are simply divided by the resonant transition rotational line strengths $S(J_0, J_1)$, as calculated by a computer program explained elsewhere.[6] The temperature uncertainty represents twice the standard deviation. The best-fit lines for the various branches do not coincide.

Having recorded a spectrum under iso-power conditions, the problem still remains of correcting for the interdependent effects of saturation and intermediate state alignment. Saturation will affect each branch differently in that the transition strengths for the resonant step are branch dependent, and thus each branch will saturate to a different extent. The intermediate state alignment effect will also show branch dependent variations in the overall ionization efficiency. This latter phenomena can be thought of in the following manner. Suppose a beam of linearly polarized light is incident on an isotropic distribution of ground state molecules (i.e. all M sublevels within a given J are equally populated). The intermediate state will be effectively aligned through the preferential excitation of those molecules having larger projections of the transition dipole moment on the electric field vector of the linearly polarized light beam. Ionization efficiency of the intermediate state can also be M dependent, and thus the degree of anisotropy created in the intermediate state can affect the overall MPI ion production.

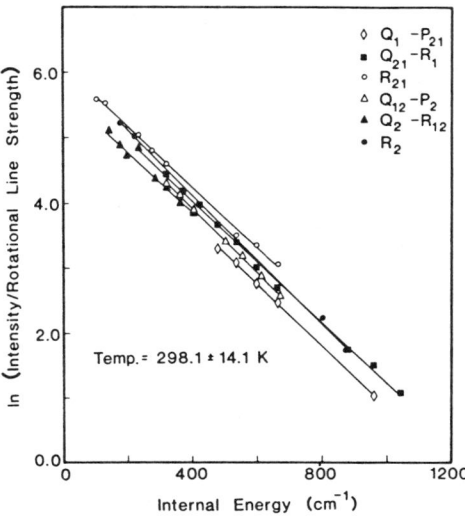

Fig. 6. A Boltzmann plot for the different branch contributions from a 1+1 REMPI spectrum of NO taken under isopower conditions (20 mJ/cm^2).

In order to accomodate these two effects, we choose to follow a quantum treatment, although a classical treatment would have also been satisfactory. Linearly polarized light restricts transitions to follow a $\Delta M=0$ selection rule. This reduces a given transition into a sum of 2J+1 independent transitions. In general, the complete reduction of spectral intensities into relative ground state populations treats each spectral transition probability as a sum over the independent transition probabilities for each M sublevel. Given the M-dependent rate constants, we may utilize Eq. (1) to calculate the saturable ionization efficiencies for each M channel. The total ionization efficiency for a particular transition is then just the sum of the contributions from each M channel.

The M-dependent rate constant for resonant absorption can be written as[7]

$$k_{01}(M) = 3 C_{01} S(J_0,J_1) \begin{pmatrix} J_1 & 1 & J_0 \\ M & 0 & -M \end{pmatrix}^2 , \qquad (4)$$

where $S(J_0,J_1)$ is the normalized rotational line strength factor, and C_{01} is a constant proportional to the Einstein B coefficient.

The M-dependent rate constant for photoionization of an intermediate state, having Σ symmetry, can be written as[7]

$$k_{12}(M) = 3(2N_1+1)(2J_1+1)\frac{\sigma}{h\nu} \sum_{M_S} \begin{pmatrix} N_1 & S & J_1 \\ M-M_S & M_S & -M \end{pmatrix}^2 \sum_{N_2} (2N_2+1) \begin{pmatrix} N_2 & 1 & N_1 \\ M-M_S & 0 & M_S-M \end{pmatrix}^2$$

$$\times \left[\Gamma \begin{pmatrix} N_2 & 1 & N_1 \\ 0 & 0 & 0 \end{pmatrix}^2 + (1-\Gamma) \begin{pmatrix} N_2 & 1 & N_1 \\ 1 & -1 & 0 \end{pmatrix}^2 \right] \quad (5)$$

Here, σ is the overall cross section for ionization and Γ is the fraction of parallel character. The quantum number N_2 represents the total angular momentum of the (NO+ + e-) ionization state, excluding spin. Figure 7 illustrates the limiting cases for $k_{01}(M)$ and $k_{12}(M)$.

The fraction of parallel character Γ can be approximated from ab initio calculations:[9] it can also be determined experimentally as a fitting parameter. The fraction of parallel character corresponds to the degree of curvature in the $k_{12}(M)$ distribution. This curvature will dramatically affect the relative ionization probability of P and R compared to Q branches.

A series of iso-power spectra recorded at various total integrated laser intensities can be analyzed together. For each iterative approximation, a trial resonant transition scaling parameter C_{01} and trial Γ parameter are used to produce Boltzmann plots for the entire spectral series. The mean of the uncertainties for the slopes of each Boltzmann plot is minimized by independently varying the values within the parameter set. The final parameter set represents the values of C_{01} and Γ for the system.

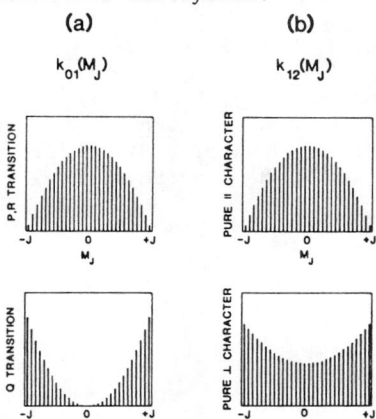

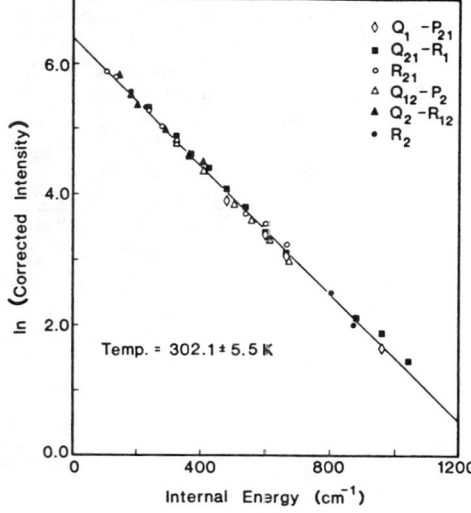

Fig. 7. The M-dependent rate constants for the limiting cases of (a) the resonant transition and (b) the ionization transition.

Fig. 8. A corrected (inclusion of intermediate state alignment and saturation) Boltzmann plot for data recorded at an integrated laser intensity of 20 mJ/cm².

RESULTS AND DISCUSSION

Figure 8 illustrates the corrected Boltzmann plot produced from the same spectral data as that used in Fig. 6. The temperature is found to agree well with the experimental 299 K temperature, and the data points scatter around a common straight line. The proposed methodology improved the precision of measuring a rotational temperature by as much as a factor of 7. The success of the method was demonstrated for laser fluences covering two orders of magnitude. The extracted fraction of parallel character is 44.6% which is in general agreement with the 26.7% calculated by Dixit et al.[9]

ACKNOWLEDGEMENTS

This work was supported in part by Amoco, the National Science Foundation (NSF CHE 85-05926) and the Department of Energy (DE-AT03-79ER10490).

REFERENCES

1. J. Morellec, D. Normand and G. Petite, Adv. At. Mol. Phys. 18, 97 (1982).
2. D. C. Jacobs, R. J. Madix and R. N. Zare, J. Chem. Phys. 85, 5469 (1986).
3. H. Zacharias, J. B. Halpern and K. H. Welge, Chem. Phys. Lett. 43, 41 (1976).
4. H. Zacharias, R. Schmiedl and K. H. Welge, Appl. Phys. 21, 127 (1980).
5. M. Asscher, W. L. Guthrie, T.-H. Lin and G. A. Somorjai, Phys. Rev. Lett. 49, 76 (1982).
6. R. Engleman, Jr. and P. E. Rouse, J. Mol. Spectrosc. 37, 240 (1971).
7. D. C. Jacobs and R. N. Zare, J. Chem. Phys. 85, 5457 (1986).
8. R. N. Zare in *Molecular Spectroscopy: Modern Research*, K. N. Rao and C. W. Mathews, Eds., (Academic Press, New York, NY, 1972).
9. S. N. Dixit, D. L. Lynch, V. McKoy and W. M. Huo, Phys. Rev. A 32, 1267 (1985).

TRANSIENT EXCITED SINGLET STATE ABSORPTION IN THE LASER DYE α-NPO

Putcha Venkateswarlu, M.C.George, Yerneni V Rao, H.Jagannath and G.Chakrapani

Depatment of Physics, Alabama A&M University, Normal, AL 35762

ABSTACT

Excited singlet state absorption of the laser dye α-NPO solution in toluene recorded in the region 4460-6260 Å shows nine submaxima and four shoulders whose origin can be traced to transitions from different vibrational levels of the lowest excited singlet state S_1 to those of two other upper singlet electronic states labeled S_3 and S_4.

MEASUREMENT, ANALYSIS AND RESULTS

Transient excited singlet state absorption of α-NPO solution (1×10^{-4} M/l) in toluene was measured in the region 4460-6260 Å using a nitrogen laser and a nitrogen laser pumped dye laser as pump and probe beams respectively. The plot of the relative differential absorption cross-section $(\sigma_e - \sigma_g)_\lambda / (\sigma_e - \sigma_g)_{max}$ versus wavelength, which essentially represents the excited singlet state absorption spectrum, shows extensive absorption with a structure consisting of nine submaxima and four shoulders. As an aid to the explanation of these spectral features a simplified energy level diagram of the molecule has been drawn up on the basis of the data obtained from the ground state absorption and the fluorescence spectra recorded for the purpose.

Eight of the nine submaxima and the four shoulders can be tentatively interpreted to be due to transitions between a single pair of electronic states labeled S_1 and S_4 while the ninth submaximum can be attributed to a transition from the lowest vibrational level of S_1 to a vibrational level of another electronic state labeled S_3. Vacuum wavenumbers of all the observed features together with the transition assignments made are presented in table I.

According to this interpretation five of the thirteen transitions originate in the first vibrational level, two each in the second and fourth vibrational levels and four in the third vibrational level of the lowest excited singlet state S_1. Further the submaxima and shoulders due

Table I. Transition Assignments to the Observed Features in the ESSA Spectrum of α-NPO Solution (1×10^{-4} M/l) in Toluene.

ESSA Maximum Submaximum or Shoulder (in cm^{-1})	Transition Assignment $T_{e,v}$ (in cm^{-1})		Electronic State	Ground State Absorption Maximum** Submaximum or Shoulder (in cm^{-1})
	Lower	Upper	Upper	
21140	26740	47870±10	S_4	47730
20530	27350	"	"	"
18690	21190	"	"	"
17980	29880	"	"	"
19640*	26750	46380±10	"	–
17180*	29190	"	"	–
20030	26740	46790±20	"	–
17600*	29190	"	"	–
16920	29880	"	"	–
20320	27350	47660±20	"	47730
18450	29190	"	"	"
21880*	26740	48620	"	–
16360	26740	43100	S_3	–

* Those marked with an asterisk are shoulders while the others are submaxima.
** Blank space in this column indicates that the upper vibrational level $T_{e,v}$ lies in the region of moderately strong $S_0 \rightarrow S_4$ or $S_0 \rightarrow S_3$ absorption but no specific maximum, submaximum or shoulder is located at the corresponding position.

to transitions from the third and fourth vibrational levels of S_1 are found to be the strongest features in the ESSA spectrum contrary to the normal expectation. Absorptive transitions from the higher vibrational levels of S_1 can be understood on the basis of the fact that the pump and the probe beams simultaneously interacted with the molecules for about 5 nS per pulse during which they were made to overlap for optimizing the excited singlet state absorption and the observed intensity distribution may be traced to the differences in the Franck-Condon factors corresponding to the transitions from the different vibrational levels of S_1.

FOUR-ATOMIC RARE GAS HALIDE EXCIPLEXES AND THEIR IMPACT ON HIGH-POWER LASER KINETICS[*]

R. Sauerbrey, F.K. Tittel, Y. Zhu and W.L. Wilson, Jr.
Dept. of Electrical & Computer Engineering, Rice University, Houston, Texas 77251-1892

ABSTRACT

The spectroscopy and kinetics of the Ar_2F / Ar_3F system with particular emphasis on the influence of electron density will be presented. The possible impact of these results on the absorption behavior of high-power lasers will be discussed.

Rg_2X exciplexes may be described as a bound state of a diatomic rare gas ion Rg_2^+ and a halogen ion X^- having triangular shape. Since the triatomic rare gas ions Rg_3^+ are bound with respect to their diatomics by about $.2 - .3\ eV$, one could expect that the four-atomic rare gas halides Rg_3X are stable at room temperature. This work presents spectroscopic evidence for the existence of the four-atomic rare gas halide exciplex Ar_3F [1]. The spectroscopic assignment is confirmed by a reinvestigation of the kinetic reactions of the trimer Ar_2F. Former investigations[2,3] revealed abnormalities in the temporal pulse shapes of Ar_2F, in particular, at high electron densities that will be explained by electron quenching.

Experiments were performed by transversely exciting a gas mixture consisting of high purity argon, F_2, and NF_3 by a pulsed electron beam ($200\ A\ cm^{-2}$, $1\ MeV$, $10\ ns$). The time-integrated fluorescence spectra were recorded by an optical multichannel analyzer using a 0.25 nm Jarrel-Ash spectrometer with .5 nm resolution. The time dependence of the fluorescence of the ArF and Ar_2F transitions at 193 nm and 285 nm, respectively, as well as a new blue continuum at 435 nm, was measured using either a fast vacuum photodiode or a photomultiplier.

Spectra of electron beam excited Ar-F_2 mixtures in the wavelength range from 200 nm to 550 nm are shown in Fig. 1. The Ar_2F emission appears as a broad band with a maximum at 285 nm. On the long wavelength side of Ar_2F, a broad continuum appears with a maximum at (435 ± 3) nm, and a spectral width of 100 nm FWHM. It was shown that the emission around 435 nm appears only when argon and a fluorine donor, either F_2 or NF_3, are together in the gas mixture. The emission is not due to gas impurities. It is apparent from Figs. 1a,b that the blue emission increases strongly in intensity relative to the Ar_2F emission when the argon pressure is raised from 2 atm (a) to 8 atm (b).

The temporal development of the dominating emissions from electron beam excited argon/fluorine mixtures is shown in Fig. 2. The $ArF(B-X)$ fluorescence at 193 nm follows the pumping pulse with a width of about 10 ns, indicating the rapid depopulation of this state by radiation and quenching processes, mainly by three-body collisions to Ar_2F. Therefore, the Ar_2F fluorescence increases on a time scale given by the width of the ArF fluorescence, followed by an exponential decay. The new blue fluorescence continuum exhibits a temporal rise-time on the order of the half-width of the Ar_2F fluorescence and decays on the same time-scale or slower than the Ar_2F fluorescence. This temporal behavior of the blue fluorescence indicates that Ar_2F is the precursor of the new species emitting the blue continuum.

---

[*] Supported in part by the National Science Foundation and the Welch Foundation.

© American Institute of Physics 1987

These experimental findings are consistent with an assignment of the blue continuum to the transition from the first electronically excited state in Ar_3F to its repulsive ground state[1].

When Ar/F_2 mixtures are excited at high electron densities, the measured temporal fluorescence pulse shape of Ar_2F exhibits two peaks (Fig. 3). In earlier papers this phenomenon was attributed to two different production mechanisms for Ar_2F. Detailed kinetic modeling shows that this behavior can, in fact, be explained by the quenching of ArF, the precursor of Ar_2F, by electrons. For the high pumping densities used for the experiments shown in Fig. 3, peak electron densities of about $2 \cdot 10^{16} cm^{-3}$ can be estimated. A quenching constant of $2 \cdot 10^{-7} cm^3 s^{-1}$ for ArF by electrons was assumed. A set of coupled rate equations for ArF and Ar_2F was solved yielding the calculated pulse shape shown in Fig. 3. Considering the uncertainties in the determination of electron density, electron concentration decay time, and some of the relevant rate coefficients, the agreement of the calculated and measured pulse shape is very good. The calculations also predict the behavior of the Ar_2F fluorescence yield as functions of F_2 and Ar pressures correctly, which cannot be achieved without including electron quenching. Quenching of Ar_2F by electrons directly seems to have only a minor influence on the pulse shape and the fluorescence yield of Ar_2F.

The fluorescence spectra obtained for optically-excited liquid Ar/F_2 mixtures by Jara et al.[4] exhibit an unassigned fluorescence around 455 nm, with a bandwidth (FWHM) of 110 nm. Considering the redshift and broadening of the liquid phase emissions relative to the gas phase emissions observed for other rare gas halide species, the 455 nm band in liquid Ar/F_2 mixtures may be assigned to the Ar_3F emission.

If Ar_3F exists as a bound state, Kr_3F and Xe_3F should also be stable due to the higher binding energies of the krypton and xenon trimer ions relative to their dimer ions.

In summary, a new broadband emission in the blue, observed from electron beam excited Ar/F_2 and Ar/NF_3 mixtures, has been assigned to Ar_3F. The kinetic behavior of the $ArF/Ar_2F/Ar_3F$ system is severely influenced by high electron densities. The existence of four-atomic rare gas halide exciplexes could have a profound impact on the scaling behavior of rare gas halide lasers.

REFERENCES

[1] Sauerbrey, R., Zhu, Y., Tittel, F.K., and Wilson, W.L., Jr., "Optical Emission and Kinetic Reactions of a Four-Atomic Rare Gas Halide Exciplex: Ar_3F," *J. Chem. Phys.*, **85**(3), pp.1299-1302 (Aug. 1986).

[2] Böwering, N., Sauerbrey, R., and Langhoff, H., "Kinetic Studies of Ar_2F^* in Electron Beam Excited $Ar-NF_3$ Mixtures," *J. Chem. Phys.*, **76**, pp.3524-3528 (1982).

[3] Marowsky, G., Glass, G.P., Tittel, F.K., Hohla, K., Wilson, W.L., Jr., and Weber, H., "Formation Kinetics of the Triatomic Excimer Ar_2F," *IEEE J. Quantum Electron.*, **QE-18**, pp.898-902 (May 1982).

[4] Jara, H., Pummer, H., Egger, H., and Rhodes, C.K., "Optical Properties of Rare-gas Fluoride Dimers and Trimers Dissolved in Liquid Rare Gases," *Phys. Rev. B*, **30**, pp.1-6 (1984).

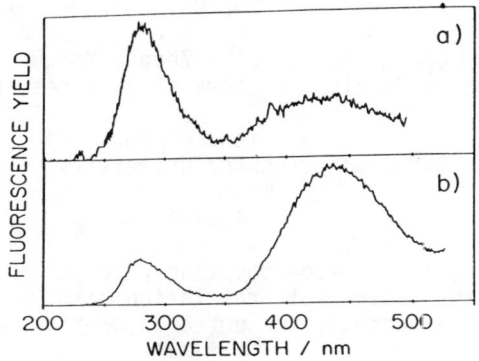

Fig. 1: Spectra from electron beam excited $Ar/5\ Torr\ F_2$ mixtures between 200 nm and 550 nm. Fig. 1a) and Fig. 1b) show the Ar_2F emission centered at 285 nm as well as a new emission in the visible centered at 435 nm.

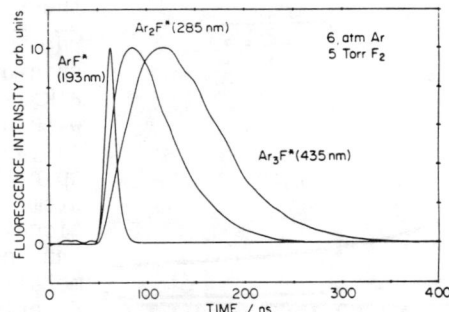

Fig. 2: Temporal dependence of normalized fluorescence signals from ArF, Ar_2F, and Ar_3F. Note the increase in the blue signal until the Ar_2F signal is well beyond its maximum.

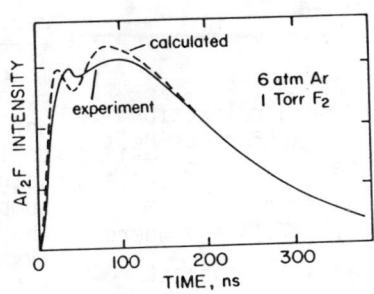

Fig. 3: Measured and calculated temporal pulse shapes for the Ar_2F fluorescence.

EXPERIMENTAL DETERMINATION OF THE SPIN-ROTATION COUPLING IN NaXe MOLECULES

Meiying Hou, Baohua Feng, Jian Zhang, YongBo She, and ZhongLu Mi
Institute of Physics, Academy of Sciences, Beijing, China

TienJie Chen
Physics Department, Peking University, Beijing, China

ABSTRACT

The mean spin-rotation coupling γN of NaXe molecules is measured by observing the sodium spin relaxation rate. It is in reasonable agreement with a recently presented theoretical expression.

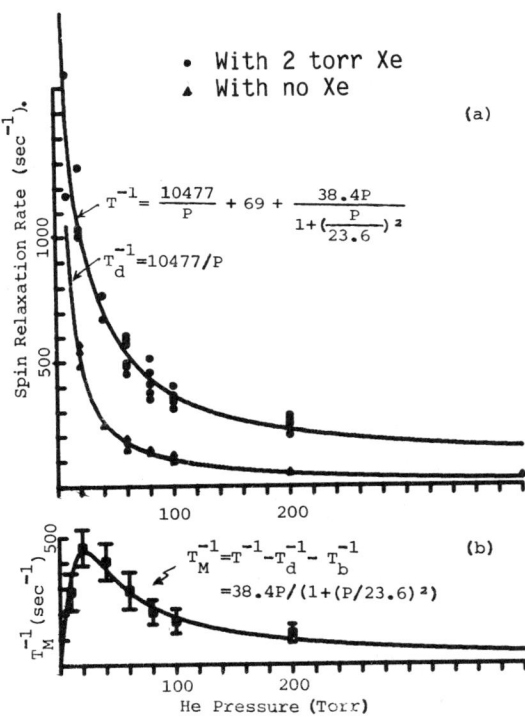

Fig. 1 (a) Measured spin relaxation rate as a function of buffer gas He's pressure. Solid line is a fit of T_1^{-1}. (b) Plot of the fitted curve of T_M^{-1} with data points ($T^{-1} - T_d^{-1} - T_b^{-1}$) where T_b^{-1} has been treated as a constant versus He pressure.

Recently Wu et al.[1] have presented a theory to calculate the spin-rotation coupling coefficient γ in alkali-metal-noble-gas van der Waals molecules. It takes the spin-rotation interaction $\gamma \vec{N} \cdot \vec{S}$ to be mainly due to the spin-orbit interaction of the alkali-metal valence electron within the core of the noble gas, where $\vec{S}$ is the electron spin of the alkali-metal atom, and $\vec{N}$ is the rotational angular momentum of the noble-gas and alkali-metal atoms about each other. Experimentally only a few of the possible pairs of different alkali-metal atoms with different noble-gas atoms have been investigated. In this paper we report our measurement of the mean interaction constant in NaXe molecules.

Since formation of van der Waals molecules is a dominant mechanism in the spin relaxation of optically pumped alkali-metal atoms in the alkali-metal-noble-gas system, we have applied the method of measuring the sodium spin relaxation rate to obtain the mean value of γN.

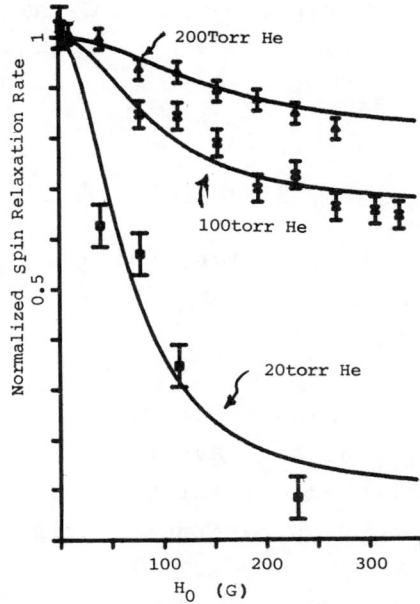

Fig.2 Dependence of sodium spin relaxation rate on the longitudinal magnetic field for different helium pressure. All cells have 2 torr xenon.

The experimental setup is similar to that described in previous papers[2,3], except that a pair of Helmholtz coils which could provide up to 350 Gauss longitudinal magnetic field were used. An Apple II computer interfaced to a signal averager was used for data acquisition and analysis.

The measured relaxation rate T^{-1} includes relaxation rate due to diffusion T_d^{-1}, rate due to binary collision T_b^{-1}, and rate due to formation of van der Waals molecules in three body collisions T_M^{-1}. The diffusion rate can be obtained by directly measuring the relaxation in a cell without xenon, as is shown in Fig.1. The binary collision rate is independent of the third body pressure and is thus a constant of He pressure. Subtracting T_d^{-1} and T_b^{-1} from the measured relaxation rate T^{-1}, we obtain the rate T_M^{-1} as a function of He pressure, as is shown in Fig. 1b. From the fitting parameters in Fig. 1b, the product of γN and the lifetime of the molecules were obtained. In order to separate γN from the lifetime, we measured the relaxation rate as a function of longitudinal magnetic field. Results are shown in Fig.2. From the half width of the magnetic relaxation rate curve and the fitting parameter of Fig.1b, we obtain

$\gamma N/h$ = (47 ± 12) MHz.

Our measured value of γN is nearly the same as was measured in KXe, RbXe, and CsXe[5]. This result shows a reasonable agreement with the theoretical predictions.

REFERENCE

1. Z. Wu, T.G. Walker, and W. Happer, Phys.Rev.Lett.54(1985)1921.
2. M. Hou, B. Cheng, and R. Ju, Chinese Physics Letters, V1, 57(1984).
3. N.D.Bhaskar, M.Hou, M.Ligare, B.Suleman and W.Happer, Phys.Rev. A22, 2710(1980).
4. M. Hou, et al., Chinese Physics Letters, V3, 397(1986).
5. Z. Wu and W. Happer, Proceedings of the Workshop on Polarized Targets in Storage Rings, Argonne National Laboratory, 17-18 May 1984.

HYPERFINE INTERACTION OF THE RYDBERG TRIPLET STATES OF Na_2

Li Li* and R.W. Field
MIT, Cambridge, Mass 02139, USA

Qingshi Zhu
Dalian Institute of Chemical Physics, Dalian, China

Abstract

The hyperfine splittings of the Rydberg triplet states, $2^3\Pi_g$, $3^3\Pi_g$, $4^3\Sigma_g^+$, and $1^3\Delta_g$, of Na_2 have been resolved by CW perturbation facilitated optical optical double resonance (PFOODR) spectroscopy. Hyperfine coupling schemes of these Rydberg states are discussed.

CW PFOODR fluorescence excitation spectroscopy via $A^1\Sigma_u^+ \sim b^3\Pi_{\Omega'u}(\Omega'=0,1)$ mixed intermediate levels has proven capable of resolving the hyperfine splittings of the Rydberg triplet states of Na_2.[1]

The two $^3\Pi_g$ states, similar to the $b^3\Pi_u$ state, are well described by the case a_β coupling scheme. The $\Omega=0$ and 2 components have resolvable hyperfine structure and the magnetic hyperfine coupling constant is positive for the $\Omega=0$ and negative for $\Omega=2$ component. The $\Omega=1$ components of both $2^3\Pi_g$ and $3^3\Pi_g$ states have much smaller hyperfine splittings than in $\Omega=0,2$ and these splittings are not resolvable in our sub-Doppler PFOODR excitation spectra.[2]

The hyperfine splitting of the $1^3\Delta_g$ state has been

*Present address: Qinghai Institute of Salt Lake, Xining, China

directly viewed via $b^3\Pi_{1u}$ intermediate levels. Every $1^3\Delta_g(v) - b^3\Pi_{1u}(v',J')$ spectrum contains five groups of lines($N=J'-2 - J'$, $N=J'-1 - J'$, $N=J' - J'$, $N=J'+1 - J'$, and $N=J'+2 - J'$). The five groups of lines from the same $b^3\Pi_{1u}v',J'$ intermediate level have identical patterns (number of components, relative intensities of components, and intervals between corresponding components). Fig. 1 (a) and (b) show the observed patterns from antisymmetric(odd-J') and symmetric(even-J') intermediate levels, respectively.

The $1^3\Delta_g$ state belongs to a case $b_{\beta S}$ coupling scheme in which the total nuclear spin angular momentum, I, first couples to the total electronic spin angular momentum, S, to yield G, which then couples to N to produce F. The G-N interaction for the $1^3\Delta_g$ state is very weak such that the F components for a given (G,N) level are unresolvable within our resolution(100 MHz). Fig. 1 (c) and (d) are predicted patterns. The calculated relative intensities and intervals satisfactorily agree with the observed.

The $4^3\Sigma_g^+$ state belongs to case $b_{\beta S}$ coupling scheme. However, under the influence of the perturbation of the $2^3\Pi_g$ and/or $3^3\Pi_g$ states, the perturbed levels tend to have a case $b_{\beta J}$ coupling.

The Rydberg states of Na_2 arise from a $3s\sigma_g$ molecular orbital and a Rydberg orbital. The Rydberg electrons are far away from the nuclei, so that the hyperfine interaction of the Rydberg triplet states of Na_2 mainly comes from the interaction between the $3s\sigma_g$ electron and the nuclei, which gives rise to a Fermi contact term b_F. The experimental value of b_F for $1^3\Delta_g$ and $4^3\Sigma_g^+$ states is about 210 MHz. Using a simple semi-quantitative LCAO molecular orbital theory,[3] we estimated the Fermi contact given by

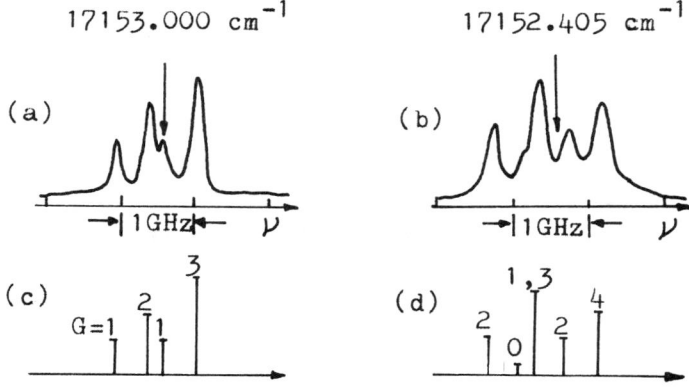

Figure 1. OODR excitation spectral patterns of the $1^3\Delta_g(v,N) - b^3\Pi_{1u}(v',J')$ lines. Probe laser frequencies are given above the observed lines. (a) observed spectrum of $1^3\Delta_g v=v_x+20, N=17 - b^3\Pi_{1u} v'=21, J'=17$, (b) observed spectrum of $1^3\Delta_g v=v_x+20, N=15 - b^3\Pi_{1u} v'=21, J'=16$, (c) calculated pattern and relative intensities for (a), (d) calculated pattern and relative intensities for (b).

the $3s\sigma_g$ electron in the Na_2 molecule. The evaluated b_F is about 150 MHz.

References

1. Li Li and R.W. Field, J. Mol. Spec. <u>117</u>, 245(1986).
2. Li Li and R.W. Field, J. Mol. Spec. (in press).
3. M.C. Curtis and P.J. Sarre, J. Chem. Soc. (in press).

Studies of the Diffuse Bands of K_2, Rb_2 and Cs_2

Wei-Tzou Luh*, John T. Bahns[†], Kenneth M. Sando,
A. Marjatta Lyyra[‡], Paul D. Kleiber[‡] and William C. Stwalley[‡]

Iowa Laser Facility and Department of Chemistry
University of Iowa
Iowa City, Iowa 52242-1294

Abstract

Direct dye laser excitations of the K_2 yellow, Rb_2 orange and Cs_2 near infrared diffuse bands have been investigated. Experimental results are shown to be consistent with free-bound $2^3\Pi_g \leftarrow 1^3\Sigma_u^+$ excitation followed by bound-free emission between the same two states. It is found that for Rb_2 and Cs_2, spin-orbit interactions become so significant that the $2^3\Pi_g$ state is split into three component states. For Rb_2, all three spin orbit components of the $2^3\Pi_g$ state are produced in direct excitation, whereas for Cs_2, most excitation wavelengths produce only one of the three spin-orbit components.

Introduction

Diffuse bands, especially of the alkali dimers, are interesting and important since they are thought to be primarily excimer-like triplet-triplet electronic transitions in which the lower triplet state is essentially repulsive. Because of inherent population inversions and relatively broad spectral ranges, they are potentially important continuously tunable laser sources[1]. The alkali diffuse bands have been observed from the UV to the IR in absorption spectra, electric discharge excitations, and laser excitations. Moreover, Bahns and Stwalley[2] have measured significant gain in the Na_2 violet band, Dinev et al.[3] have observed the triplet-triplet IR laser emission of Na_2, and Heneghan, Chakravorty and Stwalley[4] have measured significant gain in the K_2 yellow band. Here, in order to understand the characteristics of the potential curves of the electronic states involved in the diffuse bands mentioned above, we have used direct dye laser excitations.

Experimental

Three heat-pipe ovens (K, Rb, Cs) were used in this study. The dye laser (Coherent 599-21) was pumped by an argon ion laser at

---

* Also Department of Chemistry, National Kaohsiung Teachers' College, Kaohsiung 80243, Taiwan, Republic of China
† AFRPL/LKCS/Stop 24, Edwards Air Force Base, Edwards, California 93512-5000
‡ Also Department of Physics

514.5 nm. Four laser dyes, Rhodamine 560, 590 and 610, and LDS 698 (all from Exciton), were used. Their lasing spectral ranges are 540-600 nm, 570-650 nm, 601-675 nm and 688-708 nm, respectively. The former two dyes were used for K_2, the latter two for Rb_2 and Cs_2, respectively. The dye laser (usually operated multimode) was tuned at intervals of 1 nm across the wavelength ranges of the diffuse bands. For each excitation wavelength, the fluorescence spectra were taken at selected pressures (typically a few torr), resolved using a Jarrell-Ash 8200 half-meter spectrometer, detected using an RCA 4832 photomultiplier with a PAR 134 electrometer, and recorded with a Linseis LS44 strip chart recorder.

K_2 Yellow Diffuse Band

A series of fluorescence spectra is shown in Figure 1 at a pressure of 2.9 torr. The observed fluorescences are obviously unstructured continua. They can be interpreted as follows. The K_2 yellow diffuse band is presumably due to a triplet-triplet transition ($2^3\Pi_g - 1^3\Sigma_u^+$)[5] analogous to that observed for the Na_2 violet band[6]. If so, one may observe unstructured bound-free continuum fluorescence following nonselective free-bound excitations since, according to recent ab initio calculations[7], the lowest triplet state $1^3\Sigma_u^+$ of K_2 has a very shallow well of depth 498 cm^{-1} (which is approximately kT in this work), and one may expect most of the absorbing species in this triplet state are in free levels. Our observations in Figure 1 are indeed unstructured fluorescence continua and suggest that the $1^3\Sigma_u^+$ state is the lower electronic state involved. If the major relaxation processes of the upper electronic state bound levels are radiative, a measure of the spectrally integrated diffuse band fluorescence intensity versus excitation wavelength should resemble the absorption spectrum provided that the spectral response of the PMT detector in the range of interest is independent of wavelength. This is in fact the case if one compares our results to the reduced absorption coefficients of the same diffuse band from Johnson et al.[8] and Pichler et al.[9]. This suggests that the relaxation processes from the upper electronic state bound levels are mainly back to the lower $1^3\Sigma_u^+$ state. One expects, that bound-bound transitions are also possible, although the well depth is only about kT. To look for this, a single mode dye laser was also used through the entire diffuse band range to obtain excitation spectra. There was no clear evidence that any bound-bound triplet emission is excited. The free-free fluorescence is expected to be negligible at these wavelengths. In summary, the above observations show that one is indeed dealing with bound-free fluorescence following free-bound excitation, which energetically corresponds to the $2^3\Pi_g - 1^3\Sigma_u^+$ electronic transition.

Rb_2 Orange Diffuse Band

Similar diffuse band spectra were taken for Rb. One observes that these spectra are resolved into three peaks, located at 601.1 nm, 604.0 nm and 606.1 nm, respectively, unlike the single broad one

Figure 1. K$_2$ spectrally resolved diffuse band fluorescence in the 564-587 nm range excited by dye lasers in the range 565-578 nm at 1 nm intervals. The reference line at the right of each spectrum is Kr at 587.1 nm. The intense peak is scattering at the excitation wavelength.

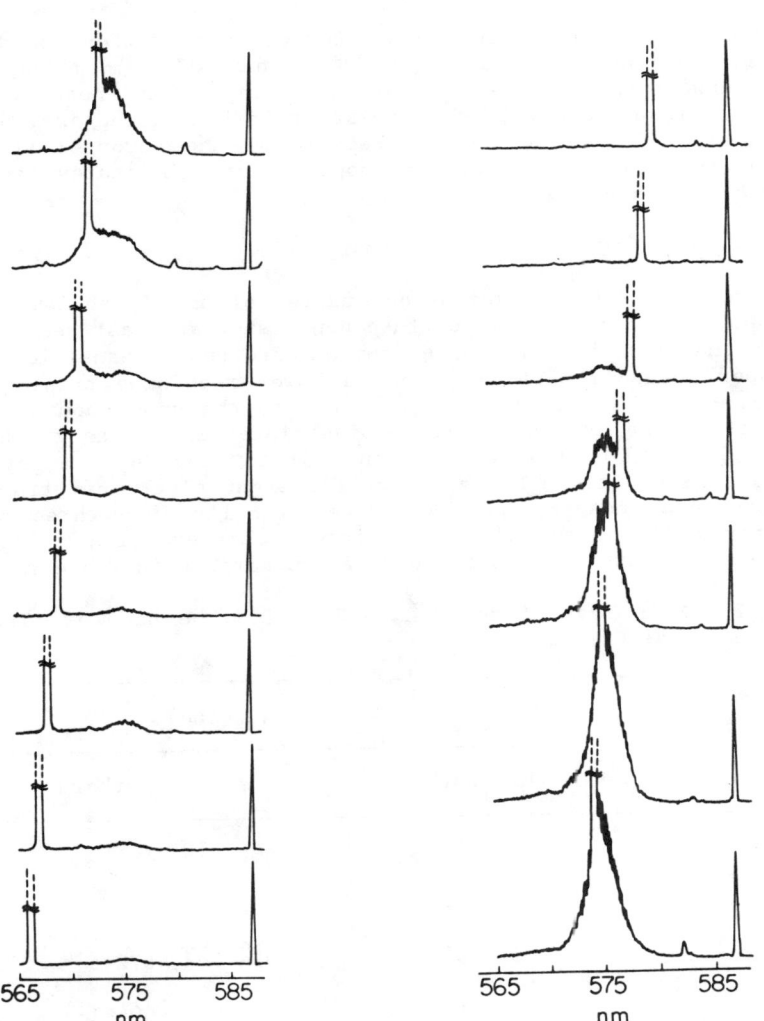

for the K$_2$ yellow diffuse band. It is well-known that spin-orbit interaction occurs in the multiplet electronic states with $\Lambda \neq 0$ and its significance increases rapidly with increasing numbers of electrons. Thus for the heavy molecules, the $2^3\Pi_g$ state is expected to be split into three different states: $2^3\Pi_{0g}$, $2^3\Pi_{1g}$ and $2^3\Pi_{2g}$. These dye laser excitation studies show that the Rb_2 orange diffuse band can be interpreted in terms of the $2^3\Pi_g - 1^3\Sigma_u^+$ transition in which the upper state is split into three component states due to spin-orbit interaction[10].

Cs_2 near infrared diffuse band

Observed fluorescence spectra for Cs_2 show diffuse band fluorescence, peaking at 718.9 nm, 712.9 nm and 707.5 nm. The separations between these three locations are larger than those of the Rb_2 orange diffuse band and are consistent with a stronger spin-orbit interaction expected for Cs_2. Because of the stronger spin-orbit interaction, the three component upper electronic states can be excited quite independently.

Summary

Dye laser direct excitation studies of the K_2 yellow, Rb_2 orange and Cs_2 near infrared diffuse bands show at least semiquantitatively that (1) they originate from an electronic transition between an upper bound triplet state and a lower repulsive triplet state; (2) from the point of view of energetics, the upper bound state is very likely the $2^3\Pi_g$ state which dissociates into a ns ground state atom and a (n-1)d excited atom; and (3) for the Rb_2 orange and Cs_2 near infrared diffuse bands, three different electronic transitions are involved because the $2^3\Pi_g$ state is split into three separate components due to the stronger spin-orbit interactions. Agreement with previous experiments is good, as summarized in Table I.

Table I. Main Peak of the $2^3\Pi_g - 1^3\Sigma_u^+$ Diffuse Band of Alkali Metal Dimers.

M_2	λ_{peak}(nm)	
	this work	others
Li_2	--	458.5[12]
Na_2	--	436.5[1,13]
K_2	575.0	574.5[13], 572.5[14], 575.0[10]
Rb_2	606.1	605.4[10], 606.1[15]
	604.0	603.2[10]
	601.1	600.7[10], 600.8[15]
Cs_2	718.9	718.5[10], 718[11], 718.7[16]
	712.9	712.9[10], 713[11], 712.7[16]
	707.5	705.4[10], 707[11], 706[16]

Acknowledgments

We would like to thank Drs. S. P. Heneghan and G. Pichler for helpful discussions and Dr. D. D. Konowalow for providing his ab initio results before publication. This research was supported by the Air Force Office of Scientific Research, the National Science Foundation and the Ministry of Education of the Republic of China.

References

1. J. P. Woerdman, Opt. Commun. 26, 216 (1978).
2. J. T. Bahns and W. C. Stwalley, Appl. Phys. Lett. 44, 826 (1984).
3. S. G. Dinev, I. G. Koproinkov and I. L. Stefanov, Opt. Commun. 52, 199 (1984).
4. S. P. Heneghan, K. P. Chakravorty and W. C. Stwalley, private communication.
5. G. Pichler, J. T. Bahns, K. M. Sando, W. C. Stwalley, D. D. Konowalow, L. Li, R. W. Field and W. Mueller, Chem. Phys. Letters, 129 425 (1986).
6. W. T. Luh, J. T. Bahns, K. M. Sando, W. C. Stwalley, S. P. Heneghan, G. Pichler and D. D. Konowalow, Chem. Phys. Letters 131, 335 (1986).
7. D. D. Konowalow, J. L. Fish, and C. Duda, private communication, March, 1984.
8. D. E. Johnson and J. G. Eden, J. Opt. Soc. Am B 2, 721 (1985).
9. D. Veza, S. Milosevic and G. Pichler, in K. Burnett, editor, *Spectral Line Shapes*, Volume 2 (de Gruyter, New York, 1983), p. 679.
10. G. Pichler, S. Milosevic, D. Veza and R. Beuc, J. Phys. B 16, 4619 (1983).
11. P. Cavaliere, G. Ferrante and L. L. Cascio, J. Chem. Phys. 62, 4753 (1975).
12. J. T. Bahns, W. C. Stwalley and G. Pichler, J. Chem. Phys., in press.
13. Cz. Radzewicz, P. Kowalczyk and J. Krasinski, Opt. Comm. 44, 139 (1983).
14. M. M. Rebbeck and J. M. Vaughan, J. Phys. B 4, 258 (1971).
15. J. M. Brom, Jr. and H. P. Broida, J. Chem. Phys. 61, 982 (1974).
16. W. Finkelnburg and O. Th. Hahn, Phys. Z. 39, 98 (1938).

THE Li_2 $b^3\Sigma_g^+$ - $X^3\Sigma_u^+$ TRANSITION AND A NEW UV ABSORPTION BAND

Jin Feng, Zhang Chune, and Tang Baoyin
Laser Division, Harbin Institute of Technology, China

ABSTRACT

In this paper the absorption spectrum (1.0-1.7 μm) is reported and the absorption coefficient obtained both experimentally and theoretically. We believe that it results from the $b^3\Sigma_g^+$ - $X^3\Sigma_u^+$ transition. The possibility of laser action in this region in lithium vapor is discussed.

INTRODUCTION

During the last decade, many experimentalists and theoreticians have been interested in the alkali dimer molecule Li_2, which, next to H_2, is the lightest stable homonuclear molecule. For the applied sciences Li_2 is of interest as an active laser medium.

EXPERIMENT AND DISCUSSION

Figure 1. shows the observed absorption spectrum originated from $b^3\Sigma_g^+$ - $X^3\Sigma_u^+$ transition over the region of 1.0-1.7 μm at P_{He} = 400 Torr and T = 1000°C. The experimental absorption coefficient calculated from the curves (A) and (C) of Figure 1 is given in Figure 2. curve A, and the calculated absorption coefficient versus wavelength is also obtained via the formula[1] (see Figure 2. curve B):

$$K_\nu(T) = \frac{4\pi^4 \nu D^2 g^4 n^2 R^2(\nu)}{|d\nu/dR|} e^{-V/kT}$$

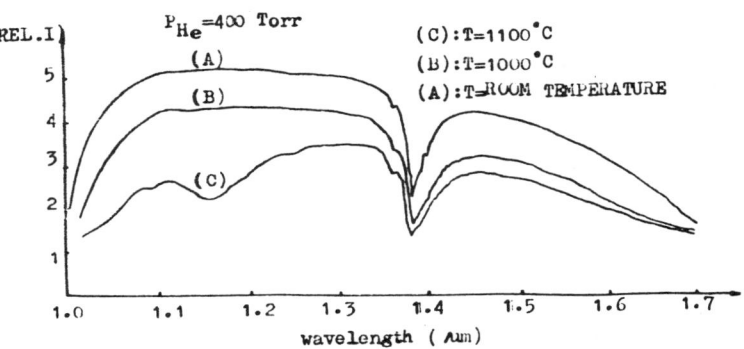

Figure 1. Absorption spectra of the first triplet state in Li_2.

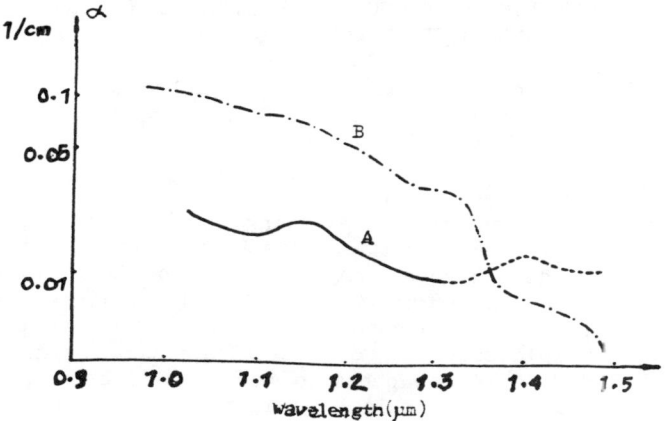

Figure 2. The absorption coefficient of the first triplet states.

Both curves at wavelengths shorter than 1.35 μm are similar, but they are quite different in the region of $\lambda > 1.35$ μm, possibly because of the low signal to noise ratio in the experiment.

The main reason for the big difference between the absorption curves A and B is that we have not taken any steps to decrease the noise which may result from the index fluctuation at the contact surfaces between lithium vapor and buffer gas.

In the experimental curve, there is an absorption peak centered around 1.154 μm which extends for nearly 700 Å. Calculations suggest that this absorption may correspond to an internuclear separation $R = 6.9\ a_0$.

CONCLUSION

We argue that the absorption spectra that we have shown here comes from the $b^3\Sigma_g^+ - X^3\Sigma_u^+$ transition. The peak of the absorption from $b^3\Sigma_g^+ - X^3\Sigma_u^+$ is at 1.154 μm, but the peak of emission from the same electronic transition may appear at 1.3 μm[2]. Since they do not overlap in wavelength, this is hopefully a lasing system, and if there were an effective way to excite the upper state $b^3\Sigma_g^+$, the possibility of a tunable laser with high conversion efficiency and low threshold might be anticipated in the near-infrared region.

REFERENCES

1. J. Huennekens, S. Schaefer, M. Ligare, and W. Happer, J. Chem. Phys. **80**, 4794 (1984).

2. D. D. Konowalow and P. S. Julienne, J. Chem. Phys. **72**, 5815 (1980).

PRECISE MULTIPHOTON SPECTROSCOPY OF EXCITED STATES OF H_2

E.E. Eyler, J.M. Gilligan and E. McCormack
Yale University Physics Department

ABSTRACT

Precise wavelength measurements involving the ground state of H_2 are difficult because the transitions lie in the vacuum ultraviolet region. We have circumvented this problem by exciting three-photon transitions to the $B(2p\sigma)$ $^1\Sigma_u$ and $C(2p\pi)$ $^1\Pi_u$ states, a technique whose potential for accurate wavelength measurements has gone largely unexplored. Resonant multiphoton ionization is observed in a collisionless supersonic beam, collimated to reduce the Doppler width. The principal limitations are the AC Stark effect, which can be largely eliminated at low laser powers, and a blue shift that depends strongly on number density. This unusual effect apparently arises from the interference of three-photon excitation and third harmonic generation. Preliminary measurements of C←X and B←X intervals agree well with the best conventional measurements, and are about twice as accurate. We will also report a preliminary investigation of the photodissociation of H_2 to $H(1s) + H(2s,2p)$ using stepwise laser excitation.

INTRODUCTION

Since H_2 is the simplest neutral molecule, accurate measurements of its excited state energies play an extremely important role in testing molecular theory. Unfortunately all transitions to these states from the ground state lie in the vacuum ultraviolet, and as a result the precision of measurements involving the ground state has never exceeded 0.1 cm^{-1}, an order of magnitude less accurate than the best laser measurements of optical intervals between excited states. One way to circumvent this problem is to excite a three-photon transition in the near ultraviolet. Although three-photon spectra have been observed in numerous atoms and molecules with moderately high resolution, the applicability of this technique for accurate wavelength measurements has gone virtually unexplored. We report preliminary results from an experiment in which we have investigated this possibility by performing precise three-photon spectroscopy on the $B(2p\sigma)$ $^1\Sigma_u$ and $C(2p\pi)$ $^1\Pi_u$ states. We also describe a related investigation of photodissociation in H_2 using an alternative, multi-step method.

EXPERIMENTAL

For the three-photon experiment, an excimer-pumped pulsed amplifier chain is used to amplify the output of a cw laser operating near 600 nm into 10 nsec pulses with an energy of up to 15

© American Institute of Physics 1987

mJ. These pulses are frequency doubled in a KDP crystal, yielding UV pulses with an energy of 0.1-1.5 mJ and a linewidth of approximately 0.005 cm^{-1}. The UV light is focused with a 15 cm lens into a collision-free pulsed supersonic beam. A collimation ratio for the molecular beam of roughly 10:1 reduces the Doppler width correspondingly, to about 0.01 cm^{-1} for the visible fundamental. Both the B and C states are detected by 3-photon resonant, 4-photon ionization, in a fashion similar to recent work by Pratt, Dehmer and Dehmer at lower resolution.[1]

To study possible systematic effects, we have made a careful examination of three lines, the Q(1) and R(1) branches of the (2-0) band of the C←X transition, and the R(1) branch of the (3-0) band of B←X. Two sources of line shifts were encountered. The first is the AC Stark effect. Since there is a distribution of power levels experienced by different atoms in the focused beam, this effect shows up principally as a power-dependent asymmetric broadening. For the strong C←X transitions it was relatively easy to reduce this effect to unobservable levels by reducing the energy of the UV beam and slightly defocusing it, but for the weaker B←X transition it could not be eliminated completely. By extrapolating the line position to zero power, we were nevertheless able to determine the B←X interval to an accuracy of about 0.06 cm^{-1}.

The second systematic effect is a blue shift that depends strongly on number density in the molecular beam. These shifts are far in excess of normal pressure shifts, and are prominent even in the tenuous molecular beam environment. This effect was only observed clearly for the C←X transition, and appears to arise from the interference of 3rd harmonic generation and 3-photon excitation. Similar effects have been observed in atomic xenon and in acetylene molecules, although at higher pressures and with lower laser resolution.[2] Fortunately, by operating near the lowest pressures for which a signal can be observed, the linewidth was reduced to the Doppler-limited width and no further pressure dependence was observed. Further studies of this curious phenomenon are underway, using circular polarization to suppress the 3rd harmonic generation.

The results of the three preliminary measurements are given below. They are in excellent agreement with measurements by Dabrowski using a conventional VUV spectrometer,[3] and about twice as accurate. We believe that the accuracy can be improved at least to the level of 0.01 cm^{-1} before the systematic effects become intractable. It will also be possible to measure certain of these intervals using single-photon transitions, by generating 3rd harmonic generation in a separate cell of Xe or Kr gas.

Branch	This measurement	Dabrowski, 1984
C←X, 2-0, Q(1)	103620.00 ± 0.06 cm^{-1}	103620.05 ± 0.10 cm^{-1}
C←X, 2-0, R(1)	103509.39 ± 0.06 cm^{-1}	103509.45 ± 0.10 cm^{-1}
B←X, 3-0, R(1)	94032.75 ± 0.06 cm^{-1}	94032.69 ± 0.10 cm^{-1}

It is especially interesting to consider the application of multiphoton techniques to the two most fundamental energy spacings of H_2, the ionization potential (IP) and dissociation energy. Recent measurements of the Rydberg nf states by Eyler, Short and Pipkin[4] have allowed the determination of the IP to an accuracy of 0.03 cm^{-1} relative to the metastable c (2pπ) state, and the determination of appropriate multiphoton intervals will allow this accuracy to be transferred to the ground state as well. One particularly promising possibility is to use use 3-photon spectroscopy to excite the lower nf Rydberg states directly from the ground state, yielding the IP directly from an extrapolation to n=∞.

Experiments are also underway at Yale to study the photodissociation of H_2. The dissociation energy has been measured only to an accuracy of 0.5 cm^{-1}, far worse than the claimed accuracy of the best calculations. We are studying the dissociation limit to H(1s) + H(2s,2p) at high resolution using stepwise excitation. An excimer laser at 193 nm excites a two-photon transition to the long-lived E,F (2sσ) state, and a pulse-amplified cw laser subsequently excites the phtotodissociation continuum adjoining the B' (3pσ) state. A Lyman-α detector is used to detect the photodissociation products. One advantage of this arrangement is that by using an electric field to quench the metastable 2s atoms, the H(2s):H(2p) branching ratio can be determined. Preliminary results suggest the dissociation limit lies slightly below the previous experimental value. Surprisingly, a large photodissociation signal is also observed arising from two-photon absorption from the E,F state to dissociative states lying well above the ionization limit. In measurements now underway, a delayed electric field is used to selectively detect the slow H(2s) atoms produced just above threshold, by waiting until the faster H atoms produced by two-photon absorption have left the field of view of the detector.

[1] S.T. Pratt, P.M. Dehmer and J.L. Dehmer, Chem. Phys. Lett. <u>105</u>, 28 (1986).

[2] T.M. Orlando, L. Li, S.L. Anderson and M.G. White, Chem. Phys. Lett. <u>129</u>, 31 (1986), and references therein.

[3] I. Dabrowski, Can. J. Phys. <u>62</u>, 1639 (1984).

[4] E.E. Eyler, R.C. Short and F.M. Pipkin, Phys. Rev. Lett. <u>56</u>, 1602 (1986).

KINETIC SPECTROSCOPY USING A COLOR CENTER LASER

J. W. Stephens, J. L. Hall, W. B. Yan, H. Solka, M. L. Richnow,
R. F. Curl, Jr., G. P. Glass, and F. K. Tittel
Rice University, Houston, TX. 77251

Infrared kinetic spectroscopy has been used to study the ethynyl radical, C_2H. In this work, three new C_2H absorption bands in the infrared have been discovered and the reaction rate constants of C_2H with O_2, H_2, and NO have been measured.

The kinetic spectroscopy experimental arrangement has been described in detail elsewhere[1]. Briefly, the C_2H radicals are produced by the photolysis of C_2H_2 or CF_3C_2H in a 1 meter cell. The photolysis is accomplished by a 10 nsec laser pulse from the 193 nm ArF line of an excimer laser while the the resulting species are probed with a scanning color center laser spectrometer[2] operating between 2.3 and 3.3 µm. The 3 MHz line width of the color center laser along with the ability to make continuous high resolution scans of up to 10 cm^{-1} provide the ability to observe easily the rotational structure of C_2H.

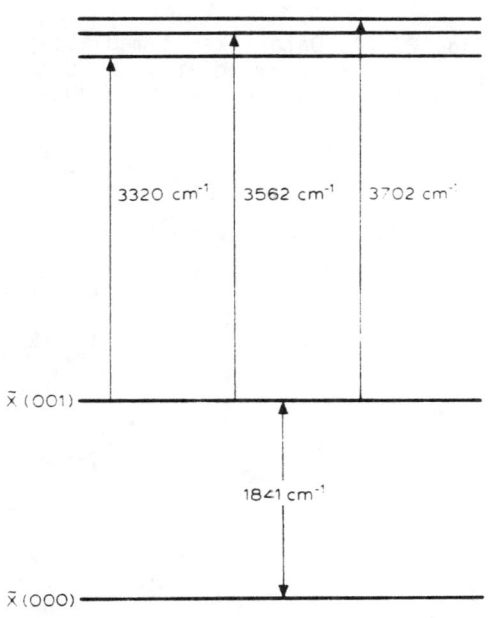

Fig. 1. The three new bands observed by excimer laser flash photolysis. The transition at 1841 cm^{-1} has been observed previously by Kanamori and Hirota.

© American Institute of Physics 1987

Recently, three new bands of C_2H have been observed upon the photolysis of C_2H_2 in our cell[3]. The spectrum was recorded by subtracting data immediately before the excimer laser flash from data taken immediately after the flash. By setting these two channels 200 nsecs apart, the noise is greatly reduced. The three

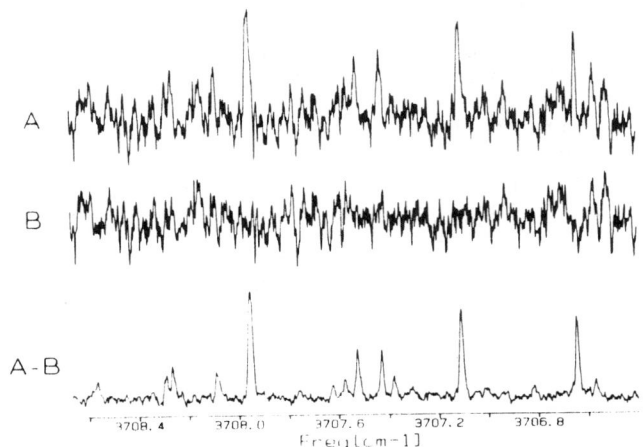

Fig. 2. (A) Data taken immediately after the excimer laser flash. (B) Data taken immediately before the flash. (A-B) The two channels are subtracted to reduce noise.

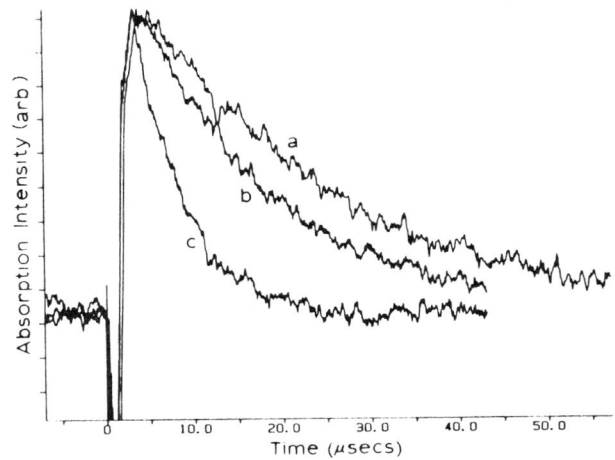

Fig. 3. Decay in C_2H absorption at several O_2 pressures. (a) No O_2 (b) 18.8 mTorr O_2 (c) 111 mTorr O_2. The spike at time = 0 is due to electrical interference from the excimer laser.

transitions lie at 3320, 3562, and 3702 cm^{-1} and are all of $^2\Sigma - ^2\Sigma$ symmetry. These transitions all share a common lower state, the $X^2\Sigma^+(001)$ state observed by Kanamori and Hirota[4].

For the rate constant measurements, C_2H was produced by the photolysis of CF_3C_2H. The decay in the C_2H absorption due to reaction of C_2H with O_2, H_2, or NO was monitored by using liquid nitrogen cooled InSb infrared detectors with fast preamplifiers giving a system time contant of less than 1 μsec. The transient digitizer signal was fitted to an exponential decay. The decay constants obtained were corrected for reaction of C_2H with the precursor and the resulting constants plotted versus the reactant pressure to yield the reaction rate. The rates measured for C_2H with O_2, H_2, and NO were 4.2×10^{-11}, 4.8×10^{-13}, and 3.5×10^{-11} cm^3 molecule^{-1} sec^{-1} respectively.

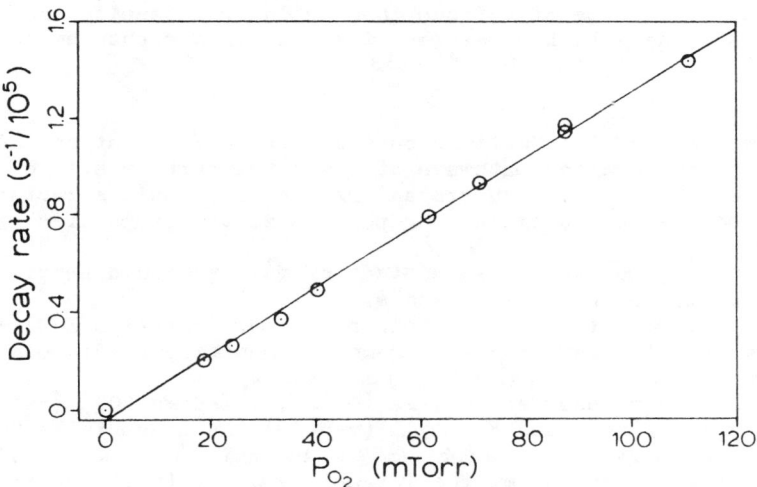

Fig. 4. Decay constants for various pressures of O_2.

1. J. L. Hall, D. Zeitz, J. W. Stephens, J. V. V. Kasper, G. P. Glass, R. F. Curl, and F. K. Tittel, J. Phys. Chem. 90, 2501 (1986)
2. J. V. V. Kasper, C. R. Pollock, R. F. Curl, Jr., and F. K. Tittel, Appl. Opt. 21, 236 (1982)
3. W.-B. Yan, J. L. Hall, J. W. Stephens, M. L. Richnow, and R. F. Curl, J. Chem. Phys., in press.
4. H. Kanamori and E. Hirota, to be published.

This work was supported by the Department of Energy under grant DE-FG05-85ER 13439 and the Robert A. Welch Foundation under grant C-071.

$Fe(CO)_5$ MULTIPHOTON IONIZATION MASS SPECTRA

Han Jingcheng, Li Shutao, Shi Jiliang[*], Liu Houxianf,
Li Fanglin[*], Gu Jianping, Wu Cunkai

Laboratory of Laser Spectroscopy, Anhui Institute of Optics
and Fine Mechanics, Academia Sinica, P.O.Box 25, Hefei, China

ABSTRACT

The multiphoton ionization (MFI) of $Fe(CO)_5$ under the action of an XeCl excimer laser beam has been investigated by using both diffusive molecular beam and modified quadrupole mass spectrometer. We not only observed CO^+, Fe^+, $Fe_2(CO)_4^+$ and $Fe(CO)_n^+$ (n=1—5) mass peaks, but also found the mass peaks of $Fe_2(CO)_m^+$ (m=5—9) in MPI mass spectrum induced by XeCl excimer laser. In addition, the dependences of the yields of various ions on laser intensity and sample pressure have been meassured. The formation mechanism for $Fe_2(CO)_m^+$ (m=5—9) ions is discussed.

The study of the multiphoton dissociation/ionization and the excited state reaction pathways of the metal carbonyls is of considerable importance in understanding the electronic structure of such complexes and explaining photocatalytic properties of the complexes.

In this paper we report a study of multiphoton dissociation and ionization of $Fe(CO)_5$ molecule.

The experimental setup consists of four parts: a XeCl excimer laser, a molecular beam arrangement, a modifild quadrupole mass spectrometer, and a signal processing system.

We not only observed CO^+, $Fe_2(CO)_n^+$, Fe^+ and $Fe(CO)_n^+$ (n=1—5) ion peaks, but also found $Fe_2(CO)_m^+$ (m=5—9) ion peaks in $Fe(CO)_5$ multiphoton ionization mass spectrum. The intensity of the Fe^+ signal is the largest among them. The intensities of $Fe(CO)_n^+$ and $Fe_2(CO)_4^+$ are typically lower than that of Fe^+ signal by a factor of about 10^2. The intensities of $Fe_2(CO)_m^+$ are typically lower than that of Fe^+ signal by a factor of about 10^4.

The laser intensity and sample pressure dependences of various ion signals at low power are shown in the table I and table II.

Table I. Experiments of laser intensity dependences of molecular ions at low power.

Fe^+	CO^+	$Fe(CO)^+$	$Fe(CO)_2^+$	$Fe(CO)_3^+$	$Fe(CO)_4^+$	$Fe(CO)_5^+$	$Fe_2(CO)_4^+$
4.0	3.9	3.4	4.0	3.9	4.0	4.1	4.1

$Fe_2(CO)_5^+$	$Fe_2(CO)_6^+$	$Fe_2(CO)_7^+$	$Fe_2(CO)_8^+$	$Fe_2(CO)_9^+$
4.12	3.94	3.84	3.89	4.13

Table II. Experiments of sample pressure dependences of molecular ions.

$Fe(CO)^+$	$Fe(CO)_2^+$	$Fe(CO)_3^+$	$Fe(CO)_4^+$	$Fe(CO)_5^+$	$Fe_2(CO)_4^+$
1.75	1.90	2.10	2.00	1.90	1.90
$Fe_2(CO)_5^+$	$Fe_2(CO)_6^+$	$Fe_2(CO)_7^+$	$Fe_2(CO)_8^+$	$Fe_2(CO)_9^+$	
2.88	2.75	2.89	2.90	3.07	

As shown in Table I and Table II, the dependences on laser intensity for Fe^+, $Fe(CO)_n$ (n=1-5) and $Fe_2(CO)_4$ are all about fourth order while pressure dependences of $Fe(CO)_n$ (n=1-5) are about second order. These experimental results support the analysis on photodissociation dynamics of $Fe(CO)_5$ molecule presented by Grant[1]:

$$Fe(CO)_5 \xrightarrow{4h\nu} Fe(I) + 5CO$$

$$Fe(I) \longrightarrow Fe(II)$$

$$Fe(II) + Fe(CO)_5 \longrightarrow [Fe_2(CO)_5^+]^* \longrightarrow Fe_2(CO)_4^+, Fe(CO)^+ \ldots\ldots$$

The formation of Fe^+ ion is a 4 photon process; $Fe_2(CO)_4^+$, $Fe(CO)_n^+$ (n=1—5) come mainly from the ion-molecule reaction of Fe^+ and $Fe(CO)_5$.
It is supposed that $Fe_2(CO)_m^+$ (m=5—9) ions may come from two channels. These ions not only may be produced by the second ion-molecule reaction of $Fe(CO)_n^+$ ions (n=1—5) and the parent molecule $Fe(CO)_5$ or neutral fragment CO, but also may come from the photofragmentation of the cluster produced by the supersonic expansion.
But taking into account that our molecular beam apparatus is diffusive, the sample pressure dependence of $Fe_2(CO)_m^+$ ion signal is third order, and the laser intensity dependence of $Fe_2(CO)_m$ ion signal is fourth order, it is infered that front channel is dominant.

REFERENCES

1. E.R.Grant. J.Chem. Phys. <u>79</u>, 4899 (1983).

* Present address: Shanghai Institute of Organic Chemistry, Academia Sinica, Shanghai, China.

Light-Induced Drift in a Discharge Tube

Lin Fucheng, Li Shifang, Hu Qiquan, and Huang Zhenjiang
(Shanghai Institute of Optics and Fine Mechanics; Academia Sinica, Shanghai, China)

Abstract

Recently the light-induced drift(LID) effect has been studied theoretically and experimentally /1-6/. The evolution equations of this effect are nonlinear and with a soliton-type solution. The LID effect in discharge condition will allow using more kinds of atoms(using sputtering effect, for example)and more atomic states(metastable state) so that some possible new applications may be found. When we deal with LID in a discharge tube, we have to consider a multi-level syatem instead of two-level system as the non-discharge case does and have to consider the Langmuir mobile effect.

Let us take the neon atom as an example. The simplified energy levels are shown in Fig.1. The ground state(g), metastable state(m), a nearby state(a) and an upper state (u) are involved. The processes involved are laser coupling(m and u), electron collision excitation and de-excitation, and spontaneous emission (from a to g). We treat m and u semiclassically as in /6/ and add two rate equations for states g and a. In this case, we also consider the Langmuir mobile velocity. The result of this model is shown in Fig.2 where we assume the atomic density at one end of the tube is constant n(0) and the other end is closed. The x-axis represents the discharge current normalized to the saturation current of the metastable level, i.e. when the electron impacting de-excitation rate is equal to the de-excitation rate by other effects then the current is equal to 1. The y-axis represents the ratio of densities at two ends. The parameter appearing in the Fig.2 is the laser intensity J. Fig.2(a) is for infinite length and Fig.2(b) is for a tube two absorption length long.

The metastable state m can be excited to u and then decay to a and at last to g. This means the m state will be exhausted which causes the saturation intensity to become very small. Because of this saturation, the light can not excite more atoms. In other words, the force of the LID is also saturated. This effect makes the slope of density profile decrease, as shown in Fig.3. Other physics process has to be considered is the Langmuir mobile. The atoms are subjected the impact from the electron and get longitudinal velocity. This makes the atoms crowd to the end of the anode and a pressure difference is formed. A diffusion from anode to cathode will form to reach a steady state. The longitudinal velocity caused by these two effects and the total velocity, i.e. the Langmuir mobile, distrbution along the radius of the tube corresponds to the upper, lower and middle curve of Fig.4(a), respectively, where a is the radius of the tube. This Langmuir mobile can be used to overcome the wall effect of LID, i.e. the velocity caused by the LID effect is a Bessel function of radius r. Fig.4(b) shows this effect, the curve labbeled 1,2, and 3 corresponds the LID effect, Langmuir mobile and the total drift, respectively.

Using the experimental data /4/ we estimate that for observing the optical piston effect, the density of Ne is about 10 cm. If one observes the drift of Ne, this is the partial pressure of this isotope and the total pressure of neon then is several torrs.

Reference:
/1/ F.Kh.Gel'mukanov and A.M.Shalagin, JETP Lett., Vol.29.711(1979)
/2/ A.D.Antsigin et al., Opt.Comm., Vol.32,237(1980)
/3/ P.J.Chapovsky and A.M.Shaiagin, Opt.Comm., Vol.0,129(1981)
/4/ H.G.C.Werij et al., Phys.Rev.Lett., Vol.237(1984)
/5/ J.P.Woerdman et al., invited paper on IQEC'86, San Francisco, 1986
/6/ Gerard Hichuis, Phys.Rev.A Vol.31, 1636(1985)

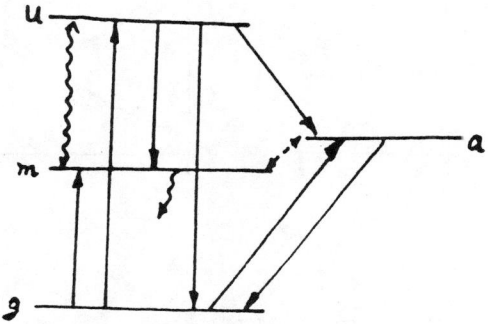

Fig. 1: The simplified energy levels.

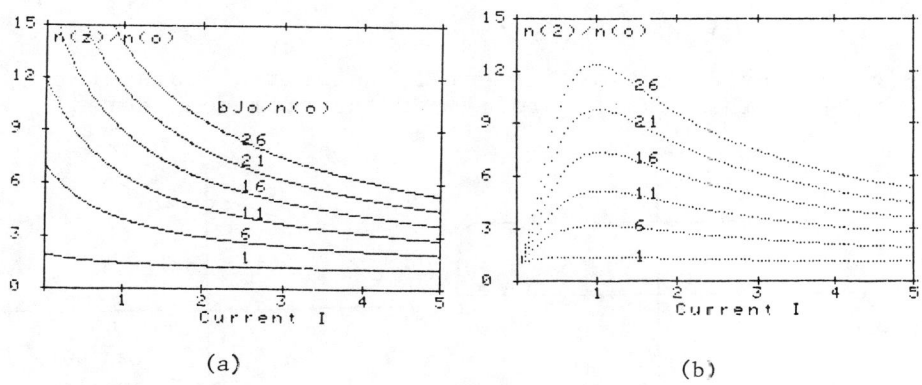

Fig. 2: The ratio of the stationary densities at two ends for a tube with infinite length **a.** and two absorption length with homogeneous density n(0) **b.** b is a constant depending on the collision parameter.

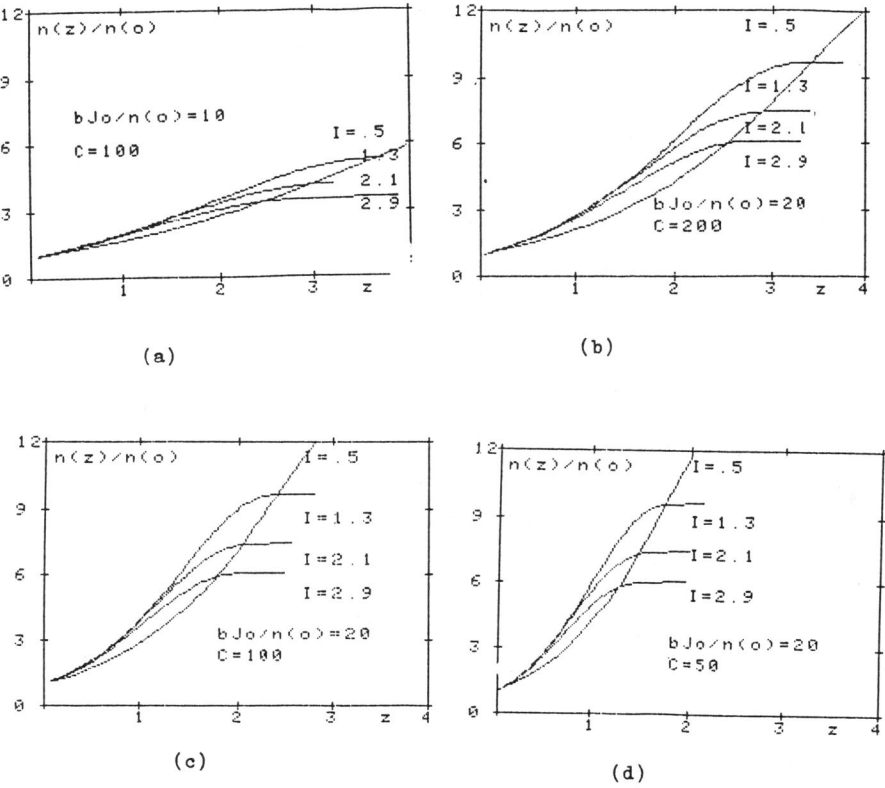

Fig. 3: The effect of current I, saturation parameter C, J/n(0) on the density profile of the atom considered.

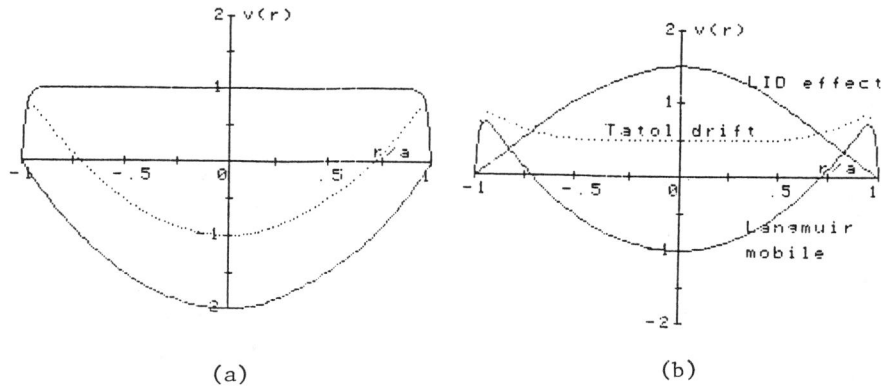

Fig. 4: The Langmuir mobile **a.** and using it to overcome the wall effect **b.**

ns´ AUTOIONIZING RYDBERG SERIES LINES OF Xe

Kiyoshi Ueda*
University of Maryland, College Park, MD 20742

ABSTRACT

Complex quantum defects are shown to be convenient for comparisons of theoretical studies (which lead to a set of eigen quantum defects μ_α and transformation matrix elements $U_{i\alpha}$) with experimental determinations of resonance energies and line widths using the ns´ autoionizing Rydberg series of Xe as a specific example.

INTRODUCTION

The $J=1^o$ autoionizing Rydberg series of Xe have received extensive experimental investigations which include VUV laser spectroscopy by Bonin et al.[1] and two-photon laser spectroscopy by Wang and Knight.[2] Theoretical investigations of the $J=1^o$ channel-structure based on multichannel quantum defect theory (MQDT)[3] include semiempirical determinations of MQDT parameters by Lu[4] and Geiger[5] and ab initio calculations of MQDT parameters by Johnson et al.[6] Parameters obtained normally from experimental studies are line-shape parameters such as resonance energy ω_R, line width Γ, and, in Ref. 1, the asymmetry parameter q defined by Fano's resonance formula.[7] On the other hand, available theoretical studies[4-6] provide MQDT parameters in the eigenchannel representation such as eigen quantum defects μ_α and transformation matrix elements $U_{i\alpha}$. Thus, to compare experimental and theoretical data requires transformation of eigenchannel parameters to alternative MQDT parameters which are more closely correlated to the line-shape parameters. This paper demonstrates the use of complex quantum defects[6] to make such comparisons.

COMPLEX QUANTUM DEFECTS

Note that we are interested in the region between the first and second thresholds where the ns and nd $J=1^o$ channel structure consists of three open channels and two closed and degenerate channels, a situation which leads to the well known ns´ and nd´ autoionizing series. In this region, complex quantum defects

---

* Permanent address: Research Institute for Scientific Measurements, Tohoku University, Katahira, Sendai 980, Japan.

$\mu_{ck}=\alpha_k+i\beta_k$ are defined in terms of eigen values of a 2×2 submatrix χ_{cc} which expresses the degenerate closed-closed-channel part of the 5×5 global scattering matrix χ.[3] Thus,

$$\chi_{cc} \mathbf{X} = \mathbf{X} \exp(2\pi i \mu_c) \quad , \quad (1)$$

where $X^t X = 1$. The real and imaginary parts of the complex quantum defect, α_k and β_k, are related to the resonance energy ω_{kn} and the resonance width Γ_{kn} by[3]

$$\omega_{kn} + E_0 = I - Ry/(n_k-\alpha_k)^2 \quad (2a)$$

and

$$\Gamma_{kn}/2 = 2Ry\beta_k/(n - \alpha_k)^3 \quad . \quad (2b)$$

Eq. (2b) is valid for relatively narrow lines with $\pi\beta_k \ll 1$ such as the ns´ series of Xe under consideration. Ry is the Rydberg constant, I is the second threshold energy, E_0 is the energy of the lower level of the transition, and k and n specify the channel (or the series) and the line, respectively.

Note that $\bar{\mu}_\alpha$ and $U_{i\alpha}$ are defined in terms of the eigen values and eigen vectors of the global scattering matrix χ. Thus, elements of the submatrix χ_{cc} can be calculated from the eigenchannel parameters $\bar{\mu}_\alpha$ and $U_{i\alpha}$ using the relation[3]

$$\chi_{k,k´} = \sum_{\alpha=1}^{5} U_{k\alpha} \exp(2\pi i \bar{\mu}_\alpha) U_{k´\alpha} \quad (3)$$

and complex quantum defects can be calculated by diagonalization of the 2×2 complex matrix constructed from Eq. (3).

COMPARISON BETWEEN EXPERIMENTAL AND THEORETICAL STUDIES

The complex quantum defect for the ns´ series calculated from various sets of eigenchannel MQDT parameters available in the literature[4-6] are compared in Table I with those obtained from two recent experiments[1,2] through use of Eq. (2). This work was supported in part by the National Science Foundation under Grant PHY-83-14060.

Table I Values of the complex quantum defect of the ns´ channel derived from experimental and theoretical data.

	Experimental		Theoretical		
	BMY[a]	WK[b]	LU[c]	GR[d]	JCHL[e]
α	0.017	0.013(2)	0.041	−0.013	−0.019
β	0.0020(2)	0.0030(9)	0.0050	0.0022	0.0031

a. For one line 11s´ in Ref. 1.
b. For ns´ with n ⩾ 13 in Ref. 2.
c. From energy independent semiempirical MQDT parameters in Ref. 4.
d. From energy dependent semiempirical MQDT parameters at the second threshold in Ref. 5.
e. From ab initio MQDT parameters at the second threshold in Ref. 6.

REFERENCES

1. K. D. Bonin, T. J. McIlrath, and K. Yoshino, J. Opt. Soc. Am. B$\underline{2}$, 1275 (1985).
2. L. G. Wang and R. D. Knight, Phys. Rev. A (in press).
3. M. J. Seaton, Rep. Prog. Phys. $\underline{46}$, 167 (1983).
4. K. T. Lu, Phys. Rev. A$\underline{4}$, 579 (1971).
5. J. Geiger, Z. Physik A$\underline{282}$, 129 (1977).
6. W. R. Johnson, K. T. Cheng, K.-N. Huang, and M. Le Dourneuf, Phys. Rev. A$\underline{22}$, 989 (1980).
7. U. Fano, Phys. Rev. $\underline{124}$, 1866 (1961); U. Fano and J. W. Cooper, Phys. Rev. $\underline{137}$, A1364 (1965).

REMEASUREMENT OF THE RYDBERG CONSTANT BY A CROSSED-BEAM LASER

PING ZHAO, WILLIAM LICHTEN,*
Yale U., Box 6666, New Haven, CT 06511

HOWARD LAYER AND JAMES BERGQUIST
The National Bureau of Standards,
Gaithersburg MD 20899 and Boulder CO 80303

ABSTRACT

We measure the Rydberg constant, via a single photon determination of the balmer-alpha wavelength. The result is:
$R = 109\ 737.315\ 69(7)\ cm^{-1}$, where $c = 299\ 792\ 458$ m/s by definition. The precision is 6 parts in 10^{10}. Preliminary results for the Balmer-β transition are: $R = 109\ 737.315\ 73(3)\ cm^{-1}$ (error 3×10^{-10}).

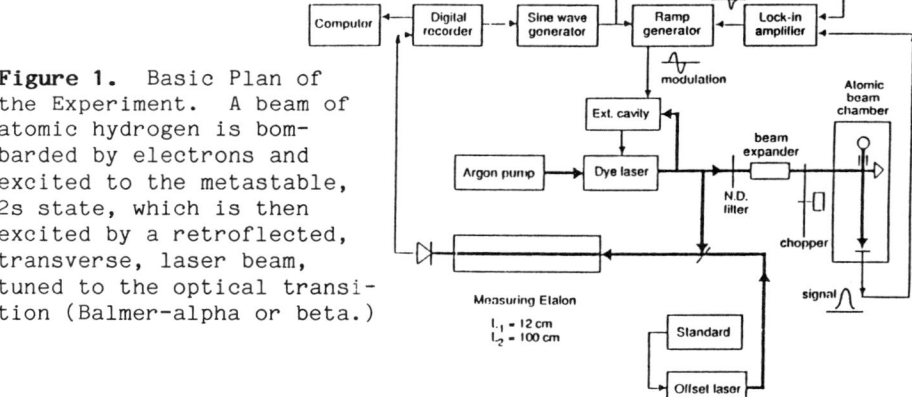

Figure 1. Basic Plan of the Experiment. A beam of atomic hydrogen is bombarded by electrons and excited to the metastable, 2s state, which is then excited by a retroflected, transverse, laser beam, tuned to the optical transition (Balmer-alpha or beta.)

Table I. Corrections and Errors for R (parts in 10^{10}).

Effect	Correction	Error
2s hfs	636	0
3p hfs	0	0
Optical Pumping, Light Shifts	-1	2
2nd Order Doppler	5	0
Photon Recoil	-8	0
2nd Order Doppler, Photon Recoil, hfs, Index of Refraction, Minor Effects		1
Statistical		5
Recording and Pressure Scan		3
Wavelength Standard		1.6
Non-uniform Mirror Coatings		1
Total Corrections and rms Error	632	6.5

Table II. Comparison of the Present Result with Other Recent Values

Reference	$R - 109\ 737\ \text{cm}^{-1}$
Goldsmith et al.	0.315 04(32)
Amin et al.	0.315 44(11)
Hildum et al.	0.314 92(22)
Biraben et al.	0.315 69(6)
Present Result	0.315 69(7)
Balmer-β (prelim.)	0.315 73(3)

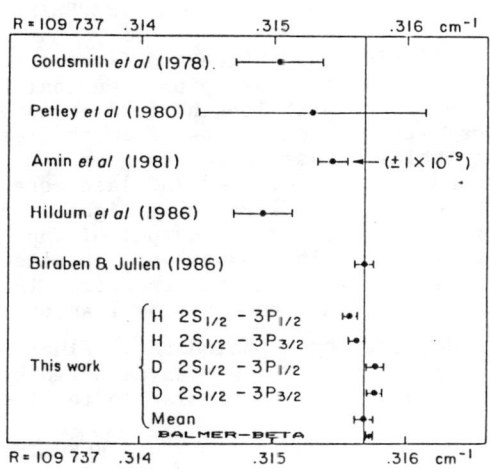

Figure 2.

Comparison of the present groups, including two other measurements by us (Amin et al. and Balmer-β).

The present value does not disagree significantly with the less accurate, previous result of the Yale group (Amin et al.). The agreement is excellent with that of the University of Paris group (Biraben et al.) and with our Balmer beta measurement. The two measurements of the Stanford group, (Goldsmith, Hildum et al.), both agree with each other and disagree significantly with our values. The Stanford measurements suffered from pressure shifts or "chirping". However, in paper THA1, T. Hänsch now reports agreement with our more precise values.

CONCLUSIONS

Our measurements of the Rydberg constant, via the Balmer beta line, are the most precise determination of any physical constant. There are no longer discrepancies in values of this constant. Our errors approach the limits for accuracy of optical measurements. Further progress in this area must lie outside of the visible region.

*Aided by grants from the National Science Foundation. A more detailed report is published in Phys. Rev. A, Dec. 1, 1986 Issue.

LASER ABSORPTION AND FLUORESCENCE STUDIES OF THE LITHIUM 2S-3D TRANSITION

G. C. Tisone and P. J. Hargis, Jr.
Sandia National Laboratories, Albuquerque, NM 87185

ABSTRACT

Laser absorption and fluorescence spectroscopy were used to study the single-photon dipole-forbidden $2s^2S$ to $3d^2D$ transition in lithium at 319.6 nm and the two-photon allowed transition between the two same levels at a wavelength of 639.2 nm.

EXPERIMENT

Lithium absorption and fluorescence measurements were made in a 1.27-cm diameter heat pipe oven that was built in the form of a cross with a path length of 10 cm in each arm. Lithium vapor densities were determined from the temperature of the heat pipe. The incident laser beam was generated by a dye laser pumped with the second harmonic of a Nd:YAG laser operating at a pulse repetition rate of 10 Hz. Ultraviolet light at 319.6 nm was generated by frequency doubling the output of the dye laser. Absorption and fluorescence line shapes were obtained by tuning the dye laser wavelength through the transition that was being studied. For absorption measurements, the transmitted laser light was attenuated and detected by a photodiode. Fluorescence from the $3d^2D$ to $2p^2P$ transition at 610.3 nm was detected by a filtered photomultiplier tube that was perpendicular to the incident dye laser beam.

RESULTS

A 6% absorption signal was measured at the center wavelength of the single-photon transition when the heat pipe was operated at its maximum temperature of 725°C--corresponding to a lithium vapor number density of 6.0×10^{15} cm^{-3}. Using this lithium density and a laser linewidth of 0.02 nm, an absorption cross section of 0.6×10^{-17} cm^2 is obtained. This value is in good agreement with the theoretical value of 0.6×10^{-17} cm^2 that can be calculated from the oscillator strengths of Caves[1] or Beck.[2] The single-photon excited 3D to 2P fluorescence signal was measured as a function of laser intensity and lithium density. The fluorescence signal was found to be nearly linear with intensity while the fluorescence signal was found to be proportional to $(density)^{0.74}$ over a density range of 5×10^{13} to 5×10^{15} cm^{-3}.

The 2S to 3D two-photon transition was studied as a function of laser wavelength, laser intensity and lithium density. Figure 1 shows the absorption line shape at a laser intensity of 1.8×10^6 Wcm^{-2} for two different lithium densities. Two-photon-

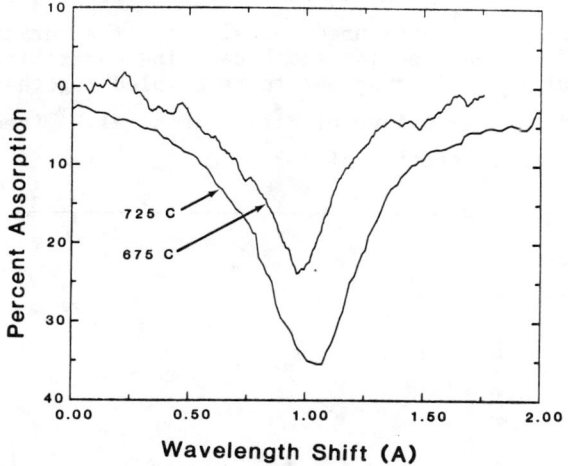

Fig. 1. Two-photon absorption as a function of wavelength

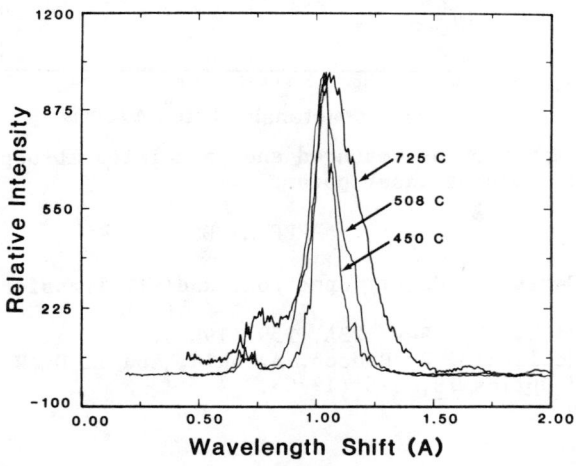

Fig. 2. The two-photon fluorescence line shape

excited fluorescence line shapes are shown in Fig. 2. The laser linewidth for the data in this figure was narrowed to 0.01 nm with an intracavity etalon. An unidentified peak is observed on the short wavelength side of the main 2S-3D fluorescence peak. Figure 3 shows the dependence of the two-photon absorption signal on input laser intensity at three different lithium densities. A simple two-photon absorption model that included a Lorentzian absorption cross section and a Gaussian laser line shape as a function of wavelength was used to extract a two-photon cross section and a broadening coefficient from the data in Fig. 3. A cross section of

6.25×10^{-31} cm^4 sec and a broadening coefficient of 3×10^{-7} cm^3 sec^{-1} were used to calculate the curves shown in Fig. 3. The cross section and broadening coefficient are not unique, but the values appear to be coupled together. For comparison, the technique of McIllrath et al.[3] gives an estimated absorption cross section of 2.2×10^{-31} cm^4 sec.

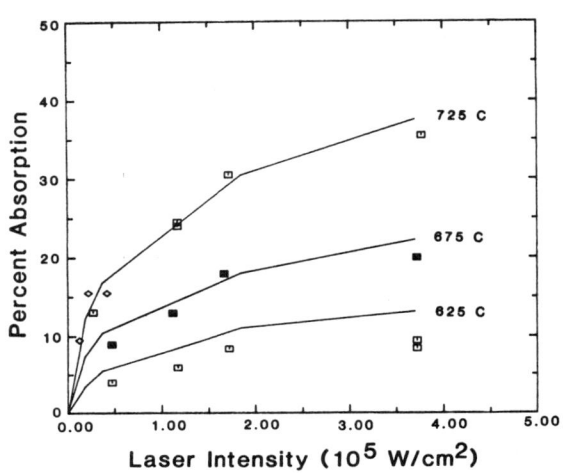

Fig. 3. Comparison of measured and calculated absorption as a function of laser power

REFERENCES

1. T. C. Caves, J. Quant. Spectroc. Radiat. Transfer 15, 439 (1974).
2. D. R. Beck, Phy. Rev. 23A, 159 (1981).
3. T. J. McIllrath, R. Hudson, A. Akin, and T. D. Wilkerson, Applied Optics 18, 316 (1979).

This work was supported by the U.S. Department of Energy under Contract DE AC04-76-DP 00789.

V. Condensed Matter, Surface, and Particle Spectroscopy

- A. Laser Raman 1: Coherent Phenomena 408
- B. Clusters 2: Metals 439
- C. Clusters 1: Non-Metals 458
- D. Time-Resolved Laser Probes of Surface Dynamics 475
- E. Atomic and Molecular Surface Imaging 496
- F. Laser Particle Interactions 509
- G. Ultrahigh-Speed Photodetectors 543

PRELIMINARY OBSERVATION OF NONRELAXATIONAL INERTIAL MOTION IN CS_2 LIQUID BY FEMTOSECOND TIME-RESOLVED IMPULSIVE STIMULATED SCATTERING

Leah R. Williams[*], S. Ruhman[†], Alan G. Joly,
Bern Kohler, and Keith A. Nelson[‡]
Department of Chemistry, Massachusetts Institute of Technology
Cambridge, Massachusetts 02139

ABSTRACT

Preliminary femtosecond time-resolved impulsive stimulated scattering data from CS_2 liquid are shown which clearly indicate an inertial component of the short-time motion which cannot be described in terms of Debye relaxational dynamics. A discussion of time-domain and frequency-domain light scattering techniques is given to illustrate the comparative difficulty of characterizing this type of motion by conventional methods.

INTRODUCTION

The dynamics of orientational and intermolecular motion in molecular liquids has long been an area of intense theoretical and experimental study[1]. Much of the experimental work has involved light scattering spectroscopy[2], which in principle can provide most of the dynamical information of primary interest. Depolarized quasielastic light scattering spectra of simple fluids often show an approximately Lorentzian central peak "riding" on top of another approximately Lorentzian, broader feature with both features centered at zero frequency. Both features are attributed primarily to orientational and "collision-induced" fluctuations in the fluid. The narrower linewidth gives in some cases a collective molecular reorientation time which is associated with entropy-driven return to the isotropic state following a fluctuation-induced partial alignment of molecules. The broader linewidth gives a faster "local" relaxation time which is generally associated with an individual molecule coming into local equilibrium with its immediate neighbors. This kind of "interaction-induced" local motion leads to subpicosecond relaxation times in simple liquids[1-3]. Recent subpicosecond time-resolved optical Kerr effect experiments on CS_2 clearly resolved two lifetimes[4] which were compatible with light-scattering (LS) results[3].

Although a "two-parameter" model can in many cases yield reasonable fits to LS spectra of simple molecular liquids, it is clear that the underlying dynamics are usually more complicated. Various models with three or more variables (one for the collective reorientation and two or more for the "local" motion) have been discussed theoretically at great length[1,2]. Perhaps the simplest of these recognizes that the intermolecular interactions in the fluid

[*]AT&T Bell Laboratories Ph.D. Scholar.
[†]Weizmann Postdoctoral Fellow.
[‡]Presidential Young Investigator Awardee.

may produce local potential minima with finite restoring forces against molecular libration and translation. In this case, "local" motion must have nonrelaxational (i.e. underdamped or overdamped vibrational) character described by two parameters: a natural undamped frequency (coming from the configuration-averaged intermolecular potential) and a damping rate. Other models emphasize the "free-rotational" character of local orientational motion which should also yield nonrelaxational dynamics and which requires at least a two-parameter description when intermolecular interactions are considered.

Here we present preliminary femtosecond time-resolved "impulsive" stimulated scattering (ISS) data on CS_2 liquid which clearly indicate nonrelaxational short-time motion. The data are fit very roughly by a simple "local overdamped oscillator" model. We also provide some background on the ISS experiment. We show that, although the information content of time-domain ISS data is in principle identical to that of frequency-domain LS data, in some cases nonrelaxational dynamics may in practice be far more easily resolved in the time domain than in the frequency domain.

ISS EXPERIMENT

The impulsive stimulated scattering experiment is illustrated schematically in Fig. 1. Two ultrashort "excitation" laser pulses derived from the same laser with central frequency and wave vectors $(\omega_L, \vec{k}_1)$ and $(\omega_L, \vec{k}_2)$, are overlapped spatially and temporally inside the sample. If the pulse duration, τ_L, is sufficiently short, then the intersecting pulses can exert a spatially periodic, temporally impulsive force on a LS-active mode of the sample through stimulated scattering[5]. This "impulse" driving force produces in the medium a coherent spatially periodic, standing-wave response whose time-dependence is given directly by the impulse response function, $G(t)$, of the LS-active mode[6]. The time-dependent motion, $Q(\vec{r},t) \sim G(t)\cos\vec{q}\cdot\vec{r}$, where $\vec{q} = \vec{k}_1 - \vec{k}_2$, is monitored by coherent scattering of variably delayed "probe" pulses which are phase-matched for optimum "diffraction" from the dynamic standing-wave "volume grating" of wave vectors $\pm\vec{q}$. The diffraction efficiency and therefore the time-dependent signal, $I(\vec{q},t)$, is given by

$$I(\vec{q},t) \propto |G(\vec{q},t)|^2. \tag{1}$$

Thus the time-dependent response of the mode is observed directly in time-resolved ISS data. In general, the response function in (1) is that of the dielectric constant (i.e., $G^{\epsilon\epsilon}(\vec{q},t)$). In cases where only one LS-active mode is excited and probed the response function of the mode is proportional to $G^{\epsilon\epsilon}$ and can be used in Eq. (1). A general treatment of ISS, including scattering from many modes and involving different tensor components of $G^{\epsilon\epsilon}$, has been presented[6].

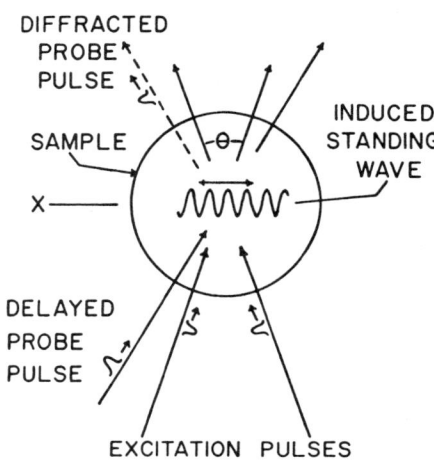

Fig. 1. Schematic diagram of the impulsive stimulated scattering experiment. The ultrashort, crossed excitation pulses "impulsively" excite a standing-wave material response which is monitored by coherent scattering of variably delayed probe pulses.

For underdamped vibrational modes characterized by a natural frequency, ω_0, and a damping rate, γ, the response function is

$$G(\vec{q},t>0) \propto e^{-\gamma t}\sin\omega_\alpha t, \qquad (2)$$

where $\omega_\alpha = (\omega_0^2 - \gamma^2)^{1/2}$, and the "impulse" force is applied at $t = 0$. For overdamped vibrational modes ($\gamma^2 > \omega_0^2$),

$$G(\vec{q},t>0) \propto e^{-\gamma_1 t} - e^{-\gamma_2 t}, \qquad (3)$$

where $\gamma_{2,1} = \gamma \pm (\gamma^2 - \omega_0^2)^{1/2}$. For Debye relaxational modes, characterized by a relaxation rate, Γ,

$$G(\vec{q},t>0) \propto e^{-\Gamma t}. \qquad (4)$$

Simulated ISS data from these three types of modes is shown in the right-hand side of Fig. 2. ISS data from underdamped vibrational modes (Fig. 2a) shows damped oscillations. ISS data from overdamped vibrational modes (Fig. 2b) shows a gradual rise after $t = 0$, followed by monotonic, nonoscillatory decay. The dynamical parameters, ω_0 and γ, can be extracted readily from data like that in Figs. 2a and 2b. ISS data from relaxational modes (Fig. 2b, dashed curve) shows an <u>instantaneous</u> rise followed by exponential decay whose time-dependence yields the dynamical parameter Γ. Fig. 2b shows that relaxational and overdamped vibrational responses are clearly distinguishable in ISS data if the time resolution is sufficient to resolve the initial rise of the overdamped response (i.e., laser pulse duration $\tau_L < \gamma_2^{-1}$).

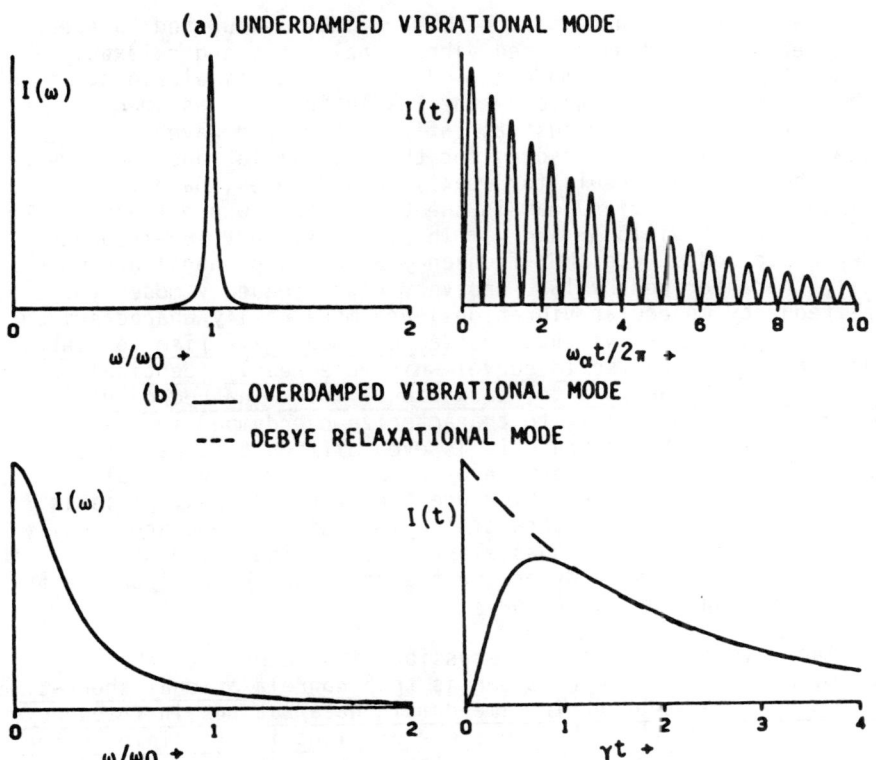

Fig. 2. Simulated LS spectra (left-hand side) and ISS data (right-hand side) from vibrational and relaxational modes. Only the Stokes side of the LS spectra are shown. (a) Underdamped vibrational mode with $\gamma = \omega_0/50$. Both LS and ISS data can be analyzed accurately. (b) Overdamped vibrational mode (solid curve) with $\gamma = 2\omega_0$; relaxational mode (dashed curve) yielding a similar LS spectrum. The LS spectra are indistinguishable even though the material dynamics are very different. Analysis of ISS data is straightforward in either case.

In frequency-domain quasielastic light scattering from a single mode, the spectrum can be described by

$$I(\vec{q},\omega) \propto \frac{k_B T}{\omega} \text{Im}[G(\vec{q},\omega)], \tag{5}$$

where k_B is the Boltzmann constant and $G(\vec{q},\omega)$, the Fourier transform of $G(\vec{q},t)$, is related to thermal fluctuations of the mode through the fluctuation-dissipation theorem. Equations (1) and (5) show that, <u>in principle, the information content of ISS and LS data is identical</u>. However, the extraction of dynamical information from the time- and frequency-domain data differs considerably in

practice. The left-hand side of Fig. 2 shows simulated LS spectra
for underdamped and overdamped vibrational modes and relaxational
modes with the same parameters used for the ISS simulations.
Infinite frequency and wave vector resolution were assumed
for the LS simulations just as infinite time and wave
vector resolution were assumed for the ISS simulations. Figure 2a
shows the familiar result for weakly damped modes, namely a
well-defined Lorentzian peak in the LS spectrum whose frequency
shift and width yield ω_0 and γ. In practice, very low-frequency
underdamped modes (e.g. MHz-frequency acoustic phonons) are more
easily characterized by ISS, and very high-frequency modes (e.g. >10
THz-frequency molecular vibrations) are more easily characterized
by LS. Figure 2b shows that overdamped modes give rise to central
peaks in LS spectra (solid curve) which are nearly identical in form
to the central peaks due to relaxational modes (dashed curve). Thus
it is extremely difficult to characterize overdamped modes
accurately from LS spectra. It is even difficult to make the
qualitative distinction between overdamped and relaxational
responses even for modes which are barely overdamped. This is
especially true in LS spectra of liquids, since there are usually
several contributions to quasielastic scattering which make it
impossible to precisely determine the complete shape of the central
feature arising from each mode.

The main point of this discussion with respect to the
preliminary ISS data to be shown is that nonrelaxational short-time
dynamics may be apparent in time-domain data but not in frequency-
domain data, even when the two are equivalent in principle in terms
of information content. More complete comparison between ISS and LS
data has been presented[6].

CS_2 ISS DATA

ISS data were collected from CS_2 using 65 fs, 620 nm pulses as
illustrated schematically in Fig. 1. The excitation pulses were V
and H polarized relative to the scattering plane, and were crossed
with a 5° angle between them. The phase-matched probe pulse was V
polarized, and the coherently scattered light was H polarized. Thus
the data collected are analogous to V-H LS data. A 2 mm spectrophotometer cuvette was used to hold the CS_2 liquid.

The laser system used was similar to one which has been
described in detail earlier[7]. Briefly, a Nd:YAG laser is cw
mode-locked, and its frequency-doubled output synchronously pumps a
dye laser (rhodamine 6G gain dye) with an anti-resonant ring which
contains a saturable absorber dye jet (DODCI absorber dye). The
output of the dye laser is a stream of 65 fs, 620 nm pulses. A
second Nd:YAG laser regeneratively amplifies a pulse from the
first, and its synchronized output (a 1.2 mJ, 100 ps, 1.06 µm pulse)
is frequency-doubled and used to longitudinally pump a homebuilt
3-stage amplifier chain. The amplified output after grating-pair
compression is a 65 fs, 620 nm, 6 µJ pulse with a 500 Hz repetition

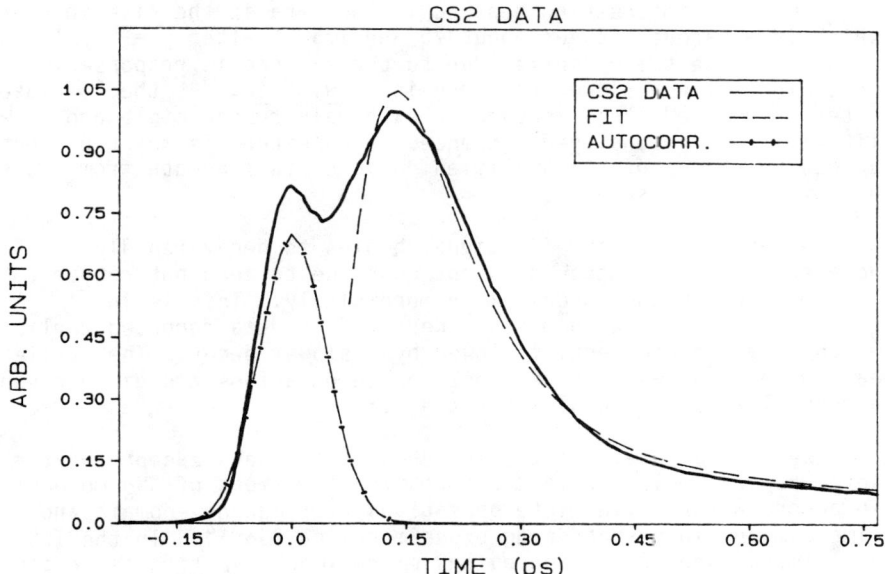

<u>Fig. 3</u>. ISS data from CS_2 liquid. The signal after t = 0 continues to rise, indicating the <u>nonrelaxational</u> character of short-time motion in the fluid. After reaching a maximum, the signal decreases rapidly and then more slowly.

rate. The pulse is split three ways to yield the excitation and probe pulses which are crossed in the sample. For the CS_2 experiments, the excitation and probe pulse energies were 150 nJ and 30 nJ, respectively, after extensive filtering to avoid power-dependent effects. The spot size of each pulse at the sample was approximately 250 μm. The variable delay of the probe pulse is controlled by a 1-μm stepping-motor delay line.

ISS data from CS_2 liquid is shown in Fig. 3. Centered near t = 0 is a "spike" which is due to the nearly-instantaneous <u>electronic</u> response of the CS_2 to the nonresonant light fields. This feature appears only when the excitation and probe pulses are all inside the sample simultaneously. It is a familiar feature of coherent four-wave mixing experiments in almost any type of sample including liquid, glass, crystal, etc. We have defined t = 0 by replacing the CS_2 with H_2O liquid and recording ISS data, which showed <u>only</u> a symmetric "spike". The peak of the H_2O signal defines t = 0. The H_2O data was nearly identical to the autocorrelation trace shown in Fig. 3 (broken curve).

The most interesting feature in the data is the rise in signal which is apparent between about 75 and 175 fs after $t = 0$. It is clear that the $t = 0$ "spike" due to the electronic response is declining while a slower response is growing in. At the low laser intensities used, this feature is extremely reproducible and shows little if any intensity-dependence. The feature is never present in autocorrelations of the amplified pulse or in ISS data from glasses or several other samples.

After ~175 fs, the ISS signal begins to decay rapidly. However, the rapid decay does not continue to zero but rather to a low level which then decays much more slowly. This is in qualitative accord with optical Kerr effect data reported earlier[4], which show a rapid decay followed by a slower decay. The earlier data were taken with longer laser pulse durations and did not show a a gradual rise in signal after $t = 0$.

We attribute all of the features in the data except the $t = 0$ spike to intermolecular and orientational motions of CS_2 molecules, in accord with earlier interpretations of frequency-domain and time-domain light-scattering experiments on CS_2[3,4]. In the ISS experiment, the excitation pulses exert impulsive torques on the molecules through the single-particle molecular polarizability. The excitation pulses also exert impulsive forces on nearest-neighbor molecule pairs (i.e., intermolecular or collision-induced stimulated scattering), causing intermolecular librational and translational motion. Molecular positions and orientations (on which ISS signal depends) do not change instantaneously, however, and so the signal (other than the $t = 0$ spike) rises gradually after $t = 0$. The gradual rise in ISS signal reflects the nonrelaxational, inertial nature of intermolecular and/or orientational motion in the fluid. It has been suggested that in CS_2 the short-time signal (or equivalently, the high-frequency part of the LS spectrum) arises predominantly from collision-induced scattering, and the long-time signal is due mainly to single-molecule scattering[3,4]. The slow decay, whose lifetime, $\Gamma^{-1} = 1.4$ ps, was characterized accurately through ISS experiments run out to 6 ps delay, gives a collective orientational relaxation time. The short-time dynamics, previously characterized as relaxational, reflect intermolecular motions which are essentially vibrational in character. These motions may be thought of as local librational and translational "phonons" which are LS-active. The rapid decay of the short-time ISS signal probably reflects very rapid dissipation of energy in these modes ("homogeneous dephasing" due to T_1 processes) as well as the wide range of local vibrational frequencies arising from many different intermolecular configurations ("inhomogeneous dephasing").

An approximate fit to the data, intended only as a qualitative guide to its interpretation, is shown in Fig. 3. The fit was generated by assuming an overdamped oscillator response at short times ("local" intermolecular motion) and a relaxational response at long times ("collective" orientational relaxation). From Eqs. (1),

(3), and (4), the functional form used to generate the fit is

$$I(t) = [A(e^{-\gamma_1 t} - e^{-\gamma_2 t}) + Be^{-\Gamma t}]^2, \qquad (5)$$

with $\gamma_1 = 10.2 \times 10^{12} s^{-1}$, $\gamma_2 = 15.1 \times 10^{12} s^{-1}$, $\Gamma = 0.73 \times 10^{12} s^{-1}$, $A = 40$, and $B = 4.78$ (arbitrary units).

Although the nonrelaxational aspect of the short-time data is unmistakable and reproducible, we emphasize the preliminary nature of the fit and the qualitative nature of its interpretation here for several reasons. First, the fit was generated with no convolution carried out to account for the 65 fs excitation and probe pulse durations. The electronic contribution to t = 0 was not subtracted in any way, but rather the first part of the data was ignored during fitting. The second limitation on our interpretation of the data is the simplicity of the "local oscillator" model used. The wide range of local vibrational frequencies (i.e., inhomogeneous dephasing) has not been accounted for. Moreover, the more slowly relaxing orientational motion may also show "nonrelaxational" short-time behavior since an impulsive torque should produce a delayed, not instantaneous, orientational response. More systematic experimental work currently under way should suggest the proper theoretical description.

These reservations notwithstanding, we close by discussing the important implications of these results. They support basic notions about short-time dynamics in molecular fluids which have been discussed at length but which are difficult to observe directly by conventional experimental methods. As we have indicated, it is difficult to extract these effects, even qualitatively, from frequency-domain LS data. This applies to both spontaneous and stimulated frequency-domain LS techniques. Time-domain methods with sufficient resolution will provide the most direct means for observation of these short-time nonrelaxational dynamics. Systematic examination of a variety of pure and mixed fluid systems under different experimental conditions (sample temperature and pressure, laser power, etc.) should further clarify the nature of the short-time motion and may yield configuration-averaged intermolecular potentials, dissipation rates, and additional information fundamental to understanding the forces felt by molecules in a fluid. We note finally that the "vibrational" nature of intermolecular motion in liquids may play a major role in electronic excited-state relaxation (e.g. Stokes-shifting dynamics) in fluids[8] and in some liquid-state chemical reactions[8] whose rates are influenced by the solvent.

This work was supported in part by NSF Grant No. DMR-8306701.

REFERENCES

1. See, e.g., P. Madden and D. Kivelson, Adv. Chem. Phys. $\underline{16}$, 467 (1984); W.A. Steele, Adv. Chem. Phys. 34, 1 (1976); D. Kivelson and T. Keyes, J. Chem. Phys. $\underline{57}$, 4599 (1972); R.G. Gordon, J. Chem. Phys. $\underline{44}$, 1830 (1966); N.D. Gershon and I. Oppenheim, Physica $\underline{62}$, 198 (1972); and references therein.
2. See G.D. Patterson and P.J. Carroll, J. Phys. Chem. $\underline{89}$, 1344 (1985); D. Kivelson and P.A. Madden, Ann. Rev. Phys. Chem. $\underline{31}$, 523 (1980); T. Keyes and D. Kivelson, J. Chem. Phys. $\underline{56}$, 1057 (1972); N.D. Gershon and I. Oppenheim, J. Chem. Phys. $\underline{59}$, 1337 (1973); and references therein.
3. P.A. Madden, in Ultrafast Phenomena IV, ed. by D.H. Auston and K.B. Eisenthal (Springer-Verlag, Berlin 1984), p. 244; G.A. Kenney-Wallace, in Applications of Picosecond Spectroscopy to Chemistry, ed. by K.B. Eisenthal (D. Reidel, 1984), p. 139.
4. B.I. Greene and R.C. Farrow, Chem. Phys. Lett. $\underline{98}$, 273 (1983).
5. M.M. Robinson et al., Chem. Phys. Lett. $\underline{112}$, 491 (1984); S. DeSilvestri et al., Chem. Phys. Lett. $\underline{116}$, 146 (1985).
6. Y.-X. Yan, L.-T. Cheng, and K.A. Nelson, Adv. IR and Raman Spectrosc. $\underline{16}$ (1987), in press; Y.-X. Yan and K.A. Nelson, J. Chem. Phys., submitted.
7. T. Sizer et al., IEEE J. Quant. Electron. QE-19, 506 (1983).
8. G. van der Zwan and J.T. Hynes, J. Chem. Phys. $\underline{89}$, 4181 (1985); J.T. Hynes, Ann. Rev. Phys. Chem. $\underline{36}$, 573 (1985).

DEVELOPMENT OF RIKE TECHNIQUES USING PICOSECOND LASERS*

M. W. Schauer, M. J. Pellin, B. M. Biwer, D. M. Gruen
Argonne National Laboratory, Argonne, IL 60439

ABSTRACT

The sensitivity of the Raman-induced Kerr effect is greatly enhanced through the use of picosecond lasers. Experiments in dilute solutions of benzene indicate that sensitivity at the monolayer level is achievable. Applications to transparent media, to fluorescing samples, and to in situ measurements of electrode surfaces are discussed.

INTRODUCTION

With the advent of tunable picosecond lasers, coherent Raman experiments have attained the sensitivity necessary for the in situ study of electrochemical processes.[1] Our primary interest is in the study of corrosion processes at metal surfaces, which can be controlled in an electrochemical cell. Raman data from an electrode surface can help to ascertain the molecular nature of the passive oxide layer[2] and any intermediate corrosion products formed during oxidation of the bare metal.

It soon became clear that, in the development of a coherent Raman technique with the sensitivity necessary to study electrode surfaces, the technique would also be sensitive enough for other applications such as the study of monolayers and thin films on mirrors and transparent substrates, catalytic systems, dilute solutions of highly fluorescent species, systems involving very low frequency Raman modes, and systems for which vibrationally selective microscopy would be useful.

As demonstration experiments, dilute solutions of benzene and pyridine were observed in transparent and electrochemical cells. Current experiments approach monolayer sensitivity in an experimental configuration amenable to surface studies. Future directions and improvements are outlined.

EXPERIMENTAL APPARATUS

The coherent Raman technique being developed is derived from the Raman-induced Kerr effect (RIKE) experiment described by Eesley.[3] In this experiment, the large nonresonant background in a Raman gain experiment is greatly reduced by analyzing the vertically polarized probe beam through a horizontal polarizer (see Fig. 1). The

---

*Work supported by the U.S. Department of Energy, BES-Materials Sciences, under Contract No. W-31-109-ENG-38.

circularly polarized pump beam induces ellipticity in the probe beam through the RIKE effect.

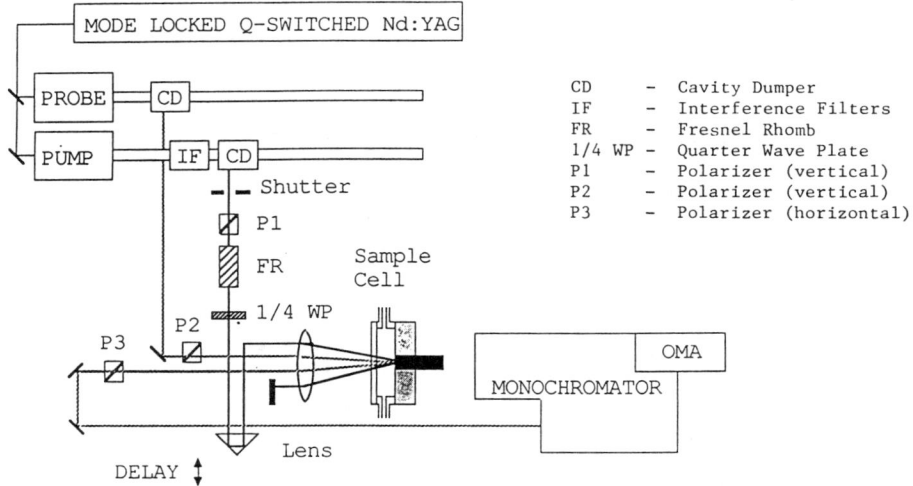

Fig. 1. Schematic representation of the coherent Raman apparatus.

Several modifications have improved the sensitivity of the experiment and made it amenable to the study of electrode surfaces. Two synchronously pumped, picosecond, cavity-dumped dye lasers are pumped by a mode locked and Q-switched Nd:YAG laser. The pump and probe lasers produce ~20 mW each at 1 kHz repetition rate. These beams are focused to ~50 µ on the electrode surface.

Significant reduction in noise is obtained by operating the probe laser broad band (~200 cm^{-1}) and using multichannel detection. One can then signal average to obtain the desired signal-to-noise ratio (S/N). The signal is divided by the background (signal without pump) and Raman gain is plotted as a function of the frequency difference between the lasers.

RESULTS AND DISCUSSION

The sensitivity of the current apparatus was determined by observing the 992 cm^{-1} symmetric stretch of benzene in a dilute solution (Fig. 2). A S/N of ~4 for a 10^{-3} M solution of benzene indicates a limit of sensitivity of ~ 2.5 x 10^{-4} M (~ 5 x 10^{10} benzene molecules). This sensitivity compares well with that reported for a CARS experiment.[4] With this sensitivity, a monolayer of benzene should have a S/N of .5 for 2 minutes of signal averaging. Higher sensitivity is possible for dilute solutions by increasing the laser intensities, but the intensities used in this experiment are amenable to studies of electrode surfaces.

The system was further characterized by observing signals from an aqueous solution of pyridine above a polished aluminum mirror (Fig. 3). Although the signal levels in this experiment are quite satisfactory, the noise level is unacceptably high due to impurity and focusing problems. However, this experiment demonstrates signal levels adequate for monolayer detection with the lasers focused onto a metal electrode. Of course to actually observe acsorbed monolayers, one must interact with the electrode at ~60° from the surface normal. This geometry presents polarization problems which are currently being addressed.

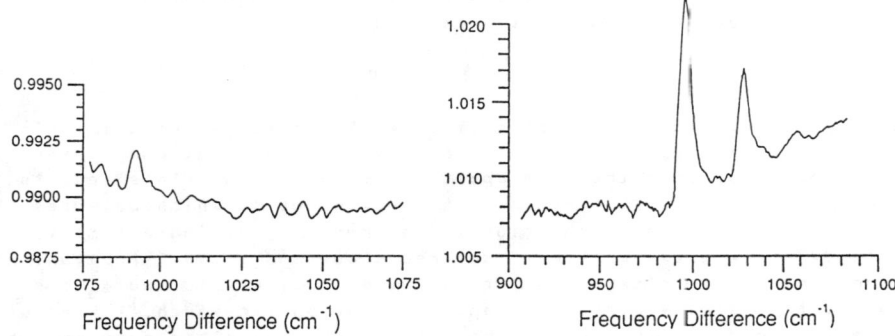

Fig. 2. Coherent Raman spectrum of a 10^{-3} M solution of benzene in carbon tetrachloride in a transparent sample cell.

Fig. 3. Coherent Raman spectrum of .1 M pyridine in water above an aluminum mirror. Two modes of pyridine are shown.

Through the use of picosecond lasers, advanced polarization techniques, and multichannel detection, it is possible to design coherent Raman experiments with the sensitivity necessary to study thin films and monolayers on electrode surfaces. Further work is focused on improving the sensitivity and adapting the technique to the study of electrode surfaces.

REFERENCES

1. (a) M. W. Schauer, M. J. Pellin, B. M. Biwer, and D. M. Gruen, Appl. Phys. Let., in press.
   (b) J. P. Heritage, D. L. Allara, Chem. Phys. Let. 74, 507 (1980).
2. (a) K. Carr, S. Baer, and C. A. Melendres, Corrosion 42, 307 (1986).
   (b) A. Hugot-Le Goff, C. Pallotta, J. Electrochem. Soc. 132, 2805 (1985).
3. (a) G. L. Eesley, Coherent Raman Spectroscopy, Pergamon Press, Inc., New York, 1981, pp. 40-51.
   (b) Gary L. Eesley, M. D. Levenson, and William M. Tolles, IEEE J. Quant. Elect. QE-14, 45 (1978).
4. Jean-Louis Oudar, Robert W. Smith, and Y. R. Shen, Appl. Phys. Let. 34, 758 (1979).

STEADY STATE LIGHT PULSES IN STIMULATED BACKWARD SCATTERING

D.N. Ghosh Roy and D.V.G.L.N. Rao
University of Massachusetts, Boston, Ma. 02125

ABSTRACT

Steady state backward SRS and SBS light pulses are analyzed in the limit where rate equation approximation fails and the dynamics of the medium fluctuations must be considered. If the pump and the medium have different attenuations, then it is shown that under certain assumptions, the parametrically coupled equations can be reduced to a one-dimensional Dirac system.

INTRODUCTION

In backward, transient stimulated Raman (SRS) and Brillouin (SBS) scattering, the scattered beam may exceed peak pump intensity and become broader than the spontaneous linewidth. This results in an effective gain g_e the nonlinear growth due to which balances the loss due to finite attenuation in the medium yielding a limiting steady state pulse with well defined shape and width. For vanishing pump and medium attenuation, this pulse is symmetric and g_e a simple product of steady state gain and fluctuation linewidth.[1] Here we present formal solutions to the problem when both the pump and the medium have finite attenuations.

BASIC EQUATIONS

The parametric equations for the pump laser (l) and the medium (m) are:

$$(X_l, X_m)^T \underline{Y} = \underline{Z} \qquad (1)$$

where, $X_{l,m} = (\partial/\partial t) + \Gamma_{l,m}$, $\underline{Y}$ is a column vector with elements N_l, N_m, and $\underline{Z}$ is another column vector with elements $A_l N_l N_m$ and $A_m N_l N_m$. T denotes the transpose operation and N the flux (#/cm^2-sec). The stoke's flux N_s is given by $N_s = C_s N_l N_m$. $A_{l,m}$ and C_s are constants depending upon the medium characteristics and are different for SRS and SBS processes.

Difficulties of solving solutions for arbitrary $\Gamma_{l,m}$ are well known.[2] Exact solutions are obtained for $\Gamma_l = \Gamma_m = 0$ and $\Gamma_l = \Gamma_m = \Gamma$.

Case I: $\Gamma_l = \Gamma_m = \Gamma$

With the transformation $\tau = (1/\Gamma)(1-\exp\{-\Gamma t\})$, eqs. (1) reduces to the undamped form $\dot{N}_{l,m} = A_{l,m} N_l(\tau) N_m(\tau)$. (2)

"." denotes differentiation with respect to τ. Equivalent Manley-Rowe relation is:

$$N_m = (A_m/A_l)N_l + (N_{m_0} - (A_m/A_l)N_{l_0}),$$

© American Institute of Physics 1987

$N_{\ell o} = N_\ell(\tau = \tau_o)$, an initial condition. We assume $N_{mo} = 0$, i.e., the intensity of the medium fluctuation flux is negligible compared to that of the light fields. Then solving (2), we obtain for N_s the following expression

$$(N_s/N_{\ell o}) = (C_s \Gamma / A_\ell)[4(\mu/\Gamma)e^{-\Gamma t} \text{sech}^2 \{(\mu/2\Gamma)(1-e^{-\Gamma t})+(\phi/2)\}$$
$$+ \tanh\{(\mu/2\Gamma)(1-e^{-\Gamma t})+(\phi/2)\} + \{1+e^{(\mu t + \phi)}\}^{-1}]. \quad (3)$$

The constant ϕ can be eliminated by shifting the time scale. As $t \to 0$ $N_s \to 0$ as expected. For $\Gamma \to 0$, (3) reduces to the result of Maier et.al.[1], namely,

$$(N_s/N_{\ell o}) \longrightarrow \text{sech}^2 \{(\mu t/2) + \phi/2\}$$

where $\mu = g_e \propto G\Gamma$. However, in the presence of finite attenuation, such simple expression for g_e is not obtained, pulse becomes asymmetric and relatively more broadened.

Case II. $\Gamma_\ell \neq \Gamma_m$, but $\Gamma_{\ell,m} = \Gamma + \Delta_{\ell,m}$, $\Delta_{\ell,m} \ll \Gamma$, the mean attenuation. Let $n_{\ell,m}$ denote the solution for $\Delta = 0$, i.e., solutions to eq. (2). Equations (1) are rewritten as:

$$(\tilde{X}_\ell, \tilde{X}_m)^T \underline{Y} = \underline{\tilde{Z}} \quad (4)$$

where $\underline{\tilde{Z}}$ is now a column vector with elements $Q(t)N_m$ and $R(t)N_\ell$, $Q(t)$ is $A_\ell n_\ell$ and $R(t)$ is $A_m n_m$. Let $N_{\ell,m} = \tilde{N}_{\ell,m} \exp(-\Gamma_{\ell,m} t)$. Then $\tilde{X} = (\partial/\partial t)$ If it is possible to write $\tilde{Q} = \tilde{\tilde{Q}} + \lambda_\ell$ and $\tilde{R} = \tilde{\tilde{R}} + \lambda_m$, then eq.(4) is:

$$(\underline{B}\,\tilde{X} + \underline{M})\,\underline{\tilde{Y}} = \lambda \underline{\tilde{Z}} \quad (5)$$

where,

$$\underline{B} = \begin{pmatrix} 0 & 1 \\ -1 & 0 \end{pmatrix}, \quad \underline{M} = \begin{pmatrix} \tilde{\tilde{Q}}(t) & 0 \\ 0 & \tilde{\tilde{R}}(t) \end{pmatrix}, \quad \underline{\lambda} = \begin{pmatrix} \lambda_\ell \\ \lambda_m \end{pmatrix}.$$

Equation (5) is the one-dimensional Dirac system[3]. Equation (5) is equivalent to the fact that $(N_{\ell,m} - n_{\ell,m})\exp(-\delta t)$ remain unchanged in first order, δ being $(\Gamma_\ell - \Gamma_m)$. Application of Levitan translation operator[3] to the solution of (5) is under progress.

REFERENCES

1. M. Maier, W. Kaiser, and J. A. Giordmaine, Phys. Rev. <u>177</u>, (1969).
2. A. D. D. Craik, Wave Interactions And Fluid Flows (Cambridge University Press, Cambridge, 1985).
3. B. M. Levitan and I. S. Sragsjan, Introduction to Spectral Theory (Transl. of Math. Monogr., Amer. Math. Soc., 1975).

THE RAMAN SOLITONS

Farres P. Mattar
Department of Physics, New York University, NY, NY 10003
and G.R. Harrison Spectroscopy Laboratory,
Massachusetts Institute of Technology, Cambridge, MA 02139

The observation by Carlsten et al [1] of a pump soliton in stimulated Raman scattering (SRS) [2], initiated by an optical phase shift imposed on the *Stokes* beam, is a vivid example of the richness of this two-field nonlinear interaction. The importance of understanding the result has motivated a number of theoretical investigations, beginning with the founding group's own original and later analyses based on solution of the SRS equations formulated to simulate collisional damping of coherence. (The two fields are coupled in a reduced, effective two-level system). Kaup has provided indirect support for this interpretation by showing in general the possibility of soliton creation from dissipation [3]. The question of Stokes initiation by four-wave mixing has been treated by Ackerhalt et al [4] and through quantum fluctuation by Englund et al [5].

Our own contribution to this theoretical effort has been to explore the detailed mechanisms of the propagation through computational analysis [6] of the equations of motion for the field-matter interaction. We hope in this way to free the problem from most approximations or *ad hoc* assumptions, revealing the essential physics directly.

Since there is some complexity in the computer results, we shall first present the qualitative picture that emerges from the calculations before describing the specific calculations. To help in this presentation we shall refer occasionally to propagation in a two-level medium.

Viewed very broadly, and allowing for arbitrarily long interaction lengths in the beam, the calculations show a clear sequence of events. The Stokes beam experiences a buildup, gain saturation, and then a stabilization which implies the creation of a soliton, during which time the pump shows a continuing depletion, as expected. Further propagation results in decay of the Stokes and destruction of the soliton and a concurrent recovery of the pump (i.e., the "anomalous pump-depletion reversal" reported by Drühl et al [7]) followed by pump stabilization to a 'pump soliton'. We stress particularly the orderly evolution of these events which is rooted in the two-beam SRS interaction modulated by transverse effects. Looking ahead to the numerical experiments, we define the soliton as a pulse for which the time-integrated area, time-integrated energy (fluence), and effective temporal width do not vary appreciably over several optical thicknesses in the asymptotic regime where gain saturation takes place.

The physical model is a collection of identical three-level atoms, each having an energy-level scheme where the 1↔3 transition is induced by a coherent injection pulse of frequency ω_p and the transition 3↔2 evolves from the Stokes seed at a much lower frequency ω_{st}. The spacing is such that $\varepsilon_1 < \varepsilon_2 \ll \varepsilon_3$, so that the fields at frequencies ω_p and ω_{st} can be treated by separate wave equations even though the medium mixes the two fields. The energy levels ε_1 and ε_2 are not coupled radiatively because of parity considerations. The input pump pulse is specified by boundary and initial conditions which are consistent with having nearly all the population in the ground state; the area of the Stokes seed is a small fraction, δ (with $\delta \ll 1$), of that of the pump. Once part of the on-axis 2π pump creates a population inversion in both the 1↔3 and 3↔2 transitions, the Stokes begins to build up, which incidentally explains why the Stokes arises after the pump peak occurs, and is narrower in time.

It is important to realize that the polarization P_a in the pump transition is driven not only by the pump field E_a but also by the Stokes field E_b, since, with Q denoting a pseudo-quadrupole, the product $E_b Q$ corresponds to an oscillation at the pump frequency also. Similarly, the product $E_a Q^*$ contributes to the temporal rate of change of the P_b polarization as in parametric oscillation [8].

During the early linear regime ($N_1 = 1$, $N_{2,3} = 0$), the pump evolves as a self-induced transparency (SIT) [9] pulse, to zeroth order in δ. The Stokes emission is similar to superfluorescence [10] and builds up as first order in δ, whereas the feedback of the Stokes field into the pump is of second order δ^2. There is no cross-correlation between the pump and the Stokes pulses in this regime, but as soon as the nonlinearities become effective, the two pulses lose their independence.

Since it is essential that there be an efficient exchange of energy between pump and Stokes, the two pulses should overlap as much as possible throughout their motion. This requirement is not guaranteed, however, and one must examine the ways in which various properties of the system affect the velocities. First of all, because the delay experienced by a pulse in a resonant two-level atomic system (also applicable to the three-level case) is proportional to the reciprocal area, the pulse length, and the characteristic Beer's length α^{-1} (with $\alpha = \pi\mu^2\omega/n\hbar c$), a large-area pump pulse can propagate with the same group velocity as a weaker Stokes pulse only if the Stokes Beer's length is considerably larger than the pump's. (This condition is achieved by selecting a metallic vapor with different oscillator strengths in its three-level structure.)

Furthermore, group velocity depends upon whether the medium is absorptive or emissive, a fact which has a more subtle effect on the overlap requirement since the initial population of the three-level atom dictates that the pump meet an absorbing transition while the Stokes interacts with an amplifying one. As soon as a π-part of the pump area is absorbed and establishes a population in the upper level, each pulse encounters a different medium: the pump absorptive; the Stokes emissive. The pump thus experiences delay with respect to an observer moving with the velocity of light, ($v_{group}^{absorption} < c$), while the Stokes pulse advances in time

($v_{group}^{amplifier} > c$). This peak acceleration of the growing Stokes and peak retardation of the depleting pump with respect to the moving frame effectively reduces the required ratio of pump-to-Stokes Beer's length so that the overlap can always be near optimum. The Beer's length for the pump and for the Stokes can now be as little as $\alpha_p \ell > 5$ and $\alpha_{st} \ell > 55$ respectively.

All of the previous discussion is consistent with no more than a uniform wave approximation. However, the experimental observations are made with finite beams and therefore often depart from these predictions. The reason is located in the competition between coherent transient and self-focusing effects. To visualize this, we recall that the more intense a coherent pulse is, the faster it travels and therefore the group velocity of the pulse peak at the center pencil (one pencil for each beam radius) thus exceeds the corresponding off-axis group velocity. After a few absorption lengths, the most intense central part of the beam is ahead of the outer sections, creating non-gaussian beam profiles with depression near the center at the tail of the pulse while maintaining gaussian profiles at the front of the pulse. As a result there is temporal and transverse reshaping which distorts the beam and a differential temporal beam narrowing or spatial pulse compression may occur. This reshaping is one of the complex but essential elements in the overall analysis which favors a direct computational treatment. These effects depend upon properties of the atomic system and the transitions, most importantly the oscillator strengths, and therefore the output field may be tailored with some control. Under certain conditions one obtains coherent self-focusing while under others self-trapping (or beam channeling) may result.

Turning now to the Raman solitons we will invoke the general principle that a soliton in any medium is formed as the resolution of competing mechanisms. The soliton in an optical fiber is an example wherein pulse broadening caused by dispersion alone is counterbalanced by phase modulation arising from the nonlinear medium. *In the Raman case the balance for each transition occurs between the nonlinearity and the linear diffraction.* This may be seen by considering at first some features of the much simpler propagation of one field in one dimension and a two-level medium: the pump is analogous to a self-induced-transparency (SIT), while the counterpart of the Stokes pulse is a superfluorescent (SF) emission. For example, a superfluorescence pulse with ringing approaches asymptotically (in a nonlinear swept-gain amplifier) an undistorted pulse, *provided* a linear loss is imposed [11], reflecting how a soliton-like wave appears to evolve out of dissipation in an amplifier. With this in mind, consider the more realistic three-dimensional (still two-level) case.

The medium provides amplification while the transverse communication across the beam imposes a diffractive loss. The beam reshapes longitudinally and transversely so that its on-axis area, which first reaches a maximum corresponding to self-focusing, oscillates in a steady-state decay towards a constant π value in the saturation region, the total process resulting in a sustained shape-preserving propagation [12]; again a soliton state is created. A measure of the gain/loss condition would be a Fresnel number F_g defined in terms of $\alpha_{p,st}$ the Beer's length transition, i.e., $F_{gp,st} = \pi r_p^2/\lambda_{p,st}\alpha_{p,st}^{-1}$. F_g is the ratio of the nonlinear effect (Beer's length) to the Rayleigh diffraction length (linear loss $\kappa^- = \pi r_p^2/\lambda_{p,st}$). The nonlinear interaction throughout the entire propagation of the beam is thus characterized by the Beer's length Fresnel number associated with a specific profile of the pre-excited atomic density.

In the three-level case, there remain some useful lessons from the above if we understand that the pump experiences a greater absorption than a self-induced-transparency pulse in a two-level medium because of its leakage into the Stokes pulse, and conversely, the Stokes buildup will not be as strong as that of a two-level superfluorescence.

We require now that the Stokes beam experience a gain large enough to compensate the losses. To achieve the appropriate balance, one needs a Raman cell considerably longer than that used for conventional amplifiers, and the Beer's length seen by the pump increases. Even though the pump sees the large optical thickness which would cause a coherent SIT on-resonant self-focusing, there is no on-axis enhancement due to energy leakage to the Stokes pulse. But in a sufficiently long propagation length the Stokes pulse does display the same z-independent, on-axis area stabilization as its two-level counterpart. Also, we find that its effective pulse length defined in terms of its output power (radially-integrated Stokes energy) becomes saturated. In all, the features which characterize a solitary pulse are observed.

To probe the details of the Stokes soliton problem, we have calculated the effect of changing the ratio of relaxation time to pump duration in both plane-wave and transverse models. The results show that, depending upon this ratio, the on-axis Stokes area stabilization over a significant propagation distance will be under-, over-, and/or critically damped. Moreover, to see the importance of gain/loss values in the creation of the Stokes solitons, calculation of gain and appropriate Fresnel number dependence have been carried out. Also the effect of input radial shape of the Raman pump has been studied.

Our calculations account for the *pump soliton* too since for a still longer interaction region, a shape-invariant solitary wave signature appears at the pump frequency signifying that the total nonlinear action experienced by the pump compensates its diffraction in the same way as for the Stokes. The Stokes now incurs depletion while the pump shows a depletion reversal! *The conditions which support a pump soliton will destroy those for the Stokes wave and vice versa; we cannot have both.* Thus our "three-level density matrix" calculation, which includes diffraction and phase variation, shows a direct connection with the experiment of Carlsten *et al*, the planar wave "reduced effective two-level" computation by Drühl *et al*, and the analytical theory by Kaup.

We have performed preliminary calculations which show the effect of dispersion in the presence of detuning for the cases in which either one of the beams is detuned while the other is resonant or both are off-resonant. The resulting self-action phenomena (self-phase modulation, self-focusing, self-trapping, etc.), which play a role in soliton formation, now change threshold. Specifically, the phase buildup due to the off-resonant condition, (independent of the sign of the detuning), can either interfere constructively with or be cancelled by the diffraction-induced phase. For example, the Stokes detuning reduces the Stokes gain and diminishes the coherent transfer of energy from the pump to the Stokes seed with the Stokes growing at a smaller rate. This causes the pump to deplete less, and reduces the gain-to-loss ratio required for soliton formation.

It is interesting to note that a certain influence on the Stokes evolution by the pump can be recognized even in the asymptotic regime where saturation occurs, and the evolution of a Stokes soliton can be readily accepted as a typical process of beam control in a Raman amplifier. However, the concept that the controller (Raman pump) itself, through nonlinear feedback, can become stabilized while the Stokes depletes in its turn is rather striking. The interchange of role between the pump and the probe insures that the two beams would not stabilize concurrently over an extended optical thickess. [13]

In summary, rigorous calculations have been described for (i) Stokes generation, build up, and gain saturation; (ii) Stokes area stabilization and soliton formation; and (iii) the Stokes decay, pump renewal, pump area stabilization, and Raman pump soliton formation. These *in toto* account for experimental observations and more generally we have attempted to extract the central physical processes underlying soliton propagation in stimulated Raman scattering.

The table below is a distillation of the essence of our report.

* For convenience an energy diagram is shown.

This work was supported in part by NSF, ARO/AFOSR and ONR.

REFERENCES

1. J.L. Carlsten, R.G. Wenzel, and K. Drühl, SPIE $\underline{380}$, 201 (1983) and R.G. Wenzel, J.L. Carlsten and K.J. Drühl, J. of Stat. Phys. $\underline{39}$, 615 (1986).
2. M.G. Raymer and J. Mostowski, Phys. Rev. $\underline{A24}$, 198 (1981).
3. D.J. Kaup, Physica $\underline{19D}$, 621 (1986).
4. J.R. Ackerhalt and P.W. Milonni, Phys. Rev. $\underline{A33}$, 3185 (1986).
5. J.C. Englund and C.M. Bowden, Phys. Rev. Lett. $\underline{57}$, 266 (1986).
6. F.P. Mattar, Appl. Phys. $\underline{17}$, 53 (1978); F.P. Mattar and M.C. Newstein, Comp. Phys. Commun. $\underline{20}$, 139 (1980); F.P. Mattar and J.H. Eberly, in *Laser-Induced Processes in Molecules*, ed. K.L. Kompa and S.D. Smith (Springer-Verlag, 1979) p. 61; and B.R. Suydam and F.P. Mattar, SPIE $\underline{380}$, 439 (1983).
7. K.J. Drühl, R.G. Wenzel and J.L. Carlsten, Phys. Rev. Lett. 51,

1171 (1983) and K.J. Drühl, J.L. Carlsten and R.G. Wenzel, J. of Stat. Phys. 39, 615 (1986).
8. F.P. Mattar, P.R. Berman, A.W. Matos, Y. Claude, C. Goutier and C.M. Bowden, in *Multiple-Photon Excitation and Dissociation of Polyatomic Molecules,* ed. by C.D. Cantrell, Topics in Current Physics, Vol. 35 (Springer-Verlag, 1986) p. 223-283
9. S.L. McCall and E.L. Hahn, Phys. Rev. Lett. 28, 3-8 (1967) and Phys. Rev. 183, 457 (1969) and Phys. Rev. A2, 861 (1970); and S.L. McCall, Ph.D. thesis in Physics, University of California at Berkeley (1968), Section II, p. 5-53, "The Self-Induced Transparency Phenomenon".
10. M.S. Feld and J.C. MacGillivray in *Coherent Nonlinear Optics,* ed. M.S. Feld and V.S. Letokhov, Topics in Current Physics Vol. 21, Springer-Verlag (1980) p. 7; Q.H.F. Vrehen and H.M. Gibbs in *Dissipative Systems in Quantum Optics,* ed. by R. Bonifacio, Topics in Current Physics Vol. 27, Springer-Verlag (1982) p. 111; M. Gross, S. Haroche, Phys. Rep. 93, 301-396 (1982); and M.F.H. Schuurmans, Q.H.F. Vrehen, D. Polder, H.M. Gibbs, in *Advances in Atomic and Molecular Physics,* Vol. 17, ed. by D.R. Bates, B. Bederson (Academic, New York, 1981) pp. 168-228.
11. R. Bonifacio, F.A. Hopf, P. Meystre and M.O. Scully, Phys. Rev. A12, 2568 (1975).
12. C.M. Bowden and F.P. Mattar, SPIE 288, 364 (1981), SPIE 369, 151 (1983) and in *Coherence and Quantum Optics V,* ed. by L. Mandel and E. Wolf (Plenum, 1984) p. 507.
13. In order to forestall any concern that this result appears to be in conflict with the interesting earlier study of Konopnicki and Eberly, [Phys. Rev. A24, 2567 (1981)], who showed the possibility of two *simul*taneous soli*tons* (simultons) occurring as the result of a double SIT, we should like to point out that their investigation is not simply comparable with ours. Since we have, in fact, reexamined the relationship between the two and will present the details in the future, we note here only certain pertinent distinctions. Primarily, while diffraction is the fundamental mechanism in establishing the Raman solitons, it will destroy simulton formation. [Cf. Mattar *et al* in *Advances in Laser Science-I,* ed. by W.C. Stwalley & M. Lapp (A.I.P., 1986) p. 324] In addition, the creation of simultons is restricted to a special set of conditions, fixing the initial populations, oscillator strengths, and Rabi frequencies, which are different from those which apply for general Raman amplification.

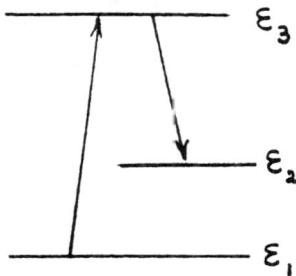

THE RAMAN SOLITONS

	Raman Pump	Stokes	Relative Magnitudes
Transition frequency ω	large	small	$\omega_p > \omega_{st}$
wavelength λ	short	long	$\lambda_p < \lambda_{st}$
Rayleigh diffraction length $z = \pi r_p^2/\lambda$	large	small	$z_p > z_{st}$
Fresnel number $F = z/\ell$	large	small	$F_p < F_{st}$
Diffraction coupling $\kappa = z^{-1}$	weak	strong	$\kappa_p < \kappa_{st}$
Phase evolution due to free space diffraction ϕ^{dif}	small	larger	$\phi_p^{dif} < \phi_{st}^{dif}$
Beer's nonlinear length (τ = characteristic time, e.g., T_1, T_2) $\alpha = \frac{\mu^2 \omega}{n \hbar c} N\tau$	small (but medium still optically thick, $d_p \ell > 1$)	larger	$\alpha_p < \alpha_{st}$
Transition dipole moment	small	larger	$\mu_p < \mu_{st}$
phase evolution due to the transition nonlinearity ϕ^{NL}	small (longer threshold)	larger	$\phi_p^{NL} < \phi_{st}^{NL}$
Distance for linear diffraction to be compensated by transition nonlinearity	longer	shorter (sooner)	
When balance occurs leading to soliton formation	subsequent	first	
Beer's length Fresnel number $F_g = \pi r_p^2/\lambda \alpha^{-1}$ $= \kappa^{-1} \alpha^{-1}$	larger	smaller	$F_{gp} \quad F_{gst}$

CRITICAL PARAMETRIC RATIO

Name	Definition	Equivalence	Order of Magnitude
Nonlinear gain ratio	$\beta = \alpha_{st}/\alpha_p$	$(\mu_{st}/\mu_p)^2/(\lambda_s/\lambda_p)$	Larger than unity
Relative free space diffraction	$\gamma = F_{st}/F_p$	$\kappa_p/\kappa_{st} = (\lambda_p/\lambda_{st})$	Smaller than unity
Relative gain-to-loss ratio	$\delta = F_{gst}/F_{gp}$	$(\lambda_p \alpha_p^1)/(\lambda_{st} \alpha_{st}^1) = (\lambda_p/\lambda_{st})(\alpha_{st}/\alpha_p) = \gamma\beta$	Larger than unity

TECHNIQUES FOR FAR ULTRAVIOLET RESONANCE RAMAN SPECTROSCOPY

P. B. Kelly[*], S. Li, G. D. Strahan, and B. Hudson
Department of Chemistry and Chemical Physics Institute,
University of Oregon, Eugene, Oregon 97403

ABSTRACT

Improvements in apparatus used for far ultraviolet resonance Raman spectroscopy are discussed.

INTRODUCTION

Unique information about the symmetry and geometry of excited electronic states may be obtained by the use of resonance Raman spectroscopy.[1-6] This is due to the intrinsically high resolution of Raman spectra and the fact that the vibrational pattern is analyzed with respect to the ground state potential surface where the molecular symmetry and normal modes are well known. This technique is particularly useful when resolved absorption spectra cannot be obtained. Recent advances in laser technology permit generation of ultraviolet radiation at a great variety of wavelengths up to, and beyond, the air transmission cut-off near 180 nm. This paper discusses some aspects of the apparatus used in our laboratory for this experimental method.

GENERATION OF FAR UV RADIATION

The basic component of this device is a Q-switched Nd:YAG laser of unstable resonator design. Most of our experiments have been performed with an oscillator-only laser. The fundamental output of this laser is 200 mJ per pulse at 1064 nm with a repetition rate of 30 Hz. Conversion of the IR radiation to higher harmonics in conventional angle tuned KD*P crystals produces radiation at 532 nm with a pulse energy of 100 mJ, at 355 nm with a pulse energy of 40 mJ and at 266 nm with a pulse energy of 20 mJ. The repetition rate of 30 Hz results in an average power at 266 nm of 600 mW. Our earliest UV Raman experiments[7] were performed with the "fifth" harmonic radiation of 212.8 nm. This is generated by summation of the 1064 nm and 266 nm beams in a KDP (non-deuterated) crystal held at -40° C where it 90° phase matches.[8] The resulting pulse energy at 212.8 nm is about 0.5 mJ (15 mW average power at 30 Hz). In recent experiments an amplified laser has been used. This produces 1 J pulses at 1064 nm and 80 - 100 mJ pulse energies at 266 nm. The repetition rate is 20 Hz. The performance of this laser for far ultraviolet studies is still being evaluated.

[*]Present address: Department of Chemistry, University of California, Davis, CA 95616.

A variety of other wavelengths can be generated using stimulated Raman shifting in molecular gasses.[10] For Raman shifting in diatomic hydrogen with a vibrational frequency of 4155 cm^{-1} the following wavelengths (in nm) are generated from the initial radiation indicated at the left:

532:	320	282	253	229	209	192	178	166	
355:		309	273	246	223	204	188	175	163
266:		299	266	240	218	200	184	171	160	150
213:				233	213	196	181	168	157	148

The production of radiation shifted in frequency from the input value is due to stimulated Raman production of the first Stokes beam followed by 4-wave mixing of this radiation with the pump beam to produce second Stokes and first anti-Stokes radiation. The stimulated Raman process is automatically phase matched so that it is (roughly) colinear with the pump beam. There is a threshold value for stimulated Raman gain. The 4-wave mixing processes require phase matching. This is facilitated by focusing the input beam and by minimizing the dispersion of the medium by keeping the gas pressure low. There is an optimum pressure for each anti-Stokes wavelength.[10]

The use of deuterium with a vibrational frequency of 2991 cm^{-1} results in additional lines. HD with a frequency of 3630 cm^{-1} can also be used. Both D_2 and HD appear to have higher thresholds for Raman shifting than H_2. Birefringence in the entrance window of the cell can result in unwanted generation of rotational Raman scattering due to introduction of circular polarization. This unwanted radiation also appears to be produced when the fundamental (1064 nm) radiation is not completely eliminated from the input beam.

The production of fifth harmonic radiation (212.8 nm) with enough pulse energy to exceed the threshold for stimulated Raman scattering is not possible with the method in use.[7-9] However, this and several other frequency generation processes will be greatly facilitated by the pending availability of β-barium borate nonlinear optical crystals.[11]

THE SAMPLE

The radiation generated by these processes is then directed to the sample. Either backscattering or 90° geometry may be used. Backscattering has the distinct advantage that the Raman scattered intensity does not depend critically on the optical density of the sample. The samples that we have used in our studies have been streams of gases or dilute solutions. Several liquid and vapor sampling systems have been devised. For liquids, a guided flow system provides a flat liquid sheet several mm thick. A shielded gas flow device has also been constructed for studies of low vapor pressure samples.

COLLECTION OPTICS, DISPERSION AND DETECTION

The Raman scattered radiation is collected by large S1-UV quartz lenses and directed into a monochromator. Both half-meter double and 1-meter single monochromators have been used. Ultraviolet Raman studies require higher wavelength dispersion than is commonly used in the visible because of the reciprocal nature of the relationship between wavelength and wavenumber.

The dispersed radiation is detected with an S-5 solar blind photomultiplier. A boxcar amplifier processes the resulting signal. The output of the boxcar is sent to the ADC of a minicomputer for collection. The minicomputer enables multiscan averaging of several such experiments. Computer scaling and plotting greatly simplifies data analysis, as this is often performed by comparison of spectral intensities of spectra obtained at different excitation wavelengths. Wavelength calibration is achieved by simultaneous recording of Raman spectra and mercury spectra.

In a new system which we are developing, the S1-UV quartz optics are replaced by CaF_2 and MgF_2 optics which have lower wavelength cutoffs (S1-UV is transparent above 180 nm, whereas CaF_2 and MgF_2 are transparent above 130 nm). Purging the surrounding atmosphere with N_2 or Ar permits the study of transitions in the VUV region. This system utilizes reflective rather than refractive collection optics with a considerable increase in collection efficiency.

REFERENCES

1. **Proceedings of the Tenth International Conference on Raman Spectroscopy**, W. L. Peticolas and B. Hudson, eds. (University of Oregon, Eugene, 1986) pgs. 17-8, 17-10, 17-24 and 20-11.
2. B. Hudson, P. B. Kelly, R. R. Chadwick, R. A. Desiderio and L.D. Ziegler in **Advances in Laser Science - I**, W. C. Stwalley and M. Lapp, eds. (Amer. Inst. Phys., New York, 1986) pgs 690; 706.
3. B. Hudson, Spectroscopy 1, 22 (1986).
4. B. Hudson, P. B. Kelly, L. D. Ziegler, R. A. Desiderio, D. P. Gerrity, W. Hess and R. Bates in **Advances in Laser Spectroscopy**, Vol. 3, B. A. Garetz and J. R. Lombardi, eds. (J. Wiley & Sons, New York, 1986) pgs 1-32.
5. B. Hudson and L. Mayne in **Biological Applications of Raman Spectroscopy**, T. G. Spiro, ed. (Wiley, New York, 1986) in press.
6. B. Hudson and L. Mayne, Methods in Enzymology 130, 331 (1986).
7. L. D. Ziegler and B. Hudson, J. Chem. Phys. 74, 982 (1981).
8. L. D. Ziegler and B. Hudson, J. Phys. Chem. 88, 1110 (1984).
9. G. H. Lesch, J. C. Johnson and G. A. Massey, IEEE J. Quantum Electron. 12, 83 (1976).
10. Recent references include K. G. H. Baldwin, J. P. Marangos, D. D. Burgess and M. C. Gower, Optics Commun. 52, 351 (1985); B. Bobbs and C. Warner, Optics Lett. 73, 88 (1986); V. Wilke and W. Schmidt, Appl. Phys. 18, 177 (1979); D. J. Brink and D. Proch, Opt. Lett. 7, 494 (1982).
11. K. Kato, IEEE J. Quantum Electron. 22, 1013 (1986).

APPLICATIONS OF ULTRAVIOLET RESONANCE RAMAN SCATTERING IN MOLECULAR ELECTRONIC SPECTROSCOPY

Peter B. Kelly,[*] Shijian Li and Bruce Hudson
Department of Chemistry and Chemical Physics Institute
University of Oregon, Eugene, Oregon 97403

Resonance Raman spectroscopy can provide unique information about the nature of excited electronic states that is often not available from other techniques. The features of a Raman spectrum that are strongly enhanced under resonance conditions can determine the electronic symmetry or the geometry of an excited electronic state and may characterize a photodissociation pathway even when the absorption transitions to that state are so diffuse or so complex that analysis is not possible. The application of this technique to a variety of species has been described recently.[1-4] Here we present the results of several unpublished studies.

Methylethylenes. The 3S Rydberg state of the dimethylethylenes lies lower than the $\pi\pi*$ state. The absorption spectra of these Rydberg states (240 nm-200 nm) show some sharp features. However, precise analysis of the absorption spectra is difficult without knowledge of the excited state vibrational frequencies. Resonance Raman spectroscopy has the advantage that displacements in the upper state normal modes are made apparent by increased intensity of the corresponding ground state modes in the Raman spectrum.

The resonance Raman spectra of the butenes show strong intensity in the C=C stretch motion and its overtone. The olefinic C-H in-plane bend, CH_3 asymmetric and symmetric deformation, CH_3 asymmetric stretch, and in-plane CH_3 rock all show enhancement as the laser approaches resonance with the 3S Rydberg state. There is no intensity in the overtone of the C=C torsion mode for any of the butenes with excitation in the Rydberg region. This is in contrast to the case of ethylene with excitation in the $\pi\pi*$ band.[5]

This spectral pattern demonstrates that the geometry change associated with excitation involves lengthening of the C=C bond and deformation of the HCH bond angles. This is similar to the CH_2 in-plane scissors motion observed in ethylene.[5] The most surprising feature of the methyl substituted olefin Raman spectra is the lack of activity in the torsion about C=C.[5] This lack of intensity indicates that the molecule remains planar in the 3S Rydberg state for the isomeric butenes. Recent theoretical calculations have shown that the 3S Rydberg state is planar in ethylene.

Enhancement of the symmetric and asymmetric CH_3 deformation and asymmetric stretch of the methyl C-H bonds is consistent with dissociation of a methyl C-H bond on the Rydberg excited state potential surface for the butenes studied.

[*] Present address: Dept. of Chemistry, University of California, Davis, CA 95616

Chloroethylenes. Resonance Raman spectra of chloroethylenes have been obtained with 240nm to 184nm excitation. The N→V ππ* transition in chloroethylenes is broad and structureless at energies below 50,000 cm^{-1}. The very broad nature of this absorption suggests a significant lengthening of the C=C distance in the excited state, although no structural information can be determined from the absorption spectra. Qualitatively, normal modes that show fundamental and overtone intensity in a resonance Raman spectrum are displaced in the excited state. Normal modes with small or no displacement or force constant change are not enhanced. We have used resonance Raman scattering to probe the geometry of the ππ* excited state of dichloroethylenes.

The N→V transitions of dichlroethylenes extend from 235 nm to shorter wavelengths. The resonance Raman spectra show enhancement of the C=C stretch, the overtone of antisymmetric CCl_2 stretching, the overtone of torsion about the C=C axis, symmetric CCl_2 stretch, CCl in-plane bending, and combination bands. This indicates that the excited state geometry has a lengthened C=C bond and is twisted about this bond as in ethylene.[5] There are also changes in the C-Cl bond distance and the Cl-C-Cl bond angle.

In conclusion, the resonance Raman spectra of the dichloroethylenes show that the ππ* states of these molecules are similar to that of ethylene with lengthened olefin C=C bond and displacement in the twisting coordinate. The enhanced activity of the torsional mode in the **cis** compound relative to that of the **trans** compound indicates that there is a significant steric component to the twisting motion but the presence of this distortion in the upper state of 1,1-dichloroethylene indicates that steric effects are not entirely dominating the upper state potential. The significant activity of antisymmetric CCl stretch and symmetric CCl stretch may indicate a dissociation pathway that would contribute to the diffuse nature of the absorption spectrum.

Benzene Derivatives. Previously reported Raman studies of benzene[6,7] and its methyl-substituted derivatives[8] using 212.8 nm excitation have been extended to other excitation wavelengths, other substitution patterns and to include fluoro- and chlorobenzene derivatives. Interest in these systems stems from the fact that lowering the symmetry of the D_{6h} benzene system can result in induction of an allowed transition component in the electronic transition from the A_{1g} ground state to the B_{1u} ($1L_a$) transition of the benzene ring. There is, however, still a significant vibronic component to this transition due to mixing with the allowed E_{1u} excited state by e_{2g} vibrations, primarily ν_8. In the absence of substituents, or with symmetric 1,3,5 substitution, the only mechanism for resonance enhancement due to proximity of the excitation frequency with the $1L_a$ transition is purely vibronic C-type activity.[2,3,6-8] In the linear vibronic coupling scheme with harmonic vibrations and only small force constant changes, this results in strong activity in the two-quantum transition of e_{2g} vibrations. This, in fact, is what is observed for symmetric systems. In the lower symmetry species, the cross term between the allowed and vibronic transition mechanisms results in activity in the fundamental vibrations of these modes via the B-type mechanism.[8] The relative intensity of the fundamental to overtone transition is a measure of the allowed to vibronic intensity in the transition.

The present studies have shown that mono- and para dimethyl and chloro-benzene derivatives have significant intensity in the fundamental vibrational transition of the mode corresponding to ν_8 of benzene. This indicates that substituent-induced allowed intensity is comparable to that of the vibronic component in these cases. On the other hand, fluoro substitution results in very weak intensity in the fundamental transition relative to that of the C-type overtone indicating that the electronic transition is still primarily vibronic.

Another interesting observation concerns the effect of the geometrical pattern of the substitutions. As noted above, symmetric 1,3,5-tri-substitution results in an effectively D_{3h} system so that the $1L_a$ transition is of B' symmetry and is forbidden. This leads to the expectation of no intensity in the fundamental transition of ν_8 as is observed. Considerable intensity in the first overtone remains. On the other hand, a study of the asymmetric 1,2,3-trimethylbenzene also shows no activity in the corresponding fundamental band. This result is in agreement with early studies on the effects of such a substitution pattern on the induced absorption intensity of the vibronic states of benzene.

The spectra of chloro- and 1,3-dichlorobenzene exhibit a number of binary overtone and combination transitions involving out-of-plane motion of the Cl atoms (b' in C_{2v} and b_{3u} in D_{2h}). The activity of these modes clearly demonstrates that the chlorine atom(s) is (are) displaced out of the benzene plane in the excited ^1L_a excited state.

REFERENCES

1. **Proceedings of the Tenth International Conference on Raman Spectroscopy**, W. L. Peticolas and B. Hudson, eds. (University of Oregon, Eugene, 1986) pgs. 17-8, 17-10, 17-24 and 20-11.
2. B. Hudson, P. B. Kelly, R. R. Chadwick, R. A. Desiderio and L.D. Ziegler in **Advances in Laser Science - I**, W. C. Stwalley and M. Lapp, eds. (Amer. Inst. Phys., New York, 1986) pg 706.
3. B. Hudson, Spectroscopy 1, 22 (1986).
4. B. Hudson, P. B. Kelly, L. D. Ziegler, R. A. Desiderio, D. P. Gerrity, W. Hess and R. Bates in **Advances in Laser Spectroscopy**, Vol. 3, B. A. Garetz and J. R. Lombardi, eds. (J. Wiley & Sons, New York, 1986) pgs 1-32.
5. L. D. Ziegler and B. Hudson, J. Chem. Phys. **79**, 1197 (1983).
6. L. D. Ziegler and B. Hudson, J. Chem. Phys. **74**, 982 (1981).
7. D. P. Gerrity, L. D. Ziegler, P. B. Kelly, R. A. Desiderio and B. Hudson, J. Chem. Phys. **83**, 3209 (1985).
8. L. D. Ziegler and B. Hudson, J. Chem. Phys. **79**, 1134 (1983).

STIMULATED ELECTRONIC RAMAN SCATTERING IN SODIUM VAPOR

Han Xiaofeng Lü Zhenguo Ma Zuguang Cheng Yongkang
Laser Division, Harbin Institute of Technology, China

ABSTRACT

The observation of a stimulated electronic Raman scattering(SERS) on 3s-4s transition in sodium vapor pumped by a PTP dye laser is reported. Infrared output(Stokes) tunable from 2.38 μm to 2.65 μm has been obtained as the dye laser is tuned from 334.0 nm to 338.8 nm. Maximum photon conversion efficiency is about 30%.

INTRODUCTION

Stimulated electronic Raman scattering(SERS) was first observed by Rokni and Yatsiv [1] and Sorokin et al. [2] in 1967. Since Raman shift of SERS is large, frequency conversion from ultraviolet and visible regions into infrared region can be realized directly with SERS process. For simplicity of energy level structure of metal atoms and familiarity with those atomic transition, SERS process in metal atom vapor has been investigated widely. This paper reports the observation of 3s-4s SERS tunable from 2.38 μm to 2.65 μm in sodium vapor excited by an excimer-pumped dye laser(334.0-338.8 nm). Analysis and discussion on the relation between the SERS intensity and the pump laser tuning are made. We believe that if the dye laser output exceeds the pumping power threshold on the two wings of the dye tuning curve, the tuned range of the SERS may be extended further and the intensity of the SERS may be increased considerably by resonant enhancement.

THEORETICAL CONSIDERATION

The SERS scheme is illustrated in the electronic energy level diagram of sodium atom(Fig.1). An intense dye laser is used to excite a Raman transition between the electronic ground state 3s and the excited state 4s. In this way, Raman shift(Stokes) radiation is produced at a frequency $\nu_s = \nu_1 - \Omega_{4s3s}$, where ν_1 is the pumping frequency and $h\Omega_{4s3s}$ is the energy of the electronic Raman

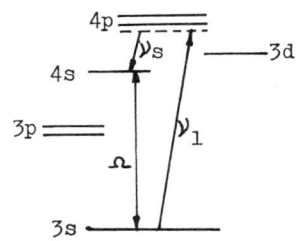

Fig.1. Partial level diagram of sodium atom showing the 3s-4s SERS.

transition. The Raman shift Ω_{4s3s} in sodium vapor is 25 739.86 cm^{-1}.

With the coupling wave method, the Stokes gain factor is shown as follows [3]

$$G_s = (4\pi\omega_s/c)\left|6\chi^{(3)}_{xxxx}(\omega_s)\right|\left|E(\omega_1)\right|^2 \quad \cdots\cdots (1)$$

where $\left|\chi^{(3)}_{xxxx}(\omega_s)\right| = 3.145\times 10^7 \dfrac{N}{\Gamma_{4s3s}}\Big[2(\gamma_{(n+1)p_{3/2}} - \gamma_1)^{-1} +$

$(\gamma_{(n+1)p_{1/2}} - \gamma_1)^{-1}\Big]\times\left|<n+1p0\tfrac{1}{2}|x|n+1s0\tfrac{1}{2}><ns0\tfrac{1}{2}|x|n+1p0\tfrac{1}{2}>\right|^2$ esu

$\cdots\cdots (2)$

N is the atom density and Γ_{4s3s} is a damping parameter. For sodium, it is assumed that N is 1.59×10^{17} cm^{-3} at the heat pipe temperature T=560 °C and $\gamma_{4p_{3/2}}$=30 272.51 cm^{-1}, $\gamma_{4p_{1/2}}$=30 266.88 cm^{-1}, $<4p0\tfrac{1}{2}|x|4s0\tfrac{1}{2}>$ =-3.12×10^{-8} cm, $<3s0\tfrac{1}{2}|x|4p0\tfrac{1}{2}>$ =-0.124×10^{-8} cm. Γ_{4s3s} is approximately 0.5 cm^{-1} taken as the dye laser linewidth. We have a curve for the power gain factor $G_s(\gamma_s)/I(\gamma_1)$ as a function of γ_1 (see Fig.2.). This shows the infrared output tunable in wide range can be generated only if an ultraviolet tunable dye laser is intense enough.

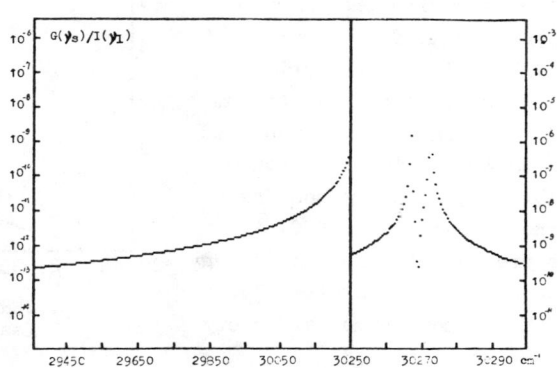

Fig.2. Power gain factor of the 3s-4s SERS in sodium vapor.

EXPERIMENTAL SETUP

The sodium vapor, at 4.5-31 torr, is produced in a heat pipe oven which is totally 90 cm long with a heated zone of ~35 cm. The buffer gas is argon. The radiation of an excimer-pumped PTP dye laser tunable from 334 nm to 347 nm with output power of 60 kW and linewidth of 0.5 cm^{-1} is focused into the heat pipe. Passing through a germanium plate as a filter and a scanning monochromator(WDG 30), the forward Stokes radiation is measured with a PbS infrared detector from which the electric signal is processed by a Boxcar integrator and recorded by a chart recorder(K200).

RESULTS AND DISCUSSION

Using the experimental setup as described above, we have achieved an infrared output tunable from 2.38 μm to 2.65 μm as the dye laser is tuned from 334.0 nm to 338.8 nm. It is found that the infrared radiation always has a certain sharp pump power threshold for each pumping wavelength and generated infrared frequency has a constant shift $\Omega = 25739$ cm^{-1}, corresponding to all tuned dye laser frequency. This frequency shift is consistent with the Raman shift on 3s-4s transition of sodium atom. It has been theoretically and experimentally proved that the infrared radiation is Stokes radiation based on the 3s-4s SERS process in sodium vapor. The dependence of the Stokes intensity on the pump laser tuning was measured(see Fig.3.). It is shown that the Stokes intensity increases as y_1 increases toward to 4p energy level. This agrees with the calculated power gain factor curve of 3s-4s stimulated electronic Raman scattering in sodium vapor(see Fig.2.). The measured optimum sodium vapor pressure is 13.7 torr corresponding to the temperature 560 °C. At this pressure the pumping energy threshold was lowest(250 μJ). The photon conversion efficiecy of 30% was obtained.

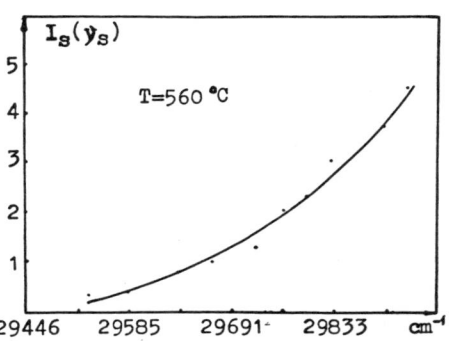

Fig.3, SERS intensity vs. dye laser frequency y_1.

REFERENCES

1. M.Rokni, S.Yatsiv, IEEE J, QE-3, 329-330 (1967)
   Phys. Lett. 24A, 277-278 (1967)
2. P.P.Sorokin, N.S.Shiren, J.R.Lankard, E.C.Hammond, T.G.Kazyaka,
   Appl. Phys. Lett. 10, 44-46 (1967)
3. Y-R. Shen, Nonlinear Infrared Generation, Berlin, Springer, 1977.

QUANTUM LEVEL PROBES OF SMALL METAL CLUSTERS AND THEIR OXIDATION

James L. Gole
School of Physics, Georgia Institute of Technology,
Atlanta, GA 30332

INTRODUCTION

Recently, we have witnessed widespread interest and a growing effort to understand several aspects of the structure and properties of atomic, molecular, and ionic clusters.[1] Within this grouping, metal clusters represent unique intermediate states of matter, the analysis of whose properties and reactivity should reveal much about the growth of atoms into small metal particles,[2] and the development of features inherent in the bulk metallic phase.[3] Further, sufficient evidence now exists to demonstrate that metal clusters are of importance to the fundamental mechanisms of catalysis and numerous chemical conversions.[4] Thus, the basic properties, (geometry, bond strength, reactivity) of small metal clusters, M_n ($2 \leq n < 6$), and their compounds have become the subject of intense, theoretical and experimental study.[5]

While it is clear that the electronic and reactive properties of small metal clusters lie intermediate to those of the atom and those of the bulk metallic phase, the detailed aspects of this picture especially with respect to internal mode structure have not yet been colored. Quantum level probes which have already indicated the need for models including vibronic coupling[6] as well as demonstrating the unique dynamic behavior of small metal clusters as a function of temperature are slowly emerging.[7] It has now become apparent that it will be difficult to probe resultant quantum levels even with extremely sensitive laser spectroscopic techniques unless high flux cluster beam sources ($\approx 10^{10} - 10^{12}$/cc) are developed. These cluster sources must be designed to overcome the substantial loss mechanisms including predissociation and rapid intramolecular conversion and relaxation which deplete excited state populations at a rate competing effectively with those spectroscopic probes that can be applied to the analysis of quantum level structure.

Quantum level probes of metal cluster oxidation,[8] which must overcome similar constraints to those imposed on the study of naked metal clusters, afford an opportunity to parameterize the intermediate region bordered on the one side by the gas phase oxidation of metallic atoms and dimers and on the other by the oxidation of the bulk metallic phase. It has also been suggested[9] that these studies may yield information useful for the assessment of short and long range interactions which effect the bulk oxidation process, providing at least a means of evaluating the local environments which characterize surface oxidation.

Outline of the Experimental Approach

In order to characterize small metal clusters, metal cluster oxidation, and the products of metal cluster oxidation, we have

chosen to develop high intensity oven based cluster sources which
allow not only for the production of the requisite clusters but also
provide for their formation primarily with low internal temperatures
(T $\leq$ 300K). This requirement is necessary since the analysis of
features which map energy levels and structure, and allow a reasonable assessment of dynamiic behavior may be prohibitive if not
impossible at the elevated temperatures required to produce these
species through vaporization of the bulk metal. These sources
require a careful mesh of the appropriate material science and (1)
the techniques of rarefied gas dynamics to produce either an isentropic supersonically expanded "free jet" of cold metal clusters or
(2) an appropriate entrainment and agglomeration of an intense
metallic flux into cluster flow in a well controlled cooling bath
gas.

Once formed in sufficient concentration, naked metal clusters
are probed using a combination of laser induced fluorescence
techniques (primarily excitation spectroscopy) and photoionization-
quadrupole or time-of-flight mass spectrometry. To approach the
study of the optical signatures and internal mode structure of metal
cluster oxides and halides and the dynamics of metal cluster oxidation, we extrapolate experience gained in the study of chemiluminescent metal atom oxidation reactions across a wide pressure range.[10]
A high flux continuous metal flow is used to create <u>large concentrations of small metal clusters</u> ($\approx 10^{11} - 10^{12}/cm^3$) by forming an
environment which is intermediate to that of a low "source" pressure
effusive device, producing primarily atoms, a small percent diatomics
and, in some cases, small percentages of polyatomics,[11] and those
conditions which prevail subsequent to the agglomeration of the
metallic plasma formed in laser vaporization as it is entrained in a
continuous or pulsed rare gas flow at high pressures. The latter
technique, especially when operated in the pulsed supersonic expansion mode, produces a wide diversity of much larger clusters (vs.
effusive source) although at small ($\approx 10^7/cm^3$) concentration. The
seed for the initial phases of a cluster forming environment is
created using a high metal flux which, within itself, can lead to
some agglomeration to form clusters. This seed flux is further
agglomerated to larger concentrations of small clusters by an entraining argon or helium flow at room to liquid nitrogen temperature.
The agglomeration-entrainment device is depicted schematically in
Figure 1. Here a metal is heated in a particularly designed crucible
to temperatures producing a vapor pressure between one and three
orders of magnitude greater than that employed for effusive operation. The resulting clusters are exposed to an oxidizing environment
where primarily multicentered highly exothermic reactions can lead to
chemiluminescent emission from the products of oxidation. Using this
source we have successfully obtained the first quantal information on
the energy levels and optical signatures of several metal cluster
oxides and select halides ($M_n O$, $M_n X$),[12] the qualitative nature of
which we will outline in following discussion. Thusfar, the metal
cluster oxides and halides are being studied using a combination of
chemiluminescence and mass spectrometry. We suggest that these
studies outline the potential for chemiluminescent probes of metal
cluster oxide quantum levels, not only within themselves but also as

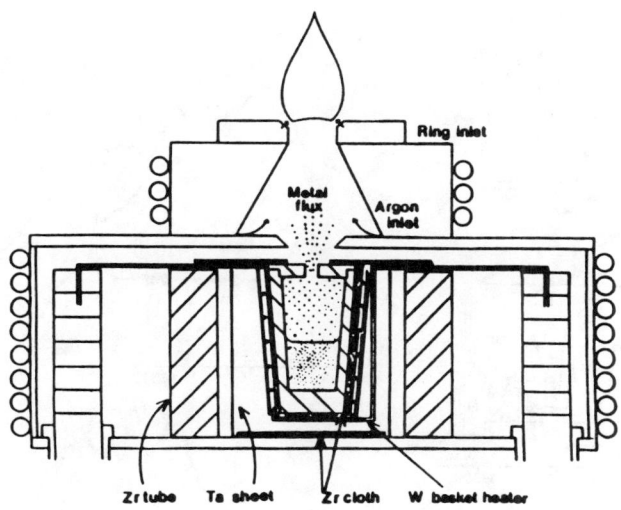

Fig. 1. Schematic of metal entrainment - agglomeration oxidation device showing tungsten basket heater, insulation, entrainment region, and oxidation region.

a means of suggesting future laser fluorescent probes of the metal cluster oxides.

"The Spectroscopy of Copper and Nickel Trimer"

Using a high flux supersonic cluster beam source outlined in detail elsewhere,[7] we have successfully obtained extensive single photon laser induced excitation spectra for copper and nickel trimer under a variety of experimental conditions.[7,8] The observed system positions and peak separations for copper trimer correlate well with both of the recent studies of Smalley,[13] Moskovits[14] and coworkers. Using laser vaporization pulsed expansion techniques, Smalley et al.[13] obtain the two-photon photo-ionization and depletion spectra in Fig. 2(a) with which we compare in Fig. 2(b). In the limit of the coldest spectrum obtained thusfar with our supersonic source ($T_{Rot} <$ 50K), the observed spectral features correlate well with those observed by Smalley et al. with several important additions. These additions involve (1) a significant number of ground state hot bands and (2) the observation of several spectral features associated with strongly predissociating excited state levels of the copper trimer molecule (Smalley depletion spectrum).

There now exists some controversy regarding the assignment of the Cu_3 transition responsible for the spectra depicted in Figure 2. An assignment to a $A^2E"-X^2E'$ transition[13,14] in which both the ground and excited state undergo Jahn-Teller distortion has now been brought into question by recent quantum chemical calculations which indicate that the upper A state may in fact be of A_1 symmetry ($^2A_1$)

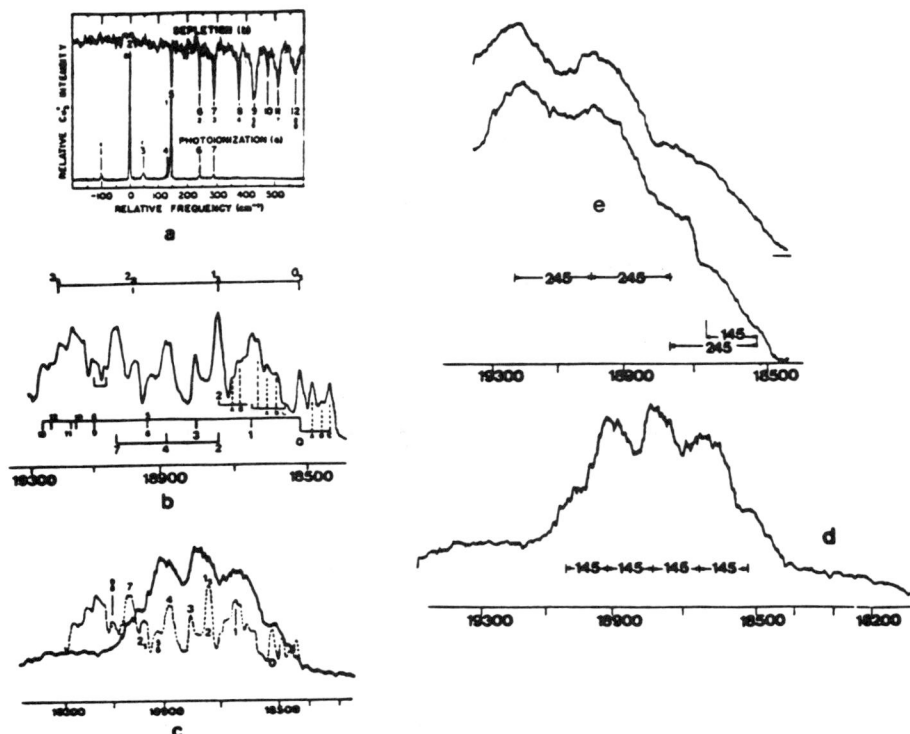

Fig. 2. (a) Copper trimer ion intensity profiles recorded in the photo-ionization (R2PI) and depletion spectroscopic studies of Morse et al.[13] The frequency shifts are given relative to the origin band at 18 525 cm^{-1}. The larger numbers indicate the original numbering scheme of Morse et al. while the smaller numerals correspond to the numbering scheme from the vibronic calculations of Thomson et al.[16] Letters denote hot bands.
(b) Laser induced excitation spectrum for copper trimer in the region 18400-19500 cm^{-1} (Coumarin 7 (535)) produced in a "strong" supersonic expansion of the pure metal at a stagnation pressure > 100 Torr. Estimated rotational temperature is T_{Rot} < 50K < T_{Vib}. Vibronic structure is labeled below using the numbering of Thomson et al. A short progression in the excited state symmetric stretch is labeled above (1_s...). Letters denote hot bands. The dye laser cutoffs are at 18400 and 19300 cm^{-1} respectively. The correlation with the R2PI (photoionization) and depletion spectroscopic studies of Morse et al. is excellent and several additional features are also noted. (c) Cooled laser induced excitation spectrum for copper trimer in the region 18200-19500 cm^{-1} (Coumarin 7 (535)) produced in a milder supersonic expansion than the spectrum (b). The effects of heating lead to a merging of the features in (b) and the burning out of higher frequency spectral features (vibronic bands 8-12, symmetric stretch feature 3). The estimated rotational temperature is T_{Rot} ~ 100K. The 145 cm^{-1} frequency separations corresponding to vibronic features 0(2), 1(5), 3(7), 5,6(9) in Figures (a) and (b) are still apparent in the figure. The dye laser cutoffs are at 18200 (Fig. 4(d)) and 19300 cm^{-1} respectively.
(d,e) Laser induced excitation spectra obtained in progressively milder stages of supersonic expansion showing a change of those excited state features dominating the spectrum. (d) Laser induced excitation spectrum in the region 18200-19400 cm^{-1} (T_{Rot} ~ 100K < T_{Vib}). (e) At higher effective temperature (100K < T_{Rot} < 200K) the dominant spectral features in the region 18500 - 19400 cm^{-1} correspond to peak separations (~ 245 cm^{-1}) indicative of the dominance of the trimer excited state symmetric stretch.

undergoing no vibronic distortion.[15] It is apparent that the upper state spectral region is complex with several possible overlapping states in a narrow energy region.

Recently, assuming that the upper state results from the Jahn-Teller distortion of a $^2E''$ conformation, Thomson, Trular and Mead[16] have carried out an elegant theoretical analysis of the Smalley spectrum, obtaining a detailed vibronic analysis of the Cu_3 excited state level structure. Not only do their calculations predict the Smalley spectrum precisely, but also there is a one-to-one correspondence between the additional excited state levels which we observe at frequencies higher than 19000 cm^{-1} and their predicted level structure. In an alternate assignment based on an upper $^2A_1$ state, Morse[17] assigns all observed features to vibronic levels correlating with short progressions and combination tones in symmetric stretch and bending modes, this assignment requiring the assessment of very large anharmonicities. While the assignment of the symmetric stretch manifold is unaffected, those features associated by Trular et al.[18] with strongly vibronically coupled levels are now assigned to a progression in the upper state bending mode and stretch-bend combination tones. An extrapolation of this assignment also appears consistent with the data in Figure 2(b). The resolution of these two assignments[7(b)] must thus await a high resolution study of the observed band systems.

Those levels indicated in the Smalley depletion spectrum are strongly predissociating, therefore; using the continuous source developed in our laboratory, we observe excitation spectra from several levels which are lost to the two photon ionization technique (features 4-13 and symmetric stretch bands 2_S, 3_S in Figure 2(b)). Based on this result and a comparison of the relative sensitivities of the single photon laser induced excitation and resonant two photon ionization techniques, it is apparent that our source of copper trimer is between three and six orders of magnitude more intense than others currently available.

The trimer spectrum undergoes significant changes as a function of experimental conditions, the most profound effect resulting from a combination of temperature change and the dynamic nature of the trimer ground state potential. As the efficiency of expansion and cooling decreases one observes (Fig. 2(d,e)) the merging and at higher excited state energies burning out of the spectral features depicted in Fig. 2(b). The modifications of those features dominating the spectrum is believed to be a direct result of the dynamic Jahn-Teller effect.

Cu_3 represents an example of a class of molecules thought not to have a definite shape in their ground states.[18] The Cu_3 molecule, if it adopts an equilateral triangle configuration (D_{3h}), corresponds to a $^2E'$ ground electronic state which will distort according to the Jahn-Teller theorem along a coordinate which belongs to the direct product representation E x e. The ground state potential surface which represents the nature of this behavior is usually plotted (Fig. 3) as a contour map in Q_x, Q_y space, where Q_x and Q_y represent normal coordinates describing the e' vibration of an X_3 molecule.[18] These coordinates would correlate with a bending and asymmetric stretching vibration if the molecule were reduced to C_{2v} symmetry. The surface

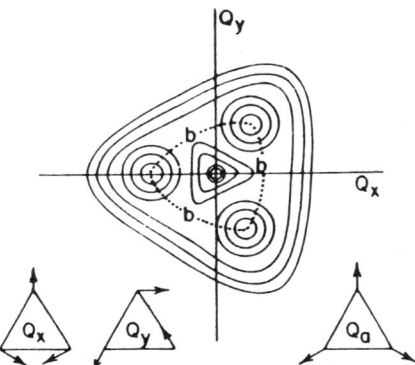

Fig. 3. Contour map of the potential energy surface of a triatomic molecule, X_3, capable of exhibiting the dynamic Jahn-Teller effect.

consists of three minima which correspond to equivalent "obtuse" forms ($> 60°$) of the X_3 molecule separated by three barriers representing saddle points at the top of which the molecule has an "acute" form ($< 60°$). The obtuse and acute configurations are defined with respect to the $60°$ D_{3h} configuration. At the D_{3h} geometry, the surface displays a maximum, higher in energy than the barriers between the obtuse minima.

The surface described above and in Fig. 3[14] is classified in terms of two parameters. The Jahn-Teller stabilization energy is defined as the difference in energy between the D_{3h} form and the minimum energy obtuse angled configuration. The barrier to pseudorotation, which corresponds to the height of the acute angled saddle point above the obtuse angled minimum, is of fundamental importance in determining the dynamics of the copper trimer molecule. When the barrier is high, the molecule will be localized in one of the three obtuse angled minima and the observed energy level pattern should correspond to that of an ordinary bent triatomic. When the barrier is low, the molecule can easily translate from one minimum to another, does not adopt a definite shape, and is able to access a much wider region of configuration space.

A low barrier to pseudorotation may explain the characteristic changes in the copper trimer excitation spectrum as a function of a varying expansion source and increasing effective temperature. Specifically we consider the changes associated with the spectra in Figs. 2(d) and 2(e) where the cooler spectrum in Fig. 2(d) is dominated by a 145 cm^{-1} progression and the spectra in Fig. 2(e) are apparently dominated by the Cu_3 excited state symmetric stretch. A small barrier to pseudorotation may lead to profound temperature dependent effects in the observed emission spectra for a fluxional triatomic molecule. The behavior of the molecule may alter with increasing temperature as those levels lying above the pseudorotation barrier but within the Jahn-Teller well are given to an increasingly higher molecular population. We suggest that this variation in the population distribution with temperature can lead to an effective variation of the configuration space available to the molecule as a function of temperature such that the dominant spectral features associated with the transition are altered.[7]

Because the form of the vibronic wave functions and the configuration space available to the copper trimer molecules varies considerably above and below the pseudorotation barrier, the nature of optical pumping from vibronic levels below and above this barrier is expected to be quite different. We suggest that as the rovibronic levels accessible to optical pumping are spread over a much wider range of the ground and excited state as temperature and the density of available states above the pseudorotation barrier increases. When the trimer molecule translates from one region which would correspond to an obtuse configuration to another, refusing to adopt a definite shape, this promotes a spreading over a wider region of configuration space.[17] Although the configuration space associated with the pumped vibronic configurations is expanded, the effect on the symmetric stretch level structure is negligible. As a result, the spectrum [Fig. 2(e)] is eventually dominated by the "less diluted" symmetric stretch progression.

The lack of an R2PI signal at frequencies in excess of 18800 cm^{-1} indicated in the lower portion of Figure 2(a) results when the efficient excited state dissociative channels deplete a sufficient portion of the excited state population well before a second ionizing photon is able to sample a requisite number of trimers. In many instances, the presence of substantial loss mechanisms including predissociation and rapid intramolecular conversion and relaxation deplete the excited state population at a rate which competes effectively with two-photon spectroscopy. It appears, however, that a substantial increase in cluster beam flux offers a plausible means of overcoming these loss mechanisms. In order to further exploit this capability, we have pursued studies of the nickel trimer molecule and have obtained the first trimer excitation spectrum.

A portion of the nickel trimer excitation spectrum is displayed in Figure 4. This excitation spectrum was obtained using an argon ion pumped stilbene dye laser, operating at 0.5 cm^{-1} resolution, to excite Ni_3 produced in a moderate supersonic expansion of the pure metal at a stagnation pressure of 30 Torr (T_{oven} = 2600K). The cooling obtained ($<T_{Rot} < 40K < T_{Vib}$) is such that vibrational features which can be correlated with (1) ground ($v_1'' \approx 230 \pm 5$ cm^{-1}, $v_2'' \approx 105 \pm 5$ cm^{-1}) and excited state ($v_1' \approx 200 \pm 5$ cm^{-1}, $v_2' \approx 95 \pm 5$ cm^{-1}) stretching and bending modes can be identified. The spectrum appears to consist of a combination of short progressions in all the modes indicated and sequence structure ($\Delta v \approx 30$ cm^{-1}) involving the ground and excited state symmetric stretch mode. It is characterized by a considerably reduced rotational temperature (relative to the oven source) while maintaining substantial vibrational excitation. This spectrum (which will be extended to longer wavelength) correlates well with the 480 nm absorption system observed by Moskovits and Hulse[19] in rare gas matrices and with the laser Raman spectrum obtained by Moskovits' and DeLella.[20] It is significant that this gas phase Ni_3 spectrum is lost to the resonant two photon ionization (R2PI) technique as loss mechanisms apparently deplete the excited state population before the requisite R2PI two-photon pump-probe scheme can be made operative. It is apparent that the <u>single photon</u>

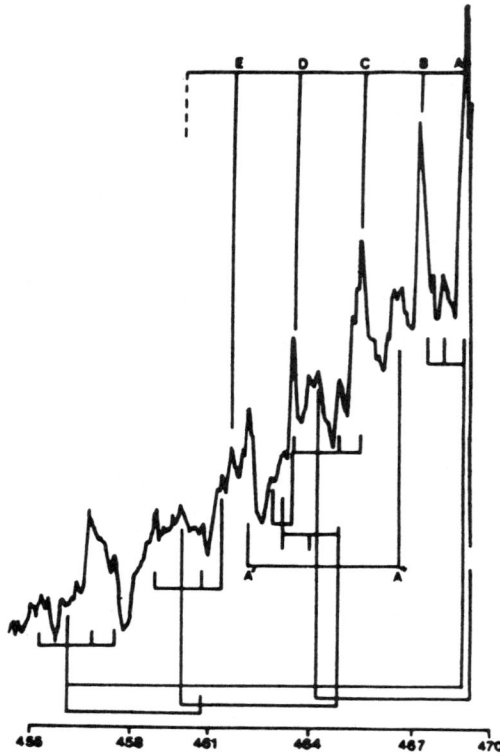

Fig. 4. Portion of the Ni$_3$ excitation spectrum showing short progressions which can be correlated with ground and excited state symmetric stretch and bending modes. See text for discussion.

nickel trimer excitation spectrum is obtained as a result of overcoming those loss mechanisms which have plagued other extremely sensitive spectroscopic techniques. Thus an initial aspect of the promise of our copper trimer excitation spectral studies,[7] which have yielded spectra for strongly predissociating excited state levels of the copper trimer molecule, is now undergoing the extension envisioned when this oven-based supersonic expansion approach was formulated. We would be remiss if we did not mention the possible correlation of our observation of fluorescence from Cu$_3$ and Ni$_3$ excited states and the observations of Imre et al.[21] in their elegant study of the dynamics and resonance Raman spectra of dissociating excited states.

Metal Cluster Oxidation

"Ag$_x$ + O$_3$"

Using the agglomeration-entrainment source described previously, we are studying ozone metatheses with copper and silver fluxes[8] entrained in argon from room to near liquid nitrogen temperature. A sampling of the spectra obtained for silver is shown in Fig. 5 as a

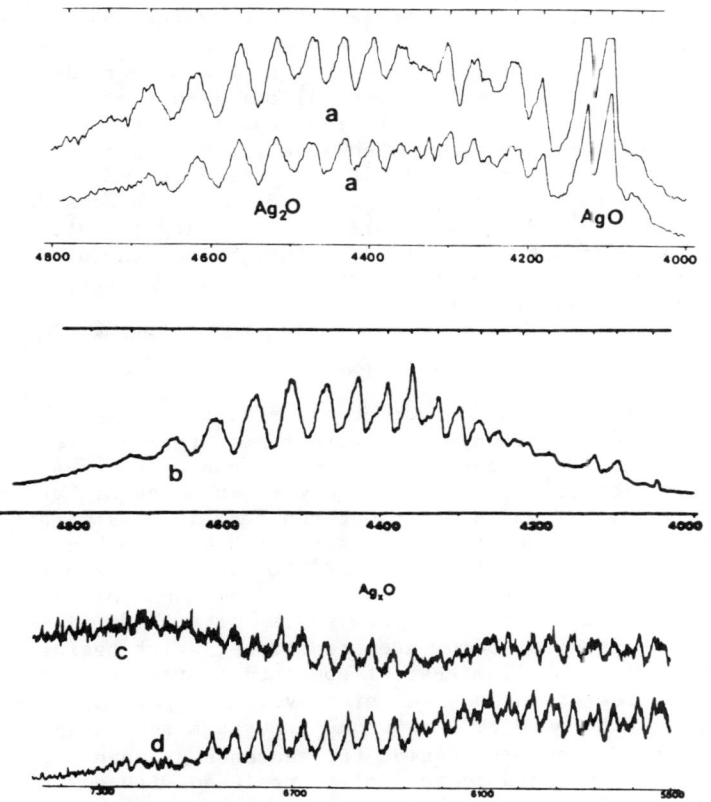

Fig. 5. (a) Chemiluminescent spectra resulting from the ozone oxidation of a moderate silver flux ($K_{Knudsen} < 1$) entrained in room temperature argon showing emission corresponding to the AgO $A^2\Pi - X^2\Sigma$ band system (4000-4300 Å) and what is believed to be an Ag_2O emission system 4400-4800 Å). (b) With increased silver flux the AgO features (4000-4300 Å) are quenched and the spectrum is dominated by Ag_2O fluorescence. (c,d) Closeup of the observed emission in the region 5500-7300 Å believed to correlate with the metal cluster oxides Ag_xO_y (x≥2). The upper spectrum (c) is obtained at considerably higher silver flux vs. spectrum (d). Spectral resolution is 5 Å. See text for discussion.

function of increasing silver flux. The spectrum in Figure 5(a), obtained with a moderately high silver flux, corresponds primarily to the overlap of AgO (4000 - 4200 Å) and Ag_2O (4400 - 4800 Å) features. It is to be noted that both the silver atom and dimer reactions with ozone are not sufficiently exothermic to produce the excited states of the metal monoxide.[8] We require at least the multicentered reaction of Ag_3 to populate the excited states of AgO and higher metal cluster oxides Ag_xO.[8] As the silver flux is increased, the AgO features are quenched (Figure 5(b)) and the Ag_2O features are much more pronounced. Further increase of the silver flux leads to some quenching of the initial Ag_2O features and the subsequent buildup of spectral features in the range 5000 - 7500 Å. The observed frequency separations in the range 5500 - 7000 Å are indicative of silver-silver stretch, bending, and torsional modes, however more definitive statements will require considerable further evaluation under a wider variety of conditions leading to the generation of the observed spectral features. Finally we note that as the silver flux is again increased (Figure 5(c)) further spectral complication arises with the definite emergence of a new red shifted feature at $\lambda > 6800$ Å.

The results outlined here in conjunction with previous studies of the 5000 - 6000 Å region[8] suggest that the observed features result, at least in part, from Ag_xO, $x \geq 3$, although considerable further experiments and analysis will be necessary to establish definitive correlations. As one might anticipate, the observed emission features appear to red shift as the size of the metal cluster oxide increases.

"$B_x + NO_2$, N_2O"

The results obtained for copper and silver represent specific examples of a general trend. In Figure 6, we provide a further comparison of the significant changes which can be observed using the agglomeration technique in conjunction with the much higher temperature boron system. When an intense boron beam is agglomerated in dry ice cooled argon and subsequently reacts with NO_2, the spectrum is found to consist largely of a modified "BO_2" emission system[22] which overlaps BO emission features (BO $A^2\Pi - X^2\Sigma^+$) emanating from v' = 0, 1, 2, $A^2\Pi$. There also appears a new system at $\lambda > 5900$ Å, not observed in the study of boron atom reactions, which begins to dominate the total chemiluminescent spectrum as the boron flux is increased. A view of the new emission system is presented in Figure 6. At least two $\Delta\nu = 40$ cm^{-1} sequence groupings separated by ~ 440 cm^{-1} and a second long progression (or sequence grouping) with $\Delta\nu \sim 142$ cm^{-1} are observed and tentatively correlated with emission from the asymmetric BBO molecule.[22]

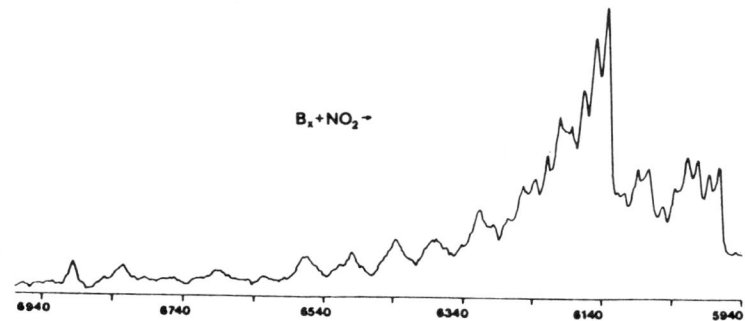

Fig. 6. Closeup of $\lambda > 5900$ Å emission region associated with the reaction of boron agglomerates with NO_2. The spectrum appears to be a combination of both sequential structure and a long progression or sequence structure corresponding to a higher frequency mode or a low frequency (~ 142 cm^{-1}) vibrational mode. Spectral resolution is 5 Å.

"$Mn_x + O_3$"

In Figure 7 we summarize one further comparison, an extension to the transition metals which again emphasizes the significant changes

449

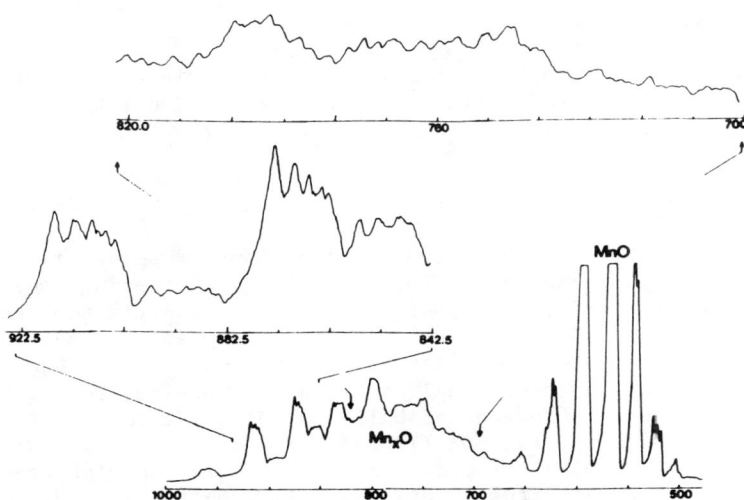

Fig. 7. Chemiluminescent spectra resulting from the multiple collision ozone oxidation of manganese atoms to form MnO* and manganese molecules to form Mn_xO where x is most likely 2. Spectral resolution is 3 Å.

induced using the agglomeration technique. We have studied the manganese-ozone reaction ($Mn + O_3 \rightarrow MnO + O_2$) across a wide pressure range from single to multiple collision conditions. Under both single and multiple collision conditions ($Mn + O_3 + Ar$, He), the $Mn-O_3$ reaction leads to the formation of the lowest-lying MnO* $A'^7\Sigma^+$ excited electronic state and the observation of a strong chemiluminescent signal corresponding to the MnO $A'\Sigma - X'\Sigma$ band system. When an intense manganese beam is agglomerated in LN2 cooled helium and subsequently interacts with O_3, the intense MnO A-X emission system which extends from 500 to 700 nm is accompanied by a new system extending from 670 to at least 1000 nm which grows relative to the MnO A-X system with increased entrainment gas cooling. This new system, whose first order spectrum is depicted in Figure 7, is believed to result from the interaction of manganese clusters and is tentatively associated with moderate progressions in two modes of the ground electronic state of Mn_2O ($Mn_2^+O^-$).

Acknowledgements

It is a pleasure to acknowledge the National Science Foundation, the Petroleum Research Fund of the American Chemical Society and the Silver Institute for their partial support of the research effort described herein.

REFERENCES

1. Symposium on the Structure, Spectroscopy, and Dynamics of Atomic, Molecular and Ionic Clusters, 191st ACS Meeting, New York, New York (1986). Symposium on Clusters and Cluster Ions, 20th Mid Atlantic ACS Meeting, Baltimore, Maryland (1986).
2. S. H. Bauer, private communication. See also, R. T. V. Kung and S. H. Bauer, 8th International Shock Tube Symposium, ed. J. L. Stollery, London: Chapman and Hall (1971); J. Keifer and B. Lutz, J. Chem. Phys. $\underline{44}$, 658 (1966); H. G. Stever, Condensation Phenomena in High Speed Flows. Princeton Series on High Speed Aerodynamics and Jet Propulsion. Vol. III, Sect. F; D. J. Frurip, Light Scattering and Absorption by Nucleating Metal Vapors. H. J. Freund and S. H. Bauer, "Homogeneous Nucleation in Metal Vapors. II. Dependence of the Heat of Condensation on Cluster Size"; S. H. Bauer and D. J. Frurip, "Homogeneous Nucleation in Metal Vapors. III. A Self Consistent Kinetic Model".
3. C. Kittel, Introduction to Solid State Physics (Wiley, New York, 1971), 4th ed.; L. Solymar and D. Walsh, Lectures on the Electrical Properties of Materials (Oxford U. P., London, 1975); W. Harrison, Solid State Theory (McGraw-Hill, New York, 1970).
4. J. H. Sinfelt, "Catalysis by Alloys and Bimetallic Clusters," Acc. Chem. Res. $\underline{10}$, 15 (1977). J. H. Sinfelt, "Heterogeneous Catalysis: Some Recent Developments," Sci. $\underline{195}$, 641 (1977). M. E. Geusic, M. D. Morse, and R. E. Smalley, J. Chem. Phys. $\underline{82}$, 5901 (1985). S. C. Richtsmeier, E. K. Parks, K. Liu, L. G. Pobo, and S. J. Riley, J. Chem. Phys. $\underline{82}$, 3659 (1985). R. L. Whetton, D. M. Cox, D. J. Trevor, and A. Kaldor, J. Chem. Phys. $\underline{89}$, 566 (1985); Phys. Rev. Lett. $\underline{54}$, 1494 (1985).
5. J. L. Gole, "The Gas Phase Characterization of the Molecular Electronic Structure of Small Metal Clusters and Cluster Oxidation," in "Metal Clusters," edited by M. Moskovits, John Wiley and Sons. J. L. Gole and W. C. Stwalley, "Characterization of Alkali Metal Aggregation from Dimer to Bulk," in Advances in Atomic and Molecular Physics, in press.
6. W. H. Gerber, "Theorie des dynamischem Jahn-Teller-Effekts in Li_3 and Untersuchung von Lithium-Molekularstrahlen", Ph.D. Thesis, Bern University, Switzerland, 1980. W. H. Gerber and E. Schumacher, J. Chem. Phys. $\underline{69}$, 1692 (1978). G. Delacretaz, E. R. Grant, R. L. Whetton, L. Woste, and J. W. Zwanziger, Phys. Rev. Lett. $\underline{56}$, 2598 (1986).
7. (a) W. H. Crumley, J. S. Hayden and J. L. Gole, J. Chem. Phys. $\underline{84}$, 5250 (1986). (b) R. Woodward, S. H. Cobb, and J. L. Gole, work in progress. (c) "Formation and Oxidation of Intense

Metal Cluster Beams and Flows", in Proceedings of the International Workshop on Ionized Cluster Beam Techniques (ICBT) '86, pg. 85.
8. W. H. Crumley, J. L. Gole, and D. A. Dixon, J. Chem. Phys. 76, 6439 (1982). J. L. Gole, R. Woodward, J. S. Hayden, and D. A. Dixon, J. Phys. Chem. 89, 4905 (1985). R. Woodward, P. N. Le, M. Temmen, M. McQuaid and J. L. Gole, "Potential Probes of the Metal Cluster Oxide Quantum Levels - Optical Signatures for the Oxidation of Small Metal Clusters M_x(M = Cu,Ag,Mn,B)", J. Phys. Chem., in press.
9. T. N. Taylor, C. T. Campbell, J. W. Rogers, Jr., W. P. Ellis, and J. M. White, Surface Science 134, 529 (1983).
10. See, for example, references in J. L. Gole, and G. J. Green, Chem. Phys. 100, 133 (1985); R. Woodward, J. S. Hayden, and J. L. Gole, Chem. Phys. 100, 153 (1985).
11. See reference 5 for a complete Table of mass spectrometric studies.
12. In a given system, there will be a variety of conditions under which cluster distributions optimal for oxidation to form strongly emitting metal cluster oxides will be produced. As experimental conditions are varied, shifting cluster distributions, we proceed through optimal regions for the excitation of a given metal cluster oxide fluorescence spectrum; we thus engage in a search for those experimental conditions optimizing the metal cluster oxide spectrum of interest. In effect, we refer to results obtained upon the oxidation of effusive metal atom flows and determine those changes which occur as agglomeration ensues.
13. M. D. Morse, J. B. Hopkins, P. R. Langridge-Smith, and R. E. Smalley, J. Chem. Phys. 79, 5316 (1983).
14. M. Moskovits and J. E. Hulse, J. Chem. Phys. 67, 4271 (1977). D. P. Dilella, K. V. Taylor, and M. Moskovits, J. Phys. Chem. 87, 524 (1983). M. Moskovits, Chem. Phys. Lett. 118, 111 (1985).
15. B. Laskowski and S. P. Walch, J. Chem. Phys. 34, 2734 (1986).
16. T. C. Thompson, D. G. Trular, and C. A. Mead, J. Chem. Phys. 82, 2392 (1985).
17. M. Morse, private communication.
18. G. Herzberg, Electronic Spectra and Electronic Structure of Polyatomic Molecules (Van Nostrand, New York, 1966); H. C. Longuet-Higgens, U. Opik, M. H. L. Pryce, and R. A. Sack, Proc. Roy. Soc. London. Ser. A244, 1 (1958).
19. M. Moskovits and J. E. Hulse, J. Chem. Phys. 66, 3988 (1977).
20. M. Moskovits and D. P. DiLella, J. Chem. Phys. 72, 2267 (1980).
21. D. Imre, J. L. Kinsey, A. Sinha, and J. Krencs, J. Phys. Chem. 88, 3956 (1984).
22. R. Woodward, T. C. Devore, and J. L. Gole, work in progress.

SILVER CLUSTERS AS THE ACTIVE SITES FOR SURFACE ENHANCED RAMAN SCATTERING

T. E. Furtak
Colorado School of Mines
Golden, Colorado 80401

D. Roy
Rensselaer Polytechnic Institute
Troy, New York 12181

ABSTRACT

We propose that the active sites for resonant Raman scattering on Ag are small positively charged Ag clusters. These are stabilized by the "special" surface treatments that have been shown to produce the largest efficiencies for surface enhanced Raman scattering (SERS). We present a summary of our arguments which are based on low energy data and a dynamical calculation for vibrational energies of the clusters.

We have identified the long debated origin of the short-range component of SERS as Ag_4^+ clusters that are stabilized by co-adsorption or by low temperature on rough Ag substrates [1]. Our position is presented in detail in a recent article [2]. Here we merely summarize our conclusions.

Figure 1 shows the low energy region of a SERS spectrum from an electrochemically roughened Ag electrode in Cl^- solution together with our calculation for the vibrational energies of the cluster [2]. The largest feature is due to Ag-Cl vibrations and is associated with anions co-adsorbed on or next to the active cluster. The smaller features at lower energy are the cluster vibrations. These are observed in any SERS spectrum from Ag. The model employs force constants from bulk Ag that have been properly scaled for charge, surface [3] and field [4] effects. Also predicted is a mode at 75 cm^{-1} that is not in our data but which is observed elsewhere [5]. Theoretical energies are close to those in the experiments, particularly if broadening, due to inhomogeneous environments, would be included. The two features at 100 cm^{-1} and 118 cm^{-1} would merge into one, under those circumstances.

Additional evidence in favor of the positively charged, four-atom silver cluster follows:
  1. The cluster modes are observed in a wide variety of environments independent of the adsorbate [5,6].
  2. There is a strong correlation between the cluster modes in SERS and Ag cluster vibrations

seen in matrix isolation resonant Raman spectra [7].
3. The stability of SERS depends on temperature [8] and applied voltage [9] in ways consistent with the expected stability of positively charged clusters.
4. There are indications that the active sites are Lewis acid electron acceptors [10], as would be the case with positively charged clusters.
5. There is a strong association between the active sites in photographic emulsions and Ag_4^+ [11]. This has been verified by independent ion beam dosing experiments. It has also been shown [12] that very large levels of enhancement of SERS can be achieved if Ag is exposed to light during electrochemical oxidation and reduction in halide solutions. It is likely that photoactivation causes nucleation of the small clusters.

This work was supported by the U. S. Department of Energy under contract DE-AC02-82ER12032.

REFERENCES

1. T. E. Furtak and D. Roy, Surf. Sci. $\underline{158}$, 126 (1985). W. J. Plieth, J. Phys. Chem. $\underline{86}$, 3166 (1982).
2. D. Roy and T. E. Furtak, Phys. Rev. $\underline{34}$, 5111 (1986).
3. J. H. Rose, J. R. Smith, and J. Ferrante, Phys. Rev. $\underline{B28}$, 1835 (1983).
4. C. A. Barlow and J. R. MacDonald, Adv. Elect. and Electrochem. Eng. $\underline{6}$, 94 (1967).
5. I. Pockrand and A. Otto, Solid State Commun. $\underline{38}$, 1159 (1981).
6. A. Regis, P. Dumas, and J. Corset, Chem. Phys. Lett. $\underline{107}$, 502 (1984). M. Fleischmann, P. J. Hendra, I. R. Hill, and M. E. Pemble, J. Electroanal. Chem. $\underline{117}$, 243 (1981).
7. U. L. Kettler, P. S. Bechthold, and W. Krasser, in The Physics and Chemistry of Small Clusters (NATO ASI Series B, P. Jena, B. K. Rao and S. N. Khanna (eds.), Plenum, New York, 1987).
8. S. H. Macomber and T. E. Furtak, Solid State Commun. $\underline{45}$, 267 (1983).
9. J. F. Owen, T. T. Chen, R. K. Chang, and B. L. Laube, Surf. Sci. $\underline{131}$, 195 (1983).
10. T. Watanabe, O. Kawanamai, K. Honda, and B. Pettinger, Chem. Phys. Lett. $\underline{102}$, 565 (1983).
11. P. Fayet, F. Granzer, G. Hegenbart, E. Moisar, B. Pishel, and L. Woste, Phys. Rev. Lett. $\underline{55}$, 3002 (1985).
12. S. H. Macomber, T. E. Furtak, and T. M. Devine, Chem. Phys. Lett. $\underline{90}$, 439 (1982).

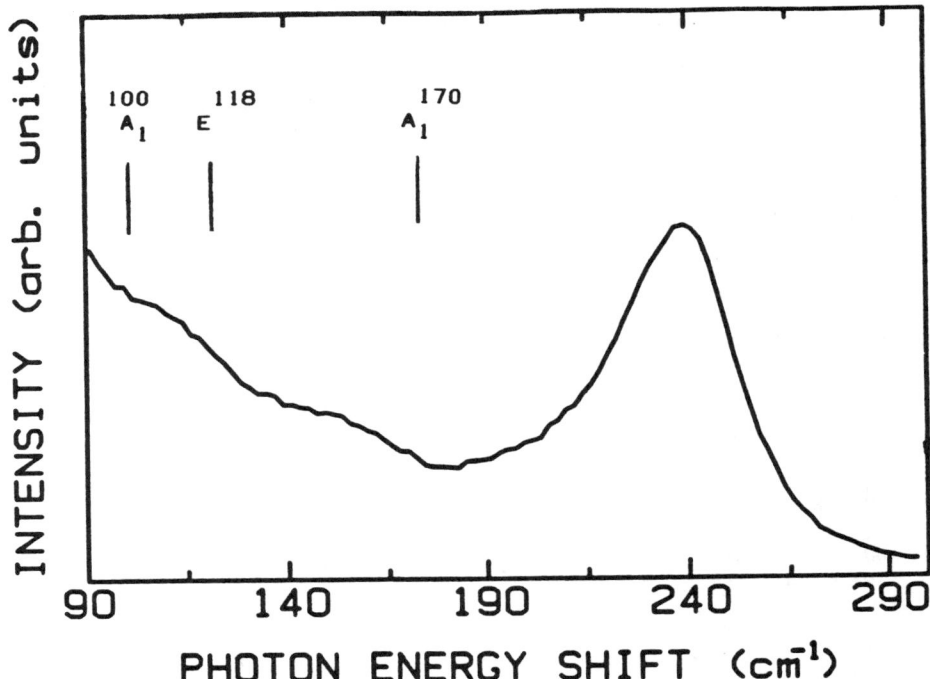

Figure 1

SERS from electrochemically roughened Ag in 1M KCl at −0.3 volts versus the SCE reference electrode. The solutions of the dynamical calculation are shown. The large feature is due to Ag-Cl vibrations.

OPTICAL PROPERTIES OF FRACTAL CLUSTERS

Zhe Chen*, Ping Sheng,
D. A. Weitz, H. M. Lindsay, M. Y. Lin
Corporate Research Science Laboratories
Exxon Research and Engineering Company
Route 22 East, Clinton Township
Annandale, NJ 08801
and
P. Meakin
E. I. DuPont de Nemours and Company, Inc.
Wilmington, DE 19898

ABSTRACT

Optical scattering and absorption characteristics of fractal clusters are calculated with a self-consistent scheme in which the multiple scattering is explicitly taken into account. Our results indicate that for optical scattering the gold colloidal aggregates with fractal dimension D=1.75 behave essentially as low dimensional objects and therefore can be accurately described by the single scattering Born approximation. However, the absorption spectrum of the clusters is shown to be significantly affected by the proximity of the gold particles to their nearest neighbors in the aggregate structure.

INTRODUCTION

Fractal is a term which denotes objects with dilatational symmetry. The density-density correlation function of a fractal structure is generally given by $C(r) \sim r^{d-D}$, which insures that the Fourier transform of $C(r)$, denoted as the structure factor $S(q)$, will also have a power-law behavior, i.e. $S(q) \sim q^{-D}$, where D is the fractal dimension and d is the Euclidean dimension of the embedding space. Within the single-scattering Born approximation, the intensity of the scattered wave is proportional to the structure factor. Within the past few years, this fact was employed extensively in light-scattering experiments to extract intrinsic structural information, such as the fractal-dimension, from fractal objects [1]. However, this approach is based on the assumption that the effects of multiple scattering are negligible. In this short note we examine the legitimacy of this assumption and report the main results of our study on optical scattering and absorption characteristics of gold colloidal aggregates.

CALCULATION AND RESULTS

Our calculation is based on Ewald's self-consistent field method [2] in which the local exciting field at each particle site, including multiple-scattering effects to all orders, is obtained exactly. The scattered field outside the cluster is then the sum of radiation emitted by all particles in the cluster; and the absorption spectrum is similarly obtained by summing the energy absorbed by each particle. Implicit in our approach is that the polarizability of each gold particle is not given by the expression for a sphere in vacuum. Rather, due to the proximity of each gold particle to its neighbors, the polarizability is calculated by taking into account all the near-field effects arising from higher-order-multipole interactions. The effect of this near-field interaction will be commented on later.

Calculation is performed on fractal clusters generated by an off-lattice cluster-cluster aggregation algorithm. Its structure closely resembles the gold colloidal aggregates obtained in the laboratory. Fig. 1 shows the log-log plot of the q-dependence of the polarized scattering intensity at $\lambda=488$ nm. The solid line is obtained by averaging over different clusters and over different incident directions for each cluster. In the small q region, the wave cannot resolve the detailed structure of the cluster and the curve is therefore flat. In the region $R_G^{-1} < q < a^{-1}$, where R_G is the cluster radius and a the particle radius, an excellent linear behavior is obtained as would have been expected for fractal clusters. A Born approximation calculation gives an almost identical curve (although the magnitude is somewhat shifted). The coincidence of the two sets of results indicates that the multiple-scattering effect is not important for gold aggregates with fractal dimension D=1.75.

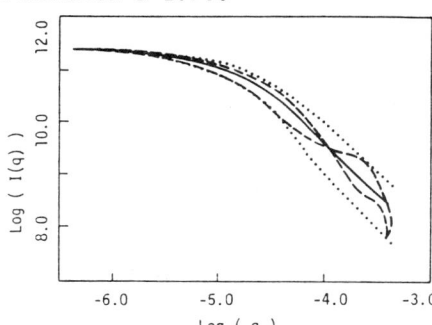

Fig. 1. Log-lot plots of the scattering intensity versus wave-vector q. Solid line is obtained by averaging over different clusters and over different directions of the same cluster; dashed lines are examples of the behavior of two individual clusters averaged over different directions; dotted lines indicate the standard deviation.

What is also shown in Fig. 1 is the scattering behavior of individual clusters averaged over different incident directions as denoted by dashed curves. It is clear that they are not fractal-like at all. The power-law behavior emerges only after averaging, and the standard deviation is indicated by the dotted curves.

While multiple scattering does not bring a measurable difference to the optical scattering, the absorption spectrum is greatly affected by the proximity of the gold particles to their neighbors (in spite of the low average density of the clusters). Since in an aggregate a particle always touches its neighbor(s), it could be asked whether the particle-particle interaction is resistive or capacitive in nature. We performed order-of-magnitude estimates of the RC time constant between the touching particles and found that, at optical frequencies, $\omega \gg 1/RC$ by at least an order of magnitude. Therefore, the near-field coupling is basically capacitive and has the same physics as that underlying the dielectric anomaly in the Maxwell-Garnett theory [3]. An exact calculation for a periodic chain of particles has been performed [4]. It is shown that the effective polarizability of a particle, renormalized by all the higher-order-multipole interaction with its neighbors, has a red-shifted resonance peak as compared with that of a Mie-resonance for a single particle in vacuum. Since the particles in fractal clusters basically form chain-like structures, the shift in the resonance peak of the polarizability is indeed manifested experimentally as seen in Fig. 2. Two absorption peaks are shown. The smaller one at left is centered at the Mie-resonance frequency and arises from (1) the single particles in the solution that are not part of the aggregated cluster and (2) particles situated at the peripheral of a cluster. The larger one at right is red-shifted and arises from the particles inside the clusters. As a verification for the origin of these peaks, experiments have been performed on an aggregate cluster whose distribution is much more biased towards the single-particle end. As expected, the relative weight of the two peaks is shifted as a result.

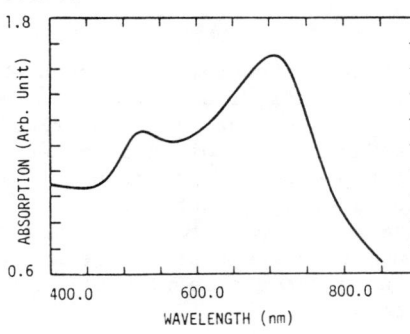

Fig. 2. Experimental absorption spectrum.

SUMMARY

To summarize, we have shown that (1) multiple-scattering will not bring any measurable difference to the Born-approximation description of scattering intensity; (2) the power-law behavior in scattering intensity is an average behavior; and (3) the polarizability of each gold particle is significantly renormalized by its near-field interaction with its neighbors, resulting in the red-shift of the absorption (resonance) peak. A more detailed presentation of these results will be published elsewhere.

REFERENCES

*Also at the Physics Department of CCNY.
1. See, for examples D. A. Weitz, J. S. Huang, M. Y. Lin, and J. Sung, Phys. Rev. Lett. 54, 1416 (1985); P. Mangin, B. Rodmacq, and A. Chamberod, Phys. Rev. Lett. 55, 2899 (1985).
2. M. Lax, Rev. Mod. Phys. 23, 287 (1951).
3. R. Landauer, in Electrical Transport and Optical Properties of Inhomogeneous Media (American Institute of Physics, N. Y., 1978), p. 2.
4. P. Sheng and Z. Chen, to be published.

DYNAMICS OF CLUSTER DISSOCIATION

R. G. Keesee and A. W. Castleman, Jr.
The Pennsylvania State University, University Park, PA 16802

ABSTRACT

The dynamics of dissociation of clusters induced by multiphoton ionization (MPI) is examined by time-of-flight mass spectrometry with the aid of a reflecting electric field. The systems discussed include ammonia, methanol, xenon, and p-xylene$(Ar)_n$ clusters. Ammonia and methanol clusters undergo rapid intracluster reactions to yield protonated clusters. Much of the excess energy which leads to dissociation in ammonia, methanol, and xenon clusters results from the energy differences in the ground states of the neutral and ionic systems. On the other hand, in the case of p-xylene$(Ar)_n$ the energetic differences are much smaller, so that the excess absorbed photon energy may be an important contribution.

INTRODUCTION

A subject of interest in chemical kinetics concerns the avenues and rates by which energy is transferred and distributed within a system following an excitation process. Clusters are particularly valuable in such studies. The high frequency intramolecular modes (within molecules of a cluster) versus the much lower frequency intermolecular modes between the molecules of a cluster offer a convenient system in which to assess the coupling of modes of widely different energies. Furthermore, the number of degrees of freedom in the system can be increased by increasing cluster size without changing the chemical units of the system.

Energy redistribution and dissociation processes are particularly relevant to nucleation phenomena in which energy transfer and redistribution of the heat of condensation occurs as the cluster attempts to grow. Clusters are also often used as models for investigating processes in liquids and solids.

Studies on neutral clusters commonly involve an ionization step for mass spectrometric detection. Consequently, an understanding of the effect of ionization in regard to the extent and time-scale of fragmentation is important in relating the cluster size distribution observed in mass spectra to the original neutral cluster size distribution. This paper presents results on the dynamics of cluster dissociation following multiphoton ionization as obtained with time-of-flight mass spectrometry.

EXPERIMENTAL

Although a relatively old technique in mass spectrometry, the time-of-flight mass spectrometer (TOFMS) is experiencing a revival in popularity. The advent of improved data acquisition such as transient digitizers and laboratory computers are partly responsible. Technological improvements, such as pulsed nozzles which can produce clusters and cold molecules with reduced pumping loads on vacuum systems, complement the pulsed nature of TOFMS. A major advance relevant to this conference is the introduction of pulsed lasers which have enhanced mass resolution by supplying efficient and short durations of ionization in a small volume. The great advantage of TOFMS over spectrometers based on mass filtering is that, in principle, every ion produced can be detected.

The apparatus employed in these studies has been described in detail elsewhere[1] and only a brief description is given here. Clusters are formed via adiabatic cooling of a gas exiting a pulsed nozzle. The clusters are intercepted by a laser beam and undergo multiphoton ionization (MPI) in an electrostatic field of the TOFMS. The resulting ions are accelerated, typically to 1 or 2 keV, into a field-free region. At the end of this region, another electric field (the reflectron) reflects the ions before they are finally detected whereupon a time-of-flight (or arrival) spectrum is obtained.

The reflectron is used in our laboratory to investigate dissociation in the field-free region, although this modification to TOFMS was originally developed to improve resolution.[2] The separation of nondissociating parent ions and dissociated daughter ions occurs as a result of the loss of kinetic energy (due to mass loss) with essentially no change in the velocity of the ion upon dissociation. Species with greater kinetic energy penetrate deeper into the reflectron and hence arrive later at the detector (see Figure 1). In addition to affording TOF separation, the reflectron may also be used as an energy analyzer. By setting U_K less than U_T, only daughter ions of kinetic energy less than U_T are reflected and other ions pass through the reflectron. Typically, ions enter the field-free region about a microsecond after ionization and reach the reflectron 10 to 100 μsec later, so that dissociation processes on the order of 10^4 to 10^6 sec^{-1} are observable. The dissociation rate as the cluster enters the field-free region can be determined by changing the extraction voltages in such a way as to vary the acceleration time (the time interval from ionization to entry into the field-free region) but maintain a constant birth potential, U_o. Faster dissociation processes in the neighborhood of 10^6 to 10^8 sec^{-1} are often discernible in conventional TOFMS (i.e., no reflectron) because of the effect of dissociation in the accelerating fields on the peak shapes.[3,4]

DISCUSSION AND RESULTS

Interaction with the laser beam leads to ionization of one of the molecules in the cluster with ejection of an electron. Subsequently, the ionized molecule may react with a neighboring molecule in the cluster and molecules in the cluster will rearrange into an orientation dictated by the presence of the charge (due to ion-dipole interactions, for instance). The relaxation processes may include

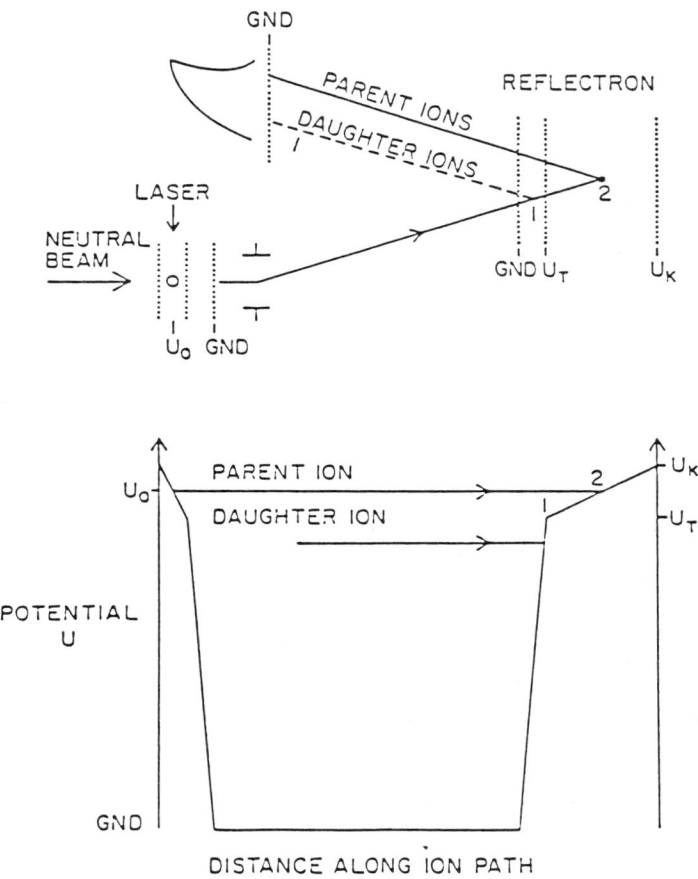

Fig. 1. (top) Schematic of the different trajectories of parent and daughter ions in the reflectron. Daughter ions are reflected at point 1 and parent ions, with higher kinetic energy, at point 2. (bottom) The electric field potential as seen by ions. The turning points depend on the final kinetic energy of the ion and the electric potential used.

dissociation as well as radiative relaxation. An excited cluster loses internal energy via the energy required to break a cluster bond along with any kinetic and internal energy that the departing molecule may carry. An example of these processes is shown below.

$$(NH_3)_m + nh\nu \rightarrow [NH_3^+(NH_3)_{m-1}]^* + e^- \qquad (1)$$

$$[NH_3^+(NH_3)_{m-1}]^* \rightarrow [NH_4^+(NH_3)_{m-2}]^* + NH_2 \qquad (2)$$

$$[NH_4^+(NH_3)_{m-2}]^* \rightarrow NH_4^+(NH_3)_{m-\ell-2} + \ell\, NH_3 \qquad (3)$$

The study of dissociation of ammonia clusters following MPI at 266 nm has been reported previously.[1] Briefly, it was found that the $NH_4^+(NH_3)_n$ clusters are formed rapidly (in less than 10 ns) since the fragmentation of reaction (2) was not observed in the experimentally available time-scales. However, the evaporative loss of NH_3 molecular units was apparent in the time-scale of the reflectron experiments. Up to six NH_3 molecules were found to dissociate from NH_4^+-$(NH_3)_8$ after entering the field-free region. Since fragmentation may occur as the result of either unimolecular (unicluster) dissociation of metastables or collision-induced dissociation, an investigation of dissociation as a function of background pressure in the field-free region is required to assess the relative contribution of these processes. Unicluster dissociation was shown to have a significant contribution for clusters larger than $NH_4^+(NH_3)_3$.

More recently, methanol clusters have been examined.[5,6] As with ammonia, methanol clusters upon MPI undergo a rapid internal (intracluster) proton transfer rection to produce $H^+(CH_3OH)_n$ followed by the slower evaporative processes. Up to five methanol units are lost from the protonated septamer and octamer. The dissociation rate coefficients for loss of the first methanol molecule, as determined by variation of the acceleration time, are shown in Figure 2. In contrast to the ammonia system, a slow intracluster reaction is also observed; specifically the elimination of H_2O from the protonated dimer. The gas-phase analog of this reaction is also known.[7] The H_2O elimination reaction between $CH_3OH_2^+$ and CH_3OH occurs on a time-scale which is accessed by the reflectron technique. On the other hand, elimination of H_2O from the larger cluster ions was not observed. This result is also in accord with the outcome of the gas-phase ion-molecule reaction between $CH_3OH_2^+ \cdot CH_3OH$ and CH_3OH in which only the three-body association reaction is seen.[7] One would expect the loss of H_2O to be at least as energetically favorable as the loss of CH_3OH. The indication is that a structural rearrangement may be necessary for H_2O loss and this rearrangement process may be hindered in the larger clusters. Some evidence that there may be a delay or induction time for the H_2O elimination process also supports the idea that a rearrangement is prerequisite.

The multiphoton ionization (by 266 nm photons) of xenon clusters is also being explored. The mass spectrum is virtually identical with that obtained by electron impact ionization.[3] Under single field acceleration, the peaks for Xe_2^+, Xe_3^+, and Xe_4^+ exhibit tails

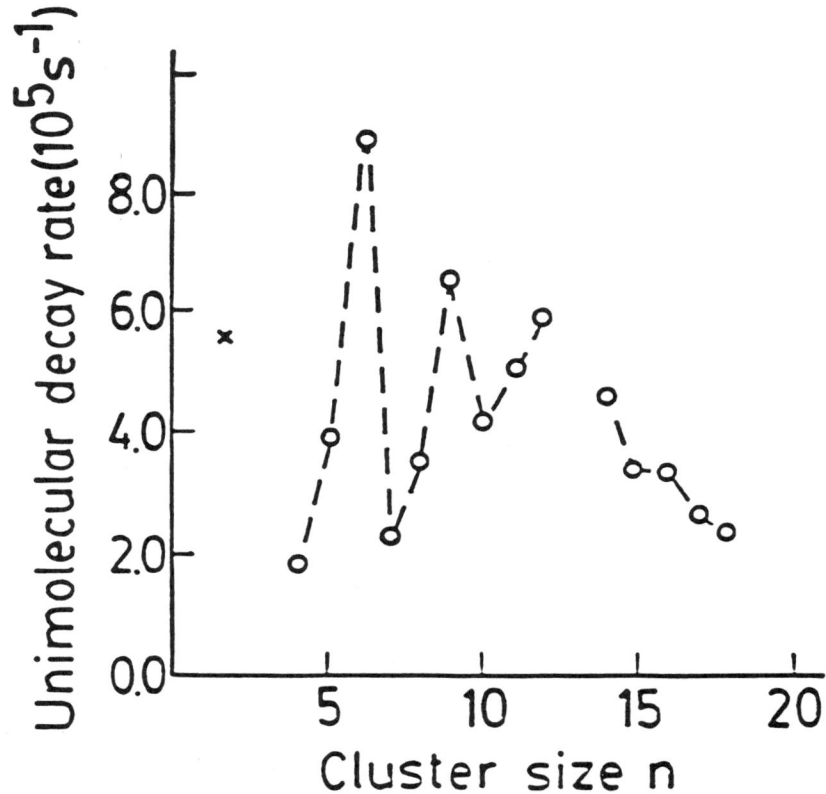

Fig. 2. Unimolecular (or unicluster) decay rate for loss of a methanol molecule (o) or elimination of water (x) from $H^+(CH_3OH)_n$.

toward longer flight times as the result of rapid dissociation (on the order of 5×10^7 sec^{-1}) of Xe_3^+, Xe_4^+, and Xe_5^+. No tails are apparent in the Xe^+ peak or in the peaks for clusters larger than Xe_4^+. The dissociation rates for clusters of ten or more atoms have been reported to be much slower.[9]

Finally, the implementation of resonant multiphoton ionization promises to expand one's ability to study dissociation processes. When a chromophore, such as p-xylene, is clustered or solvated by a species, such as argon, the frequencies of electronic transitions of the chromophore are shifted. The initial size of the neutral species is identifiable by the spectral shift, as shown in Figure 3.

Consequently, dissociation, even on very short time-scales, is evident by the appearance of these shifted peaks in the MPI spectrum of the product ion. Furthermore, with two-color methods in which one

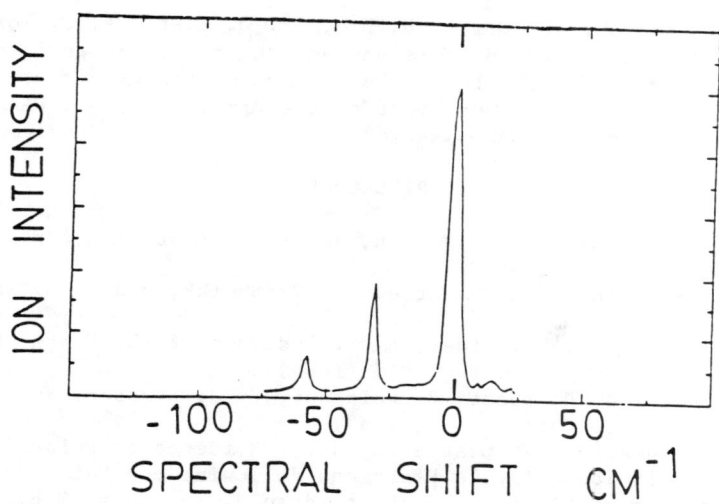

Fig. 3. The ion intensity of p-xylene$^+$ from the one-color multiphoton ionization of p-xylene(Ar)$_n$. The peak at zero spectral shift is from the resonance ($S_0 \rightarrow S_1$) enhanced MPI of bare p-xylene at 36740 cm^{-1}. The peaks to the left at -30 and -60 cm^{-1} are due to $S_0 \rightarrow S_1$ resonances of p-xylene in p-xylene(Ar) and p-xylene(Ar)$_2$, respectively, with subsequent ionization and fragmentation into p-xylene$^+$.

color excites the chromophore and the other supplies the additional energy required for ionization, one can probe the dissociation dynamics from the ionization threshold into the continuum. More dissociation is seen in the one-color MPI of p-xylene(Ar)$_n$ for which ionization occurs about 0.7 eV above the threshold than in two-color experiments where ionization is near threshold.[10]

Another interest in a system such as p-xylene(Ar)$_n$ is that the ionized and neutral systems are much more similar energetically and structurally than are those systems are for ammonia, methanol, and xenon clusters. The ionization threshold shifts by only about 120 cm^{-1} (15 meV) for each argon atom on p-xylene.[10] For the other systems discussed here, the ionized clusters are more strongly bound (by well over an eV) than the original neutral system. Thus dissociation of p-xylene $(Ar)_n$ should reflect the influence of the absorption of energy above threshold as opposed to the relaxation necessitated by large differences in the binding energy and equilibrium geometries of the neutral and ionic clusters.

ACKNOWLEDGMENTS

The authors thank Dr. P. D. Dao, Dr. O. Echt, Ms. S. Morgan, and Mr. M. Cook for helpful discussions and their part in performing the work. Financial support by the Department of Energy, Grant No. DE-AC02-ER-60055, and the Department of the Army, Grant No. DAAG-29-85-K-0215, is gratefully acknowledged.

REFERENCES

1. O. Echt, P. D. Dao, S. Morgan, and A. W. Castleman, Jr., J. Chem. Phys. 82, 4076 (1985).
2. V. A. Mamyrin, V. I. Karataev, D. V. Shmikk, and V. A. Zauglin, Sov. Phys. JETP 37, 45 (1973).
3. J. L. Durant, D. M. Rider, S. L. Anderson, F. D. Proch, and R. N. Zare, J. Chem. Phys. 80, 1817 (1984).
4. H. Kuhlewind, U. Boesl, R. Weinkauf, H. J. Neusser, and E. W. Schlag, Laser Chem. 3, 3 (1983).
5. S. Morgan and A. W. Castleman, Jr., "Evidence of Delayed 'Internal' Ion-Molecule Reactions Following the Multiphoton Ionization of Clusters," submitted to J. Am. Chem. Soc.
6. S. Morgan and A. W. Castleman, Jr., "Unimolecular Dissociation of Protonated Methanol Clusters," to be submitted.
7. L. M. Bass, R. D. Cates, M. F. Jarrold, N. J. Kirchner, and M. T. Bowers, J. Am. Chem. Soc. 105, 7024 (1983).
8. O. Echt, K. Sattler, and E. Recknagel, Phys. Rev. Lett. 47, 1121 (1981).
9. D. Kreisle, O. Echt, M. Knapp, and E. Recknagel, Phys. Rev. A 33, 768 (1986).
10. P. D. Dao, S. Morgan, and A. W. Castleman, Jr., Chem. Phys. Lett. 113, 219 (1985).

The Chemistry of Size-Selected Silicon Clusters as Studied by Fourier Transform Mass Spectrometry

William D. Reents, Jr.
Mary L. Mandich
Vladimir E. Bondybey

AT&T Bell Laboratories
Murray Hill, New Jersey 07974

INTRODUCTION

The manufacture of silicon devices relies heavily upon chemical reactions to both deposit onto and etch material from the device. In spite of the importance these reactions have, an understanding of the reaction mechanisms on the surface is largely unknown. Yet this is the area where many reactions, especially etching, occur. Silicon surfaces have dangling bonds which are the basis for reactivity. Depending upon the crystal orientation, these dangling bonds could consist of either radical electrons or silylene (paired) electrons. The reactivity of small organosilicon molecules containing these groups has been studied and each group has unique reaction products. The question arises as to whether the dangling bonds on a silicon surface react similarly to the small molecules with the corresponding type of dangling bonds. This question can be addressed by studying the reactions of small silicon clusters. We have initiated studies involving the reactions of ionic silicon clusters with specific reagents. These studies include growth of silicon clusters by reaction with CH_3SiH_3,[1] etching of silicon clusters by reaction with NO_2,[2] and deposition/fluorination by reaction with WF_6.[3] The clusters examined are either positively or negatively charged and contain from one to eight silicon atoms. Although these clusters are small and have geometries which differ significantly from bulk silicon, their reactions are applicable to specific reactive sites expected on a silicon surface and thus serve as a simple model for understanding the details of surface reactions. This paper will present a portion of our findings on cluster reactions with a discussion on relevance to reaction mechanisms for bulk silicon.

EXPERIMENTAL

Silicon cluster ions are created by laser evaporation of a stationary bulk silicon target located just outside the trapped ion cell of a modified Nicolet FT/MS-1000 Fourier transform mass spectrometer (FTMS).[4] The resulting trapped negative or positive ion populations consist of clusters containing up to seven or nine atoms, respectively.[1] The ion cell and method of laser vaporization are depicted in Figure 1. In this configuration, the cell consists of two sides which are separately pumped. The enormous advantage of the dual cell is that it prevents contamination of the target surface by the reagent gas. When the target and reactant gas are left in contact, the initial cluster population is altered: the larger clusters decrease in intensity and product-like functionalized clusters appear directly from laser vaporization.

© American Institute of Physics 1987

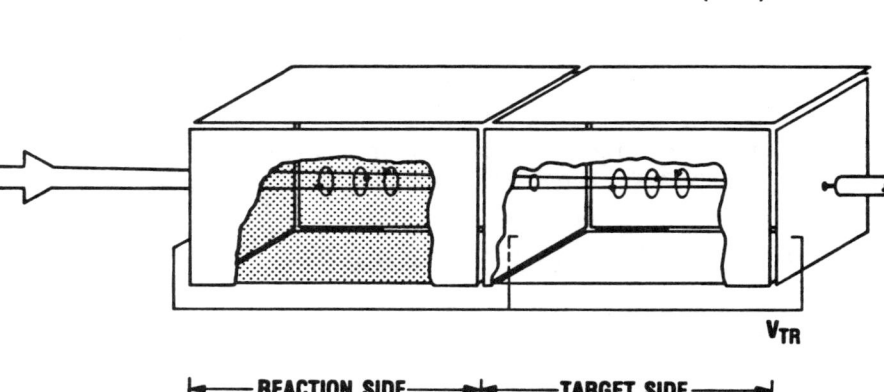

Figure 1. FTMS dual cell depicting sample target for laser evaporation

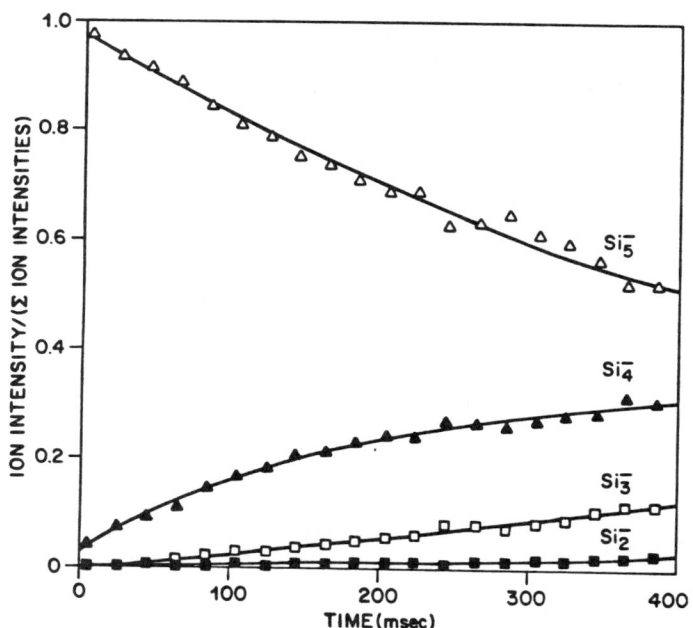

Figure 2. Time evolution of Si_{5-n}^- product formation from reaction of Si_5^- with NO_2.

The exothermic ion-molecule reactions of the clusters are studied by first isolating a single cluster size from the total ion population using standard double resonance techniques. [4] The bimolecular reaction rates and products of these selected clusters are then monitored by recording the mass spectra of the ions in the cell at a sequence of reaction times. The time dependencies of the normalized reactant and product ion intensities afford a means for determining reaction rates and for differentiating between primary versus secondary and higher order products (Figure 2). All reactant ion intensities are single exponential and scale linearly with the pressures ($\sim 10^{-6}$ - 10^{-7} torr) in these experiments. The reaction times are sufficiently long to allow for collisional and radiative relaxation of any excess translational and internal energy that may be present initially in the cluster ions.

RESULTS AND DISCUSSION

There are certain differences between the reactions of a silicon surface and the reactions of small silicon clusters. First, small ions such as the silicon clusters studied here do not have the large lattice which a bulk material has to absorb the energy released in a reaction. As a result most ion/molecule reactions rely on releasing a neutral molecule in order to remove this excess internal energy. Second, because of the charge present on these ionic clusters, the attraction of a neutral to the charged cluster enhances the reaction rate over that of conventional neutral-neutral reactions. Third, surface reactions are typically examined at pressures of 10^{-5} to 1 torr. In these cluster studies, the low pressure of the reactant gas (10^{-7} to 10^{-6} torr) combined with the short reaction time (0.01 to 5 seconds) permits our examination of individual collisions of the cluster with reactant molecules and allows us to discern single collision reactions from multiple collision reactions. Fourth, the nature of the cluster surface, i.e. the presence of contaminants is readily observed. Since we are dealing with clusters containing several atoms, the presence of any atom, including hydrogen, would dramatically increase the mass of the cluster. Therefore, it is easy to unambiguously observe reactions for the bare cluster ions. Surface cleaning, e.g. by ion bombardment with possible destruction of the surface structure, is unnecessary.

Theoretical calculations [5] show that the small silicon clusters contain both radical electrons and silylene (lone pair) electrons. These structures show a trend important to our discussion: the smaller clusters exhibit significant silylene character whereas the larger clusters (> 3 atoms) exhibit increasingly greater radical character. The implication is that silylene-based reactions will occur predominantly for the smaller clusters. Since the charged clusters have a radical electron at the site of the charge, it is expected that radical-based reactions will occur for clusters of all sizes.

The first indication of reaction mechanism will therefore come from examining the reaction rate with a specific reactant as a function of cluster size. Figure 3 plots the log of the reaction rates vs. cluster size for the reactants studied thus far. Those reactants which show only slight variation of reaction rate with the clusters as a function of cluster size, e.g. NO_2, would be predicted to react via a radical-based mechanism. CH_3SiH_3 shows a significant decrease in reaction rate for Si_6^+ and Si_7^+ (no observed

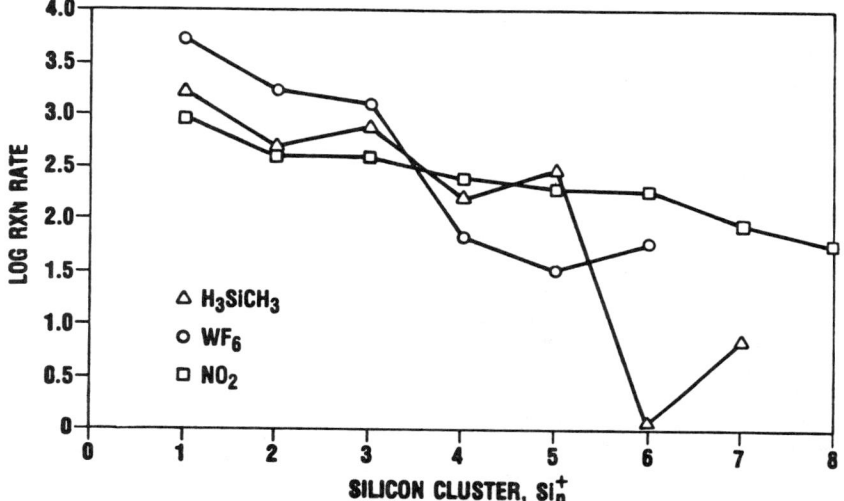

Figure 3. Si_n^+ reaction rate constants.

reaction for them) relative to the smaller clusters and therefore is predicted to react via a silylene-based mechanism. WF_6 is an intermediate case wherein the reaction rate drops for Si_{4-6}^+ but is nevertheless measurable. This is accounted for by the blockage of the silylene mechanism for clusters with more than 3 atoms; the reactivity of these larger clusters is totally due to a radical mechanism. These observations are in agreement with known reactions of small organosilicon compounds. Silylene insertion into Si-H bonds, e.g. in CH_3SiH_3, occurs readily. [6] Silicon radicals abstract halogens or react with radicals. [7] This is expected for halogen-containing compounds, e.g. WF_6, and for radical compounds, e.g. NO_2, as well.

The reaction products are consistent with the reaction mechanisms proposed above. Methylsilane reacts with Si_{1-4}^+ predominantly by loss of dihydrogen whereas Si_5^+ reacts by loss of a hydrogen or methyl radical. The anionic silicon clusters do not react with CH_3SiH_3.

$$Si_{1-4}^+ + CH_3SiH_3 \rightarrow Si_{2-5}CH_4^+ + H_2 \qquad (1)$$
$$Si_{1-4}^+ + CH_3SiH_3 \rightarrow Si_{2-5}CH_2^+ + 2H_2 \qquad (2)$$
$$Si_5^+ + CH_3SiH_3 \rightarrow Si_6CH_5^+ + H \qquad (3)$$
$$Si_5^+ + CH_3SiH_3 \rightarrow Si_6H_3^+ + CH_3 \qquad (4)$$

Loss of H_2 in the CH_3SiH_3 reaction occurs readily after silylene insertion. The higher energy products (radicals) produced in reactions 3 and 4 must result from a more exothermic reaction. This contrasts with the smaller clusters which have greater reaction rates but are less exothermic.

Table I. Reaction of cationic silicon clusters with WF_6.

	Reaction	Product Fraction (%)
$Si^+ + WF_6$	$\rightarrow SiF^+ + WF_5$	14
	$\rightarrow WF_3^+ + SiF_3$	1
	$\rightarrow WF_4^+ + SiF_2$	67
	$\rightarrow WF_5^+ + SiF$	18
$Si_2^+ + WF_6$	$\rightarrow SiF^+ + WF_3 + SiF_2$	7
	$\rightarrow Si_2F^+ + WF_5$	8
	$\rightarrow Si_2F_2^+ + WF_4$	8
	$\rightarrow WF_2^+ + 2SiF_2$	27
	$\rightarrow WF_3^+ + SiF_2 + SiF$	15
	$\rightarrow WF_4^+ + Si_2F_2$	22
	$\rightarrow WF_5^+ + Si_2F$	13
$Si_3^+ + WF_6$	$\rightarrow Si_3F^+ + WF_5$	7
	$\rightarrow Si_3F_2^+ + WF_4$	4
	$\rightarrow WF_2^+ + SiF_2 + Si_2F_2$	2
	$\rightarrow WF_3^+ + SiF_2 + Si_2F$	5
	$\rightarrow WF_4^+ + Si_3F_2$	17
	$\rightarrow WSiF_2^+ + 2SiF_2$	55
	$\rightarrow WSiF_3^+ + SiF_2 + SiF$	5
	$\rightarrow WSiF_4^+ + Si_2F_2$	5
$Si_4^+ + WF_6$	$\rightarrow Si_4F^+ + WF_5$	100
$Si_5^+ + WF_6$	$\rightarrow Si_5F^+ + WF_5$	100
$Si_6^+ + WF_6$	$\rightarrow Si_6F^+ + WF_5$	100

Abstraction of a fluorine from WF_6 occurs for the larger clusters as would be expected for a radical mechanism. These details are shown in Table I for the cationic clusters. Similar reactions occur for the anionic clusters.[3] Atom abstraction also occurs for NO_2, producing a neutral SiO.

$$\text{Si}_n^{+/-} + \text{NO}_2 \rightarrow \text{Si}_{n-1}^{+/-} + \text{SiO} + \text{NO} \qquad (5)$$

This process is analogous to etching bulk silicon without leaving a residue. Loss of SiO is the mechanism used by the ion to remove the excess energy.

How do these reaction mechanisms pertain to the silicon surface? The mechanism is tied to the reactive sites available. The reactive sites for the clusters were either radical electrons or lone pair electrons. A silicon 111 surface has radical electrons whereas silicon 100 has a silylene surface. Recent work has indicated that reconstruction of the silicon 100 surface occurs to form a radical surface. The radical reactions appear most useful in relating to silicon surfaces since the silylene mechanism is unimportant for larger clusters and the fact that the presence of silylene groups on bulk silicon surfaces is uncertain. Defect sites could, however, be either radical or silylene in nature with reactivities similar to those of small clusters.

The simplest radical type reaction that we have observed for silicon clusters has been etching by NO_2 which yields essentially one set of products. The reactions of NO_2 with silicon surfaces, however, are not well characterized. Based on the cluster reactions, NO_2 would be expected to transfer a single oxygen atom to bulk silicon to form an oxidized surface and NO via a radical mechanism. Formation of SiO in the cluster reactions was due to the need for the clusters to remove excess energy. Due to their small size and to the 0-1 eV exothermicity of the reaction, they effectively have a very high internal temperature during the reaction. Thus, this production of SiO may be analogous to the formation of SiO as a dominant species during high temperature fabrication of silica.

WF_6 is used to deposit tungsten on silicon surfaces. The ultimate products from WF_6 processing are W(s) deposited as a surface film accompanied by evolution of SiF_4(g). No mechanism has been proposed for the bulk process but, based upon these studies, a radical mechanism involving the initial transfer of a single fluorine atom from WF_6 (forming an intermediate WF_5) to the silicon surface is plausible. The resulting WF_5 probably remains bound to the surface which allows for further reaction with the silicon surface, perhaps by multiple single fluorine atom transfers via radical mechanisms.

These cluster atoms contain two to three silicon-silicon bonds, making them far more unsaturated than normal surface silicon atoms which contain three silicon-silicon bonds typically. As a result, they appear to be far more reactive, especially for silylene-based mechanisms which should be nonexistent for silicon 111. However, this has serious implications for the reactivity of damaged surfaces, e.g. by ion bombardment. Ion bombardment is known to enhance the reactivity of a silicon surface, presumably by exposing fresh surface to the reactant gas. We propose that ion bombardment may also result in the formation of less saturated silicon atoms such as we describe here. Not only are they more reactive, but also new reaction mechanisms can occur due to the formation of silylene groups on the less saturated silicon atoms.

CONCLUSIONS

Bare silicon clusters exhibit reaction rates and product distributions which are ascribed to two different reaction sites: a silylene (lone electron pair) site and a radical electron site. Silylene chemistry is observed for Si_{1-5}^+ with CH_3SiH_3 and for small ionic silicon clusters with WF_6. Radical chemistry is observed for all ionic silicon clusters with WF_6 and NO_2. Extrapolating to bulk silicon surface, radical chemistry should dominate. Ion bombardment of the surface could create highly unsaturated silicon sites which exhibit enhanced reactivity similar to these clusters.

REFERENCES

1. M. L. Mandich, W. D. Reents, Jr., and V. E. Bondybey, J. Phys. Chem. , 90 , 2315 (1986).
2. M. L. Mandich, V. E, Bondybey, and W. D. Reents, Jr., J. Chem. Phys. , in press.
3. W. D. Reents, Jr., M. L. Mandich, and V. E. Bondybey, Chem. Phys. Lett. , 131 , 1(1986).
4. a) J. L. Beauchamp, Ann. Rev. Phys. Chem. , 22 , 527(1971). b) A. G. Marshall, Acc. Chem. Res. , 18 , 316(1985).
5. K. Raghavachari and V. Logovinsky, Phys. Rev. Lett. , 55 , 2853(1985).
6. Y. Tang, in Reactive Intermediates, Vol. 2 , R. A. Abramovitch, ed., (Plenum Press, N. Y., 1982).
7. J. Wilt, in Reactive Intermediates, Vol. 3 , R. A. Abramovitch, ed., (Plenum Press, N. Y., 1983).

PHOTODETACHMENT AND PHOTODISSOCIATION STUDIES OF SEMICONDUCTOR CLUSTER IONS

Y. Liu, Q. Zhang, S. C. O'Brien, J. R. Heath,
R. F. Curl, F. K. Tittel, and R. E. Smalley
Rice University, Houston, TX 77251

A considerable advance in the study of clusters has been possible due to the development of techniques for producing both cold ionic and neutral cluster beams. Si, Ge and GaAs negative clusters ions have been produced recently in direct laser vaporization[1] and cooled by a subsequent free supersonic expansion. Ions of a particular mass are selected and studied by laser photodetachment and photodissociation followed by time-of-flight mass analysis. Figure 1 depicts schematically the apparatus used in these studies.

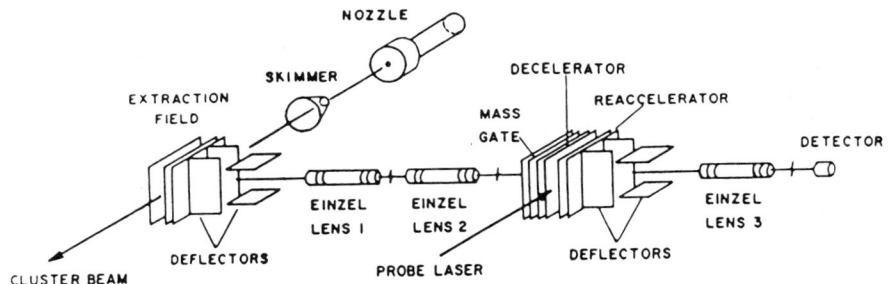

Fig. 1. Schematic of the experimental arrangement. A 532 nm laser (30 mj/pulse, 5 ns pulse width) is used to evaporate a semiconductor disc placed at a supersonic nozzle throat.

Electron affinities (EA) of the clusters are explored by measuring the dependence of the detached electron intensity upon the probing laser fluence. Using several discrete laser wavelengths, we have been able to roughly determine the EA's of various GaAs clusters up to 30 atoms in size. The evolution of the EA as a function of the cluster size for GaAs is shown in Fig. 2. In these measurements the clusters studied are those with approximately an equal number of Ga and As atoms. The EA's of Ga rich or As rich clusters have also been studied. As expected for Ga_xAs_y with x+y constant, the EA increases with increasing ratio of y to x. It is clear in Fig. 2 that GaAs clusters with an even number of atoms have lower EA's than their neighboring odd ones. A similar even/odd alternation in the ionization potentials (IP) of

GaAs clusters has been observed previously[2] where even clusters have higher IP's than the odd ones. This suggests that clusters with an even total number of atoms have fully paired singlet ground states with no dangling bonds on the surface while the odd clusters have the unpaired electron in a non-bonding frontier orbital.

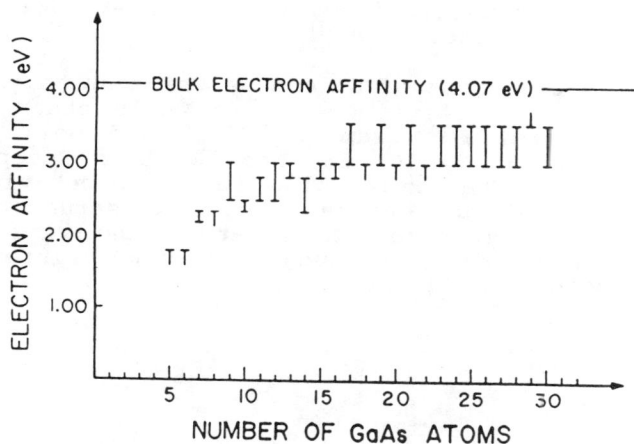

Fig. 2. Estimates of the electron affinity of clusters with nearly stoichiometric GaAs.

Photodissociation processes are found to be quite competitive with photodetachment for GaAs negative cluster ions. Both electrons and fragmentation products are observed with the electron signals always much more intense. For Si and Ge anions it is found that the relative importance of detachment and fragmentation depends strongly upon the laser wavelength. Generally speaking, the bonding energies of Si_x^- and Ge_x^- are lower than the electron affinities. One-photon detachment and one-photon dissociation can be observed simultaneously when the photon energy is just above the detachment threshold of the clusters with the fragmentation signals dominating the electron signals. As the photon energy is increased, electron signals become larger than those of the fragmentation daughters. Presumably, if the photon energy is much larger than both D_e and EA, electron detachment will be totally dominant.

The fragmentation channels of Si and Ge negative cluster ions observed are nearly the same (although the relative intensities of the daughters are not equal) between Si and Ge, implying that Si and Ge clusters have almost identical structures. Both Si and Ge anions fragment by fissioning into only a few channels, mainly in the 5 to 10 atom size range with the 6-atom and, especially, the 10-atom negative ions being the favored daughters. In contrast, GaAs anions are found to fragment like metals by a non-fissioning process: they can lose one, two, or more atoms, indicating a fundamental difference in chemical bonding between GaAs clusters and Si or Ge clusters. Fig. 3 shows the fragmentation patterns of Si_{20}^- and Ge_{20}^-. No dependence of the fragmentation patterns upon the photon energy is observed in the range of our study. For example, Si_{20}^- only produces Si_{10}^- when irradiated with 2.48 eV, 2.81 eV, 2.99 eV and 3.51 eV photons.

Some negative ions of Si and Ge fragment only into ions of 10 atoms, e.g. Si_{17}^- and Si_{20}^-, while other ions have several daughters. For example, Si_{18}^- and Ge_{18}^- produce daughters of 5, 6, 9, 10, 11, and 12 atoms independent of laser frequency. The explanation for this behavior may be that there are large changes in the structure of the cluster with cluster size. The 10-atom negative ion appears to be a favorite daughter suggesting the existence of a special structure for 10 atoms.

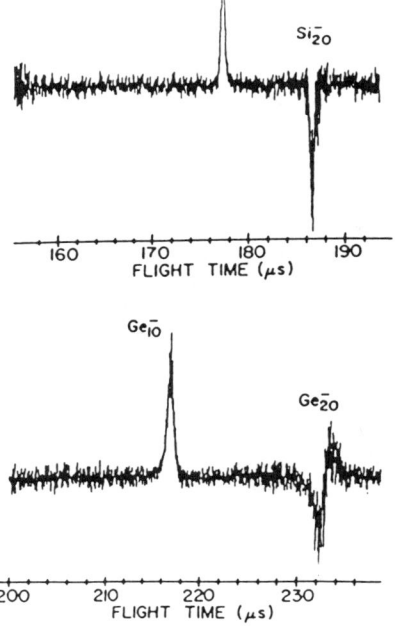

Fig. 3. Fragmentation patterns of Si_{20}^- and Ge_{20}^-. The horizontal axis is the flight time from the extraction region to the detector. The vertical axis is the difference between the ion signals with the probing laser on and off.

1. Y. Liu, Q-L. Zhang, F. K. Tittel, R. F. Curl, and R. E. Smalley, to be published in J. Chem. Phys.
2. S. C. O'Brien, Y. Liu, Q. Zhang, J. R. Heath, F. K. Tittel, R. F. Curl, and R. E. Smalley, J. Chem. Phys. 84, 4074(1986).

STUDIES OF SURFACE DYNAMICS USING SECOND-HARMONIC GENERATION

Harry W. K. Tom
AT&T Bell Laboratories, Holmdel, New Jersey 07733

ABSTRACT

Recently, optical second-harmonic generation (SHG) has been shown to be a sensitive and versatile probe of adsorbate coverage, composition, binding site, and molecular orientation on surfaces. In addition SHG has been shown to reveal the symmetry of atomic order on surfaces. It is thus able to monitor symmetry-changing surface phase transitions. Because of the great sensitivity and the time-resolution obtainable with pump-probe techniques, SHG is a unique tool for monitoring surface dynamics. Results from SHG measurements with time resolutions of 100 msec and 100 fsec are presented.

TIME-RESOLVED SURFACE STUDIES

One of the ultimate goals of surface science is to be able to understand and predict reaction rates. The standard approximation is to assume that reactions occur in equilibrium and that a detailed balance approach is valid. One then measures reaction rate constant, k, vs. temperature, T, to arrive at a relation of the form $k = \nu_0 \exp(E_A/k_B T)$, where the preexponential factor ν_0 is related to the change in entropy, E_A is the activation energy of the reaction, and k_B is the Boltzmann constant. For such a relation to have predictive value, the conditions under which equilibrium holds must be verified. Non-equilibrium effects become important when the reaction rates are limited by the rate at which states of a single degree of freedom reach statistical equilibrium (for example the relaxation of an excited vibrational state takes less than 100 psec, CAVANAGH, these proceedings) or the rate at which several degrees of freedom reach equilibrium (for example the exchange of kinetic to vibrational and rotational energy during molecular adsorption). Thus two kinds of measurements appear necessary. First one must measure reaction rates over many orders of magnitude of rate to determine when non-equilibrium effects become important. Second, one must probe non-equilibrium effects directly in the time-domain at subnanosecond time-scales. A further motivation for time-resolved surface studies is to identify the primary events involved in surface reactions, such as adatom diffusion and reaction intermediates. Again subnanosecond time-resolution is desired.

For the most part conventional surface probes are restricted to time-scales greater than a second. Several techniques have been developed recently to access these faster time scales. Time-resolved electron energy loss has been able to probe the millisecond time-scale.[1] Molecular beam experiments can limit the time of particle-surface interaction to the particle's residence time,[2] which may be on the order of 10^{-9} or less seconds. However, direct measurement of the residence time is a formidable problem even if the incident particles are prepared with a short laser pulse due to the spread in particle velocities after the scattering event.

Surface techniques that incorporate lasers to initiate a reaction or surface response can obtain time-resolution under 10^{-9} sec. By using a laser pulse to heat the sample, temperature ramps in excess of 10^5 K/sec can be obtained and thermal desorption spectroscopy and reaction kinetics have been studied in this way.[3] In such experiments however, the time or temperature at which the particles desorb or react cannot be measured directly. Low-Energy Electron Diffraction following an intense laser pulse that initiates structural change has been reported with 10 nsec time-resolution.[4] Photoemission induced by a laser pulse following excitation of the electronic surface states by another laser pulse

© American Institute of Physics 1987

has been reported with 70 psec resolution.[5] Better time-resolution is possible with these pulsed electron spectroscopy techniques, however, there is a tradeoff between time-resolution and signal due to space charge effects.

Second-harmonic generation (SHG) is an *insitu* probe so molecular scattering or desorption events can be time-resolved directly. In addition, intermediate states that may not be stable or distinguishable in the gas phase may be probed with SHG. SHG is also an all optical technique. It is therefore not as drastically effected by space charge limitations as are electron spectroscopies. By using high repetition rate pulse lasers like the 8KHz Cu vapor laser, 0.12 msec resolution is easily obtainable for kinetic measurements. By using pump-probe techniques, the ultimate time resolution is limited only by the time-duration of a single pulse. With time-resolution as low as a few fsec, SHG makes possible direct measurement of non- equilibrium effects of surface reactions.

SECOND-HARMONIC GENERATION: EXPERIMENTAL CONSIDERATIONS

In SHG experiments, a laser pulse at the excitation frequency induces a polarization response at twice that frequency in the surface and near-surface bulk. This polarization radiates at the SH frequency and is detected in reflection with gated photon-detecting electronics. From clean metal or semiconductor surfaces, several 100's of SH photons are generated for an input pulse of 10 mJ focused to 1 cm^2 with 10 nsec duration. These photons are detected with typically 5 to 8% efficiency. The SH yield scales as the square of the input fluence and inversely with the area and pulse time duration.

The extreme sensitivity to surface (vs. bulk) electronic states obtained in SHG comes from the fact that SHG is dipole-forbidden in the bulk of centrosymmetric media but always dipole-allowed at the surface where centrosymmetry is broken normal to the surface. In addition, certain surface states may be intrinsically more highly polarizable than corresponding bulk states. Finally, the large gradient in the electric field at the surface may make the radiation from even surface quadrupole-allowed SH as large as that from bulk quadrupole-allowed SH.[6]

The polarization for the surface and bulk regions may be written as:[6-8]

$$P_i^s(2\omega) = X_{ijk}^s(2\omega)E_j(\omega)E_k(\omega) \qquad (1)$$

$$P_i^b(2\omega) = X_{ijkl}^b(2\omega)E_j(\omega)\nabla_k E_l(\omega) \qquad (2)$$

where the superscripts s and b denote surface and bulk, subscripts i,j,k and l denote vector components, the sum over repeated subindices is implied, $\hat{E}(\omega)$ is the fundamental field amplitude and $\tilde{X}(2\omega)$ is the SH-nonlinear susceptibility. All the material specific information is in the tensor elements of $\tilde{X}^s(2\omega)$ and $\tilde{X}^b(2\omega)$ tensors. The second-harmonic field $\hat{E}(2\omega)$ is calculated from vector $\vec{P}(2\omega)$ by Maxwell's Equations. For a given excitation geometry (angle of incidence and field polarizations) the SH signal is proportional to the square modulus of some linear combination of the elements of $\tilde{X}(2\omega)$ which we will denote as $|X_{eff}|^2$ in this text. We note here that the surface dipole and quadrupole contribution to SH are incorporated into $\tilde{X}^s$ in (1).

SH STUDIES OF ADSORPTION KINETICS

The first figure demonstrates the sensitivity and time-resolution available with SHG in a typical kinetic rate measurement of the adsorption of O_2 on Rh(111).[9] Starting with an atomically clean and well-ordered Rh(111) surface at 315K, the SHG signal was monitored as the UHV chamber was backfilled with 5×10^{-8} torr O_2. The signal, shown in Fig. 1, decreases smoothly to a saturation value of 0.12 times the bare metal value at about 1.8 Langmuirs (1 Langmuir= 10^{-6} torr-sec corresponds to 20 seconds). A sharp 2X2 LEED pattern was not obtained until about 20 L. The time resolution is limited here by the pulse repetition rate of the laser which was 0.1 sec.

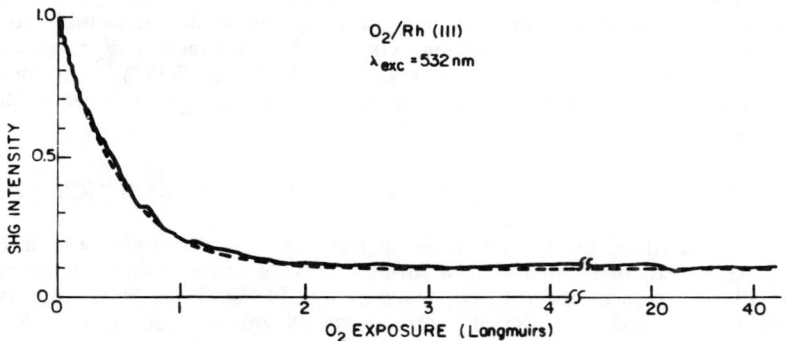

Fig. 1. Adsorption of O_2 on Rh(111). SH intensity normalized to signal for clean Rh(111). Data (solid line) obtained as a continuous function of O_2 exposure (1 L in 20 sec). Theory (dashed line).

One sees immediately that SHG is extremely sensitive to the oxygen overlayer. The dashed line shows a fit to the data that is consistent with a model of Langmuir adsorption kinetics (sticking coefficient is coverage independent). The kinetics were previously established with Auger by YATES, et al.[10] Our model is that the change in surface nonlinear response per adsorbate is constant with coverage (consistent with all empty sites being equivalent) and that therefore the nonlinear susceptibility of the surface may be given by the expression:

$$X_{eff} = A + BD/D_s ,$$

where D and D_s are the fractional and saturation surface coverage with respect to surface Rh atoms, and A and B are coefficients associated with the nonlinearity of the bare metal and the induced change in surface nonlinearity per adsorbate.

The coverage as a function of exposure is just given by Langmuir adsorption kinetics in the limit of negligible desorption: $D(t)/D_s = 1-\exp(-Kpt)$ where K is a constant accounting for the sticking coefficient and p is the oxygen pressure. The SH signal is proportional to $|X_{eff}|^2$. The fit gives $B/A = 1.03 \exp(i160°)$ and $K/D_s = 0.93$/layer. The latter is in close agreement with the value 0.78/layer obtained by YATES, et al.,[10] considering possible differences in pressure calibration.

SHG from Rh(111) shows similar high sensitivity to other adsorbates, among them CO, benzene and pyridine and their hydrocarbon fragments upon dehydrogenation,[11] and the alkali metals.[9,12] Similar results have been obtained on Cu(111),[13] Ni(111),[13,14] Pt(111),[14] Ag(110),[15] and Si(111)-7X7.[16,17,18] SHG then promises to be generally applicable to metal and semiconductor-adsorbate systems. Readers should also be aware of the great sensitivity of SH to adsorbates on surfaces in electrochemical environments.[19,20]

In general, adsorbates that reduce the free-electron-like polarizability of the surface by either shifting the Fermi level or directly eliminating surface states, will reduce the SHG. Adsorbates such as the alkali that increase the free electron response of the surface by shifting the Fermi level or by forming a metallic overlayer, increase the surface SHG. In some cases, interband transitions can dominate the SH response and significantly modify this simple free electron picture.

For O_2/Rh(111), the sticking coefficient was independent of coverage and the SH took a particularly convenient form. By calibrating the SH signal to coverage by an independent technique such as thermal desorption yield or Auger it would be possible to extract coverage as a function of time or sticking coefficient as a function of coverage. Such studies have been performed for CO on Rh(111),[9] Cu(111) and Ni(111),[13] and for CO, O_2 and H_2 on Pt(111) and Ni(111),[14] and departure from Langmuir kinetics has been distinguished.

SH STUDIES OF DESORPTION AND REACTION KINETICS

While the details of the SHG function of time during a reaction may be difficult to interpret without extensive calibration of surface coverage or composition at intermediate points, relative reaction rates can be compared easily. In Fig. 2, the SHG signal is shown during the thermal decomposition of a thin thermally grown oxide layer on Si(111) at 900° C.[16] The time origin in Fig. 2 begins immediately after dosing the 900° C sample with 40 L of O_2 at 10^{-6} torr Immediately the SHG signal begins to increase toward the clean Si(111) level, indicating the oxide is decomposing. After approximately 2 minutes, Auger showed that the oxygen coverage was equivalent to that of a saturated room temperature chemisorbed layer. The fit to the data shown as the dashed curve is obtained with the same model of the surface SH susceptibility described in (3), except the fit was made with

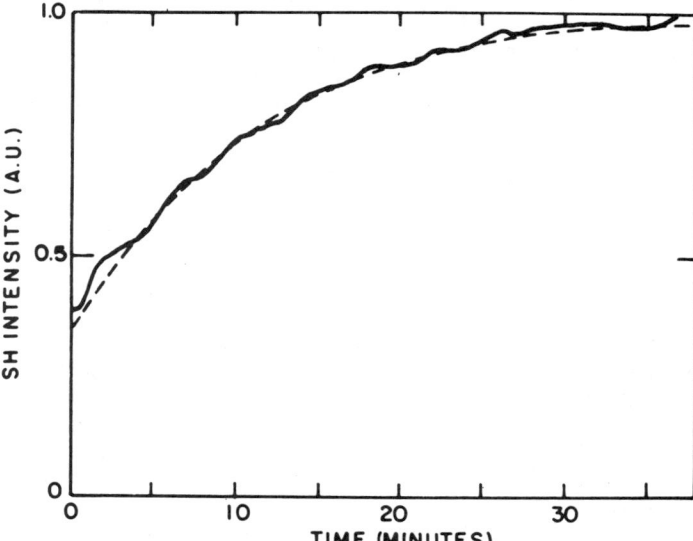

Fig. 2. Isothermal decomposition of oxide on Si(111) surface formed at 900°C by exposure to 40 Langmuirs of O_2. Time axis begins immediately after O_2 exposure. SH data (solid line) normalized to signal for clean Si(111) at 900° C. Theory (dashed line).

$B/A = -0.412$ and $D(t)/D_s = \exp(-kt)$ where $k = 1.7 \times 10^{-3}/s$. The same experiment was performed at 1100°C and the shape of the SH data was the same except it was fit with a faster rate constant, $k = 5.3 \times 10^{-2}/s$. If we assume there is only a single activation energy such that $k = \nu_0 \exp(-E_D/k_BT)$, E_D, the energy of desorption, is 2.4 eV and $\nu_0 = 3.1 \times 10^7/s$. Our value of E_D may be compared to the value of 3.1eV obtained for desorption from Si(100) between 700 and 790°C.[21] A similar SHG study of the effect of coadsorbed sulfur on methoxy decomposition on Ni(111) has been performed.[22]

It is easy to imagine extending this kind of measurement to much shorter time scales by using a laser pulse to rapidly heat the sample and then using a time-delayed probe pulse to time-resolve the evolution of the reaction. Recently, the laser-induced desorption of Rhodamine 610 dye molecules from a fused silica substrate was studied with a time-delayed SH probe with 6 psec time-resolution.[23] For a pump fluence of 0.5 J/cm^2, the change in the SH probe was so rapid that it could not be resolved in 6 ps. The results indicate that either the molecules photofragmented or that they were desorbed in a few psec with a thermal velocity distribution in excess of 5000K.

SH STUDIES OF PHASE TRANSITIONS

SHG is also highly sensitive to surface order as manifested in the symmetry of the surface electronic susceptibility. As shown in Fig. 3, the SH intensity for a given excitation geometry depends on the orientation of the sample as it is rotated about the surface

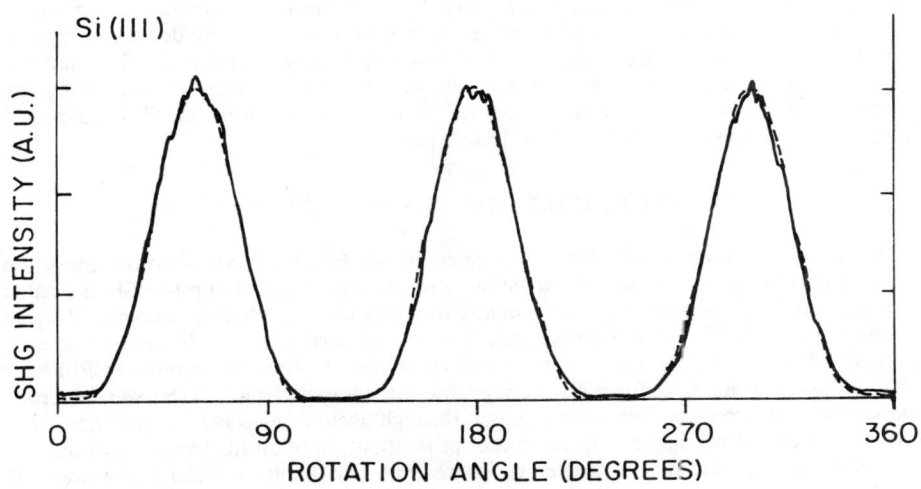

Fig. 3. SH from Si(111) in arbitrary units as a function of sample rotation angle about its normal. The sample was cleaned and atomically-ordered in UHV before exposing to air. The rotation angle is measured between the surface mirror plane and the optical plane of incidence. Data (points). Theory (solid line).

normal.[7] In the case shown, the SH from a Si(111) surface shows the 3m symmetry of the surface. For 3m symmetric surfaces, the SH susceptibility is generally given by:

$$X_{eff} = L + Mf(3R) \qquad (4)$$

where R is the rotation angle and coefficients L and M are the angle-independent and angle-dependent contributions to SH, respectively. It can be shown that for p̂-polarized input and ŝ-polarized SH, the angular function f(3R) is simply sin(3R) and the angle-independent coefficient L is zero. It is important to realize that the value of M may be used as a monitor of surface order because M is only non-zero if the surface and near-surface bulk have 3m symmetry. On the other hand for p̂-polarized input and p̂-polarized SH, L is not zero and f(3R) is cos(3R). The data are fit in Fig. 3 with L = −M.

Rotational anisotropy is predicted from the form of the tensors $\tilde{X}^s(2\omega)$ and $\tilde{X}^b(2\omega)$ in (1) and (2), for all surface symmetries 4m and lower.[7] The dependence appropriate for the 4m symmetry (Si(100))[7] and 2m symmetry (Si(111)-2X1)[24] have also been observed. Rotational anisotropy is also observed on metal surfaces, specifically the Cu(111).[25]

Phase-transitions that change symmetry may then be studied with SHG. The transformation of a Si(111)-2X1 to Si(111)-7X7 (2m to 3m symmetry change) as the sample temperature was ramped at 3K/s was studied with SHG.[24] The transition occurs rapidly in 10's of seconds around 285° C. Those authors also studied ion-beam-induced disordering,[17] and annealing of an amorphous Si overlayer on Si(111)-7X7,[26] all as continuous functions of time on the scale of seconds which is suitable for many processing applications.

The time-evolution of surface order of a Si(111) surface irradiated with a laser pulse intense enough to induce melting was recently measured with 100 femtosecond time-resolution.[27] The rotational anisotropy of the SH signal was measured as a function of delay time with respect to the intense melting pulse. The rotational anisotropy decreased rapidly within the first 150 fs but had a slow decay component lasting 1 to 3 psec. The authors interpreted the SH signal decrease as consistent with the transformation from crystalline solid to a disordered liquid. Similar studies[28,29] of optical melting of GaAs surfaces indicated that disorder was induced in less than 2 psec.

SH STUDY OF OPTICALLY-INDUCED DISORDER OF Si(111)

In Fig. 4 we present results from a more recent study of the order-disorder transition of Si(111) (held in air) after optical excitation. In this experiment, the probe SH is excited with p-polarized light and the sample is rotated to a position such that the ŝ-polarized signal is proportional to $|M|^2$ and the p̂-polarized signal is proportional to $|L|^2$ where L and M were defined in (4). Pulses were 100 fsec in duration derived from an amplified CPM laser at 610 nm and had an Airy function beam profile after a spatial filter. The weaker probe pulse was split off from the laser beam, passed through a variable delay line, and focused to a spot size of 25 micron diameter at the center of the pump spot on the sample surface. The pump spot was 80 microns in diameter to insure even excitation over the probe area. A new spot on the sample was used for each shot of the laser. The pump and probe were incident at 20° and -25° with respect to the surface normal in order to reduce the detection of SH induced by the intense pump.

The top register of Fig. 4 shows the cross-correlation of the pump and probe on the surface obtained from the second-harmonic generated by the product of the pump and probe. This signal marks the arrival time of the probe with respect to the pump and the time-resolution of the measurement. The lower two registers of Fig. 4 show the ŝ- and p̂-polarized SH probe signals. All signals are shown as a function of probe delay with respect to the pump.

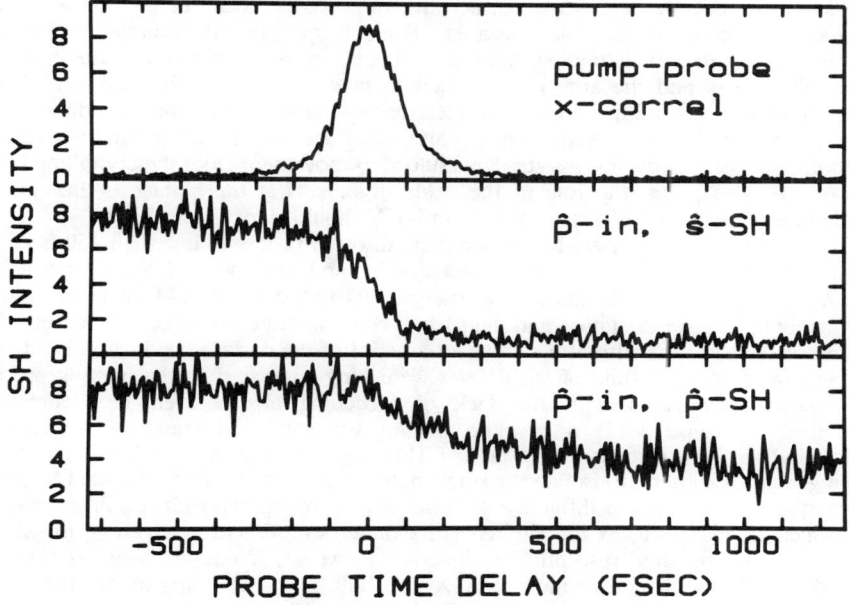

Fig. 4. SH from Si(111) as a function of probe delay with respect to pump arrival at surface. Pump intensity is 2 times the threshold for creating a melted spot in the Si(111) surface.

We see in the figure that the ŝ-SH signal that depends on order is reduced within 150 fsec or the resolution of the experiment. The p̂-SH changes much more slowly, on the time-scale of 500 fsec. Both changes in SH are threshold dependent. At 0.5 X threshold no change in M is observed with delay time. This indicates that the rapid decrease in M depends critically on the density of excited electronic states and is not due to bleaching of the nonlinear susceptibility which should not depend on threshold. The SH signal does not recover from its value at 250 fsec indicating that whatever changes occur do so permanently. We note that the threshold (around 0.1 J/cm^2) for the rapid SH change corresponds to the melting threshold of the bulk as well: melt spots in the Si wafer are observed with an optical microscope only for pump intensities exceeding threshold. The slow decrease of the p̂-SH is still under investigation, but is thought to arise from melting of the bulk. Quadrupole-allowed SH may arise from the topmost 100Å of the bulk and may be contributing to this p̂-SH signal. Previous studies using linear reflection have shown that the liquid melt front propagates into the bulk at rates in excess of 10^6 cm/sec[30] which corresponds to 100Å in 1 psec.

The less than 150 fsec change in surface order is most surprising. These new results are consistent with results published earlier[27] if one ignores the longer decay feature, takes into account a possible shift in the origin of time, and the fact that the previous study measured a signal proportional to $|L+M|^2$ where L/M was roughly 1 whereas we measure $|L|^2$ and $|M|^2$ directly. However, in contrast to the previous work which deduced a disorder time around 1 psec which would be consistent with thermal melting, this work is consistent with a complete loss of order within 1 or 2 electron-phonon relaxation times, a time much too short to have phonon equilibrium or thermal melting. In this case, disorder would be electronically-induced.

Such rapid disordering of the surface Si atoms could occur in an optical process analogous to photodissociation for molecules. Roughly speaking, the bonding orbitals that hold the Si atoms in their lattice positions are depopulated so efficiently that the bonds are temporarily broken and the atoms are pushed to new positions by the remaining lattice electric fields and new local fields due to electrons in highly excited "anti-bonding" states. This may be more likely to occur at the Si-SiO$_2$ interface rather than in the bulk because the amorphous oxide provides less steric resistance to movement from the crystalline lattice positions. At this point, the role of the oxide or defects at the surface is unclear, but repeating this experiment on a clean sample in UHV should clarify such effects.

The results do not necessarily imply atomic disorder occurs in less than 150 fsec: SH measures the instantaneous electronic susceptibility and not the real atomic periodicity directly. For example, the bonding state susceptibility that dominates M might be reduced immediately upon laser excitation and the surface could melt before M could recover. This could occur through bleaching if a significant amount of the surface conduction band were filled and the relaxation time of such states were long compared to the melting process. This appears inconsistent with the lack of bleaching effects even near threshold. Alternatively, because M is a measure of 3m symmetry, M could be reduced by transforming the 3m symmetry to 6m order. This might be possible if sufficient electronic screening were established to isolate the single outermost layer of bulk terminated Si (which has 6m symmetry) from the influence of lower layers in registry with the 3m symmetric bulk (which normally makes the surface 3m symmetric). Electron screening might also weaken the bonding state susceptibility directly. However, if one extrapolates from the carrier density induced by excitation at 0.2X threshold, e.g., assuming no saturation, the maximum number of excited states at threshold is 3.2×10^{21}/cm^3 or 6% of the Si atoms.[31] Even at such density it is unlikely that screening alone could dramatically reduce the nearest neighbor bond susceptibility. We can also dismiss the possibility that the Si surface and native oxide could be evaporated in 100 fsec considering that even if the atoms instantaneously acquired a velocity of 10^5 cm/s, they could only move 1Å or a half of the interatomic distance in 100 fsec.

If the Si atoms are indeed disordered by a photodissociation-like process, it would be the first such observation for a solid surface. This fast surface structural change may prove to be an essential step (in analogy to nucleation) in the slower bulk structural change and related advance of the liquid melt front away from the surface into the bulk. Similar experiments on metal surfaces are in progress.

CONCLUSION

Second-harmonic generation has the sensitivity to adsorbates and surface order and also the time-resolution to be a useful probe of surface reaction kinetics. Certainly, several problems remain with SHG as a surface probe, among them how to relate the changes in electronic susceptibility as measured by SHG to the changes in adsorbate composition in studies of reaction intermediates or to changes in atomic structure in structural phase transition studies. However, the theory of the electronic structure of surfaces is in itself a growing field and there is no reason that the SHG experiments should not contribute to that understanding as well as benefit by its development. Even under the present lack of microscopic theory, SHG promises to be a most useful tool for investigating a host of surface science questions in a time-regime hard to access by any other means. Even with millisecond kinetic studies it may be possible to detect non-equilibrium effects on reactions at high temperature and reactant pressures. At the very shortest time-scales, it may be possible to study non-equilibrium effects directly. In addition, new effects due to high densities of highly excited surface electronic states may reveal new and interesting physics and photochemistry.

REFERENCES

1. T. H. Ellis, L. H. Dubois, S. D. Kevan, Science 230, 256 (1985).
2. J. Misewich, H. Zacharias, and M. M. T. Loy, Phys. Rev. Lett. 55, 1919 (1985); C. T. Rettner, F. Fabre, J. Kimman, and D. J. Auerbach, Phys. Rev. Lett. 55, 1904 (1985).
3. R. B. Hall and A. M. DeSantolo, Surf. Sci. 37, 421 (1984).
4. R. S. Becker, G. S. Higashi, and J. A. Golovchenko, Phys. Rev. Lett. 52, 307 (1984).
5. R. Haight, J. Bokor. J. Stark, R. H. Storz, R. R. Freeman, and P. H. Bucksbaum, Phys. Rev. Lett. 54, 1302 (1985); R. Haight and J. Bokor, Phys. Rev. Lett. 56, 2846.
6. P. Guyot-Sionnest and Y. R. Shen, Phys. Rev. B33, 8254 (1986); and submitted to Phys. Rev. B.
7. H. W. K. Tom, T. F. Heinz, and Y. R. Shen, Phys. Rev. Lett. 51, 1983 (1983).
8. N. Bloembergen, R. K. Chang, S. S. Jha, and C. H. Lee, Phys. Rev. 174, 813 (1968); 178, 1528(E) (1969).
9. H. W. K. Tom, C. M. Mate, X. D. Zhu, J. E. Crowell, T. F. Heinz, G. A. Somorjai and Y. R. Shen, Phys. Rev. Lett. 52, 348 (1984).
10. J. T. Yates, P. A. Thiel, and W. H. Weinberg, Surf. Sci. 82, 45 (1979).
11. H. W. K. Tom, Ph. D. Thesis, Univ. Cal. Berkeley, (1984).
12. H. W. K. Tom, C. M. Mate, X. D. Zhu, J. E. Crowell, Y. R. Shen, and G. A. Somorjai, Surf. Sci. 172, 466 (1986).
13. X. D. Zhu, Y. R. Shen, and R. Carr, Surf. Sci. 163, 114 (1985).
14. S. G. Grubb, A. M. DeSantolo, and R. B. Hall, submitted to J. Phys. Chem.
15. A. Burns, H. -L. Dai, D. Heskett, E. W. Plummer, and K. -J. Song, to be published in J. Chem. Phys.
16. H. W. K. Tom, X. D. Zhu, Y. R. Shen, and G. A. Somorjai, Surf. Sci. 167, 167 (1986).
17. T. F. Heinz, M. M. T. Loy, and W. A. Thompson, J. Vac. Sci. Technol. B 3, 1467 (1985).
18. H. W. K. Tom and G. D. Aumiller, submitted to Phys. Rev. B.
19. C. K. Chen, T. F. Heinz, D. Ricard, and Y. R. Shen, Phys. Rev. Lett. 46, 1010 (1981).
20. G. L. Richmond, Chem. Phys. Lett. 110, 571 (1984); H. M. Rojhantalab and G. L. Richmond, these proceedings.
21. M. P. D'Evelyn, M. M. Nelson, T. Engel, submitted to Surf. Sci.
22. R. B. Hall, A. M. DeSantolo, and S. G. Grubb, submitted to J. Vac. Sci. Technol.
23. G. Arjavalingam, T. F. Heinz, J. H. Glownia, in Ultrafast Phenomena, V, eds. G. R. Fleming and A. E. Siegman, (Springer-Verlag, 1986).
24. T. F. Heinz, M. M. T. Loy, and W. A. Thompson, Phys. Rev. Lett. 54, 63 (1985).
25. H. W. K. Tom and G. D. Aumiller, Phys. Rev. B 33, 8818 (1986).
26. T. F. Heinz, G. Arjavalingam, M. M. T. Loy, J. H. Glownia, paper THII1, International Quantum Electronics Conference, June 9-13, 1986, San Francisco, CA.
27. C. V. Shank, R. Yen, and C. Hirlimann, Phys. Rev. Lett. 51, 900 (1983).
28. S. A. Akhmanov, N. I. Koroteev, G. A. Paition, I. L. Shumay, M. F. Guljaudinov, and E. I. Shtyrkov, Opt. Commun. 47, 202 (1983).
29. A. M. Malvezzi, J. M. Liu, and N. Bloembergen, Appl. Phys. Lett. 45, 1019 (1984).
30. C. V. Shank, R. Yen, and C. Hirlimann, Phys. Rev. Lett. 50, 454 (1983).
31. M. C. Downer and C. V. Shank, Phys. Rev. Lett. 56, 761 (1986).

INFRARED LASER-INDUCED DESORPTION OF NO AND CO
FROM ALUMINA SUBSTRATES

W. H. Weber and B. D. Poindexter
Physics Dept., Research Staff, Ford Motor Co., Dearborn, MI 48121

ABSTRACT

We present results of laser-induced desorption (LID) experiments using a CO laser on layers of CO and NO physisorbed at low temperature (6-40K) on fire-polished alumina substrates. Resonant LID is observed for NO but not CO. The time-of-flight (TOF) spectra of both molecules agree with Maxwell-Boltzmann distributions and show no additional structure.

INTRODUCTION

Resonant infrared LID, a process in which a vibrational mode of an adsorbed molecule is excited thereby promoting its desorption, has been reported for a variety of molecules.[1] Theoretical predictions by Kreuzer et al.[2] suggest that the TOF spectra of the desorbed molecules should show structure and should deviate from a Boltzmann distribution. The purpose of this paper is to examine both resonant and non-resonant LID in a new physical system, i.e., CO and NO on alumina, and measure the TOF spectra.

EXPERIMENTAL

The experiments were done in a bakeable UHV chamber with a Helitran cryostat providing substrate temperatures down to 6K. A BaF_2 window allowed access for the ir beam. The substrate was kept at a temperature low enough to assure a sticking probability near unity (below 30K for CO and 50K for NO) and dosed through a capillary exiting 2 cm away. Assuming a $\cos^2\theta$ distribution a dose of 0.8 mTorr-liter would yield 1 monolayer ($\approx 3 \times 10^{15}$ molecules/cm^2). Dosing was done from a 1-liter flask in which the pressure was monitored continuously with a capacitance manometer. This allowed accurate and reproducible doses to be made.

The radiation source was a line-tunable cw CO laser capable of operating on over 400 lines in the 1500-1900 cm^{-1} region with a typical output power of 0.3-0.5 W.[3] We used a fast intracavity rotating chopper synchronized with a slower mechanical shutter to obtain single laser pulses of 0.1-0.5 ms duration. The laser beam was p-polarized and incident on the sample at 65° from the normal after being focused with a 22-cm focal length lens.

Desorbed molecules were detected with a Granville-Phillips SPECTRA-SCAN 400 quadrupole mass analyzer (QMA) located in a differentially pumped side arm of the chamber, oriented along the normal to the sample surface. The entrance to the side arm was a 3-mm diameter aperture, 5 cm from the sample. The ionizer for the QMA was 14 cm from the sample and the aperture accepted only those molecules that could reach the ionizer on a straight-line trajectory. The TOF spectra were recorded using a Data Precision D-6000 data acquisition system. To decrease the QMA response time the usual electronics were bypassed with a fast amplifier directly across the photomultiplier.

RESULTS

The CO measurements were done at temperatures between 6 and 25K, since CO begins to desorb above 30K. To eliminate contamination from the background residual of $^{12}C^{16}O$ present in our chamber, we used the $^{13}C^{18}O$ isotope. No evidence for resonant desorption of CO was found -- for the same fluence, all laser lines were equally effective at desorbing CO. This is not unexpected, since the highest laser frequency (1950 cm^{-1}) is about 150 cm^{-1} below the expected CO vibrational mode. Presumably the laser pulse is heating the substrate directly, which is weakly absorbing at these frequencies, leading to thermal desorption from the surface. A typical TOF spectrum is shown in Fig. 1. The data points are an average of 45 single shots of the laser and the sample was cleaned and redosed after each shot. One laser pulse removed 70% of the molecules in the irradiated area (0.65 mm^2). The solid line is a calculated Maxwell-Boltzmann distribution, convoluted with a square-wave laser pulse function:

$$I(t) = \int dt' F(t-t')(t_0/t')^4 \exp\{-(t_0/t')^2\}, \quad (1)$$

where $t_0 = d(m/2kT)^{1/2}$, d is the distance to the detector, F is a constant for $0<t-t'<w$ and zero otherwise, and w is the laser pulse width (FWHM). The observed and calculated spectra agree well except for the

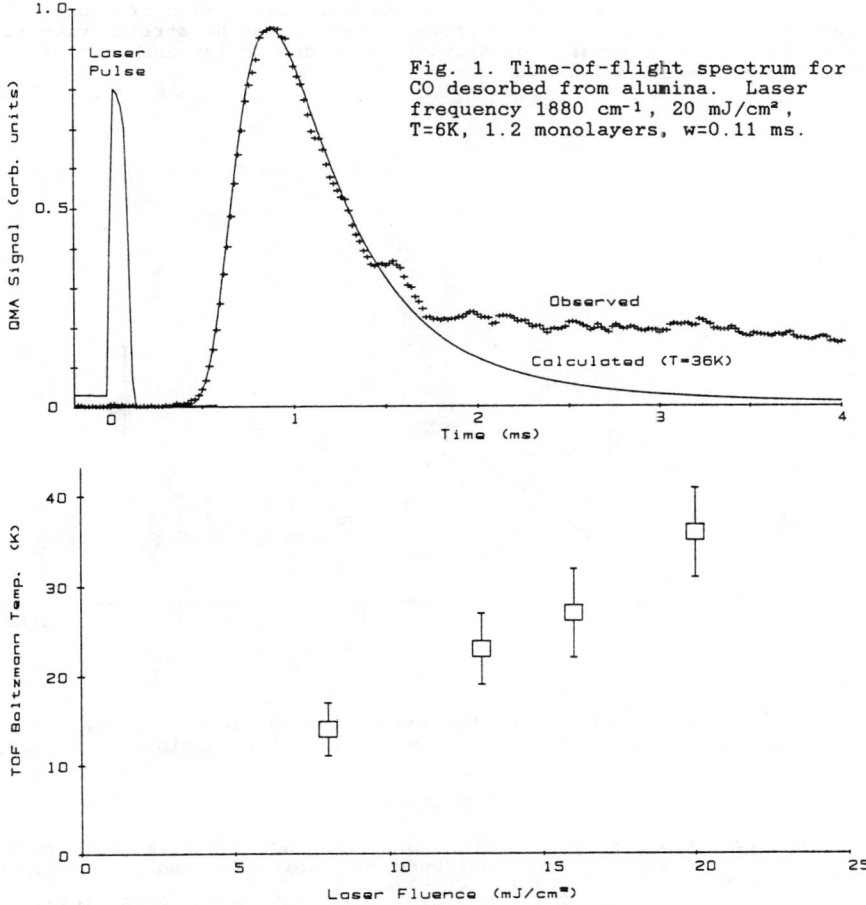

Fig. 1. Time-of-flight spectrum for CO desorbed from alumina. Laser frequency 1880 cm^{-1}, 20 mJ/cm^2, T=6K, 1.2 monolayers, w=0.11 ms.

Fig. 2. Effective TOF Boltzmann temperature versus laser fluence for CO desorbed from alumina; same experimental conditions as in Fig. 1.

excess signal at long times, which we believe is due to molecules entering the ionizer after one or more collisions with the walls. This excess signal was much larger without the aperture.

The dependence of the effective Boltzmann temperature T_{eff} of the desorbed molecules on the laser fluence is shown in Fig. 2. It is rather surprising that for low fluences these temperatures are well below those at which normal thermal desorption is observed.

The TOF spectra for NO were very similar to those for CO. No unusual structure was observed and the T_{eff} were at or 10-30K above the substrate T. The primary difference between the results for the two molecules is that NO shows a resonant dependence on the laser frequency, as shown in Fig. 3. These data were obtained at T=45K (NO begins to thermally desorb above 50K) with a fluence of 25 mJ/cm^2. At higher fluences the nonresonant behavior dominates, which is similar to the effect observed by Chuang and Hussla[4] for NH$_3$ on Cu. The peak signal, occurring at 1843 cm^{-1}, corresponds to desorption of about 80% of a monolayer. This peak is probably due to the NO stretch vibration, which has been down-shifted about 60 cm^{-1} due to interactions with the substrate and other molecules.

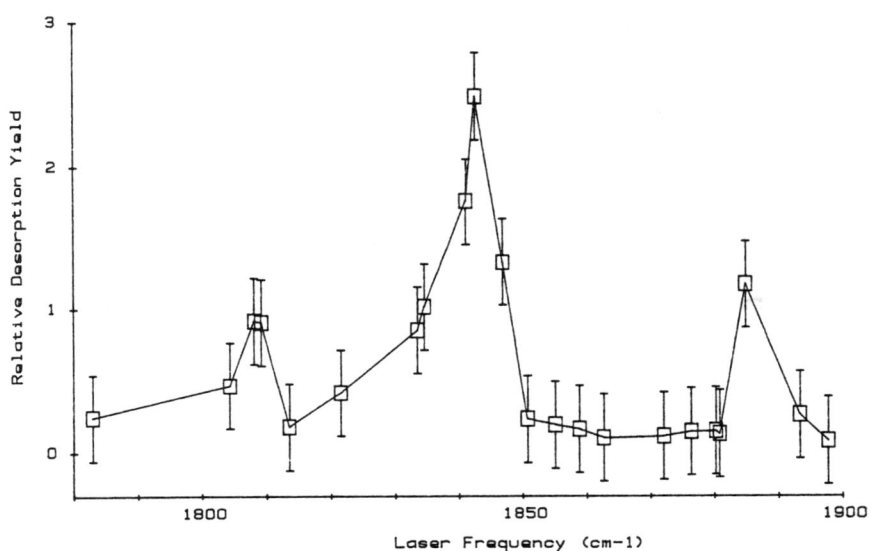

Fig. 3. Frequency dependence of the laser-induced desorption yield for NO on alumina. Laser fluence 25 mJ/cm^2, T=45K, 1.2 monolayer coverage.

REFERENCES

1. See for example T.J. Chuang, H. Seki, and I. Hussla, Surface Sci. 158, 252 (1985) and J. Heidberg, H. Stein, E. Riehl, Z. Szilagyi, and H. Weiss, Surface Sci. 158, 553 (1985).
2. H.J. Kreuzer and Z.W. Gortel, Phys. Rev. B 29, 6926 (1984); Z.W. Gortel, H.J. Kreuzer, P. Piercy, and R. Teshima, Phys. Rev. B 27, 5066 (1983).
3. W.H. Weber and R.W. Terhune, J. Chem. Phys. 78, 6422 (1983).
4. T.J. Chuang and I. Hussla, Phys. Rev. Lett. 52, 2045 (1984).

SECOND HARMONIC GENERATION AND DIFFERENTIAL CAPACITANCE STUDIES OF SMOOTH SILVER ELECTRODE-AQUEOUS ELECTROLYTES

Hossein M. Rojhantalab and Geraldine L. Richmond
Chemistry, University of Oregon, Eugene, OR 97403

ABSTRACT

Optical second harmonic generation and differential capacitance have been measured for electrochemical interfaces of electropolished polycrystalline silver electrode-halide ions. A close correlation is observed between the SHG signal and the surface excess charge density q_m on the metal electrode for different concentrations of electrolytes.

INTRODUCTION

In-situ characterization of a solid surface in contact with a solution requires interface-selective probes which are insensitive to bulk properties. Recent studies have shown that second harmonic generation (SHG) can be a fast and useful technique in studying the double layer structure of a metal-liquid interface[1,2]. Earlier work on evaporated films of silver had suggested that the second harmonic signal for Si and Ag in KCl is related quadratically to the applied bias voltage[3]. More recent work on polished silver single crystal surfaces, and silver thin films[4] suggested that for simple adsorption-desorption of ions on the surface, the SH intensity is proportional to the square of the surface charge density at metal, q_m^2.

In this paper we have tested this model by comparing the SHG and differential capacitance simultaneously. We report the results for polycrystalline smooth silver potentiostated in different mixtures of NaCl-NaClO$_4$ electrolytes at constant ionic strength.

EXPERIMENTAL

Experimental details can be found elsewhere[5]. The working electrode was a 10x13 mm silver disk (99.999% purity) press-fitted into a KELF plunger. It was mechanically polished to a mirror finish with 0.05 micron alumina grit and then electrochemically polished to remove any oxide layer. Pure and reproducible surfaces were obtained by this method as judged by the featureless cyclic voltammograms[6]. The differential capacitance versus potential was measured using a lock-in-amplifier with an internal oscillator set at a frequency of 57 Hz and 4 mv peak-to-peak AC amplitude. The SHG experiments were done similarly to the earlier studies[5].

RESULTS AND DISCUSSION

Second harmonic generation at an interface between two centrosymmetric media is proportional to the square of the nonlinear polarizability, $\vec{P}_{nl}(2\omega)$ of the interface. Under conditions where the electric dipole approximation holds, nonlinear polarizability in an external applied dc potential can be expressed as:

© American Institute of Physics 1987

$$\vec{P}_{nl}(2\omega) = \overset{\leftrightarrow}{\chi}{}^{(2)}(2\omega) : \vec{E}(\omega) \cdot \vec{E}(\omega) +$$
$$\overset{\leftrightarrow}{\chi}{}^{(3)}(2\omega) : \vec{E}_{dc} \cdot \vec{E}(\omega) \cdot \vec{E}(\omega)$$

where $\overset{\leftrightarrow}{\chi}{}^{(2)}$, and $\overset{\leftrightarrow}{\chi}{}^{(3)}$ are the second order and third order nonlinear susceptibility tensors which have contributions from both sides of the interface[5]. Although $\overset{\leftrightarrow}{\chi}{}^{(3)}$ is several orders of magnitude smaller than $\overset{\leftrightarrow}{\chi}{}^{(2)}$, the large electric field strength at the interface ($E_{dc}=10^6$ V/cm) makes the third order term of comparable or greater magnitude. Assuming a flat surface, the electric field at the surface should be proportional to q_m by Gauss's law.

The experimental results for differential capacitance, C_m, on electropolished polycrystalline silver electrode in various concentrations of NaCl-NaClO$_4$ mixture at constant ionic strength are displayed in Fig. 1. The scan was initiated at a potential positive relative to hydrogen evolution and terminated prior to oxidation of the metal. Extreme care was undertaken to avoid oxidative processes at the surface. $C_m(E)$ curves, which showed little or no hysteresis upon repeated potential sweep, are very similar to those reported earlier[6]. The maximum in $C_m(E)$ curve corresponding to ½ coverage, is found to shift cathodically with increasing anion concentration.

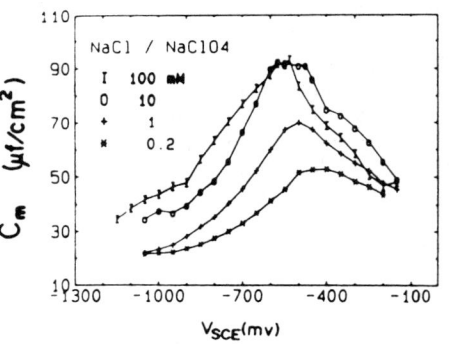

Fig. 1. Differential Capacitance vs potential for electropolished polycrystalline silver electrode in (0.5-x)M NaClO$_4$ + x M NaCl mixtures, x=0.2, 1, 10, 100mM.

As the interface is polarized, the charge density on the solution side of the interface, q_s, increases and is distributed throughout the double layer. By electroneutrality, an equal and opposite amount of charge builds on the metal side. Therefore, q_m is a measure of the excess charge density on the metal at a given potential brought about by adsorbing anions. The magnitude of q_m can be found by back integrating the $C_m(E)$ curves. In Fig. 2 the resultant q_m^2 vs E is shown for our $C_m(E)$ data. For comparison in Fig. 3 the SHG vs potential is displayed for the same electrode. The SH signal in Fig. 3 is dependent on the potential and anion concentration. The results in Figures 2 and 3 clearly show a close correlation between q_m^2 and SH signal and support the q_m^2 model. In all the solutions the onset of the signal is near the potential of zero charge, PZC. The reported value of PZC for polycrystalline silver in NaClO$_4$ is -950 mv[2,6]. Our concentration studies show values of -900 to -1050 mv for 0.2 to 100mM concentration of NaCl in NaClO$_4$. This negative shift in potential is expected for an electrolyte containing specifically adsorbing anions such as Cl$^-$. Additional studies with other electrolytes give similar indications.

This study provides further evidence that for SHG from silver electrodes polarized in simple electrolytes, the dc electric field component of the polarizability is a dominating effect. Further work is in progress with the goal of making quantitative measurements of adsorbing ions and molecules at electrode surfaces.

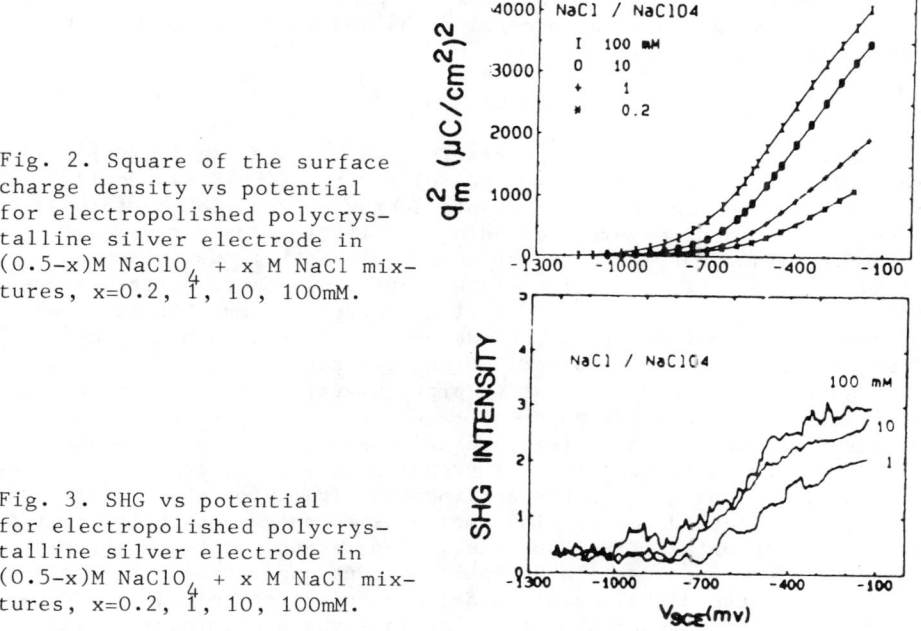

Fig. 2. Square of the surface charge density vs potential for electropolished polycrystalline silver electrode in $(0.5-x)$M NaClO$_4$ + x M NaCl mixtures, x=0.2, 1, 10, 100mM.

Fig. 3. SHG vs potential for electropolished polycrystalline silver electrode in $(0.5-x)$M NaClO$_4$ + x M NaCl mixtures, x=0.2, 1, 10, 100mM.

ACKNOWLEDGEMENT

Financial support for this work from the National Science Foundation (NSF#8513008) and ACS Petroleum Research Fund (PRF-ACS16905) is appreciated.

REFERENCES

1. For reviews, see: Y.R.Shen in "Chemistry and Structure at Inter-Faces", R.B.Hall and A.B.Ellis (VCH publishers, Deerfield Beach, FL, 1986), chapter 4.
2. G.L.Richmond, Chem. Phys. Lett. 106, 26 (1984); 110, 571 (1984); 113, 359 (1984).
3. C.H.Lee, R.K.Chang, and N. Bloembergen, Phys. Rev. Lett. 18, 167 (1967).
4. R.M.Corn, M.Romagnoli, M.D.Levenson, and M.R.Philpott; Chem. Phys. Lett. 106, 30 (1984); J. Chem. Phys. 81 4127 (1984).
5. G.L.Richmond, H.M.Rojhantalab, J.M.Robinson, and V.L.Shannon, J. Opt. Soc. Am., B 4 (1987), in press.
6. J.T.Hupp, D.Larkin, M.J.Weaver, Surf. Sci. 125, 429 (1983).

NONLINEAR OPTICAL STUDIES OF SEMICONDUCTOR INTERFACIAL PROPERTIES

J. M. Robinson and G. L. Richmond
University of Oregon, Eugene, OR 97403

ABSTRACT

Second harmonic generation from amorphous selenium thin films and single crystal transition metal dichalcogenides under ambient conditions is reported.

INTRODUCTION

Optical second harmonic generation[1] (SHG) is a highly sensitive probe of the surface morphology and reactivity of bulk isotropic materials. Under the electric dipole approximation, SHG is forbidden in the bulk of centrosymmetric media.[1,2] Its intrinsic sensitivity to the interface between two such media arises from the symmetry breaking properties of the interface. Surface symmetry[3], the orientation of adsorbed species[4] and surface charge accumulation at the solid/liquid interface[5] have been determined by static measurements of SH intensity. Time resolved SHG measurements using a fast pulsed laser system have the potential to probe kinetics and dynamics of interfacial processes on nanosecond and faster timescales. These processes include photocarrier generation and recombination as well as electron transfer reactions or trapping in surface defects. We have chosen to investigate the nonlinear optical properties of two types of chalcogenide materials, amorphous selenium and the transition metal dichalcogenides $ZrSe_2$ and WSe_2. Understanding the photogenerated carrier dynamics of a-Se has important application in Xerography whereas the dichalcogenides are photoelectrochemical cell components. The work described here comprises the preliminary ambient measurements of signal intensity and damage threshold under steady state conditions and provides the groundwork for time-resolved SHG experiments.

EXPERIMENTAL APPROACH

Figure 1 shows the experimental apparatus for the investigation of the Se films on quartz substrates. The focussed p-polarized 1.06 um light (~3 W) from a cw mode-locked Nd:YAG laser (120 psec pulses at 82 MHz) strikes the sample at ~45°. Single photon counting detects the total SH light (532 nm) generated in reflectance following rejection of the fundamental beam by color filters and a monochromator. The SHG measurements of the dichalcogenides employ the collimated output of a Q-switched Nd:YAG laser (10 ns pulses at 10 Hz) and gating detection in a similar configuration.

© American Institute of Physics 1987

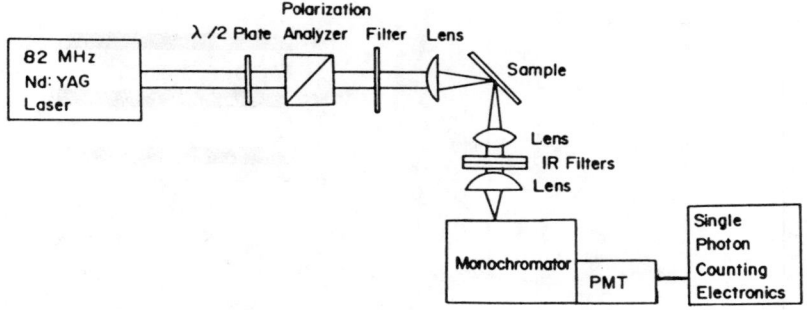

Fig. 1. Second harmonic generation experimental apparatus.

RESULTS AND DISCUSSION

Second harmonic generation from the surface of the isotropic Se films is readily observed using the picosecond laser system described above. The measured harmonic signal at 532 nm is monochromatic. The bandgap of a-Se is 2 eV, well above the energy of the fundamental beam.[6] By employing a subbandgap probe, these experiments successfully avoid the broad infrared photoluminescence which has been observed from a-Se using visible photoexcitation at low temperature.[7] The ease with which the signal is obtained, the lack of interfering luminescence and the known sensitivity of SHG to electric field effects make SHG a potentially ideal probe of kinetic processes in a-Se.

The SH signal from these films can be enhanced by controlling the thickness of the film. Figure 2 displays the relationship between the depth of the film and the observed SH intensity. SHG increases with sample thickness and attains a maximum for films of bulk thickness. Similar behavior was observed by Wokaun et al.[8] when SHG was monitored from gold island films of 0 to 200 Å thickness. The authors attribute the increase to the launching of extended surface plasmons.

Both of the transition metal dichalcogenides studied here, WSe_2 and $ZrSe_2$, exhibit optical SH behavior. A single crystal, oxide free surface is easily prepared by pealing the top layer of the material from the bulk sample (Fig.3).[9] No surface damage or accompanying photoluminescence is observed at low laser fluence. Photoexcitation of many of the dichalcogenides results in efficient photocarrier producation but the excitation wavelengths[9,10] are in the visible (Fig.3), well beyond the 1.06 um probe wavelength used here. The unique combination of simple surface preparation, resistance to oxidation and high photocurrent yields make these layered compounds attractive solar cell[11] candidates. With their strong nonlinear response, they are also good model compounds for monitoring photoinduced dynamics by SHG.

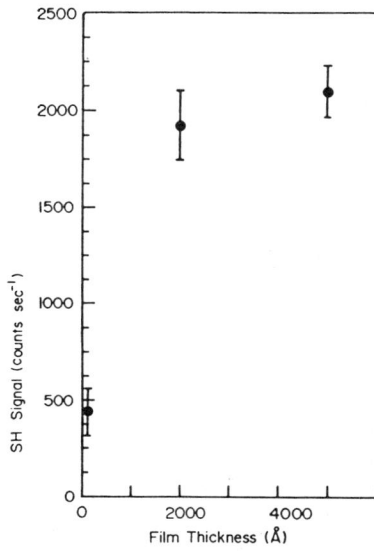

Fig. 2. Depth dependence of SH signal from a-Se films.

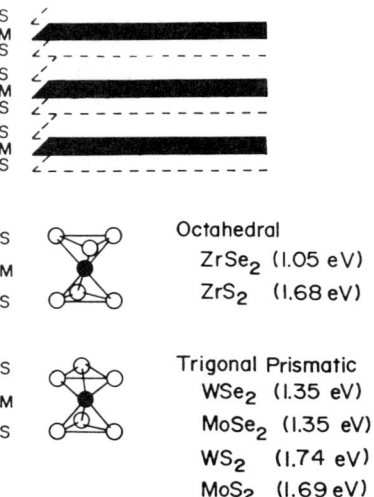

Fig. 3. Structure of the transition metal dichalcogenides. Bandgap energy listed. After ref. 10.

ACKNOWLEDGEMENTS

We acknowledge the support of the Department of Energy (Grant DE-FG0G-86ER45273). We thank B. A. Parkinson and M. Crawford for the materials and several useful discussions.

REFERENCES

1. Y.R. Shen, The Principles of Nonlinear Optics (Wiley-Interscience N.Y., 1984).
2. N. Bloembergen and P.S. Pershan, Phys. Rev. 128, 606 (1962).
3. See for example: H.W.K. Tom, T.F. Heinz and Y.R. Shen, Phys. Rev. Lett. 51, 1983 (1983).
4. T.F. Heinz, H.W.K Tom, and Y.R. Shen, Phys. Rev. A 28, 1883 (1983).
5. See for example: R.M. Corn, M. Romagnoli, M.D. Levenson and M.R. Philpott, J. Chem. Phys. 81, 4127 (1984). G.L. Richmond, Langmuir 2, 132 (1986).
6. A.K. Bhatnagar and K. Venugopala Reddy, J. Non-Cryst. Solids 76, 409 (1985).
7. R.A. Street, T.M. Searle and I.G. Austin, Philos. Mag. 29, 1157 (1974).
8. A Wokaun, J.G. Bergman, J.P. Heritage, A.M. Glass, P.F. Liao and D.H. Olson, Phys. Rev. B 24, 849 (1981).
9. K.K. Kam and B.A. Parkinson, J. Phys. Chem. 86, 463 (1982).
10. H. Tributsch, Discuss. Faraday Soc. 70, 189 (1980).
11. B.A. Parkinson, Acc. Chem. Res. 17, 431 (1984).

VIBRATIONAL EXCITATION OF AN ADBOND BY A SHORT-PULSED LASER

Sander van Smaalen and Thomas F. George
Departments of Physics & Astronomy and Chemistry, 239 Fronczak Hall,
State University of New York at Buffalo, Buffalo, New York 14260

ABSTRACT

A system of a phonon-damped adbond, coherently excited by a laser, is considered. The effect of a series of laser pulses is compared with the effect of a continuous-wave laser.

INTRODUCTION

Consider an adsorbed atom in a vibrational bond, driven by a laser, and damped due to the lattice vibrations of the substrate. We approximate the vibrating adatom by a one-dimensional oscillator, for the motion pependicular to the surface.

The interaction between the adatom and the substrate is given by the potential between the atom and the nearest surface atom. Choose the origin on the average position of the surface atom . Let z be the position of the adatom and denote the z-component of the displacement of the surface atom by u_z. Then the problem can be separated in a vibration of the adatom in the potential $V(z-z_0)$, and the interaction with the lattice vibrations, in first order given by

$$H_{ap} = -u_z \frac{dV(z-z_0)}{dz}. \tag{1}$$

A master equation for the time evolution of the reduced density operator of the adbond is then[1-3]

$$\frac{d\sigma(t)}{dt} = \frac{1}{i\hbar} [H_a, \sigma(t)] - \Gamma \sigma(t), \tag{2}$$

where Γ is the Liouville relaxation operator. Using that the adbond potential is strongly anharmonic then leads to a set of equations for the level populations $\sigma_n(t)$,[1-3]

$$\frac{d\sigma_n(t)}{dt} = \sum_k \{a_{kn} \sigma_k(t) - a_{nk} \sigma_n(t)\}. \tag{3}$$

Expressions for the transition rate constants a_{kn} can be found in refs. 1-3.

PULSED-LASER EXCITATION

The laser couples the two levels $|g\rangle$ and $|e\rangle$ of the adbond. Then the interaction of the ad-bond with the laser is given by,

$$H_{ar}(t) = \Omega(t) \{|e\rangle\langle g| \exp(-i\omega t) + |g\rangle\langle e| \exp(i\omega t)\} \qquad (4)$$

where $\Omega(t) = -\vec{\mu}\cdot\vec{E}_0(t)/\hbar$ is the Rabi frequency, $\vec{\mu} = \langle g|\vec{\mu}|e\rangle$ is the transition dipole moment, and $\vec{E}(t) = \vec{E}_0(t) \cos(\omega t)$ is the electric field of the laser beam. For a cw laser $\vec{E}_0(t)$ is independent of time. For a pulsed laser we assume that $\vec{E}_0(t)$ is slowly varying compared with the laser frequency ω.

An expression for the time evolution of the reduced density operator when the laser is present is obtained by adding $H_{ar}(t)$ to H_a in Eq.2. It is easily seen that the master equation (Eq.3) then also involves the coherences of σ between $|g\rangle$ and $|e\rangle$.

Assume the pulse duration Δt to be much shorter than the inverse of the rate constants in Eq.3. Then the relaxation can be neglected during the pulse, and the effect of the pulse is found from the equations for a two-level system. On resonance it depends only on the pulse area:

$$\theta = \int_{pulse} dt\, \Omega(t) \qquad (5)$$

The solution is

$$R_2(\Delta t) = R_2^0 \cos(\theta) + R_3^0 \sin(\theta),$$
$$R_3(\Delta t) = R_3^0 \cos(\theta) - R_2^0 \sin(\theta), \qquad (6)$$

where $R_3 = \sigma_e - \sigma_g$ is the population inversion, and $R_2 = i(\tilde{\sigma}_{eg} - \tilde{\sigma}_{ge})$ is the imaginary part of the coherence in the rotating frame.

A series of equally-spaced π-pulses ($\theta = \pi$) is considered. From Eq.6 it follows that the effect of a π-pulse is to change R_3^0 into $-R_3^0$. If we assume that $R_2^0 = 0$, then it remains zero throughout. Between the pulses (occurring at intervals t_p), the adbond evolves in time according to the master equation (Eq.3). After several pulses the system will reach a quasi steady state in which the time evolution of $\sigma(t)$ is identical in each interval t_p. Then we can express the populations entirely in the rate constants a_{kn} and the time t_p.[4]

COMPARISON WITH CW LASER EXCITATION

An adbond, irradiated by a cw laser, reaches a steady state in which the populations are independent of time. The steady-state values, $\sigma_n(\infty)$, are obtained by solving the master equation for

$d\bar{\sigma}(t)/dt = 0$. The result is given in table I. For the pulsed laser, with the adbond in the quasi steady state, an average value for the populations can be defined as

$$\sigma_n(av) = \frac{1}{t_p} \int_0^{t_p} dt\, \sigma_n(t). \qquad (7)$$

It is then possible to compare the average "exciting power" of the pulsed laser with that of the cw laser by calculating $\eta = \sigma_e(av)/\sigma_e(\infty)$, and using some criterion to compare both laser.

The first criterion is to require that both lasers have the same average power. In the low-intensity limit, $\Omega_{cw} \ll \gamma$, we find $\eta \ll 1$, that is, the cw laser is much more effective in exciting the adbond then the pulsed laser. In the high-intensity limit $\sigma(\infty)$ and $\sigma(av)$ aquire the same limiting values, and $\eta = 1$.

The second criterion is to require that the average power absorption from the laser is equal in both cases. It follows that $\eta = 1$, independent of the laser power. Conversely, this expresses the fact that the energy flow into the substrate is proportional to the excitation of the adbond and does not depend on the details of this excitation.

Table I Comparison of a cw laser and a pulsed laser

	continuous wave laser	pulsed laser
average laser power	Ω_{cw}^2	$\dfrac{\pi^2}{\Delta t\, t_p}$
average absorbed power	$\hbar\omega \dfrac{\Omega_{cw}^2}{\gamma}(\sigma_g - \sigma_e)$	$\hbar\omega \dfrac{1}{t_p} R_3(\Delta t)$
average excited-level population	$\sigma_e(\infty) = \dfrac{\Omega_{cw}^2}{\gamma^2 + 2\Omega_{cw}^2}$	$\sigma_e(av) = \dfrac{(1 - \exp[-\gamma t_p])}{t_p(1 + \exp[-\gamma t_p])}$

This work was supported by ONR and AFOSR.

REFERENCES

1. H.F. Arnoldus, S. van Smaalen and T.F. George, Phys.Rev. B34 (1986).
2. S. Efrima, L. Jedrzejek, K.F. Freed, E. Hood and H. Metiu, J. Chem. Phys. 79, 2436 (1983).
3. Z.W. Gortel, H.J. Kreuzer, P. Piercy and R. Teshima, Phys. Rev. B28, 2119 (1983).
4. S. van Smaalen and T.F. George, Surf. Sci., submitted.

ATOMIC AND ELECTRONIC IMAGING OF SEMICONDUCTOR SURFACES WITH SCANNING TUNNELING MICROSCOPY

J. E. Demuth, R. J. Hamers, and R. M. Tromp
IBM Watson Research Center, Yorktown Heights, NY 10598

ABSTRACT

The principles of scanning tunneling microscopy and its application to study silicon surfaces are briefly reviewed. Scanning tunneling microscopy "topographs" contain both geometric information about the locations of atoms at the surface as well as about the charge densities of surface localized states. We describe procedures by which these two components can be distinguished so as to produce images of the surface electronic states with atomic resolution. This ability to spatially resolve the surface electronic structure provides new information to understand the local structure and nature of bonding, and in some cases can be used as a means to chemically image specific features of the surface.

INTRODUCTION

The scanning tunneling microscope (STM) was invented less than five years ago by Gerd Binnig and Heinrich Rohrer at the IBM Research Laboratory in Zurich, Switzerland.[1] Their STM image of the Si(111) 7×7 surface three years ago[2] provided a major breakthrough in our understanding of the 7×7 surface and has since stimulated much more STM work on semiconductor surfaces. On October 16, 1986, Binnig and Rohrer were honored for their work with the Nobel Prize in Physics together with Ernst Ruska, the inventor of the electron microscope in 1933.

The STM with its ability to study surfaces in real space has from the beginning changed our perception of surfaces. We can now see not only the periodic part of the surface but also the local irregularities, defects and imperfections at surfaces that have long eluded other methods. (Most of these previous surface science methods measure properties averaged over large regions of the surface.) Another important aspect of STM is that since the electron tunneling process depends on the density of surface electronic states, it ultimately conveys information about the local surface electronic structure. These two properties of tunneling make it an ideal probe for semiconductor surface studies. In this paper we will show the types of local structural and electrical information that can be obtained from STM measurements. In particular, we will focus on our own studies of Si(111) and Si(100) surfaces.

© American Institute of Physics 1987

EXPERIMENTAL PROCEDURES

In scanning tunneling microscopy a fine metal tip, usually an electrochemically etched Tungsten wire, is brought very close to the sample to allow tunneling. The tip-sample distance required is typically 5-10Å and a tunnel current of 1 nA is typically used. At these very small distances the wavefunctions of the atoms in the surface and on the tip overlap, and by applying a voltage difference between them (anywhere from 1 mV to 2V) the electrons can tunnel from the tip to sample or vice versa. In scanning tunneling microscopy, the tip is mounted to three orthogonal piezoelectric transducers and is scanned parallel to the surface. Changes in the tunnel current reflect changes in the distances between the tip and surface and can be used to monitor this separation. More typically, the variations in tunnel current are monitored by a feedback circuit which applies a feedback voltage to the z-piezo transducer (normal to the surface) so as to maintain a constant tunnel current while scanning. This correction voltage applied to the z-piezo is measured during an x-y raster scan and presented as a real space image of the surface corrugations and called a 'topograph'.

Most scanning tunneling microscopy is performed in a vacuum chamber which allows the preparation and preservation of clean, well-defined surfaces. For our studies of Si(111) and (100) surfaces we have prepared clean surfaces by a thermal oxide removal method which when properly executed produces extremely clean, uniform silicon surfaces.[3,4] Other surfaces such as graphite and some layered compounds such as MoS_2 are physically inert and do not necessarily require a vacuum chamber for tunneling studies.[5] However, the vacuum chamber also serves a dual function in providing acoustic and electrical shielding which may be necessary for performing some tunneling measurements. The vibration isolation of our STM is provided by an air bearing table upon which the chamber rests, as well as by a stack of vibrationally damped isolation plates inside the chamber which support the STM. Our STM has novel design features to enable a variety of semiconductor surface studies and is described in more detail elsewhere.[3]

The tips we prepare by electrochemical etching usually always provide atomic scale lateral resolution so long as they appear sharp under inspection at 400× in an optical microscope. Our SEM examinations of working tips show a rather large tip radius of up to 1000Å. Presumably there is one microstructure or cluster of atoms at the end of the tip (closest to the surface under study) which has atomic scale features so as to provide the high lateral resolution. Sputtering or heating of the tip prior to tunneling is useful in removing oxides and impurities left on the tip after the etching process.

STM TOPOGRAPHS

In Fig. 1, we show linescans of a Si(111) 7×7 surface and its corresponding grey scale image. In the grey scale image the protrusions are only ~1Å high and correspond to the lighter colors. We readily observe the periodic structure or unit cell of the Si(111) 7×7 surface which is 27Å per side and is outlined in Fig. 1. In addition some protrusions are missing which would normally be expected to exist on the ideal surface. Surprisingly, the long range order of these 7×7 unit cells is preserved even though many of the protrusions are missing. Binnig and Rohrer, who first saw this 7×7 STM pattern, identified these 12 protrusions as three-fold coordinated Si ad-atoms on the top layer.[2] Local arrangements of these ad-atoms structure can also arise under different processing conditions to produce locally different ordered arrays as has been found for example on laser annealed Si(111).[6] Some other STM images of Si(111) 7×7 are also shown in Fig. 2. for a positive and negative bias of 2V on the sample and tip, respectively. While both images are from the same Si(111) 7×7 sample a dramatic asymmetry occurs between the two halves of the unit cell. This asymmetry at negative bias is now known to arise from the enhanced tunneling from surface states near E_F of the sample.[7] The need to distinguish the structurally derived STM features from those associated with tunneling from localized electronic states at the surface has been the focus of much recent work. Recent theoretical work has been directed at understanding the imaging process and the physical meaning of STM topographs.[8-10]

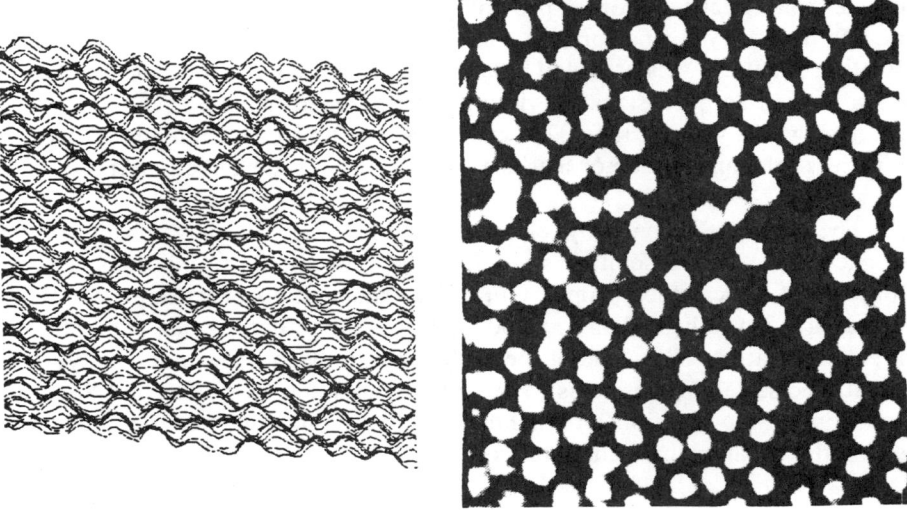

Figure 1. A linescan of a Si(111) 7×7 surface and a corresponding grey scale image. The unit cell is indicated and the height variations correspond to ~1.5Å A bias voltage of +2 V is applied to the sample.

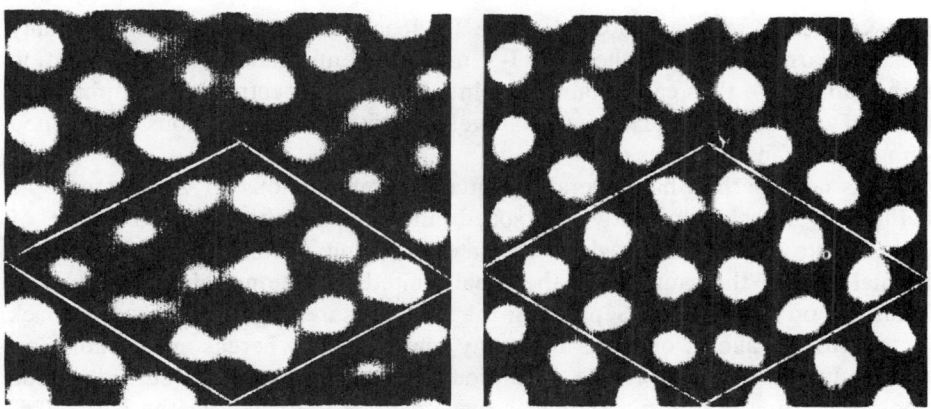

Figure 2. Bias dependent images of Si(111) 7×7 for (a) +2 V on the sample and (b) −2 V on the sample. Under these bias conditions electrons tunnel into the sample or tip, respectively.

IMAGING OF SURFACE ELECTRONIC STATES

As shown in Fig. 2 bias dependent STM images also contain information about the surface electronic states. Other bias dependent measurements of cleaved Si(111) 2×1 surfaces[11] have been used to selectively image defect states lying in the surface state band gap. Here, for sufficiently low bias voltages one tunnels predominantly into the states in the gap and observes large enhancements of the topographic features near certain types of irregularities. Several other approaches have been utilized in an effort to obtain more detailed spectroscopic information about surfaces including small signal modulation of the bias voltage[12,13] and measuring I-V data.[14] Both of these to date have been limited by different problems. The modulation technique applied in a constant current scanning mode introduces topographic features into dI/dV[16] while I-V measurements are difficult to perform in a scanning mode or at well defined positions on the surface.[12]

In order to obtain I/V type of data in a scanning mode we have developed a new method, called CITS, for operating the feedback control while scanning.[17-18] Essentially CITS introduces a timing sequence to the tunneling measurement so that the feedback control can operate independent of and during I-V measurements.[7] This is achieved by repetitively ramping the sample bias over some energy range at 2.2 kHz and sampling the tunneling current at particular bias voltages or in reality at different times during the voltage ramp. One of the bias voltages is selected to control the feedback and maintains an essentially constant barrier height for the other I-V measurements made during the duration of the ramp. We thus can obtain 50 point, 12 bit I-V spectra for each pixel element of a 100×100 pixel image together with a simultaneously obtained topograph *all* in 5 minutes.

This new method has several advantages over previous STM spectroscopic methods. Most importantly, its speed and the simultaneous collection of topographic and electronic data alleviates tracking problems associated with thermal drift of the sample and the usual limited operational lifetime of exceptionally good probe tips. Separating the I-V measurements from the feedback control further allows one to sample any bias voltage whereas most studies on semiconductors[3,12] have found it difficult to obtain constant current mode topographs for the low bias voltages needed to probe states in the gap. This method also allows us to reduce our I-V data into 2-D images of the various surface states contributing at the different bias voltages. This type of presentation of the data is simpler to visualize in many cases than I-V spectra.

Si(111) 7×7

In Fig. 3 we show I-V data obtained by this method and plotted as I/V versus V for selected positions in the unit cell of the Si(111) 7×7 surface. For comparison we show photoemission (PES) measurements by Himpsel and coworkers of the filled and empty states.[19] As discussed elsewhere[14] one expects onsets in the tunneling conductance when new states can contribute to the tunneling with increasing bias voltages. Such onsets near E_F at -1 V below or at $+0.5$ V above E_F correspond to surface states seen in PES and appear to be localized at certain atomic positions in the surface unit cell. In order to spatially image these states we have stabilized the feedback at a particular bias of $+2$ V which fortuitously allows the tip to follow a contour of atomic charge density for the dimer adatom stacking fault model of the Si(111) 7×7 surface.[7] Allowing the tip to follow this contour of the total atomic charge density greatly simplifies separating geometric and electronic contributions since it maintains a constant barrier height at each point above the surface during the scan. Plotting the tunneling current obtained at these other biases, then directly reflect the atomic locations of these surface states.

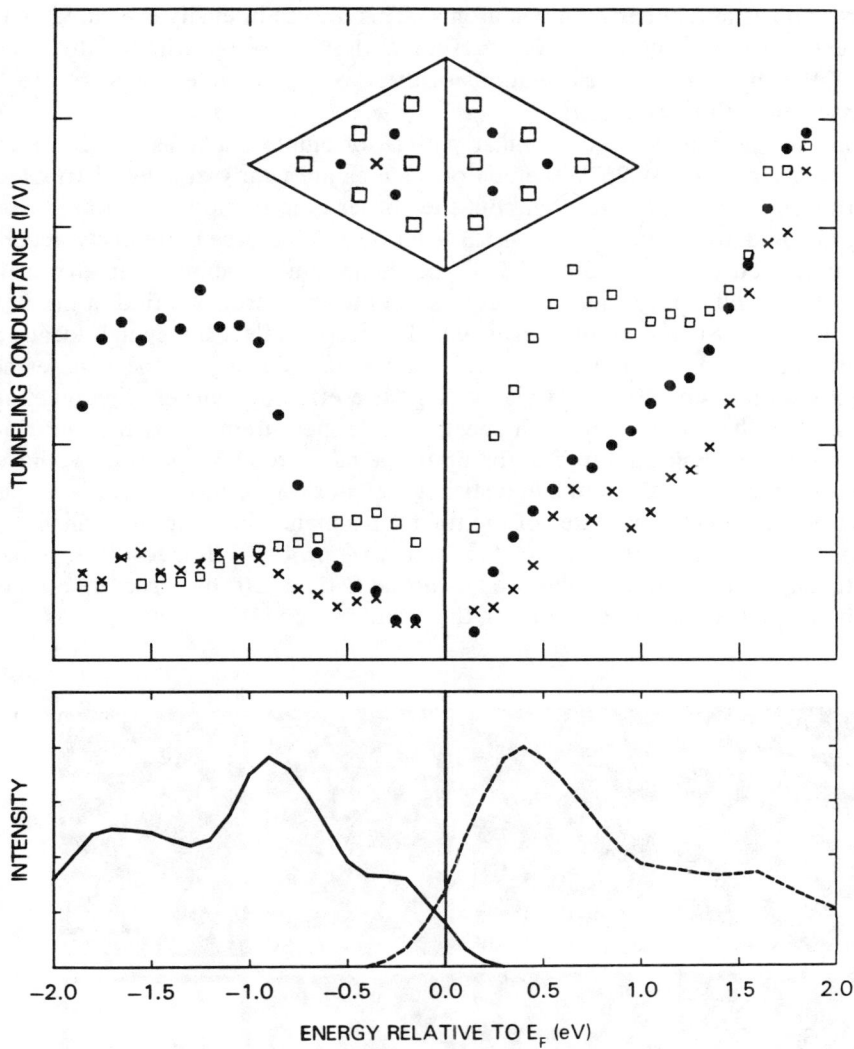

Figure 3. Conductance (I/V) spectra as a function of bias voltage for selected atomic positions in the Si(111) 7×7 unit cell as indicated in the top of the figure.

Current images obtained in this manner as a function of different bias voltages are shown in Fig. 4. At −.25 V we see the highest lying surface states located on adatoms on the faulted half of the unit cell. At −0.70 V we see a superposition of the adatom derived states and states located on atoms in between the adatoms, called "rest atoms". At −1.45 V the tunneling from rest atom states now dominate and at −2 V we see additional diffuse intensity near the corner holes and in between rest atoms. This diffuse intensity is located near the edges of atoms in the 7×7 structure and has been attributed to back bonds.[12] Recent theoretical calculations of these occupied states by Northrup[20] are consistent with these results.

For +.15 V bias, we see a similar pattern of empty states as for the filled states. Starting above +.25 V the adatoms states are nearly equally distributed but with a slight 3 fold symmetry about the corner hole rotated 180° to that for the occupied adatom state. Above +1.3 V an overall increase in intensity occurs for the unfaulted side of the unit cell. These changes in tunneling symmetry seen for both filled and empty states also correspond to the onsets we find in the I-V curves in Fig. 3. Such current images, which directly reflect the spatial location of the surface states, do not reflect the relative state density of these states at different energies since for each bias voltage the effective barrier for tunneling changes. Another concern in such spectroscopic measurements is the contributions of the electronic structure of the tip to the measured I-V spectra. We have not found evidence for tip dominated structure in our I-V results, and believe this occurs since the electronic states of the tip are resonance-like features which are much broader in energy than the surface states we probe. The two-dimensional current image of the surface states are less prone to tip electronic structure effects since the tip states are the same during the scan.

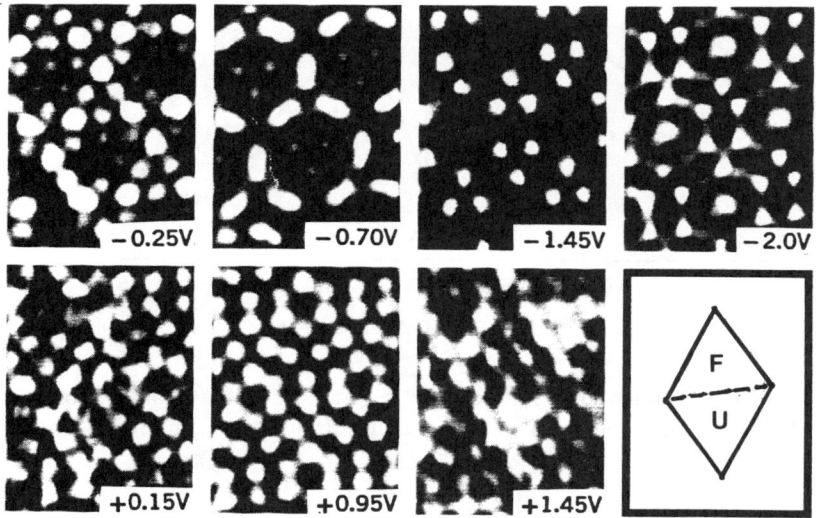

Figure 4. Current images of Si(111) 7×7 at selected negative and positive bias as indicated with the feedback stabilized at +2 V. The orientation and location of the faulted (F) and unfaulted (U) halves of the unit cell are indicated.

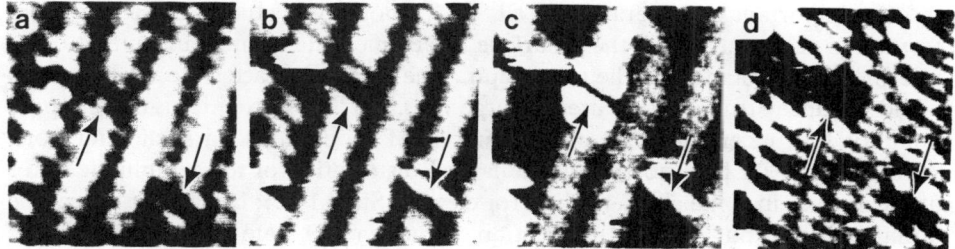

Figure 7. Grey scale images of (a) the surface topography of Si(111) 2×1 obtained with a −2 V bias and simultaneously acquired current images for a bias of (b) −1.2 V, (c) −0.8 V and (d) +0.8 V. The arrows denote two missing dimer defects discussed in the text. The distance between rows of dimers is 7.5Å.

Si(100) 2×1

In Fig. 5a we show a topographic image of the Si(100)-2×1 surface. In general this surface has more atomic scale defects than on Si(111) 7×7. As discussed in our previous studies of Si(100) 2×1 we find both buckled (asymmetric) and non-buckled (symmetric) dimers.[4] The symmetric dimers appear as the uniform rows and the buckled dimer as the zig-zag structure along the rows as seen, for example, in the left side of Fig. 5. The filled and empty surface states of the symmetric and buckled dimers have been imaged and can also be seen in 5c and 5d respectively. These states are discussed in more detail elsewhere.[18] Briefly, the occupied states on the dimers have a broad surface state localized between them while the unoccupied states have charge lobes over each of the dimers with a node in between.[21] The similarities in the energy of the symmetric and asymmetric dimers suggest that the symmetric dimer is a time average of dynamically buckling dimers.[18]

SUMMARY AND CONCLUSIONS

Scanning tunneling microscopy contains both geometric and electronic information. Here we discuss some of the procedures that allow us to separate these two contributions and image surface electronic states with atomic scale resolution. This ability to obtain both topographic and local electronic structure information in one experiment is particularly novel and can potentially allow one to understand the bonding and structure in one experiment. Any structural model proposed to explain the local geometry must also account for the local surface states one observes. In a sense it forces one to derive a "self-consistant" model of the surface.

An important consideration in all STM topographic studies of semiconductors is the degree to which these surface electronic states can contribute to or be interpreted as distances in the topograph. The largest surface state we have seen could produce structural features as large as 0.8Å. This provides a useful upper limit for the uncertainties in interpreting STM topographs on semiconductors. Also Feenstra[11,15] has shown that the lateral distribution of topographic features at the surface in some cases, such as for cleaved Si(111) 2×1, may not reflect the atom positions, but instead the location of the surface state charge distribution. In general this makes it more important to know the relation between electronic and geometric structure when interpreting STM images.

The ability to image specific surface state features should further provide a means of chemical imaging in STM. Valence orbitals of different surface compounds can be used as a more specific signature of these different materials than the changes in effective barriers or work functions would. This type of surface state imaging is also limited as these states are the most sensitive to the details of bonding and would sensitively dependent on their environment. Despite these present limitations, electronic imaging with the STM promises to provide important new information about bonding and structure at surfaces.

ACKNOWLEDGEMENT

The authors wish to acknowledge the Office of Naval Research for partial support of this work.

REFERENCES

1. G. Binnig and H. Rohrer, Helv. Phys. Acta, **55**, 726 (1982).
2. G. Binnig, H. Rohrer, Ch. Gerber, and E. Weibel, Phys. Rev. Lett. **50**, 120 (1983).
3. J. E. Demuth, R. J. Hamers, R. M. Tromp, and M. E. Welland, IBM Jour. of Res. and Dev. **30**, 396 (1986).
4. R. J. Hamers, R. M. Tromp, and J. E. Demuth, Phys. Rev. B **34**, 1388 (1986).
5. J. Moreland, S. Alexander, M. Cox, R. Sonnenfeld, and P. K. Hansma, Appl. Phys. Lett. **43**, 387 (1983).
6. R. S. Becker and J. A. Golorchenko, to be published.
7. R. M. Tromp, R. J. Hamers, and J. E. Demuth, Phys. Rev. B **34**, 1388 (1986).
8. A. Baratoff, Physica **127B**, 143 (1984).
9. J. Tersoff, and D. R. Hamann, Phys. Rev. B **31**, 805 (1985).
10. N. D. Lang, Phys. Rev. Lett. **56**, 1164 (1986).

11. R. M. Feenstra, W. A. Thompson, and A. P. Fein, Phys. Rev. Lett. **56**, 608 (1986).
12. R. S. Becker, J. A. Golovchenko, and B. S. Swartzentraber, Phys. Rev. Lett. **55**, 987 (1985).
13. G. Binnig, K. H. Frank, H. Fuchs, J. Kübler, N. Garcia, B. Riehl, H. Rohrer, F. Salvan, and A. R. Williams, Phys. Rev. Lett. **55**, 991 (1985).
14. W. J. Kaiser, and R. C. Jaklevic, IBM Jour. of Res. and Dev. **30**, 411 (1986).
15. R. M. Feenstra, J. Stroscio, and A. P. Fein, to be published.
16. G. Binnig and H. Rohrer, IBM Jour. of Res. and Dev. **30**, 355 (1986).
17. R. J. Hamers, R. M. Tromp, and J. E. Demuth, Phys. Rev. Lett. **56**, 1972 (1986).
18. R. M. Tromp, R. J. Hamers and J. E. Demuth, Science **234**, 304 (1986).
19. F. J. Himpsel, D. Straub, and Th. Fauster, *Proc. 17th Int'l Conf. Phys. Semiconductors,*, Ed. J. Chadi and W. A. Harrison, Springer Verlag, NY, p. 39 (1985).
20. J. E. Northrup, Phys. Rev. Lett. **57**, 154 (1986).
21. J. Ihm, M. L. Cohen, and D. J. Chadi, Phys. Rev. B **21**, 4592 (1980).

SECOND HARMONIC AND SUM FREQUENCY GENERATION ON DYE-COATED SURFACES USING COLLINEAR AND NON- COLLINEAR EXCITATION GEOMETRIES

R. E. Muenchausen, D. C. Nguyen, R. A. Keller, and N. S. Nogar
Chemistry and Laser Science Division, Los Alamos National Laboratory,
Los Alamos, NM. 87545

ABSTRACT

Doubly resonantly enhanced sum frequency generation from rhodamine 6G monolayers adsorbed on glass substates is compared with resonantly enhanced second harmonic generation using a collinear excitation geometry. Second harmonic and sum frequency generation with a non-collinear excitation geometry is also reported where spatial filtering of the non-collinear output is shown to increase the scattered light rejection by more than 4 orders of magnitude.

SUMMARY

Second harmonic generation (SHG) with a collinear excitation geometry is becoming an important tool for surface studies.[1] Phase-matching on the surface is inherently satisfied since the nonlinear generation occurs over distances corresponding to only a few monolayers. We report resonantly enhanced sum frequency generation (SFG) for both collinear[2] and non-collinear[3] excitation geometries.

A pulsed dye laser probed the $S_1 \leftarrow S_0$ transition of rhodamine 6G, (ca. 525 nm) collinearly with the Nd:YAG fundamental (1064 nm) such that the sum frequency output was also resonant with the $S_2 \leftarrow S_0$ transition as shown in Fig. 1 which also shows the pumping scheme for resonantly enhanced SHG.

The results show that doubly resonant SFG is enhanced by more than 2 orders of magnitude relative to resonant SHG for submonolayer coverages of rhodamine 6G. In addition, the measured SFG excitation spectrum of rhodamine 6G is in good agreement with the calculated excitation spectrum shown in Fig. 2.

In a separate experiment, results for the non-collinear excitation of second harmonic and sum frequency generation from rhodamine 6G coated substrates demonstrate that the efficiency of nonlinear generation is not significantly reduced for crossing angles less than 15°. The output beam coherence can be exploited to achieve spatial separation from the incident beams, as shown

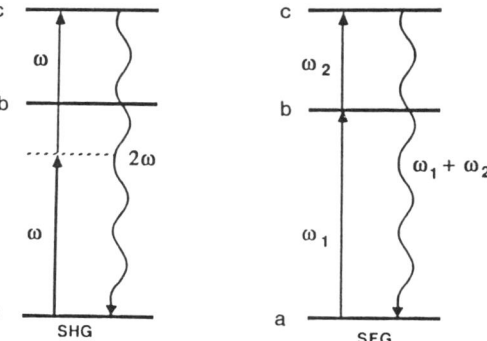

Fig. 1. Pumping schemes for resonantly enhanced SHG and SFG in rhodamine 6G.

for SHG and SFG in Fig. 3. Note that the noncollinear SHG output bisects the collinear outputs and that the relative signal intensities are in good agreement with the 2:1 ratio of noncollinear to collinear expected for equal input beam irradiances as shown in Fig. 3a. In Fig. 3b the noncollinear SFG output is resonantly enhanced angularly separated from the reflected visible beam by 2°.

The resulting spatial filtering of the nonlinear output possible with the noncollinear excitation geometry can increase the scattered light rejection by more than four orders of magnitude for crossing angles small as 6° as shown in Fig. 4 for SHG.

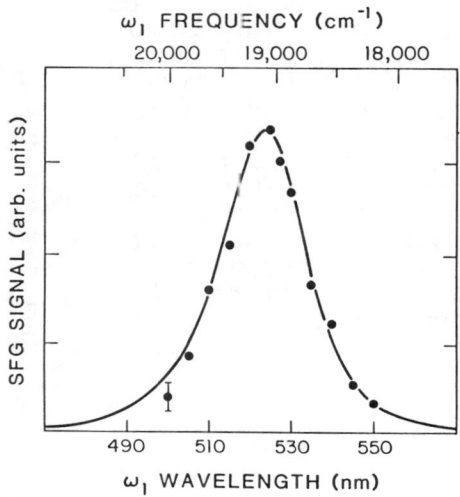

Fig. 2. SFG excitation spectrum, dots are the experimental data with the theory (solid curve) overlayed.

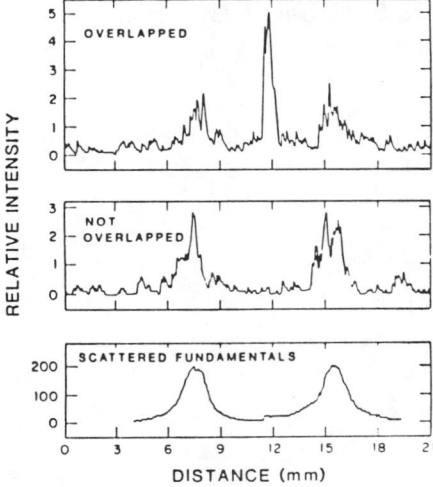

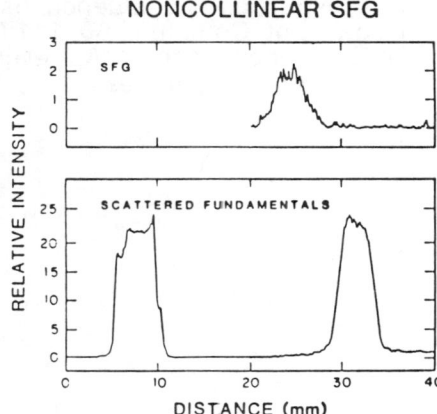

Fig. 3. a) Spatial location of the reflected input beams at 695 nm (bottom), SHG output at 347 nm for no input beam overlap on the surface (middle), and SHG output when the input beams are overlapped on the surface (top). The middle peak is the noncollinear SHG output. b) SFG output (top) at 351 nm with respect to the spatial location of the reflected input beams (bottom) at 1064 nm (left) and 525 nm (right).

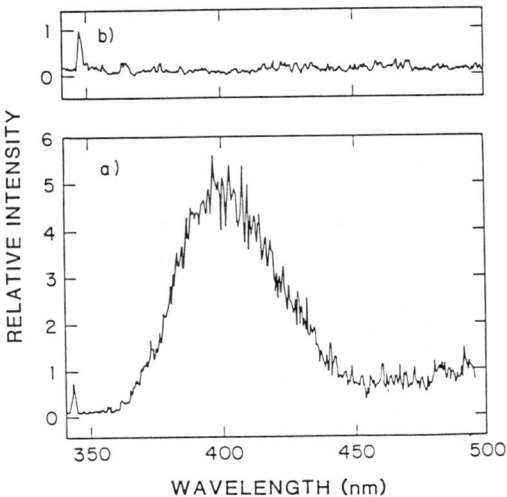

Fig. 4. a) Monochromator scan showing collinear SHG output and the two-photon excited fluorescence of rhodamine 6G. The increasing background is due to the reflected fundamental at 695 nm. b) Same scan conditions except that the spatial filter passes only the noncollinear SHG ouput. The color filter used for collinear detection has also been removed.

REFERENCES

1. Y. R. Shen, J. Vac. Sci. Technol. B, **3**, 1464 (1985).
2. D. C. Nguyen, R. E. Muenchausen, R. A. Keller and N. S. Nogar, Opt. Commun., **60**, 111 (1986).
3. R. E. Muenchausen, R. A. Keller and N. S. Nogar, J. Opt. Soc. Am. B; **4** (1987); in press.

MICROMETER-SIZE DROPLETS AS OPTICAL CAVITIES:
LASING AND OTHER NONLINEAR EFFECTS

Richard K. Chang
Yale University
Section of Applied Physics and Center for Laser Diagnostics
New Haven, Connecticut 06520

ABSTRACT

The morphology of a droplet enhances the internal and external intensities at specific locations and produces high Q-factor feedback for specific wavelengths of emission generated inside a droplet. Coherent nonlinear emissions have been observed from micrometer-size droplets along with other nonlinear effects such as wavelength broadening due to the intensity-dependent refractive index. Physical and chemical properties of the droplet can be deduced from such coherent emissions. Laser-induced droplet explosion and a laser-supported detonation wave result at input intensity levels higher than those necessary to observe such nonlinear effects. The use of spatially resolved spectroscopy to analyze emissions resulting from microexplosion has provided some preliminary details about droplet and air breakdown mechanisms.

DISCUSSION

For micrometer-size liquid droplets (with radius a) surrounded by a medium with lower index of refraction, a droplet can be envisioned as a lens to concentrate the incident laser intensity (with wavelength $\lambda < a$) and as an optical cavity to provide feedback for the internally generated radiation. These two droplet roles will be reviewed in the context of coherent emissions such as lasing, stimulated Raman scattering (SRS), coherent anti-Stokes Raman scattering (CARS), coherent Raman mixing (CRM), and phase-modulation broadening. In addition, the effects of laser-induced explosion of droplets, when the incident laser intensity exceeds the breakdown threshold, will also be briefly reviewed.

The results of a Lorenz-Mie calculation[1] for the internal field and the external near field are schematically shown in Fig. 1. Three regions along the principal diameter, which is parallel to the incident beam direction, have enhanced intensities: (1) outside the shadow face, located at ~1.5a, with 10^3 enhancement; (2) inside the shadow face, located at ~0.75a, with 10^2 enhancement; and (3) inside the illuminated face, located at ~ -0.75a, with ~1/3 of the intensity at ~0.75a. Lorenz-Mie calculations can also specify the wavelengths[2] which are commensurate with the numerous morphology-dependent resonances (MDR's, specified by mode numbers and mode orders) of a sphere with a and with ratio of refractive index m. The number of MDR's within a wave-number interval (e.g., 10 cm^{-1}) has been calculated,[3] depending on the specific range of Q-factors or linewidths to be included. Lorenz-Mie calculations therefore provide detailed

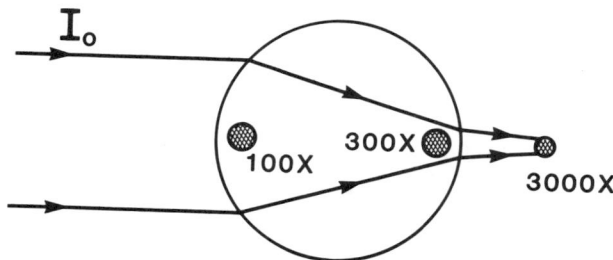

Fig. 1. Summary of Lorenz-Mie calculations of enhancements and locations of internal and external (near-field) intensity for a micrometer-size droplet upon illumination by an input beam with I_0 intensity and with wavelength in the visible range.

information as to where the pumping intensity will be dominant within and outside the droplets and at what wavelengths in the emission spectrum the internally generated radiation will experience feedback with specific Q-factors.

Geometric optics can provide a useful and qualitative picture of the intensity enhancement and optical feedback associated with droplet morphology. The curved droplet interface refracts and reflects the incident rays to form the three focal volumes shown in Fig. 1. Fluorescence and spontaneous Raman scattering are amplified within the two internal focal volumes. While most of these amplified rays emerge from the droplet, some are trapped within the droplet by total internal reflection. The angle at which the rays strike the interface is related to the MDR mode order, and the number of total internal reflections during one round trip around the circumference is related to the MDR mode number. The droplet acts as an optical cavity when the wave front at the beginning is the same as that at the end of a round trip around the droplet circumference. Lasing and SRS are achieved when the round-trip gain, located in the two internal high-intensity regions shown in Fig. 1, exceeds the round-trip loss due to absorption and/or the nonzero leakage from the droplet. Depending on the Q-factor of the cavity, the internal rays can make a large number of round trips, i.e., the cavity lifetime can be long and can even exceed the input laser pulse duration.

Lasing and SRS spectra consisting of nearly equally spaced wavelength peaks have been observed when the input intensity is adjusted to be near the threshold values for the lasing and SRS processes.[4,5] The wavelengths of each emission peak correspond to MDR's with different mode numbers but the same mode order. Several physical and chemical properties pertaining to the droplet can be deduced from these discrete emission peaks in the lasing and SRS spectra: (1) radius, by fitting the measured spacing of the peaks

with the Lorenz-Mie calculations for a sphere with a and m;
(2) evaporation and condensation rates, by measuring the peak shifts with time;[6] (3) dynamic surface tension, by measuring the oscillation frequency of the peak resulting from the droplet shape oscillation after a perturbation which distorts the droplet;[7] (4) bulk viscosity, by measuring the decay of the peak shift as the shape oscillations dampen;[7] and (5) chemical content, by noting the energy loss (i.e., the Raman frequency shifts) of the emission peaks relative to the incident frequency.[8] As the pumping intensity increases beyond the threshold values, the emission spectra no longer consist of discrete equally spaced wavelength peaks belonging to one set of MDR's. In fact, the spectra evolve into several sets of equally spaced wavelength peaks, and eventually the peaks merge into a continuum within the fluorescence or Raman gain profile as more and more sets of MDR's reach the threshold condition.

Four-wave mixing processes (CARS and CRM) have also been observed in a single droplet.[9,10] Compared to the large sample volume case (e.g., a cell 1 cm long), the phase-matching requirement for the droplet is relaxed. The angle-tuning curve was noted to be broad and asymmetric. The breadth is consistent with the short interaction length within the droplet (shown in Fig. 1). The asymmetry is associated with the spatial overlap of the two high-intensity regions of the pump and Stokes radiation as the angle between these two beams is changed from the optimum phase-matched angle. No MDR associated peaks were observed in the CARS and CRM spectra, implying that the conditions for phase-matching and MDR's cannot be simultaneously satisfied within the droplet.

Even though the droplet size is many orders of magnitude smaller than the length of optical fibers used in nonlinear optical experiments, several nonlinear optical processes normally observed from meter-long fibers have been observed from micrometer-size droplets. Multiorder Stokes shifts [up to (n=14)th order] have been observed in CCl_4 droplets, as a result of successive first-order stimulated Raman pumping of the nth order by the (n-1)th order.[11] The nonmonotonic decrease of the observed multiorder SRS peaks is related to which MDR's happen to fall within the Raman linewidth of the nth order.[3] Should the Q-factor for the (n-1)th mode be large, the leakage radiation of the (n-1)th mode from the droplet would be less (hence, the detected signal would be less). The internal radiation remaining to pump the nth order would therefore be more [hence, the nth order would be more intense than the (n-1)th order].

In addition, wavelength broadening due to the intensity-dependent index of refraction effect (phase-modulation broadening), well known in propagation through optical fibers, has also been observed when an intense beam is incident on a micrometer-size liquid droplet.[12] The elastic scattering from a CS_2 droplet was noted to be asymmetrically broadened by as much as 1000 cm^{-1} on the longer wavelength side. Similarly, all the CS_2 multiorder SRS peaks were asymmetrically broadened on the longer wavelength side. Such asymmetric phase-modulation effects in droplets imply that internal intensity can substantially change the CS_2 low-intensity value n_0 to $n(t) = n_0 + n_2 I$ via molecular orientation alignment. Such large phase-modulation effects are consistent with the fact that the rays pass through the two high-intensity regions many times, thereby accumulating phase shifts during each passage through the two regions with n(t).

The demonstration of the intensity-dependent refractive index n(t) in a droplet implies that the standard Lorenz-Mie formalism must be modified to treat the field and the material parameter in a self-consistent manner, i.e., the nonuniform internal intensity distribution can alter the refractive index, and the resultant spatial variation of the refractive index can alter the internal intensity distribution. Because the optical path length for a wave propagating within the droplet can be intensity dependent via n(t), the MDR's can also be intensity dependent. Therefore, intensity-dependent mode pulling and pushing and optical bistability can result via the n(t) effect. To date, optical bistability in a levitated droplet has been demonstrated through thermal heating of the droplet which causes its radius to change.[13]

Laser-induced droplet vaporization and explosion occur when the input intensity is increased beyond the values needed to observe the above-mentioned nonlinear effects. Spatially resolved emission spectra of the droplet and the accompanying forward and backward plumes can be detected[14] by imaging the droplet and the plumes onto the entrance slit of a spectrograph (see Fig. 2). At the exit plane of the spectrograph, a vidicon camera detects the spatially resolved wavelength-dispersed emission spectra along a line passing through the forward plume, the droplet, and the backward plume.

Figure 3 shows the spatially resolved emission spectra from an ethanol (ETOH) droplet (a ≃ 35 μm) containing rhodamine dye (R-590) and being irradiated by ~6 GW/cm^2 input intensity. The vidicon tracks pertaining to the droplet itself are designated by shaded bars, and the input laser direction is designated by an arrow. Four different wavelength regions are displayed to illustrate the various

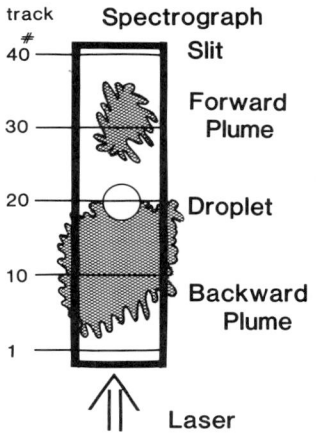

Fig. 2. Schematic of the focused image on the entrance slit of a spectrograph equipped with a two-dimensional vidicon camera at the exit plane. The forward and backward plumes and the droplet are spatially resolved along one dimension by different track numbers of the vidicon camera. [From Ref. 15]

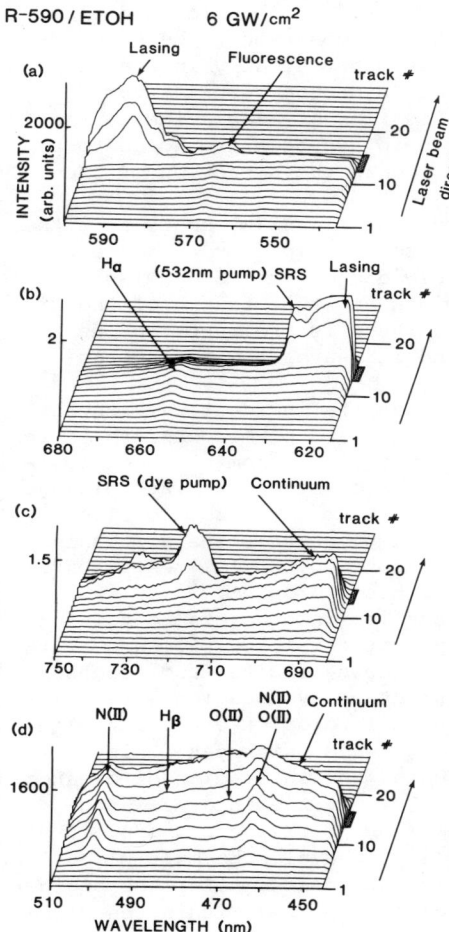

Fig. 3. Spatially resolved emission spectra from within the droplet and the backward plume. Track numbers assigned to the region within the droplet are shaded and the direction of the laser relative to the droplet is indicated. Four different wavelength regions detected by the vidicon camera are shown (a)-(d). The input laser wavelength is 532 nm, and the radius of the ethanol (ETOH) droplets containing rhodamine dye (R-590) is ~35 μm. Note that the lasing and SRS radiation is confined within the droplet. The plasma continuum and fluorescence are found both inside and outside the droplet. The hydrogen Balmer lines and the O(II) and N(II) lines are detected only in the backward plume. No emission is detected in the forward plume. [From Ref. 14]

emission processes occurring when a lasing droplet explodes.

Figure 3(a) shows that lasing is confined within the droplet. At this high-intensity level, the lasing spectra are devoid of discrete MDR peaks as many different sets of MDR's with various mode numbers and mode orders can exceed the lasing threshold. However, the fluorescence emission exists both inside the droplet and in the backward plume, the latter coming from the dye molecules streaming from the illuminated face of the droplet. Because of the dye absorption, the illuminated face of the droplet is heated more than the other portions of the droplet.

Figure 3(b) shows the laser emission for a wavelength portion which is longer than that shown in Fig. 3(a), as well as the ethanol SRS pumped by the incident laser (at 0.532 μm). Both these coherent emissions are <u>confined</u> within the droplet. In the backward plume, the hydrogen Balmer line emission (H_α) is observed to be superimposed upon a plasma continuum which is monotonically increasing toward the shorter wavelength. Atomic hydrogen results from the decomposition of ethanol after the creation of plasma inside the droplet. Both the plasma and the atomic hydrogen are noted to be streaming from the illuminated face toward the direction of the laser beam.

In an even longer wavelength region, Fig. 3(c) shows that the ethanol SRS within the droplet can also be pumped by the internal dye lasing intensity. The broad linewidth of this SRS emission is a convolution of the Raman gain linewidth and the breadth of the droplet lasing spectrum. The plasma continuum is prominent inside the droplet and in the backward plume and increases toward shorter wavelengths.

In a wavelength region shorter than the incident wavelength [see Fig. 3(d)], the emission in the backward plume consists of the plasma continuum, the hydrogen Balmer line (H_β from the decomposition of ethanol), and the emission lines of singly ionized nitrogen and oxygen [N(II) and O(II)]. The continuum continues to rise toward the UV. The presence of N(II) implies that the air behind the droplet illuminated face has been ionized. O(II) results from the decomposition of air and of the ethanol droplet.

The laser-induced breakdown mechanism in droplets containing dye is different from that in transparent droplets. Because of the dye absorption, the initiation of breakdown results from superheating the droplet at the internal high-intensity location within the illuminated face. Superheating results in droplet explosion with an accompanying backward propagating shock wave which can ionize the gas and vapor. In the transparent case, the initiation of breakdown results from multiphoton ionization (MPI) at the internal high-intensity location within the shadow face.[15] MPI can produce the priming electrons necessary for subsequent avalanche multiplication and breakdown within the droplet. In both cases, after the beginning part of the laser pulse has created a plasma within the droplet, the subsequent part of the laser pulse is substantially absorbed by the plasma and further feeds the heating process. This process forms a laser-supported detonation wave which propagates toward the direction of the laser beam. The plasma density and temperature should be extractable from emission spectra such as those shown in Fig. 3 from the Stark shifts and broadening of the emission lines and from the intensity ratio of the emission lines with the continuum.

In conclusion, the morphology of a droplet provides an interesting case for studying nonlinear optical effects in 10^{-8} cm^3 of liquid (some tens of femto-liter). Many nonlinear optical effects which have been reported to date in micrometer-size droplets are normally associated with centimeter-long liquid samples or meter-long optical fibers. From nonlinear effects such as SRS and CARS, it is possible to determine several physical parameters of a droplet (size, evaporation and condensation rates, dynamic surface tension, and bulk viscosity) and chemical speciation of the content within the droplet as the droplets flow past the laser beam. Experimental results can be qualitatively explained by combining nonlinear optics and Lorenz-Mie formalisms. However, to be more quantitative, the latter must be expanded to include the nonuniform perturbation of the complex refractive index by the internal intensity distribution. The study of spatially and temporally resolved emission spectra of microexplosions has furthered our understanding of laser-induced air breakdown and laser-supported detonation waves when aerosols are present in the air.

This work was done in close collaboration with the following colleagues: W.-F. Hsieh, J. H. Eickmans, J. B. Snow, S.-X. Qian, P. W. Barber, and H.-M. Tzeng. Partial support of this work by the U. S. Air Force Office of Scientific Research (Contract No. F49620-85-K-0002) and the U. S. Army Research Office (Contract No. DAAG29-85-K-0063) is also gratefully acknowledged.

REFERENCES

1. D. S. Benincasa, P. W. Barber, J.-Z. Zhang, W.-F. Hsieh, and R. K. Chang, "Spatial Distribution of the Internal and Near-Field Intensity of Large Cylindrical and Spherical Scatterers," submitted to Appl. Opt.
2. P. R. Conwell, P. W. Barber, and C. K. Rushforth, J. Opt. Soc. Am. A $\underline{1}$, 62 (1984).
3. S. C. Hill and R. E. Benner, J. Opt. Soc. Am. B $\underline{3}$, 1509 (1986).
4. H.-M. Tzeng, K. F. Wall, M. B. Long, and R. K. Chang, Opt. Lett. $\underline{9}$, 499 (1984); H.-B. Lin, A. L. Huston, B. L. Justus, and A. J. Campillo, Opt. Lett. $\underline{11}$, 614 (1986).
5. J.B. Snow, S.-X. Qian, and R.K. Chang, Opt. Lett. $\underline{10}$, 37 (1985).
6. H.-M. Tzeng, K. F. Wall, M. B. Long, and R. K. Chang, Opt. Lett. $\underline{9}$, 273 (1984).
7. H.-M. Tzeng, M. B. Long, R. K. Chang, and P. W. Barber, Opt. Lett. $\underline{10}$, 209 (1985).
8. J. Eickmans, S.-X. Qian, and R. K. Chang, in *1. World Congress on Particle Technology, Part I. Particle Characterization*, K. Leschonski, ed. (NMA, Nuremberg, 1986), p. 125.
9. S.-X. Qian, J. B. Snow, and R. K. Chang, Opt. Lett. $\underline{10}$, 499 (1985).
10. S.-X. Qian, J. B. Snow, and R. K. Chang, in *Laser Spectroscopy VII*, T. W. Hänsch and Y. R. Shen, eds. (Springer-Verlag, Berlin, 1985), p. 204.
11. S.-X. Qian and R. K. Chang, Phys. Rev. Lett. $\underline{56}$, 926 (1986).
12. S.-X. Qian and R. K. Chang, Opt. Lett. $\underline{11}$, 371 (1986).
13. S. Arnold, in *International Laser Science Conference Technical Digest* (OSA, Washington, DC, 1986), p. 47.
14. W.-F. Hsieh, H.-M. Tzeng, and R. K. Chang, in *Special Issue of the Annual Report of the Institute of Physics, Academia Sinica (Taiwan)*, Vol. 16, 1986 in honor of Prof. Ta-You Wu's 80th birthday, p. 1.
15. J. H. Eickmans, W.-F. Hsieh, and R. K. Chang, "Laser-Induced Explosion of H_2O Droplets: Spatially Resolved Spectra," Opt. Lett., in press.

RAMAN-MIE SCATTERING FROM OPTICALLY LEVITATED SINGLE PARTICLES

W. Kiefer
Karl-Franzens-Universität Graz, Institut für Experimentalphysik,
A-8010 Graz, Austria

ABSTRACT

We report on a Raman microprobe technique where micron-sized solid particles or liquid droplets are trapped in stable optical potential wells using only the force of radiation pressure from a continuous gas laser. We demonstrate this technique with Raman spectra from single particles of sizes ranging between ca. 5 and 40 μm. The light scattering spectra of spherical particles of these sizes showed pronounced structural resonances. The observed resonances could be assigned by using the well-known Lorenz-Mie formalism, and a good correlation is found between these experimental and theoretically predicted Raman-Mie spectra.

INTRODUCTION

Recently there has been a good deal of interest in investigations of the light-scattering properties of dielectric spheres. Studies of the wavelength-dependent features of this type of scattering are of particular interest because of the sharp structural resonances that occur as the Mie size parameter $x = 2\pi a/\lambda$ (a = radius of sphere, λ = wavelength in the surrounding medium) is varied. These structural resonances have been predicted for incident and/or inelastically scattered wavelengths which correspond to the natural modes of dielectric spheres. The first observation of morphology-dependent resonances in Raman scattering from dielectric microparticles has been published by Owen et al.[1] Recently, we reported on the observation of such structural resonances in the Raman spectra of single microspheres of glass[2] and liquid droplets[3]. The experiments were performed applying a straightforward sample arrangement[4] where the particle is excited by laser radiation while free in space without any supporting sample holder. This was achieved by employing the technique of optical levitation by radiation pressure, which has been pioneered by Ashkin[5] some years ago. In this paper we summarize these results and also give a tutorial description of the optical levitation technique.

OPTICAL LEVITATION BY RADIATION PRESSURE

It has been shown by Ashkin[5] that optical levitation is based on the ability of laser light to trap nonabsorbing particles stably by the force of radiation pressure. In this technique a continuous wave vertically directed focused TEM_{00} Gaussian-mode laser beam supports

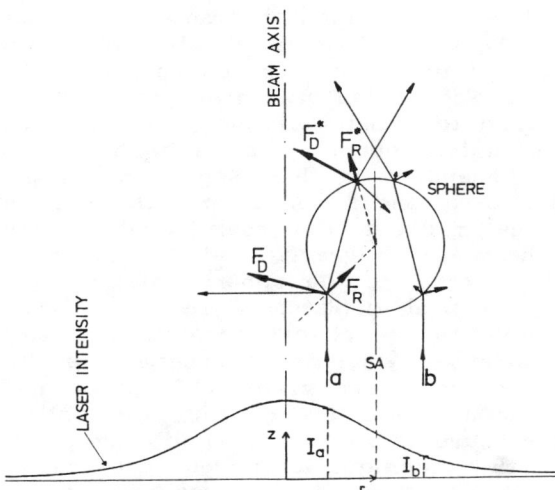
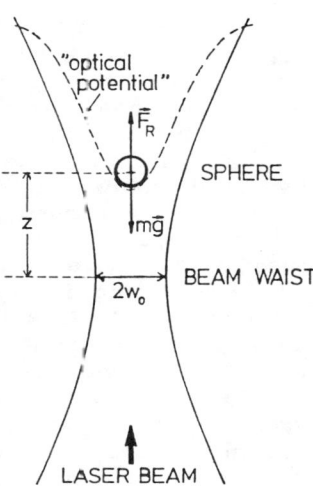

Fig. 1. Forces acting on a dielectric sphere situated off axis of a TEM_{00}-mode Gaussian laser beam traveling along the z - direction. a, b = light rays; F_D, F_D^* = deflection forces; F_R and F_R^* = reflection forces; * denotes output face, I_a, I_b = intensity of rays a and b, respectively; SA = sphere axis; r = radial distance of sphere center from beam axis. For further explanation see text.

Fig. 2. A schematic diagram of a spherical particle sitting at the equilibrium position above the focus of a TEM_{00}-mode Gaussian beam, where gravity (mg) and the axial light force in the upward direction (F_R) balance. Dashed line: schematically drawn optical potential due to the radial inward force.

the particle's weight and simultaneously pulls the particle transversely into the region of high light intensity on the beam axis. Ashkin[5] showed how the transverse component of the radiation force acts on a dielectric sphere located off axis in a Gaussian beam (see Fig. 1): consider a pair of light rays, a and b, situated symmetrically with respect to the center of the sphere, which is off axis by distance r. Both rays undergo reflection and deflection at the input and output faces. These result in radiation pressure forces F_R, F_R^* and F_D, F_D^* for reflection (suscript R) and deflection (suscript D), respectively (the star denotes the forces at the output face). All forces are located along the direction of the momentum changes for the rays and give accelerations in the $+z$ direction. The radial components of F_R and F_R^* add radially in the $-r$ direction for ray a and in the $+r$ direction for ray b. However, since ray a due to the Gaussian beam profile is more intense (intensity I_a, see Fig. 1) than ray b (intensity I_b), the radial component of F_D^a (F_D^a = vectorial sum of F_D and F_D^*

of ray a, not shown in Fig. 1) is greater than the radial component of $F_D{}^b$ ($F_D{}^b$ = vectorial sum of F_D and $F_D{}^*$ of ray b, also not shown in Fig. 1). The total radial force thus is directed towards the beam axis and the sphere as a whole is accelerated inward and forward.

The radial inward force and the axial radiation pressure force allow us therefore to construct a true optical potential well. This is schematically shown in Fig. 2. Dependent on the laser intensity, the particle size, and the focal length of the focusing lens, there is an equilibrium point some distance z above the beam waist of the focused TEM_{00}-mode Gaussian beam, where gravity and the total axial light force acting in upward direction to the sphere balance. This equilibrium is stable since any vertical displacement from this point results in a restoring force due to the change in light intensity caused by the beam divergence, and any lateral displacement results in a restoring force due to the transverse gradient force as discussed above. If spherical particles are brought in this optical potential well they will rest for hours stably at the equilibrium point, as shown by Ashkin[5]. Thus, if this technique is applied such that at the sample position of a usual Raman spectrometer a vertically directed continuous wave laser beam stably traps and simultaneously excites a micronsized particle, Raman-Mie-spectra of spherical particles can ideally be obtained.

Several techniques can be employed to bring micronsized particles to the stable minimum of the optical potential well, allowing to take inelastic light-scattering spectra from the trapped particle. Various techniques for levitating solid particles were descibed in reference 4. The technique to trap liquid droplets is outlined in reference 3. For details concerning Raman spectrometer, laser and laser power used, detection system, etc., we refer to our earlier publications[3,4].

THEORY OF MORPHOLOGY DEPENDENT RESONANCES

For dielectric spheres in the Mie-size range, the Lorenz-Mie formalism can be used to calculate the precise positions of the peaks associated with the morphology-dependent resonances[6-8].

As reviewed by Barber[8], the scattering efficiency for a dielectric sphere, is given by

$$Q_s = \frac{2}{x^2} \sum_{n=1}^{\infty} (2n+1)(|a_n|^2 + |b_n|^2), \tag{1}$$

where x is the Mie size parameter and a_n and b_n are the expansion coefficients given by

$$a_n = \frac{j_n(x)[mxj_n(mx)]' - m^2 j_n(mx)[xj_n(x)]'}{h_n^{(2)}(x)[mxj_n(mx)]' - m^2 j_n(mx)[xh_n^{(2)}(x)]'} \tag{2}$$

and

$$b_n = \frac{j_n(x)[mxj_n(mx)]' - j_n(mx)[xj_n(x)]'}{h_n^{(2)}(x)[mxj_n(mx)]' - j_n(mx)[xh_n^{(2)}(x)]'}, \tag{3}$$

where m is the relative refractive index and j_n and $h_n^{(2)}$ are the spherical Bessel functions and Hankel functions of the second kind, respectively. The primes denote differentiation with respect to the argument of the function. The coefficients a_n and b_n show resonance behavor and each of them can be associated with a structural mode of electromagnetic vibration of the sphere[8].

Similarly to the a_n and b_n coefficients, which correspond to the TM and TE modes of the scattered field, respectively, the c_n and d_n coefficients (TM and TE modes of the internal field, respectively), defined by[7]

$$c_n = \frac{mi/x}{h_n^{(2)}(x)[mxj_n(mx)]' - m^2 j_n(mx)[xh_n^{(2)}(x)]'}, \quad (4)$$

$$d_n = \frac{i/x}{h_n^{(2)}(x)[mxj_n(mx)]' - j_n(mx)[xh_n^{(2)}(x)]'}, \quad (5)$$

exhibit also sharp peaks at certain values of the size parameter. The c_n and d_n coefficients can further be used to calculate the total (volume averaged) internal electric field intensity as a function of x:

$$I_{\text{total}}(x) = \frac{3}{a^3} \int_0^a dr r^2 I(r), \quad (6)$$

where

$$I(r) = \frac{1}{4\pi} \int_0^{2\pi} d\varphi \int_0^{\pi} \sin\theta d\theta \mathbf{E}^i(r,\theta,\varphi) \mathbf{E}^{i*}(r,\theta,\varphi)$$

$$= \frac{E_0^2}{2} \cdot \sum_{n=1}^{\infty} \{(2n+1) j_n^2(m 2\pi r/\lambda) |d_n|^2$$

$$+ [(n+1) j_{n-1}^2(m 2\pi r/\lambda) + n j_{n+1}^2(m 2\pi r/\lambda)] |c_n|^2 \} \quad (7)$$

is the internal electric field intensity $\vec{E}\vec{E}*$ averaged over all spherical angles. $\mathbf{E}^i$ is the vector of the internal electric field and E_O is the incident field strength. For further details see references 3 and 6-8.

RAMAN-MIE SPECTRA OF SOLID MICROSPHERES

Fig. 3 shows the experimentally observed light-scattering spectrum (upper trace) from an optically levitated glass sphere with a diameter of ca. 27 μm. The lower part of Fig. 3 shows a Raman spectrum of a pellet pressed from spheres of the same glass material but with random diameters ranging between 0.5 and 35 μm. Comparison of the upper and lower spectra in Fig. 3 reveals that the general features of the pellet Raman spectrum are well reproduced in the spectrum of the levitated glass sphere. However, in addition, a regular structure of sharp peaks with spacing of about 40 cm^{-1} is observed in the spectrum of the micron-sized sphere. The observed ripple structure is interpreted as structural resonance initiated by the Raman light inside the glass sphere[2]. Since the Raman spectrum of glass is essentially continuous in the region of interest (see the lower spectrum of Fig. 3), there are always frequencies which coincide with the frequencies of the structural resonances of the particular sphere. Since at these frequencies the wavelength of the Raman emission corresponds to a resonance condition of the microparticle, the Raman emission will be enhanced at that wavelength. We

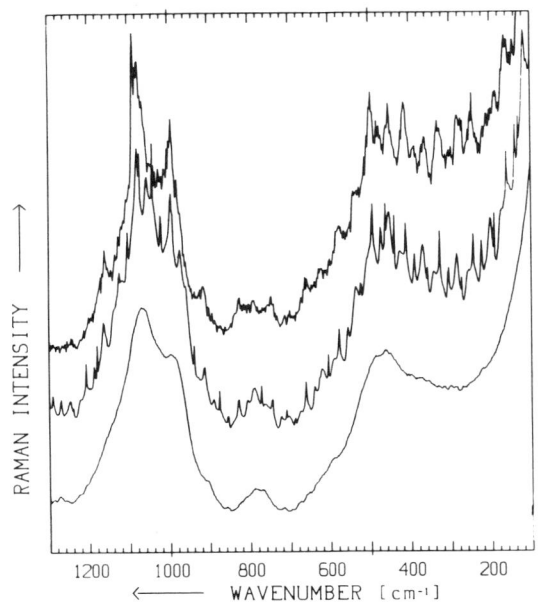

Fig. 3. Experimentally observed Raman-Mie-spectrum of an optically levitated glass sphere of diameter ca. 27 μm (upper trace). The middle trace is a theoretically calculated inelastic light scattering spectrum (see text) and the lower trace is a Raman spectrum of a pressed pellet of glass spheres with diameters ranging between 0.5 and 35 μm. Excitation wavelength λ = 514.5 nm; 0.5 watt. Slit width 3.5 cm^{-1}

have calculated[2] a synthetic spectrum of a single glass sphere (radius 13.831 μm, refractive index m = 1.577) by modulating the Raman spectrum of the glass material with the scattering cross section of the sphere (see Equ. 1). The middle spectrum of Fig. 3 shows this synthetic Raman-Mie spectrum. The observed and calculated spectra of the microsphere show fairly good agreement in the position and intensity of the observed structural resonances and hence prove the correct interpretation as a combined Raman-Mie scattering mechanism. For further details see reference 2.

RAMAN-MIE SPECTRA OF LEVITATED LIQUID DROPLETS

Morphology dependent resonances were also observed in the light scattering spectra of levitated single droplets[3]. In order to avoid rapid evaporation of pure water droplets, we investigated the scattering spectra of mixtures of water with glycerol. In Fig. 4, we display in the lower field A part of the Raman spectrum (3200 - 3500 cm^{-1}) of an 1:6 water-glycerol mixture (refractive index m = 1.45), which was obtained from a large cell filled with this mixture. Curve B of Fig. 4 represents an observed Raman-Mie spectrum of an optically levitated single droplet of the same mixture as for A. The radius a of this dielectric liquid sphere is ca. 15±1 μm as measured with a microscope. In the upper field C of Fig. 4 we show a theoretical calculation of a synthetical Raman-Mie spectrum which one expects for a dielectric sphere of refractive index m = 1.45 and a radius of a = 14.422 μm, if the above outlined theory is applied. It should be mentioned that in this case we derived the resonances using the c_n and d_n coefficients[3]. Again, good agreement between the observed Raman-Mie spectrum and the calculated spectrum is obtained. Assumed

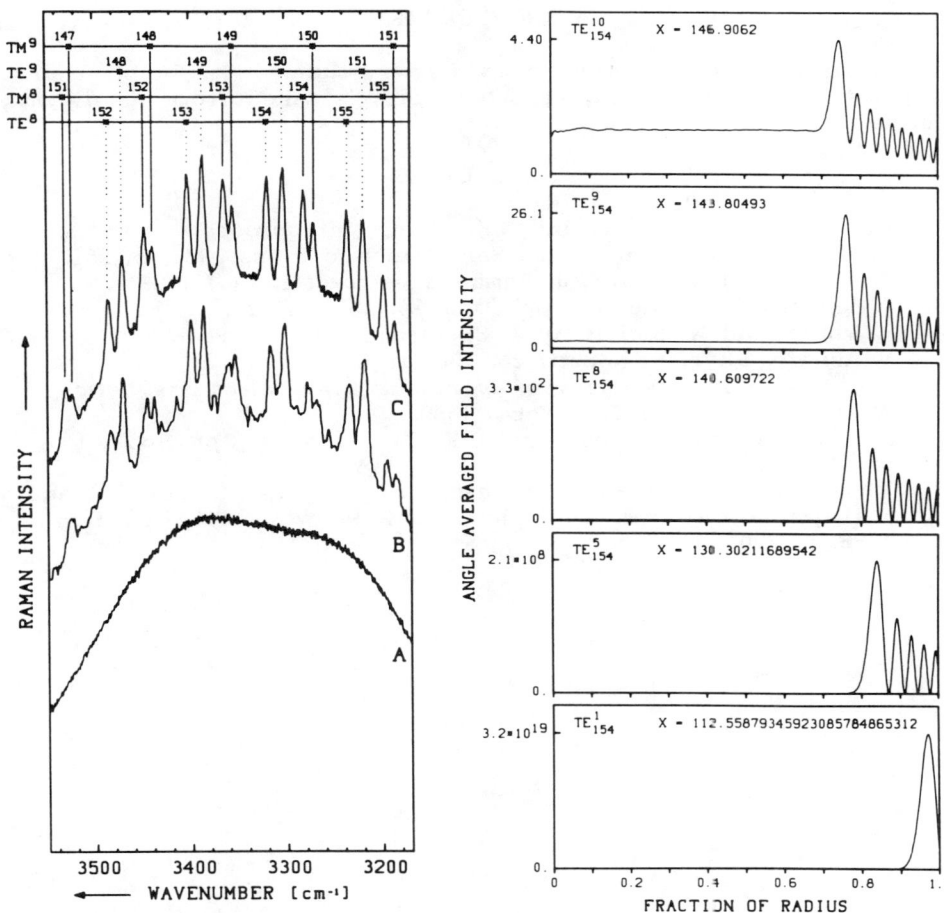

Fig. 4. A, O-H stretching region of the Raman spectrum of a liquid 1:6 water-glycerol mixture (refractive index $m = 1.45$); B, observed Raman-Mie spectrum of an optically levitated liquid droplet of the same mixture as for A (radius of sphere $a = 15$ μm; resolution = 7 cm^{-1}); C, calculated Raman-Mie spectrum.

Fig. 5. Calculated angle averaged field intensity within a lossless dielectric sphere as a function of the fraction of radius. Calculations are shown for $m = 1.45$ and x satisfying resonance conditions for TE$_{154}$ modes.

explanations, why only high order morphology dependent resonances are observed in this case, are given in reference 4, where we also have calculated the distribution of the internal electric field intensity as a function of sphere radius. Part of these calculations are reproduced in Fig. 5, where we display the angle averaged field intensity within a lossless dielectric sphere for a particular resonance condition. For further details we refer to reference 3.

ACKNOWLEDGMENT

Acknowledgment is made to my former student, Mr. Rudolf Thurn. A great part of the work, reported here, is taken from his diploma thesis.

REFERENCES

1. J.F. Owen, P.W. Barber, and R.K. Chang, in *Microbeam Analysis 1982*, K.F.J. Heinrich, editor, San Francisco Press 1982, p. 255.
2. R. Thurn and W. Kiefer, J. Raman Spectrosc. **15**, 411 (1984).
3. R. Thurn and W. Kiefer, Appl. Opt. **24**, 1515 (1985).
4. R. Thurn and W. Kiefer, Appl. Spectrosc. **38**, 78 (1984).
5. A. Ashkin, Phys. Rev. Lett. **24**, 156 (1970).
6. M. Kerker, *The Scattering of Light and Other Electromagnetic Radiation*, Academic, New York, 1969.
7. P.R. Conwell, P.W. Barber, and C.K. Rushforth, J. Opt. Soc. **A1**, 62 (1984).
8. P.W. Barber, in *Advances in Laser Science - I*, W.C. Stwalley and M. Lapp, editors, Am. Inst. Phys. Conf. Proc. No. 146, New York, 1986, p. 720.

RESONANCE LIGHT SCATTERING FROM A SUSPENSION OF MICROSPHERES

Thomas R. Lettieri and Egon Marx
National Bureau of Standards, Gaithersburg, MD 20899

ABSTRACT

Resonance light scattering spectra have been obtained from a liquid suspension of dielectric microspheres and then used to determine the mean diameter and width of the size distribution of the spheres.

INTRODUCTION

Light scattering cross sections of dielectric microspheres within a certain size range contain sharp resonances when plotted as a function of wavelength[1]. These resonance light scattering (RLS) spectra have been used by several experimenters to measure accurately the size and refractive index of single microspheres, either levitated[2], falling[3], or resting on substrates[4]. One could, in principle, determine the mean diameter of a size-distributed sample by measuring each sphere individually using RLS; this could take a prohibitive amount of time if thousands of spheres were to be measured. We show in this paper that it is easier and faster to measure the RLS spectrum of a collection of microspheres and then use this collective spectrum to get the mean diameter and the width of the size distibution[5].

EXPERIMENTAL APPARATUS

A schematic diagram of the experimental apparatus is shown in Fig. 1. The wavelength range of the dye laser was 570 to 620 nm with the Rhodamine 590 dye used. The beam was vertically incident into the sample cell, and both the perpendicular-polarized (I_v) and the parallel-polarized (I_h) components of the light scattered at 90° could be detected by proper orientation of the collection arm. Within the cell was a water suspension of nominal 10-μm-diameter polystyrene microspheres at a weight concentration of about 15 ppm.

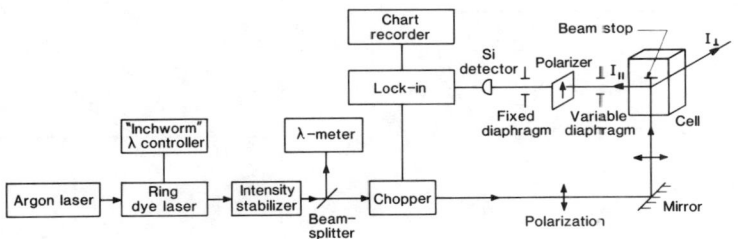

Fig. 1 Schematic diagram of experimental apparatus.

Several experimental factors should be considered to get usable RLS spectra from a collection of microspheres. First, the coefficient of variation, CV, of the size distribution must be small (CV = standard deviation/mean diameter); in the present case, it was less than 1 percent. Second, the acceptance angle of the collection optics must also be small (here, about 0.4° half-angle). Third, the scattering angle must be accurate to less than 1°. Fourth, the liquid suspension of particles must be kept free of foreign debris to prevent spurious peaks from appearing in the spectra. Fifth, the particle concentration must be adjusted to minimize multiple scattering while maintaining an adequate signal-to-noise ratio[5].

COMPUTER ANALYSIS

Calculated RLS spectra were generated on a CYBER 205 computer using a vectorized program based on Wiscombe's Mie-scattering code[6]. The spectra from a collection of microspheres were then calculated (assuming independent scattering) by integrating over diameter, which was weighted by a Gaussian distribution. This was done for about 10,000 diameters in each spectrum.

RESULTS

Figures 2 and 3 show calculated and experimental spectra for I_v and I_h. The top spectrum in each Fig. is a calculation for a single microsphere of 9.91 μm diameter, while the middle spectra are those calculated for a collection of microspheres having a size distribution standard deviation of 0.035 μm. The bottom spectra are experimental results. Good agreement between the calculated and the experimental spectra for the collection of microspheres can be seen in each Figure.

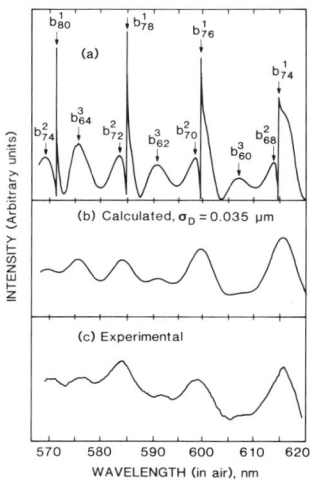

Fig. 2 I_v for 9.91-μm spheres.

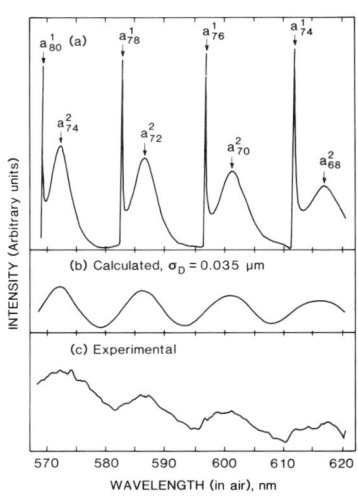

Fig. 3 I_h for 9.91-μm spheres.

The mean diameter of 9.91 µm was determined by a least-squares fit between the calculated and the experimental peak wavelengths, starting with a known particle diameter of 9.89 ± 0.04 µm. The fits yielded a series of nearly equispaced minimizing diameters[7], all but one of which could be ruled out due to either a poor match between the experimental and the calculated RLS spectra shapes, or a poor fit between the experimental and the calculated peak wavelengths. The mean diameter of 9.91 µm measured with RLS agrees very well with the values of 9.89 and 9.886 µm determined by two other techniques, one based on optical microscopy and the other on scanning electron microscopy.

The standard deviation of the size distribution was determined by qualitatively comparing the experimental RLS spectra for I_v and I_h with spectra calculated for a range of standard deviations (see Figures 2 and 3). The value of 0.035 µm measured this way was much smaller than that from other methods, which gave standard deviations of 0.09 and 0.1 µm. The origin of this disagreement is not known.

CONCLUSION

The resonance light scattering technique can be used to find an accurate mean diameter for a collection of dielectric microspheres in liquid suspension, provided that an approximate diameter is known a priori and that several experimental precautions are taken. The method is less time consuming than alternative single-particle methods, and will work with a range of diameters from about 5 to 50 µm, depending on the standard deviation of the size distribution.

REFERENCES

1. See for example: A. Ashkin and J. Dziedzic, Appl. Opt. 20, 1803 (1981).
2. A. Ashkin and J. Dziedzic, Phys. Rev. Lett. 38, 1351 (1977).
3. H.-M. Tzeng, K. F. Wall, M. B. Long, and R. K. Chang, Opt. Lett. 9, 273 (1984).
4. S. C. Hill, R. E. Benner, C. K. Rushforth, and P. R. Conwell, Appl. Opt. 23, 1680 (1984).
5. T. R. Lettieri and E. Marx, Appl. Opt. (in press).
6. W. J. Wiscombe, Appl. Opt. 19, 1505 (1980).
7. S. C. Hill, C. K. Rushforth, R. E. Benner, and P. R. Conwell, Appl. Opt. 24, 2380 (1985).

TWO-WAVE MIXING IN LIQUID
SUSPENSIONS OF MICROPARTICLES

R. McGraw and D. Rogovin
Rockwell International Science Center
Thousand Oaks, California 91360

ABSTRACT

The response of an artificial Kerr medium to moving electromagnetic gratings is examined. Such gratings arise in nondegenerate phase conjugation, signal processing, and two-wave mixing applications. The field equations for two-wave mixing in microparticle suspensions are presented and examined from the standpoint of energy conversion and gain. In particular, the out-of-phase component of the frequency-response function is related to the power dissipation.

INTRODUCTION

In this paper we examine the response of a liquid suspension of microparticles (in the Rayleigh size regime) to time varying fields of the kind that occur in nondegenerate four-wave mixing, signal processing, and two-wave mixing applications. The polarizable particles experience an electrostrictive force which is proportional to the gradient of the field intensity. The corresponding electrostrictive potential for a pair of incident plane waves takes the form:

$$U(r,t) = U_0 \cos(\vec{Q} \cdot \vec{r} - \Omega t) \qquad (1)$$

In Eq. (1), $U_0 = -1/2\, \alpha(\omega)\, \varepsilon_1 \varepsilon_2 (\hat{e}_1, \hat{e}_2)$ where $\alpha(\omega)$ is the particle polarizability and ε_j and $\hat{e}_j$ are the amplitude and unit polarization vector for wave j. For $\vec{k}_j$ and ω_j the wavevector and frequency of wave j, we have in Eq. (1), $\vec{Q} = \vec{k}_1 - \vec{k}_2$ and $\Omega = \omega_1 - \omega_2$ and consider only the case $\Omega \ll \omega_1, \omega_2$ as the frequency range over which the medium can effectively respond. In the following sections we examine the medium response to the potential of Eq. (1) including energy conversion and two-wave mixing gain.

MEDIUM RESPONSE

Since Eq. (1) describes a uniformly moving field grating, it is convenient to shift analysis to the frame of reference depicted in Fig. 1. Viewed in this frame the electrostrictive potential is independent of time. For $\alpha<0$ the direction of the electrostrictive force is indicated by the lower arrows and the particles tend to bunch in the low field regions of the grating. This tendency is countered by Brownian diffusion and, for moving gratings, by the drag force (open arrows) encountered as particles are driven through the fluid medium.

© American Institute of Physics 1987

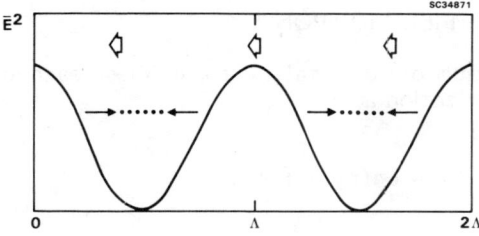

Fig. 1 Formation of a particle density grating in the frame of the moving electrostrictive field.

The time evolution of particle density is governed by the Planck-Nernst equation [1,2]

$$\frac{\partial}{\partial t}n(\vec{r},t) = D\vec{\nabla}\cdot[\vec{\nabla}n(\vec{r},t) - \frac{\vec{F}(\vec{r},t)}{kT} n(\vec{r},t)] \qquad (2)$$

where D is the diffusion constant and $\vec{F} = -\nabla U$ is the electrostrictive force. The solution to Eq. (2) for the moving grating has been obtained in expanded form:[2]

$$n(r,t) = \sum_{\ell=0}^{\infty} A_\ell \cos[\ell(\vec{Q}\cdot\vec{r} - \Omega t)] + B_\ell \sin[\ell(\vec{Q}\cdot\vec{r} - \Omega t)] \quad . \qquad (3)$$

A number of important physical quantities are found to depend only on the in-phase (A_1) and out-of-phase (B_1) components of the first order grating. These include the energy density $<U> = U_0 A_1/2$; the particle current density $<J_r> = g\ D Q B_1/2$; and the power dissipation per unit volume $<Power> = U_0 \Omega B_1/2$ where $g = U_0/kT$ and the angular brackets indicate averaging over a grating period. The coefficients A_1 and B_1 characterize the frequency-response properties of the medium and are shown in Fig. 2.

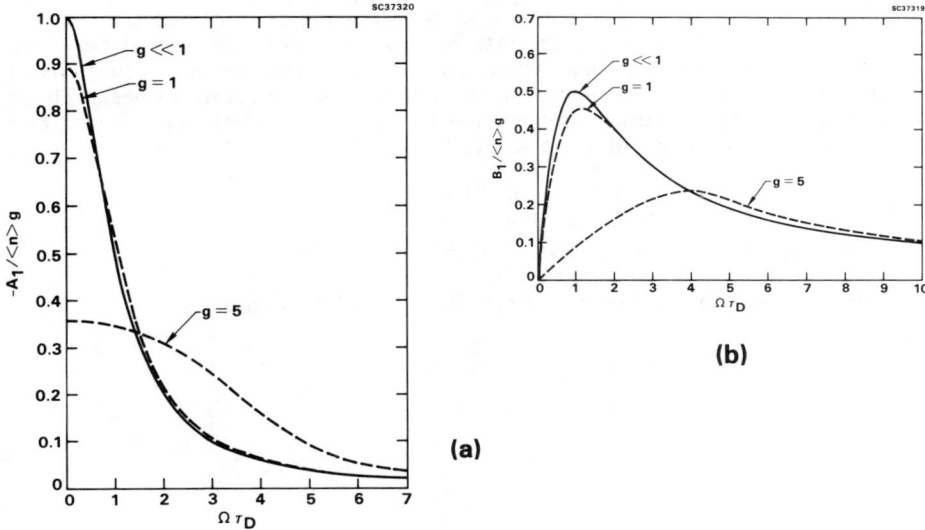

Fig. 2 Frequency dependence of the in-phase and out-of-phase grating coefficients.

FIELD RESPONSE

For a medium consisting of identical microparticles, each of whose polarizability is α, the polarization is [1,2]

$$\vec{P}(r,t) = \alpha n(r,t)\, \vec{E}(r,t) \tag{4}$$

where $n(r,t)$ is given by Eq. (3). Substitution into the electromagnetic wave equation, and use of the slowly varying envelope and phase approximation, results in the coupled field amplitude equations for two-wave mixing.[2]

$$(\vec{K}_1 \cdot \vec{v})\varepsilon_1 = \pi \alpha K^2 B_1 (\hat{e}_1 \cdot \hat{e}_2)\varepsilon_2 \tag{5a}$$

$$(\vec{K}_2 \cdot \vec{v})\varepsilon_2 = \pi \alpha K^2 B_1 (\hat{e}_1 \cdot \hat{e}_2)\varepsilon_1 \tag{5b}$$

Equations 5 imply a net transfer of energy from the reference beam ε_1 to the lower frequency (stokes shifted) beam ε_2, dependent on the out-of-phase grating coefficient B_1. Since the total number of photons is conserved this transfer results in a net loss of field energy from the two beams. This loss represents the power expended on the medium electric polarization and is given precisely by our expression for the power dissipation which also contains the $\pi/2$ out-of-phase component of the medium response.

In summary, we have treated here the response of an artificial Kerr medium to an imposed time varying electromagnetic field. This response included the formation of a particle density grating characterized by its in-phase and out-of-phase components A_1 and B_1. Not considered here are effects due to thermal noise which give rise to spontaneous fluctuations in the grating coefficients about their average field-induced values. These fluctuations are of utmost importance to signal processing applications and will be examined in a following study.

REFERENCES

1. D. Rogovin and S. Sari, Phys. Rev. A31, 2375 (1985).

2. R. McGraw and D. Rogovin, Phys. Rev. A35, forthcoming.

LASER DOPPLER VELOCIMETRY
FOR SUB-MICROMETER PARTICLE SIZE DETERMINATION

Norman J. Dovichi
University of Alberta, Edmonton, Alberta, T6G 2G2 CANADA

Fahimeh Zarrin
Colorado State University, Fort Collins, Colorado, 80523

ABSTRACT

The modulated light scatter signal produced in laser Doppler velocimetry may be used to measure particle size with excellent noise immunity. However, high precision particle size measurement is only possible if particles are constrained to flow through the center of the interference fringe region at a constant velocity. The sheath flow cuvette is used in the biomedical technique of flow cytometry to constrain particles to flow in a thin stream through a laser beam. This cuvette is combined with laser Doppler velocimetry for particle size determination. Polystyrene spheres ranging in size from 45-nm to over 1-μm radius may be analyzed with a resolution limited by the inherent size distribution of the standards.

INTRODUCTION

Laser Doppler velocimetry, LDV, has been employed to measure the velocity of particles embedded in a flowing stream [1]. In LDV, a laser beam is split into two equal intensity beams which are crossed in the flowing stream. Interference between the two beams generates a set of light and dark fringes within their intersection region. As particles pass through the fringe region, a burst of light scatter is generated. The frequency of the light scatter is given by V/Ω where V is the velocity of the particles and Ω is the fringe spacing. Since the fringe spacing is fixed by the intersection angle of the two laser beams, the modulation frequency may be used to determine flow velocity within the stream.

LDV also has been used to measure both flow velocity and particle size within the flow stream [2,3]. Essentially, the depth of modulation of the light scatter signal is related to particle size whereas the frequency of the signal is related to velocity. Small particles will produce a light scatter signal which is proportional to the laser intensity profile; the light scatter signal is completely modulated. To first approximation, larger particles produce a light scatter signal which is the integral over the particle area of laser intensity. The light scatter signal generated by larger particles is incompletely modulated. The depth of modulation scheme for particle size determination is interesting because it is relatively imune to variations in the refractive index of the particle.

Unfortunately, measurements of light scatter modulation depth in LDV have not produced high resolution measurements of particle size. The modulation depth is related to particle size only if particles pass through the center of the interference fringe region. Particles which pass far from the fringe center generate a distorted light scatter signal. For example, particles which pass on axis but far from the intersection region will generate a light scatter signal which consists of two Gaussian humps as the particle travels through the non-interfering laser beams.

© American Institute of Physics 1987

A deconvolution routine has been employed to correct partially for this artifact; however, quite broad particle size distributions are produced [4].

Rather than use the depth of modulation for particle size determination, we utilize the amplitude of the modulated component of the signal. The accuracy of this amplitude measurement suffers since the signal depends upon both particle size and refractive index. However, the precision of the measurement is quite high and limited by the inherent size distribution of particle standards. This high precision is a result of an important property of LDV: the light scatter signal is modulated while background signals, such as light scattered from the windows, are not modulated. Low frequency sources of noise, such as 1/f and line frequency laser noise, may be eliminated in light scatter measurements with a bandpass filter centered at the modulation frequency. The modulation frequency typically falls in the radio frequency band. For example a fringe spacing of 3-μm and a particle velocity of 1-m/s produces a modulation frequency of 300-kHz. However, it is necessary to constrain particles to traverse the center of the fringe region to produce high precision light scatter distributions.

The sheath flow cuvette may be employed to constrain particles to flow through the center of the LDV fringe region. In this cuvette, a sample stream is introduced into the center of a flowing sheath stream under laminar flow conditions. The size of the sample stream is adjusted by varying the relative flow rates of the sample and sheath streams [5]. The sample flows as a narrow, 3-μm radius, stream in the center of the cuvette which may be located in the center of the fringe region.

EXPERIMENTAL

A 5-mW beam is produced at 442-nm with an unpolarized helium-cadmium laser. The beam is split into two parallel beams with a cube beam splitter. These beams are focused with a 25-mm focal length biconvex lens. A sheath flow cuvette with 250-μm square flow chamber and 1.5-mm thick windows is located at the beam intersection region.

Light scatter is collected with a microscope at two locations: at right angles to the fringe region and at a shallow forward angle, 12° to the plane formed by the beams. A 0.45 numerical aperture microscope objective is employed for right angle collection whereas a 0.20 numerical aperture objective is used for forward angle collection. A 20X microscope eye piece is fitted with a small, 0.4 mm diameter, pinhole to restrict the field of view of the detector to the fringe intersection region. A 1P28 photomultiplier tube detects the light scatter signal. A locally constructed circuit is used to demodulate the light scatter signal.

Suspensions of polystyrene particle standards are prepared in filtered and deionized water. The particle concentration is quite low to prevent more than one particle being present in the fringe region. The sheath fluid is also filtered water. Flow is generated with syringe pumps.

RESULTS

The sheath flow cuvette was used to investigate the effect of sample stream size upon the light scatter distribution generated by LDV. To produce light scatter distributions with high precision, it is necessary to utilize a sample stream radius which is smaller than the spot-size of the laser beam which produces the LDV fringes. However, if the sample stream is less than 3-μm in radius for a 10-μm spot-size laser beam, then the light scatter distribution is limited by the inherent distribution of particle size within the sample [6].

The relationship between LDV signal intensity and particle size was investigated for both forward and right angle light scatter. The data followed the expected Mie scatter dependence for particles much smaller than the fringe spacing. The forward light scatter intensity was monotonically related to particle size over the range of 175 to 545-nm radius. Smaller particles produced a light scatter signal which was buried in the background signal. This intense background signal was due to light scatter from the cuvette windows.

Right angle light scatter produced much lower background signals. Particles as small as 45-nm radius produced a light scatter signal which was well resolved from the background signal. The width of the light scatter distribution was dominated by the inherent size distribution for the particle standards. This inherent particle size distribution dominated the light scatter distribution for all particle sizes observed and was less than 1% for good standards [6].

REFERENCES

1. H.D. Thompson and W.H. Stevenson, Ed. Laser Velocimetry and Particle Sizing (Hemisphere Publishing Co., Washington, 1979).

2. W.M. Farmer, Appl. Opt. 11, 2603 (1972).

3. W.M. Farmer, Appl. Opt. 15, 1984 (1976).

4. D. Holve and S.A. Self, Pg. 397 of reference 1.

5. F. Zarrin and N.J. Dovichi, Anal. Chem. 57, 2690 (1985).

6. F. Zarrin, D.J. Bornhop, and N.J. Dovichi, Anal. Chem., in press

Diffusive and Convective Evaporation of Irradiated Droplets
by
R. L. Armstrong
New Mexico State University, Physics Department
Las Cruces, NM 88003
and
A. Zardecki
Los Alamos National Laboratory, Theoretical Division
Los Alamos, NM 87545

The evaporation of a liquid droplet uniformly heated by a high-irradiance laser beam is investigated on the basis of a hydrodynamic description of the system composed of the ejected vapor and the ambient gas. For low irradiance beams, diffusive mass transport and conductive energy transport are the dominant interactions between the droplet and it's environment.[1,2] In this, the isobaric case, changes in the thermodynamic state of the ambient medium are small. Alternatively, for high-flux beams, convective transport processes become significant, and droplet vaporization is generally accompanied by the production of strong shock waves in the surrounding gas. Following Knight,[3] conservation of mass, momentum, and energy at the droplet surface leads to the establishment of jump conditions in the thermodynamic variables at the droplet surface. These jump conditions permit a complete hydrodynamic description of the evaporating droplet to be obtained in the high-flux regime.[4]

For either moderate-flux beams or high-flux beams at sufficiently early times, neither of the two limiting regimes discussed in the proceeding paragraph are adequate to describe the droplet vaporization dynamics. Rather, a correct description must include both diffusive and convective transport processes. In this paper, we describe an extension of Knight's analysis to include diffusion. Diffusion arises naturally in this analysis as a result of the constraints placed on the mean velocities of the vapor and ambient gas species. Specifically, the vapor mean velocity is constrained by the vapor mass balance condition at the droplet surface. The ambient gas mean velocity is constrained by the assumption of immiscibility, and by the specification of a thermal accommodation coefficient (a) for ambient molecules reflected from the droplet surface. Numerical calculations are presented for a = 0 in this paper. The effect of a non-zero value of a will be deferred to a later paper.

Figures 1-2 illustrate representative results for the case of a moderate-flux (F_{max} = 1.52×10^6 W/cm^2), 1.06 µm laser beam incident on a 20 µm (diameter) water droplet. The high-flux case has been treated in an earlier paper[4] and here we emphasize the regime where diffusive transport must be taken into account. Figure 1 illustrates the behavior of the bulk flow Mach number and the vapor mass fraction on the leading edge of the laser beam (the peak intensity occurs at 30×10^{-8} s). Figure 2 is a three-dimensional plot of the bulk flow velocity distribution in the neighborhood of the droplet. The remaining thermodynamic variables are not plotted since they remain virtually identical to their ambient values.

In summary, with the addition of diffusive transport terms to the droplet boundary conditions, a numerical computer code now exists that will successfully model aerosol-bean interactions over a range of beam irradiances from very weak beams to high-flux beams where explosive droplet vaporization occurs.

References

1. R. L. Armstrong, Appl. Opt. 23, 148 (1984).
2. R. L. Armstrong, S.A.W. Gerstl, and A. Zardecki, J. Opt. Soc. Am. A2, 1739 (1985).
3. C.J. Knight, AIAA Journ. 17, 519 (1979).
4. R. L. Armstrong, P. J. O'Rourke, and A. Zardecki, Phys. Fluids 29, 3573 (1986).

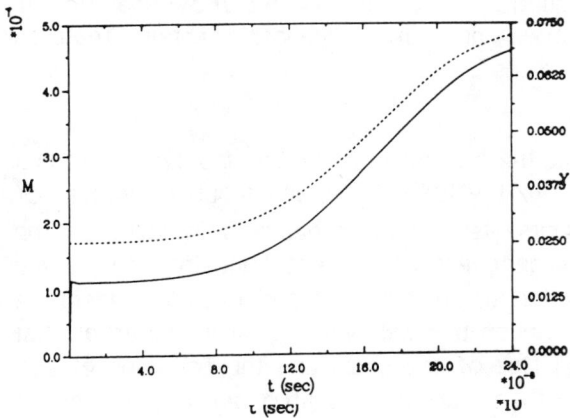

Fig. 1
Bulk flow Mach number (M) and vapor mass fraction (Y) for water droplets (10 μm radius) irradiated by 10.6μm Gassian beam with peak irradiance 1.52×10^6 W/cm^2, pulse width 5×10^{-8} s.

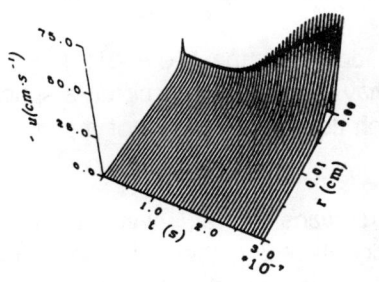

Fig. 2
Three dimensional plot of bulk velocity for system of Fig. 1.

A self-similar approach to the explosion of droplets by a high energy laser beam.

Shirish M. Chitanvis
Los Alamos National Laboratory, Theoretical Division, P-371
Los Alamos, New Mexico 87544.

ABSTRACT.

We have constructed a model in which a small droplet is exploded by the absorption of energy from a high energy laser beam. The beam flux is so high that we assume the formation of a plasma. We have a single-fluid model of a plasma droplet interacting with laser radiation. Selfsimilarity is invoked to reduce the spherically symmetric problem involving hydrodynamics and Maxwell's equations to quadrature. We show analytically that our model reproduces in a qualitative manner certain features observed experimentally by Eickmans et al.

I. Introduction.

Eickmans et al[1] have reported the explosion of water droplets of radius 35µm, by a laser beam with a flux $I \sim 10^9$ W/cm^2, and a pulse width τ_p of 20 nanoseconds. In these experiments, the laser beam is incident on the droplet, which acts as a convergent lens and causes focusing toward the rear of the droplet. The wavelength of the incident (Nd:YAG) laser is 0.532µm. One would not expect much thermal absorption in water at that wavelength. But due to the high flux of the beam and the focusing effect, the droplet ionizes.[1] The plasma then absorbs quite strongly from the beam, and the drop explodes with the formation of a layer of vapor. In the vapor,[1] we clearly see a decrease in the density followed by an increase. We theorize that the hot spot created within the droplet expels a mass of vapor. Far from the droplet, the temperature drops and the vapor becomes denser and then appears as a plume.

We have distilled the fundamental idea that the blow-off followed by cooling causes the non-monotonic behavior,[1] and constructed a spherically symmetric model of the explosion which does indeed predict a plume.

II. The single-fluid model.

In the explosions reported by Eickmans et al,[1] we basically have ionized *fluid* streaming out of the droplet. But there is no compelling reason to treat the two components of this plasma separately with great

---

[1] J.H. Eickmans, W.F. Hsieh and R.K. Chang, submitted to *Opt. Lett.*

care. We therefore simply consider the plasma to be a hydrodynamic fluid with a certain local velocity (v), mass density (ρ), temperature (T), pressure (P), a specific heat (C_v) and an absorption coefficient (α). For simplicity we shall consider the plasma to be perfect gas. We shall take C_v and α to be phenomenological parameters to be fitted using experimental data.

As discussed in a previous paper on explosions[2] we shall use the following self similar variable to simplify the coupled equations of hydrodynamics and electromagnetism:[2]

$$\xi = r/c(t+t_0) \quad (2.1)$$

where c is the surface speed given approximately by the conservation of energy:[3]

$$c \approx \sqrt{\{2\alpha' I \tau_p / \rho_0\}} \quad (2.2)$$

where $\alpha' = 3\alpha/4a_0$, α being a dimensionless absorption efficiency, a_0 is the unperturbed droplet radius, I is the flux of the beam (W/cm^2), τ_p is the pulse length, and ρ_0 is the unperturbed density.

If we make the ansatz[2,3] that $\rho(r,t) = \rho'(\xi)$, $v(r,t) = v'(\xi)$, $T(r,t) = T'(\xi)$, $E(r,t) = F'(\xi)$, and we assume a perfect gas law for ease of computations:

$$P(\xi)\rho^{-1}(\xi) = R_g T(\xi) \quad (2.3)$$

we get the following set of coupled ordinary differential equations, using $F(\xi) = 1/\xi$ (please see references 2 and 3 for details):

$$dv(\xi)/d\xi = 2\kappa_2[(v(\xi)-\xi)^2 - \kappa_2(\kappa_3-1)T(\xi)]^{-1}$$

$$[(\kappa_3+1)v(\xi)T(\xi)/\xi + \kappa_1 \xi (d/d\xi |F(\xi)|^2)/(2\rho(\xi))] \quad (2.4)$$

$$dT(\xi)/d\xi = -\kappa_3 T(\xi)(dv(\xi)/d\xi + 2v(\xi)/\xi)/(v(\xi)-\xi)$$

$$-\kappa_1 \xi(d/d\xi |F(\xi)|^2)/(\rho(\xi)(v(\xi)-\xi)) \quad (2.5)$$

$$d\rho(\xi)/d\xi = -\rho(\xi)(dv(\xi)/d\xi + 2v(\xi)/\xi)/(v(\xi)-\xi) \quad (2.6)$$

$$d^2F(\xi)/d^2\xi^2 = -2F(\xi)/\xi \quad (2.7)$$

where ρ,v,T,F are normalized functions as follows: $\rho(\xi) = \rho'(\xi)/\rho_{(0)}$, $v(\xi) = v'(\xi)/c$, $T(\xi) = T'(\xi)/T_{(0)}$, $F(\xi) = F'(\xi)/F'(\xi=1)$, where $\rho_{(0)}$ is the density of the droplet at the surface, $T_{(0)}$ is the temperature of the drop at

---
[2] S.M. Chitanvis, *Physica* 137A ,271, (1986)
[3] S.M. Chitanvis, *CRDEC Conference Proceedings* (1986)

the surface, and c is the speed of the surface. We also need the following definitions: $\kappa_1 = \alpha |F'(\xi=1)|^2 / (8\pi C v T_{(0)}\rho_{(0)} c)$, $\kappa_2 = R g T_{(0)}/c^2$, $\kappa_3 = Rg/Cv$.

III. The analysis.

Notice that in Eqn (2.4)-(2.6), there is a potential source of singularity when $\rho(\xi) \to 0$ or $v(\xi) \to \xi$ for some $\xi = \xi^*$.

We shall perform a local analysis of the solution for $\xi_1=\xi^*+\delta\xi \leq \xi \leq \xi_2=\xi^*+\delta\xi$, $\delta\xi \to 0^+$ when $v(\xi) = v^* \approx \xi^*$. We shall take $dv/d\xi \equiv 0$ here. This assumption was suggested by our earlier[2] numerical solution of the problem. This leads to the identity:

$$T(\xi) = \kappa_1/(v^*(\kappa_3+1)) \, (1/\rho(\xi)) \, (1/\xi) \qquad (3.1)$$

From Eqn. (2.13) we get:

$$\rho(\xi) \approx \rho(\xi_1) [\, (v^*-\xi)/(v^*-\xi_1)/\xi \,]^2 \qquad (3.2)$$

This yields:

$$T(\xi) \approx \kappa_1/(\rho(\xi_1)(\kappa_3+1)) \, [(v^*-\xi_1)/(v^*-\xi)]^2 \qquad (3.3)$$

It is easy to show that Eqn (3.3) satisfies Eqn (2.5) approximately for $\xi \approx \xi^*$.

Thus we see that at the singular point $\xi = \xi^*$, the density dips to zero, then starts to rise. The temperature on the other hand, has a second order pole at $\xi = \xi^*$. This is not unphysical behaviour, since the total energy of the fluid viz. $1/2\rho v^2 + \rho C_v T$ remains finite.

Far beyond this dip, for some $\xi \geq \xi_a$ we can get the asymptotic behaviour of the hydrodynamic variables. We again take $dv/d\xi = 0$, and $v(\xi \to \infty) = v_0$ so that:

$$\rho(\xi) \approx \rho(\xi_a) \exp[-2v_0 \, (1/\xi - 1/\xi_a) \,] \qquad (3.4)$$

$$T(\xi) \approx [\, \kappa_1 v_0/\rho(\xi_a)(\kappa_3+1)] \, (1/\xi) \exp[2v_0 \, (1/\xi - 1/\xi_a) \,] \qquad (3.5)$$

Eqn. (3.5) satisfies Eqn. (2.5) approximately, as long as κ_1 (which is dimensionless) $<< 1$. For $I \leq 10^{12}$ W/cm^2, this condition is satisfied.

We therefore have the following picture of the density; it starts off at some value $\rho(\xi=1)$ at the surface, decreases to zero, and then tends to an asymptotic value if there is no ambient medium around the spherically exploding droplet. In the presence of an ambient medium, the motion stops at $\xi = 1 + ct'/$(initial radius), where t' is the time past the explosion.

We therefore see analytically in our model that a spherical *plume* or *blowoff* is formed around the exploding droplet. This is analogous to the plume seen in experiments.[1]

NEAR-FIELD LIGHT SCATTERING BY PARALLEL GLASS FIBERS

Daniel S. Benincasa, Tak-Goa Tsuei, and Peter W. Barber
Clarkson University, Potsdam, NY. 13676

ABSTRACT

Single glass fibers whose radius-to-wavelength ratio is large, for example, $r/\lambda = 100$, strongly focus normally incident laser light to give greatly enhanced intensities both within and in the near field of the fiber, even when the fiber is not at resonance. Calculated results show that the enhancement is dependent on the radius-to-wavelength ratio and on the fiber index of refraction. Even greater enhancements are observed for spherical particles. The near-field intensity for a single fiber is greatly modified when a second fiber is moved into its near field. Calculations of the near-field intensities for two coupled fibers provide insight into the coupling mechanism, especially the dependence of the coupling on the fiber separation and on the orientation of the two fibers relative to the incident wave.

INTRODUCTION

Multiple scattering interactions between closely-spaced particles have recently been studied for two spheres[1] and two circular cylinders.[2] Both investigations have shown the dependence of the far-field scattered intensity on the polarization of the incident wave, and on the particle separation and orientation.

Particle-particle interactions are strongly influenced by the spatial distribution of the near-field intensity around an isolated particle. Calculated intensity distributions will first be shown for single fibers. Near-field and internal intensity calculations for two parallel fibers are then shown to be remarkably similar when the two fibers are widely spaced. However, the interaction between the fibers results in significant perturbations of the near-field intensity as the separation decreases.

ISOLATED FIBERS

Calculations for single spheres and cylindrical fibers[3] show a complicated spatial distribution of near-field intensity, with a prominent focal peak located in the forward direction (the shadow side of the particle) just outside the surface. This peak occurs for particles with size parameters (defined by $x = 2\pi a/\lambda$, where a is the radius and λ is the incident wavelength) which are much greater than unity. For these large particles, the curved surface of the particles together with the index of refraction discontinuity at the interface result in a lens-like structure. A spot (sphere) or line (cylinder) focus is located external to such a thick lens, with a focal length which is dependent on the size parameter and the index of refraction ratio m between the lens and the surrounding medium.

Figure 1 shows the calculated intensity on a line through the center of a fiber, perpendicular to the axis, and parallel to the direction of the incident wave. The size parameter of 100 corresponds to a fiber of ~ 8μm radius at an incident wavelength of 0.5145 μm. The index of refraction of 1.5 is typical of glass in this wavelength range. The most prominent feature in Fig. 1 is the intensity maximum which occurs outside the shadow side of the fiber. For unit incident intensity, this peak of nearly 30 units located at r/a $\simeq$ 1.3 illustrates the focusing effect of a cylindrical lens.

Other features of the near-field intensity can be seen more clearly in Fig. 2, which shows the intensity distribution over a plane around the outside of the fiber. The size parameter of x = 488.5 corresponds to a 40 μm radius fiber illuminated at a wavelength of 0.5145 μm. In addition to the strong focused intensity peak, the shadow cast by the fiber and the specular reflection behind the illuminated side of the fiber are clearly discernible. The interference pattern between the incident and scattered fields is evident in all directions.

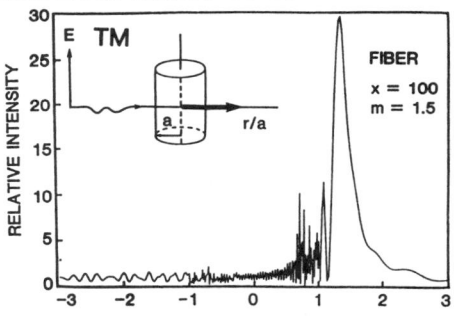
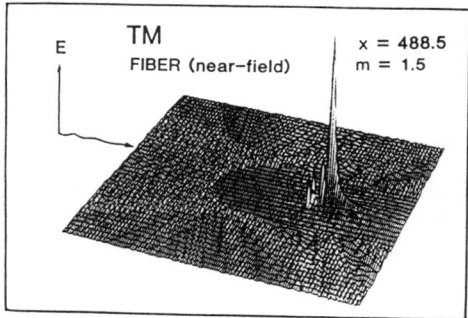

Fig. 1. Relative intensity vs r/a on a line through the center of a fiber. r/a = $\pm$ 1 defines the boundary of the fiber. Incident intensity is unity.

Fig. 2. Near-field intensity distribution of a fiber located at the center of the plot with the internal intensity set equal to zero. The maximum intensity at the peak is 68.1 times the incident intensity.

TWO FIBERS

The solution for the scattering of an electromagnetic wave by a pair of non-overlapping parallel circular cylinders has been given in a convenient form for numerical calculation.[4] Of particular interest is the solution for two identical cylinders separated by a distance 2d and illuminated along a line joining the cylinder centers by a TM plane wave as shown in Fig. 3.

Scattering calculations using the two cylinder formulation are shown in Fig. 4 for two identical fibers with x = 50 separated by a distance $2\pi(2d)/\lambda = \delta = 1000$. With the fibers widely separated and for end-on illumination as shown, the intensity distribution around each of the individual fibers is similar to that for an isolated fiber, even though the second fiber is in the shadow of the first fiber. The magnitude of the focal peak is less than that shown for the fiber in Fig. 1 because the fiber size and the index of refraction are both less in Fig. 4. When the separation decreases, the intensity distribution both within and in the near field of each of the fibers becomes significantly different from that for a single isolated fiber.

Geometrical optics ray-tracing is currently being used to develop a physical understanding of the dependence of the multiple scattering interaction on the fiber size and index of refraction, the separation of the fibers, and the orientation and polarization of the incident wave.

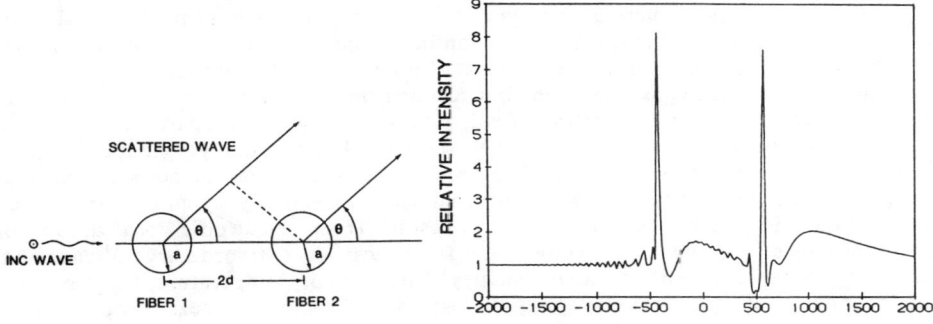

Fig. 3. Two cylinders illuminated by a TM incident wave. The scattered intensity is calculated at an angle θ.

Fig. 4. Relative intensity on a line through the center of two dielectric fibers for a TM polarized wave incident from the left, x = 50, δ = 1000 and m = 1.33.

REFERENCES

1. K.A. Fuller, G.W. Kattawar, and R.T. Wang, Appl. Opt. 25, 2521 (1986).
2. B. Schlicht, K.F. Wall, R.K. Chang and P.W. Barber, J. Opt. Soc. Am. A in press (1986).
3. D.S. Benincasa, P.W. Barber, J-Z. Zhang, W-F. Hsieh, and R.K Chang, Appl. Opt. submitted (1986)
4. G.D. Olaofe, Radio Sci. 5, 1351 (1970).

Optical Bistable Interaction of Laser Radiation with Microparticles

K. M. Leung

Polytechnic University
Department of Physics, Brooklyn, NY, 11201

ABSTRACT

Intrinsic optical bistability in the scattering and absorption of light from Rayleigh-sized microparticles having an intensity-dependent refractive index is theoretically investigated. For particles near plasmon resonances optical switching occurs with several orders of magnitude reduction in the switching intensity. Besides acting as an optical memory element whose size is small compared with the incident wavelength, the microparticle also exhibits an optical transistor mode with a sizable differential gain.

1. INTRODUCTION

We are interested here in studying the intrinsically nonlinear interaction of laser radiation with microparticles. Nonlinear optical effects in microparticles can be expected to be enhanced above that found in bulk materials owing to two separate physical mechanisms. The first mechanism is due to the local field effect which increases the local electric field intensity inside the particle when the frequency of the incident light is close to the particle's plasmon resonance. This is basically a purely classical effect. The second mechanism has to do with the fact that the optical nonlinearities of a microparticle can actually be much larger than that found in bulk materials. This results from electronic localization or quantum-size effect, which is responsible for a host of extraordinary phenomena, such as enhanced optical nonlinearities at room temperature, found in semiconductor superlattice structures.[1] Instead of being squeezed into a two-dimensional space in a superlattice, the electrons and holes inside a microparticle are compressed on all three sides, and the electronic behavior is expected to be more like that of a zero-dimensional object, although at present, the physics of such an object is still very far from being understood.

2. OPTICAL BISTABILITY IN RAYLEIGH PARTICLES

We contend ourselves here with a classical analysis. The particle will be treated as a sphere whose diameter is small compared with the wavelength of light, and therefore a theoretical treatment based on electrostatics suffices. The particle is also assumed to have a dielectric constant having the same Drude form as in the bulk. The response to incident radiation is supposed to be instantaneous and local, and a Kerr-like nonlinear dielectric constant is assumed, thus the sphere has a dielectric constant of the form

$$\frac{\varepsilon_s}{\varepsilon_h} = \varepsilon_\infty \left[1 - \frac{\omega_p^2}{\omega(\omega + i\tau^{-1})} \right] + \alpha |\vec{E}_s|^2 \qquad (1)$$

where ε_h is the dielectric constant of the host medium outside the particle.

The equation that determines the electric field is given by [2] $\nabla \cdot (\varepsilon \vec{E}) = 0$. The solution in the linear medium outside the particle is well known. However, we have not yet been able to obtain the complete solution inside the particle, but (as in the linear case) the solution for which $|\vec{E}_s|^2$ is uniform is an acceptable one. With that solution, the field inside the particle can be determined from the neccessary boundary conditions. We find that it obeys the following equation [2]

$$x = y/(2\delta + x)^2 + 1, \qquad (2)$$

for frequency close to the plasmon frequency of the particle. Equation (3) has been written in dimensionless form with $y = 9\alpha E_h^2(\omega_0 \tau)^3/(\varepsilon_\infty + 2\varepsilon_h)^2$ as the reduced incident intensity, $x = \alpha E_s^2(\omega_0 \tau)/(\varepsilon_\infty + 2\varepsilon_h)$ as the reduced internal intensity, $\delta = (\omega - \omega_0)\tau$ as the reduced frequency detuning,

$$\omega_0 = \omega_p \left(\frac{\varepsilon_s}{\varepsilon_\infty + 2\varepsilon_h} \right)^{1/2} \qquad (3)$$

as the plasmon frequency in the linear limit, and ω_p as the plasmon frequency of the bulk material.

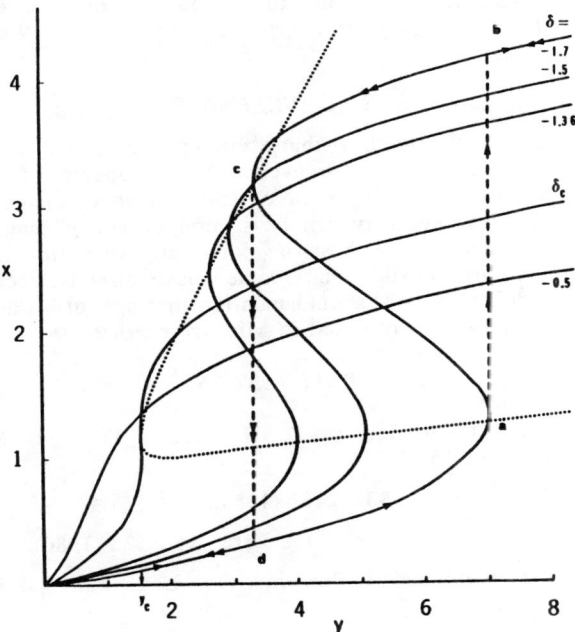

Figure 1. The reduced internal intensity, x, as a function of the reduced incident intensity, y, for various values of the frequency detuning, δ. δ_c is the critical detuning, and y_c is the critical incident intensity. At $\delta = \delta_c$ the microparticle acts as an optical transistor.

Once the internal intensity is determined, the nonlinear polarizability, α_s, can then be computed as

$$\alpha_s = \frac{3\left(\varepsilon_s/\varepsilon_h - 1\right)}{4\pi\left(\varepsilon_s/\varepsilon_h + 2\right)}. \tag{4}$$

The differential scattering cross-section and the absorption cross-section are then given, respectively, by the square modulus and the imaginary part of α_s in the usual way.

Since x and y have the same sign as that of α, we see from eq.(4) that only the self-focusing ($\alpha > 0$) case needs be considered. The self-defocusing ($\alpha < 0$) case can be obtained by simultaneously changing the signs of x, y, and δ. Also because of the factor $(\omega_0 \tau)^3$ in the definition of y, we see that near sharp resonance where $(\omega_0 \tau) \gg 1$, the minimum threshold for the on set of optical bistability will be reduced by several orders of magnitude compared with the non-resonance case.

Our result for the internal intensity, x, as a function of the incident intensity, y, at fixed values of the frequency detuning, δ, is shown in Fig. 1. The critical values for the incident intensity and the detuning for the on set of optical bistability can be shown to be given respectively by $y_c = 8\sqrt{3}/9$ and $\delta_c = -\sqrt{3}/2$.

3. CONCLUSION

Our above results clearly show that a microparticle, whose size can be quite small compared with the incident wavelength, is capable of exhibiting optical bistability and therefore can act as an optical memory element. The threshold intensity for the incident intensity can be several orders of magnitude lower than that found in bulk materials. A host of potentially very important applications based such a system are possible and some have already been pointed out by Chemla and Miller.[3] Finally we should mention that optical bistability based on a similar effect has recently been reported in Mie-size particles.[4]

REFERENCES

1. See, for example, the review by D. S. Chemla, D. A. B. Miller, and P. W. Smith, Opt. Eng. 24, 556 (1985).

2. K. M. Leung, Phys. Rev. B, 33, 2461 (1986).

3. D. S. Chemla and D. A. B. Miller, Opt. Lett. 11, 522 (1986).

4. S. Arnold, K. M. Leung, and A. Pluchino, Opt. Lett. 11, 800, (1986).

HIGH GAIN-BANDWIDTH-PRODUCT AVALANCHE PHOTODIODES FOR MULTIGIGABIT/S DATA RATES

J. C. Campbell
AT&T Bell Laboratories
Crawford Hill Laboratory
Holmdel, NJ 07733

ABSTRACT

For long-wavelength (1.3 µm and 1.5µm) high-bit-rate (>400 Mbit/s) lightwave systems the highest receiver sensitivities have been achieved with III-V compound avalanche photodiodes with separate absorption and multiplication regions (SAM-APDs). Initial APDs of this type exhibited poor frequency response owing to charge accumulation at the heterojunction interfaces. A significant improvement in the bandwidth was achieved by introducing a transition region between the multiplication and absorption layers (SAGM-APDs). Early SAGM-APDs exhibited bandwidths in the range 1 GHz to 3 GHz and gain-bandwidth products as high as 18 GHz. Recently, the progression of new lightwave systems to higher and higher bit rates has stimulated efforts to further increase the bandwidth of these SAGM-APDs. By optimizing the design parameters bandwidths as high as 5 GHz and 7 GHz and gain-bandwidth products of 50 GHz and 60 GHz have been reported for planar and mesa structures, respectively. The improved frequency response of these APDs has contributed to the success of lightwave systems experiments at 2 Gbit/s, 4 Gbit/s, and 8 Gbit/s.

INTRODUCTION

For lightwave systems one must usually choose between a p-i-n photodiode or an avalanche photodiode as the photodetector in the front end of the receiver. This choice is usually predicated on a number of interrelated factors such as cost, circuit complexity, bit rate, and system margin. It has been shown theoretically and experiments confirm that APD's provide a few decibels higher sensitivity than p-i-n's.[1] The margin depends on several factors including the device parameters of the photodetectors, the characteristics of the preamplifiers which constitute the first gain stage, and the bit rate. For example, the margin increases with bit rate. In fact, for most systems operating at 400 Mbit/s or higher the increased sensitivity provided by the internal gain of the APD justifies its higher cost and more complex biasing circuitry. Therefore it is important that APD's be able to operate at high frequencies or, more importantly, that they have high gain-bandwidth products.

FREQUENCY RESPONSE - THEORY

Many early lightwave transmission experiments utilized Ge APD's[2] but as a result of their high dark current and unfavorable ratio of ionization coefficients, it is generally felt that the highest performance can be achieved with APD's fabricated from III-V compounds. The structure that has achieved the best performance to date is the SAM-APD[3] (<u>s</u>eparate <u>a</u>bsorption and <u>m</u>ultiplication regions) which was first developed to eliminate the tunneling component of the dark current in $In_{0.53}Ga_{0.47}As$ (referred to henceforth as InGaAs) avalanche photodiodes. In the SAM-APD the p-n junction and thus the highest field region are located in a wide-bandgap material such as InP where tunneling is negligible and absorption occurs in an adjacent narrow-bandgap material such as InGaAs. It has been shown that the frequency response of an SAM-APD is accurately characterized by the following expression[4]:

$$\frac{M(\omega)}{M_o} = T(\omega) \left[[1+(\omega RC)^2] [1+(\omega/e_h)^2] [1+(\omega \tau_e M_o)^2] \right]^{-1/2} \quad (1)$$

where R is the total series resistance, C is the diode capacitance, e_h is the emission rate for holes trapped at the heterojunction interfaces, τ_e is the effective transit time through the multiplication region, and M_o is the dc avalanche gain. The transit-time factor $T(\omega)$ can be written in terms of the contributions of the primary holes and electrons and secondary electrons, $P(\omega)$, $N(\omega)$, and $N_s(\omega)$, respectively

$$T(\omega) = | P(\omega) + N(\omega) + N_s(\omega) | \quad (2)$$

where

$$P(\omega) = [1-\exp(-\alpha x_a)]\{1-\exp(\omega x_m/v_p)\}$$
$$+ \exp(\alpha x_t - j\omega x_m/v_p) \left[\frac{1-\exp(j\omega x_a/v_p + \alpha x_a)}{j\omega + \alpha v_p}\right]$$
$$- \exp(-\alpha x_a - j\omega x_t/v_p) \left[\frac{1-\exp(-j\omega x_a/v_p)}{j\omega}\right] \quad (3)$$

and

$$N(\omega) = \left[\frac{1-\exp(-j\omega x_a/v_n)}{j\omega}\right] - \exp(-\alpha x_a)\left[\frac{1-\exp(-j\omega x_a/v_n+\alpha x_a)}{j\omega - \alpha v_p}\right] \quad (4')$$

and

$$N_s(\omega) = (M_o-1)[1-\exp(-j\omega x_a/v_p)]\left[\frac{1-\exp(-j\omega x_a/v_n)}{j\omega}\right.$$

$$\left. - \frac{1-\exp(\alpha x_a v_p/v_n - j\omega x_a/v_n)}{j\omega + \alpha v_p}\right]. \quad (5)$$

In these equations ω is the angular frequency, x_a is the width of the absorption region, x_m is the width of the multiplication region, x_t-x_m is the width of the transition region, α is the absorption coefficient, , and v_n and v_p are the saturated electron and hole drift velocities, respectively.

The term $T(\omega)$ in Eq. (1) is due to the finite transit time of carriers through the depletion region[5]. The transit time for the SAGM-APD structure will usually be longer than that of a standard homojunction p-i-n photodiode because the absorption and multiplication regions are not coincident. This adds length to the drift region. In addition, there will be a transit time associated with secondary electrons which are injected from the multiplication region into the absorbing layer. In fact, for significant multiplication these secondary electrons will dominate the transit time. The expression for the transit time given in Eq. (2)-(5) take into account the added drift region for the holes and the effect of the secondary electrons. For typical SAGM-APD parameters the transit time should not be a limiting factor for bandwidths less than 9 GHz.

The RC time constant, which is represented by the second term in Eq.(1), is usually somewhat more restrictive than the transit-time term. The capacitance of these APDs is typically less than 0.25 pFd at the operating voltage. Assuming a load resistance of 50 Ω we find that the RC term is not significant for bandwidths less than 8 GHz.

The third term in Eq.(1) is due to carrier trapping at the heterojunction interfaces[6]. This effect is illustrated by the valence band diagram in Fig. 1. As the photogenerated holes are swept from the narrow-bandgap absorbing layer toward the wide-bandgap multiplication region some of them become trapped in the potential well at the heterojunction interface. The number of holes that accumulate in the well and the rate at which they are emitted, e_h, depend on the height and width of the barrier, E_b and W, respectively.

Both of these parameters depend, in turn, on the magnitude of the valence band discontinuity E_v and the amount of grading at the heterojunction interface. It has been found that for InP/InGaAs SAM-APDs this effect gives rise to a long tail ($\tau > 10$ ns) in the pulse response and essentially renders this type of APD useless for high-bit-rate lightwave systems.

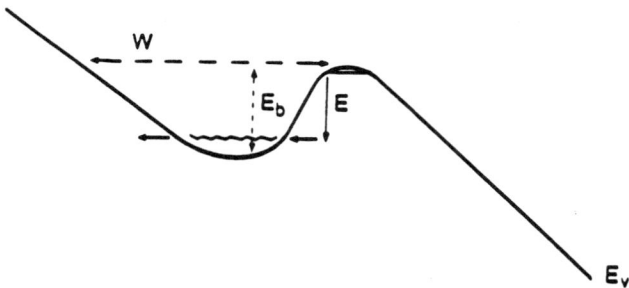

Fig. 1 Valence Band at Heterojunction Interface

There have been several solutions proposed to eliminate the problem of charge accumulation at the heterojunction interface. Each involves introducing a transition region between the multiplication and absorption regions. The optimum approach would be to continuously grade the bandgap energy between that of InP and $In_{0.53}Ga_{0.47}As$[6]. However, it is extremely difficult to grade the crystal composition to provide a uniform change in the bandgap and simultaneously maintain the lattice constant. The most sophisticated approach, proposed and demonstrated by Capasso et al.[7], utilized a chirped period superlattice to mimic grading of the bandgap.

The simplest and most successful solution, to date, has been to do the grading step-wise with a small number of intermediate-bandgap layers[8-11]. We refer to this structure as an SAGM-APD, the "G" denoting the presence of the intermediate-bandgap "grading" layers. The effect of adding these layers is to reduce the valence band discontinuity. Also at each heterojunction interface there is a certain amount of compositional grading. Both of these effects reduce the barrier height and this has resulted in a significant improvement in the frequency response. Most of the early SAGM-APDs utilized a single transition layer. The bandwidths of these APDs were typically in the range 1GHz to 2 GHz with gain-bandwidth products < 18 GHz[4,10-12]. As lightwave systems push toward multi-gigabit transmission rates it is clear that even higher bandwidths will be required and this has necessitated a closer examination of interface hole trapping.

There is still some uncertainty regarding the exact nature of the emission process. Originally, it was proposed that the trapped holes escaped solely by thermionic emission[6]. However, Fowler and Nordheim[13] have shown in another context that for sufficiently high potential barriers or low temperature tunneling is the dominant emission process. At higher temperatures the strong field emission becomes sensitive to temperature and finally blends into thermionic emission. For SAM-APDs the data appear to be fit best by a thermionic-field emission model which exhibits characteristics of thermionic emission and tunneling.[14] In this model carriers with energy E (Fig. 1) tunnel through the barrier. Higher values of E correspond to narrower barriers and consequently higher tunneling rates. We have used the WKB approximation to calculate the emission rate for holes having energy E relative to the bottom of the potential well. Average emission rates and time constants were then obtained by averaging over the energy distribution (assuming thermal equilibrium in the well).

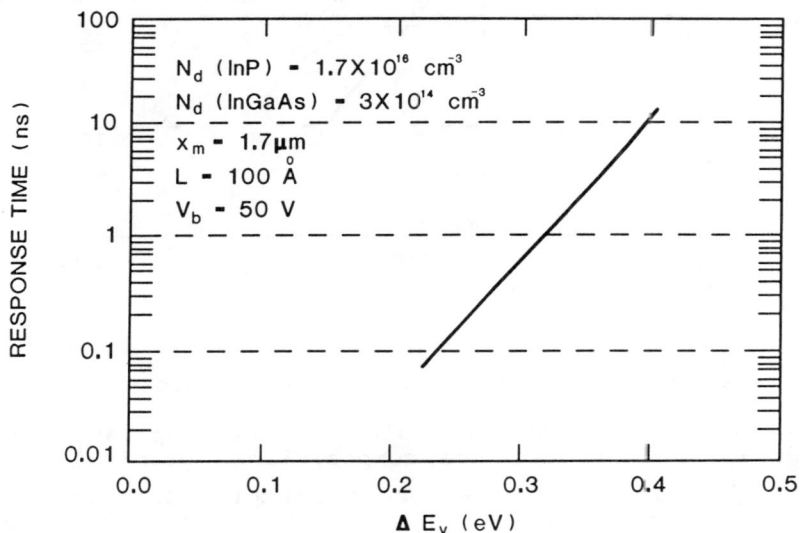

Fig. 2 Time Constant Versus Valence Band Discontinuity

Figure 2 shows the calculated time constant as a function of the valence band discontinuity. The parameters in the the figure are defined as follows: N_d is the donor carrier concentration, x_m is the width of the InP multiplication region, L is the compositional grading parameter, and V_b is the bias voltage. The value $\Delta E_v = 0.4$ eV corresponds to an abrupt InP/InGaAs heterojunction (SAM-APD). The point $\Delta E_v = 0.24$ eV corresponds to the valence band step between InP and $In_{0.7}Ga_{0.3}As_{0.65}P_{0.35}$, an intermediate-bandgap composition which has been used in the transition region of SAGM-APDs. This reduction in ΔE_v results in a decrease in the time constant by almost two orders of magnitude which is consistent with experimental observations.

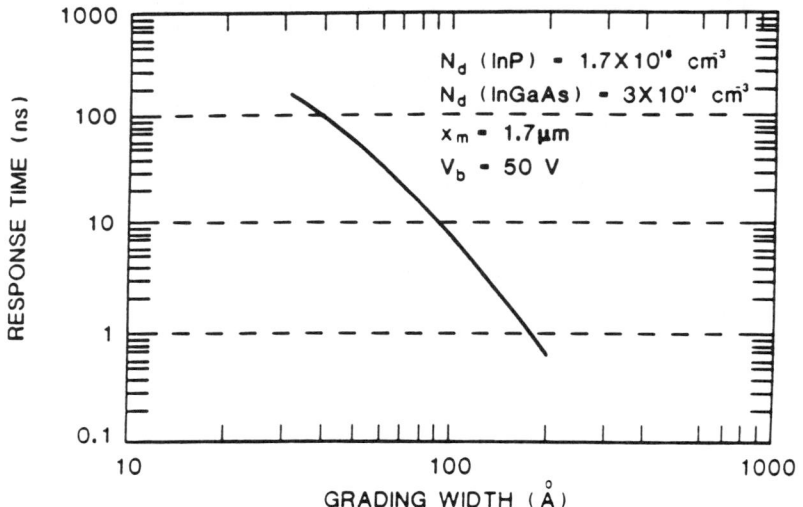

Fig. 3 Time Constant Versus Grading Width

Figure 3 shows the calculated time constant as a function of the width of the compositional grading, L, at the heterojunction interface. The model of Cheung et al.[15] for the variation of the band gap with L has been assumed for this calculation. We observe that the time constant drops by over two orders of magnitude as the grading parameter is increased from 30 Å to 200 Å. The degree of compositional grading depends on the crystal growth technique and the growth parameters used to fabricate the APD wafers. For example, heterojunction interfaces grown by vapor phase techniques such as molecular beam epitaxy (MBE) or metal organic chemical vapor deposition (MOCVD) tend to be more abrupt (L<50 Å) than those grown by liquid phase epitaxy (LPE) (L>100 Å).

Figure 4 shows the upper limits on the bandwith for interface emission rates of $1/e_h$=10 ps and $1/e_h$=20 ps. We conclude from these calculations that barrier lowering in the transition region of an SAGM-APD is essential for the realization of high frequencies. It also clear that even if the heterojunction interfaces are very abrupt, a few (2 to 4) discrete transition layers should be sufficient to achieve bandwidths of several GHz.

The fourth term in Eq.(1) is due to the avalanche buildup time. This term gives rise to a constant gain-bandwidth product and is often the most restrictive on the bandwidth. One of the factors that determine the avalanche buildup time and hence the gain-bandwidth product is the ratio of the electron (α) and hole (β) ionization coefficients. Emmons[16] has shown that for a given avalanche gain the more α and β differ the higher the bandwidth that can be attained. This has been one of the driving forces behind the APD structures such as the "staircase APD"[17] that have been designed to achieve artifically high β/α ratios by bandgap engineering. For the SAGM-APDs the values of α and β are determined by the crystal composition of the multiplication region. There are, however, structural modifications that provide some improvement in the gain-bandwidth product.

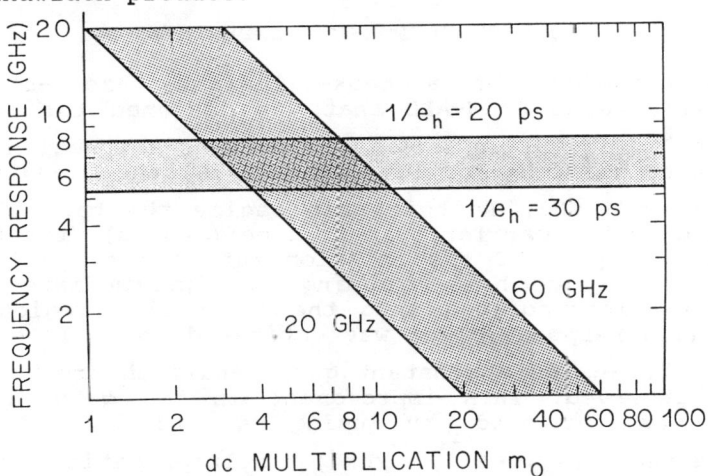

Fig. 4 Bandwidth Versus Avalanche Gain

Since the gain-bandwith product arises from the regenerative nature of the multiplication process, the effects of carrier feedback can be reduced by decreasing the width of the multiplication region. In the SAGM-APD structure this must be accompanied by an increase in the carrier concentration in order to insure that the performance will not be degraded by carrier diffusion into the drift region or excess dark current.

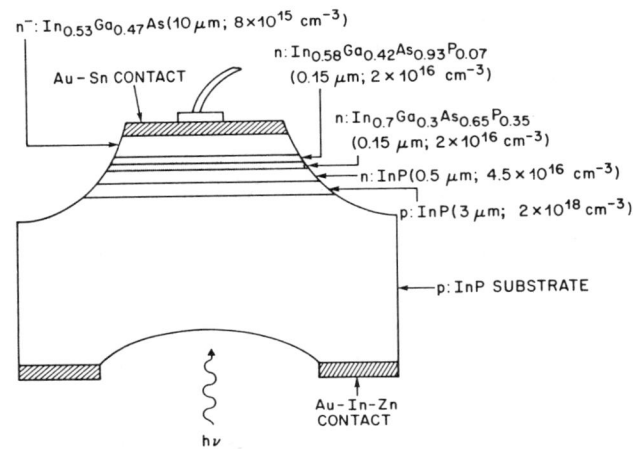

Fig. 5 Back-Illuminated Mesa-Structure SAGM-APD

FREQUENCY RESPONSE - EXPERIMENT

The bandwidth of a back-illuminated mesa-structure InP/InGaAsP/InGaAs SAGM-APD that was designed for improved bandwidth[18] (Fig. 5) is shown in Fig. 6 as a function of the DC avalanche gain, M_o. For low gains (M_o<9) the bandwidth was greater than 6 GHz. In this gain regime the bandwidth was determined by the carrier transit time ($\simeq$45 ps), the RC time constant ($\simeq$35 ps) and the emission rate for trapped holes. In order to reduce hole trapping two intermediate-bandgap layers were incorporated into the transition region. From the low-gain response, $1/e_h$ was estimated to be less than 5 ps. At higher gains a constant gain-bandwidth product of 60 GHz was observed. This improvement in the gain-bandwidth product was achieved by using a thin ($\simeq$ 0.5 µm) heavily-doped ($N_d \simeq 5 \times 10^{16}$ /cm^3) multiplication region. Recently Sugimoto et al.[19] have achieved a gain-bandwidth product of 50 GHz in a planar SAGM-APD with a similar multiplication region.

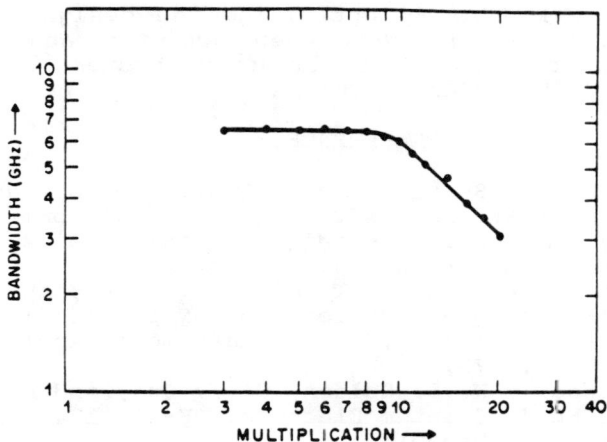

Fig. 6 Bandwidth Versus Avalanche Gain

The ultimate test for a photodetector is how well it performs in an optical receiver. Table I lists "champion" receiver sensitivities at several bit rates in terms of average received optical power and photons per bit. It can be seen that the InP/InGaAsP/InGaAs SAGM-APDs provide a significant margin over Ge APDs and InGaAs p-i-n photodiodes. In addition, the improvement in frequency response has contributed to the success of systems experiments at 2 Gbit/s, 4 Gbit/s, and 8 Gbit/s.

Table I Champion Receiver Sensitivities

Bit Rate (Mbit/s)	λ (μm)	N (Ph/bit)	P (dBm)	Det.	Ref.
420	1.52	437	-46.2	SAGM-APD	20
450	1.52	840	-43.1	Ge-APD	21
565	1.54	703	-42.9	SAGM-APD	22
565	1.3	1635	-38.5	InGaAs-PIN	23
1000	1.52	472	-42.1	SAGM-APD	20
1200	1.54	646	-40.0	SAGM-APD	22
1200	1.55	2359	-34.4	GE-APD	2
1200	1.53	1436	-36.5	InGaAs-PIN	24
2000	1.51	847	-36.6	SAGM-APD	25
2000	1.54	705	-37.4	SAGM-APD	22
4000	1.51	1440	-31.2	SAGM-APD	25
8000	1.3	2150	-25.8	SAGM-APD	26

CONCLUSION

We have examined the factors that limit the frequency

response of SAGM-APDs. By optimizing the design parameters high frequency response has been achieved and this has contributed to the success of lightwave systems experiments at 2 Gbit/s, 4 Gbit/s and 8 Gbit/s.

REFERENCES

1. S. R. Forrest, "Sensitivity of Avalanche Photodetectors for High-Bit-Rate Long-Wavelength Optical Communication Systems", Semiconductors and Semimetals, Vol.22 (Academic Press, Orlando, 1985), p.329.
2. J. I. Yamada, A. Kawana, T. Miya, H. Nagai, and T. Kimura, IEEE J. Quantum Electron. QE-18, 1537(1982).
3. K. Nishida, K. Taguchi, and Y. Matsumoto, Appl. Phys. Lett. 35, 251(1979).
4. J. C. Campbell, W. S. Holden, G. J. Qua, and A. G. Dentai, IEEE J. Quantum Electron. QE-21, 1743(1985).
5. J. C. Campbell, unpublished.
6. S. R. Forest, O. K. Kim, and R. G. Smith, Appl. Phys. Lett. 41, 95(1982).
7. F. Capasso, H. M. Cox, A. L. Hutchinson, N. A. Olsson, and S. G. Hummel, Appl. Phys. Lett. 45, 1193(1984).
8. Y. Matsushima, A. Akiba, K. Sakai, Y. Kushiro, Y. Noda, and K. Utaka, Electron. Lett. 18, 945(1982).
9. J. C. Campbell, A. G. Dentai, W. S. Holden, and B. L. Kasper, Electron. Lett. 19, 818(1983).
10. K. Yasuda, T. Mikawa, Y. Kishi, and T. Kaneda, Electron. Lett. 20, 373(1984).
11. Y. Sugimoto, T. Torikai, K. Makita, H. Ishihara, K. Minemura, and K. Taguchi, Electron. Lett. 20,653(1984).
12. R. Trommer and L. Hoffmann, Proc. 11th Int. Symp. on GaAs and Related Compounds, Bioritz, France (1984).
13. R. H. Fowler and L. Nordheim, Proc. Roy. Soc.(London) 119, 173(1928).
14. F. A. Padovani and R. Stratton, Solid-State Electron. 9, 695(1966).
15. D. T. Cheung, S. Y. Chiang, and G. L. Pearson, Solid-State Electron. 18, 263(1975).
16. E. B. Emmons, J. Appl. Phys. 38, 3705(1967).
17. G. F. Williams, F. Capasso, and W. T. Tsang, IEEE Electron Dev. Lett. EDL-3, 71(1982).
18. J. C. Campbell, W. S. Holden, J. F. Ferguson, A. G. Dentai, and Y. K. Jhee, 5th Int. Conf. on Integrated Opt. and Opt. Fiber Comm., Venice, Italy (1985).
19. Y. Sugimoto, T. Torikai, K. Makita, H. Ishihara, and K. Taguchi, 5th Int. Conf. on Integrated Opt. and Opt. Fiber Comm., Venice, Italy (1985).
20. B. L. Kasper, unpublished.
21. H. Toba, Y. Kobayashi, K. Yanagimoto, H. Nagai, and M. Nakahara, Electron. Lett. 20, 370(1984).
22. M. Shikada, S. Fujita, I. Takano, N. Henmi, I. Mito, K. Taguchi, and K. Minemura, 5th Int. Conf. on Integrated Opt. and Opt. Fiber Comm., Venice, Italy (1985).

23. D. R. Smith, R. C. Hooper, P. P. Smyth, and D. Wake, Electron. Lett. <u>18</u>, 453(1982).
24. M. C. Brain, P. P. Smyth, D. R. Smith, B. R. White, and P. J. Chidgey, Electron. Lett. <u>20</u>, 894(1984).
25. B. L. Kasper, J. C. Campbell, A. H. Gnauck, A. G. Dentai, and J. R. Talman, Electron. Lett. <u>21</u>, 982 (1985).
26. B. L. Kasper, J. C. Campbell, J. R. Talman, A. H. Gnauck, J. E. Bowers, and W. S. Holden, IEEE J. Lightwave Tech. LT-5, to be published.

VI. Laser Photochemistry and Photophysics
A. Laser Challenges in Photochemistry and Photophysics 556
B. Laser Raman Spectroscopy 2: Chemical and Biochemical Reactions 571
C. Laser-Induced Surface Reactions 586
D. Laser Photophysics 608

USING INCOHERENT LIGHT TO GENERATE COHERENT EXCITATIONS

S. R. Hartmann
Physics Department, Columbia University, New York, N.Y.

ABSTRACT

Photon echo generation by incoherent light is both practical and efficient. In the regime where the noisy excitation pulses overlap fast transient response is obtained and in addition psec beating is observed in Na vapor, the latter arising from coherent superposition states associated with the D line transitions. Only the fast response requires broadband noisy light: ultrafast modulated echoes are produced with either broadband noisy light spanning the D line transitions or with a pair of narrow band lasers resonant with each of the D line transitions.

INTRODUCTION

The photon echo effect is normally thought of as a coherent effect brought about by the interaction of an ensemble of atoms or molecules with a sequence of laser pulses which must be coherent.[1] This notion is not altogether correct as was demonstrated in '84 when an echo experiment was performed with the usual coherent laser sources replaced by broadband amplified spontaneous emission sources.[2] Not only was it possible to generate echoes in this unorthodox manner but the echoes so produced were larger than those obtained using coherent laser pulses.

Although usually not explicitly stated it is understood that the laser excitation pulses must be resonant with the optical transition to be excited. A fuller appreciation of this obvious requirement leads to an understanding of the unexpected effectiveness of echo generation with noise. With this motivation we ask the question: What constitutes resonant radiation? For the case in which all homogeneous relaxation times are long compared to the laser pulse width and the lasers are weak so that power broadening is negligible, then all atoms within a bandwidth $\Delta\omega = 1/\tau_p$, where τ_p is the duration of a coherent laser pulse at frequency ω, interact strongly with the radiation field. Outside this regime the interaction falls quickly and can therefore be called nonresonant. A stronger statement than this can in fact be made which is that the interaction strength within this bandwidth is uniform. That is, an atom at exact resonance and one removed by $\Delta\omega$ from exact resonance experience a pulse with the same area θ. But our interest is with incoherent radiation pulses. We now note that if such pulses have a duration τ_p then any radiation sampled for that duration and lying within a bandwidth no greater than $\Delta\omega = 1/\tau_p$ from exact resonance must appear coherent. This is an uncertainty argument result. It follows that any particular atom sees the applied incoherent laser field as a coherent one. It also follows that all atoms within the common bandwidth $\Delta\omega$ see the same coherent field and behave as a coherent group. With these results it is clear that echo production with incoherent light is to be expected. Each group of atoms within a bandwidth $\Delta\omega$ acts as a unit. If P_{echo} is the coherent dipole moment density of one such unit the radiated intensity I_{echo} of the atom ensemble varies as $N P_{echo}^2$ where N is $\Delta\Omega/\Delta\omega$, the number of units within the linewidth $\Delta\Omega$.

Echo experiments on the Na D lines in Na vapor using single mode coherent 7 nanosecond laser pulses can only excite the fraction $\Delta\omega/\Delta\Omega \approx 1/30$ of the resonance line. Incoherent echo generation gains us a factor of 30 by utilizing all the atoms in the resonance line. The statistical nature of the incoherent light leads to a reduction to 33% in the average value of P_{echo}^2 but the overall advantage of using incoherent light in this situation still corresponds to a factor of (1/3)(30) or 10. In the echo experiments we had performed using coherent light the lasers ran in several modes and so we only achieved an enhancement factor of 3 when incoherent sources were used. We constructed the incoherent sources by taking ordinary dye lasers and removing all elements which might define a cavity. Over the region where these incoherent sources lased, their output was effectively thermal.

FOUR WAVE MIXING

The object of these experiments was to establish that echo generation by incoherent light was both possible and practical. As an adjunct to this work a four wave mixing signal was developed in Na vapor at a temperature well above 445 K with temporally overlapping pulses from a single intense incoherent laser (thermal) source. The source output was split to produce two separate pulses we label 1 and 2 and then recombined after introducing an optical delay. The result is shown in figure 1.[1] When the delay between pulse 1 along **k** and 2 along **k** + **K** exceeds the 7nsec duration of these pulses the resulting four wave mixing signal along **k** + 2**K** is just the ordinary photon echo. The data we show here goes only to the 7nsec limit. The purpose of this experiment was to show that no well defined laser modes existed in our amplified spontaneous emission laser source. Had they been present there would have been a beat structure on the four wave mixing signal. The rapid fall in intensity for negative τ occurs in the order of a 100 psec. For positive τ there is a signal fall off with a time constant of the order of a few hundred psec followed by a slower fall off due to the diminishing of the temporal overlap of the excitation pulses.

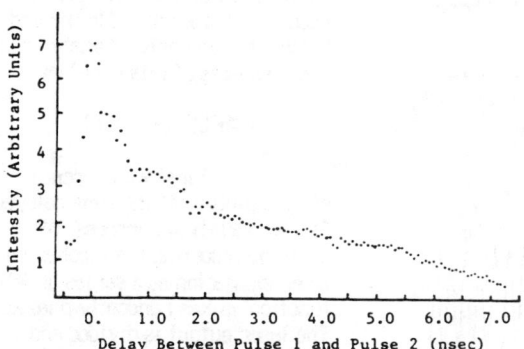

FIG. 1. Four-wave mixed signal, generated by use of one thermal source, as a function of pulse separation. The absence of any large-scale beating demonstrates that there is no stable mode structure in the thermal excitation pulses.

At the time of this work parallel experiments were being carried out by Yajima et al[3] and Asaka et al[4]. The time delayed four wave mixing experiments of Yajima et al. used picosecond noise pulses and the experiments of Asaka et al. used a noisy cw laser. In both these cases the laser excitations were weak and a perturbation theory was suitable.

In an effort to understand better the interaction of noisy light with matter we proceeded to study four wave mixing with intense radiation fields.[5] We further broadened the spectral width of our incoherent source, and reduced the temperature of the Na vapor to 445 K so that collisional broadening effects were negligable. We found that as had been reported by Yajima we obtained beats in the four wave mixing signal when we varied the time delay τ. These beats correspond to the 6 Å splitting of the Na D lines.[6] Whereas our incoherent echo experiments were performed using light with a bandwidth of only a few GHz our four wave mixing experiments with intense fields had a bandwidth of 12 Å, sufficient to excite both D lines. Our results are shown in figure 2.[5] These results are similar to those of figure 1 except for the dramatic 1.9 psec beating. The fast fall off in intensity S for positive τ wasn't expected as we now worked in a collision free regime where the result of Morita and Yajima[7], valid when the longitudinal and transverse relaxation times, T_1 and T_2, are large compared to the inverse 1/e half width of the inhomogeneous line, $\delta\omega^{-1}$, applies, as was the case in our

$$S = \int_{-\delta\omega\tau/\sqrt{2}}^{\infty} dy \exp(-y^2)$$

experiment. Our observed degradation in S for large τ is due to laser noise induced relaxation effects and is a subject of continuing interest.[8] Our discussion here will be limited to explaining why coherent signals should be produced by broadband noise and to what extent the broadband noise is essential in effecting a particular result. We limit ourselves to the weak excitation limit and we refer the reader to the work of Morita and Yajima for an analytic analysis of many aspects of this problem.

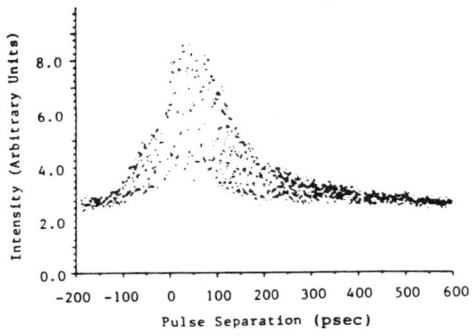

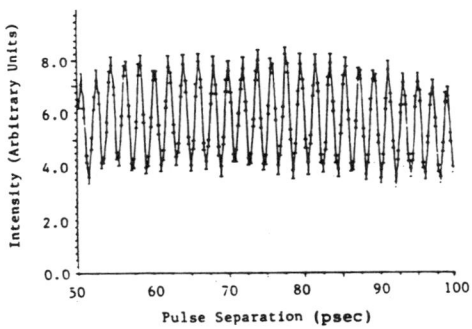

FIG. 2. (a) Experimentally measured signal intensity vs excitation pulse separation in a 10-cm-long Na cell at 445 K. (b) The same data for pulse separations of 50–100 psec with error bars and lines connecting the data. The 1.9-psec modulation is due to the Na $3P$ fine-structure beating.

SIMPLE ANALYSIS

First we present a simple picture of the time delayed four wave mixing process. We begin by modeling the incoherent laser excitation as a series of delta function spikes randomly phased. The laser output is divided and recombined so that the sample is irradiated with two identical laser pulses directed along **k** and **k** + **K** with relative delay τ. These two spike trains are pictured in the upper part of figure 3. The lower part of this figure shows two recoil diagrams[9]. In the upper recoil diagram the spike along **k** preceeds the spike along **k** + **K** while in the lower recoil diagram it follows it. In all parts of the figure the time variable is synchronized and increases to the right. Let's consider the upper recoil diagram. It shows the effect of the first four noise spikes along **k** separately generating excited state amplitudes which recoil until deexcited by the corresponding noise spikes along **k** + **K**. In this manner a macroscopic ground state amplitude with momentum corresponding to −**K** is built up. The essential thing here is that although the noise spikes in each directed pulse train are randomly phased there is a fixed phase difference between noise spikes separated by τ in the two pulse trains. When we consider the effect of a noise spike along **k** + **K** acting on the ground state amplitude the effect is as shown in the central section of the upper recoil diagram. An excited state amplitude is generated which recoils along **k** + **K** and collides τ later with the ground state amplitude recoiling along −**K**. The result is the formation of a macroscopic dipole moment with wave vector **k** + **K** −(−**K**) =

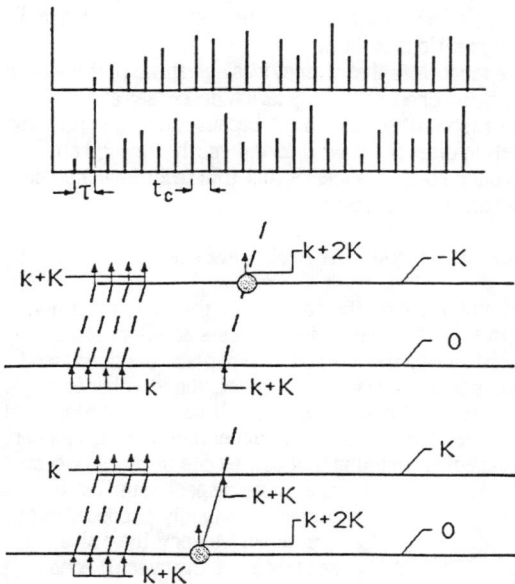

FIG. 3. Upper section: Two identical laser spike trains with relative delay τ along k+K and k. Lower section: Recoil diagrams showing signal generation along k+2K for positive τ (above) and for negative τ (below). Time increases to the right in all cases.

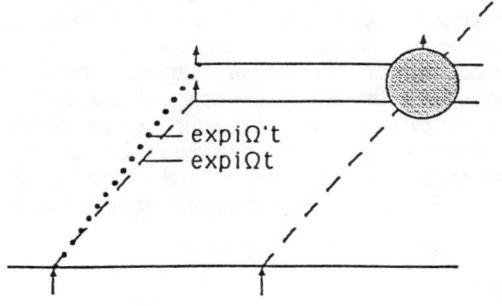

FIG. 4. Recoil diagram showing origin of ground superposition state modulation.

k + 2K in the phase matched direction in which we look. Each noise spike along k + K therefore generates a macroscopic dipole moment which radiates. Since these noise spikes are randomly phased the signal we see is the sum of the intensities which would be radiated separately by each of the induced dipole moment densities. The picture we have presented corresponds to a series of accumulated photon echoes which combine incoherently to yield the result expressed above in Eq(1). The signal falls to half its assymptotic limit at τ = 0 since the echo amplitude is fully formed on application of the noise spike along k + K and the contribution to the time integral of the echo signal during its formation is not obtained. For negative τ a similar analysis using the lower recoil diagram obtains and the resulting signal is seen to be best interpreted as a series of incompleted free decays.

The 1.9 psec beats we observed can be explained in terms of the macroscopic ground state population recoiling along −K being modulated at the difference frequency of the sodium D lines. Consider figure 4 which shows the recoil diagram appropriate when two separate transitions are excited by the same delta function spike. Each excited state amplitude evolves according to its energy eigenvalue with the result that the ground state formed τ later has an amplitude proportional to [exp(−iΩτ) + exp(−iΩ'τ)] ~ cos[(Ω−Ω')τ/2] where Ω and Ω' are the eigenfrequencies associated with the two excited states. As before, the effect of a later spike along k+K is to generate a recoiling state which collides with the reformed ground state to produce an echo. This ground state having an amplitude which is modulated at half the difference frequency of the two excited states gives rise to an echo intensity modulated by $\cos[(\Omega-\Omega')\tau]$. As can be seen from the figure this is true whether the interrogation spike which produces the recoiling excited state (states) is resonant with either of the two transitions or both.

The virtue of the recoil diagram approach is that it provides a graphic picture of the echo formation or time delayed four wave mixing process. If relaxation processes are introduced then the radiated signal will diminish with increasing τ because the superposition states will dephase reducing the echo formation directly as well as indirectly through an incomplete formation of the macroscopic ground state. It appears that this technique may be well suited for the measurement of fast relaxation phenomena.[10]

ULTRAFAST SPECTROSCOPY WITH NARROW BAND LASERS

The incoherent echo analysis was motivated by the realization that only radiation resonant with a transition will interact with it. This observation suggests an alternative method for generating the psec beats observed above. The method is to replace the broadband incoherent laser whose spectral output covers both lines with two narrowband coherent lasers each resonant with one of the transitions.[11] The output of these lasers is combined and treated as a single pulse. We performed a regular two pulse photon experiment using regular dye lasers with 7 nsec duration pulses. As expected we found that echoes are produced which are in all respects similar to echoes previously produced except that when we vary the pulse separation τ the overall echo intensity varies according to $\cos[(\Omega-\Omega')\tau]$. Thus the psec beating is obtained with narrow band excitation. This beating is observed even when a spectrometer is introduced in front of the detector so that radiation is detected from only one of the two transitions excited. This demonstrates that the beating arises from an interference in the common ground state population.

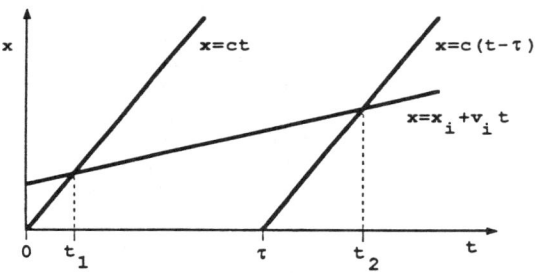

FIG. 5. Space-time diagram showing excitation pulses and an atom traveling at v_i.

This result holds independent of the frequency splitting of the two optical transitions excited. By proper selection of materials narrow band excitation of beats in the femtosecond and attosecond regimes is possible. These beats can be observed with slow detectors as the beat is present in the integrated echo intensity signal not in the individual echo signal envelope.

The echo intensity is calculated by following a moving atom which is irradiated at t_1 and t_2. The lasers are not at exact resonance but must both excite the same velocity subgroup as we are interested only in the situation where beats occur. We use a + and − subscript to indicate time just before and after the pulse arrival and we list the wavefunction development as follows:

$t_{1-}>=|g>$, $t_{1+}>=|g>+|e>+|e'>$,

$t>t_{1+}>=|g>+\exp[-i\Omega(t-t_1)]|e>+\exp[-i\Omega'(t-t_1)]|e'>$,

$t_{2-}>=|g>+\exp[-i\Omega(t_2-t_1)]|e>+\exp[-i\Omega'(t_2-t_1)]|e'>$,

$t_{2+}\rangle = |e\rangle + |e'\rangle + \{\exp[-i\Omega(t_2-t_1)]|g\rangle + \exp[-i\Omega'(t_2-t_1)]|g'\rangle\}$,

$t > t_{2+}\rangle = \exp[-i\Omega(t-t_2)]|e\rangle + \exp[-i\Omega'(t-t_2)]|e'\rangle$
$+\{\exp[-i\Omega(t_2-t_1)] + \exp[-i\Omega'(t_2-t_1)]\}|g\rangle$.

From the wavefunction at $t > t_{2+}$ we calculate the dipole moment of the atom being followed and label it with the subscript i:

$P_i = \{\exp i\Omega[(t-t_2)-(t_2-t_1)]\} \bullet [1 + \exp -i(\Omega'-\Omega)(t_2-t_1)] + \text{c.c.}$.

The dipole density is then obtained by summing over all atoms, yielding

$P(x,t) = \sum_i P_i(t)\delta(x-x_i-v_i t)$ where $\sum_i \to \int dv g(v)$. The lineshape is given by

$g(v) = (1/\sqrt{2\pi})(1/v_p)\exp-(v-v_s)^2/v_p^2$ where $v_s = (\Delta\Omega/\Omega)c$ is the average velocity of the interacting atoms as determined by the detuning of the laser from the resonance frequency of an atom at rest, the quantity $v_p = (\tau_D/\tau_p)v_D$ is the spread in velocities covered by the laser pulse whose duration is τ_p. The radiated field from those atoms moving at v_i is obtained from

$E \sim \int dx P(x, t+x/c) \sim \sum_i P_i[t+(x_i+v_i t)/(c-v_i)]$.

The relationship between the various parameters is displayed in figure 5 and used to evaluate P_i to order v_i/c. This done we can obtain E and the corresponding integrated intensity which we write as

$\int dt\, I_{echo} \sim 1 + \exp\{(-1/4)[\tau v_p(\Omega-\Omega')/c]^2 \cos[(\Omega-\Omega')\tau(1 + v_s/c)]\}$

Since the modulation is observed only when both lasers are simultaneously resonant on the same atom the Doppler shift occurs only in the difference frequency. In the limit that the lasers are multimode and the mode density is dense then the response becomes Doppler free and is given by

$\int dt\, I_{echo} \sim 1 + \exp\{-(1/4)[\tau(\Omega'-\Omega)/c]^2(v_D^2+v_p^2/2)\}\cos(\Omega-\Omega')\tau$

where v_D is the Doppler velocity. The term involving v_D degrades the echo in Na with a time constant of 400 nsec when the Na temperature is 410 K. This is the dominant relaxation term after lifetime relaxation and only becomes important when the time between the first excitation pulse and the echo exceeds 50 fluorescence lifetimes.

CONCLUSION

It has been demonstrated that incoherent or broadband light is suitable for generating photon echoes and for performing time delayed four wave mixing experiments. The broadband character of the light makes it possible to excite a wide spectrum of resonances. This is useful when relaxation broadening is important. The simplicity of the technique makes it a promising tool for the study of fast relaxation processes and wide band spectroscopy. Insight into the generation of coherent beats with broadband light was obtained by performing a parallel experiment with a pair of narrowband lasers. Doppler free spectroscopy should be possible with this technique.

ACKNOWLEDGMENTS

We thank B. Brody, D. DeBeer, and F. Moshary for their imformed comments. This work was supported by the U. S. Office of Naval Research and by the Joint Services Electronics Program (U. S. Army, U. S. Navy, U. S. Air Force) under contract No. DAAG29-85-K-0049.

REFERENCES

1. N. A. Kurnit, I. D. Abella, and S. R. Hartmann, Phys. Rev. Lett. **13**, 567 (1964).
2. R. Beach and S. R. Hartmann, Phys. Rev. Lett. **53**, 663 (1984).
3. Tatsuo Yajima, Norio Morita, and Yuzo Ishida, in XIII International Quantum Electronics Conference Technical Digest, Anaheim, California, 1984 (unpublished), p. 112.
4. S. Asaka. H. Nakatsuka, M. Fujiwara, and M. Matsuoka, Phys. Rev. A **29**, 2286 (1984).
5. R. Beach, D. DeBeer, and S. R. Hartmann, Phys. Rev. A **32**, 3467 (1985)
6. The correponding experiment in Rb. vapor produces a 139 fsec beat. J. E. Golub and T. W. Mossberg, Opt. Lett. **11**, 431(1986)
7. N. Morita and T. Yajima, Phys. Rev. A **30**, 2525 (1984).
8. M. Defour, J. C. Keller, and J. L. LeGouët, JOSA B **3**, 544 (1986)
9. R. Beach, S. R. Hartmann, and R. Friedberg, Phys. Rev. A **25**, 2658 (1982)
10. Masahairo Fujiwara, Ryo Kuroda, and Hiroki Nakatsuka, JOSA B **2** 1634 (1986)
11. D. DeBeer, L. G. Van Wagenen, R. Beach, and S. R. Hartmann, Phys. Rev. Lett. **56**, 1128 (1986).

A THEORY OF COHERENT MULTI-COLOR LASER EXCITATION OF LOCALIZED STATES

John S. Hutchinson
Department of Chemistry, Rice University
Houston, Texas 77251

ABSTRACT

A proposal is presented for preparation of photochemically interesting coherent superpositions of molecular eigenstates using more than one laser. This superposition continues to be the excited molecular state as long as the lasers are on. The proposal is applied to excitations of local mode vibrational overtones and to separation of singlet and triplet electronic excitations.

INTRODUCTION

Considerable interest exists in the preparation of state-selective and mode-selective photochemical excitations.[1] If created, such excitations hold the promise of laser control of reaction rates non-thermally. For example, high energy vibrational excitations of local HC bond overtones have been observed, and several studies have been reported of photochemistry induced by such excitations. These studies have found, with only slight deviation, that the rates of such photochemical reactions are in agreement with statistical (thermal) rate constants, thus revealing that single laser excitation of overtone states does not provide a measure of photochemical control.[1]

We have recently studied[2] the dynamics of the preparation of overtone states in a polyatomic molecule, and have shown that the excitation process is actually orders of magnitude slower than intramolecular vibrational relaxation. As a direct consequence, overtone excitations (at least of bound states) are extremely eigenstate specific. Thus, the extent to which the excitation is localized is governed by the extent of localization of the single eigenstate prepared, and is not determined by the method of state preparation.

In the remainder of this paper, we will propose that multi-color excitation of these states may provide the desired mode-specific local excitation.

THEORY OF TWO LASER COHERENT EXCITATION

We have presented the general theory of excitation of a superposition of n eigenstates with n lasers.[3,4] In this paper, we will consider only the n=2 special case. The Hamiltonian of interest in this study describes the interaction between a molecular Hamiltonian, H_M, and a quantized radiation field, H_F, via[2]

$$H = H_M + H_F - \mu(4\pi\omega^2/V)^{1/2}Q_F \, , \qquad (1)$$

From our earlier study of single laser excitations, we know that each laser may be considered to interact with only one eigenstate. Thus, in a quantized field dress-states basis, the dynamics of the two laser excitation can be described by only three states: $|0\rangle$, the dressed ground molecular state, and $|1\rangle$ and $|2\rangle$, the two dressed molecular eigenstates. The Hamiltonian matrix in this three state model is

$$H = \begin{bmatrix} E_0 & \gamma & \gamma \\ \gamma & E_0 & 0 \\ \gamma & 0 & E_0 \end{bmatrix} \quad (2)$$

where all three dressed basis states are made degenerate by precise tuning of the two lasers to the excited eigenstates. Each dressed molecular eigenstate is coupled only to the ground state, and by scaling the laser intensities, these couplings may be made equal.

The eigenstates of the Hamiltonian in Eq.(2) are

$$\Phi_1 = 2^{-\frac{1}{2}}|0\rangle + \tfrac{1}{2}|1\rangle + \tfrac{1}{2}|2\rangle \qquad E_1 = E_0 - 2^{\frac{1}{2}}\gamma$$

$$\Phi_2 = 2^{-\frac{1}{2}}|0\rangle - \tfrac{1}{2}|1\rangle - \tfrac{1}{2}|2\rangle \qquad E_2 = E_0 + 2^{\frac{1}{2}}\gamma \quad (3)$$

$$\Phi_3 = 2^{-\frac{1}{2}}(|1\rangle - |2\rangle) \qquad E_3 = E_0.$$

Only two of the eigenstates project onto the initial state at t=0, which is just $|0\rangle$. Therefore, the time-dependence of the excitation is given by the wave function

$$\Psi(t) = 2^{-\frac{1}{2}}(\Phi_1 e^{-iE_1 t/\hbar} + \Phi_2 e^{-iE_2 t/\hbar}) \quad (4)$$

so that the dynamics of excitation are simply a quantum beating between the initial state $|0\rangle = \Phi_1 + \Phi_2$ and the excited state $\Psi_f = \Phi_1 - \Phi_2$. Examining the eigenfunctions in Eq. (3), we see then that the excitation is just a superposition of the dressed molecular eigenstates, $|1\rangle+|2\rangle$. Moreover, this is the state at all times while the lasers are on: the excited state Ψ_f does not relax.

The key to the significance of this theory is the possibility that $|1\rangle+|2\rangle$ is a chemically interesting excitation. We have applied this postulate to both electronic[3] and vibrational[4] excitations. Consider first a model for polyatomic vibrations[4] in which H_M is

$$H_M = H_y + H_x(y) \quad (5)$$

where H_y is a Morse oscillator modelling a local HC stretch, and H_x is a harmonic oscillator modelling a "bath" mode for relaxation of the local excitation. The molecular zeroth-order states $|v_y,v_x\rangle$ are heavily mixed in the molecular eigenstates; for example, near the third overtone of the HC mode, we have the following states,

$$|1\rangle = .731|4,0\rangle - .495|3,2\rangle - .105|2,4\rangle + .406|1,6\rangle$$
$$|2\rangle = .654|4,0\rangle + .544|3,2\rangle - .141|2,4\rangle - .455|1,6\rangle , \quad (6)$$

creating substantial vibrational delocalization in a single laser
excitation, as seen in Fig. 1. Note that both localized and
delocalized states rise together. However, with a two laser
coherent excitation, we can superimpose the eigenstates in Eq.(6).
By inspection, this superposition is just the localized excitation,
|4,0>, as is shown in Fig. 2. Note most significantly that the
local excitation does not relax, as long as the lasers are on!

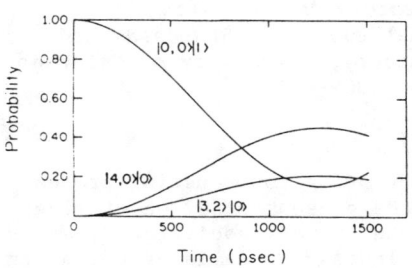
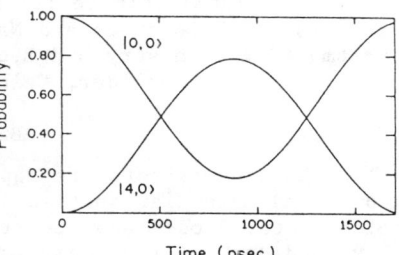

Fig.1. Excitation dynamics   Fig.2. Coherent Excitation of
of State |1> by one laser.   superposition of |1> and |2>
                                by two lasers.

We have also applied this coherent excitation effect to
electronic excitations of CS.[3] The $^1\Pi$ state in CS is mixed with
four nearby triplet states, most notably $^3\Sigma$ and $^3\Delta$. Electronic
excitation of CS with one laser thus yields a state of mixed
singlet and triplet character. However, we have shown that, with
two lasers, one may "interfere out" either the $^3\Sigma$ or the $^3\Delta$ triplet
state, creating a state of substantially enhanced singlet
character. The reader is referred to the original papers for
further details.[3,4]

ACKNOWLEDGMENTS

The author gratefully acknowledges his co-worker, Thomas A.
Holme, for his major role in this work. This work was supported in
parts by grants from the National Science Foundation, the Robert A.
Welch Foundation, and the Atlantic Richfield Foundation of Research
Corporation. Acknowledgment is made to the Donors of the Petroleum
Research Fund, administered by the American Chemical Society, for
partial support of this research.

REFERENCES

1. F.F. Crim, Ann. Rev. Phys. Chem. 35, 657 (1984).

2. T.A. Holme and J.S. Hutchinson, J. Chem. Phys. 84, 5455 (1986).

3. T.A. Holme and J.S. Hutchinson, J. Chem. Phys. 86, xx (1987).

4. T.A. Holme and J.S. Hutchinson, Chem.Phys.Lett. 124, 181 (1986).

APPLICATION OF SEMICONDUCTOR DIODE LASERS TO PROBE PHOTODISSOCIATION DYNAMICS

Harold K. Haugen
Department of Physics, University of Toronto
Toronto, Ontario, Canada, M5S 1A7

Wayne P. Hess and Stephen R. Leone[*]
Joint Institute for Laboratory Astrophysics
University of Colorado and National Bureau of Standards and
Department of Chemistry and Biochemistry, University of Colorado
Boulder, Colorado 80309-0440

ABSTRACT

Tunable diode lasers are rapidly proving to be useful probes of atomic and molecular species. In the present study we utilize a semiconductor diode laser to probe the photodissociation dynamics of alkyl iodides. A room temperature InGaAsP diode laser operating at 1315 nm is used to probe the transition between $I^*(^2P_{1/2})$ and $I(^2P_{3/2})$ atoms. I^* quantum yields are obtained in the UV laser photolysis of n- and i-C_3F_7I, CH_3I and ICN by time resolved laser gain versus absorption spectroscopy. The high amplitude stability of the diode laser and the internal normalization of the gain versus absorption technique allow for a sensitive and accurate determination of the I^* quantum yields.

INTRODUCTION

The photolysis of alkyl iodides has been the subject of numerous investigations,[1] due both to applications to I^* chemical lasers and relevance to fundamental photodissociation dynamics. We have recently developed a diode-laser-based gain versus absorption spectroscopy technique to measure accurately I^* quantum yields in these systems.

EXPERIMENTAL TECHNIQUE

The experimental approach (Fig. 1)[2,3] consists of a two laser, pulse and probe technique. A pulsed UV photolysis laser creates the initial I/I^* population and a cw diode laser probes the subsequent I atom dynamics. Following photolysis, the gain or absorption of the I^*/I system on the diode probe laser is monitored by a fast (65 MHz) Ge photodiode. The transient signal is amplified, digitized and signal averaged. The initial population inversion following pulsed laser photolysis and the total absorption after quenching of all excited iodine atoms are measured on a short (~10 μs) time scale. Thus the gain versus absorption method has the advantage that it obtains the quantum yield from a single time resolved transient. The internal normalization inherent in probing the coupled pair of I atom states provides an accurate measurement that is insensitive to

---

[*]Staff Member, Quantum Physics Division, NBS.

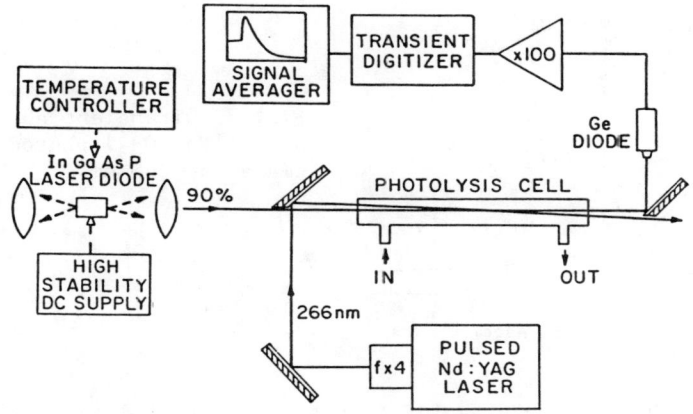

Fig. 1. Schematic of the experimental apparatus.

several experimental parameters: these include laser and atomic linewidths, the pressures of the precursor and buffer gases, and the powers of the photolysis and probe lasers.

The diode laser is tuned to the $I^*(^2P_{1/2}) \rightarrow I(^2P_{3/2})$ transition (~1315 nm) by a combination of diode temperature and current control. For the C_3F_7I and CH_3I quantum yields the diode is operated in a "free running" multimode configuration; no optical feedback is employed. A recent improvement in the experimental technique[4] is the use of optical feedback to control the mode of the diode laser. Light emitted from the rear facet of the diode is collimated and imaged onto a Littrow configuration grating. The dispersed back-reflected light enhances a particular mode at the expense of the other modes (~95% of emitted power in the desired mode).

RESULTS AND DISCUSSION

A typical transient absorption signal is presented in Fig. 2. Following the photolysis pulse, at t=0, prompt gain is observed. The I^* atoms are then quenched by molecular oxygen on a short time scale to yield ground state iodine atoms with unit efficiency. The quantum yield is obtained by the simple formula[2,3]

$$\Phi_{I^*} = \frac{1}{3}\left(\frac{S_i}{S_f} + 1\right) .$$

Here S_i is the back extrapolated time zero signal and S_f is the total absorption after all the I^* atoms have been quenched. The factor of 1/3 accounts for the hyperfine degeneracies of the ground and excited states. The quantum yields for n- and i-C_3F_7I, CH_3I, and ICN are determined to be 102±4, 102±7, 73±4, and 66±2%, respectively, for 266 nm photolysis. The reported errors are one standard

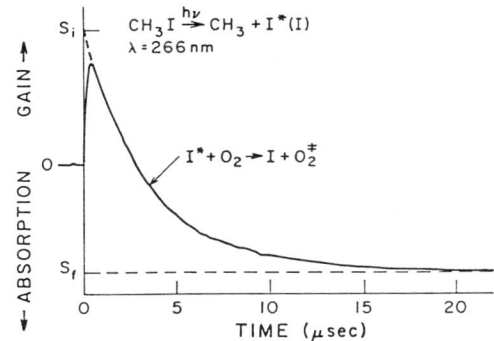

Fig. 2. Transient absorption signal for CH_3I photodissociated at 266 nm.

deviation assuming no systematic errors. The quantum yields for C_3F_7I may not be greater than 1.0, however, by the nature of the measurement, statistical fluctuations may carry the result above 1.0.

The unity quantum yields for the C_3F_7I isomers indicate that the 266 nm excitation accesses a single excited state which correlates with I^* products. No significant curve crossing occurs during dissociation. The 73% quantum yield for CH_3I could be due to a curve crossing from a state that correlates with I^* products to a state that correlates with I ground state products, or there could be two electronic states that are directly accessed by the 266 nm excitation. The "purity" of the transition moment determined in molecular beam experiments suggests that a curve crossing explanation is more likely. Recently we have evaluated quantum yields for ICN at seven wavelengths between 248 and 284 nm.[4] The ICN quantum yield peaks at 266 nm with a value of 66% and falls off to 44% and 53%, at 248 and 284 nm, respectively.

CONCLUSION

The absorption versus gain technique has been shown to be a powerful new means of obtaining accurate quantum yields. The tunable diode laser is a sensitive probe of I and I^* populations and is readily generalized to probing the dynamics of I and I^* atoms. The diode laser shows great promise as a probe of photodissociation product states, numerous kinetic phenomena, and as a diagnostic for various I^* laser media.

This work supported by NSF, NASA and AFWL.

References

1. S. R. Leone, Adv. Chem. Phys. 50, 255 (1982).
2. H. K. Haugen, E. Weitz and S. R. Leone, J. Chem. Phys. 83, 3402 (1985).
3. W. P. Hess, S. J. Kohler, H. K. Haugen and S. R. Leone, J. Chem. Phys. 84, 2143 (1986).
4. W. P. Hess and S. R. Leone, to be published.

SELECTIVE IR PHOTOISOMERIZATIONS IN SOLIDS

James S. Shirk[#] and C.L. Marquardt
Naval Research Laboratory
Washington, D.C. 20375

ABSTRACT

Quantum yield measurements show vibrational relaxation to the host is important in the IR photochemistry of 2-fluoroethanol in solid Ar. Spectral linewidths imply these relaxation rates depend upon the conformer and specific vibrational mode. Selective vibrational relaxation will produce nonthermal and mode selective photochemistry in matrices.

INTRODUCTION

IR excitation of a variety of molecules trapped in solid rare gases leads to photoisomerization, a photo-induced conformational change.
In previous papers [1,2] it was shown that: 1) the quantum yields for these reactions in the 2-haloethanols depend upon the vibrational mode excited; 2) the quantum yields for the formation of thermodynamically unstable conformers are often higher than those for the photoreaction to form the stable molecule; and 3) the quantum yield for isomerization has no obvious dependence on the density of states of the molecule excited. Result 2) means that broad band IR irradiation (with a glower) converts a sample almost completly to the thermodynamically unstable form, a process quite impossible thermally.

This paper presents quantum yields for the photoiscmerization of 2-fluoroethanol from the thermodynamically stable Gg' form to the less stable Tt form and back. These and spectral linewidths provide evidence that vibrational relaxation to the solid matrix is important and depends upon the conformer and vibrational mode.

RESULTS AND DISCUSSION

The experimental apparatus has been described[1]. The sample of 2-fluoroethanol in Ar (1/1000) was irradiated at different CH and OH(OD) stretching modes with a tunable F-center laser or a Raman shifted ND/YAG pumped dye laser. Initial quantum yields were calculated from FTIR spectra recorded before and after photolysis.

Table I gives quantum yields for the Gg' to Tt isomerization at different wavelengths and the ratio of the forward to reverse quantum yields.

TABLE I

Photolysis Frequency		ϕ	ϕ_f/ϕ_r
1035 cm^{-1}	(CF)	0	-
2688	(OD)	[.1]	>5
2992	(CH)	.02	10
3644	(OH)	.16	20
7128	(OH)	>.2	1

It is clear that the quantum yields are not a monotonic function of energy. Exciting the OH(OD) stretch is more efficient than exciting the CH stretch. The small quantum yields imply relaxation is fast compared to reaction.
If the reaction were from thermalized molecules the relative forward to reverse quantum yields should be related to the equilibrium constant since

the quantum yield, ϕ, is given by:

$$\phi = k/(k+k_R) = k/k_R \qquad (1)$$

where k is the isomerization rate and k_R is the rate of vibrational relaxation. The quantum yield ratio is then:

$$\phi_f/\phi_r = (k_f/k_f+k_R)/(k_r/k_r+k_R) = k_f/k_r \qquad (2)$$

where k_f, k_r are the forward and reverse rate constants. The last expression (equivalent to the equilibrium constant) holds when k_R is the same for reagent and product and fast compared to the reaction rates. Since the Gg' conformer is at least 4kJ/mole more stable than any other conformer, the maximum value a thermodynamic ϕ_f/ϕ_r can have is 0.5. At low excitations the experimental ratio is well above the largest possible thermodynamic value. The photochemical reaction is not thermal.

Part of the reason for the small Tt to Gg' quantum yield may be found in the spectrum of the Tt conformer. Table II shows a striking variation in the linewidth (FWHM) of the various vibrations of the Tt. In comparison the linewidths of the Gg' are all in the range 1.5-3 cm^{-1}.

TABLE II

BAND	FWHM	ASSIGNMENT
888 cm^{-1}	>9 cm^{-1}	CH$_2$ rock
1015	1.6	CF stretch
1214	5	COH bend
1492	1.5	CH$_2$ bend
3668	3	OH stretch
7168	3	OH overtone

If we assume that the linewidths of the broader lines is due to lifetime broadening, then some modes of the Tt conformer show a shorter lifetime than the Gg'. Apparently some modes of the Tt relax rapidly to the lattice. This can account for the large ratios for the forward to reverse quantum yields: the Tt to Gg' reaction is quenched due to the rapid vibrational relaxation of the Tt conformer.

The wide variation in linewidth implies that the vibrational relaxation rate depends upon the mode which is excited. Some modes of the Tt form communicate faster with the lattice than with other modes of the same molecule.

CONCLUSION

The IR induced photochemistry of 2-fluoroethanol in Ar is very different from simply heating the sample. Vibrational relaxation competes with the reaction. The linewidths of the Tt conformer imply the vibrational relaxation rates depend upon the conformer and the specific vibrational mode. In such a case the reactions can be "mode selective".

REFERENCES

1. W.F. Hoffman and J.S. Shirk, J. Phys. Chem. **89**, 1715 (1985)
2. W.F. Hoffman and J.S. Shirk, J. Phys. Chem. **90**, 5706 (1986)

\# permanent address: Dept. of Chemistry, Illinois Institute of Technology, Chicago, Il 60616

INTERFERENCES IN THE RAMAN EXCITATION PROFILE FOR THE INTENSITY OF
NORMAL MODES OF AGGREGATED CHLOROPHYLL a

by

L.V. Haley, T.L. Collier, T.A. Mattioli, D.L. Thibodeau and J.A.
Koningstein
Department of Chemistry, Carleton University, Ottawa, Ont. K1S 5B6

ABSTRACT

The Raman excitation profile for the intensity of normal modes of aggregated Chlorophyll a for wavelength resonant with the Soret absorption band centered around 430 nm, shows two minima at 434.1 nm and 435.4 nm. These minima have a width of about 10 cm^{-1} in an excitation profile characterized by a full width of half height of about 100 cm^{-1}. The minima are explained in terms of interference effects between the terms of the Raman scattering tensor arising from contributions of close lying resonating electronic states.

RESULTS AND DISCUSSION

Raman excitation profiles can be investigated as function of the wavelength of the exciting line and in time. In this article we discuss aspects of the former while studying the wavelength dependent intensity of vibrational scattering of solutions of aggregated Chlorophyll in hexane.

Aggregated solutions of Chlorophyll a in dry hexane can exist of dimers and higher aggregates. The energy level diagram of such aggregates is more complicated than that of the monomer. Whereas the latter could exist of a series of states well separated from each other, that of the higher aggregates may exist of a series of levels which are closely together and centered around the position of a level of the monomer. As a result, the absorption spectrum of the aggregates can be broader if compared to that of the monomer. The excitation profile for the intensity of vibrational Raman lines of the ground state should go through a maximum if the wavelength of the radiation of a tunable laser (Raman probe) is in resonance with the position of the zero-zero line of an electronic band system. The profile for an aggregate should in principle exhibit several of such maxima because of close lying electronic states.

A complication can occur, if the profile due to one electronic state overlaps with that of another, as can be expected for solutions containing Chlorophyll aggregates. In that case interference effect can take place resulting in abrupt changes in the resonance Raman excitation profile in the spectral region where the overlap occurs. Such interference can be explained in the following way. Suppose one considers an electronic energy level diagram existing of a ground state and two close lying electronic states with the position of the tunable laser (the Raman probe) located at the cross over of the absorption bands of the two states. The summation which occurs in the well known expression for the Raman tensor for resonance enhanced scattering in the ground state is then thought to

© American Institute of Physics 1987

extend over these two states only. We are in particular interested in the form of the denominators of the truncated terms which appear in the expression for an element of the tensor. The energy denominator in the summation is negative for one of the states but positive for the other state. The Raman intensity is proportional to the absolute square of an element of the tensor and in the square appears besides the positive contribution from two diagonal terms also that of the off-diagonal terms (the cross products). The denominator of these cross products is negative, hence to possible occurence of a reduction (interference) in the Raman excitation profile if the laser is tuned through the absorption band system. The magnitude if such interferences depends to a large degree on i) the frequency difference of laser and the position of the electronic excited state, ii) on the damping factor and iii) on the magnitude of electric dipole matrix elements which appear in the numerator of the expression for an element of the tensor. We have observed interferences which appear as holes with a width of 10 cm^{-1} in the resonance Raman excitation profile of a normal modes of the ground state having an overall width of 100 cm^{-1}. We believe that such a phenomenon may well be the finger print of any system which contains aggregated molecules and the interference effect allows an accurate determination of the positions of origins in the absorption band systems of these solutions.

The simplest form of a Chlorophyll aggregated is a dimer. In dry hexane such a molecule exists where the binding of the two individual Chlorophyll molecules takes place via the C=O keto group on ring V of monomer to the M_g atom of the other. The electronic energy level diagram of such a donor-acceptor molecule has been obtained from one and two-photon induced fluorescence, their time dependence and from linear and non-linear absorption spectroscopy.

In the Soret spectral region we find the origin of an electronic band system of the acceptor molecule at 436 nm and that of the donor molecule at 449 nm. The resonance Raman profile for normal modes of the ground state shows

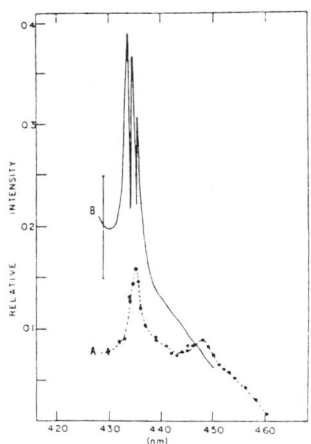

Fig. 1

maxima at these positions. The energy difference between these two electronic state is much larger than the width of the excitation profiles and from the above follows that interference effects are not anticipated. The experimental data, lower trace A of Fig. 1 are in agreement with the above. The profile of the solutions with higher concentrations of Chlorophyll a containing other aggregates are characterized by sharp interferences, upper trace B of Fig. 1 belongs to a Chlorophyll cluster which contains more than two molecules because in order to explain the sharp interferences there must be at least three electronic levels with origins between 430 nm and 436 nm. In addition, there should be a large difference in the intensity of the electric dipole transitions and trimeric or tetrameric clusters to exhibit an energy level diagram which satisfies such a requirement although such is also the case for a dimer where the rings of the Chlorophyll molecules are parallel to each other. However, it seems doubtful that such a complex could exist in solution because a similar dimer made up of bacteriochlorophyll molecules, found in the reaction center of certain bacteria exist by virtue of the fact that the cluster is only a stable chemical entity because it is surrounded by a protein molecule. We are now studying these solutions with the laser tuned to the vibronic side bands belonging to specific vibrations (like the C=O stretch). For the donor-acceptor complex we anticipate resonance enhancement profiles which could provide an inside in the bonding between the Chlorophyll molecules. The intensity of the Raman line due to the C=O group on ring V of the donor molecule should show an increase in intensity if the laser is tuned some 1700 cm^{-1} away from the origin at 449 nm while such is not the case for the intensity of the C=O stretch of the acceptor molecules and vice versa. In addition, little is known of the excited state spectra of monomeric and aggregated Chlorophyll a in solution. Some of the excited states in the red part of the spectrum can be populated with nanosecond pulsed lasers and selective pumping of levels of the donor or acceptor molecule can then provide us with an additional experimental tool to pin down the energy level diagram and intermolecular energy transfer.

STUDIES OF ENZYMES BY RESONANCE RAMAN SPECTROSCOPY

P.R. Carey and A.C. Storer
Division of Biological Sciences, National Research Council of
Canada, Ottawa, Canada K1A 0R6

ABSTRACT

A recent approach to the study of enzyme-substrate intermediates by resonance Raman (RR) spectroscopy involves the creation of a RR probe, in the form of a dithioester, at the point where the substrate is covalently linked for a short time to the enzyme. By this means it is possible to obtain the vibrational spectrum associated with those bonds undergoing catalytic transformation. Consequently, a wealth of detail becomes available on the substrate during catalysis; the conformation of the substrate in the active site can be monitored and characterised, structure-rate constant relationships developed, reaction pathways mapped, and evidence sought for geometric distortions. A recent development is the use of the RR data to detect 'action at a distance', i.e. the effect of enzyme-substrate contacts far from the point of catalysis on substrate conformation at that point.

INTRODUCTION

Where suitable resonance Raman (RR) probes are available, RR spectroscopy provides a powerful means of eliciting structural and dynamical information for enzyme-substrate transients[1-6]. At the same time the study of enzyme-substrate reactions offers a great challenge to the spectroscopist; the constantly changing nature of an enzyme reaction mixture and the uncertainties associated with the chemistry of enzyme active sites mean that spectral data are sometimes difficult to obtain and interpret. The difficulties have now been overcome for several enzyme-substrate systems and this summary will emphasise the unique information obtained from the RR data.

The initial RR studies of enzyme-substrate complexes[4-6] involved the use of chromophoric substrates based on e.g. cinnamic or furylacrylic acids. While these studies yield unique information on factors such as the electrical properties of active sites[1] recent attention has focussed on the use of dithioester RR probes which enable us to study 'natural' enzyme-substrate complexes under turnover conditions.

USING DITHIOESTERS TO MONITOR THE BONDS UNDERGOING CATALYTIC TRANSFORMATION

By creating a dithioester, $-C(=S)S-$, as a RR probe it is possible to obtain the vibrational spectrum of that part of a transient enzyme-substrate complex undergoing catalytic transformation[2]. The probe is generated by the use of a substrate which is a peptide sequence terminating in a thionoester,

© American Institute of Physics 1987

peptide-C(=S)OCH$_3$, and a cysteine proteinase where the active site -SH group of the enzyme forms a link to the substrate. With this combination the acyl enzyme intermediate is a dithioacyl enzyme, peptide-C(=S)S-enzyme. A prototype reaction is shown in Fig. 1:

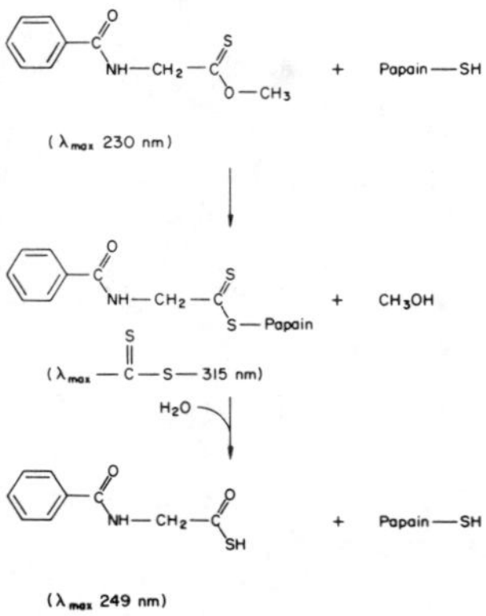

Fig. 1

From the point of view of introducing spectroscopic selectivity the intermediate has a λ_{max} at 315 nm, whereas the substrate and product both absorb below 250 nm. As can be seen in Figure 2 by excitation with the 324-nm Kr$^+$ line, the intermediate gives rise to a RR spectrum with many bands in the 500-1200 cm^{-1} range[7]. The peaks in Figure 2 contain contributions from the stretching and other motions of the C=S and C-S-C bonds, and thus intense modes from the bonds modified during catalysis are observed in the RR spectrum. Moreover, the reaction scheme can be generalised to use, for example, a polypeptide sequence as substrate and any enzyme forming a transient ester linkage involving an active-site thiol.

In order to interpret the RR spectra of dithioacyl papains it has been necessary to carry out extensive studies on model compounds. Vibrational and theoretical investigations were carried out for simple dithioesters, such as CH$_3$C(=S)SCH$_3$[8]. However, the key to interpreting the RR spectra of dithioacyl papains has come from model compounds which bear a closer chemical resemblance to the enzyme substrate compounds. For example, ethyl dithioesters of N-acylglycine have been invaluable for understanding the RR spectra of N-acylglycine dithioacyl papains. Joint spectroscopic and X-ray crystallographic studies have been crucial in understanding the relationships between the RR spectra and dithioester conformations[9,10]. Approximately 15 N-acylglycine ethyl dithioesters, RC(=O)NHCH$_2$C(=S)SC$_2$H$_5$, have been analysed to date. They give rise to intense RR bands in the 500-700 and 1000-1200 cm^{-1} regions of the spectrum. In solution the relative intensities of the bands were found to be very sensitive to temperature and solvent. These facts, taken with other considerations, indicated

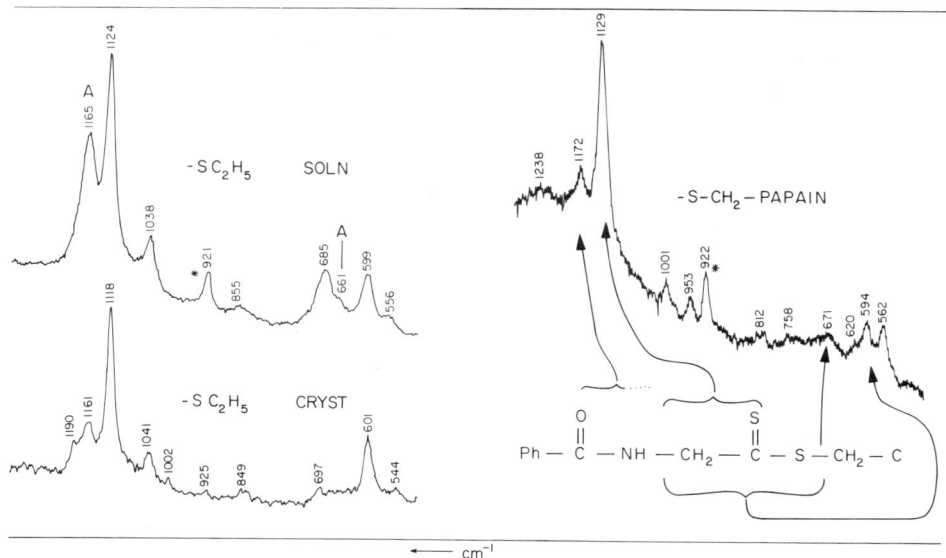

Fig. 2. The 324 nm excited RR spectra of N-benzoylglycine dithioester-ethyl dithioester, left hand side; the dithioacyl papain and an indication of conformationally sensitive peaks, right hand side. Adapted from ref. (1) with permission.

the presence of more than one conformer. It was found that in aqueous or acetonitrile solutions there are two major conformational states, designated conformers A and B, and that each conformer has a characteristic and separate RR spectrum in both the 600 and 1100 cm^{-1} regions. In order to form an exact description of conformers A and B, combined X-ray crystallographic and Raman analyses were undertaken on single crystals of N-acylglycine ethyl dithioesters. Conformers A and B are depicted in Fig. 3. The combined X-ray crystallographic RR approach could, in turn, be used to understand the RR spectra of the N-acylglycine dithioesters in solution and the RR spectra of the dithioacyl papains. In keeping with the higher thermodynamic stability of conformer B, most of the N-acylglycine dithioesters crystallize in this form.

Figure 2 compares the RR spectrum of crystalline $PhC(=O)NHCH_2C(=S)SC_2H_5$ with the RR spectrum of this molecule in H_2O. In solution (top spectrum) the spectral signatures of forms A and B are present, but for the crystals only the signature of conformer B is found, e.g. in the upper trace in Figure 2 most of the intensity of the 1165 cm^{-1} peak is due to a conformer A mode that is absent in the spectrum of the crystalline material. An X-ray diffraction analysis on the crystal provides an accurate structure of conformer B, and this is shown in Figure 3. To date, for glycine-based dithioesters only $pNO_2PhC(=O)NHCH_2C(=S)SC_2H_5$ has crystallized in a form giving rise to a conformer A type RR

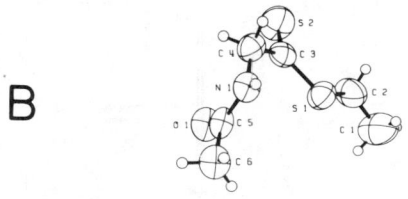

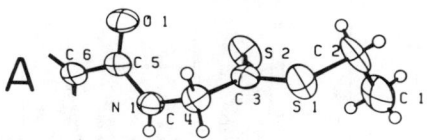

Fig. 3. Conformers A and B

signature; its structure is compared to conformer B in Figure 3. The major conformational difference between conformers A and B is a rotation of ~150° about the C(3)-C(4) bond. The rotation about the C-C linkage changes the vibrational coupling in and about the dithioester moiety and accounts, at least in part, for the different Raman spectral signatures of conformers A and B.

The RR spectra of N-acylglycine dithioacyl papains, e.g. the RR spectrum of PhC(=O)NHCH$_2$C(=S)S-papain seen in Figure 2, demonstrate unequivocally that between pH 4 and 9 the great majority of acyl groups assume a B-type conformation. In Figure 2 the intense peak at 1129 and the peak at 594 cm^{-1} are modes characteristic of a B conformer - the corresponding peaks are seen in the RR spectrum of the crystalline ethyl dithioester. Thus, the enzyme's active site exerts conformational selection; it binds just one of the conformational states available to N-acylglycine dithioesters. Moreover, detached comparison of model and enzyme-substrate RR spectra, facilitated by isotopic substitutions, demonstrates that the enzyme binds the substrate without causing major geometric strain or distortion[11].

Papain has an extended binding site which can accomodate up to a tetrapeptide on the acyl-binding side. Since it has often been proposed that energy from enzyme substrate contacts removed from the point of catalytic attack may be funnelled to this point to produce rate acceleration, it is of considerable interest to see if extending the substrate into the full binding site affects the geometry about the scissile linkages. To this end, the substrates MeOC(=O)PheGlyC(=S)OCH$_3$, MeOC(=O)GlyPheGlyC(=S)OCH$_3$ and MeOC(=O)GlyGlyPheGlyC(=S)OCH$_3$ have been synthesised. The RR spectra of the resulting dithioacyl enzyme intermediates involving these multi-peptide substrates are compared in Fig. 4 to that based on the single amino acid substrate C$_6$H$_5$(CH$_2$)$_2$C(=O)NHCH$_2$C(=S)S-papain. As can be seen in Figure 4 the intermediates based on the di-, tri-, and tetrapeptide substrates have very similar RR spectra

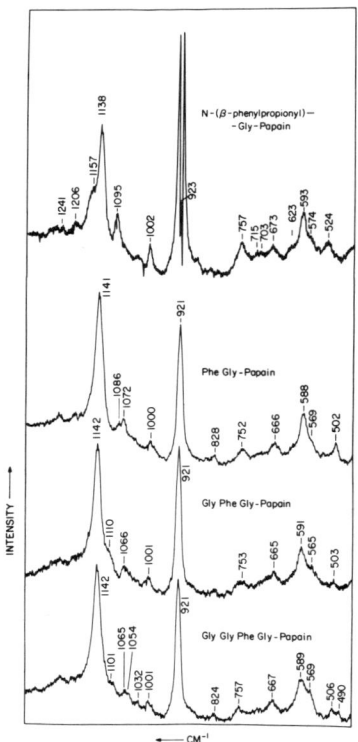

Fig. 4. The 324 nm excited RR spectra of four dithioacyl papains containing one, two, three and four substrate amino acid residues in the active site (top to bottom). Reproduced for ref. (12) with permission.

and this similarity is mirrored in the kinetic constants for breakdown of these dithioacyl papains. However, both the kinetic and spectral properties of the intermediate based on the single amino acid substrate differ markedly from the class formed by the di-, tri- and tetrapeptides. The RR and kinetic data taken together provide for the first time a molecular explanation for the phenomenon of 'action at a distance', in this case pointing to the importance of an enzyme substrate contact involving the -C(=O)NH-PheGly substrate linkage for fine tuning the deacylation kinetics[12].

ACKNOWLEDGEMENT

We would like to express our thanks to our colleagues and co-workers whose names appear in the list of references. Without their help the work described in this review could not have been accomplished. This article is NRCC number 26634.

REFERENCES

1. P.R. Carey and A.C. Storer, Ann. Rev. Biphysics Bioeng. 13, 25-49 (1984).
2. P.R. Carey and A.C. Storer, Acc. Chem. Res. 16, 455-460 (1983).
3. P.R. Carey, Biochemical Applications of Raman and Resonance Raman Spectroscopies, Academic Press, New York (1982), Chapter 6.
4. P.R. Carey and H. Schneider, Acc. Chem. Res. 11, 122-128 (1978).
5. B.A.E. MacClement, R.G. Carriere, D.J. Phelps and P.R. Carey, Biochemistry 20, 3438-3447 (1981).
6. D.J. Phelps, H. Schneider and P.R. Carey, Biochemistry 20, 3447-3454 (1981).
7. A.C. Storer, W.F. Murphy and P.R. Carey, J. Bicl. Chem. 254, 3163-3165 (1979).
8. J.J.C. Teixeira-Dias, V.M. Jardim-Barreto, Y. Ozaki, A.C. Storer and P.R. Carey, Can. J. Chem. 60, 174-189 (1982).
9. C.P. Huber, Y. Ozaki, D.H. Pliura, A.C. Storer and P.R Carey, Biochemistry 21, 3109-3115 (1982).
10. K.I. Varughese, A.C. Storer and P.R. Carey, J. Am. Chem. Soc. 106, 8252-8257 (1984).
11. A.C. Storer, H. Lee and P.R. Carey, Biochemistry 22, 4789-4796 (1983).
12. R.H. Angus, P.R. Carey, H. Lee and A.C. Storer, Biochemistry 25, 3304-3310 (1986).

LASER RAMAN PHONON SPECTROSCOPY
OF SOLID STATE PHOTOREACTION :
PHOTODIMERIZATION OF O-METHOXY
TRANS CINNAMIC ACID

Urmi Ghosh and T. N. Misra
Optics Department, Indian Association for the
Cultivation of Science, Calcutta 700 032
India

ABSTRACT

Laser Raman spectroscopy shows that the photodimerization reaction of o-methoxy trans cinnamic acid crystal is mediated by a lattice phonon which softens with reaction progress.

INTRODUCTION

Strong exciton-phonon coupling has been shown to be involved in many crystalline state photoreaction[1,2]. O-methoxy trans cinnamic acid (MCA) when irradiated in the crystalline state dimerizes to give 2,2' dimethoxy — α — truxillic acid, a centrosymmetric cyclobutane derivative[3]. Our laser Raman phonon spectroscopic study suggests that this photoreaction is phonon soft mode mediated.

RESULTS AND DISCUSSION

Purified MCA was irradiated by a 500 W Xenon lamp. Samples were taken at different stages of reaction and studied.

Characterization of the reactant and the product by IR spectroscopy :—

The progress of reaction was monitored by electronic absorption spectroscopy. With reaction progress the intensity of monomer absorption at 315 nm decreases and in completely dimerized sample it disappears (Fig 1).

The infra-red spectra of MCA at different stages of reaction show that the intense aliphatic C = C stretching mode at 1612 cm^{-1} and the HC = CH trans vibration at 992 cm^{-1} in the monomer decrease in intensity with the extent of dimerization and disappear in the dimer crystal These observations indicate cyclobutane ring formation.

Phonon spectral change during reaction progress :—

In fig 2 the phonon spectra of MCA at different stages of reaction are shown. In the monomer spectrum bands are observed at 32.5, 57.5, 65, 89, 95.5 and 160 cm^{-1}. With reaction progress the sharp phonon band at 32.5 cm^{-1} shifts to lower frequency and finally disappears in dimer

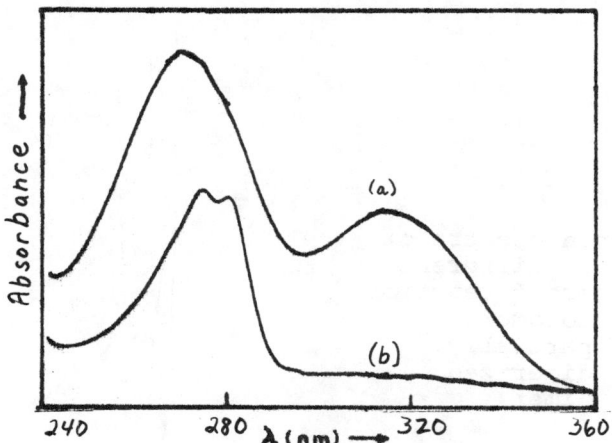

Fig.1. UV absorption spectra of MCA in methanol
(a) monomer (b) dimer

spectrum. We believe that this is not due to rise in Rayleigh background. This has been checked using some chemically stable systems as was done by Dwarkanath and Prasad[4] We, thus, conclude that the experimental observation suggests softening of the 32.5 cm^{-1} phonon mode with reaction progress. The softening of this mode causes the photo reaction to occur.

The basic idea of mode softening is that one of the normal modes becomes unstable due to phonon-phonon interaction and its frequency decreases with temperature. The restoring force softens unless the phonon condenses into the lattice to cause a structural phase transition. Mode softening also produces large amplitude displacements which are analogs of molecular collisions in a gas phase and thus can be expected to assist in reactivity. Mode softening was also observed in thermal rearrangement reaction[4]. Decrease of electrostatic interaction between regions of unbalanced and the Van der Waals interaction between atoms cause phonon mode to soften.

The other features of phonon spectra show that in the partially dimerized crystal (Fig 2b), the monomer phonon bands loose their intensity and the new phonon bands appear at 49, 68 and 110 cm^{-1}. When the reaction is almost complete, the monomer phonon bands disappear and a new phonon spectrum is observed (Fig 2c) with bands at 44, 49, 68 and 110 cm^{-1}. The appearance of seggregated phonon spectra in the partially dimerized crystal suggests that the reaction is heterogeneous in character and the reactant and the product form separate lattices.

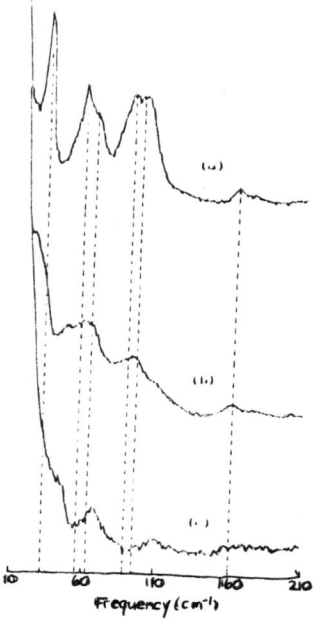

Fig.2. Phonon spectra of MCA at different stages of reaction
(a) monomer
(b) partially dimerized
(c) dimer

ACKNOWLEDGEMENT

This work was supported by the Department of Science and Technology, Govt. of India, Grant No.14(7)/83-STP II.

REFERENCES

3. M.D. Cohen, G. M. J. Schmidt and F. I. Sonntag, J. Chem. Soc., 2000, 1964.
1. T. N. Misra and P. N. Prasad., Chem. Phys. Lett., 85, 381, 1982.
2. Urmi Ghosh and T. N. Misra, Bull. Chem. Soc. Jpn., 58, 2403, 1985.
4. K. Dwarkanath and P. N. Prasad, J. Am. Chem. Soc., 102, 4254, 1980.

APPLICATIONS OF ULTRAVIOLET RESONANCE RAMAN SPECTROSCOPY TO PROTEIN STRUCTURE

Leland Mayne and Bruce Hudson
Department of Chemistry and Institute of Molecular Biology
University of Oregon, Eugene, Oregon 97403

ABSTRACT

Ultraviolet resonance Raman spectroscopy can be a useful tool for the study of proteins and their constituents. Excitation with wavelengths shorter than about 250 nm results in resonance enhancement from electronic excitations of the peptide bond. This can be used to obtain information concerning the nature of the electronic transitions of this group and as a method for obtaining selective structural information. The present report concerns an investigation of the use of this technique to probe the state of isomerization of the linkage to proline residues. This is of interest because of the proposed role of isomerization about this linkage in the folding of proteins to form a native structure.

INTRODUCTION

Far UV resonance Raman spectroscopy of protein components has been of considerable recent interest[1-7]. The aromatic amino acids have been studied and analyzed in terms of benzene modes. N-methylacetamide has been investigated as a model for the peptide bond, leading to some further understanding of the nature of its vibrations and electronic states.[6] Strong resonance enhancement is observed for the amide II and III in-plane vibrations. A very large effect of deuterium substitution at the amide nitrogen is observed. Only the amide II' band has significant intensity in the deuterated species. This is interpreted as due to uncoupling of the C-N stretch and C-N-H bending motions. Isotopic labeling with C-13 and N-15 at the amide C-N bond results in a shift to lower frequency consistent with the normal mode of amide II' being purely C-N stretch in character. This is interpreted in terms of a change in the geometry of the excited state that is primarily C-N elongation.[6]

The resonance Raman spectroscopy of X-proline peptide linkages is of particular interest because the structure around the nitrogen in this group makes it different from the normal secondary amide peptide bond in terms of its electronic and vibrational spectra. Proline bonds are also of interest because of their implication in studies of protein renaturation kinetics. In particular, it has been proposed that a "slow-folding" form of proteins results from the presence of the incorrect conformer of this linkage in the unfolded protein. Figure 1 shows the conformational equilibria for typical peptide bonds and for the X-proline linkage. In the present work, selective excitation of the resonance Raman spectrum of the X-proline peptide linkage is demonstrated.

584

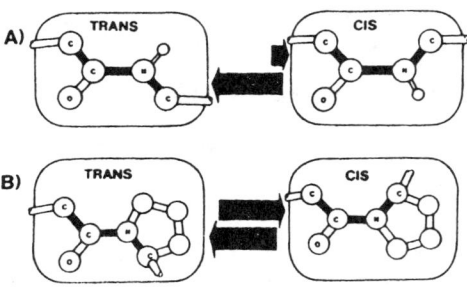

Figure 1. The equilibria of normal peptide bonds and the X-proline linkage. For normal peptide bonds, less than 1% of the cis isomer is present due to steric interactions. For the X-proline bond about 30% of the cis isomer is present.

RESULTS AND DISCUSSION

We have studied several small model compounds containing proline residues. Figure 2 shows the Raman spectra of pro-gly (dashed) and gly-pro (solid) taken with 218nm radiation. We see that in the gly-pro case the absence of the amide proton causes the C-N stretch motion to be decoupled from the N-H bending. This leads to an almost pure C-N stretch amide II'-like mode at 1485 cm^{-1}. In a normal peptide these motions are coupled and give rise to the amide II and III modes. This is analogous to the case of N-methylacetamide and N-deuterated N-methylacetamide. Preliminary work indicates that the frequency of this band is sensitive to the state of cis-trans isomerization of the X-pro linkage[7].

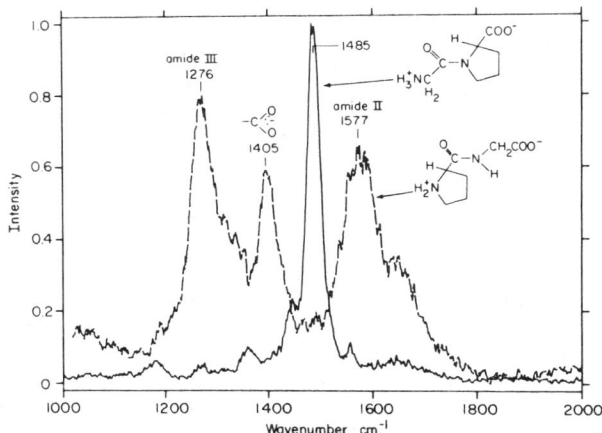

Figure 2. The resonance Raman spectra of prolylglycine (dashed line) and glycylproline (solid line). Both spectra obtained with 218 nm excitation of 10 mM solutions of the dipeptides.

In order to apply this technique to protein conformational equilibria it will be necessary to enhance the vibrations of the few X-pro bonds relative to that of normal peptide bonds. To this end we have measured the intensity of the amide II mode of pro-gly and the amide II'-like mode of gly-pro relative to a sodium sulfate internal standard at a variety of excitation wavelengths. These results are shown in figure 3. The amide II'-like mode is enhanced relative to the amide II mode by a factor of about 30 at excitations near 230 nm. This enhancement is consistent with the red-shifted absorption of this linkage and indicates that it will be possible to locate the X-proline bond signal in proteins.

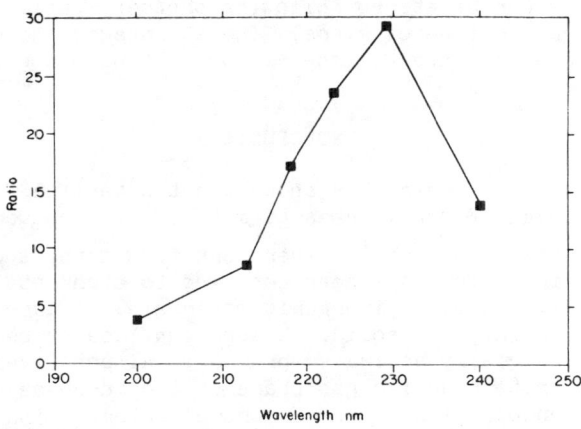

Figure 3. The intensity of the amide II'-like mode of gly-pro relative to the amide II mode of pro-gly. Intensities are based on peak height.

REFERENCES

1. B. Hudson in **Advances in Laser Science - I**, W. C. Stwalley and M. Lapp, eds (American Institute of Physics, New York, 1986) pg 690.
2. B. Hudson and L. Mayne in **Biological Applications of Raman Spectroscopy**, T. G. Spiro, ed. (Wiley, New York, 1986) in press.
3. B. Hudson and L. Mayne, Methods in Enzymology **130**, 331 (1986).
4. R. A. Copeland and T. G. Spiro, Biochem., **24**, 4960 (1985).
5. J. M. Dudik, C. R. Johnson, S. A. Asher, J. Phys. Chem., **89**, 3805 (1985).
6. L. Mayne, L. D. Ziegler and B. Hudson, J. Phys. Chem., **89**, 3395 (1985).
7. L. Mayne, T. Ramahi, T. Oas, and B. Hudson, Biophys. J., **45**, 322a (1984).

LASER-INDUCED ETCHING

Carol I. H. Ashby
Sandia National Laboratories, Laser and Atomic Physics Division,
Albuquerque, New Mexico 87185

ABSTRACT

Laser-induced etching of metals, semiconductors, and other inorganic materials can be etched by ablation, localized heating to enhance a thermally activated chemical reaction, or photogeneration of a good etchant from a less reactive precursor. In addition, semiconductors can be etched following photoexcitation of the surface to produce free carriers. The advantages and disadvantages of these four approaches to laser-induced etching are reviewed in this paper.

INTRODUCTION

Laser-induced etching has shown great potential for the selective removal of small areas of material to produce a particular pattern on a solid surface [1]. There are four methods of laser-assisted etching which have been employed to etch metals, semiconductors, and other inorganic materials. A laser can be used to physically ablate the solid. Lasers can locally heat a solid to accelerate a thermally activated process. Alternatively, the laser light can generate reactive gas-phase or liquid-phase etchants by photolytic decomposition of precursor molecules. Finally, the laser light can create free electrons and holes in semiconductors; these carriers then participate directly in the etching process. The most important characteristics of these four types of laser-induced etching provide a basis for selecting the most appropriate process for a given application.

ABLATION

Ablation occurs when high laser power densities ($>10^6$ W/cm^2) are employed to rapidly etch a solid without deliberate introduction of a chemically reactive species. Etching is generally attributed to a laser-induced phase change where the substrate is volatilized by the rapid energy deposition during a laser pulse. A threshold laser power density must be reached before any etching occurs. Ablation can be used to etch any type of material rapidly; it shows little if any selectivity between different light-absorbing materials. Some commercial processes for etching metals, such as Al, and for poly-Si link breaking use laser ablation [2]. Ablation-based processes can leave a significant amount of redeposited material, or debris, on the edge of the etched region. The debris can be greatly reduced or even eliminated if ablation occurs in the presence of a reactive gas, such as in the ablation of SiO_2 in the presence of HCl [3]. Low pressures of Cl_2 both increase the etch rate to near 1 micrometer per laser pulse and decrease debris during ablation of Al [4]. The proposed mechanism involves chemical reaction in the time between laser pulses and subsequent ablation of the reaction products. A laser may also be used to remove or crack a passivating film, exposing the more reactive underlying material to an etchant. This approach does not necessarily require power densities sufficient to ablate the passive layer. Aluminum has been etched by Cl_2 using laser power densities which are a factor of 5-10 below the ablation threshold to crack the passivation oxide surface layer and expose Al to the reactive gas [5].

THERMALLY ACTIVATED CHEMICAL REACTIONS

In thermal processes, the laser heats the solid in a highly localized region to increase the rate of a thermally activated chemical reaction. Any type of material for which there is a step in an etching process which is strongly temperature dependent, i.e., has a large energy of activation, can be etched using this approach. Material selectivity is limited only by the chemical selectivity of the particular etching reaction. Much lower laser power densities may be employed than for ablation since the temperature of the solid may need to rise only a few tens of degrees to produce substantially higher etching rates in the irradiated area. However, the fastest reactions involve melting or near-melting of the solid. Etching rates between 10 and 30 micrometers/second have been reported for Si [6] and GaAs [7] while molten. The lower melting temperature of amorphous Si compared to crystalline Si has been used to produce selective etching of the amorphous form [8]. Some inorganic materials, such as $LiNbO_3$, are very unreactive with acids or bases at room temperature but can be made highly soluble by a chemical transformation of the ion responsible for the low reactivity through reaction with hydroxide or halide ions while molten; i.e., by fusion of salts. This can be done in a highly localized manner when the surface is melted by a laser pulse [9].

Thermal processes are non-reciprocal; this means that the temperature rise depends not simply on the total energy deposited but rather on the rate at which the energy is deposited. This nonreciprocity makes it possible to etch features in a solid surface which are either larger or smaller than the actual laser beam diameter [10]. Smaller features are possible when the laser beam is pulsed or scanned rapidly [10-12]. Larger features can be produced when the beam is stationary or scanned slowly [10,11]. The principal factor determining feature size is thermal diffusion. Large thermal gradients favor the formation of features smaller than the laser beam diameter, but they also increase the potential for producing defects [13] and residual strain [14] even if the maximum temperature rise is insufficient to produce melting. Microfractures can also form if the material has both low thermal conductivity and a high thermal expansion coefficient, as is found in ZnO [15].

Several approaches different from those already mentioned have been successfully employed. Typical examples of laser-induced thermal chemical etching of metals and other inorganics are provided by the etching of W or Mo by O_2 or F_2 at 700-800°C [12,16] and of refractory materials such as TiB_2 and TiC by Cl_2 [17]. Laser heating has been used to shift the equilibrium from Al_2O_3 formation to dissolution in the reaction of Al with an aqueous solution of H_3PO_4, HNO_3, and $K_2Cr_2O_7$ [18]. Selective etching of Fe-garnet, which absorbs blue and green light, off a garnet substrate, which is transparent at these wavelengths, occurs because only the absorbing layer becomes hot enough to react rapidly with H_3PO_4 [19]. A material which sublimes readily, such as CdSe, can be etched in the absence of a reactant by photosublimation [20].

PHOTOCHEMICAL REACTANT GENERATION

In these processes, the laser produces a chemical species which is much more reactive than the precursor, usually by photolysis, but sometimes by nondissociative photoexcitation of the precursor. Photodissociation can occur in the gas or liquid phase or even when the precursor is adsorbed on the substrate surface. The laser beam can be either parallel or perpendicular to the surface, but rates are generally faster with a perpendicular beam since this heats the surface as well as generates the reactive species responsible for etching. The laser wavelength required for this process is dictated by the absorption spectrum of the precursor. If a range of wavelengths in the UV region can produce the reactant, use of longer-wavelength light may reduce the possibility of semiconductor substrate damage [21]. As in all light-induced processes, shorter wavelengths permit smaller features to be etched if the resolution is ultimately determined by the diffraction limit of the light. However, other diffusion-related effects may be more important in determining process resolution. Reactant diffusion seldom limits resolution with liquid-phase reactants but can be a serious limitation with gas-phase reactants. Since the minimum feature size is determined by the sum of the beam diameter and twice the mean free path of the reactive species, reaction conditions which decrease the time between molecular collisions, such as increasing the pressure or the temperature, permit the etching of smaller

features. Material selectivity is determined by the chemical
selectivity of a particular etching reaction. Preferential etching
of certain crystallographic planes is a common phenomenon, just as
it is in nonphotochemical etching.

The generation of the highly reactive species can involve
either multiphoton or single photon excitation of the precursor.
Single-photon processes require absorption of only one photon to
generate the reactant. Consequently, a high intensity light source
is not required. Etching rates from these processes are generally
linearly dependent on photon flux. Most visible- or ultraviolet-
driven reactions involve single-photon processes. Multiphoton
processes, in contrast, require simultaneous absorption of several
photons to produce a sufficiently excited precursor to produce
enhanced etching. Etch rates are higher than first order in photon
flux. Multiphoton excitation requires a high intensity light
source. Most infrared-driven reactions require multiphoton
absorption [22,23].

Many processes involve one-step photodecomposition of a
precursor molecule [11,21,24-26]. However, more complicated schemes
have been employed. For example, photons have been used to produce
the strong oxidant Ce^{4+} from Ce^{+3} in an aqueous solution containing
the Cl^- ion. The oxidant produces Cl atoms from the Cl^- ions; these
Cl atoms then etch Cu metal [27]. A different approach first employs
a UV laser to deposit Zn on an Al surface by photodecomposition of a
volatile organozinc compound. A cw visible laser is then used to
locally heat the surface, which causes the Zn to diffuse into the Al
in the heated region. The resulting Al-Zn alloy is more soluble in
acid than pure Al and can be preferentially dissolved [28].

PHOTOEXCITATION OF CARRIERS IN SOLID

When a semiconductor surface is irradiated with photons with
energies greater than the semiconductor band gap, additional free
electrons and holes are produced in the near-surface region. These
photogenerated carriers can participate directly in a chemical
reaction, producing enhanced etching of the semiconductor. There
are two general types of these reactions: photoelectrochemical
(PEC) and photochemical.

The PEC processes require ohmic contacts to the semiconductor
and an external power supply. Etching is controlled by the applied
voltage while photons create a plentiful supply of both electrons
and holes in the surface region. The chemical properties of the
electrolyte solution in which the semiconductor is immersed are of
secondary importance. These processes provide very precise control
of the amount of material removed, since this is directly related to
the time integrated current. Etching rates are determined by both
light intensity and applied bias voltage. These processes are
generally used for n-type material with an acid or base as the
electrolyte. However, very chemically mild electrolytes, such as

KCl, have been used successfully to etch n-CdSe [29]. Bias potential cycling permits etching of p-type material [30].

The photochemical type of process does not require an external current source for etching to occur. In contrast to PEC etching, the chemical properties of the solution or gas mixture are very important in determining whether etching occurs and at what rate. The etching rate generally depends linearly on light intensity. With liquid etchants, the illuminated and unilluminated regions of the semiconductor surface serve as anode and cathode for a galvanic reaction. Material selectivity is determined by both the chemical selectivity of the particular etching reaction and the electronic properties of the semiconductor.

An etching process based on electronic excitation of the solid can display a high level of selectivity between materials with very similar chemical reactivity by using the electronic properties of the semiconductor to control the generation or subsequent behavior of the photogenerated carriers. Since photogenerated carriers participate directly in the etching process, the effect of various processes such as surface or bulk recombination and lateral carrier diffusion can play a vital role in determining the etching rates and resolution of these reactions. For carrier-driven reactions, the ultimate resolution of the etching process is determined either by optical considerations or by the lateral diffusion of the carriers before recombination. When carrier diffusion limits feature size, resolution can be improved by decreasing the time available for lateral diffusion. This can be achieved by increasing the chemical reaction rate or by increasing the impurity dopant concentration [31] if the application permits this change.

Etching based on photoexcitation of the semiconductor surface has exhibited the greatest potential for highly selective discrimination between chemically similar materials. Since photocarriers are generated only by photons with energies in excess of the semiconductor band gap, it is possible to achieve an extremely high degree of etching selectivity between materials which are chemically very similar but which have different band gaps. This is achieved by using photons with energies intermediate between the two band gaps. This has been demonstrated by the selective etching of $GaAs_{0.8}P_{0.2}$ versus $GaAs_{0.63}P_{0.37}$ [32]. Selective etching based on dopant type has been demonstrated using both wet and dry etching processes. Selectivity using wet processes arises from galvanic effects where the light and dark regions of the surface provide anodic and cathodic regions for reaction. In general, illumination enhances wet etching of n-type relative to p-type materials [33, 34, 35]. In the reaction of GaAs with Cl atoms, application of an appropriate negative bias can suppress the etching of n^+-GaAs almost completely, while p-GaAs and n-GaAs etch at unimpeded rates under the same conditions [36].

SUMMARY

A variety of laser-induced etching processes have been developed for metals, semiconductors, and other inorganic materials. Each approach has inherent advantages and disadvantages; clearly the most appropriate choice depends on the specific application. However, the work done to date has shown the great potential of laser-induced etching to provide spatially localized, high-resolution, material-selective etching of solid surfaces.

ACKNOWLEDGMENT

This work performed at Sandia National Laboratories supported by the U. S. Department of Energy under contract number DE-AC04-76DP00789 for the Office of Basic Energy Sciences.

REFERENCES

1. C. I. H. Ashby, "Laser-Induced Etching," in Physics of Thin Films, Vol. 13, M. H. Francombe and J. L. Vossen, eds., Academic Press, Orlando, Florida, 1987.

2. R. T. Smith, J. D. Chlipala, J. F. M. Bindels, R. G. Nelson, F. H. Fisher, and T. F. Mantz, IEEE J. Solid-State Circuits SC-20, 399 (1985).

3. B. T. Dai, B. S. Agrawalla, and S. D. Allen, Proc. Symp. on Beam-Induced Chemical Processes, Ext. Abstr., 1985 Fall Mtg. Materials Research Soc. (R. J. von Gutfeld, J. E. Greene, and H. Schlossberg, eds.), Dec. 2-6, 1985, Boston, MA, p. 143.

4. G. Koren, F. Ho, and J. J. Ritsko, Appl. Phys. Lett. $\underline{46}$, 1006 (1985).

5. K. E. Greenberg, A. W. Johnson, J. W. Medernach, and K. Jungling, Proc. Symp. on Beam-Induced Chemical Processes, Ext. Abstr., 1985 Fall Mtg. Materials Research Soc. (R. J. von Gutfeld, J. E. Greene, and H. Schlossberg, eds.), Dec. 2-6, 1985, Boston, MA, p. 59.

6. R. J. von Gutfeld and R. T. Hodgson, Appl. Phys. Lett. $\underline{40}$, 352 (1982).

7. A. W. Tucker and M. Birnbaum, Pro. Soc. Photo-optical Instrum. Engineers $\underline{385}$, 131 (1983).

8. E. F. Krimmel, A. G. K. Lutsch, R. Swanepoel, and J. Brink, Appl. Phys. A $\underline{38}$, 1099 (1985).

9. C. I. H. Ashby and P. J. Brannon, Appl. Phys. Lett. $\underline{49}$, 475 (1986).

10. M. Takai, J. Tokuda, H. Nakai, K. Gamo, and S. Namba, Jpn. J. App. Phys. 22, L757 (1983).

11. D. J. Ehrlich, R. M. Osgood, Jr., and T. F. Deutsch, Appl. Phys. Lett. 38, 1018 (1981).

12. G. Koren, Appl. Phys. Lett. 47, 1012 (1982).

13. D. L. Parker in "Energy Beam-Solid Interactions and Transient Thermal Processing" (J.C.C. Fan and N.M. Johnson, eds.), p. 359. Elsevier Science Publishing Co., Inc., New York, 1984.

14. M. Takai, H. Nakai, S. Nakashima, T. Minamisono, K. Gamo and S. Namba, Jpn. J. Appl. Phys. 24, L755 (1985).

15. R. F. Wood and D. H. Lowndes, Cryst. Latt. Def. and Amorph. Mat. 12, 475 (1985).

16. G. Koren, J. Appl. Phys. 59, 1667 (1986).

17. A. W. Johnson and R. V. Smilgys, Proc. Symp. on Laser Chemical Processing of Semiconductor Devices, Ext. Abstr., 1984 Fall Mtg. Materials Research Soc. (F. A. Houle, T. F. Deutsch, and R. M. Osgood, Jr., eds.), Nov. 27-29, 1984, Boston, MA, p. 108.

18. J. Y. Tsao and D. J. Ehrlich, Appl. Phys. Lett. 43, 146 (1983).

19. K. Ando and S. Tsukahara, Jpn. J. Appl. Phys. 21, L347 (1982).

20. C. Uzan, R. Legros, Y. Marfaing, and R. Triboulet, Appl. Phys. Lett. 45 879 (1984).

21. Y. Horiike, M. Sekine, K. Horioka, T. Arikado, M. Nakase, and H. Okano, Proc. Syump. on Laser Chemical Processing of Semiconductor Devices, Ext. Abstr., 1984 Fall Meeting Materials Research Soc. (F. A. Houle, T. F. Deutsch, and R. M. Osgood, Jr., eds.), Nov. 27-29, 1984, Boston, MA, p. 99.

22. T. J. Chuang, J. Vac. Sci. Technol. 18, 638 (1981).

23. F. A. Houle and T. J. Chuang, J. Vac. Sci. Technol. 20, 790 (1982).

24. G. L. Loper and M. D. Tabat, J. Appl. Phys. 58, 3649 (1985).

25. D. J. Ehrlich, R. M. Osgood, Jr., and T. F. Deutsch, Appl. Phys. Lett. 36, 698 (1980).

26. R. W. Haynes, G. M. Metze, V. G. Kreismanis, and L. F. Eastman, Appl. Phys. Lett. 37, 344 (1980).

27. T. Donohue, Proc. Symp. on Beam-Induced Chemical Processes, Ext. Abstr., 1985 Fall Mtg. Materials Research Soc. (R. J. von Gutfeld, J. E. Greene, and H. Schlossberg, eds.) Dec. 2-6, 1985, Boston, MA, p. 139.

28. D. J. Ehrlich, R. M. Osgood, Jr., and T. F. Deutsch, Appl. Phys. Lett. $\underline{38}$, 399 (1981).

29. C. A. Kavassalis, D. H. Longendorfer, R. A. LeLievre, and R. D. Rauh in "Laser-Controlled Chemical Processing of Surfaces" (A. W. Johnson, D. J. Ehrlich, and H. R. Schlossberg, eds.) p. 151, Elsevier Science Publishing Co., Inc., New York, 1984.

30. F. W. Ostermayer, Jr., and P. A. Kohl, Appl. Phys. Lett. $\underline{39}$ 76 (1981).

31. R. M. Lum, A. M. Glass, F. W. Ostermayer, Jr., P. A. Kohl, A. A. Ballman, and R. A. Logan, J. Appl. Phys. $\underline{57}$, 39 (1985).

32. C. I. H. Ashby and R. M. Biefeld, Appl. Phys. Lett. $\underline{47}$, 62 (1985).

33. D. V. Podlesnik, H. H. Gilgen, and R. M. Osgood, Jr., Appl. Phys. Lett. $\underline{45}$, 563 (1984).

34. A. W. Johnson and G. C. Tisone in "Laser-Controlled Chemical Processing of Surfaces" (A. W. Johnson, D. J. Ehrlich, and H. R. Schlossberg, eds.), p. 145, Elsevier Science Publishing Co., Inc., New York, 1984.

35. F. Kuhn-Kuhnenfeld, J. Electrochem. Soc. $\underline{119}$, 1063 (1972).

36. C. I. H. Ashby, Appl. Phys. Lett. $\underline{46}$, 752 (1985).

UV LASER INDUCED THIN FILM DEPOSITION

R. Solanki
Department of Applied Physics and Electrical Engineering
Oregon Graduate Center, Beaverton, OR 97006-1999

ABSTRACT

This paper briefly reviews the process of UV laser induced thin film deposition, with emphasis on microelectronic applications. The unique aspects of this technique include low process temperatures and spatially selective deposits. Both large area and localized depositions are examined.

INTRODUCTION

The importance of high purity and defect-free thin films is well recognized for a wide variety of applications, ranging from optics to microelectronics. As these technologies evolve, they demand more stringent film properties which stretch the limitations of conventional deposition methods. To meet these demands, new film deposition techniques are being explored. One area of research that has attracted considerable interest over the past few years is the use of lasers for materials processing, including deposition and etching. Here, the process of laser deposition, especially using UV lasers, will be briefly reviewed. Due to space limitations, in-depth review or theoretical discussion is not possible, however interested readers are referred to review articles.[1-3]

LASER DEPOSITION

There are several different laser deposition techniques, some of which are modifications of conventional processes. One of these is laser evaporation or sputtering, where a high flux of photons is directed onto a solid target, which is close to a substrate. The interaction of the photons with the target leads to the removal of its atoms, which then condense on the substrate.

Infrared lasers (generally CO_2 lasers) have been utilized to deposit powdery metallic and ceramic films using a cold wall 'resonance' process. The IR radiation is absorbed by the vibrational modes of the donor molecules, which in turn dissociate. An off-resonance process called 'dielectric breakdown' has also been explored for deposition of powdery films. In this process, a short, high energy photon pulse is focused into a deposition chamber containing appropriate parent gases. The high electromagnetic field induces a spark inside the chamber which leads to deposition of a film consisting of fine particles.

The deposition process that has attracted the most interest utilizes focused laser beams to directly 'write' diffraction limited structures. The ability to deposit films with both spatial and temporal selectivity will allow considerable flexibility in processing microelectronic integrated circuits, as well as fabrication of new device structures. The direct write process can be achieved by either a pyrolytic or photolytic approach. In the pyrolytic process, the focused laser (e.g. argon ion, CO_2) beam heats a localized area on a substrate which is kept at ambient temperature. Reactant gases in the vicinity of this hot spot undergo thermal decomposition analogous to that in a conventional chemical vapor deposition (CVD) process. By moving either the focused spot or the substrate, patterned structures can be grown, eliminating the need for a lithographic mask. The selection of the laser wavelength is such that it is absorbed by the substrate and is transparent to the reactant gases. A wide range of materials, extending from metals to semiconductors have been deposited using this technique.

UV LASER DEPOSITION

The subject of this review is the use of a UV laser to induce a photochemical reaction which results in thin film deposition. Here, the laser beam is absorbed in a single- or multi-photon process by the reactant molecules which are in gas phase or adsorbed to the substrate. The UV absorption raises the electronic energy of the molecule to an excited state, which if above the dissociation limit, can cause the molecule to fragment. Products of the photo-fragmentation then either react chemically with other species or, if in gas phase, will condense onto nearby surfaces. The fragmentation of molecules occurs in a well defined and repeatable fashion, hence, in principle, better control of film stoichiometry is possible compared to plasma CVD. Also, by changing the wavelength of light, one may define a different reaction channel. It should be noted that the effectiveness of this process depends on the absorption spectra of the reactant molecules and the wavelength of the available laser radiation. This requirement does limit the choice of starting gases.

EXPERIMENTAL ARRANGEMENT

Experimental arrangement for UV laser deposition is sketched in Figure 1. Two different laser beam geometries are shown, one at

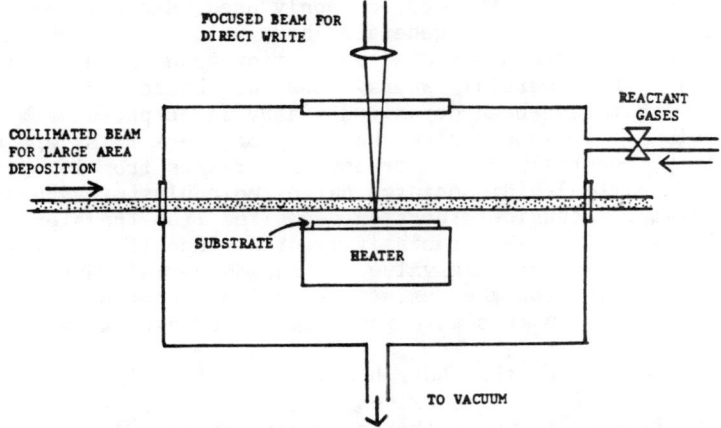

Fig.1. Experimental arrangement for UV laser deposition.

normal incidence to the substrate for direct 'write' and the other at parallel incidence for deposition over large areas. The substrate sits on a heater, which is heated only for large area deposition. The deposition pressures of the reactant gases range from 10^{-1} to 10^2 torr.

The spot-size of the focused beam determines the width of the structures that can be deposited. In principle, the finest (diffraction limited) spot-size one can achieve with $f1$ optics is approximately equal to the wavelength of the laser. At typical deposition pressures there exists a reasonably thick adsorbed layer of the reactant molecules on the substrate. The photochemical reaction between the UV photons at the focused spot and the adsorbed layer produces surface nucleation which helps to localize the condensation. The nucleation occurs after overcoming certain barriers, namely surface tension and reaction activation. The subsequent growth rate is dependent on the diffusion rate of the reactant molecules to the reaction site and the dissociation process. The interaction between the laser beam and the substrate leads to several surface reactions (e.g. electron-hole creation) which can help or hinder the deposition process, as well as influence the structure (e.g. ripple formation) of the deposit.

The parallel incidence beam is employed when deposition over large areas is desired. The UV source used is generally an excimer laser whose output is collimated to produce a beam of approximately 3 mm × 20 mm cross section. This beam creates a dissociation zone a few millimeters above the substrate. The deposition process is dominated by gas phase reactions. The non-volatile by-products of the photo-reaction diffuse out of the reaction volume and settle on the substrate, which is generally heated to temperatures between 200°C to 400°C. Without substrate heating the films are powdery, probably due to lack of surface mobility of the deposited species. By scanning the substrate under the laser beam or vice versa, photodepositions can be obtained over relatively large areas.

In the photolysis based CVD technique, it is the optical absorption properties of the gaseous donor molecules which govern the ability to deposit films. The most commonly used reactant gases are organometallic compounds which generally have relatively high vapor pressures at room temperature and an absorption band in the UV that can be reached with commercial lasers. Some of the metals and semiconductors that have been deposited via UV laser photolysis are listed in Table I. We should note that many of these films are not of high quality, generally due to trapped impurities from low purity starting gases, partially dissociated parent molecules or vacuum leaks. For example, consider tungsten deposited from tungsten hexacarbonyl (Table II). The resistivity of as-deposited film is approximately 20 times the bulk value. A rapid thermal anneal (RTA) cycle was required to drop the resistivity to an acceptable level. The dominant impurtiy in this case was oxygen from vacuum leaks in the system.

Among the insulators that have been photodeposited, the best results have been achieved with oxides (e.g. SiO_2, Al_2O_3). This is illustrated in Table III, where the properties of photodeposited

Table I. Metals and Semiconductors Deposited via UV Laser Photolysis

Film	Donor Gas
Al	Trimethyl Aluminum (TMA)
Au	Dimethyl Gold Acetylacetonate
Bi	Trimethyl Bismuth
Cd	Dimethyl Cadmium
Cr	Chromium Hexacarbonyl
Cu	Hexafluoropentanedionate Copper
Fe	Iron Carbonyl
In	Trimethyl Indium
Mn	Manganese Carbonyl
Mo	Molybdenum Hexacarbonyl
	Molybdenum Hexafluoride
Ni	Nickel Carbonyl
Pb	Tetraethyl Lead
Pt	Platinum-bis-hexafluoroacetyl acetonate $Pt(PF_3)_4$
Sn	Tetramethyl Tin
Ti	Titanium Tetrachloride
W	Tungsten Hexacarbonyl
	Tungsten Hexafluoride
Zn	Dimethyl Zinc
Ga	Trimethyl Gallium
GaAs	$Cl_3Ga\ As(C_6H_5)_3$
Ge	Germane
InP	$(CH_3)_3\ InP(CH_3)_3$
Si	Silane, Disilane

Al_2O_3 are compared to those obtained via RF magnetron sputtering. It can be seen that the quality of the photodeposited alumina film is similar to that obtained via a conventional process. Success with other insulators is limited. For example, nitride films deposited using ammonia consist of large amounts of hydrogen, which adversely affect their physical and electrical properties. In general, the quality of the UV laser induced CVD films have been improving as new reactant gases are being synthesized which are more suited to this process.
 The use of UV lasers for depositing thin films is a relatively new process and its full potential either as a research tool or in a manufacturing environment has yet to be realized. Its major drawback in an industrial setting is its through-put limitation. However, its direct 'write' capability is being recognized as an ideal tool for

Table II. Laser Deposited W from $W(CO)_6$

Deposition Temperature	23°C – 35°C
Deposition Rate	1700 Å/min over 1" × 1"
Thickness Uniformity	± 3%
Laser Flux at 248 nm	3 W/cm^2
Resistivity	102 μΩ-cm (4000 Å thick) 17.2 μΩ-cm after Rapid Thermal Anneal at 700°C in H_2
Impurities: Oxygen	5%
Carbon	0.7%
Adhesion (Si, SiO_2)	> 6.5 × 10^8 dynes/cm^2
Stress	< 2 × 10^9 dynes/cm^2 Tensile

Table III. Comparison Between Laser Deposited and Magnetron Sputtered Al_2O_3

	Laser Deposited	rf Sputtered
Deposition Rate (Å/min)	2000	350
Adhesion (dynes/cm^2)	> 6.5 × 10^8	Strongly Adherent
Pinhole Defects	< 1 in 5 cm^2 (1100-Å-thick film)	31/cm^2 (2500-Å-thick film)
Stress (dynes/cm^2)	< 6 × 10^9 (tensile)	2.8 × 10^9 (compressive)
Refractive Index	1.63	1.66
Stoichiometry	Al_2O_3	Al_2O_3
Impurities	C < 1%	Ar ~ 5%
Resistivity (Ω-cm)	10^{11}	10^{12}
Dielectric Constant at 1 MHz	9.74	9.96

custom fabrication of ICs at production or prototype level, and for repair of complex devices late in the fabrication process to increase the yield. Another application that is already a commercial product is the use of focused laser beams for mask repair.

In this review, we have examined only the practical aspects of this rapidly evolving technology. Several questions remain to be addressed regarding the full potential of this approach, as well as the basic science involved.

REFERENCES

1. D. J. Ehrlich, R. M. Osgood, Jr., and T. F. Deutsch, J. Vac. Sci. Technol. $\underline{2}$, 23 (1982).
2. D. J. Ehrlich and J. Y. Tsao, J. Vac. Sci. Technol. B.$\underline{1}$, 969 (1983).
3. R. Solanki, C. Moore, and G. Collins, Solid State Technol. June 1985, p. 220.

LASER-INDUCED DEPOSITION OF GOLD

Thomas H. Baum
IBM Almaden Research Center, San Jose California
95120-6099

ABSTRACT

The laser-induced, vapor-phase deposition of gold metal from dimethyl-(2,4-pentanedionato) gold (III) and two fluorinated derivatives was examined by the photopyrolytic (LCVD) method of deposition. High purity gold deposits are formed by LCVD with the rates of deposition being extremely vapor pressure dependent and thus, gold complex dependent.

INTRODUCTION

Laser chemical vapor deposition (LCVD) is a modification of conventional CVD, in which a focused laser is absorbed by the substrate and used as a localized heat source. An organometallic, adsorbed or colliding with the surface, undergoes pyrolytic decomposition to metal with the liberation of volatile reaction products. The spatially-resolved surface temperature profile, induced by laser heating of the substrate, defines the reaction zone and controls the overall reaction. The optical and thermophysical properties of the substrate and deposited metal are thus important process parameters. Another important parameter for the deposition process is the organometallic complex which is used as the metal precursor. Dimethyl-(2,4-pentandionato) gold (III) and two fluorinated derivatives[2] undergo clean thermal decomposition to gold. The incorporation of fluorine substituents results in increased volatility of the gold complex[3] at room temperature. All three gold compounds (1a, 1b and 1c) undergo exothermic decomposition beginning at ~160 °C, as determined by differential scanning calorimetry (DSC).

RESULTS

Previous studies[1] have demonstrated that rates of gold deposition by LCVD are highly vapor pressure dependent. Utilizing 1a, rates were an order of magnitude faster at elevated cell temperature, compared with rates under identical conditions at room temperature. Concurrently, use of 1b and 1c showed dramatically increased rates of gold deposition, as expected. Rates were compared for one laser power, focal spot size and substrate (SiO_2 on silicon) and found to scale with the increasing vapor pressure of the gold complex. Vertical growth rates were found to be 0.1 μm/sec for 1a, 0.45 μm/sec for 1b and 2.5 μm/sec for 1c, under identical conditions. Experimentally determined rates showed good agreement with a theoretical model[4] and are predominantly controlled by the gas-phase concentration of the reactant gold complex[5].

The purity of the gold deposits was found to be ≥ 95% by XPS and scanning Auger. Electrical resistivities of gold lines deposited from 1a and 1b were observed to be 4 - 20 times bulk gold (2.44 μΩ-cm) prior to thermal annealing. However, thermal annealing (350 °C for 30 minutes) consistently reduced the resistivities to 3 - 4 times bulk gold. TEM plates of gold[1] deposited by LCVD were polycrystalline and of high purity. After thermal annealing, the thin, gold line had coalesced into larger, discrete islands of metal. The morphological behavior upon annealing explains the observed decreases in resistivity. Gold lines produced from 1c were found to be 2 - 5 times bulk gold, were large grained initially and little change was observed with annealing.

CONCLUSIONS

The laser-induced pyrolytic deposition of high purity gold films from three organogold complexes has been highlighted. The incorporation of fluorine substituents into the molecular structure may be used to increase the precursor's vapor pressure and thus, the rates of gold deposition under ambient conditions.

REFERENCES

1. T. H. Baum and C. R. Jones, Appl. Phys. Lett., 47, 538 (1985); T. H. Baum and C. R. Jones, J. Vac. Sci. Tech. B4, 1187 (1986).
2. F. H. Brain and C. S. Gibson, J. Chem. Soc., 762 (1939). S. Komiya and J. K. Kochi, J. Amer. Chem. Soc., 99, 3695 (1978).
3. W. R. Wolf, R. E. Sievers and G. H. Brown, Inorg. Chem., 11, 1995 (1972).
4. D. J. Ehrlich and J. Y. Tsao, J. Vac. Sci. Tech B1, 969 (1983).
5. T. H. Baum, J. Electrochem. Soc., in press.

LASER-INDUCED DESORPTION FROM THE (111) SURFACE OF BaF_2

E. Matthias, H.B. Nielsen, J. Reif
Freie Universität Berlin, 1000 Berlin 33, F.R. Germany

A. Rosén, E. Westin
Chalmers University of Technology, Göteborg, Sweden

ABSTRACT

We have investigated the wavelength, polarization, and intensity dependence of the electron and ion emission from the (111) surface of BaF_2 under irradiation with pulsed tunable laser light in the green and blue spectral ranges. We also carried out cluster calculations to obtain a prediction for the electronic surface structure of BaF_2 (111). Based on these results, a qualitative model for the desorption of positive ions from this surface is proposed.

INTRODUCTION

In a study of the interaction of strong pulsed tunable laser light with transparent optical materials[1,2], at photon energies far below the bandgap and intensities well below the damage threshold, we observed the emission of electrons and positive ions (Ba^+, Ba^{++}, $(BaF)^+$, F^+) from the (111) surface of BaF_2. In the blue[1] and green spectral ranges we found a pronounced wavelength and polarization dependence of both the electron and the ion emission yields. In addition, cluster calculations were carried out to model the crystal surface of BaF_2 (111). So far, our studies indicate that the desorption of positive ions is the consequence of preceding multiphoton photoemission which is resonantly enhanced by intrinsic surface states.

CLUSTER CALCULATIONS

To model the (111) surface of BaF_2, a ten-atom planar cluster was found convenient in which a central Ba atom is surrounded by three F atoms, and six more Ba atoms successively, the F atoms lying slightly below the Ba plane (configuration $Ba_1Ba_6F_3$ in C_{3v} symmetry). The calculations were performed with a DVM-SCF method[3,4] within the local density approximation (LDA) and with the form of the exchange-correlation potential proposed by von Barth and Hedin[5]. The molecular wavefunctions were constructed as a linear combination of atomic orbitals (LCAO) with the addition of some virtual orbitals for Ba ($|Xe|6s^26p^05d^0$) and F ($1s^22p^53s^0$).

To illustrate the result of these calculations, the total density of states (DOS) for the 10 atom cluster is shown in Fig. 1a), where each discrete level is arbitrarily broadened by a Lorentzian of 0.1 eV. The full line represents the situation for a relaxed system where one electron has been removed from the last occupied state at the Fermi energy. Ionization from the F 2p band results in a shift of this band as indicated by the dashed peak while the excitonic levels at higher energies are hardly affected. We notice that there exist occupied surface states near the middle of the

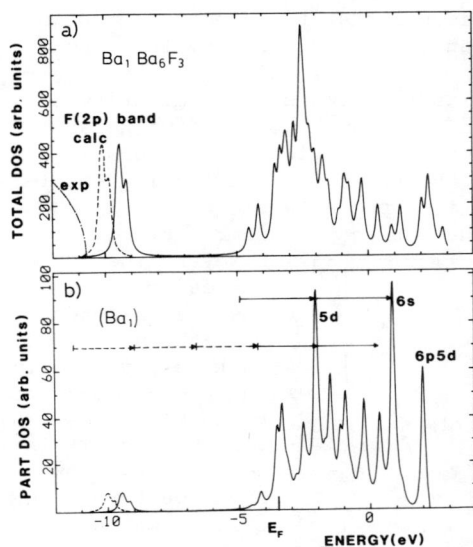

Fig. 1. Density of states for BaF$_2$ (111) surface from cluster calculations.
a) Total density of states.
b) Partial density of states at center Ba atom.

bandgap (-5 to -4 eV). Further, the large density of unoccupied surface states in the upper half of the bandgap originates from the Ba atoms. In Fig. 1b) the contribution from the central Ba atom is shown in the projected partial density of states.

EXPERIMENT

Clean BaF$_2$ (111) surfaces were prepared in ultra-high vacuum of about 3×10^{-8} Pa by in situ cleaving of the crystals. For details of the experimental setup see Ref. 1. The yields of emitted electrons and ions as a function of wavelength in the green spectral region are shown in Fig. 2 for s- (Fig. 2a), b)) and p-polarized light (Fig. 2c)). The strong similarity between electron and ion yield spectra shown in Fig. 2a) and b) is also observed for perpendicular polarization. We take this as an indication that the emission of positive ions is a consequence of multiphoton e$^-$ emission. This was verified by the experimental results from the intensity dependence of the emission yields shown in Fig. 3. The electron yield depends on the laser intensity by the second and for higher intensities by the fifth power, indicating two- or five-photon ionization, respectively. Nevertheless, the spectrum of the excitation does not change as can be seen from Fig. 2c). The arrows in Fig. 1b) indicate possible excitation schemes for blue (upper) (cf. Ref. 1) and green (lower) laser light. From comparison of the low and high intensity results (Fig. 1c)) for green light and the spectra obtained with the blue light[1] we suspect that the same resonances determine the spectra of all processes. However, the reason for the drastic change of the emission yield spectra when turning the polarization by 90° is not yet understood.

DISCUSSION

Even though the ion emission displays the same spectral dependence as the multiphoton photoemission (cf. Fig. 2) it is observable only at higher intensities and depends on the laser intensity by approximately twice the exponent of that of the electrons ionized by a 5-photon process from the 2p-valence band (Fig. 3). We interpret the result in Fig. 3 as strong evidence that two electrons must be emitted before one positive ion can leave the surface. This process, especially remembering that we observe positive but no negative fluorine ions with the blue laser[1] shows great similarity to a model recently proposed by Itoh and Nakayama[6]. There, two holes on neighboring sites attract each other via lattice distortions, leading to emission of the atom where the two holes are finally localized (F^+). Emission of other positive species would then follow due to electrostatic instability of the surface (Coulomb explosion).

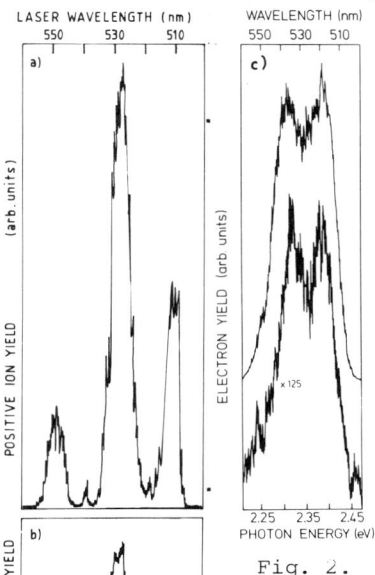

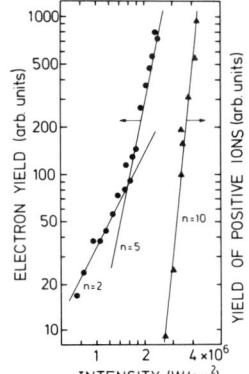

However, the identification of the observed exponents with the number of photons involved in the process must be taken with some caution because it is derived from lowest order perturbation theory, which may be no longer justified for resonant processes.

Fig. 3. Intensity dependence of electron and ion yields for p-polarized light at 518 nm.

Fig. 2. Comparison between yields of positive ions (a) and of electrons (b) for s-polarized light. c): Electron yield for p-polarized light; intensity: 4×10^6 W/cm^2 (upper curve) and 8×10^5 W/cm^2 (lower curve).

ACKNOWLEDGEMENTS

We gratefully acknowledge financial support by the Deutsche Forschungsgemeinschaft, Sfb 161, and by the Swedish Natural Science Council and the Swedish Board for Technical Development.

REFERENCES

1. J. Reif et al., Appl.Phys.Lett. 49, 770 (1986)
2. J. Reif et al., Appl.Phys.Lett. 49, no.15 (1986)
3. A. Rosén et al., J.Chem.Phys. 65, 3629 (1976)
4. B. Delley and D.E. Ellis, J.Chem.Phys. 76, 1949 (1982)
5. K. von Barth and L. Hedin, J.Phys. C5, 1629 (1972)
6. N. Itoh and T. Nakayama, Phys.Lett. 92A, 471 (1982)

LASER EXCITATION SPECTROSCOPIC STUDIES OF METAL ION BINDING IN POLYMERS

Eric K. L. Wong and Geraldine L. Richmond
U. Oregon, Chemistry Department, Eugene, OR 97403

ABSTRACT

Europium(III) ion has been used as a fluorescent probe to study the metal ion binding properties of the perfluorosulfonate membrane, Nafion (DuPont). The excitation spectrum shows multiple peaks corresponding to several different environments. Information about the pH dependent and cation substitution properties of these binding sites have been derived from the spectra.

INTRODUCTION

Nafion membrane has many different applications, several of which involve exchange of cations across a membrane that separates solutions of different ionic concentration. The membrane shows high selectivity to different metal ions with the binding predominately taking two forms: homogeneous binding of individual ions to anionic sites in the membrane, and binding in the form of ionic clusters. The ionic clusters consist of anionic exchange sites, exchangeable counterions, and sorbed water[1]. Understanding the structure of ionic clusters and the metal ions binding properties is very important for improved performance and diversification of the industrial application of Nafion. Various experimental techniques[2-6] have been used to elucidate the structure of the ionic clusters, but the exact nature of these ionic clusters is not fully resolved. We have employed the unique optical properties of europium (III) to probe the structure of ionic clusters in the Nafion membrane. The results reported here are preliminary studies aimed at providing insight into europium interaction with Nafion membrane and the dependence of the binding on different environmental parameters.

In this paper, we use laser induced excitation spectroscopy of the $^7F_0 \longrightarrow {}^5D_0$ transition of europium(III) to monitor the binding properties of metal ions in Nafion. This relatively sharp (8 cm^{-1}) non-degenerate transition near 580 nm has many unique properties which include a 'hypersensitivity' to the binding environment. The energy, fluorescence quantum yield and lifetime of the excited state are very sensitive to the number and the type of ligands in the inner coordination sphere.

EXPERIMENTAL METHOD

Nafion membrane (1100 equivalent weight) was immersed in 0.04M EuCl$_3$ solution for several hours with stirring.

After the membrane was washed with deionized water, it was dried under reduced pressure and stored in a desiccator. The excitation spectrum of the film was obtained with a Nd:YAG pumped dye laser scanned near the $^7F_0 \rightarrow {}^5D_0$ transition. The emission was collected at 614 nm using a double monochromator and a thermoelectrically cooled photomultiplier tube. The signal was amplified and averaged using a boxcar integerator. The data was fit by least-square Lorentzian and exponential curve fitting.

RESULTS AND DISCUSSION

Upon binding of the ion to the membrane, a large increase in the fluorescence is observed. Figure 1 shows the resulting excitation spectrum for a film prepared with a solution of $EuCl_3$ at pH 5.48. The spectra can be resolved into two peaks corresponding to two different binding sites. The lifetime of these peaks was measured as 130 usec.

When the film is prepared under conditions where the initial pH of the solution is higher, the excitation spectrum indicates significantly different binding. The overall fluorescence intensity increases by a factor of 4. In addition, two other peaks corresponding two new binding sites appear in the spectra as shown in Figure 2. The lifetimes of the third peak and the fourth peak were measured as 195 and 160 usec., respectively.

Different alkali metal ions were added to the europium bound Nafion in studies aimed at understanding the relative binding properties of those four binding sites. The addition of Li^+ ion produces no significant change in the excitation spectrum. Adding Na^+ ion produces an apparent decrease in fluorescence intensities of all four binding sites. The addition of K^+ ion results in disappearence of the third and fourth peak. With the addition of Cs^+ ion, the fluorescence was not detectable. The results show that the alkali ions compete with europium ion for those binding sites, and demonstrate that the third and fourth binding sites are quite distinguishable from the other two binding sites. The spectra also indicate that the binding strength of alkali ions decreases as $Cs^+ > K^+ > Na^+ > Li^+$ for Nafion.

From the above preliminary results, the europium certainly binds to Nafion in at least four different binding sites depending on the pH. The lifetime measurements indicate significantly different coordinating sites. The cation substitution study shows the high selectivity of metal ion binding in the polymer. The further studies of concentration dependence, solvent dependence, temperature dependence, and energy transfer within the polymer are in progress in order to gain a better understanding of the structure and binding properties of the ionic clusters.

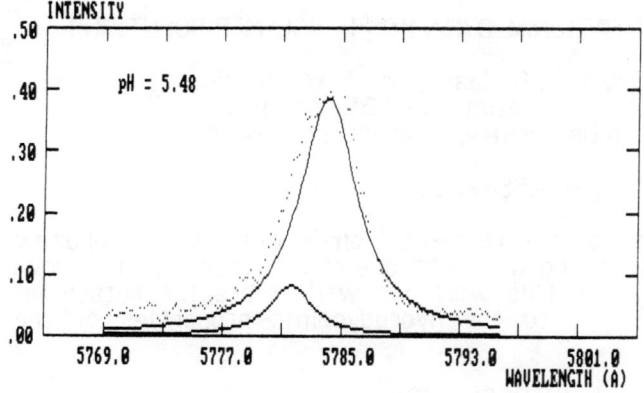

Fig. 1. Eu(III) excitation spectrum at low pH. The dotted curve is data and the solid curves are fits.

Fig. 2. Eu(III) excitation spectrum at high pH. The dotted curve is data and the solid curves are fits.

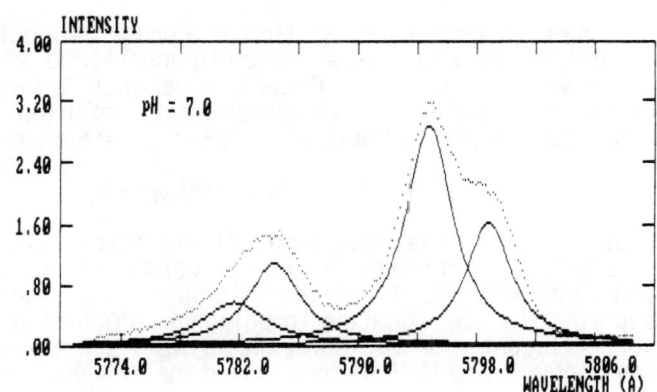

ACKNOWLEDGEMENT

Support from National Science Foundation (CHE8451346 & CHE8408340) is gratefully acknowledged.

REFERENCES

1. M. Lopez, B. Kipling and H. L. Yeager, Anal Chem, 49, 875 (1977).
2. M. Fujimura, T. Hashimoto, H. Kawai, Macromolecules, 14, 1309 (1981).
3. M. Falk, Can. J. Chem, 58, 1495 (1980).
4. C. L. Marx, D. F. Caulfiea, S. L. Cooper, Macromolecules, 6, 344 (1973).
5. E. J. Roche, M. Pineri, R. Duplessix, A. M. Levelut, J. Polym. Sci., Polym. Phys. Ed. 19, 1 (1981).
6. R. A. Komoroski, K. A. Mauritz, J. Am. Chem. Soc. 100, 7484 (1978).

PICOSECOND REORIENTATIONAL DYNAMICS IN POLYMER SOLUTIONS

E. L. Quitevis, K. G. Casey, and T. W. Sinor
Department of Chemistry and Biochemistry
Texas Tech University, Lubbock, TX 79409

ABSTRACT

We describe studies of the reorientational relaxation of oxazine dyes in polymer solutions using the technique of picosecond pump-probe absorption spectroscopy. In this work we will relate the rotational reorientation of these dyes to the hydrodynamic properties of the polymer solution.

INTRODUCTION

The rotational reorientation of a probe molecule that is not bound to the polymer has been widely used to understand structure and dynamics in polymer solution[1]. Recent experiments have shown that many molecules rotate hydrodynamically in solution[2]. The hydrodynamic assumption leads to the Debye-Stokes-Einstein (DSE) equation,

$$\tau_{rot} = \eta V_{hyd} / kT, \qquad (1)$$

where τ_{rot} is the rotational reorientation time, η is the solvent viscosity, V_{hyd} is the hydrodynamic volume of the rotor, k is Boltzmann's constant, and T is the absolute temperature. The viscosity obtained by applying the DSE equation to polymer solutions is often lower than the shear viscosity of the solution. This viscosity is a property of local environment and is therefore called the microviscosity.

EXPERIMENTAL SECTION AND RESULTS

An ideal technique to study rotational reorientation is picosecond pump-probe spectroscopy using a synchronously-pumped dye laser. In this technique, a pump pulse bleaches the ground-state absorption. To monitor the decay of the bleaching the transmission of the weak probe pulse, derived from the same laser pulse, is then measured as a function of delay time and polarization with respect to the pump pulse. In such an experiment one measures the polarization anisotropy R(t) which is given by

$$R(t) = [\Delta T_{\parallel} - \Delta T_{\perp}] / [\Delta T_{\parallel} + 2 \Delta T_{\perp}], \qquad (2)$$

where $\Delta T_{\parallel}$ and $\Delta T_{\perp}$ are the changes in the probe transmission as a function of the delay time t for the parallel and perpendicular polarizations, respectively. In this paper we describe picosecond pump-probe absorption measurements of the ground-state rotational reorientation of cresyl violet and oxazine 1 in methanol solutions containing varying concentrations of poly(ethylene oxide) (PEO, molecular weight ~ 10^5). These systems were chosen because of the well characterized reorientational dynamics of these dyes[3] and the solution properties of PEO[4]. For these dye molecules, R(t) can be described by a single exponential:

$$R(t) = R(0) \exp(-t/\tau_{rot}),\qquad(3)$$

where R(0) is the initial anisotropy.

Our pump-probe apparatus makes use of difference-frequency modulation spectroscopy[5]. In this detection scheme the pump and probe beams are acousto-optically modulated at two different radio frequencies f_1 and f_2. The beams interact in the sample to generate a signal at the difference frequency $f_1 - f_2$, which is detected in the probe beam by a MHz lock-in amplifier. The reference signal is obtained by combining signals from the oscillators at f_1 and f_2 in a double balance mixer. By using this modulation scheme the signal is obtained at a frequency where the laser noise is low and the signal is discriminated from the background due to the pump and probe. Further details can be found in Ref. 6.

The rotational reorientation time was obtained by fitting the experimentally derived R(t) to the exponential function in Eq. 3. The rotational reorientation times are plotted as a function of polymer concentration in Fig. 1.

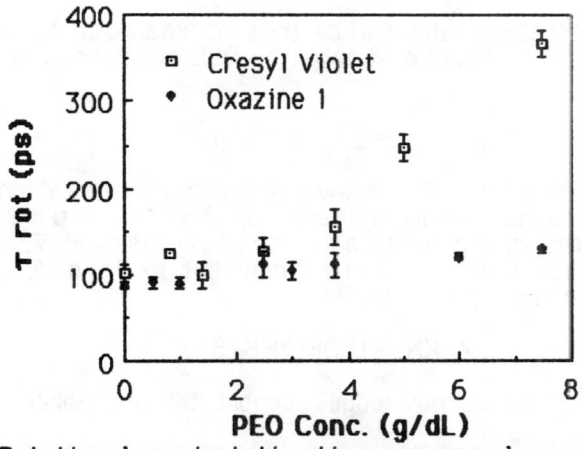

Figure 1. Rotational reorientation time versus polymer concentration.

DISCUSSION

The dependence of the τ_{rot} on the polymer concentration c can be interpreted in the following way. In dilute polymer solutions, where polymer chains are isolated, the dye molecules are primarily in a solvent environment. As the polymer concentration increases, chains become entangled and the dye molecules become encaged in a polymer network. If ϕ is the polymer volume fraction, the dilute concentration regime is characterized by $\phi \ll 1$. For dilute solutions ϕ is proportional to the product $c[\eta]^7$, where $[\eta]$ is the intrinsic viscosity of the polymer which is a measure of the polymer coil size. The onset of substantial coil overlap is given by $c[\eta] \geq 1$. Viscosities measured at 25°C with a Cannon-Fenske calibrated viscometer yielded a value of $[\eta] = 0.995$ dL/g Hence, coil overlap is prevalent for $c[\eta] \geq 1.2$ g/dL, which is consistent with the marked change of the rotational reorientation time for cresyl violet.

The striking difference between cresyl violet and oxazine 1 in Fig. 1 can be attributed to the degree of solvent attachment that occurs for each dye[3]. Strong hydrogen-bonding can occur between methanol and the amino groups on cresyl violet. Thus, the rotational reorientation of cresyl violet is strongly coupled to the-long range hydrodynamic modes of the solution as reflected in the shear viscosity. In the case of oxazine 1, hydrogen-bonding cannot occur readily to the diethylamino groups because of steric hindrance. Hence, the rotational reorientation of the dye is not as strongly coupled to the hydrodynamic modes of the polymer solution.

The microviscosity that is obtained by using the DSE equation can be associated with the dye-polymer hydrodynamic interaction. In a theory proposed by Mazo[8], the solvent is treated as a structureless continuous fluid with viscosity η_0 and the polymer as a wall. For a sphere of radius r_0 rotating in a concentric cavity of radius ρ, the effective viscosity is given by

$$\eta_{eff} = \eta_0 [1 - (r_0/\rho)^3]^{-1} \qquad (4)$$

If the polymer-encaged dye can be treated as a rotating sphere in a cavity, then by substituting Eq. 4 into the DSE equation, the following ratio results,

$$\tau_{rot}^0/\tau_{rot} = [1 - (r_0/\rho)^3] \qquad (5)$$

where τ_{rot}^0 and r_0 are taken to be the rotational reorientation time and the hydrodynamic radius in pure solvent. For example, in pure methanol cresyl violet rotates as if it were a sphere of radius $r_0 = 5.73$ Å based in the DSE equation. By using Eq. 5 one finds that in 7.5 g/dL, the dye is trapped in a microcage of radius 6.27Å.

ACKNOWLEDGEMENTS

We are pleased to acknowledge support by the Robert A. Welch Foundation.

REFERENCES

1. K. Hirota and Y. Nishijima, Rep. Prog. Polym. Sci. Jpn. 23, 55 (1980).
2. D. P. Millar, R. Shah, and A. H. Zewail, Chem Phys. Lett. 66, 435 (1979).
3. G. S. Beddard, T. Doust, and G. Porter, Chem. Phys. Lett. 61, 17 (1981).
4. L. E. Pierre, in: Polyethers, Part I, ed. N. G. Gaylord (Interscience, New York, 1963) pp. 83-168.
5. E. L. Quitevis, E. F. Gudgin Templeton, and G. A. Kenney-Wallace, Appl. Opt. 24, 318 (1985).
6. E. L. Quitevis, K. G. Casey, and T. W. Sinor, Chem. Phys. Lett., in press.
7. R. Simha and J. L. Zakin, J. Chem. Phys. 33, 1791 (1960).
8. R. Mazo, Biopolym. 15, 507 (1976).

VII. Diagnostic and Analytical Applications of Lasers
   A. Laser Diagnostics for Large Molecules 612
   B. Laser Microprobes and Microscopy 624
   C. Optical Aspects of Laser Diagnostics 644
   D. Imaging in Multiphase Media 676
   E. Atomic and Molecular Diagnostics 683

SUPERCRITICAL FLUID INJECTION OF NONVOLATILES WITH RESONANT TWO PHOTON IONIZATION DETECTION IN SUPERSONIC BEAM MASS SPECTROMETRY

D. M. Lubman, C. H. Sin, H. M. Pang
Department of Chemistry, University of Michigan
Ann Arbor, MI 48109

ABSTRACT

Supercritical fluids of N_2O and CO_2 are used to solubilize nonvolatile compounds for supersonic jet expansions. R2PI is then used to ionize these compounds in a time of flight mass spectrometer. By tuning the dye laser wavelength a cold wavelength ionization spectrum can be obtained that is characteristic of that species.

INTRODUCTION

In our work resonant two-photon ionization (R2PI) in supersonic jets is used as a means of optically selecting ions for mass spectrometry. The use of supersonic beam injection provides internally ultracold molecules with sharp spectral features for chemical analysis for molecules whose spectra would otherwise exhibit broad, unresolvable contours at room temperature. Our present work focuses on the use of this method for analysis of small biological species based upon the combination of wavelength selectivity and the high efficiency for producing molecular ions by the R2PI technique. One method investigated for use in solubilizing nonvolatile and thermally labile molecules involves expansions of supercritical fluids of CO_2 and N_2O at up to 400 atm backpressure into a mass spectrometer at 10^{-5} torr. Direct liquid introduction (DLI) techniques developed for HPLC/MS could also be used for introduction of nonvolatiles into vacuum. However, the much higher volatility of most gases that might be used in SFI (CO_2, N_2O, NH_3, etc.) provides a significant advantage over HPLC or liquid introduction into a mass spectrometer since the supercritical fluid immediately converts into a beam of molecules after introduction into vacuum, whereas DLI may require significant heating of the nozzle in order to prevent large clusters or crystals of material from condensing out upon expansion. In addition, SFI can utilize carrier gases which have a low enough number of internal modes to form a molecular beam so that significant cooling of the seeded material can be achieved. The expansion of this ultradense jet can thus provide sharp spectral features for analysis of nonvolatiles in the reservoir at room temperature which would normally require significant heating for volatilization into the gas phase.

EXPERIMENTAL

The experimental set-up is described elsewhere (1). It consists of a differentially pumped vacuum system with a TOF mass spectrometer sitting vertically on top of the chamber. The ultraviolet laser beam enters the chamber through quartz windows and R2PI is produced in the acceleration region of the TOFMS. The novel feature of this experiment is that a 150-200 μm orifice can be used to obtain high on-axis density in the jet even with 400 atm backpressure using a specially designed high pressure pulsed injection valve. Using this valve in conjunction with efficient LN_2 cryopumping, a pressure of ~10^{-5} torr can be maintained in the mass spectrometer. The laser source is a Quanta-Ray Nd:YAG pumped laser which is frequency doubled in KD*P to produce tunable near-UV light. The supercritical fluid pressure of CO_2 and N_2O is provided by a Varian 8500 syringe pump.

RESULTS AND DISCUSSION

In Figure 1a is shown a spectrum of the $S_0 \to S_1$ transition of carbazole expanded from supercritical CO_2 at 190 atm back pressure and 40°C. The melting point of carbazole is 245°C and at 40°C essentially no signal is observed from a 1 atm back pressure of Ar. In Figure 1b is a spectrum of carbazole obtained using 1 atm of CO_2 at 180°C. This spectrum was taken by use of a special pulsed solenoid valve capable of heating to at least 300°C. However, using supercritical fluid expansions of CO_2, a low temperature can be used to seed carbazole into a molecular beam, though it has a fairly high melting point.

Both spectra in Figure 1 have absorptions that are broader than are observed with Ar since the rotational population is not as effectively cooled by CO_2 as by Ar. This result is expected since the terminal Mach number (M_T) depends on the heat capacity ratio, $\gamma = C_p/C_v$, which is larger for Ar than for CO_2. In essence, the monatomic gas has no vibrational and rotational degrees of freedom in which to store energy and thus reaches its final translational temperature more quickly than in a polyatomic gas. This incomplete cooling of the beam leaves a significant number of molecules in excited rovibronic states, so that resonant ionization of these molecules makes a substantial contribution to the signal seen as the laser is swept over a relatively broad spectral range which is why the background baseline never goes to zero.

The mass spectra obtained for the laser photoionization of various PNAH's in a time-of-flight mass spec. are shown in Figure 2, which is a mass spectrum of acenaphthene, carbazole, phenanthrene, pyrene, and tetracene obtained from a supercritical expansion of 350 atm CO_2 at T=40°C using 266 nm as the ionization source. These compounds all dissolve strongly in the supercritical fluid even at 40°C even though tetracene has a melting point of 300°C. We have also explored a number of small biologicals including tryptophan, tyramine, phenoxazine and benzimidazole which can be solubilized by supercritical CO_2 or N_2O at essentially room temperature and have obtained wavelength dependent spectra for benzimidazole. The unique features of the mass spectra obtained are 1) an operating temperature only slightly higher than room temperature can be used to dissolve the nonvolatile or thermally labile compounds and 2) with the use of laser ionization, molecular ions are produced without fragmentation.

Our most recent work utilizes derivatization as a means of increasing the solubility of polar biological species in supercritical CO_2 or N_2O by decreasing the polarity of the -COOH group. We have been able to solubilize various catecholeamines and their metabolites by converting the -COOH to a -COOCH$_3$ group. These derivatized compounds readily dissolve in the fluid and can be ionized at 280 nm. This is expected since the derivatization does not strongly change the absorption or ionization properties of the original molecule. We have recently obtained a wavelength ionization spectrum of the methyl ester derivative of indole-3-acetic acid using supercritical CO_2 expansion and its spectrum compares favorably with that of the underivatized compound.

REFERENCES

1. C. H. Sin, H. M. Pang, D. M. Lubman, and J. Zorn, Anal. Chem., 58, 487 (1986).
2. H. M. Pang, C. H. Sin, D. M. Lubman, and J. Zorn, Anal. Chem., 58, 1581 (1986).

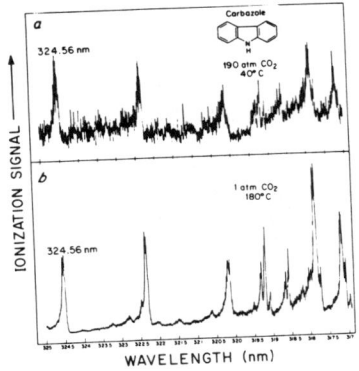

Fig.1 R2PI spectrum of carbazole expanded in a supersonic jet of (a) supercritical CO_2 at 190 atm and 40°C and (b) 1 atm CO_2 at 180°C.

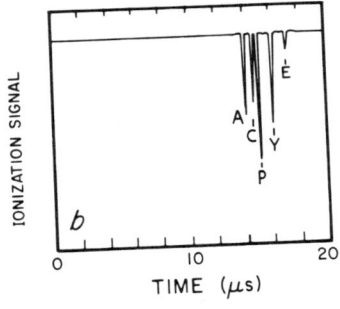

Fig.2 Time-of-flight mass spectra of acenaphthene(A), carbazole(C), phenanthrene(P), pyrene(Y) and tetracene(E) expanded from a jet at 350 atm CO_2 and 40°C.

LASER-BASED CIRCULAR DICHROISM DETECTION OF MOLECULES IN FLOWING LIQUID SYSTEMS USING HIGH FREQUENCY POLARIZATION MODULATION

Robert E. Synovec
University of Washington, Seattle, WA 98195

Edward S. Yeung
Iowa State University, Ames, IA 50011

ABSTRACT

A highly sensitive laser-based circular dichroism detector for flowing liquid systems, specifically liquid chromatography, is presented. Both transmission and fluorescence detected circular dichroism were developed. High frequency polarization modulation was used to reduce laser amplitude noise. Detection selectivity was demonstrated since optically inactive, yet absorbing, molecules do not produce an appreciable detector response. Riboflavin was detected at sub-ng levels, with optical activity information.

INTRODUCTION

Selective detection and study of optically active species in biological samples is an important area of chemical research.[1,2] Development of optical detectors that perform acceptably with liquid chromatography (LC) systems are critically needed. Circular dichroism (CD) can fill this need, which can be transmission detected (TDCD)[3] or fluorescence detected (FDCD)[4]. Since CD is inherently a "second order" absorbance measurement, problems related to obtaining the correct signal are addressed. Further, reducing much of the background noise is accomplished, since noise limits detector performance.

Fig. 1. CD System: LA-laser, M-electro-optic modulator, C-LC cell, LC-LC system, D-detector (either TD or FD), W-waveform generator, L-lock-in amplifier, R-recording device.

EXPERIMENTAL

The LC-CD system is shown in simplified form in Fig. 1. Both TDCD and FDCD were studied. Briefly, for TDCD[3], an argon ion laser at 488 nm and 20 mW was used to produce, alternately, left and right circularly polarized light. Use of a high modulation frequency, 500 kHz, provided optimum results by stabilizing the argon ion laser noise from 1 part in 10^2 to nearly 1 part in 10^6 peak-to-peak noise (PPN) to modulated light intensity.

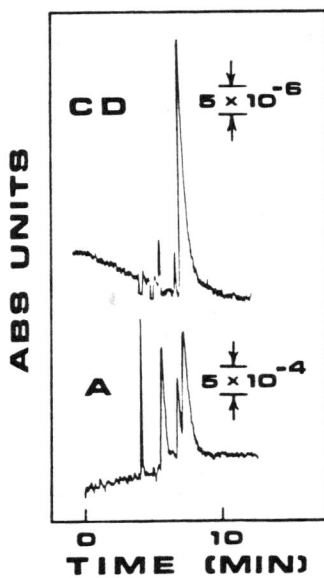

Fig. 2. LC separation and detection of $Co(NH_3)_5Cl^{2+}$ (5.7 min), $Cr(NH_3)_6^{3+}$ (6.8 min), and $(+)-Co(en)_3^{3+}$ (7.3 min); CD: TDCD, A: Absorbance; 10 μl injected volume, 5×10^{-4}M $(+)-Co(en)_3^{3+}$, 0.67 mL/min flow.

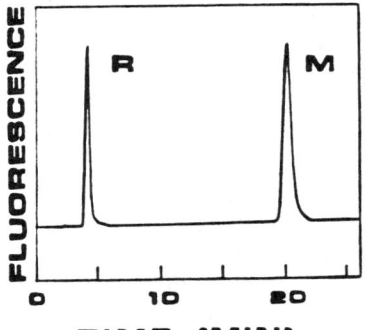

Fig. 3. Fluorescence chromatogram. R-(-)-riboflavin (1.05×10^{-5}M); M-4-methylumbelliferone (4.25×10^{-6}M); 1μl injected volume, 200μL/min.

Similarly, for FDCD, a helium cadmium laser was used at 325 nm and 8 mW, and at 150 kHz modulation frequency[4].

Reversed-phase LC systems were employed under typical operating conditions for both conventional and microbore high performance LC. Inorganic complexes were studied for LC-TDCD: $(+)-Co(en)_3^{3+}$, $(-)-Co(en)_3^{3+}$, $Co(NH_3)_5Cl^{2+}$, and $Cr(NH_3)_6^{3+}$. For $(+)-Co(en)_3^{3+}$, $\varepsilon_{max}(469nm) = 84$ L $cm^{-1}mol^{-1}$, and $\Delta\varepsilon_{max}(493nm) = +1.89$ L $cm^{-1} mol^{-1}$. All of the complexes absorb at 488 nm, while only the $Co(en)_3^{3+}$ complexes are optically active. For LC-FDCD, (-)-riboflavin and 4-methylumbelliferone were studied. Both absorb at 325 nm, and strongly fluoresce, while only (-)-riboflavin is optically active with $\Delta\varepsilon$ ca $+2.0$ L $cm^{-1}mol^{-1}$.

RESULTS AND DISCUSSION

Selectivity of detection is demonstrated for LC-TDCD in Fig. 2. The absorbance chromatogram (Fig. 2, A) contains three peaks for the three absorbing complexes, while the TDCD chromatogram (Fig. 2, CD) contains predominately the $(+)-Co(en)_3^{3+}$ peak. Note that apparent signals are shown at roughly 5 minutes and before. These are LC injection disturbances due to large refractive index (RI) changes caused by mixing the sample in a solvent of different RI than the LC eluent. Note also that the optically inactive complexes do show some TDCD response. This is due to an imbalance in intensities in the modulated beams. The imbalance produces a signal as a function of sample absorbance[3].

In LC-FDCD, the contribution due to the imbalance could not be neglected for organic molecules of biological interest[4]. By measuring both the fluorescence (Fig. 3) and FDCD (Fig. 4) chromatograms, successively, an on-

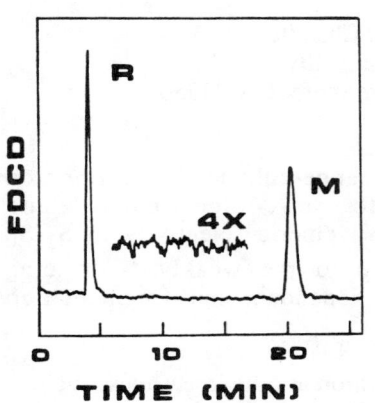

Fig. 4. FDCD chromatogram. Same sample, etc., as Fig. 3.

line procedure was developed to measure the relative magnitude of the imbalance in the modulated beams using optically inactive 4-methylumbelliferone. It was then possible to measure the contribution to the total signal of (-)-riboflavin due to optical activity. Thus, the optical activity at sub-ng injected quantities of (-)-riboflavin was measured. A summary of the performance values in terms of the limit-of-detection (LOD) are given in Table I. The LC-TDCD and LC-FDCD systems should find wide applicability in the study of biologically-related chemical systems.

Table I. LC-TDCD and LC-FDCD System Performance Values

LC System	Detection System	Path Length b, cm	ΔA LOD[a], au	Mass LOD, ng
Conventional	TDCD	2	7.5×10^{-7}	19
Microbore	TDCD	1	2.5×10^{-6}	2.4
Conventional	FDCD	1.8	1.7×10^{-6}	0.72

[a] $\Delta A = \Delta\varepsilon bC$, with LOD = 1 x PPN, and 1 s time constant.

REFERENCES

1. J. C. Kuo and E. S. Yeung, J. Chromatogr., 223, 321 (1981).
2. J. C. Kuo and E. S. Yeung, J. Chromatogr., 229, 293 (1982).
3. R. E. Synovec and E. S. Yeung, Anal. Chem., 57, 2606 (1985).
4. R. E. Synovec and E. S. Yeung, J. Chromatogr., accepted (1986).

LASER SPECTROSCOPY OF JET-COOLED CHLORINATED AROMATIC HYDROCARBONS*

E. A. Rohlfing and D. W. Chandler
Combustion Research Facility
Sandia National Laboratories, Livermore, CA 94550

ABSTRACT

Laser-induced fluorescence and resonance-enhanced multiphoton ionization spectroscopies are applied to jet-cooled mono- and dichloronaphthalenes. Both techniques provide isomeric selectivity in S_1-S_0 excitation, however REMPI is more sensitive to the weakly fluorescing dichloronaphthalenes and shows isomerically dependent ion fragmentation patterns.

The ultrasensitive detection of chlorinated aromatic hydrocarbons, such as polychlorinated biphenyls (PCBs) and polychlorinated dibenzo-p-dioxins (PCDDs), is currently a problem of intense concern due to the toxic and/or carcinogenic nature of these species and their widespread presence in the environment.[1] The requirements for a gas-phase detection technique are: (1) It must be ultrasensitive (ppb level or less). (2) It must be isomerically selective, since different positional isomers often have dramatically different toxicities. (3) It should be real-time so that the release of toxics to the environment can be mimimized.

In this paper, we report on the use of laser induced fluorescence (LIF) and resonance-enhanced multiphoton ionization (REMPI) for the detection of mono- and dichloronaphthalenes. These molecules are similar in nature to PCBs and are good surrogates for more complicated species such as PCDDs. In our experiments, chloronaphthalene vapor is diluted to approximately 10 ppm in 3-4 atm helium and allowed to expand freely into vacuum through the 0.5 mm orifice of a pulsed valve . This free jet expansion cools the molecules to very low rotational temperatures (typically < 10 K) and dramatically simplifies the LIF or REMPI excitation spectrum by narrowing rotational contours and eliminating vibrational hot bands. The expansion is crossed at approximately 50 orifice diameters downstream by a frequency-doubled dye-laser beam. For LIF studies, the fluorescence is collected, filtered, and imaged onto a photomultiplier tube; the photomultiplier signal is amplified and processed with a boxcar averager. For REMPI studies, the jet expands into the ionization region of a small time-of-flight (TOF) mass spectrometer. Photoions generated by the UV laser are accelerated, allowed to drift, and detected with a dual microchannel plate; this signal is amplified and processed with a boxcar averager. In all cases, the initial one photon excitation step is the S_1-S_0 transition. Since the S_1 state lies less than halfway to the ionization potential, additional absorption of two photons is necessary to produce parent ions in a 1+2 REMPI process.

In Figure 1, the LIF spectrum of jet-cooled 2-chloronaphthalene (2CN) is displayed. Also indicated in Figure 1 are the origin and first vibronic band of

*work supported by the U. S. Department of Energy.

naphthalene which was present at low concentrations in the pulsed valve. This spectrum clearly demonstrates that 2CN can be differentiated from naphthalene based on the red shift of its origin band. In naphthalene, the purely electronic S_1-S_0 transition is weak and the spectrum derives much of its intensity from vibronic interaction with the second excited singlet state, S_2. Thus, the origin is weak and the most intense bands are the vibronic bands arising from the antisymmetric vibrational modes that couple S_1 and S_2. In 2CN however, the origin is by far the strongest band and the vibronic bands are now dominated by progressions in totally symmetric vibrational modes. Clearly, the addition of a Cl substituent greatly increases the S_1-S_0 oscillator strength. The addition of Cl also enhances the rate of intersytem crossing to the triplet manifold, thereby reducing the lifetime of the S_1 state (from ~300 ns in naphthalene to ~30 ns in 2CN) and fluoresence quantum yield (from ~0.24 in naphthalene to ~0.007 in 2CN)[2]. The reduced quantum yield makes LIF far less sensitive toward 2CN than it is toward naphthalene, nonetheless, we estimate a detection limit of ~30 ppb for 2CN.

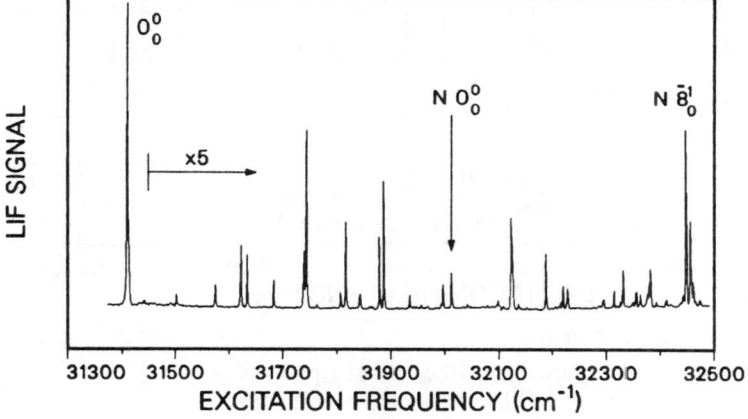

Fig. 1. The jet-cooled LIF spectrum of 2CN seeded at approximately 10 ppm in 3 atm helium. The S_1-S_0 origin band is indicated, as are the origin and first vibronic band of naphthalene (N) which was present at low concentrations in the valve.

The addition of a second chlorine substituent to naphthalene even further reduces the S_1 lifetime and fluorescence quantum yield -- so much so, in fact, that we were unable to observe fluorescence from any of the dichloronaphthalenes (DCNs). However, REMPI was found to be much more sensitive under similar conditions than LIF and we have obtained jet-cooled REMFI spectra of a series of DCNs. In Figure 2, we display the parent ion REMPI spectra of 1,8 and 1,4 DCN. Unlike the CN excitation spectra, the origin bands are not dominant in the DCN spectra, indicating that the purely electronic transition is weaker in the DCNs and that vibronically induced transition strength may again be important. This could explain the dense vibronic structure, especially in the 1,8 spectrum, where there may be a combination of progressions in totally symmetric modes and false origins due to vibronically active modes. At the laser intensities necessary to effect 1+2

REMPI there is substantial fragmentation of the DCN parent ion into small fragments. For all the DCNs the relative intensities of the C_1^+-C_{10}^+ fragments are similar. However, the DCNs in which the two chlorines are on adjacent carbons (1,2 and 2,3 DCN) show a larger ratio of fragment-to-parent ion intensities. We speculate that in these DCNs $C_{10}H_6^+$ is formed via the concerted loss of Cl_2 from the parent ion, a process that is energetically possible when the parent ion absorbs one UV photon. Conversely, in DCNs that have the chlorines separated by more than one carbon, $C_{10}H_6^+$ is formed by sequential loss of two Cl atoms from the parent ion, a process that requires the absorption of two photons. Subsequent fragmentation to smaller fragments then occurs independently of how $C_{10}H_6^+$ was initially formed.

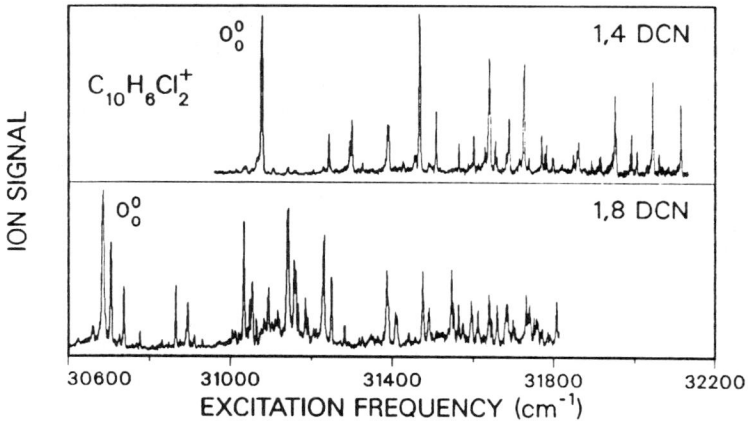

Fig. 2. The jet-cooled REMPI spectra of 1,4 and 1,8 DCN. Expansion conditions are similar to those in Figure 1 and the REMPI process is a one photon excitation of the S_1-S_0 transition followed by two photon ionization.

The jet-cooled REMPI spectra in Figure 2 clearly show that the 1,4 and 1,8 DCN isomers can be distinguished based on the positions of their S_1-S_0 origins. From similar spectra, we have measured the S_1-S_0 origin positions as red shifts relative to the naphthalene origin for a series of DCNs (1,2; 1,4; 1,5; 1,8; 2,3; and 2,7). These data demonstrate that isomeric selectivity for the DCNs is possible via jet-cooled S_1-S_0 spectroscopy. Jet-cooling is absolutely essential for this selectivity; the differences in origin position, which are as small as 3 cm^{-1}, could never be resolved for room temperature DCNs.

References

1. Dioxins in the Environment, M. A. Kamrin and P. W. Rodgers, eds., (Hemisphere, New York, 1985).

2. H. Saigusa, T. Asumi, M. Sumitani, and K. Yoshihara, J. Chem. Phys. 72, 1713 (1980).

PHOTO-THERMAL DEFLECTION VELOCIMETRY IN LAMINAR AND TURBULENT FLOWS USING A TRANSIENT GRATING

Cameron J. Dasch and Jeffrey A. Sell
General Motors Research Laboratories, Warren, MI 48090

ABSTRACT

The precision, sensitivity, and directionality of photo-thermal deflection velocimetry (PDV) have been greatly improved by pumping with a grating pattern. The first PDV measurements in a turbulent flow are presented.

SUMMARY

There is currently a widespread effort to develop velocity measuring techniques which will improve upon the widely used Laser Doppler Velocimetry (LDV) technique. These research efforts center on elimination of seed particles, increased data rates, and whole-field measurements. For velocity measurements at a point[1-4] or along a line[5], photothermal deflection spectroscopy has recently been developed as a very attractive "marker" velocimetric method. Here, we report a new LDV-like photothermal deflection method which improves PDV velocity precision by an order of magnitude, improves the signal size by an order of magnitude, and gives the sign of the local speed.

In PDV, a pulsed laser is tuned to a molecular absorber in the flowing gas. This heated gas acts as a negative lens, and its translation with the gas can be detected by the deflection of a probe laser. To date, PDV velocity was measured by time-of-flight across a gap between the pump and probe beams. The advantages of PDV over LDV include no seed particles, data rates up to available laser repetition rates, and immunity to local luminosity.

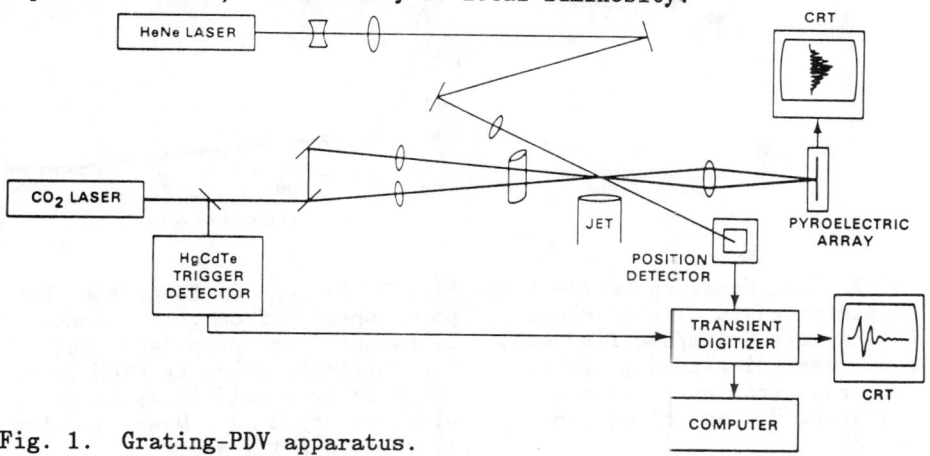

Fig. 1. Grating-PDV apparatus.

© American Institute of Physics 1987

In this work, the usual Gaussian-shaped PDV pump beam is replaced with a grating pattern[6], and the probe beam is focussed smaller than the pump fringe spacing. Using multiple fringes for the pump increases the refractive index gradients, gives multiple velocity markers, and allows the pump and probe beams to be overlapped so that flows both up and down give signals. The grating pattern is generated by splitting the pump into two beams and crossing them at a small angle in LDV fashion (see Fig. 1). The pump beam is a 0.5 mJ Q-switched CO_2 laser which vibrationally excites a few percent ethylene in an atmosphere of N_2. The fringe spacing is 245 μm. Ethylene relaxes quickly to form a transient thermal grating whose movement oscillates the probe laser across the position-sensitive detector. Velocities were measured at the end of a long straight pipe exiting into air. Fig. 2 shows the increased sensitivity. Blocking one of the grating-PDV pump beams gives a single-beam PDV trace similar to previous data.

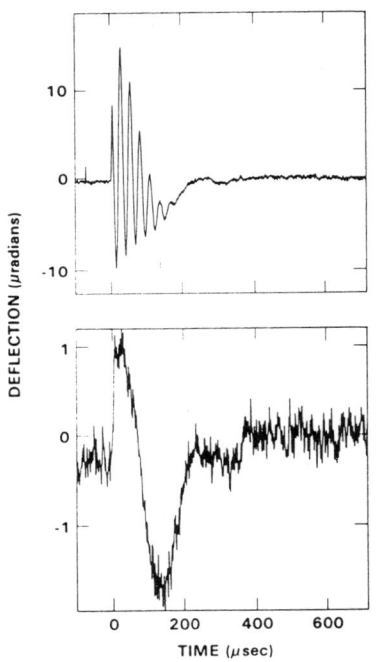

Fig. 2. Time-dependent deflection signals averaged over 64 shots in a laminar 9.6 m/sec flow with probe laser displaced slightly downstream from pump center.
a) grating-PDV and b) single-beam PDV.

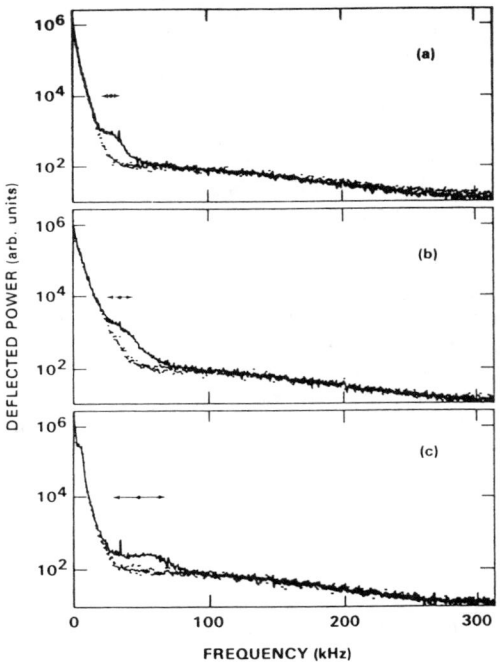

Fig. 3. Average of 64 grating-PDV power spectra from grid-induced turbulent flow: pump laser on(—), pump off(••), velocity FWHM (←→). Distance from grid is a) 28 mm, b) 15 mm, c) 2 mm. Mean velocity at grid is 7.1 m/sec.

The transient grating allows the first PDV measurements in turbulent flows. Single-beam time-of-flight PDV measurements are precluded by fluctuations in the pump-probe gap. The turbulent flows were generated by a wire mesh grid at the gas jet exit. The Fourier transform power spectra of the deflection signals is averaged and approximates the velocity PDF (Fig. 3). The mean velocity and the turbulence both decay downstream as expected. Although the noise peak is much larger than the velocity peak, the velocity is resolvable after 64 shots.

REFERENCES

1. J. A. Sell, Appl. Optics 24, 3725 (1985).
2. A. Rose and R. Gupta, Optics Lett. 10, 532 (1985).
3. H. Sontag and A. C. Tam, Optics Lett. 10, 436, (1985).
4. W. A. Weiner and N. J. Dovichi, Appl. Optics 24, 2981 (1985).
5. J. A. Sell and R. J. Cattolica, Appl. Optics 25, 1420 1986.
6. C. J. Dasch, Twentieth Symposium (International) on Combustion, 1985.

Laser Raman and Fluorescence Microprobing Techniques

F. Wallart and P. Dhamelincourt
Laboratoire de Spectrochimie Infrarouge et Raman C.N.R.S. LP 2641 - Université de Lille I Bat. C5 - 59655 Villeneuve D'Ascq Cédex FRANCE

ABSTRACT

The increasing need for non-destructive analytical methods and the developments in instrumentation during the past decade have greatly stimulated interest in the use of spontaneous Raman scattering for molecular microanalysis, but only recently has this potential been realized.

The purpose of this paper is to give a brief review of the fundamentals as well as the methodology of laser Raman micro-spectroscopy and to demonstrate the practical uses of the technique in several major application areas.

Emphasis is also put on the advantages of employing modern multi-channel detection techniques that offer a significant improvement in performance.

Lastly, more recently, these techniques have led to the development of a very sensitive laser microfluorometer which offers in biology great possibilities for investigations at the single living cell level.

I - INTRODUCTION

The entry of Raman spectroscopy into the field of instrumental microanalysis represents a significant opportunity for the field of Raman spectroscopy in that Raman scattering can provide information which up to now has not been available from any other widely used microanalytical technique such as electron or ion microprobing. It is well known that Raman spectroscopy, which is based on the study of the spectral distribution of inelastically scattered light, is a highly specific technique of investigating molecular species in all phases of matter as they are fingerprinted by their vibrational spectra. So, with lasers as sources for excitation of Raman scattering and the ongoing developments in instrumentation for optical spectroscopy, Raman microprobing techniques have matured to the point at which non-destructive chemical microanalysis has become routinely practicable for both academic research and industrial purposes.

Lastly, more recently, laser microfluorescence techniques have developed which offer in biology great possibilities for unique investigations at the single living cell level.

II - Principles of Raman and Fluorescence Spectroscopy

1. Basic Raman process.

The spectroscopic measurements performed with Raman microprobes are based upon the excitation and detection of the normal or spontaneous Raman effect. This effect provides the basis for Raman spectroscopy and has been extensively described (1). It results in the appearance of scattered light at different frequencies. These frequency shifts (from the exciting line) are identified with the frequencies of the atom oscillations in polyatomic structures contained in the sample. The intensities of the lines are determined by the Raman cross section and are directly proportional to the sample which is probed. The kind of information provided by the Raman spectrum is essentially the same as that is obtained from infrared spectra. Thus, the Raman spectra can be regarded as unique fingerprints which also contain information on the local molecular environment (e.g. amorphous or crystalline phases).

2. Basic fluorescence process.

In the spontaneous Raman effect described above, the incident photon energy is below the energies of any excited electronic levels. If the exciting wavelength is such that electronic excitations occur, then other inelastic processes like fluorescence may be induced.

In normal (nonresonant) fluorescence, the light emission, at frequencies ν_F, occurs over a broad range which corresponds to the Stokes region of the Raman spectrum.

The fluorescence emission, though offering less molecular specificity, has an intensity which is often an order of magnitude or more than that of Raman scattering.

III - LASER RAMAN MICROSPECTROMETRY

1. Advantages and limitations.

The principal advantages of the technique are the molecular specificity, the spatial resolution (the laser focal spot placed on the sample may be as small as one micrometer), the ambient conditions (samples may be examined in air under ambient conditions of temperature and relative humidity), and the range of applications (all material inorganic, organic and biological can be investigated).

The major limitations, on the other hand, are the inherent weakness of Raman effect, the fluorescence interferences and the lack of quantification.

2. Instrumentation in Raman microspectrometry.

The intensity of the signal delivered by the detector of a spectrometer analyzing a Raman line can be expressed by:
$$S\nu \; I_0.\sigma.N.\Omega.T.s$$
where I_0 is the laser irradiance at the sample, σ is the Raman cross section, N is the number of molecules inside the probed volume, Ω is the solid angle for Raman scattered light collection, T is the throughput of the instrument and s is the sensitivity of the detector. When a very small quantity of matter has to be analyzed, there are only a few parameters which can be modified in order to compensate for the large reduction in the number of molecules N, namely: I_0, Ω and s. Thus, the use of microscope objectives is the best way to increase both I_0 and Ω.

That is why all the micro-Raman instruments possess a good quality light microscope.

In the instruments the same (high N.A.) objective is used to focus the laser beam on the sample region of interest and to collect the light scattered by the sample (backscattering geometry).

A first generation of instruments conceived at the laboratory on this principle has been commercially available (Instrument S.A. Jobin Yvon) under the name of MOLE (2). It comprised a scanning spectrometer and a monochannel detector for the recording of Raman spectra from microscopic region chosen in the sample. However, this instrument had also an imaging mode of operation, permitting the localization of individual components of a multicomponent heterogeneous sample. Unfortunately, this last mode, though very interesting, has been little used because of its lack of sufficient spatial resolution and sensitivity to discriminate Raman features from fluorescence background, especially for industrial samples. This instrument is now superseded by a modular system (Ramanor U 1 000).

However, the main limitation of this first generation of instrument lies in the high laser irradiance (typically 10^5 to 10^6 watts per cm^2) which may cause degradation during the long exposure time required to record spectra. This limitation is inherent to the sequential recording of the spectral elements imposed by scanning spectrometers.

Thus, in order to overcome this difficulty, in a second generation of instrument (3), we have moved towards the use of multichannel detectors which enable the simultaneous recording of all the spectral elements with the same sensitivity. As a result one can turn the signal to noise gain thus obtained into time in order to profit by either detecting very weak Raman signals or illuminating fragile materials with very low laser power.

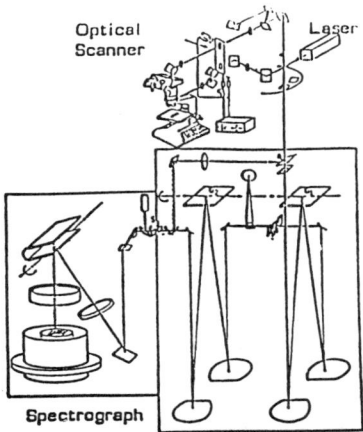

Fig. 1 Diagram of the multichannel Raman Microprobe (Microdil 28) Developped at Lille (C.N.R.S. - DILOR Compagny) FRANCE

In figure 1 is presented the optical scheme of a multichannel Raman microprobe (MICRODIL 28) which has been developed in close collaboration between our laboratory and a French company (DILOR S. A. Lille France). In addition to this unique instrument, for which all the optics has been conceived and maximized for Raman microanalysis, there exists now on the market several multichannel macro-Raman instruments on which a microscope can be adapted as a micro-Raman accessory (OMARS89 DILOR; TRIPLE SPECTROGRAPHE RAMAN - Instruments S. A., FRANCE; MICRAMATE - SPEX, USA)

3. Applications of a micro-Raman analysis.

We will try here to give some ideas on the kind of problems which are amenable to investigation by micro-Raman spectroscopy.

a - Geology - Gemology

- Point and non-destructive analysis of intramineral fluid inclusions (4).

In the course of geological phenomena, whether they be hydrothermal, magmatic or metamorphic, the role played by fluids is very important. Revealing of the composition and density of deep fluids, the intramineral inclusions (whose size varies from a few to several tens of microns) are therefore sound proof of the origin of the minerals.

Raman microspectroscopy, through its capacity to analyze "in situ" (through the host mineral) and non-destructively very small quantities of matter is particularly well adapted to the study of fluid inclusion. It enables the precise analysis (identification and determination of molar fractions) of the majority of geological fluids (CH_4, CO_2, N_2, H_2S, O_2, H_2, ...) The coupling of this technique with the microthermometry allows a correct definition of the thermobarometric conditions of fluid capture. In a like manner, Raman microspectroscopy has proven to be a perfect technique for identifying fluid or solid inclusion in gems. For the gemologist, this identification is of the utmost importance both for the identification is of the utmost importance both for the identification of the gem and for the determination of their geological origin. Moreover, this "in situ" non-destructive technique has proven to be very efficient in discriminating between natural gems and synthetic minerals made for jewelry (e.g. sapphire, ruby, emerald...) (5).

b. - Industrial material control.

Contamination or formation of defects are problems currently encountered in the commercial production of materials. The chemical identification of these defects is of particular interest because it can lead to properly selected materials and processes.

- Semi-conductor process support.

Because circuit board geometries and components are approaching micrometer dimensions, contaminants of this size often create device reliability problems. The first step towards solving these problems is contaminant identification for which Raman microspectroscopy is often the only method with a chance of success (6). Inorganic materials (like contaminants from etching, corrosion stain on connector line, ...) as well as organic materials (like contaminants from packing materials) can be readily identified "in situ" by Raman microanalysis. In addition to providing the ability to identify foreign contaminants on devices, Raman microspectroscopy can provide characterization of the materials used in fabricating the devices (7). For example, with the development of very large scale integration (VLSI) technology, considerable emphasis has been placed on devising methods of fabricating high-quality Si films on a variety of insulating substrates. A wide variety of approaches to production of such silicon overlayers has been explored and a good understanding of the crystal growth is therefore essential for the further development of device quality material. Raman microspectrometry can be used as a rapid non-destructive method of analysis for processed Si based on the position and width of the silicon band around 520 cm^{-1} which is extremely sensitive to local environment.

- Analysis of contaminants in synthetic textile filaments.

Small individual filaments (typically 5 to 20 µm in diameter) are usually used to make textile yarns. Filament breaks caused by particulate contamination of the polymer are important problems. Inclusions in the filaments arise from particulate contamination of the feed stock polymer. They can be analyzed directly by recording their Raman spectra through the textile filament. Substances present as part of the polymer recipe or produced during polymerization (internal contaminants such as antiluster, degraded polymer, carbonaceous residues, ...) as well as substances acquired during the material handling processes (external contaminants such as packing materials, workers' clothing fiber fragments, airborne particles, ...) are thus easily identified.

c - Analysis of airborne particles.

The analysis of airborne particles in the size range 1 to 10 µm is important in many fields of human endeavor including concerns related to pollution monitoring and health hazards. Raman microspectroscopy is eminently suited for the characterization of microparticles of organic and inorganic origin. In particular organic compounds give informative, diagnostically useful spectra (8). The principal species for which characterization is easy include the common minerals (silicates, oxides, carbonates, sulfates...) and some organic compounds (pesticides, insecticides, aliphatic acids, polymers, hydrocarbon films, peptides...).

d - "In situ" analysis of ancient works of art.

Historical materials or works of art are unique and cannot be replaced. Thus no injury or risk of damage proceeding from the analysis can be tolerated. Raman microspectroscopy is thus eminently suited in aiding authentification with no major damage to the historic art object being examined. Recent work has demonstrated the ability of this technique to identify pigments on various ancient works of art (Lazurite and mercury sulphide vermillon for blue and red ornamented letters on a page of a 12th century French missal, orpiment As_2S_3 as yellow paint on a fragment of an Egyptian death mask, ...)

e - Microanalysis of biological tissues

With the advent of Raman microprobe techniques, great possibilities for unique investigations appear to be opening up in biology, pathology and tissue research to obtain molecular information at the cellular level. Raman microanalysis is generally performed on standard histological sections of biological tissues. So far, because of the fragility of biological samples, studies have been devoted to problems where the material analyzed was sufficiently locally concentrated, e.g. bioaccumulations in cells and tissues (9), precipitations in vertebrate kidneys induced by pathological processes (10), hard tissue formation (11), and foreign bodies in tissue coming from the degradation of implanted prosthetis (12).

Improvements in sensitivity are required in order to extend the field of applications to the study of normal tissue in which the biological material of interest is homogeneously distributed. Raman multichannel microprobes are certainly a unique opportunity for such studies. Lastly, in vivo studies of pigments in vegetal and animal single cells are now possible using the resonance Raman effect (13).

IV - LASER MICROFLUOROMETRY

1. Advantages and limitations.

The use of laser sources for the excitation of fluorescence spectra present great advantages, especially in terms of high spatial resolution and high sensitivity (14).

However, besides its unquestionable merits, this technique has some limits connected with the nature of the analyzed material. Indeed, high irradiance induce generally rapid modification of the probed area as indicated by a fast decay of the fluorescence emission. Fragile materials are not able to sustain laser power higher than a few hundreds microwatts.

2. Instrumentation.

Most of the time, fluorescence studies require fast recordings of the whole fluorescence spectrum with both good resolution and signal to noise ratios. These requirements preclude the use of scanning spectrometers with monochannel detectors. That is why we are developing in close collaboration with the DILOR society a microfluorometer which employs spectrographic dispersion with a sensitive multichannel detection (15).

3. Applications of laser fluorescence microspectrometry.

- Biological material

Laser sources, by allowing very precise topographic investigations, are particularly well adapted to the study of biological mechanisms in single living cells (enzymology, metabolisation, drug - intracellular target interactions, intracellular microviscosity measurements using fluorescence polarization, kinetic studies, ...)

In a recent work, this technique proved to be a very efficient method to study the mechanism of human tumor cell (K 562 leukemia cells) resistance to an anticancer drug (Adriamycin) by comparing the intracytoplasmic and intranucleus fluorescence spectra obtained from living cells sensitive or made resistant to the drug (16).

- Material control.

Pigments can be analyzed in "in situ" in inks or pains deposited on any support and requiring no sampling at all, thus opening up interesting possibilities in forensic science (17).

At last, laser induced photoluminescence studies of semiconductors (InP, GaAs...) may be conducted on a microscopic scale in order to control the uniformity of electronic properties of material surface during device processing (18).

V - CONCLUSION

With routine detection limits well below nanograms and high specifity, Raman microprobing has already become in many laboratories a major microanalytical technique yielding new or more precise answers to problems left unsolved or incompletely solved by other techniques.

On the other hand, laser microfluorometry, due to its very high sensitivity and spatial resolution is a powerful tool to study intrinsic or extrinsic fluorophores (fluorescent probes) in single cells while preserving their viability so that future of this technique looks very promising.

REFERENCES

1. D. A. Long, Raman Spectroscopy (McGraw Hill Inc., London, 1977)
2. P. Dhamelincourt et al., Anal. Chem., 51, 414 A (1979)
3. J. Barbillat and M. Delhaye, Microbeam Analysis 1983 (San Francisco Press, Inc., San Francisco, 1983), p. 230
4. J. Dubessy et al., Journal de Physique, colloque C_2, 45, C_2 - 811 (1984)
5. M. L. Dele, P. Dhamelincourt, J. P. Poirot et H. J. Schubnel, J. Molecular Structure, 143, 13 (1986)
6. R. Z. Muggli and M. E. Andersen, Solid State Technology 4, 287 (1985)
7. J. B. Hopkins and L. A. Farrow, J. Appl. Phys., 59, 1103 (1986)
8. E. S. Etz et al., Environmental Pollutants (Plenum Pub. Corps., N. Y., 1978), p. 413
9. C. Ballan - DuFrancais, M. Truchet and P. Dhamelincourt, Bio. Cell., 36, 51 (1979)
10. R. Martoja and M. Truchet, Malocologia 23, 343 (1983)
11. N. D. Grympas et al., Microbeam Analysis 1982 (San Francisco Press, San Francisco, 1982, p. 333
12. J. L. Abraham and E. S. Etz, Science 206, 716 (1979)
13. J. C. Merlin, Spectroscopy of Biological Molecules (Interscience - Wiley, N. Y., 1985), p. 427
14. M. Delhaye et al., Spectroscopy of Biological Molecules (Interscience - Wiley, N. Y., 1985), p. 51
15. P. Dhamelincourt et al. Spectroscopy of Biological Molecules (Interscience - Wiley, N. Y., 1985), p. 234
16. P. Jeannesson et al., Spectroscopy of Biological Molecules (Interscience - Wiley, N. Y., 1985), p. 234
17. M. E. Andersen Microbeam Analysis 1982 (San Francisco Press, San Francisco, 1982), p. 197
18. S. K. Krawczyk and G. Hollinger, Appl. Phys. Lett., 45, 870 (1984)

A REVIEW OF THE NRL CARS MICROSCOPE

M. D. Duncan, J. Reintjes, T. J. Manuccia
Naval Research Laboratory, Optical Sciences Division,
Laser Physics Branch, Washington, D. C. 20375

ABSTRACT

This paper describes a technique that permits molecular specificity in microscopy while retaining good spatial resolution. Specificity is achieved using Raman scattering from characteristic molecular vibrations, and high resolution is obtained by imaging the distribution of visible anti-Stokes radiation. Images have been obtained from a variety of pure organic liquids, from deuterated water in onion-skin cells, deuterated liposomes and other samples. Thermal and dielectric breakdown damage to even fragile biological materials is made negligible by choice of duty cycle and average power. Sample fluorescence is avoided by the choice of anti-Stokes imaging. Imaging through (usually astigmatic) spectrometers is not needed. Deuterium substitution can be used as a general purpose and artifact-free "stain". The combination of high spatial resolution, excellent molecular discrimination, and digital image processing (background subtraction) provide the CARS microscope with capabilities not found in any other current microscopic imaging technique.

INTRODUCTION

It is particularly difficult to obtain good spatial resolution of specific molecular species in a microscopic sample. Electron microscopes produce detailed maps of elemental composition, but few techniques allow the imaging of individual molecular species to the extent that a map of their spatial distribution can be made on the scale of 1 µm or less. Certain techniques, such as infrared spectroscopy, produce good chemical selectivity but lack the high spatial resolution of visible wavelengths. Stains and fluorescing agents can give very good spatial resolution, but they are limited to certain molecules, take elaborate preparation and contaminate the sample under study. Spontaneous Raman scattering with visible light overcomes some of these problems by providing good spatial resolution and good chemical selectivity, but can usually only be used to probe single microscopic spots in a sample due to the overheating of a sample, low signal levels, and the difficulties in imaging through a spectrometer.[1-3]

The scanning coherent anti-Stokes Raman scattering (CARS) microscope developed at NRL can overcome or sidestep all of the above limitations and has been used for the past 5 years to investigate several molecular specific imaging problems in biological or biologically related samples.[4-6] The CARS microscope uses the good molecular discrimination shared by all Raman techniques and combines it with the high spatial resolution obtained by imaging the visible anti-Stokes light produced by the sample. No spectrometer is needed since in the CARS technique the anti-Stokes signal is well separated from the pump light both spatially and spectrally. The sample is not required to be stained or altered in any way. The signal levels provided by the nonlinear CARS technique usually exceed those produced by spontaneous Raman processes, so heating, even of fragile biological samples, is not a problem.

We present here two examples the demonstrate the ability of the CARS microscope to image biological compounds in a microscopic sample. They are CARS images of onion skin cells showing the spatial distribution of deuterated water and CARS images of deuterated and nondeuterated liposomes showing high spatial resolution and excellent molecular discrimination. We will discuss the use of background subtraction and contrast enhancement to improve the information content of the CARS images dramatically. We will also discuss some of the limitations of the CARS technique as it applies to the study of microscopic samples.

TECHNIQUE

The CARS process is discussed in detail in review papers by Tolles et al.[7] and by Druet and Taran.[8] The CARS microscope operates by scanning a sample region of interest with two focused laser beams. Focused laser beams are used to increase the signal produced by the nonlinear CARS interaction, and scanning is used to increase the field of view. The laser frequencies are tuned so that their difference matches that of an active Raman vibrational mode in a specific species of interest. When the CARS signal light is then collected, spectrally isolated, and imaged, those regions that contain the species of interest produce bright areas on a dark background. A schematic diagram of this process is shown in Fig. 1.

The CARS microscope is capable of identifying a number of different molecules and molecular groups within the 1700 cm^{-1} to 3300 cm^{-1} range. The examples presented here use deuterated C-D or O-D bonds and have characteristic frequencies in the 2100 cm^{-1} to 2500 cm^{-1} range. This region was chosen because of the lack of interference with naturally occurring C-H and O-H Raman bands.

The spatial resolution of the CARS microscope is determined entirely by the wavelength of the signal light and the quality of the collection optics and does not depend on the focal spot size of the input laser beams or on the size of the illumination aperture. The resolution of our instrument is limited to approximately 0.7 µm.

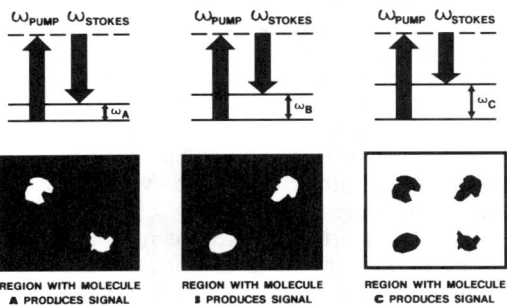

Fig. 1 - Schematic diagram of a CARS image from a multicomponent microscopic sample. When the incident lasers are tuned to a Raman resonance in molecular species **A**, only regions of the sample containing **A** appear bright. The incident lasers can also be tuned to species **B** and **C**. Laser tunings are shown in the top row; resultant images are below.

APPARATUS

The experimental apparatus is shown schematically in Fig. 2. The separate wavelengths for the CARS process are produced by cw mode-locked dye lasers pumped by a cw mode-locked argon-ion laser. One dye laser is kept at a fixed frequency (typically 570 nm) while the other is tuned. Both dye lasers operate at 150 mW average power and 300 W peak power with pulses of 6 to 8 ps duration. Both beams are overlapped in time and space and are focused onto the sample. The final turning mirror scans the beams across the sample in two orthogonal directions using piezoelectric stacks and can produce scan dimensions of 30 to 100 µm. After

Fig. 2 - Schematic diagram of the CARS microscope apparatus. A conventional microscope is used to collect light emerging from the sample and is represented by a single lens to the right of the sample. Beam separation and focusing angles are exaggerated for clarity.

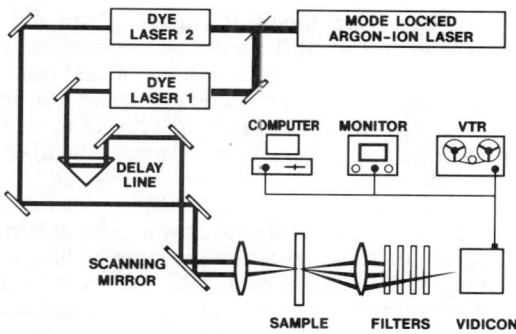

interacting with the sample, both input laser beams are blocked with dichroic filters while the anti-Stokes signal light is passed and focused onto an image intensified vidicon. The resulting video signal is digitized and captured in a video frame buffer. Signal images can then be viewed in real time, integrated, or stored for later viewing and processing.

EXPERIMENTAL RESULTS

One of the earliest systems studied with the CARS microscope was onion skin cells soaked in deuterated water. Figure 3 shows these onion skin cells imaged using the anti-Stokes light produced by the O-D vibration at 2450 cm^{-1}. Ordinary white-light images of the same regions are shown for comparison. There are many similarities between the white light and the CARS images, mainly due to the physical morphology of the sample. Wherever a white light image suffers scattering in the

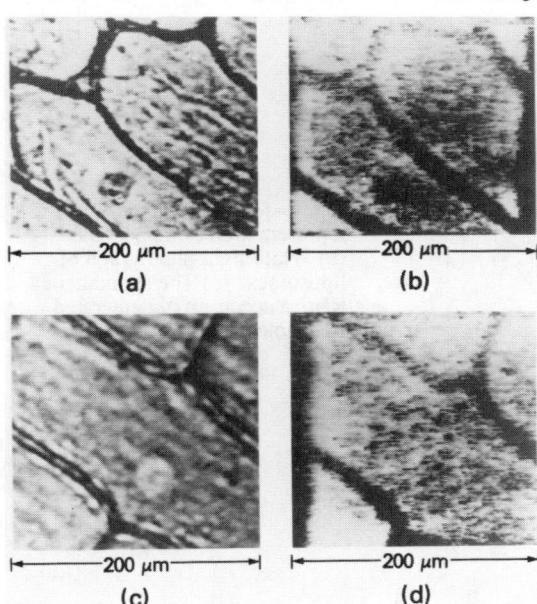

Fig. 3 - White-light and CARS images of onion skin cells that have been soaked in D_2O.
(a), (c) White-light images with slightly different focusing conditions.
(b), (d) CARS images of the same regions as (a) and (c) when the lasers are tuned to the 2450 cm^{-1} band of D_2O. The CARS images were obtained in 2 seconds. The onion skin cells were soaked in D_2O for 4 hours.

sample, so does the CARS image. Even with this similarity, the CARS images show some interesting differences when compared to the white-light images. One of these differences is the apparent lack of signal from the cell nuclei in the CARS images. This suggests that the deuterated water was not absorbed nearly as quickly by the cell nuclei as it was absorbed by the rest of the cell. This is exactly the sort of information that would have been hard or impossible to obtain using any other conventional technique.

The pictures of the onion skin cells were taken without any digital image storage or processing capabilities. It was evident during the acquisition of that data that sample morphology was a severe impediment to the interpretation of the CARS images. In addition, all CARS images in biological systems produced a large amount of nonresonant background signal due to the interference of other molecular resonances. Only by using image processing techniques could both of these problems be solved at once. The simplest technique would consist of taking one image of the sample region while the lasers were tuned to be on-resonance with the species of interest, taking one image of the sample region while the lasers were tuned to be off-resonance, and then subtracting the two images digitally. As a result of this subtraction, the purely morphological aspects of the image and the CARS nonresonant background signal could be completely suppressed. This technique does indeed work when applied to real CARS images, but normally many images are integrated in each step to increase the signal-to-noise ratio. In addition, the spectral tuning of the lasers is kept to a small value so that the nonresonant CARS background signal stays constant.

Deuterated and nondeuterated liposomes were also studied with the CARS microscope. Liposomes are microscopic organic structures that have importance in the study of artificial encapsulation of such materials as hemoglobin.[9] Figure 4(a) shows a schematic representation of their structure and illustrates the bilayers that form and

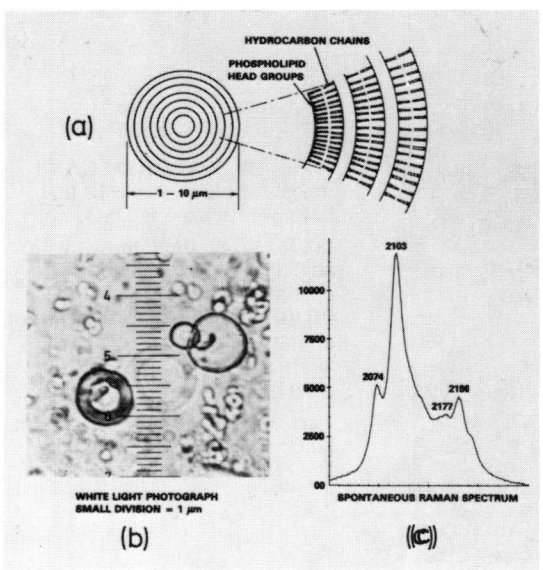

Fig. 4 - (a) A schematic representation of a liposome. (b) White-light photograph of liposomes. (c) The spontaneous Raman spectrum of deuterated liposomes.

mimic natural cell membranes. Figure 4(b) is a photomicrograph of liposomes taken in white light and Fig. 4(c) shows a region of the spontaneous Raman spectrum of liposomes that have had 50% of their C-H bonds deuterated. For study with the CARS microscope the liposomes were suspended in water and placed in 100 μm thick flat capillary tubes.

Figure 5 demonstrates the ability of the CARS microscope to produce signal from the microscopic liposomes as well as the ability of the image processing technique to eliminate the CARS nonresonant background signal. Figure 5(a) shows the CARS signal produced when the input lasers were tuned so that the 2100 cm^{-1} C-D bond of the deuterated material was in resonance. Figure 5(b) shows the signal produced when the lasers were tuned off-resonance. Both images were produced by integrating the CARS signal for 8 seconds, or, equivalently, for 255 video frames. The nonresonant signal is clearly a large part of the total signal produced by the liposome and the surrounding water. Figure 5(c) is an image produced by digitally subtracting Fig. 5(b) from Fig. 5(a) and shows a significant improvement in useful signal.

The full extent to which the CARS nonresonant background signal and the physical morphology of the material can be eliminated is illustrated in Fig. 6. Parts (a) and (c) of Fig. 6 show the signal produced by a deuterated and nondeuterated liposome, respectively, when the lasers are tuned to the 2100 cm^{-1} band of the C-D vibration. Figure 6(b) shows a processed image where the nonresonant background signal has been subtracted. It is identical to Fig. 5(c) and shows the spatial extent of the deuterated liposome. Figure 6(d) shows a processed image with its nonresonant background signal subtracted and clearly shows that there is no deuterated material

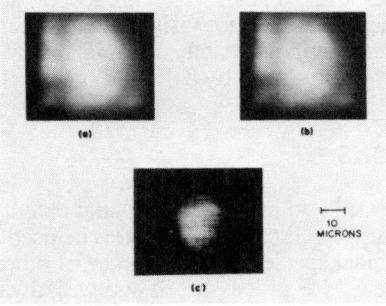
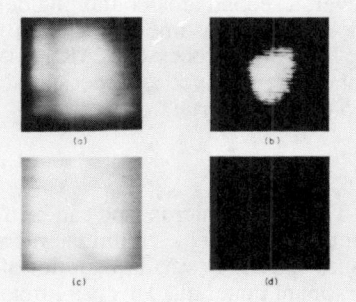

Fig. 5 - (a) Unprocessed CARS image of a deuterated liposome taken with the lasers tuned to the 2100 cm^{-1} band of the C-D vibration. (b) Unprocessed CARS image of the same region with the lasers tuned off-resonance. (c) Processed image resulting from subtracting Fig. 5(b) from Fig. 5(a).

Fig. 6 - (a) Unprocessed CARS image of a deuterated liposome taken with the lasers tuned to the 2100 cm^{-1} band of the C-D vibration. (b) Processed image with all CARS nonresonant background signal subtracted. (c) Unprocessed CARS image of a nondeuterated liposome, again tuned to the 2100 cm^{-1} band. (d) Image processed as in (b), showing complete suppression of background signal

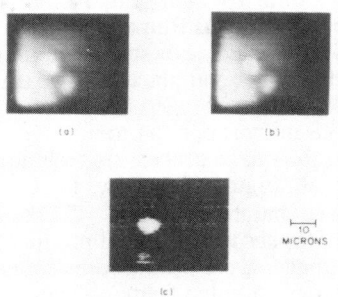

Fig. 7 - (a) CARS image of both a deuterated and nondeuterated liposome taken with the lasers tuned to the 2100 cm^{-1} band of the C-D vibration. (b) Same as (a), but with the lasers tuned to 2150 cm^{-1}. (c) Processed image showing regions containing deuterated material. Image noise has been enhanced to illustrate the complete suppression of signal from regions containing nondeuterated material.

present in that liposome at all. The physical morphology and the nonresonant background signal have been suppressed completely, leaving no signal for the case of a nondeuterated liposome.

An example of the ultimate resolution and sensitivity of the CARS microscope is show in Fig. 7. For this sample, both deuterated and nondeuterated liposomes were mixed together. Figure 7(a) is an on-resonance CARS image with two visible liposomes within the scan area. Figure 7(b) is an image of the same region with the lasers tuned off-resonance, and Fig. 7(c) is the digitally subtracted image that has been enhanced to the point that video scan line noise is made visible. It is clearly seen that only one of the originally visible liposomes contains deuterated material. In addition, however, a second area containing deuterated material can be seen. This second area is a deuterated liposome less than 5 µm in diameter and was not visible in either the white light or unprocessed CARS images. Features approximately 1 µm in spatial extent can be resolved and measureable signal was produced by a thickness of less than 5 µm of material.

LIMITATIONS

The CARS microscope can be used successfully to measure the spatial extent of major chemical species within a microscopic sample. It cannot, however, be used to measure extremely small quantities of any chemical species in the presence of other, nonresonant material. This is a limitation of the CARS process itself and will always limit the application of the CARS microscope in real biological systems. A useful guideline would be that any strong Raman scatterer that is 3% or more of a microscopic system under study would be a good candidate for study with the CARS microscope. This could include ubiquitous material such as water or any major species that could, for example, be deuterated to a large extent. Deuteration is not necessary except that it does allow the chemical under study to be taken out of more crowded regions of the Raman spectrum and put into a relatively uncrowded area.

SUMMARY

In summary, we have presented results that illustrate some of the unique capabilities of the CARS microscope for imaging distinct chemical species within a microscopic region. Images of deuterated and nondeuterated liposomes have demonstrated that the spatial resolution of the CARS microscope is 1 µm or better.

Digital image processing has been used to increase the signal-to-noise ratio of CARS microscope images by subtracting the CARS nonresonant background signal and by eliminating constant morphological features of the sample.

1. J. L. Abraham and E. S. Etz., Science **206**, 716 (1979).
2. E. S. Etz, *Scanning Electron Microscopy*, I, p. 67, IIT Research Institute, Chicago, IL (1979).
3. P. Dhamelincourt and P. Bisson, Microsc. Acta **79**, 267 (1977).
4. M. D. Duncan, J. Reintjes, and T. J. Manuccia, Opt. Lett. **7**, 350 (1982).
5. J. Reintjes, M. D. Duncan, and T. J. Manuccia, in *Picosecond Lasers and Applications*, L. S. Goldberg, ed., Proc. SPIE **322**, 87 (1982).
6. M. D. Duncan, Opt. Commun. **50**(5), 307 (1984).
7. W. M. Tolles, J. W. Nibler, J. R. McDonald, and A. B. Harvey, Appl. Spectrosc. **31**, 253 (1977).
8. S. Druet and J. P. Taran, *Chemical and Biochemical Applications of Lasers*, Vol. 4, p. 187, Academic Press, New York (1978).
9. B. P. Gaber, P. Yager, J. P. Sheridan, and E. L. Chang, Federation of Experimental Biologists Lett. **153**, 285 (1983).

CROSSED-BEAM THERMAL LENS AS A SCANNING LASER MICROSCOPE

Dean S. Burgi and Norman J. Dovichi
University of Alberta, Edmonton, Alberta T6G 2G2 CANADA

ABSTRACT

The crossed-beam thermal lens is used for the measurement of minute optical absorbance within very small probe volumes. In this technique, a cylindrically-symmetric temperature rise is produced within a sample from absorbance of a modulated pump laser beam. This heated region is probed at right angles with a second, cw laser beam. Defocusing of the probe beam caused by the heated sample is detected as a periodic change in the far-field probe beam-center intensity. Since the two beams only interact in their intersection volume, very small regions may be probed with tightly focused beams. By recording the thermal lens signal as an inhomogeneous sample is scanned through the probe volume, an image is generated based upon the absorbance of the sample. Furthermore, by recording the phase of the signal, it is possible to generate an image based upon thermal diffusivity. Images of histopathological and geological samples are presented.

INTRODUCTION

The thermal lens, first reported in 1964, is a powerful tool for the analysis of highly transparent materials [1-6]. In this technique, a lens-like optical element is formed within a sample from a temperature rise produced by absorption of a pump laser beam. This thermal lens defocuses the laser beam and is quantitated as a change in the far-field beam-center intensity. Since the temperature rise within the sample is proportional to the pump laser power, highly transparent materials may be studied by utilizing a high power laser.

A number of variants of the thermal lens experiment have been devised. The thermal lens is probed coaxially in the conventional thermal lens [7], off-axis in photothermal deflection [8], and at right angles in the crossed-beam thermal lens [9,10]. The crossed-beam technique provides good spatial resolution since the signal is generated only where the two beams intersect; tightly focused beams may be used to probe very small volume samples.

If the crossed-beam thermal lens experiment is performed within a homogeneous sample, the heated region acts as a cylindrical lens-like optical element to defocus the probe beam out of the plane containing the pump and probe laser beams. The analysis of homogeneous materials has produced spectacular absorbance sensitivity; detection limits of 120 iron-1,10-phenanthroline molecules were obtained within a 200-μm^3 probe volume [11], and sub-femtomole quantities of amino acids have been detected after chromatographic separation [12-13]. Furthermore, the temporal behavior of the crossed-beam thermal lens signal has been employed in regression analysis to measure both flow velocity and thermal diffusivity [14-15]. Thermal diffusivity is inversely proportional to the time constant of the signal and may be measured with 1% precision.

Two different phenomena can produce a crossed-beam thermal lens signal in inhomogeneous materials and are related to the normal and transverse deflection signals observed in photothermal deflection imaging of surfaces [16]. One phenomenon was described above for homogeneous materials; the cylindrical temperature rise generated by the pump beam intensity profile acts to defocus the probe beam out of the plane containing the two beams. A slit oriented in the plane of the two beams followed by a lens to focus the transmitted light onto a photodetector will measure changes in intensity associated with the conventional signal. A second phenomenon produces defocusing in a direction orthogonal to the conventional signal. The thermal gradient near an interface can defocus the probe beam perpendicularly to the plane of the interface. If the interface is perpendicular to the pump beam axis, then the thermal lens signal is associated with defocusing of the probe beam in the plane formed by the two laser beams. A slit oriented orthogonal to the plane formed by the beams followed by a lens to focus the beams onto a photodetector will be sensitive to defocusing associated with an interface.

EXPERIMENTAL DESIGN

A scanning laser microscope has been constructed using the crossed-beam thermal lens [17-18]. In this experiment, figure 1, the crossed-beam thermal lens signal is recorded as an inhomogeneous sample is translated through the intersection region of the pump and probe laser beams. A low power pump beam is employed for most of our experiments, P = 1.0 mW, and a computer controlled translation stage is used to move the sample through the laser beam intersection region. Tightly focused beams are produced by microscope objectives; a 18-mm focal length lens focuses the pump beam to a 2-μm spot and a 16-mm focal length lens focused the probe beam to a slightly larger spot. To date, we have used an argon ion laser to produced the pump beam at 488 or 514.5 nm. The probe beam is produced by a helium-neon laser at 632.8 nm. A thin slit, followed by a large diameter lens, is used to isolate the probe beam center. The slit is oriented in either the plane of the two beams or perpendicular to the plane. The transmitted intensity is focused onto a photodiode.

A dual phase lock-in amplifier is used to demodulate the thermal lens signal. The amplitude of the lock-in signal is related to the sample absorbance whereas the phase is related to both the thermal diffusivity of the sample and the sign of the signal. Data are normalized to the dc probe laser intensity. The digitized data are stored in a computer for later display. Data collection rate is relatively slow; a 1-s time constant is employed on the lock-in amplifier so that 3-s are required per data point. Over 8 hours would be required to generate the 10^4 data points in an image with 100 by 100 pixels.

FIGURE 1-Instrument

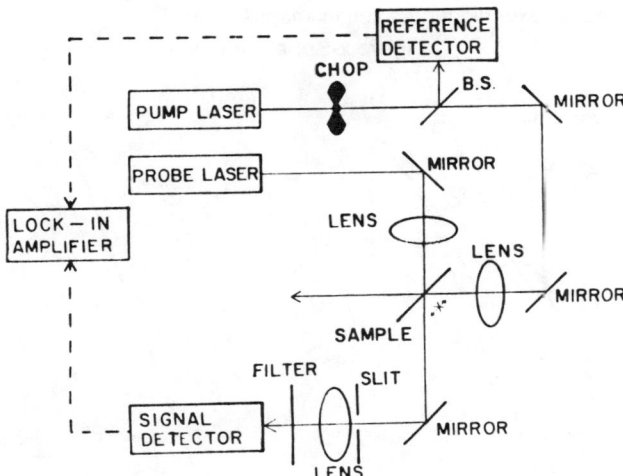

RESULTS

Figure 2 presents both amplitude and phase images generated within a 1-mm thick piece of the mineral sphalerite. This sample comes from the Creede ore deposit in southern Colorado and is produced by hydrothermal deposition wherein super-saturated fluids dissolve a complex mineral mixture within the earth's crust. Upon reaching atmospheric conditions, the mineral precipitates from the mother liquid. Sphalerite consists of a zinc sulfide matrix with various impurities, particularly iron and lesser amounts of other transition elements. In the formation of the mineral, small fluid-filled inclusions were created. These inclusions represent some of the original liquid from which the mineral was formed.

The change in refractive index with temperature is an interesting property of sphalerite. In most materials and almost all liquids, dn/dT is negative. On the other hand, dn/dT for sphalerite is positive. The thermal lens signal is directly proportional to dn/dT so that a positive signal is expected in the bulk mineral whereas a negative signal is expected in the fluid inclusion; the signal should go through zero near the interface.

The tear drop shaped hump in the center of the amplitude plot, figure 2a, is the image of a 20- by 40-μm fluid inclusion. Recall that the amplitude of the lock-in amplifier presents the absolute value of the thermal lens signal. The valley, or region of zero signal, corresponds to the interface. The plateau observed on the periphery of the image corresponds to the bulk sphalerite matrix. The phase image of the sphalerite sample is shown in figure 2b. Positive phase is shown with solid lines and negative phase is shown with dashed lines. Since the change in refractive index with temperature is positive for sphalerite but negative for water, the phase is relatively constant and positive in the homogeneous solid and negative in the inclusion. The phase goes through a rapid change at the interface which is characteristic of interfaces and may be used for their identification.

The spatial resolution for this the image appears to be about 5-μm which is similar to the minimum step-size of the micrometer. The resolution probably is limited by thermal diffusion. Assuming a thermal diffusivity of 0.1 $cm^2 s^{-1}$ in the mineral, the thermal diffusion distance is about 5-μm, the observed resolution, for a 200 Hz modulation frequency. Higher modulation frequency will be required for higher resolution images.

Figure 2-Sphalerite

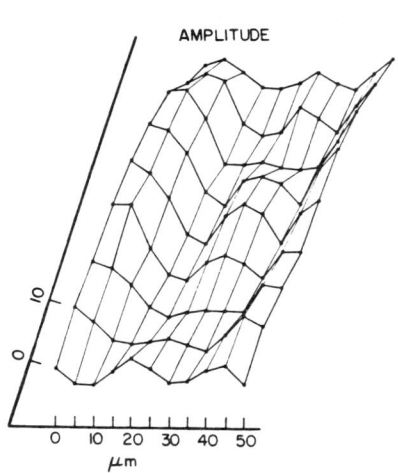
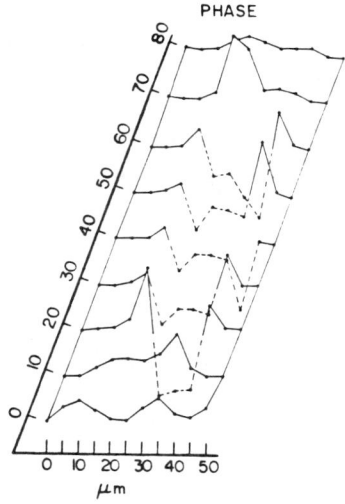

Sphalerite is a highly inhomogeneous material; fluid inclusions, opaque solid inclusions, cracks, and voids all act to distort the beam profiles. Three features of the microscope are used to minimize the effect of these inhomogeneities upon the image. First, the modulated component of the signal is used to construct the image. Any perturbation in the probe beam profile induced by inhomogeneities in the sample will not be modulated and will not generate a signal. Second, before plotting the data, the lock-in amplifier amplitude is divided by the average probe beam intensity. Since the amplitude of the modulated component of the probe beam intensity is proportional to the unmodulated component; the ratio of the two values is immune to intensity variations. Third, the probe beam intensity is averaged over the area of the slit, reducing the effect of spatial inhomogeneities upon the thermal lens signal.

Figure 3a presents an amplitude image taken with 100-μm step size through a 8-μm thick piece of human epidermis taken from a tuberculosis patient. The sample is a pathological standard and has been stained with the dye acid fast bacteria, AFB, and counterstained with methylene blue. AFB is a complex natural product which has the interesting property of staining only tuberculosis bacteria while leaving normal tissue unaffected. AFB strongly absorbs the argon ion laser 514.5 nm line; regions of high amplitude in figure 3 correspond to high concentrations of the AFB stain and, hence, localized tuberculosis bacteria infections. Figure 3b presents an image of the tissue sample taken with 5-μm step-size. The bacteria appear to colonize the tissue in long striations about 50-μm in width.

Figure 3-AFB Stained Epidermus

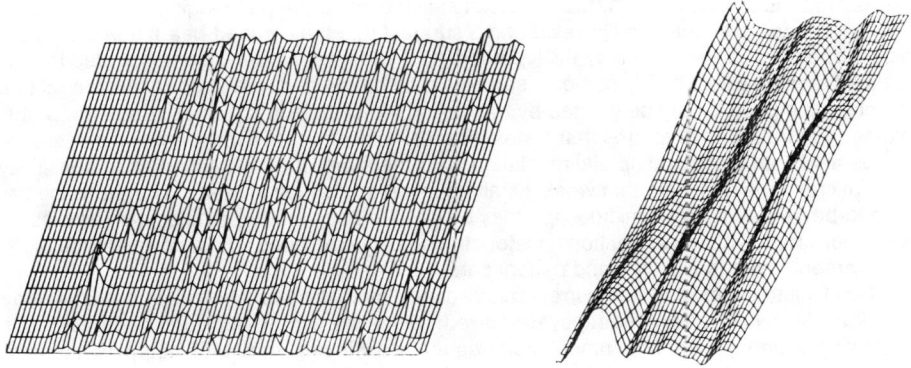

CONCLUSIONS

The crossed-beam thermal lens is a relatively new microscopy technique. The microscope presents three important and unique capabilities. First, the microscope is highly sensitive; absorbance as small as 10^{-8} may be studied within a few micrometer probe volume. This sensitivity will be of particular value in the study and analysis of highly transparent materials, such as high purity optical components or weakly stained tissue samples. Second, an image of the sample may be formed based upon thermal diffusivity. Images related to the thermal diffusivity of the sample should prove useful in the study of high power laser damage of optical components. Damage of the components may result from variations of thermal diffusion within the component which produce non-uniform expansion and structural failure.

The third, and most important, property of the microscope results from the highly localized absorbance measurement. In conventional transmission measurements, the sample absorbance is integrated over the optical path; no information is generated on the distribution of analyte along the beam path. The crossed-beam thermal lens microscope directly measures the sample absorbance at a point-like region at the intersection of the two laser beams. The data of figure 3 demonstrates that a localized region of a complex, natural sample may be imaged with good resolution. A tunable pump laser could be used to obtain a spectrum of the contents of the inclusion without interference from the surrounding matrix. Furthermore, by scanning the sample in a controlled fashion, it is possible to isolate specific features without interference from other regions of the sample.

Conventional histopathology utilizes thin, 5- to 10-μm, tissue samples for two reasons. First, the thin sample may be stained uniformly with a high concentration of dye; a Beer's law measurement of sample transmission requires a very high concentration of dye to produce an observable absorbance over the 5-μm thick path length of the tissue sample. Second, a single plane of cells is observed; the pathologist is not confused by imaging through multiple layers of cells. Unfortunately, a minimum of about 24 hours are required to prepare the tissue sample for observation. The tissue is trimmed to a convenient size and then immersed in a series of baths of steadily increasing alcohol concentration to dehydrate the tissue. Next, the tissue is embedded in hot paraffin. The dehydration steps are required since paraffin is soluble in alcohol but not water. After cooling, a 5-μm thick tissue sample is sliced with a microtome. Then, the sliced tissue sample is immersed in a series of baths of steadily increasing water concentration to rehydrate the sample. Last, the sample is stained and mounted. Most stains are water soluble which necessitates the rehydration step. When time is of the essence, frozen tissue samples may be cut on a microtome; however, the tissues are tens of micrometers thick and less satisfactory to the pathologist.

The images presented in figures 2 and 3 suggest that the crossed-beam thermal lens microscope may be used to study thick pathological samples. Figure 2 demonstrates that localized regions within highly complex samples may be imaged with good resolution so that a single plane of cells may be imaged by translating the sample in a plane through the probe volume. Figure 3 demonstrates that histological samples prepared with conventional stains may be analyzed with good spatial resolution. Furthermore, the high absorbance sensitivity of the microscope suggests that weakly stained samples may be imaged. Therefore,the crossed-beam thermal lens microscope may be used to image thick pathological samples much more rapidly than conventional histopathological techniques with a corresponding improvement in diagnosis time and patient care.

Two limitations exist for the current microscope. First, the spatial resolution is limited by the relatively low chopping frequency employed. Since the signal amplitude decreases with chopping frequency, very high modulation frequencies will require use of lasers with tens of milliwatts of power rather than sub-milliwatt powers currently employed. Second, image acquisition is slow, primarily due to the long time constant employed in the instrument. Faster electronics, higher modulation frequency, and larger signal amplitudes will improve signal acquisition time significantly.

A multichannel thermal lens microscope has recently been developed by our research group. Use of a photodiode array and pulsed pump laser allows images to be generated from 128 to 1024 times faster than the conventional microscope. For example, an image with 100 by 100 pixels may be generated in one minute compared with eight hours required by the conventional microscope.

ACKNOWLEDGEMENT

This work was funded by the Natural Sciences and Engineering Research Council of Canada.

REFERENCES

1. R.C.C. Leite, R.S. Moore, J.R. Whinnery, Appl. Phys. Lett. 5, 141 (1964).
2. J.R. Whinnery, Acc. Chem. Res. 7, 225 (1974).
3. C. Hu, J.R. Whinnery, Appl. Opt. 12, 74 (1973).
4. D.S. Kliger, Acc. Chem. Res. 13, 129, (1980).
5. J.M. Harris and N.J. Dovichi, Anal. Chem. 52, 695A (1980).
6. N.J. Dovichi, CRC Crit. Rev. Anal. Chem., in press.
7. M.E. Long, R.L. Swofford, and A.C. Albrecht, Science 191, 183 (1976).
8. W.B. Jackson, N.M. Amer, A.C. Boccara, and D. Fournier, Appl. Opt. 20, 1333 (1981).
9. N.J. Dovichi, T.G. Nolan and W.A. Weimer, Anal. Chem. 56, 1700 (1984).
10. T.G. Nolan, W.A. Weimer, and N.J. Dovichi, Anal. Chem. 56, 1704 (1984).
11. T.G. Nolan and N.J. Dovichi, IEEE Circuits and Devices Magazine, 2 54 (1986).
12. T.G. Nolan, B.K. Hart, and N.J. Dovichi, Anal. Chem. 57, 2703 (1985).
13. T.G. Nolan and N.J. Dovichi, submitted.
14. W.A. Weimer and N.J. Dovichi, Appl. Opt. 24, 2981 (1985).
15. W.A. Weimer and N.J. Dovichi, J. Appl. Phys. 59, 225 (1986).
16. J.C. Murphy and L.C. Aamodt, J. Appl. Phys. 51, 4580 (1980).
17. D.S. Burgi, T.G. Nolan, J.A. Risfelt, and N.J. Dovichi, Opt. Engin. 23, 756 (1984).
18. D.S. Burgi, W.A. Weimer, T.G. Nolan, and N.J. Dovichi, American Institute of Physics Conference Proceedings 146, 664 (1986).

FOLK WISDOM IN OPTICAL DESIGN

Anthony E. Smart
SPECTRON, Costa Mesa, Ca. 92626

ABSTRACT

This paper augments text-book design rules with lessons learned from building optical systems that actually meet specifications. Implementing optical systems requires more than a knowledge of optics and use of contemporary design codes. Achievement of excellence also requires a kind of "street" knowledge, learned not from textbooks but from experience, indeed, often through failures. These hard-won lessons are here colloquially described as "folk wisdom". Valuable experience is often hidden in traditional methodology. Sometimes such hints may be disregarded in favor of apparent improvements without serious consequences; nevertheless, raising a question before proceeding with what appears to be an obvious improvement may prevent later problems. This paper outlines techniques which, if borne in mind during the system conceptualization and design phases, can improve efficiency, effectiveness and ultimate performance. Costs, mechanical constraints, parts choice, specification and procurement, available adjustments, alignment sequence and retention, deterioration of performance with aging, and others are considered.

INTRODUCTION

These notes and tables may be useful in the design of optical systems and their necessary mechanical support structures. While minimal functionality may sometimes be achieved by competent application of textbook optical and mechanical design, excellence depends also on the intelligent applications of lessons from experience. Some of the comments here were learned painfully, and extraordinarily slowly, by designing and fielding many laser based instruments of various types and operating principles, over more than two decades.
The lessons also apply to coherent and incoherent imaging and holographic systems, radiation sources and detectors, active and passive devices, and many other components and systems. Whether the application be at the esoteric edges of research or an industrial device that must be simple, reliable and cheap enough to be profitable, careful analysis and optimization always pays off.
Advantages accrue from an early emphasis on systems analysis, with careful verification of principle, specification of performance, cost, interfaces, testing, and anticipated reliability. It is also beneficial to retain a sensitivity to partly remembered experiences, frequently available via the medium of "intuition".
The concepts and caveats described here are probably available elsewhere, spread thinly over publications and years. In the following tables are some considerations and mnemonics that may prevent some problems and offer possible solutions to others.

CREATING AN OPTICAL SYSTEM

There are many ways to obtain a working optical system. Table I suggests a possible sequence of actions that may lead to such a realization. By "optical system" here I mean not only the optical components but also the mechanical supporting structures, alignment mechanisms, and even the documentation that specifies the system and permits testing and qualification.

<u>Table I</u>   <u>Approach to a Successful Optical System</u>

1] Obtain and agree all necessary detailed specifications for performance and interfacing. Acknowledge that a modern optical system will probably involve not only optical components and mechanical design, but also electronics, computer hardware and software and user communication. Start at system level.

2] Specify the physical enclosure, its external and internal environments and and all necessary services.

3] Lay out a preliminary optical design, bearing in mind that standard optics are always to be preferred.

4] Choose appropriate methods of component mounting and adjustment and write an alignment algorithm which permits performance verification.

5] Produce an Interface Control Document with all system interfaces and connectivity accurately specified. Assure that this includes functional and logical architecture as well as hardware.

6] Produce a full scale General Assembly Drawing for all optics and mechanical components.

7] Confirm parts availability, specifications, quality, price, delivery, second sources, etc.

8] Procure parts, assemble, test, and document adequately.

Use Gedanken "thought" experiments to simulate your proposed system and what you will be doing while using it. These may suggest additional adjustments, identify necessary design changes, recommend special tools, improve calibration methods, etc.

While these steps may appear to be a logical sequence it is often necessary or desirable to loop back for verification, forced changes, correction of errors, etc. Certain activities, such as documentation and acceptance tests, permeate the entire process.

At preselected intervals in the schedule for the above process peers and colleagues should be involved in design reviews. Find the most critical detractor, who thinks it cannot be done, and listen carefully to his objections. He will help to identify obstacles while it is still early enough to surmount them.

Leave the ego outside the design review. Think of participants as "devil's advocates" rather than adversaries. Supplementary reviews handle any matters where there exist doubt, confusion, or difference of opinion. Any such must be pursued until agreement is reached: it will take less time and cost less in the long run.

For organizational, as well as technical, purposes a good project plan is required. This need not be elaborate but will usually expose any potential problem areas that may thus be accorded further attention in time for solutions to be found.

Figure 1 shows a generic experiment with a morphology typical of many active optical systems. The major interest here is in the stimulus, implying irradiation, and the response, deriving from properties of the examined entity. Although it is essential to treat these factors in the context of a whole system, here I shall discuss only the stimulus and response. A passive system is a subset of this, with the stimulus omitted, the response being replaced by a natural property of the system.

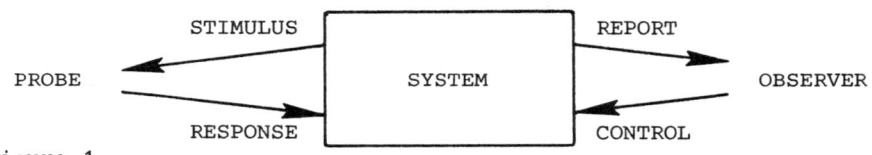

Figure 1

Table II shows some practical considerations for stimulus generation while Table III is equivalent for response. Many items in Table II are also relevant to Table III.

Table II Considerations of Optical Stimulus Generation

PROBE GEOMETRY	[Size, 3-D shape, Beam Control]
WAVELENGTH (RANGE)	[Broad Band, Monochromatic, Coherent]
SOURCE(S)	[Thermal, Laser, Flash]
INTRINSIC LUMINANCE	[Lagrange Invariant]
COHERENCE	[Source Properties]
SPATIAL	[Transverse Mode Structure]
TEMPORAL	[Resonator Stability]
POLARIZATION	["S" or "P"; Ellipticity]
ABERRATIONS	[Seidel, Chromatic, Higher Orders]
TRANSMISSION EFFICIENCY	[Materials, Components]
ALIGNMENT	[Mechanical Design]
EASE OF SET UP	[Adjustment Algorithm]
STABILITY	[Locking]
COMPONENT QUALITY	[Quality Specification]
INTEGRITY	[Testing]
COATINGS	[Efficiency]
TEST ENVIRONMENT	[Wavefront Degradation, Blooming]
	[Absorption]
	[Non-linear Effects]

Table III   Considerations of Optical Response

```
COLLECTION OPTICS    [ Pointing, Aperture, Transmission ]
STOPS                [ Placement (Field, Aperture, Hybrid) ]
FILTERS              [ Wavelength, Bandwidth, Tuning ]
DETECTORS            [ Type (PMT, APD, Others) ]
                     [ Quantum Efficiency, Gain ]
                     [ Noise (Quantum Statistics, Amplifier) ]
```

These considerations may be applied to optical components of all types, including fiber optics and special purpose devices.

ENVIRONMENT

Three levels of environment may be considered, the local conditions of the optical system, the extensive properties of the phenomena under examination, and the environmental constraints of the test facility. These extend to such physical trivia as, "Is the rig window transparent at the wavelength of interest?", "Does the building have a floor upon which to place the optical system?", "Are the necessary services available?", "Will your lenses melt?" or "How far away must I be to avoid the impact crater?". These may be thought humorous only until they happen without having been anticipated.

Table IV considers factors necessary for successful operation in the environment under which the optical system must work. Not only must the equipment survive, it must continue to perform to specification. The operating environment of a state-of-the-art optical system is rarely as benign as that of the laboratory, and many factors that cause deterioration of ultimate performance go unheeded until final installation.

Table IV   Environmental Considerations

```
MECHANICAL
    STATIC STRESSES        [Mechanical Strength, Materials]
    VIBRATION              [Rigidity, Construction]
    WEIGHT AND SIZE        [Small is Better, c.f. Cost Constraints]
THERMAL
    COMPONENT BEHAVIOR     [Temperature Sensitivity, Property Changes]
    LOCAL VARIATION        [Differential Misalignments, Distortions]
    EXTREMES               [Thermal Controls, Insulation]
RADIATION
    SPURIOUS LIGHT         [Stops, Filters, Surface Treatments]
    COMPONENT DAMAGE       [Protection Mechanisms]
AERO-OPTICS
    INHOMOGENEITY          [Refractive Distortions of Wavefront]
ELECTROMAGNETIC NOISE
    SENSITIVITY TO         [Proper Grounding, Hardening, RFI Screening]
    CREATION OF            [Screening, Isolation, Power Supplies]
```

Table V is a group of observations that I have been careless enough to disregard on one or more occasions. Always I received the inevitable retribution.

Table V   Proverbs and Aphorisms

Planning

* Demand and obtain exact specifications in writing.
* Visit the site and talk to the people who work there.
* Plan the system architecture first.
* Handle the intractable jobs at the beginning.
* Find more than one solution in critical problem areas.
* Avoid innovation where conventional solutions exist.
* Spend more resources where your understanding is less.
* Do not "sub-optimize": spread tolerances uniformly.

Design and Procurement

* Design the documentation while you design the system.
* Make a model, mathematical, computational or physical.
* Design the assembly and alignment sequence early on.
* Adjustments should be the minimum possible set, orthogonal, lockable, and with proper sensitivity.
* Everything mechanical is made of India-rubber.
* Minimize:
    numbers of non-standard or special parts
    size and complexity of parts
    parts
* Write specifications for what you want.
* If someone already makes it, buy it from them.
* Reinvented wheels often end up being square (triangular?).
* Buy the best and cry only once.
* Order the long lead items first.
* Second source critical items.

Testing

* Test all hardware as soon as it exists.
* Do not assume catalog items meet their specifications.
* Allow time to obtain replacements.
* Test at all stages; raw material, parts, subassemblies.
* Test the completed system exhaustively.
* Test it some more.
* Quantify performance to a little beyond that required.

Quality Control

* Quality is satisfying the customer.
* Rule 1: The customer is always right.
* Rule 2: If the customer ever appears not to be right, refer to Rule 1.

CONCLUSIONS

It is impossible to itemize all the various things that must be considered in the design and construction of any original optical system. What may be established is that all the obvious problems areas have been anticipated and thought about. Finding solutions to problems before they become serious makes even difficult programs proceed acceptably. Having the courage and dedication actually to think about all the facets of a complex system at the planning stage takes a great deal of mental discipline: it also avoids later pain. As with many properties of folk wisdom it is not always the most obvious nor apparently logical approach that leads to the best solution. It appears not to be well learned from written texts but more effectively by experience and apprenticeship. Table VI, a list of homilies, may make the difference between adequacy and excellence.

Table VI   Obvious Homilies

Intend the excellence of your design.
Study how others have approached or solved the problem.
Test the adequacy of this for your present purposes.
Understand why something was done the way it was done,
   before you commit to a different method.
Seek advice, comments, recommendations.
Consider them carefully.
Disregard those that don't feel right.
Never trust opinion, however informed, without evidence.
If you are being brilliantly innovative think about it harder,
   and get more people to check it.
Question everything.
Analyze and understand the answers.
Test every assumption.
Historical reasons are sometimes only unexamined prejudice.
Historical reasons are sometimes wise cautionary guidelines.
Distinguish between intelligent compromises and short-cuts.

In summary, to make an optical instrument, system or subsystem there are many things to be considered. The exercise of this analytic function, followed by synthesis of the best ideas, is the most probable way of achieving acceptable performance. The considered facets should be selected not only from standard scientific and engineering texts, but from the pool of experience available personally and from colleagues, anecdotal observations, historical preferences. Experience of making mistakes may not be the only way to learn but it is certainly effective.

PLANNING AND IMPLEMENTING A FORMAL TEST PROGRAM FOR SPACEFLIGHT INSTRUMENTS

Lemuel E. Mauldin, III
NASA Langley Research Center, Hampton, Virginia 23665-5225

ABSTRACT

Effective test management is essential in flight hardware development. The Test Engineer faces the challenging task of developing management tools uniquely suited to the organization and hardware to be tested. The Halogen Occultation Experiment (HALOE) project at the NASA Langley Research Center set up and successfully implemented an innovative management system for testing the HALOE instrument, an electro-optic remote sensor for the Upper Atmospheric Research Satellite. This paper presents the basic elements of that system. Although the details given may not be specifically applicable, they will provide the Test Engineer with helpful, general guidelines.

INTRODUCTION

The primary goal of a flight test program is to demonstrate that the instrument can meet its performance requirements under flight conditions. Planning must begin in the conceptual design phase at both the system level and the component level to meet this goal. As trade-off studies progress toward a design concept, the Test Engineer must work with the subsystem and reliability engineers to determine the level of testing required to evaluate component and subsystem performance. Since each new flight instrument will inevitably contain new or unproven technology, he must develop the necessary flight qualification test programs.

The Test Engineer must also work with system engineers to insure the fully assembled instrument can be tested adequately. Once in orbit, the only link to instrument performance is through its flight telemetry data. Therefore, use of internal test points should be minimized. If critical data becomes inaccessible due to the limited flight telemetry channels available, then this data must be correlated early in the systems test program with flight telemetry data.

In general, the Test Engineer must steer instrument design from the earliest stages toward hardware that can be tested on the ground and evaluated in orbit. In the discussion that follows, a detailed test planning cycle is given. This is followed by an effective system for implementing the test program at the system level. These techniques, which are outlined below, proved to be very effective in managing the HALOE test program.

1. Compile Test Requirements Document (TRD)
   Contains all component, subsystem, systems test requirements
2. Prepare Test Flow Chart (TFC)
   Must be responsive to TRD
3. Prepare Integrated Test Plan (ITP)
   Outlines each test in TFC
4. Conduct Test Coordination Meetings (TCM) for Each Test
   Assign Test Conductor
   Assign test preparation activities in Action Item List (AIL)
5. Conduct Test Readiness Review (TRR) for Each Test
   Close out AIL assignments

© American Institute of Physics 1987

6. Conduct Tests in Sequence given in TFC
   Document test anomalies with Test Anomaly Sheet (TAS)
   Transfer flight hardware anomalies from TAS to Test Anomaly Log (TAL
   Maintain Instrument Journal
7. Analyze Test Data
   Disseminate test results in Test Data Review (TDR)
   Assign action items using AIL to resolve test anomalies
8. Prepare Test Report
   Compare test results to test requirements
   Discuss test anomalies
   Conclusions (recommendations for retest, etc.)

PLANNING THE TEST PROGRAM

The first 3 steps are used for test planning. The initial step is to prepare a Test Requirements Document (TRD), which is a complete set of all part, subsystem, and system level tests to be performed. Test requirements come from many sources, including the instrument performance specification; flight environment specifications; subsystem and system engineering design analyses; part, subsystem, and system reliability analyses; and spacecraft interface specifications. For example, it includes all part screening, life, and flight qualification tests; all subsystem performance and environmental tests; and all system characterization, calibration, performance, and flight acceptance tests.

Compiling this TRD mandates a systems approach toward testing and resolves many issues such as conflicting subsystem specifications before they become problems. The TRD is a working document that serves as a common denominator for the multidisciplinary engineering team and valuable reference for the Test Engineer. It is placed under a document control system and is revised and reissued as the instrument evolves and test requirements change.

From the TRD and with the help of subsystem and system engineers, the Test Engineer prepares a Test Flow Chart (TFC). This is the second important step in planning the flight test program. The TFC (see reference 1) is a network grouping test requirements from the TRD and showing all serial and parallel testing, beginning with component testing and progressing through subsystem testing, system testing and flight acceptance testing.

Tests can be grouped many different ways, depending on what efficiency constraints exist within the test program. Usually, they are grouped by test setup so that the number of test setups are minimized. Other constraints that may reorder the progression are availibility of test personnel, test equipment, and test facilities. Thus, the flow chosen should be uniquely suited for a given instrument and test organization. Several project scrubbings are usually required to resolve gaps and overlaps and determine the most efficient progression.

Once the logic is established, the tests grouped into a single block are listed below the block as shown in the figures of reference 1. These later may become major paragraph headings in the test procedure. One can easily determine from the TFC what test procedures and test setups are required.

After the test sequence is established, test time durations are determined. The time durations should reflect only serial test constraints. For example, procedure preparation and test setup development normally occur in parallel with other serial test activities are are not reflected in the test durations. However, the Test Engineer is smart to provide some contingency in the test durations to cover the unexpected, especially for first time tests. Data evaluation may be either a serial or parallel path, depending on whether or not it is a constraint to starting the next test.

Test durations are best determined by the key test personnel and should be success oriented. Problems inevitably occur, but the project master schedule (which is based on a risk assessment of the whole program) should contain flexibility (such as a contingency budget) to resolve problems. The TFC should not be used for a project schedule. Adding non-test activities unnecessarily complicates the chart and defeats its overall purpose as a planning tool.

The TFC becomes the master outline for writing an Integrated Test Plan (ITP), which is the third important step in planning a flight test program. The HALOE project ITP is given in reference 1. With the help of key project personnel, the Test Engineer describes each test in the ITP in sufficient detail to allow test setup development and test procedure preparation to begin. Major headings in the ITP are General; Flight Hardware, Facility, and Test Equipment Description; Component Testing; Subsystem Testing; Subsystem Integration Testing; System Performance and Characterization Testing; Flight Acceptance Testing; and Spacecraft Integration Testing.

In the General section, the test organization is given and the responsibilities of the key test personnel are defined (for example, the role of the Test Conductor is given). Guidelines for preparing the test procedure and test report are also given. Quality assurance activities such as how to resolve test anomalies and hardware failures are described. In addition, the document control system requirements and hardware configuration requirements are given. Finally, the TFC is given and a matrix which cross references each test with the TRD is shown.

The next section of the ITP basically describes the flight hardware, major items of ground test equipment, and test facilities. Cross references provide details of test equipment and test facility requirements for each test. If major fixtures are being developed for the test program, these are also included.

Following the hardware description are paragraphs that cover each test shown in the TFC. For each test, the test objective, outline for the test procedure, proposed test setup, and data analysis requirements are presented. The most important of these is the test objective, which is a concise statement of what the test is to accomplish. If it cannot be stated clearly, the test planning cycle and test that follows is doomed to problems because of misinterpretation.

The ITP is another working document that becomes an important reference for the remainder of the test program. It is maintained as a project controlled document, and is revised and reissued if major changes in the test program occur. Issuing the ITP essentially completes the test planning cycle, and starts the test implementation cycle.

IMPLEMENTING THE TEST PROGRAM

Because of the complex, multidisciplinary nature of testing flight hardware, the Test Engineer has to efficiently organize test setup development, test procedure preparation, testing, test data analysis and evaluation, and test report preparation for each test. If these activities are organized efficiently, only the test and test data evaluation (in some cases) appear on the master project schedule as serial activities leading to spacecraft integration. This means all the other activities must be planned as parallel activities. Since the same personnel may be involved simultaneously in various stages of test preparation for several tests, the Test Engineer must have a system for effective coordination of the test team.

A system with a proven success consisting of three important meetings is presented. These meetings are (1) Test Coordination Meeting (TCM), (2) Test Readiness Review (TRR), and (3) Test Data Review (TDR). Generally, this level of formality is not required until after instrument subsystems are integrated. Subsystem testing usually does not require the level of detailed coordination as systems testing.

Test Coordination Meetings are held for each test early in the flight hardware program to allow time for test setup development, fixture design and fabrication, and test procedure preparation. These meetings, which are held in the sequence given in the TFC, are attended by key test personnel. Each meeting is held with the common agenda shown below.

<u>AGENDA TOPICS</u>

Test Requirements
Test Procedure
Flight Hardware Configuration
Test Setup Development
Fixtures, Test Equipment, and Electrical Cables
Facility
Data System Development
Data Processing and Analysis
Safety
Test Personnel Requirements

A Test Conductor is assigned overall responsibility for the test, and as each agenda topic is discussed, the Test Engineer divides the test preparation into work packages which are assigned during the meeting to key test personnel. In addition, the Test Engineer records major points of discussion. Usually, primary emphasis is placed on test objectives, test requirements and test setup. At the conclusion of this meeting, the Test Engineer prepares minutes of the meeting with an attached Action Item List (AIL) which assigns the work packages.

The AIL is maintained with a software file management system such as DBASE. The files are setup to cross reference the work packages to test, due date, responsible person, and agenda topic. After this system has been used to plan several tests, the data base becomes quite large. However, with the aid of the computer file management system, the Test Engineer sorts the information and gives each person his work package assignments. This helps both the individual and Test Engineer track progress and assign priorities.

Reports are generated weekly during critical phases of the test program. With this data base, the Test Engineer also generates reports on fixture design and fabrication progress, test procedure preparation progress, test setup development progress, etc., which are important test management aids.

For most tests, only one TCM is held. However, for some tests (such as thermal-vacuum) which require coordination of many interrelated activities over a long period of time, monthly meetings are held.

Test setup development and test procedure preparation activities are coordinated by the Test Conductor. If test setups require research and development, this must be anticipated very early in the instrument helps both the individual and Test Engineer track progress and assign priorities.

Reports are generated weekly during critical phases of the test program. With this data base, the Test Engineer also generates reports on fixture design and fabrication progress, test procedure preparation progress, test setup development progress, etc., which are important test management aids.

For most tests, only one TCM is held. However, for some tests (such as thermal-vacuum) which require coordination of many interrelated activities over a long period of time, monthly meetings are held.

Test setup development and test procedure preparation activities are coordinated by the Test Conductor. If test setups require research and development, this must be anticipated very early in the instrument development program to prevent these activities from impacting the overall instrument development schedule. If possible, tests should use established techniques to reduce program risk. Nonstandard or new setups requiring analysis and design should have documentation (in the form of engineering reports) that substantiates their ability to meet the test requirement criteria in the TRD.

The test procedure is a project controlled document that specifies the step-by-step use of the ground test equipment, facility, and test setup to provide the necessary data for comparing instrument performance to the test requirements. A detailed description of test procedure development is beyond the scope of this paper, but some of the salient features will be discussed. In addition to the step-by-step procedure, it includes the test objective; test requirement criteria; hardware configuration; test facility, equipment, and personnel requirements; precautious to be used for hardware and personnel safety; and test data analysis requirements. It must be approved by the Test Conductor, Test Engineer, Quality Engineer, and all key subsystem and system engineers prior to beginning the test. It provides a detailed documentation of test conditions and the flight hardware configuration so that the test data can be analyzed from different perspectives to help resolve hardware non-conformances or test data anomalies that occur much later in the test program.

After development of the test setup, preparation of the test procedure, and completion of all work packages necessary to perform the test, the Test Readiness Review is held. Normally, this review is held 5 days before starting the test for the purpose of closing out all the assigned work packages in the AIL and authorizing the Test Conductor to proceed.

The agenda for the TRR is the same as that used for the TCM, since last minute additions or changes may be required. The Test Engineer reviews results of each work package with the key individuals. If new work packages are assigned (or old ones modified) these are entered into the computer AIL data base. The Test Engineer again writes and distributes minutes of the meeting with open items listed on the attached AIL.

Emphasis on agenda topics during the TRR is usually placed on test equipment, fixtures, and cables. Experience has shown that the most common cause of a test delay is not having the proper electrical cables. The TRR is a valuable tool for sorting out test setup details necessary to perform the test.

After all work packages assigned in the AIL have been completed, the Test Conductor is ready to start the test. If the test preparation has been efficiently planned, the test will proceed smoothly. Since the procedure writer cannot always anticipate all details needed by the Test Conductor, changes (redlines) to the procedure are often required. Most changes will consist of minor corrections or procedural changes that do not affect the substance of the test. These changes are made without interrupting the test and are approved by the initials of the Test Conductor and Quality Engineer. The Test Engineer files the redlined test procedure after the test for future use for the "as run" version of the test. The procedure does not need to be retyped unless it will be used again.

Tests are occasionally snagged by a setup that does not deliver the necessary test requirements. Most organizations have a Quality Assurance Plan in effect to resolve this type of problem. Also, if hardware failures occur, these are resolved by the criteria in the Quality Assurance Plan. However, the Test Engineer is wise to have his own system for tracking test anomalies which are caused by either the test setup or hardware under test.

One system that works efficiently is a loose leaf notebook maintained by the Test Conductor with one page Test Anomaly Sheets (TAS). The TAS has sections for describing the anomaly, immediate action taken (if any), and recommendations to prevent future occurances. The Test Engineer reviews the TAS notebook periodically to coordinate resolution of test setup, equipment, or facility related testing problems.

If the anomaly is related to the flight hardware, it is also entered in the Test Anomaly Log (TAL) computer data base. The TAL, which contains title of test, description of anomaly, status (open or closed), and referral (to a nonconformance of failure report), is used by the Test Engineer to coordinate resolution of flight hardware anomalies.

Another important notebook used in the test program is the Instrument Journal. Each activity involving flight hardware must be documented in this journal. In many cases, the details are covered adequately in the test procedure and TAS notebook, so entries in the journal can be general and reference this documentation. However, activities not adequately covered in other documentation (such as instrument moves between facilities) should be described in detail in the journal.

Most systems flight test programs have a performance and characterization test phase, followed by an environmental (flight acceptance) test phase, followed by a spacecraft integration test phase. The performance and characterization test program will consist of several procedures. The Test Engineer reviews these procedures with the Principal Investigator to determine which parameters will be tested immediately before and after the flight environments. These tests are then represented by paragraphs in the Performance Verification Test (PVT) procedure, which will be the procedure used for all testing during the environmental test phase. The HALOE project PVT procedure is given in reference 2.

In addition to repeating critical performance and characterization tests, the PVT also contains a simulated orbit test. Here, the instrument is tested through the same operating mode sequence that the instrument sees in orbit. In many cases, performance tests can be combined with the simulated orbit test. Critical data from the PVT should be trended after each test. The trended data may indicate problems not detected in an individual test.

The Test Engineer should write the PVT procedure in a modular form with each test performed in a stand alone paragraph for three important reasons: (1) only one procedure needs to be written, (2) each test is performed several times under identical conditions which creates a large data base for trend analyses, and (3) the procedure can be adapted to fit the needs of a particular test.

This technique also allows the Test Engineer to perform different levels of testing with the same procedure. For example, a Level I test may include all the paragraphs, and is used only at the beginning and end of the environmental test program. A Level II test is less complete, takes less time to run, and is used before and after each flight environment. A Level III test is a quick test of instrument health and is run before and after each instrument move between facilities.

The PVT procedure can also be adapted to serve as the spacecraft integration test procedure. The major problem in writing a modular procedure of this type is initializing the instrument to be ready for several types of testing and test setups. However, this disadvantage is small when compared to the advantages.

Immediately after a test, the Test Data Review (TDR) is held. This meeting is the third important meeting for a given test, should be attended by the Principle Investigator and key members of the science staff. The Test Conductor reviews the test objective and compares test results to test requirements. Entries to the TAL are discussed, and the Test Engineer concludes the meeting with a summary of follow-up activities to be performed. This may include recommendations for some retesting or additional data analysis. All activities necessary to close-out the test are assigned to key individuals using the AIL.

If test anomalies were referred to nonconformance or failure reports, testing may have to stop until these are dispositioned. In many cases, workarounds can be found to minimize the schedule impact of resolving test anomalies.

Finally, a Test Report is prepared by the Test Conductor. Discussed in this report are (1) comparison of test data analysis results to test requirements, (2) test anomalies, nonconformances, and failures (for flight hardware and test equipment, setup, and facility), (3) recommendations, and (4) conclusion as to whether the test fulfilled the test objective.

A file is kept for each test that contains (1) the "as run" test procedure, (2) the TCM minutes, (3) the TRR minutes, (4) the associated AIL, and (5) the test report. If test setup engineering reports are generated, these are also contained in the test file.

For the test program presented above to be effective, the Test Engineer must be skilled at writing and holding meetings. Writing the formal and informal documents will require much of the Test Engineer's time. An assistant that writes efficiently helps to solve this problem.

Another large percentage of time is spent in meetings. Meetings can become very wasteful of project manpower if they are not held efficiently. Keys to effective meetings are (1) invite only the key people associated with the topic (2) use the agendas that were given and steer discussion away from unrelated topics, (3) prepare carefully for each agenda item and anticipate problems that may surface, (4) hold splinter sessions to resolve time consuming issues when only two or three of the attendees are involved, and (5) issue the minutes with the attached Action Item List within one day of the meeting. Meetings can be held in less than one hour using these techniques.

A different type of meeting from the ones mentioned previously is also a key element in the test management system. The Test Engineer conducts a 15 minute meeting at the beginning of each workday to outline for the test team the activities for the day. This meeting is an effective method for generating enthusiasm among the test team to accomplish the daily objectives. The Test Engineer also uses this meeting to coordinate activities for the next two or three days. However, it should not be used for long range planning since it must be conducted in 15 minutes or less to be an effective management tool.

Another effective technique the Test Engineer can use is to divide the test group into teams. One team can be conducting a test while another team is analyzing data from the last test and preparing for the next test.

Teamwork, of course, is required when more than one shift per day is used for testing. The Test Engineer should avoid using more than one shift if possible. This tends to stiffle enthusiasm and wear down the test team over a long period of time. However, some tests (such as a thermal-vacuum test) require 24 hour monitoring and shift work is unavoidable.

CONCLUSIONS

The key elements of a successful test management system used at the NASA Langley Research Center for the HALOE project were described. These concepts may not be new to successful aerospace companies that have been testing flight hardware. However, research organizations that are planning to go into flight hardware development will find the concepts presented to be very effective.

ACKNOWLEDGEMENTS

The HALOE project is managed at the NASA Langley Research Center by J. L. Raper for the Principal Investigator, J. M. Russell. Many thanks go to these individuals and the HALOE test team for their contributions to the success of this program.

REFERENCES

1. "Halogen Occultation Experiment Integrated Test Plan," L. E. Mauldin and A. J. Butterfield, NASA TM-87709, June, 1986.
2. "Halogen Occultation Experiment Performance Verification Test Procedure," L. E. Mauldin, III, NASA TM-87748, June, 1986.

Optical Ray Tracing for Crossed Beam Photothermal Deflection Spectroscopy

JEFFREY A. SELL

General Motors Research Laboratories, Warren, MI 48090

ABSTRACT

A geometrical algorithm for optical ray tracing is used to trace the trajectories of probe laser beam rays in pulsed Photothermal Deflection Spectroscopy (PDS) of a static gas. Rays are traced as a function of various parameters such as time, absorbed pump beam energy, radius, and pulse duration. The exit angle is also computed and compared to the deflection angle calculated analytically, with very good agreement for a pump pulse duration shorter than the thermal diffusion time. For a long pump pulse duration, the analytical expression is inaccurate due to the neglect of thermal diffusion during the pump pulse. Probe beam ray crossing is examined and found to be unimportant for most common PDS situations. The saturation of the PD signal is found for high absorbed pump pulse energy; this is due to excessive heating of the absorber. Finally, the maximum probe beam deflection occurs when the deflection is measured at a time corresponding to the end of the pump pulse duration.

INTRODUCTION

Photothermal Deflection Spectroscopy (PDS) has recently been demonstrated to be a very useful technique for flow diagnostics. Velocity measurements have been made at single points[1-4] and at multiple points[5] for laminar flows. More recently, a transient grating PD method was used for velocity measurements in nominally turbulent flows.[6] In order for this technique to be widely applicable, the optical physics of it must be understood. Currently, this is well beyond our capability. As a first step, however, we report here optical ray tracing for PDS in a static gaseous medium. These results should be applicable directly to concentration measurements by pulsed PDS in an absorption cell. Future calculations will consider the case of fluid flows.

RAY TRACING

We follow the approximate geometrical approach outlined in Klein[7] for optical ray tracing of rays propagating through an inhomogeneous index of refraction:

$$\vec{r}_2 = \frac{n(\vec{r}_1)}{n(\vec{r}_2)} \vec{r}_1 + \frac{\Delta s}{n(\vec{r}_2)} \nabla n. \tag{1}$$

This is the main algorithm used for a numerical computation of the ray trajectory. A FORTRAN program was written based on Equation (1) and used to trace rays through the index gradient formed by the pump laser heated medium. For the present work, a two-dimensional analysis is used where the probe beam is normal to the pump beam.

© American Institute of Physics 1987

RESULTS AND DISCUSSION

Figure 1 shows rays traced which originally are horizontal at $y_0=0.01$ cm, which is the same value as the $\frac{1}{e^2}$ intensity radius of the Gaussian pump beam. The pump pulse is assumed to last for 1 ms; the product of the absorption coefficient, the absorber concentration, and the pulse energy (hereafter called the absorbed pump beam energy) is 3.0×10^{-6}cm^{-1}J. The rays are shown as a function of time for t=0.1, 0.5, 1.0, and 2.0 ms. Initially the rays are deflected upward since the probe is initally above the pump. At t=1 ms, the pump pulse is turned off and the rays relax to the undeflected position at a rate determined by thermal conduction.

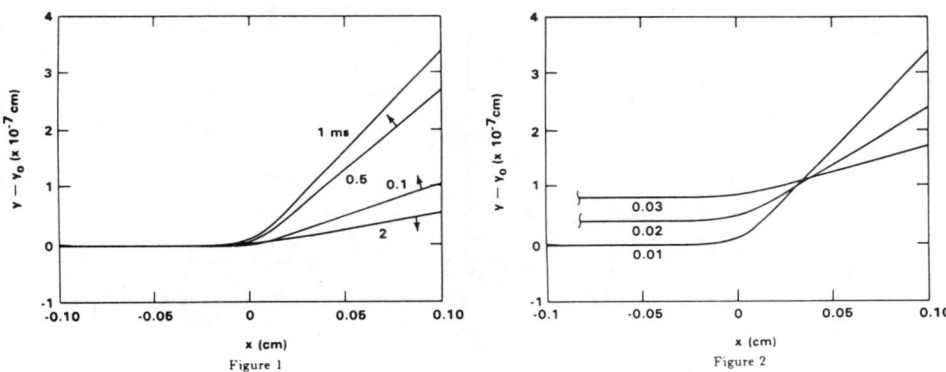

Figure 1   Figure 2

Rays traced for various initial y positions are shown in Figure 2. Here again, all initial rays are assumed to be horizontal. Figure 2 shows evidence of ray crossing as a result of the rays closer to the pump beam being deflected more strongly than those further away from it. The rays initially at $y_0=0.01$ and 0.02 cm would cross at $l=70.42$ m. This distance is much larger than the distance from the interaction region to the detector in most PDS experiments. If the conditions here (an absorbed pump beam energy of 3×10^{-6} cm^{-1}J and a pump beam waist of 0.01 cm) are considered as typical for PDS, then we conclude that ray crossing may not seriously complicate many PDS experiments. Also, position sensitive detectors are often used and such detectors typically measure the beam centroid position which would be be insensitive to ray crossing.

One of the outputs of the ray tracing program is the exit angle of the ray. This can be compared to the deflection angle computed from an analytical expression (Equation (5) of ref(2)). Both are plotted as a function of time in Figure 3 which shows the exit angle as a function of time up to 1 ms; the pump pulse duration is 1 μs, the absorbed pump energy is 3.0×10^{-6}cm^{-1}J , and the pump radius is 0.01 cm. The deflection angle computed from the analytical expression, multiplied by 2, is indistinguishable from the exit angle. (Sontag and Tam[2] apparently dropped a factor of 2 in the derivation of their Equation(5)). At t=1 ms, the agreement between the calculated exit angle and the deflection angle is within 1.5 %.

Figure 4 shows the exit angle as a function of absorbed pump beam energy up to 3.0×10^{-4} cm^{-1}J. The conditions here are $y_0 = w_0 = 0.01$ cm, and t=1 ms. Here, the exit angle calculated from ray tracing is significantly lower than the calculated deflection angle using Equation (5) of ref(2). The reason for this is at high absorber concentration, the temperature of the gas elevates significantly so that the fractional index change is smaller which reduces the deflection angle.[3,5] This is an inherent feature of the PD technique and can be thought of as saturation. The condition of high absorbed pump beam energy can occur for high absorber concentration, high absorption strength, or high pump pulse energy.

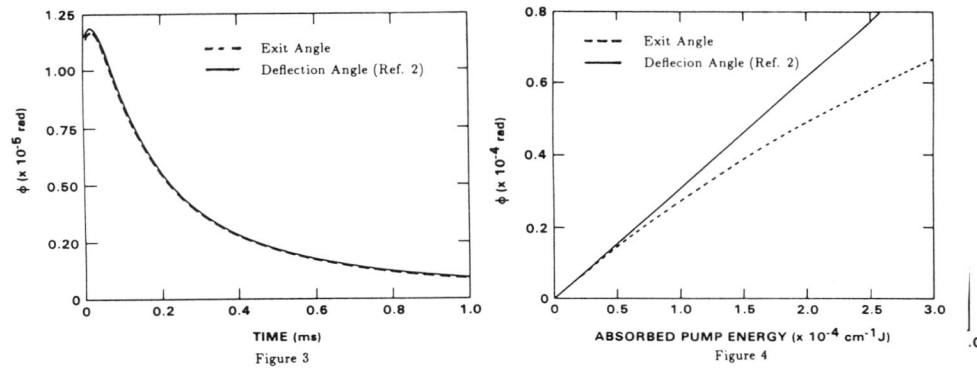

Figure 3

Figure 4

CONCLUSIONS

Optical ray tracing is very useful for visualizing the deflection of probe laser beam rays in PDS as a function of various parameters such as time, absorbed pump beam energy, radius, and duration. The exit angle computed from ray tracing agrees very well with that computed by a previously known analytical expression for a pump pulse duration shorter than the thermal diffusion time. For a long pulse duration, the analytical expression is inaccurate because it does not allow for thermal diffusion while the pump pulse is still on. The conditions for which saturation of the PD signal occurs are identified as high absorber concentration and/or high pump pulse energy. Saturation is due to excessive heating of the absorber. The probe beam deflection for a short duration pump pulse is only weakly dependent on the pump pulse duration. For a long pulse, the maximum beam deflection occurs at the end of the pump pulse (for probe rays close to the pump beam).

REFERENCES

1. W.A. Weimer and N.J. Dovichi, **Applied Optics, 24**, 2981 (1985).
2. H. Sontag and A. Tam, **Optics Letters, 10**, 436 (1985).
3. J.A. Sell, **Applied Optics, 23**, 1586 (1984).
4. J.A. Sell, **Applied Optics, 24**, 3725 (1985).
5. J.A. Sell and R.J. Cattolica, **Applied Optics, 25**, 1420 (1986).
6. C.J. Dasch and J.A. Sell, **Optics Letters, 11**, 603 (1986).
7. M.V. Klein, *Optics*, Wiley, New York, 1970.

CLASSICAL AND HOLOGRAPHIC INTERFEROMETRY SYSTEMS COMBINED TO DOCUMENT INERTIAL FUSION EXPERIMENTS

J. S. Ankney and Garland E. Busch
KMS Fusion, Inc., P.O. Box 1567, Ann Arbor, MI. 48106

ABSTRACT

An optical system has been developed which is used to document inertial fusion experiments. It records both classical interferograms of solid hydrogen isotope layers within the hollow spherical fuel container prior to the shot and holographic interferograms of the plasma during the implosion. This optical system has the advantages of visible and UV utility, 2-4 µm resolution and a low space requirement. Both systems share a catadioptric objective lens and a refractive relay lens. The visible classical interferometric beam and the UV pulsed holographic probe follow different off-axis paths within the object cone of the lenses. To document the solid layer, the beam is diverted to a shearing cube interferometer using a lens-mirror assembly on a sliding stage. This assembly then slides out of the way before the implosion to admit the UV probe to the holographic camera, which records four holographic interferograms of the plasma separated in time by 40-400 ps. The two operational modes, shearing interferometer and holographic, are serial; switching from one to the other requires only seconds.

INTRODUCTION

Inertial fusion (IF) employs laser beams to implode spherical shells containing deuterium and tritium. Freezing the DT fuel in a uniform layer within the container aids in achieving the densities required for fusion. Classical interferometry can be used to determine the degree of solid layer uniformity in the shell prior to the laser shot.

For a recent series of implosion experiments using cryogenic fuel, we wanted to make a series of holographic interferograms to diagnose the laser-produced plasma. To avoid necessitating changes in the existing holographic system previously used for non-cryogenic targets, the diagnostics for documentation of the cryogenic fuel layer and the holographic system had to share the same pair of opposing ports on the evacuated target chamber.

HOLOGRAPHIC INTERFEROMETRY

The holographic interferometry system employed at KMS[1] is capable of taking four interferograms of the plasma separated in time by 40-400 ps. The holographic probe wavelength is 263 nm (in the ultraviolet) to allow measurements at the highest possible electron density. Resolution at the target is one to two micrometers.

The holographic probe beams originate in the oscillator of the KMSF Chroma laser, operated at 1.053 µm. A pulse is selected and switched out of the oscillator pulse train using a Pockels cell switch. The timing difference between the probe pulse and the main implosion pulse is at this point determined by the interval between pulses in the oscillator cavity, 8 ns. The eventual timing differences between the probe pulses and the main pulse are all due to differences in optical path lengths.

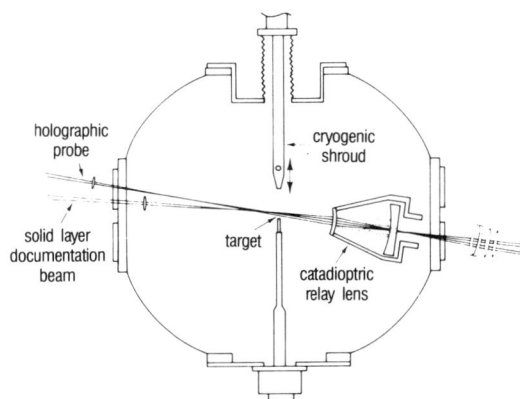

Fig. 1. Cross section of target chamber showing beam paths.

The pulse leaving the oscillator goes to a regenerative amplifier where its pulse width is shortened from 90 to 20-30 ps and its energy is increased to 200-800 J. After 27 passes, the pulse is switched out of the regenerative amplifier to a small rod amplifier operated at a gain of 10-20. After this final amplification, the pulse is frequency doubled to green in a KDP Type II crystal. The pulse then passes through a weak lens and a pinhole filter on the way to an ADP crystal where it is again frequency doubled to the UV.

All sixteen pulses required for the four holographic interferograms are generated from the same UV pulse to assure coherence. A multiplexer, using graded beam splitter coatings, splits this pulse into four pulses of equal energy and uniform spacing (40-400 ps). A double pulse generator then splits each of the four pulses into two reference and two signal pulses. The reference pulses are given an angular separation to produce the separation of holographic images upon reconstruction.

The signal pulses pass through a lens, enter the target chamber (see Figure 1), and traverse the imploding plasma. The plasma is imaged behind the holographic plate using an all-spherical two-element f/2 catadioptric relay lens and a four-element f/4.2 refractive lens assembly. The large collection aperture of the catadioptric relay lens provides a diffraction-limited resolution of 1-2 μm in the UV.

SOLID FUEL LAYER DOCUMENTATION

Interferograms of the gas filled shell and the cryogenic solid layer are formed with a Murty shearing cube interferometer using blue (488 nm) light from an argon ion laser.[2] The CW laser light is focused to a crossover about 5 mm ahead of the target at f/10 to give a diameter of 500 μm (about three target diameters) at the target.

The crossover is imaged onto the fully reflective surfaces of the interferometer; the target is imaged farther downstream, the distance depending on the relay lenses. The shearing cube is adjusted so that the image of the target from one leg of the interferometer is superimposed upon the background light surrounding the target in the beam from the other leg; an interferometric image of the target results. This image is magnified and relayed onto a camera film plane using a microscope objective.

THE TWO SYSTEMS COMBINED

The holographic and classical interferometry systems have been combined to enable both to be used in cryogenic laser fusion experiments while requiring the least amount of port space on the target chamber. The holographic signal pulses and the CW argon beam for solid layer documentation are brought in with an angular separation of about .08 radians (centers 33 mm apart on the 76 mm diameter beam port). Thus they follow different off-axis paths within the object cones of the relay lenses of the holographic system. After passing through a final focusing lens, the CW argon beam is diverted to the interferometer by a mirror. The lens and mirror are on a sliding stage which is moved out of the way prior to a target shot, allowing the UV probe to pass into the holographic camera. The sliding stage allows easy repositioning of the lens and mirror when documentation of the solid layer is again needed.

As well as allowing documentation of the solid fuel layer and holographic interferometry of the implosion on the same shot, combination of the two systems has other advantages. One of them is that, unlike the lens system used in the previous cryogenic campaign, which had to be backed out of the chamber prior to every shot, the catadioptric objective remains stationary throughout the shot. The catadioptric design gives 1-2 m resolution with the first element 115 mm from the target, far enough away to avoid significant damage to the lens.

Another advantage is that, since the holographic probe pulses are designed to arrive during the implosion, the probe is ideal for use in documenting the effects of amplified spontaneous emission (ASE) on the solid fuel layer. To check for effects of ASE, the cryogenic shroud[3] is retracted exactly as for an actual target shot, but the main driver pulse is blocked. A probe pulse, frequency doubled only to green, is used with the solid layer documentation optics to record a classical interferogram of the solid layer at the time that the main driver pulse would normally arrive.

REFERENCES

1. G. E. Busch, C. L. Shepard, L. D. Siebert, and J. A. Tarvin, Rev. Sci. Instrum. **56**, 879 (1985).
2. J. A. Tarvin, D. L. Musinski, T. R. Pattinson, R. D. Sigler, and G. E. Busch, SPIE **192** Interferometry, 239 (1979).
3. D. L. Musinski, T. M. Henderson, R. J. Simms, and T. R. Pattinson, J. Appl. Phys. **51**, 1394 (1980).

This work was supported by the U.S. Department of Energy under Contract DE-AC08-DP-40152.

VISUALIZATION OF RF ACOUSTIC WAVEFRONT BY LASER CORRELATION THEORY

Yang Xuanmin, Yi Ming
Department of Physics, Nanjing University, Nanjing, China

ABSTRACT

We provide a new theory for visualizing the rf traveling acoustic wavefront in real time, with which we have not only obtained a steady visible image of the rf acoustic traveling wavefront--a moving phase object--but have also got rid of some confused concepts of the stroboscopic method.

INTRODUCTION

Direct visual observation of traveling sound wavefront is an important technique for observing the moving phase object and research of acoustics.

For seeing a traveling sound wavefront directly, Toepler first invented a stroboscopic Schlieren method and a central dark field method. The theory is based on geometric optics and the concepts of the stroboscope. By a phase contrast method based on physical optics, the phase changes introduced by the traveling sound wave could be converted into intensity one. But some phenomena can not be explained because they have never considered the temporal spectrum in the spatial spectrum.

Now we have developed a new technique based on the correlation theory and discovered the temporal spectrum in spatial spectrum which is the foundation of filtering technique.

PRINCIPLES

To illustrate the main idea of information correlation theory and technique, let us describe the processor first.

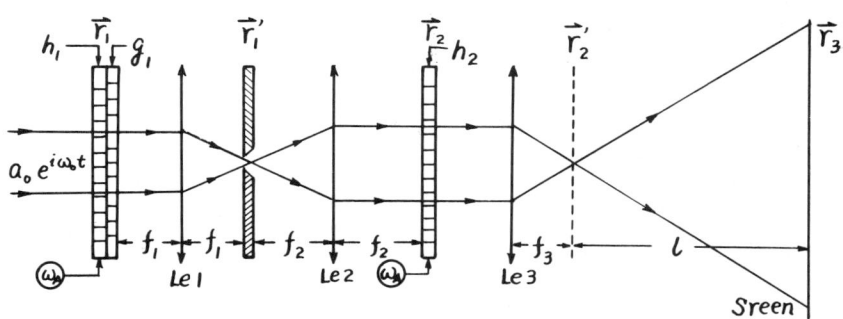

Fig. 1. Optical processor

1. Spatial correlator

Two transparencies, an acouso-optical modulator h_1 and a phase grating g_1 which are placed parallelly in direct contact in plane $\vec{r}_1$, are illuminated by a wide CW laser beam $a_0(\vec{r}_1)\exp(i\omega_0 t)$ as shown in Fig. 1. Lens Le1 Fourier transforms the amplitude transmittance $h_1 \cdot g_1$ onto plane $\vec{r}_1'$, and a stop is inserted to yield the spatial crosscorrelation function of h_1 and g_1 denoted by $S(\omega_A t)$, a double sideband modulated light wave of modulated frequency ω_A.

$$S(\omega_A t) = h(\omega_A t - \vec{k}_A \cdot \vec{r}_1) * g_1(-\vec{k}_A \cdot \vec{r}_1)$$
$$= \exp(i\omega_0 t)[A + B\exp(i\omega_A t) + C\exp(-i\omega_A t)] \quad (1)$$

2. Temporal correlator

Lens Le2 casts a parallel beam of $S(\omega_A t)$ onto the detected rf acoustic traveling wave h_2 in plane $\vec{r}_2$, then lens Le3 images the amplitude tranmittance $S(\omega_A t)\cdot h_2$ onto the detective screen $\vec{r}_3$ to get the temporal crosscorrelation function of $S(\omega_A t)$ and h_2, that is the magnified stationary phase image of rf acoustic traveling wavefront.

$$S(\omega_A t) * h_2(\omega_A t - \vec{k}_A' \cdot \vec{r}_2) = E_2(\vec{k}_A' \cdot \vec{r}_2) = E_3(\vec{k}_A' \cdot \vec{r}_3/M) \quad (2)$$

3. Phase-Amplitude conversion -- Spatial filtering

To convert the phase changed object into intensity one, we discovered that there is temporal spectrum in spatial spectrum at the back focal plane $\vec{r}_2'$ of lens Le3 shown in table, then developed a special filtering technique to get a high contrast image.

The technique is different from the central dark method, Schlieren method and phase contrast method for which the temporal spectrum in spatial spectrum had not been considered when they were used in similar situations.

TABLE Temporal spectrum in spatial spectrum

Diffracted order	Spatial spectrum	Temporal spectrum				
	$\|\vec{r}_2'\|$	ω_0	$\omega_0+\omega_A$	$\omega_0-\omega_A$	$\omega_0+2\omega_A$	$\omega_0-2\omega_A$
m	$= m\lambda f_3/\Lambda_A'$					
1	$\lambda f_3/\Lambda_A'$	$-CJ_1(\beta_3)$	$-AJ_1(\beta_3)$		$-BJ_1(\beta_3)$	
0	0	$AJ_0(\beta_3)$	$BJ_0(\beta_3)$	$CJ_0(\beta_3)$		
-1	$\lambda f_3/\Lambda_A'$	$-BJ_{-1}(\beta_3)$		$-AJ_{-1}(\beta_3)$		$-CJ_{-1}(\beta_3)$

RESULTS OF EXPERIMENTS

The results of rf acoustic traveling wavefront of magnificantion M are shown in photo 1 and 2. Photo 1 shows the case of only one of the transducers driven and photo 2 shows two transducers driven in series and out of phase.

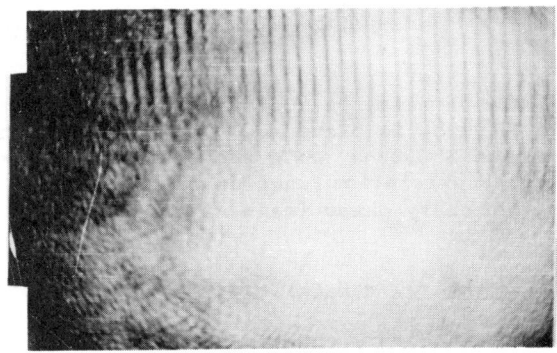

Photo 1 One of the transducers driven

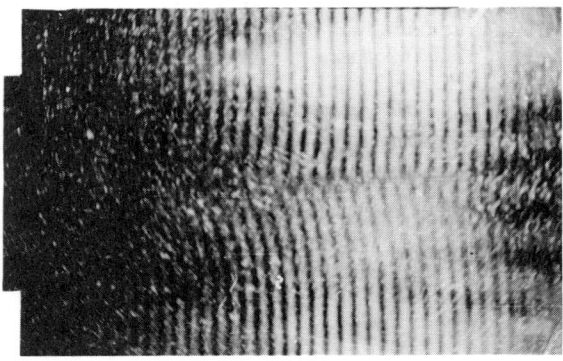

Photo 2 Two transducers driven in series and out of phase

CROSS-CORRELATION THEORY OF GRATINGS AND ITS USE FOR MEASURING THE 2-D MICROVIBRATION BY LASER BEAM

Yang Xuanmin, Yi Ming, Pan Hui
Department of Physics, Nanjing University, Nanjing, China

ABSTRACT

The cross-correlation theory of gratings is discussed and a practical application in industry is suggested.

PRINCIPLES

A moving 2-D phase grating $h(\vec{r}\cdot\vec{V}) = \exp\{j\Delta n_1 kl_1 \sin[\vec{k}_s \cdot (\vec{V}t-\vec{r})]\}$ and a fixed 2-D phase grating $g(\vec{r}) = \exp\{j\Delta n_2 kl_2 \sin[-\vec{k}_s \cdot \vec{r}]\}$ with the same spatial frequencies f_x and f_y are placed parallelly in plane $\vec{r}$. Where Δn_1, Δn_2 are the peak change of refractive index, $k = 2\pi/\lambda$; l_1, l_2 are the thickness of the gratings, $\vec{k}_s = 2\pi f_x \vec{i} + 2\pi f_y \vec{j}$, and $\vec{V} = v_x \vec{i} + v_y \vec{j}$. The gratings are illuminated by a CW laser beam:

$$a(\vec{r})\exp(j\omega_0 t) = \exp(j\omega_0 t)\mathrm{rect}(x/D_1)\mathrm{rect}(y/D_2) \qquad (1)$$

where D_1 and D_2 are the square apertures in x and y direction respectively. It may be seen in Fig.1.

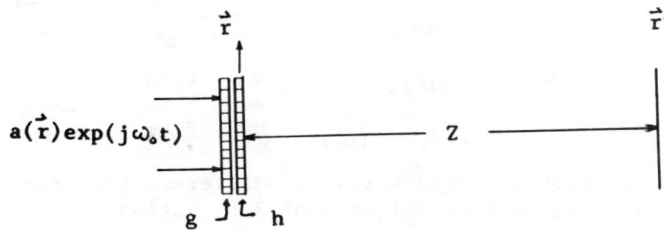

Fig. 1

The optical field distribution at far field plane $\vec{r}'$ would be the Fourier transformation of the transmitted optical field at plane $\vec{r}$. According to the cross-correlation theory it would be the cross-correlation of Fourier transformation of $h(\vec{r},\vec{V})g(\vec{r})$ and $a(\vec{r})\exp(j\omega_0 t)$. Therefore we have

$$f(\vec{r}') = \iint_{-\infty}^{\infty} a(\vec{r})\exp(j\omega_0 t)\exp(jk(\vec{r}\cdot\vec{r}')/Z)\,d\vec{r}$$
$$* \iint_{-\infty}^{\infty} h(\vec{r},\vec{V})g(\vec{r})\exp[jk(\vec{r}\cdot\vec{r}')/Z]\,d\vec{r}$$

$$= \exp(j\omega_o t) \sum_p \sum_q \exp(jp\vec{k}_s \cdot \vec{V}t) J_p(\beta_1) J_q(\beta_2)$$

$$\cdot \mathrm{sinc}\{D_1/\lambda Z[x' - (p+q)f_x\lambda Z]\} \cdot \mathrm{sinc}\{D_2/\lambda Z[y' - (p+q)f_y\lambda Z]\} \quad (2)$$

where $\beta_1 = \Delta n_1 k l_1$, $\beta_2 = \Delta n_2 k l_2$, and by use of the identity:

$$\exp(ja \sin x) = \sum_{p=-\infty}^{\infty} J_p(a) \exp(jpx) \quad (3)$$

There are some peak values at far field plane $\vec{r}'$ for which $x' = (p+q)f_x\lambda Z$ and $y' = (p+q)f_y\lambda Z$. Their frequencies and the amplitude coefficients are listed in Table 1. for a general survey.

Table 1. Frequencies and amplitude coefficients

p	q	p+q	ω	coefficients	position x	y
0	0	0	ω_o	$J_o(\beta_1) J_o(\beta_2)$	0	0
1	-1	0	$\omega_o + \vec{K}_s \cdot \vec{V}t$	$J_1(\beta_1) J_{-1}(\beta_2)$	0	0
-1	1	0	$\omega_o - \vec{K}_s \cdot \vec{V}t$	$J_{-1}(\beta_1) J_1(\beta_2)$	0	0
1	0	1	$\omega_o + 2\pi f_x V_x t$	$J_1(\beta_1) J_o(\beta_2)$	$f_x\lambda z$	0
0	1	1	ω_o	$J_o(\beta_1) J_1(\beta_2)$	$f_x\lambda z$	0
1	0	1	$\omega_o + 2\pi f_y V_y t$	$J_{-1}(\beta_1) J_o(\beta_2)$	0	$f_y\lambda z$
0	1	1	ω_o	$J_o(\beta_1) J_{-1}(\beta_2)$	0	$f_y\lambda z$

In different diffracted orders there are different frequency shift which depends on the moving velocity of 2 D grating $h(\vec{r}, \vec{V})$ at plane r. Table 1 shows that two temporal frequencies at points $(f_x\lambda z, 0)$ and $(0, f_y\lambda z)$ in plane $\vec{r}'$ produce the beating waves which may be expressed by

$$I = I_{oi}[1 + \cos(2\pi f_i v_i t)] \quad i = x, y \quad (4)$$

Measuring the frequency shift by detecting the beating waves of first order diffracted beams in x' and y' direction respectively we can determine the velocity of object which is fixed on the 2-D grating $h(\vec{r}, \vec{V})$.

However, it is not convenient to determine the components of velocity v_x and v_y and the displacements $x(t)$ and $y(t)$ by counting the wave numbers. So we have applied an integral circuit of electronics. Then the composition of 2-D vibration is obtained.

EXPERIMENTS

Two identical 2-D sinusoidal phase gratings of spatial frequencies $f_x = f_y = 100$ lines/mm are placed parallelly. A tone fork be adhered on one of the gratings acts as a vibrating source and vibrates in the direction of ϑ with x axis. The frequency of the tone fork is 260 Hz. The composition of the 2-D vibration is linear which is well desired.

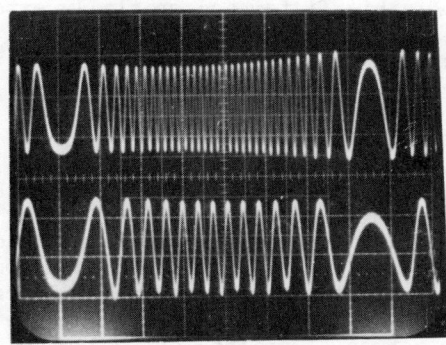

Photo 1

The Displacement Y(t) and X(t) (upper and lower) may be counted by wave numbers

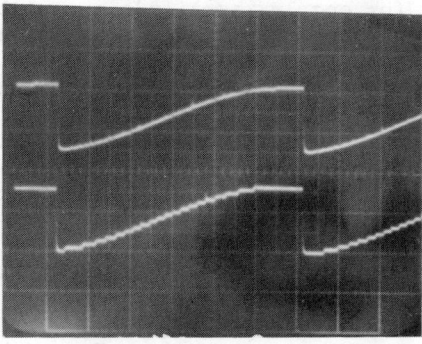

Photo 2

Displacement Y(t) and X(t)$\sim$t Pattern

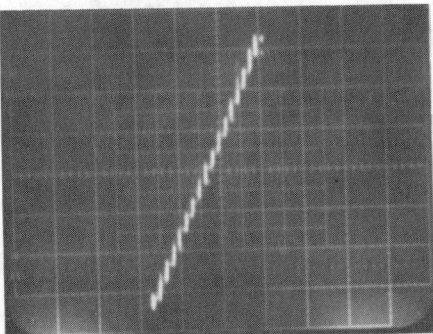

Photo 3

Composition Pattern of 2-D Vibration

A NEW TECHNIQUE FOR PATTERN RECOGNITION USING FRESNEL HOLOGRAM AND EXTENDED SOURCE

G. G. Mu, Z. Q. Wang and D. Q. Chen
(Institute of Modern Optics, Nankai University)

ABSTRACT

A new type of lensless intensity correlator for pattern recognition in which a Fresnel holographic filter and spatially extended band-limited source are used is described, and an experimental result by this technique is given. It is convenient for optoelectric hybrid system to perform real-time pattern recognition.

INTRODUCTION

Signal detection by complex spatial matched filter reported first by Vander Lugt[1] has been applied to various fields, and many successful results have been achieved. There are two main disadvantages: first, the matched filter must be reset at the exact position in Fourier plane of a coherent optical system, and this causes many troubles. Second, the objects to be correlated have to be in the form of a transparency, or a type of spatial optical modulator must be used to accomplish the transformation of incoherent light to coherent which is necessary for complex amplitude correlation[2]. We shall here present a new technique for pattern recognition which uses the Fresnel holographic filter (FHF) and a spatially extended incoherent source. By this technique the in-plane shift of the spatial filter does not influence the correlation result, the allowable out-of-plane shift extends to some millimeters, and the transformation of incoherent light to coherent is not necessary in the operation. These offer convenience in application to real time pattern recognition.

THE PRINCIPLE OF LENSLESS INTENSITY CORRELATION SYSTEM

A schematic of the FHF recording system is shown in Fig. 1. We place a characteristic transparency $f_1(x,y)$ at plane P_1, a photographic plate at plane P_2, and the converging reference beam is focused onto point B at plane P_3. After suitable exposure and developing, the desired filter is created. One of the terms of its transmittance function we are interested in is given by

$$t(\alpha,\beta) = \exp\left\{\frac{-i2\pi}{\lambda_1 d}[(\alpha-h_0)^2+\beta^2]\right\}\iint_{-\infty}^{\infty} f_1^*(x,y)\exp\left\{\frac{-i2\pi}{\lambda_1 d}[(\alpha-x)^2+(\beta-y)^2]\right\} dxdy \quad (1)$$

where λ_1 is the laser wavelength and the other parameters are noted in Fig. 1.

The schematic of correlation system is shown in Fig. 2. The object to be detected, $f_2(x,y)$, is illuminated by a spatially extended band-limited source and imaged onto input plane P_1. The FHF

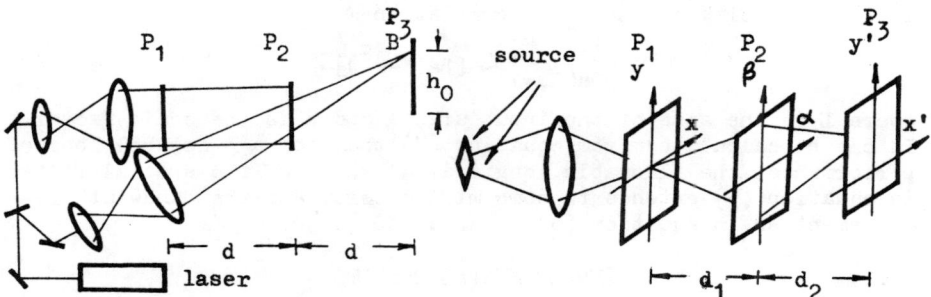

Fig. 1. The FHF making system.   Fig. 2. The correlation system.

is replaced at its original position with a small longitudinal shift d_1-d and a transverse shift α_0. It can be shown that for a certain wavelength λ_2 the irradiance at the output plane P_3 is

$$I(x',y') = \iint_{-\infty}^{\infty} |f_2(x,y)|^2 |f_1(\frac{\lambda_1 d}{\lambda_2 d_1}x+\frac{\lambda_1 d}{\lambda_2 d_2}x'-h_0-2\alpha_0, \frac{\lambda_1 d}{\lambda_2 d_1}y+\frac{\lambda_1 d}{\lambda_2 d_2}y')|^2 dxdy \quad (2)$$

which indicates that the output irradiance is the convolution of the input signal irradiance with respect to the impulse response irradiance of the system. d_1 and d_2 must satisfy

$$1/\lambda_2 d_1 + 1/\lambda_2 d_2 = 2/\lambda_1 d \quad (3)$$

To illustrate the effects of the shift of the spatial filter and the limited bandwidth of the light source, we shall discuss three different cases of equation (2). First, if λ_2 equals λ_1, d_1 and d_2 equals d, we get

$$I(x',y') = \iint_{-\infty}^{\infty} |f_2(x,y)|^2 |f_1(x'+x-h_0-2\alpha_0, y'+y)|^2 dxdy \quad (4)$$

It is the exact intensity correlation except for a displacement of $2\alpha_0$ of the correlation peak. Second, λ_2 equals λ_1, α_0 equals 0, we get

$$I(x',y') = \iint_{-\infty}^{\infty} |f_2(x,y)|^2 |f_1(\frac{d}{d_2}x'+\frac{d}{d_1}x-h_0, \frac{d}{d_2}y'+\frac{d}{d_1}y)|^2 dxdy \quad (5)$$

It is the intensity correlation of a change in scale. Third, for α_0 equal 0, d_1 equal d, and $\lambda_2-\lambda_1 = \Delta\lambda$ a small quantity, we get

$$I(x',y') = \iint_{-\infty}^{\infty} |f_2(x,y)|^2 |f_1[\frac{\lambda_1}{\lambda_2}(x+\frac{\lambda_1+2\Delta\lambda}{\lambda_1}x'-\frac{\lambda_2}{\lambda_1}h_0, y+\frac{\lambda_1+2\Delta\lambda}{\lambda_1}y')]|^2 dxdy \quad (6)$$

It is the intensity correlation of a change in wavelength-scale. Considering that any real function can be expressed as superposition of many sinusoidal functions with different frequency components, assuming that various sinusoidal components are uncorrelated, and using the fact that $\text{sinc}(x) \leq 1/x$, we shall get the relative intensity of the peak correlation

$$I_a/I_{a=1} \sim [La(1-a^2)]^{-2} \quad (7)$$

where L is the size of the input signal and a is the scale factor. It can be calculated from equation (7) that for a 3-decibel loss of performance, the allowable longitudinal shift of the spatial filter in equation (5) extends to some millimeters, and the bandwidth requirement $\Delta\lambda$ in equation (6) is about 30 Å.

EXPERIMENTAL RESULTS

In experimental demonstrations, the letters "C R T" shown in Fig. 3 are chosen to be the input object signal. The letter "C" is the detecting signal. A He-Ne laser of 60 milliwatts is used to record the FHF, and an extended source with bandwidth of 30 Å is used in the correlation system. Fig. 4 shows the experimental correlation result.

Fig. 3. The input object signal.   Fig. 4. The correlation result.

CONCLUSIONS

We emphasize that the new approach to pattern recognition is not sensitive to a change in scale as compared with the complex amplitude correlation [2]. The transverse and longitudinal shift of the spatial filter has little effect on the peak correlation. The optical system is relatively simple. Thus it offers convenience in real time applications.

REFERENCES

1. A. Vander Lugt, IEEE, IT-10, 139 (1964).
2. F. T. S. Yu and X. J. Lu, Appl. Opt., 23, 3109 (1984).
3. D. Casasent and A. Furman, Appl. Opt., 16, 1652 (1977).

COMPUTERIZATION OF AN INFRARED DIODE LASER SPECTROMETER

C. B. Dane, D. R. Lander, R. F. Curl, Jr.,
J. V. V. Kasper, F. K. Tittel
Rice University, Houston, Texas 77251
R. Brüggemann, Bonn University, F. R. Germany

The diode laser has become a useful source of tunable infrared coherent radiation between 3-20µm. Unfortunately, its small continuous frequency tuning range of typically less than 1cm^{-1} and its tendency to produce nonlinear frequency scans limit its usefulness in spectroscopy. A method for producing diode laser scans several wavenumbers long, linear in frequency, and accurately calibrated from reference spectra has been devised. Such scans are especially useful in pattern recognition during the analysis of spectra.

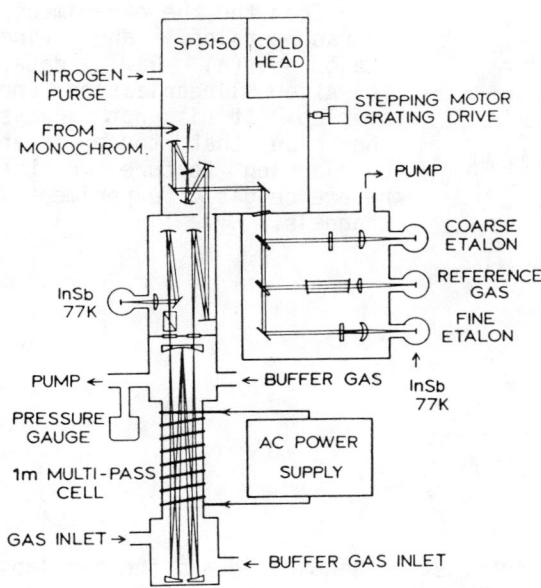

Fig. 1. Experimental set-up. The optical arrangement allows data to be collected from the experiment as well as the three diagnostic channels in a single scan.

The infrared diode spectrometer consists of a Spectra Physics diode laser source (SP-5150), an LS1-11 minicomputer, a one meter multi-pass White cell, and diagnostic instrumentation as shown in Fig. 1. The laser output beam is split by a ZnSe beamsplitter which directs 70% into the multipass cell. The remaining 30% is divided among three diagnostic channels which consist of a fine and a coarse etalon with free spectral ranges of 500MHz and 3GHz and a reference gas absorption cell.

© American Institute of Physics 1987

A particular frequency scan consists of consecutive current scans between each of which the diode temperature is adjusted. At the end of each current scan (usually about 10GHz) the computer automatically halts the scan and enters a 'ramp mode' in which the current is rapidly ramped over the region just scanned. By viewing the coarse marker cavity features on an oscilloscope, the temperature is adjusted such that the last peak on the high current side of the display is moved to the edge of the low current side. In this manner the next region to be scanned exhibits an overlap in frequency with the previous region. The user then signals the computer to continue the scan. The monochromator is adjusted during the 'ramp mode' by a stepping motor also under computer control.

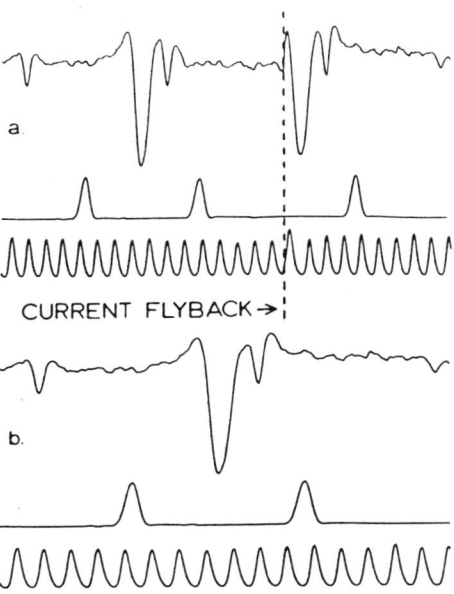

Fig. 2. A sample of the overlapping procedure. The traces from top to bottom are data from the the experiment, coarse etalon, and fine etalon. (a) Raw data. (b) After linearization and overlap. It is not always the case that there is an overlapping feature on the reference gas or experimental channels.

The resulting data set must be linearized and the overlaps removed so the adjacent current scans can be accurately joined. In order to linearize the data of a given current ramp segment, a fourth order fit to the positions of the fine markers is calculated. The marker spacings are then normalized to the maximum observed spacing and parallel corrections are made to all channels. Once linearized the overlapping segments are joined by identifying a spectral feature occurring on each side of the flyback and superimposing the centroid line centers. The spectrum can now be calibrated as one continuous scan by another set of calibration programs. Scans made in this manner are typically 3-4cm^{-1} in length although scans as long as 7cm^{-1} have been collected.

This system for producing multi-wavenumber, frequency linear, readily calibrated scans will simplify the assignment and analysis of diode laser gas phase spectra. New advances in diode fabrication technology[1] hold promise for diode lasers capable of longer range current scans. Such diodes combined with the overlapping methods presented here will significantly enhance infrared diode laser spectroscopy.

This work was supported by National Science Foundation grant CHE-8504171 and by grant C-586 of the Robert A. Welch Foundation.

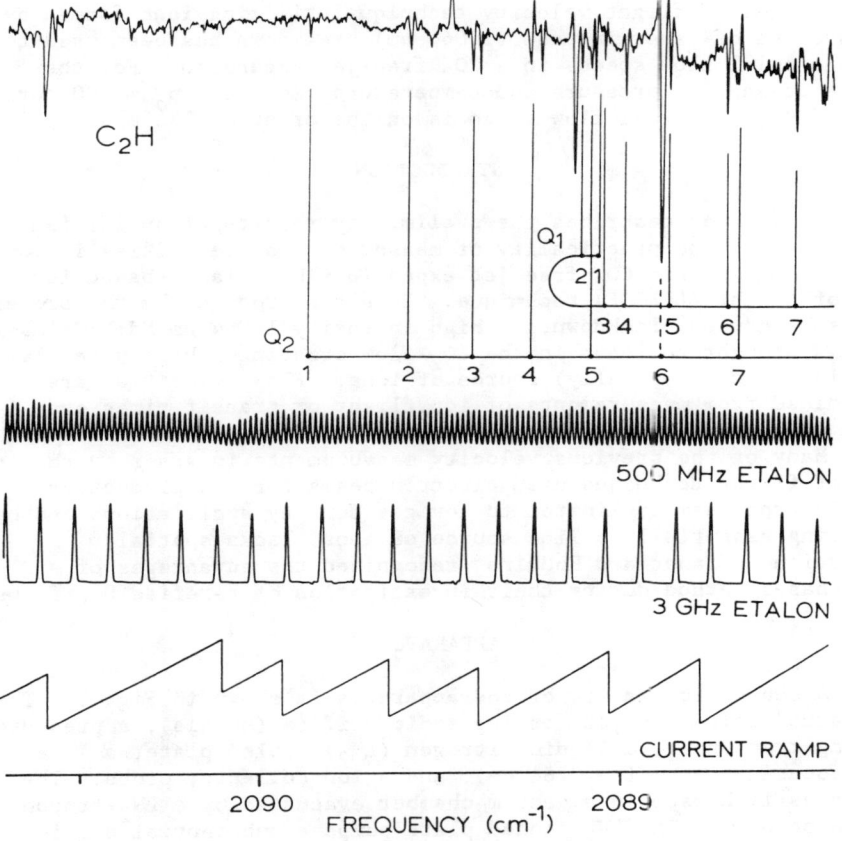

Fig. 3. A portion of a 3.5cm^{-1} continuous scan of an electrical discharge through argon over a coating of polyacetylene. A recently reported[2] C_2H band is observed using the magnetic rotation technique.

1. Y. Shani, A. Katzir, K. H. Bachem, P. Norton, M. Tacke, H. M. Preier, Appl. Phys. Lett. **48**, 1178 (1985).
2. E. Hirota, personal communication.

APPLICATION OF A LASER-INDUCED BREAKDOWN TIME-OF-FLIGHT TECHNIQUE AS A FLOW DIAGNOSTIC IN A CO_2 FREE-JET EXPANSION

P. J. Wantuck and D. E. Hof

Chemical and Laser Sciences Division
Los Alamos National Laboratory
Los Alamos, NM 87545

ABSTRACT

A time-of-flight velocity technique utilizing ions formed by the process of laser-induced (1.064 μm) breakdown has been used to measure axial flow speeds in a CO_2 free-jet expansion. For the nozzle stagnation pressure and temperature employed ($p_o \approx 200$ torr, $T_o \approx 298K$), the axial flow speed is on the order of 630 m/s.

INTRODUCTION

This paper describes the preliminary results of an initial evaluation of the practicality of measuring flow velocities in low density, supersonic CO_2 free-jet expansions by a laser-based ion time-of-flight (LIBTOF) technique. Ions are produced by the process of laser-induced breakdown. A high intensity 1.064 μm Nd:YAG laser beam is brought to focus in the CO_2 flow creating a highly resolved (spatially and temporally) source of ions. Flow velocities are determined from measurements of ion flight or transit times over known flightpath distances.

Many of the previous velocity measurements in gases by the ion time-of-flight technique used electron beams for ion production. The electron beam is limited to low gas density applications and by producing essentially a line source of ions, lacks spatial resolution. Fisher and Hodgins[1] recognized the advantages of a laser based method during their investigation of rarefied UF_6 flows.

APPARATUS

A schematic diagram of the apparatus is shown in Fig. 1. The apparatus includes a jet-forming sonic orifice (nozzle), a pressure-sensing transducer, a liquid-nitrogen (LN_2) cooled plate, a laser-beam focusing lens (f = 3.81 cm), and a ion collector probe. The apparatus is housed in a vacuum chamber evacuated by a LN_2-trapped diffusion pump. The LN_2-cooled plate pumps a substantial portion of the CO_2 flow. At the highest gas flow rates, the chamber pressure rises to approximately 5×10^{-4} torr.

The diameter of the nozzle orifice is 1.0 mm and the entire nozzle assembly is externally adjustable. The ion collector is a 1-cm diameter copper disk mounted inside a grounded shield. The portion of the shield facing the flow is a fine mesh (~90% open) steel screen. The probe can be translated parallel to the jet axis. The probe potential is held at -30 V which assures high efficiency collection of positive ions. The positive ion current is converted

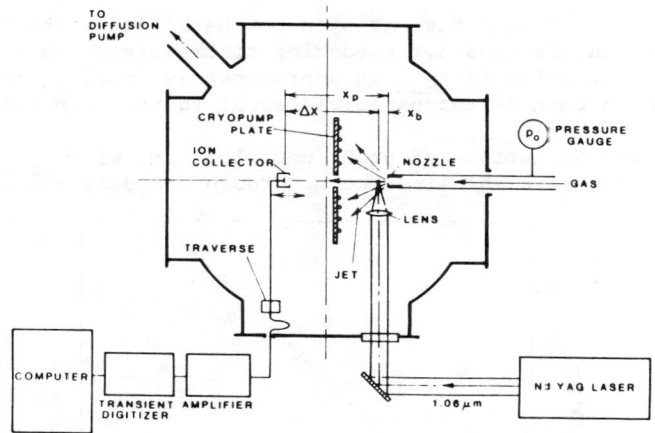

Fig. 1. Apparatus diagram.

to a potential by passing it through a 50 kΩ resistor. The resulting signal is then amplified and sent to a Tektronix R7912 transient digitizer. The processed signal is then recorded with an interfaced NOVA/Data General Eclipse computer system.

RESULTS

A typical TOF trace for fixed stagnation temperature and pressure is shown in Fig 2. The peak corresponding to zero flight time is thought to be due to (breakdown-plasma emitted) photon-induced ejection of electrons from the copper ion collector. This peak is useful in that it provides a reference time, t_o, for the TOF traces. The second peak corresponds to the arrival of the flow-related ion packet.

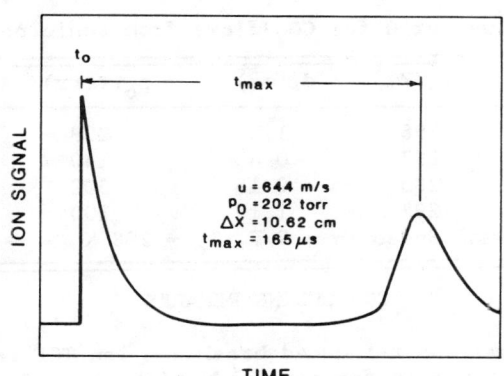

Fig. 2. Time-of-flight trace for CO_2. x_b - 1 mm, T_o - 298K, d_t - 1.0 mm, laser power density - 10^{12} W/cm^2.

The direct measurable employed to characterize the TOF distribution is the time corresponding to the signal maximum, t_{max}. Clearly, the quantity $\Delta x/t_{max}$ is approximately equal to the mean ion speed which in turn is approximately equal to the mean flow speed of the gas[2].

Observed variations of probe position, Δx, with t_{max} are shown in Fig. 3. The straight line shown through the data was fit using

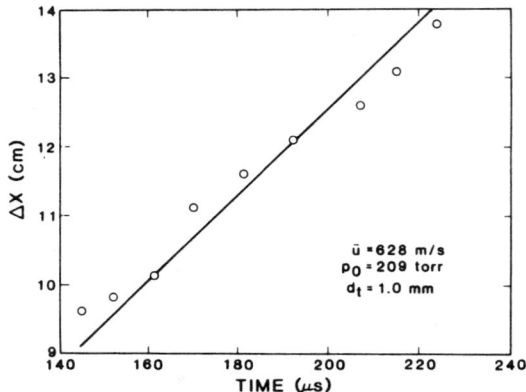

Fig. 3. Variation of probe location, Δx, vs ion peak-signal time, t_{max}. x_b = 1 mm, T_o = 298 K, d_t = 1.0 mm.

linear regression analysis. The slope, a measure of the mean flow speed, is equal to 628 (± 35) m/s. Values of mean flow speed, $\bar{u}$, for CO_2 free-jet flows are presented in Table I. The mean flow speed measured in the present experiment is in agreement with these results although tending to be somewhat higher in magnitude. This disagreement is most probably due to homogeneous condensation in the expansion[3].

TABLE I. Values of $\bar{u}$ for CO_2 flows from sonic nozzles

Source	T_o(K)	d_t(mm)	p_o(torr)	$\bar{u}$(m/s)
LIBTOF	298	1.0	209	628
Bailey[2]	297	0.4	140	610
Golomb, et al.[3]	285	0.15	300	615
Hawsey[4]	298	1.17	100	609
Limiting Thermal Speed, γ = 9/7, T_o = 298 K			------	699

CONCLUDING REMARKS

The (1.064 μm) laser-induced breakdown ion TOF technique is a viable velocity diagnostic for rarefied, high speed CO_2 flows. The primary advantage of the LIBTOF technique is the high degree of spatial resolution which can be achieved in such gas flows. The observed breakdown in other gases following irradiation with focused 1.064 μm laser light[5,6] suggests that the LIBTOF technique will be applicable to a variety gases.

REFERENCES

1. S. S. Fisher and M. G. Hodgins, AIAA 11th Aerodynamic Testing Conference, 1980, AIAA-80-0437-CP.
2. A. B. Bailey, Arnold Engineering and Development Center Report No. AEDC-TR-73-93.
3. D. Golomb, R. E. Good, A. B. Bailey, M. R. Busby, and R. Dawbarn, J. Chem. Phys. <u>57</u>, 3844 (1972).
4. R. Hawsey, M.S. thesis, Univ. of Virginia, 1978.
5. J. Stricker and J. G. Parker, J. Appl. Phys., <u>53</u>, 851 (1982).
6. T. P. Hughes, <u>Plasmas and Laser Light</u>, (Wiley, New York, 1975), pp. 182-187.

MULTISPOT LASER VIBROMETRY FOR MATERIALS AND STRUCTURE EVALUATIONS

G. L. Fitzpatrick, R. L. Skaugset, & T. J. Davis, Sigma Research, Inc.
8710 148th Avenue N.E., Redmond, Washington 98052

ABSTRACT

A portable device for performing multispot laser vibrometry is described. The device is capable of measuring arrival times, surface motion frequency, and surface velocity and displaying these quantities in various useful ways.

INTRODUCTION

The purpose of this paper is to briefly describe a new, optical measuring instrument. A tripod-mounted system for making multispot laser Doppler measurements has been constructed and has undergone initial testing. Such a system holds promise in applications ranging from the evaluation of rock structures (a purpose for which it was originally intended) to the inspection of modern aerospace composite materials. The system has a number of unique features. It employs separate send and receive telescopes (Maksutov Cassegrain design) and an optical fiber grating (25μm or larger optical fibers arranged in a grating or crossed gratings) located in the send telescope for producing an array of laser spots on a distant target (5-30m or greater). The associated electronics currently process four channels corresponding to four laser spots on the surface of interest. Measurements of the relative arrival times of an elastic wave disturbance, the peak particle velocity, and the average surface motion frequency may be displayed in numerical format. Experimental measurements on four independent vibrating targets are presented. Plans for an eight-spot, tripod-mounted system, and systems with square multispot arrays are also described.

APPARATUS

Figure 1 presents, in a highly schematic fashion, the basic optical arrangement. The current system is a four channel device and each channel records an interferometer signal of the form

$$I(t) = A_s^2 + A_o^2 + 2A_s A_o \cos\left[\frac{2v}{c} \omega_o t\right] \qquad (1)$$

when A_s is the amplitude of the light returned from the surface, A_o is the reference beam amplitude, ω_o is the laser frequency, c is the speed of light in air, and v is a constant line of sight velocity of the rock or other surface relative to the stationary interferometer, $|\Delta\omega| = (2v/c)\omega_o$ is the classical Doppler shift of the radiation reflected from the surface and is clearly proportional to the line of sight velocity. If the line of sight velocity is sinusoidal

$$V(t) = V'_o \sin(\omega'_o t) \qquad (2)$$

where ω'_o is the surface motion frequency, the interferometer signal would be a frequency modulated signal given by

$$I(t) = A_s^2 + A_o^2 + 2A_s A_o \cos\left[K_o \cos\omega'_o t\right] \qquad (3)$$

where $K_o = -\dfrac{2\omega_o V'_o}{c\omega'_o}$. $\qquad (4)$

The onset and duration of a generally sinusoidal surface motion results in a signal I(t). This signal can provide the information in Table I for any channel. The method of obtaining this information is also indicated.

TABLE I - DETECTION AND ANALYSIS ELECTRONICS

Measured Parameters	Detection Methods	Displays
1. Relative Arrival Times	-Detect analytic magnitude, perform threshold detection	-Δt digital display -Arrival sequence -Electronic bargraph
2. Max Surface Velocity V'_o	-Doppler frequency detector followed by peak detector	-Electronic bargraph -Oscilloscope
3. Surface Motion Frequency ω'_o	-Periodicity of Doppler output	-Electronic bargraph -Oscilloscope

The apparatus is illustrated in Fig. 2. The laser is a 35 milliwatt He-Ne laser and the two telescopes are 7 and 3 1/2 inch Questar telescopes, respectively. In Fig. 3, four channels of simultaneous data taken from a set of four independent vibrating targets are illustrated.

CONCLUSIONS

A multispot laser vibrometer system for measurements on remote, unprepared surfaces has been demonstrated. By using more detectors and crossed fiber gratings larger areas of a surface could be inspected. Such systems would be capable of providing information on the surface velocity, relative arrival times, and surface motion frequency of a large number of points. One can imagine numerous applications for such technology including, 1) characterization of an airfoil or other air frame component vibrations in wind tunnels, 2) panel vibration studies on any structure (auto bodies, etc.), 3) studies of the vibrations of parts of buildings or bridges subject to ambient vibrations, and 4) examination of structures in hostile environments where contact transducers are ruled out.

ACKNOWLEDGEMENTS

This work was sponsored by the U.S. Department of Interior (U.S. Bureau of Mines) under a Small Business Innovation Research (SBIR) contract J0145075. The authors would like to thank the technical project monitor, Roger McVey of the U.S. Bureau of Mines, for numerous helpful suggestions.

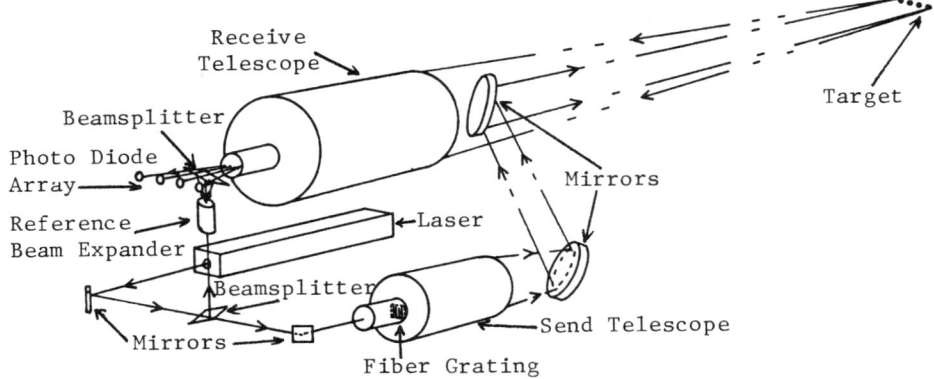

Figure 1. Optical arrangement.

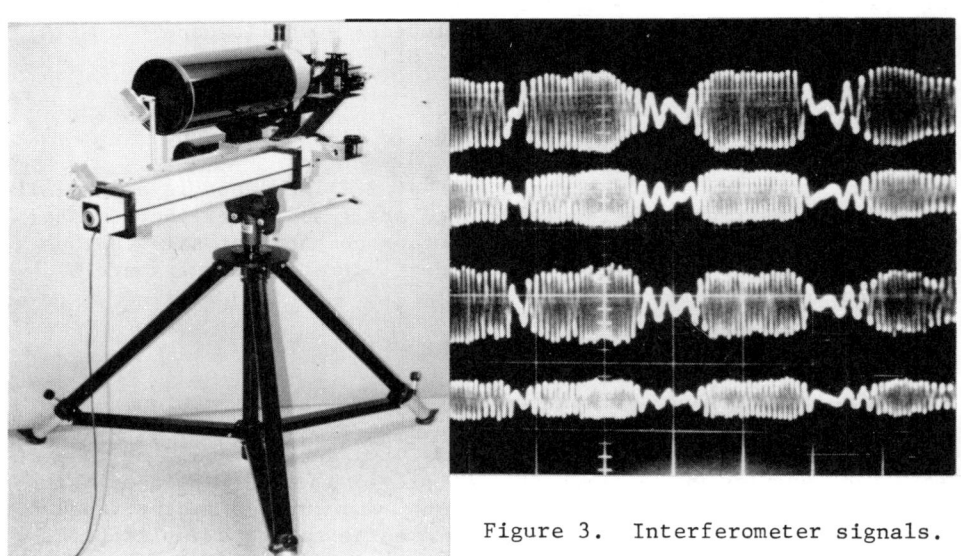

Figure 2. Apparatus.

Figure 3. Interferometer signals.

DETECTION OF TRANSIENT FLUORINE ATOMS

Gary W. Loge, Norris Nereson, and Herbert A. Fry
Chemical and Laser Sciences Division
Los Alamos National Laboratory
P. O. Box 1663, Los Alamos, NM 87545

ABSTRACT

A KrF eximer laser with a fluence of 50 mJ/cm^2 was used to photolyze either uranium hexafluoride or molecular fluorine, yielding a transient number density of fluorine atoms. The rise and decay of the atomic fluorine density was observed by transient absorption of a 25-μm Pb-salt diode laser. To prevent the diode laser wavelength from drifting out of resonance with the atomic fluorine line, part of the beam was split off and sent through a microwave discharge fluorine atom cell. This allowed a wavelength modulation-feedback technique to be used to lock the diode laser wavelength onto the atomic line. The remaining diode laser beam was made collinear with the excimer laser beam using a LiF window with a 45° angle of incidence to reflect the infrared beam while transmitting most of the uv beam. Using this setup along with a transient digitizer to average between 100 and 200 transient absorption profiles, fluorine atom number densities on the order of 10^{14} cm^{-3} in a 1.7 m pathlength were detected. The signals observed were about a factor of two less than expected from known photolysis and atomic fluorine absorption cross-sections.

INTRODUCTION

The absorption spectrum of atomic fluorine contains features only at 25 μm and at wavelengths shorter than 78 nm. The absorption at 404 cm^{-1} corresponds to the $^2P_{3/2} \rightarrow {}^2P_{1/2}$ spin-orbital transition in atomic fluorine. The detection of fluorine atoms by absorption of a 25 μm diode laser beam has previously been demonstrated[1] and used to measure the hyperfine splitting[2] for a continuous source of fluorine atoms generated in a microwave discharge cell. We report here demonstration of a technique that allows transient fluorine atoms to be detected using the same diode laser absorption.

EXPERIMENTAL APPARATUS

The demonstration of transient fluorine atom detection was accomplished using a KrF eximer laser at 248 nm to photolyze either UF$_6$ or F$_2$ forming a nearly instantaneous density of fluorine atoms (10 ns risetime), which then disappears due to recombination or diffusion out of the diode laser probe beam. The transient fluorine atoms absorbed part of a collinear diode laser beam, which was tuned to the strongest hyperfine line at 404.175 cm^{-1}. The excimer laser beam and diode laser beams were made collinear using a LiF window to transmit the excimer laser beam and reflect the diode laser beam (see figure 1). The excimer laser fluence was measured at several

© American Institute of Physics 1987

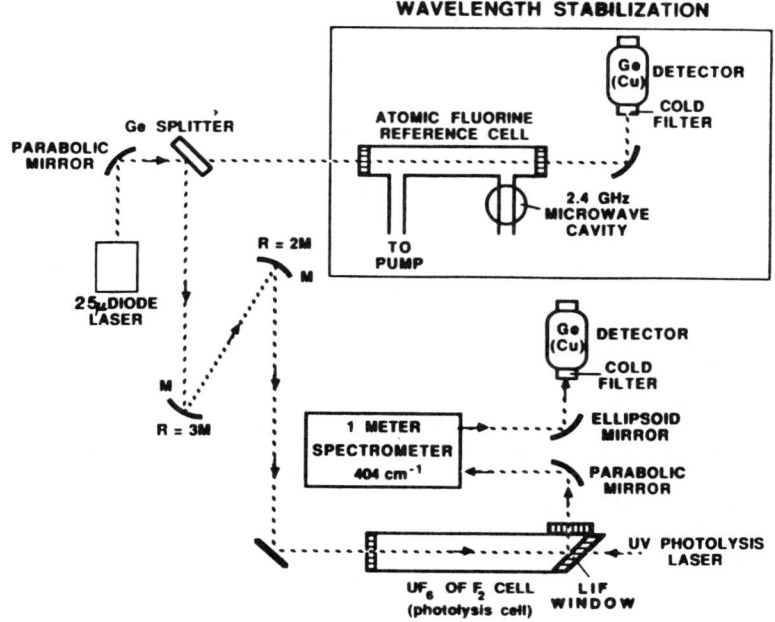

Fig. 1. Optical schematic of the equipment used to measure transient fluorine atom absorption and to wavelength stabilize the diode laser.

positions through the 1.7 m photolysis cell using a known aperture size and an energy meter. The diode laser beam diameter was kept smaller than the excimer laser beam so that the entire diode laser beam sampled the atomic fluorine number density. KBr windows were used to transmit the 25 μm diode laser beam. A monochromator was used to separate the diode laser modes and reject background IR from the LiF window produced by the excimer laser. Ge(Cu) detectors at 10K were used, and a cold filter was used to remove excess noise due to blackbody radiation. The response time of the detector was about 20 μs, which gave maximum gain. A shorter response time could have been used, as fast as 20 ns, but would have given a lower gain. The signal due to transmitted diode laser reaching the detector was sent to a transient digitizer that was triggered at the excimer laser pulse.

A Ge beamsplitter (50%T) before the photolysis cell was used to direct part of the beam into a fluorine atom absorption cell used to lock the diode laser wavelength onto the absorption line. The long term stability of the diode laser wavelength has been observed by repetitive scans across the line to be several times larger than the room temperature fluorine atom Doppler linewidth. This drift of the diode laser wavelength would cause the detection of transient fluorine atoms to be extremely difficult. For this reason, the diode laser wavelength is locked onto the fluorine atom absorption line using wavelength modulation at 5 kHz, creating a first-derivative signal. This first-derivative signal is feed into the diode current power supply, causing wavelength stabilization (see figure 2).

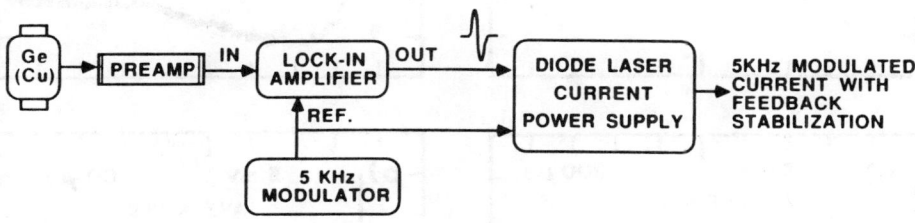

Fig. 2. Diagram of the wavelength modulation (diode laser current modulation), feedback stabilization electronics.

Although this technique does keep the diode laser wavelength in resonance, it decreases the apparent absorption cross-section because it requires that the wavelength be modulated across the absorption line rather than sitting at the peak. The wavelength modulation also causes the transient absorption signal to be modulated. However, the excimer laser is triggered randomly relative to the phase of the modulation, so signal averaging will remove modulation of the signal to a large extent.

RESULTS

Two sets of data were obtained using either UF_6 or F_2 as the photolysis precursor. Averaged signals are shown in figures 3 and 4. When using UF_6, a slower risetime for absorption is observed, and absorption is also seen when the diode laser wavelength is moved off resonance of the atomic fluorine absorption line. The suspected source of the interfering absorption is $(UF_5)_2$, which is created by collisions after photolysis of UF_6.

The sensitivity of this technique can be estimated from the data shown in figures 3 and 4. From known photolysis cross sections and excimer laser fluence, including attenuation by absorption and beam divergence, the average fluorine atom density is about 10^{14} cm^{-3}, which corresponds to an absorption of about 0.05 of the diode laser. The observed absorption was about a factor of two less than the expected value. This is probably due to the diode laser wavelength being modulated across the absorption line rather than sitting on the peak. This problem is being addressed in ongoing experiments.

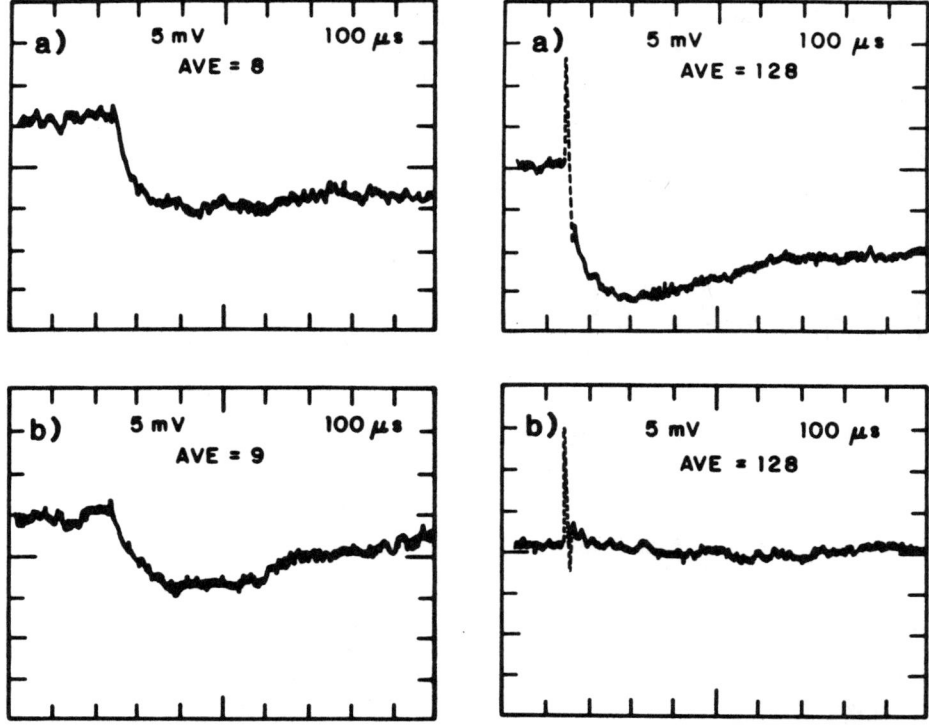

Fig. 3. Averaged transient absorption with 1.5T of UF_6 in and 5T of He in photolysis cell. a) Diode wavelength stabilized, b) 0.016 cm^{-1} off resonance.

Fig. 4. Averaged transient absorption with 5T of F_2 photolysis cell. a) Diode wavelength stabilized, b) 0.017 cm^{-1} off off resonance.

Another effect that may be occurring is that photolysis may not form a room temperature population distribution in the $^2P_{3/2}$ and $^2P_{1/2}$ states, which affects the absorption cross section. If the fluorine atoms are formed hotter, i.e., more population in the upper state, the cross section would be smaller but would increase as collisional relaxation to the room temperature distribution occurred. The data suggests this is occurring, with more occurring in the F_2 photolysis.

REFERENCES

1. A. C. Stanton and C. E. Kolb, J. Chem. Phys., 72, 6637 (1980).

2. G. A. Laguna and W. H. Beattie, Chem. Phys. Lett., 88, 439 (1982).

LASER SPECTROSCOPIC DETECTION OF OH IN
CATALYTIC REACTIONS ON PLATINUM

S. Ljungström, A. Rosén, T. Wahnström and B. Kasemo
Department of Physics, Chalmers University of Technology and
University of Göteborg, S-412 96 Göteborg, Sweden

ABSTRACT

OH radicals thermally desorbed from a polycrystalline platinum foil during the catalytic reaction between H_2 and O_2 have been studied by laser-induced fluorescence. The experiments were performed in a continuous flow system at total pressure and Pt-sample temperature ranges of 0.005-0.200 torr and 900-1200 K, respectively.

INTRODUCTION

Oxidation of hydrogen to water in gas phase only takes place at rather high temperatures due to the existence of a high activation barrier. However, in the presence of a catlyst the hydrogen and oxygen molecules can dissociate on the catalyst and recombine to form water. In both the gas and surface reactions H, O and OH are important radicals participating in the reaction. Identification and measurements of possible intermediate species and their properties in chemical reactions (including catalytic reactions) is one of the most direct ways to establish reaction routes and map out the reaction mechanisms /1/. Studies of such reaction intermediates on a catalyst can either be done in frozen situations with surface sensitive spectroscopies as infrared absorption (IR) and electron energy loss spectroscopy (EELS) or alternatively by laser induced fluorescence (LIF) detection of thermally desorbed species during dynamic reaction conditions /2,3/. The LIF technique offers unique possibilities to simultaneously identify the type of intermediate species, its internal quantum state and its (apparent) desorption energetics. In the present work we focus the study on LIF detection of intermediate OH radicals in the oxidation of hydrogen to water on a polycrystalline Pt foil.

EXPERIMENTAL

The experimental setup, which has been described in detail in earlier work /3/, consists of a vacuum chamber with a typical base pressure of 10^{-6} torr. The partial pressures of hydrogen and oxygen can be varied individually. The catalyst is a high purity Pt-foil (13x2x0.1 mm) heated resistively in the reactant gases in the pressure range 0.01-0.10 torr. The laser beam is parallell to the foil surface and its center axes lies about 4 mm from the foil. The fundamental laser wavelength, obtained from an excimer pumped dye laser, is frequency doubled to lie in the region 306.3-307.5 nm, i.e. within the absorption band of OH. For minimization of OH creation in the decomposition of water (produced in the forward reaction) liquid nitrogen cooled copper plates are mounted close to the foil.

RESULTS AND DISCUSSION

In the experiments the laser is scanned through the absorption band of OH with simultaneous detection of the emitted fluorescence light. A typical fluorescence scan obtained at a mixing ratio of $H_2/O_2=0.2$ and total pressure of 100 mtorr is shown in Fig. 1.

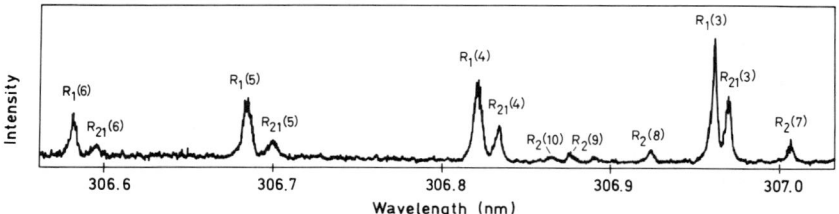

Fig. 1. Fluorescence intensity vs. laser wavelength for the $X^2\Pi(v\Pi = 0) \rightarrow A^2\Sigma(v'=0)$ transition at 1100 K and 0.05 torr.

The peaks in the spectrum reveal, by comparison with known spectroscopic constants for OH /4/, the presence of OH molecules. Since the peaks disappear when the foil temperature is decreased below 900 K it is ascertained that the detected OH originates from the surface catalyzed reaction. The most commonly adopted reaction scheme for the $H_2+O_2 \rightarrow H_2O$ reaction on Pt proceeds via dissociation of hydrogen and oxygen to adsorbed H and O atoms, which combine on the surface to OH. OH reacts further with an H atom to H_2O, which desorbs from the surface. The observed gas phase OH are produced by (thermal) desorption of surface OH radicals before they react to H_2O. This is a minority route compared with the overall water producing route.

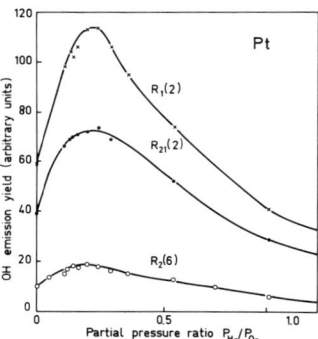 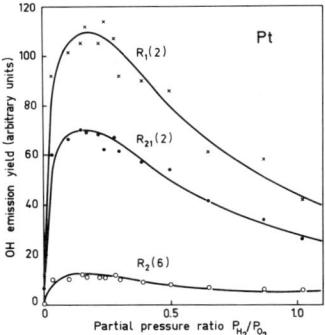

Fig. 2. OH production vs. $P(H_2)/P(O_2)$ at 1100 K and 0.1 torr.

Fig. 3. OH production vs. $P(H_2)/P(O_2)$ at 1100 K and 0.1 torr with a liq. N_2 trap.

Figure 2 shows the fluorescence intensity from the foil as a function of the partial pressure ratio at constant pressure for three peaks in the wavelength scan. Since the total pressure was held constant such a curve is a measure of the desorption flux from the surface. The form of the curve is attributed to the changes in surface reaction conditions as the coverages of H and O vary with the H_2/O_2 ratio. A maximum in the OH intensity is obtained, as expected since the production of OH should vanish when the H_2/O_2 ratio goes to infinity or zero. The intensity does not exactly reach zero in the oxygen rich end, however. The reason is that in addition to the H_2/O_2 reaction there is also a surface reaction between coadsorbed water molecules and oxygen atoms which also contributes to the OH production. In order to verify this assumption, the measurements were repeated with copper plates around the sample, cooled to liquid nitrogen temperature. Figure 3 shows the corresponding fluorescence intensity from the foil as function of the H_2/O_2 ratio at the same constant pressure. In this case the intensity at $H_2/O_2=0$ is only 5% of the peak intensity which for the uncooled conditions were 50%.

From the heights of the peaks of the type shown in Fig. 1 information can be extracted about OH production efficiency of the catalyst, the temperature and pressure dependence of reaction and apparent desorption energies /3/. Normalization of the peaks with respect to the transition probabilities makes it possible to determine also the population distribution between rotational levels of the desorbed OH molecules.

In order to obtain a measure of the OH desorption energy the intensity of a given rotational transition was measured as a function of temperature. A plot of intensity vs. the inverse temperature is shown in Fig. 4. A straight line is obtained from which an apparent desorption energy can be derived. The desorption energy derived in this way varies with the H_2/O_2 mixing ratio, which indicates that kinetic effects influence the result and that the true desorption energy is not obtained /5/.

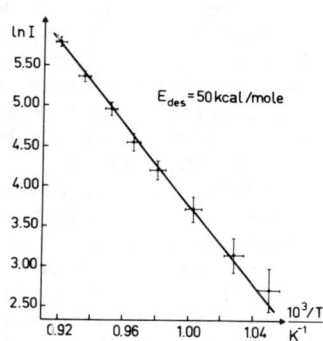

Fig. 4. An Arrhenius plot.

CONCLUSION

LIF studies of thermally desorbed intermediate reaction products in catalytic reactions promise to give valuable insight into catalytic reaction routes. This concerns both reaction kinetics and energetics of reaction. Such studies will be particularly fruitful when they are combined with the traditional methods for catalytic reaction studies.

REFERENCES

1. B. Kasemo and B.I. Lundqvist, Comments At. Mol. Phys. 14, 229 (1984).
2. L.D. Talley, W.A. Sanders, D.J. Bogan, and M.C. Lin, J. Chem. Phys. 75, 3107 (1981).
3. A. Rosén, S. Ljungström, T. Wahnström, and B. Kasemo, J. Electr. Spec. Rel. Phenomena 39, 15 (1985).
4. G.H. Dieke and H.M. Crosswhite, "The ultraviolet bands of OH", J. Quant. Spectr. Transfer. 2, 97 (1962).
5. T. Wahnström, S. Ljungström, B. Kasemo and A. Rosén, In preparation.

IR LASER ABSORPTION EDDY CORRELATION MEASUREMENT DEVICES FOR TRACE ATMOSPHERIC GASES

M. S. Zahniser, P. L. Kebabian, S. Anderson,
A. Freedman and C. E. Kolb
Center for Chemical and Environmental Physics,
Aerodyne Research, Inc., Billerica, MA 01821

ABSTRACT

The development of two open path IR absorption instruments for eddy correlation measurements of fluxes of trace atmospheric gases which impact "greenhouse" and acidic dry deposition problems is discussed. One instrument, based on a tunable lead salt diode laser, will be used to detect a variety of molecules including O_3, NO, and SO_2. The second, designed for methane measurements, uses a tunable IR HeNe laser operating at 3.39 μm.

INTRODUCTION

Over the past decade there has been increasing awareness of mankind's environmental impact on both regional and global scales. Concern about increasingly severe acid rainfall in the northeastern U.S. and Canada has generated a need to understand acid generation and deposition processes in the atmosphere.[1] One important process about which little is known is the flux of acid-related gases, such as O_3, HNO_3, and SO_2, to or from the earth's surface in the absence of precipitation, called dry deposition. Another important problem (known as the "greenhouse effect") concerns the increasing global concentrations of IR-active gases such as CH_4, N_2O, and of course CO_2. These trends are believed to be related to a variety of human activities but a quantitative understanding depends on characterization of the natural source and sink strengths for these gases. Although the eddy-correlation technique appears to be the best way to measure these fluxes, only modest progress has been made since sufficiently sensitive detectors have not been available. Several programs at ARI are aimed at developing sensitive laser based instrumentation which can measure ambient surface fluxes of atmospheric trace gases relevant to acidic dry deposition and biological greenhouse gas production processes.

Transport in the lowest few meters of the atmosphere is accomplished primarily by wind fluctuations or eddies.[2-4] The purpose of eddy correlation flux measurement devices is to study the emission or uptake of trace gases in the free air, i.e., without enclosures or traps that would disturb the local environment. This is accomplished by correlating fluctuations in the average gas concentration with fluctuations in the vertical component of the wind. When the gas source (sink) is at ground level, upward gusts will have, on the average, higher (lower) concentrations than downward gusts.

Our fluxmeters rely on the use of infrared absorption to measure trace gas concentrations. Use of lasers with the appropriate wavelength coupled with open-path (atmospheric pressure) multipass absorption cells not only provides adequate sensitivity (at 10 Hz sampling rate) and permits simultaneous measurement of gas concentration and wind and heat flux vectors in the same volume of air, but does so with minimal disruption of the sampled gas. Since both wind velocity (sonic anemometers) and heat flux (fast thermistor probes) measurement devices are commercially available, we have concentrated on developing the appropriate lasers and optics for concentration measurements.

IR LASER ABSORPTION

Tunable lead salt diode lasers provide the broadest coverage in the infrared (350 to 3500 cm^{-1}) as well as high spectral power density and narrow line width (10^{-4} cm^{-1}). Recent development of single mode, high power diodes (~1 mW) which operate at liquid nitrogen temperatures make use of this type of laser in field instrumentation more attractive. We estimate that detection limits of 1 to 5 parts per billion can be obtained (based on a 40 m path length and 10^{-4} fractional absorption) for NO, O_3, and SO_2. Figure 1 shows a measurement of ozone fluctuations due to air currents in the laboratory using a diode laser operating at 1055.351 cm^{-1}.

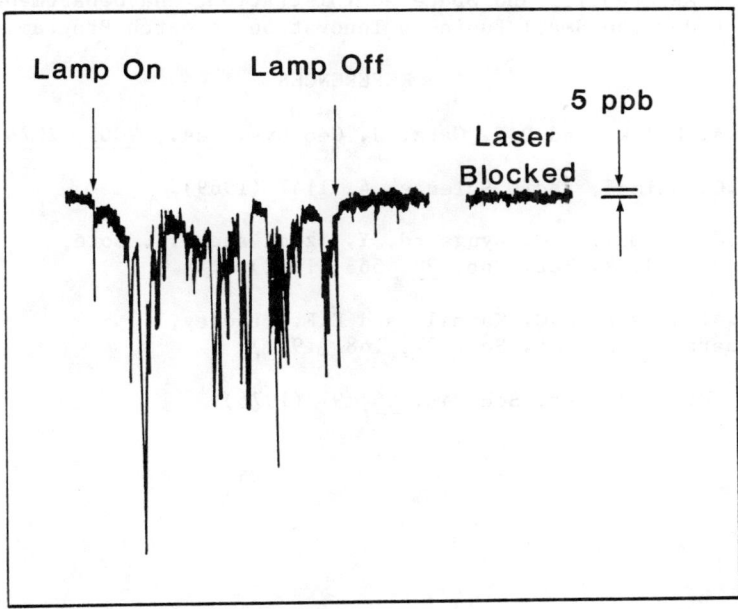

Figure 1. Ozone concentration fluctuation, in laboratory room air. Ozone is generated using a mercury lamp. The total scan time is five minutes.

An alternative approach for certain gases is to use neutral rare gas lasers where accidental coincidences between molecular absorption and laser lines occur: HeNe laser and methane (2947.9 cm^{-1}); Xe laser and nitrous oxide (2567.4 cm^{-1}); and HeNe laser and HF (4174.0 cm^{-1}), for example. We have chosen to measure methane concentrations using the 2947.9 cm^{-1} line of the helium neon laser.[5] The natural laser frequency can be Zeeman-tuned using permanent or electromagnets on a commercial laser tube, providing sensitivity to the variation in molecular absorption when the laser is tuned around line center. With an available tuning range of 1.6 GHZ, we estimate that 1% fluctuations in the ambient methane concentration (1.6 parts per million) can be measured at 10 Hz without interference from atmospheric water vapor. It should be noted that transmission of the 3.39 µm laser light to and from the multiple pass cell can be accomplished using fiber optics, simplifying the optical train.

In conclusion, advanced infrared light sources (tunable lead salt diode and neutral rare gas lasers) combined with an open-path multi-pass absorption cell provide a sensitive, general and non-intrusive method of monitoring trace gases in the free atmosphere. When coupled with appropriate windfield instrumentation, key measurements of environmentally important surface fluxes will be possible.

This work is supported by the National Science Foundation, National Aeronautics and Space Administration, and Department of Energy under the Small Business Innovation Research Program.

REFERENCES

1. D.A. Dolske and D.F. Gatz, J. Geophys. Res., 90D1, 2076 (1985).

2. J.C. Kaimal, Radio Science, 4, 1147 (1969).

3. J.C. Kaimal, J.C. Wyngaard, Y. Izumi and O.R. Coté, Quant. J. R. Met. Soc. 98, 563 (1972).

4. D.A. Haugen, J.C. Kaimal, and E.F. Bradley, Quart. J. R. Met. Soc. 97, 168 (1971).

5. A. Pine, J. Opt. Soc. Am. 66, 97 (1976).

PROPELLANT COMBUSTION STUDY BY COHERENT ANTI-STOKES RAMAN SCATTERING

Thieu H. Vu
GEO-CENTERS, Inc., Wharton, NJ 07885

Richard Field
U.S. Army ARDEC, Dover, NJ 07801-5001

ABSTRACT

We have applied the method of coherent anti-Stokes Raman scattering (CARS) in the study of propellant combustion in air using N_2 as the thermometric species. The temperatures measured in the flame are lower than the calculated equilibrium flame temperature mainly due to the unsteady flame structure. We have compared the CARS temperature in a CH_4/N_2O flame with the value obtained by Na line-reversal. There is a good agreement between the two methods. During our study of the CH_4/N_2O flame, we have also noticed a drop in the non-resonant scattering around 4717Å, possibly due to absorption by the C_2 or NH radical.

INTRODUCTION

As part of an attempt at elucidating the complex dynamics involved in ballistic events, we have applied various non-intrusive diagnostic techniques in our studies of propellant ignition and combustion. In particular, we want to determine the flame temperature. Since the ignition and combustion processes are usually very rapid and spatially ill-defined, a method with a fast response and a good spatial resolution is required. The technique of broad band BOXCARS can satisfy all the demands above.

CARS OF PROPELLANT FLAMES

The system under study is a model double-base propellant, containing 80.8% nitrocellulose and 18.2% nitroglycerine. Our intention was to use the CARS spectrum of N_2 formed in the combustion to plot a temperature profile of the propellant flame in air.

The BOXCARS setup consists of a Q-switched (10Hz) YAG-laser (532nm) to pump both the sample and broad band Stokes dye laser. The CARS beam is spatially filtered from the laser beams and directed onto a vidicon, with the spectrometer operated in the second order. The combustion in air of each propellant grain ($\cong$ 5mm high and 8mm in diameter) is completed within two seconds, with the flame being between 1 and 2 cm high. To avoid signal collection from the ambient N_2 just before and after the flame lifetime, only 10 pulses were taken for each spectrum. Figure 1 shows the N_2 CARS spectra at three different heights in the flame. A 3.2 cm^{-1} resolution limit imposed by our instrument allows us to detect the vibrational bands only.

A library of calculated spectra was generated using a procedure outlined by Hall.[1] The theoretical and experimental spectra were then matched by adjusting the temperature and the non-resonant susceptibility, while keeping the N_2-mole fraction constant. The temperatures measured at three different heights in the flame are: 2175K (11 - 12mm), 2175K (9 - 10mm), and 2200K (7 - 8mm) (Figure 1). The theoretical equilibrium flame temperature of the propellant at atmospheric pressures is 2873K. The lower temperatures measured by CARS may be due to the unsteady flame structure at the laser probe volume, leading to the averaging of spectra from both the hot flame zone and the cooler dark zone. Moreover, the regressing propellant surface during its rapid combustion led to the sampling at the flame edge.

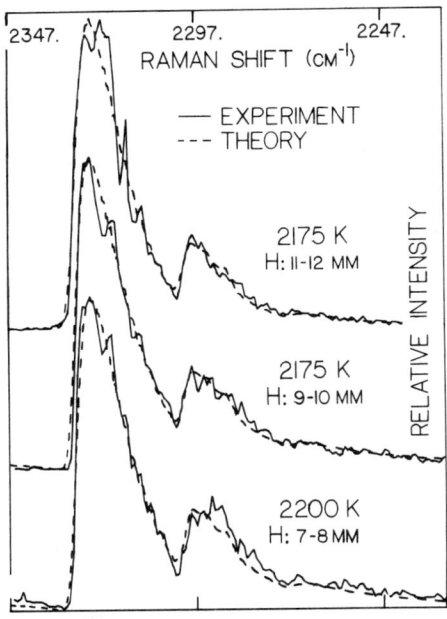

Fig. 1. N_2-CARS Spectra of Propellant Flames (H: probe position above propellant).

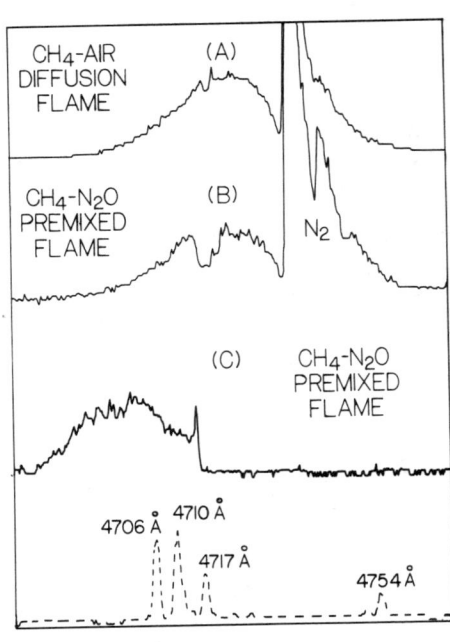

Fig. 2. N_2-CARS Spectra of CH_4/Air and CH_4/N_2O Flames (dashed curve: Neon calibration spectrum).

To assess the accuracy of our experimental setup, we have compared the temperature measured with our BOXCARS system and that by the Na line-reversal method (Na-LR). This was done on an Na-seeded CH_4/N_2O flame. The results listed in the following table indicate a reasonable agreement between the two methods:

CARS Temperature (K):	2410	2310	2425
Na-LR Temperature (K):	2448	2283	2453

CARS OF CH4/N2O FLAMES

During our studies of premixed unseeded CH_4/N_2O flames, we have noticed a sharp intensity drop around 4717Å in the non-resonant background (Figure 2). This interference is not observed in any CH_4/air diffusion flame but occurs only when N_2O is introduced. A possible cause is an absorption in the C_2 Swan bands. However, a search for another C_2 absorption of nearly the same strength at 4698Å shows no evidence (Figure 2-C). Another hypothesis to account for this intensity drop is the $X^3\Sigma^- \to b^1\Sigma^+$ transition of the NH radical formed in the flame. This singlet/triplet intercombination was often seen as a rather weak emission in the photolysis of NH_3.[2]

CONCLUSION

The temperatures of propellant flames obtained in this work should be interpreted as the spatial average temperatures due to the fluctuating flame structure. To fully take advantage of the spatial resolution of the BOXCARS method, we will try to first stabilize the flame by burning the propellant under high pressures.

As for the interference in the non-resonant scattering spectrum of the CH_4/N_2O flame, we will seek further verifications of the hypotheses by emission or laser-induced fluorescence spectroscopy.

REFERENCES

[1] R. J. Hall, Combustion Flame, <u>35</u>, 47 (1979)
[2] B. Gelernt, S. V. Filseth and T. Carrington, Chem. Phys. Lett., <u>36</u>, 238 (1975)

ACKNOWLEDGEMENTS

We would like to thank Dr. WanYee Cheung and Mr. Donald Chiu for assistance in the experimental setup and thermodynamic calculations.

VIII. Laser Research and Techniques in Medicine and Biology
A. Interactions of Laser Radiation with Biological Tissue 698
B. Biomedical Laser Applications 715

INTERACTIONS OF EXCIMER LASERS WITH POLYMERS

Y.S.Liu, H.S.Cole and H.R. Philipp
General Electric Research and Development Center
Schenectady, New York 12301

ABSTRACT

Interactions of high photon energy ($h\nu > 5$ eV) radiation from an excimer laser with polymeric materials has been studied. Photo-etching rates at both 193 nm and 248 nm were measured. These results, considered together with our recent measurements of the VUV optical properties of these polymers, give new insights into the mechanisms of the ablative photo-etching process.

INTRODUCTION

Recent observations of ablative photo-decomposition of polymers by pulsed UV lasers have generated much interest in studying the interaction of excimer lasers with these materials.[1,2] Potential use of this new process for dry-etching and lithography patterning in microelectronics was immediately recognized. Other important applications include micro-surgery in medicine and surface modifications in polymer chemistry.

Many techniques, such as laser-induced fluorescence[3], photoacoustic detection[4,5] and emission spectroscopy[6,7] among others have been applied for studying basic photo-polymer interaction. Observations generally reported in the study of ablative photodecomposition include: (a) significant ablation does not occur until the laser fluence reaches above certain threshold value which varies from polymer to polymer, (b) the products ejected during ablation are expelled away at supersonic velocities, (c) the removal of polymeric materials occurs mainly within the on-time of the laser pulse,(typically of 10 ns), and (d) patterns generated by photo-ablation generally have sharp edge definitions.

In many studies, the etch depth of a polymer irradiated at a given laser fluence level is described according to the Beer's law.[8] In this case, the etch depth per pulse, L, is given by

$$L = \frac{1}{\alpha} \ln (F/F_o) \qquad (1)$$

where α is the absorption coefficient, F is the laser fluence, and F_o the threshold fluence. The fit of observed etching rates to Eq.(1) is often not entirely satisfactory. We believe this failure may arise from the fact that the material is serially removed and/or modified during the on-time of the laser pulse. In this work, we discuss the measurements of photo-etching rates of a variety of polymers including, both aliphatic and aromatic as well as polymer blends, under various irradiation conditions. When these data are considered together with recent measurements of UVU optical properties of these polymers, some new insights into the mechanisms of the ablative photo-etching process are gained.

MEASUREMENTS

The samples were prepared by spin-coating polymer solutions on 50 mm by 50 mm quartz substrates. The smooth uniform films of several micron thickness were vacuum baked to remove the residual solvent. The optical absorbance of these polymers were measured in the spectral range 170 nm to 400 nm using a vacuum UV spectrometer. The details of VUV measurements have been discussed in a previous publication. Table 1 shows the optical absorption absorption of a selected polymers at 193, 248, and 308 nm respectively.

Table 1. Optical absorption coefficients of various polymers.

POLYMER	WAVELENGTH		
	193 nm	248 nm	308 nm
Polymide	4.2×10^5	2.8×10^5	1.2×10^5
Polysulfone	4.0×10^5	1.5×10^5	8.1×10^2
Novolac epoxy	1.0×10^5	2.1×10^4	2.4×10^3
Polycarbonate	5.5×10^5	1.0×10^4	2.2×10^1
Poly(α-methyl)styrene	8.0×10^5	6.5×10^3	8.0×10^1
Poly(methyl-methacrylate)	2.0×10^3	6.5×10^1	$< 10^1$
Poly(vinylacetate)	1.0×10^3	$< 10^2$	$< 10^1$
Polyethylene	6.3×10^2	$< 10^1$	$< 10^1$
Polypropylene	5.3×10^2	$< 10^1$	$< 10^1$
Polytetrafluoroehtylene	2.6×10^2	1.4×10^1	$< 10^1$

The etch rates at ArF (193 nm) and KrF (248 nm) laser wavelengths were measured using an excimer laser (Lambda-Physik Model EMG101) operated with a pulse width of 15 ns.[10,11] The etch depths of the polymers exposed to various laser fluence levels were measured. The etch depth was measured using a Sloan Dektak II profilometer. The etch rate given in microns per pulse is determined by the measured etch depth averaged over the total number of pulses.

RESULTS AND DISCUSSIONS

In Fig. 1, the etch rates per pulse, measured at a constant fluence level of 0.2 J/cm^2, are plotted against the absorption coefficient (cm^{-1}) of the polymer. A hypothetical etch rate curve is also drawn based on the assumption that the thickness of material ablated per pulse is solely determined by the absorption depth of the laser light which is the reciprocal of the absorption coeffcient. The data points lying below this line imply the etching depth per pulse are smaller than the optical absorption depth for a given polymer, while those data points above the straight line imply the etching depths are larger than the optical absorption depths.

As it is shown in Fig. 1, the etch depths measured for strongly aborptive polymers such as polyimide are substantially larger than the optical absorption depths of these polymers. On the other hand, for weakly absorptive polymers such as PMMA, the etch depth per pulse was significantly less than its absorption depth. For all polymers studied, we observed that etch rates vary only by less than a factor of 5 for irradiation at both 193 nm and 248 nm, in spite of the fact that optical absorption coefficients, and hence the optical aborption depths in these polymers vary by more than three orders of magnitude.

From a thermal analysis, the temperature rise at the surface of a PMMA sample is about 100 C for a fluence of 0.2 J/cm^2. For polyimide, having a much higher absorption coeffcient, this value is above 15,000 C at the same irradiation fluence level. Obviously, different etching mechanisms have to be considered.

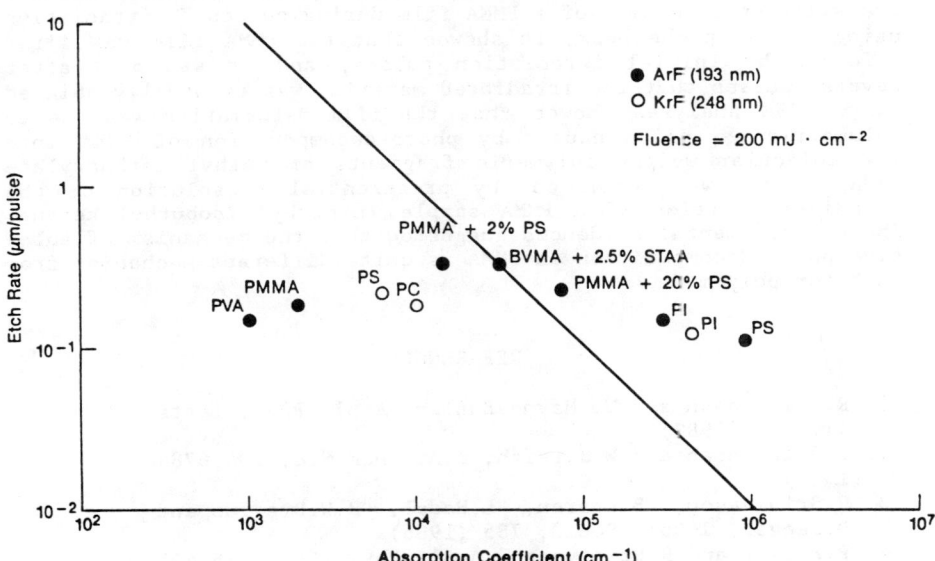

Fig. 1 Ablative photo-etching rates of various polymers plotted against the absorption coefficients at an irradiation fluence level of 0.2 J/cm^2 for 193 nm and 248 nm. The line drawn represnets the absorption depth (in microns) corresponding to the absorption coeficient.

The observed etch depth per pulse for polyimide is about 5 times the optical aborption depth. This suggests that photo-ablation takes place continuously during the intra-pulse period. Since the thermal diffusion length in a polymer is less than 0.01 micron over a period of 1 ns, rapid temperature rise takes place in and is confined to a very thin surface layer of the irradiated material. The rapid volumetric expansion due to localized heating can be considered quasi-adiabatic and rapidly leads to ruptures of the molecular bonds to cause a portion of the material to eject. This is consistent with a time-dependent etching model which has been proposed recently.[12]

On the other hand, the measured etch depth per pulse for PMMA at 0.2 J/cm^2 was almost two order of magnitude smaller than its optical absorption depth. With a temperature rise estimated to be less than 100 C in PMMA, thermal effects are insignificant at this irradiation fluence level. By measuring the in-situ optical transmission intensity of a PMMA film during pulsed UV irradiation using a HeNe probe beam, it showed that the PMMA film was first deformed by initial irradiation pulses, and it was only after several pulses that the irradiated material was eventually ablated away. SEM analyses showed that the film deformation was due to volumetric expansion caused by photo-decomposition of PMMA into low molecular weight polymeric fragments of methyl methacrylate (MMA). This was confirmed by preferential dissolution of the irradiated portion of a PMMA sample in methyl isobuthyl ketone. These experimental evidences suggested that the mechanism of ablative photo-decomposition in PMMA is quite different mechanism from that for polyimide.

REFERENCES

1. R.Srinivasan and V. Mayne-Banton, Appl. Phys. Letts. 41, 576 (1982)
2. R.Srinivasan and W.J.Leigh, J.Am.Chem.Soc.,104,6784 (1982)
3. R.Srinivason, B.Braren, L.Hadel, R.W.Dreyfus and D.Seeger, J.Opt. Soc.3, 785 (1986)
4. P.E.Dyer and R.Srinivason, Appl. Phys. Lett.,48,445 (1986)
5. G.Gorodetsky, T.G.Kazyaka, R.L.Melcher and R.Srinivason, Appl. Phys.Lett., 46, 828 (1985)
6. G.Koren and J.T.C.Yeh, Appl.Phys. Lett., 44,1112 (1984)
7. G.M.Davis, M.C.Gower, C.Fotakis, T.Efthimiopoulos and P.Argyrakis, Appl. Phys. A36, 27 (1985)
8. H.H.G.Jellinek and R. Srinivason, J. Phys. Chem.,88, 3048 (1984)
9. H.R.Philipp, H.S.Cole, Y.S.Liu and T.A.Sitnik, Appl. Phys. Lett.,48, 192 (1986)
10. H.S.Cole, Y.S.Liu and H.R.Philipp, Appl.Phys.Lett. 48,76 (1986)
11. H.S.Cole, Y.S.Liu, H.R.Philipp and R.Guida, Proc. of Materials Research Society Spring Meeting, Apr. 15-19, (1986)
12. E. Sutcliffe and R. Srinivasan (to be published)

DYNAMICS OF THE ULTRAVIOLET LASER ABLATION OF CORNEAL TISSUE

R. Srinivasan
IBM T. J. Watson Research Center
Yorktown Heights, NY 10598
U.S.A.

ABSTRACT

When pulsed, ultraviolet laser radiation falls on the surface of an organic polymer or biological tissue the material at the surface is spontaneously etched away to a depth of 0.1 to several microns. The process is characterized by the control which can be exercised over the depth of etching by controlling the temporal width of the laser pulse and its fluence (energy per unit surface area) and by the lack of detectable thermal damage to the substrate.

INTRODUCTION

Studies on the interaction of ultraviolet laser pulses with polymer films led to the discovery in 1982 of the phenomenon of "ablative photodecomposition" which results in the degradation of the structure of the polymeric solid by the photons and the expulsion of the fragments at supersonic velocities.[1] The result is an etch pattern in the solid with a geometry that is defined by the light beam. The principal advantages in using ultraviolet laser radiation rather than visible or infrared laser radiation for this purpose lie in the precision ($\pm$ 2000 Å) with which the depth of the cut can be controlled and the lack of thermal damage to the substrate to a microscopic level.

LASER ABLATION CHARACTERISTICS

Laser pulses from an excimer laser are typically about 20 nanoseconds long and can be repeated at a frequency of 0.1 to 300 pulses per second. The photons that are emitted can have wavelengths of 193nm, 248nm, or 308nm depending upon the composition of the gas mixture in the laser.[2] Fig. 1 is a schematic representation of the impact of such a pulse of light on the surface of a polymer or tissue. At the fluences used in this work, the penetration of the radiation through the polymer can be assumed to follow Beer's Law. In weak absorbers such as the cornea or polymethyl methacrylate (=PMMA) which is a good model polymer for corneal tissue, a pulse of laser radiation of 193nm wavelength will penetrate to 6.5μm depth before 95%

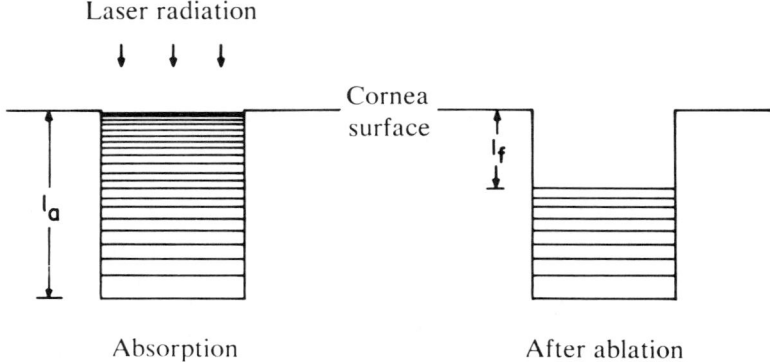

Fig. 1. Schematic representation of impact of laser pulse on polymer surface.

of it would have been absorbed. In strong absorbers, the penetration depth can be just a few thousand Angstroms. If the fluence, F of the laser beam which is defined as the energy deposited per unit area of the surface per pulse, exceeds a certain threshold value, F_o, then a depth, l_f of the material will be ablated by the pulse. If the fluence is such that $l_f < l_a$ where l_a is the depth of penetration of the radiation, then a depth $l_a - l_f$ which had been exposed to the light will be left behind. The next pulse will go through this partly irradiated material as well as through virgin material underlying it. The first pulse is therefore unique. But after the first few pulses, there is a linear relationship between the number of pulses and the depth that is etched. In practice, the depths etched by varying numbers of pulses are averaged and noted as the etch depth per pulse for that material at that wavelength and fluence. This value is reproducible within the uncertainty (± 8%) in the measurement of the etch depth and the fluence of the laser pulse. This is the reason that the etch depth can be reproduced to ± 2000 Å in most materials.

A typical plot of the etch depth/pulse as a function of fluence is shown in Fig.2 for PMMA.[3] The two wavelengths are those at which PMMA has a useful absorption. It should be emphasized that at 308nm, even though PMMA has an insignificant absorption, laser pulses will etch the sample but the thermal damage to the substrate will be obvious. The etch depth is independent of the atmosphere in which

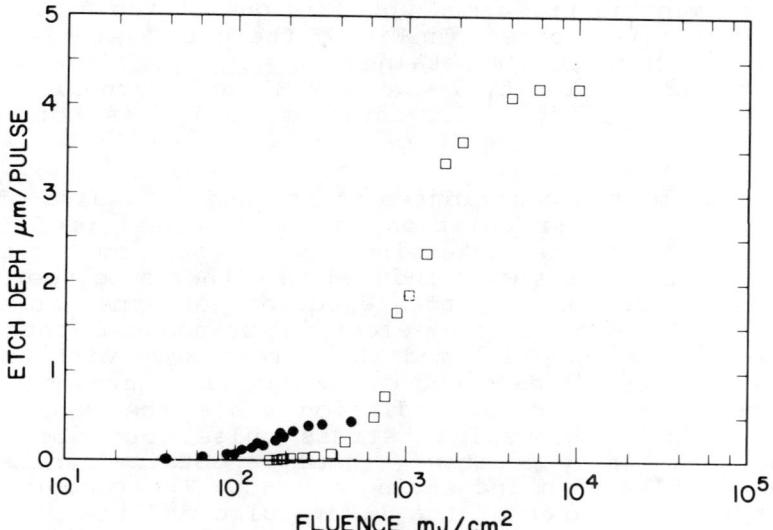

Fig. 2. Plot of etch depth/ pulse as a function of fluence for PMMA.
● 193 nm ; □ 248 nm.

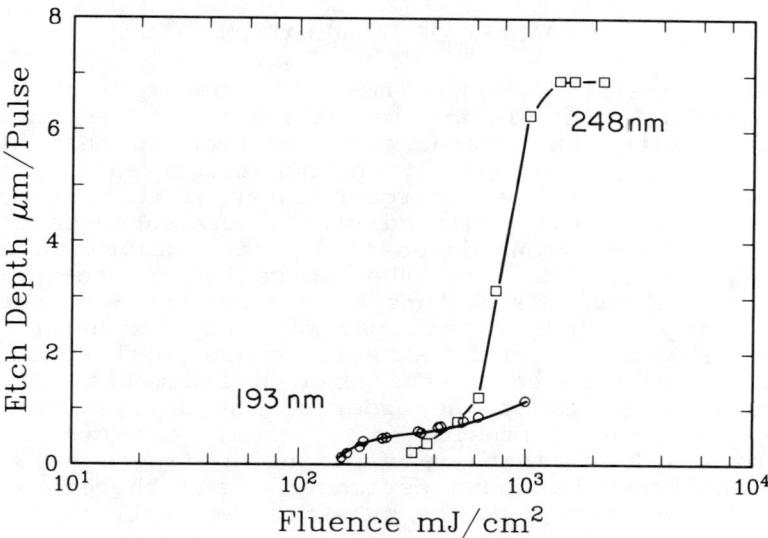

Fig. 3. Plot of etch depth/ pulse as a function of fluence for rabbit cornea.
o 193 nm ; □ 248 nm. The solid lines are for the purpose of visualization.

the experiment is performed which is one of the attractive features of this process. In Fig.3, there is a similar etch plot for rabbit cornea obtained <u>in vitro</u> by Krueger and Trokel.[4] The similarity between PMMA and corneal tissue in their sensitivity to UV laser radiation is not accidental.

A key datum that contributes to the understanding of the dynamics of UV laser ablation is the temporal profile of the process which was determined by a photoacoustic measurement of the stress wave induced in either a polymer film or corneal tissue by the ejection of the ablating material.[5] A thin, Piezo- electric transducer that was attached to the sample timed the stress wave with an uncertainty of ± 1 nanosecond. Below the threshold for ablation, a pulse of UV radiation heats the sample and gives rise to a sinusoidal stress pulse. But above the threshold, a strong positive signal is obtained which increases rapidly with increasing fluence. The temporal delay between the onset of the laser pulse and the onset of ablation is 7 nanoseconds for polymers which is well within the width of the laser pulse. The data obtained for rabbit cornea are shown in Fig. 4. The width of the ablation pulse is comparable to that of the laser pulse. With increasing fluence, the start of the ablation is seen to fall within the temporal width of the laser pulse.

MECHANISM OF UV ABLATION

It is generally accepted that the absorption of UV photons results in electronic excitation. The excited electronic state can undergo decomposition in that state itself which would be a purely photochemical reaction. Or, if the excited molecule undergoes internal conversion to a vibrationally excited ground state, any subsequent decomposition can be considered to be the equivalent of a thermal process. This is the so-called photo-thermal mechanism in which the photons merely act as a source of thermal energy. Along either pathway, any excess energy over that needed for bondbreak will remain in the products and will be dissipated in the ablated fragments. If the time for ablation is of the order of the duration of the laser pulse, the diffusion of thermal energy to the substrate would be minimal - a diffusion length of a few hundred Angstroms has been estimated - and therefore the lack of thermal damage to the substrate would be expected.

Early views of this phenomenon implicitly assumed that ablation followed the deposition of all of the photons of the laser pulse in the tissue or polymer. But the photoacoustic experiments mentioned in the previous section typically showed that when a small fraction of the

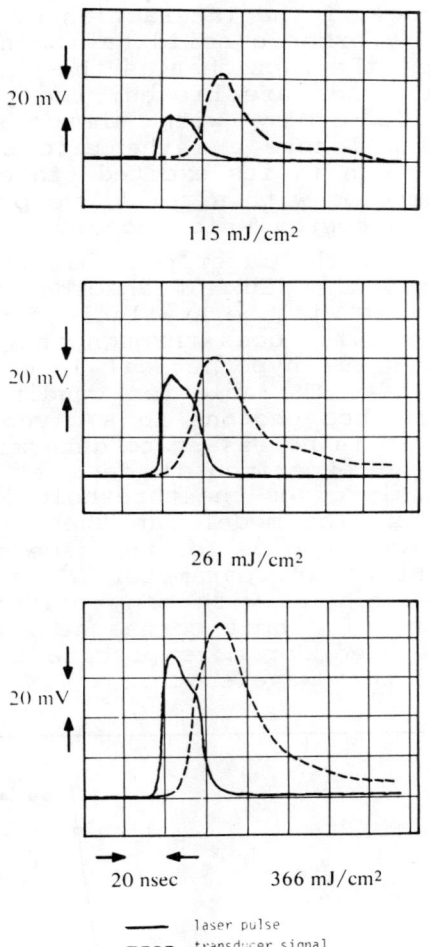

Fig. 4. Synchronized signals from laser pulse and Piezo-electric transducer at various fluences.
Sample: Cornea; 193 nm wavelength.

pulse energy had been deposited in the film, the material started to ablate. The bearing of the time profile of ablation to the mechanism lies in calculations of the temperature rise in the volume of material that ablated. These calculations which form the basis for the photothermal mechanism significantly overestimated the temperature. It can be shown that the actual rate of thermal decomposition would not be adequate to lead to ablation in the time span of the laser pulse. Detailed calculations for the ablation of the cornea have been presented elsewhere.[5]

A realistic model of the interaction of UV laser pulses with tissue or any organic solid has been proposed which takes into account the fact that as the photons penetrated the material and bonds are broken, the photon intensity at any depth will evolve with time.[6] A photochemical mechanism is assumed here and the rate of decomposition of the tissue protein in its excited state is viewed as a process which competes with alternative processes for the degradation of the energy which do not lead to decomposition. The width of the laser pulse is a critical factor because it is only the flux of absorbed photons above a threshold value that will be available for bond-breaking. As a single laser pulse goes through the tissue, its absorption in successive hypothetical layers is calculated by using Beer's Law. The ablation condition is met when the number of absorbed photons in a given volume exceeds a certain value. It is necessary to determine the ablation condition from the experimental data at one wavelength which simultaneously gives the threshold level of the absorbed photon flux. The model can then be applied to UV etching at other wavelengths for the same material without further adjustment of any parameter. This model has been found to work quite well with many polymer samples. Its fit to the etching of rabbit cornea at 248nm is shown in Fig.5. There is need for more precise data at 193nm to test the model at this wavelength.

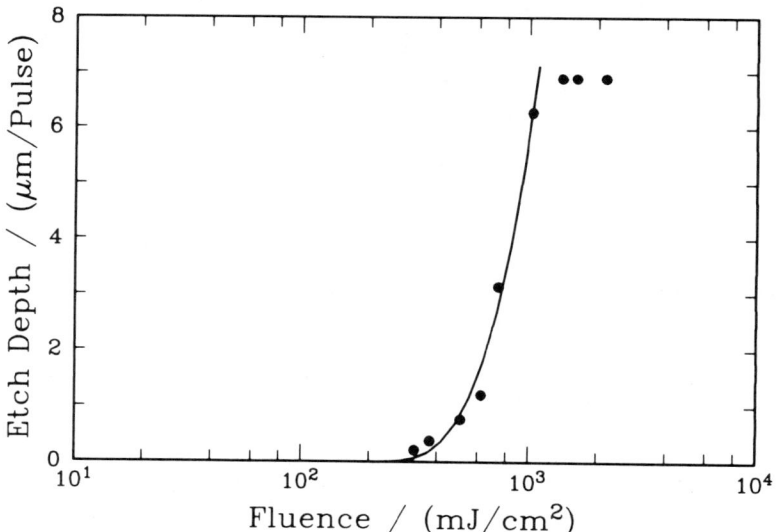

Fig. 5. Etch depth/ pulse as a function of fluence for rabbit cornea at 248nm.
The data points are from Ref. 4. The solid line is the fit according to the dynamic theory.

In summary, at present there is very little evidence to support a purely photothermal mechanism for the UV laser ablation of any polymer or biological tissue. A photochemical mechanism is more probable but it may be much more complex in nature than originally proposed. The process of excitation may include not only the first upper electronic state but, through multi-photon processes, higher states as well.

Acknowledgment: The author wishes to thank Ms. Bodil Braren for her assistance in preparing this manuscript.

REFERENCES

1. For a recent review see R. Srinivasan, Science, 234, 559 (1986).
2. H. Pummer et al., Laser Optoelektron., 17, 141 (1985).
3. R. Srinivasan et al., Macromolecules, 19, 916 (1986).
4. R. R. Krueger, S.L. Trokel, Arch. Ophthalmol., 103, 1741 (1985).
5. R. Srinivasan, P.E. Dyer, B. Braren, Lasers Surg. Med. 6, Dec (1986).
6. E. Sutcliffe, R. Srinivasan, J. Appl. Phys., 60, 3315 (1986); Lasers Ophthalm.,(in press) 1987.

INTERNAL BIOLOGICAL TISSUE TEMPERATURE MEASUREMENTS USING ZIRCONIUM FLUORIDE FIBER

Ed Sinofsky, Gary Gofstein
USCI, Div., C.R. Bard, Billerica, MA 01821

ABSTRACT

The resolution and construction of a zirconium fluoride optical waveguide coupled to a cooled PbSe detector is explored for remote radiometry in the 3-5um range. We obtain a spatial resolution of better than .5mm and thermal resolution of 3C at room temperature with immunity to high levels of laser irradiation.

INTRODUCTION

Our goal of developing a remote thermography apparatus with high spatial resolution and immunity to high power laser radiation in the near infra-red region is primarily motivated by the lack of verifying data in the modeling of laser-tissue interactions. The mechanisms of laser interaction with complex random media are still poorly understood; predictive models of absorption and temperature as a function of depth need experimental calibration.

The essential problem is the detection of temperature internal to the tissue without disturbing any of the ambient conditions such as density and uniformity. Blackbody radiation is emmitted by all objects above absolute zero. The radiation in the neighborhood of room temperature (25C) peaks in the 10um range and decays rapidly. Wavelength short of 3um are almost devoid of radiation relative to the 10um peak. Unfortunately, tissue is transparent only in the near infrared making it clear that an invasive window to the tissue interior is necessary. Optical fiber provides such a window in the wavelength range of 3-4.5um where the radiation varies greatly with temperature. The recent availability of improved fluoride glass fibers with low loss mid-infrared transmission makes radiometry of small, remote volumes possible without the absorption and noise pickup of thermocouples.

Methods which others have used to make this measurement have serious drawbacks, most notably the use of thermal cameras directly aimed at a cross-section of tissue. The need to section the tissue exposes it to different thermal boundaries than exist in the intended application. The presence of steam around the tissue is highly absorbing in the mid IR range, adding another source of uncertainty Because tissue is mostly water, we assume that it has a similar high emmissivity (greater than .9) in the 3 to 5um range.

EXPERIMENTAL SETUP

We use a section of ZrF fiber of 3cm length, 200um outer

diameter with 150um core diameter to transmit radiation in the wavelength rande 3.-4.5um from the launching end within the tissue by optical waveguiding.

We measured a loss of 1dB from the launching to the receiving end caused by two effects: fresnel reflections and scattering due to contamination at the fiber ends, and surface irregularities and fiber impurities. The fresnel loss is calculable to be 0.6 dB:

The radiation emmited by the proximal end of the fiber is optically modulated using a tuning fork chopper at 1Khz to tag it and enable it to be detected by a lockin amplifier. The chopper has polished blades in order to lower its emmissivity and increase the difference between the chopped signal on and off states. Because of the abundant background radiation, in this case at 19C, the low temperature resolution (below 30C) is decreased as the blackbody spectra of the background and the tissue approach equality.

Ideally the chopper should be of about the same size as the fiber core so that only the fiber radiation is chopped. Such geometry can be practically realized with an occulting wire of the core diameter, vibrating very close to the core. Because most choppers are quite large compared to the fiber size they also chop a considerable portion of the background radiation at 19C. This results in a large offset at the phase sensitive detector.

After chopping, the radiation is optically filtered by a longwave-pass filter with a cutoff at 3um to eliminate visible light interference and signal contamination by irradiating lasers, at or below 1.3um. The filter is a multilayer dielectric on a silicon substrate and has a pass band attenuation of 0.9dB due to fresnel reflections. At higher temperatures (65c) the 3um cutoff results in the loss of some signal (3% at 90C). A potential problem with this element is laser absorption in the silicon leading to heating and false temperature readings.

The optical signal is then spatially filtered by a 650um diameter nearly circular aperture to filter the 1Khz chopped background to an acceptable level so that its noise variations will not obscure the signal variations.

Radiation emmitted by the detector side of the aperture contributes to background noise as well. Although not chopped at 1Khz, its temperature is related to that of the detector cooler heatsink which may vary considerably depending on the quality of the thermal joints. These large temperature variations permit noise, especially low frequency drifts, to effect the output even under the tight (.025Hz) phase-frequency filtering.

Without an imaging lens, not all of the output from the fiber reaches the 1mm x 1mm detector. By assuming a uniform distribution of light throughout the solid angle, we calculate a loss of 4.1dB, due to spreading of radiation from the output of the fiber as it traveles to the detector.

A single stage thermoelectrically cooled (-25C) PbSe photodetector with a bandgap cutoff at 4.5um is used to detect the thermal radiation. The position of the detector long wavelength cutoff is nearly coincident with unrelated to the signal from entering the electrical system. We hoped to evaluate a colder detector as well, to determine if the improved detectivity and responsivity of colder PbSe would counteract the background noise increase due to the increase in cutoff wavelength. The distance between the saphire window of the detector and PbSe itself proved to be a loss mechanism because of the diverging radiation problem mentioned above.

PbSe was chosen because of its matching band-gap and high detectivity in the range of interest. The narrow bandwidth makes the photon detector the detector of choice over the energy detector since the photon detector has 100 times the detectivity of the thermal detector at the cutoff wavelength.

Following the manufacturers data sheet for 500K blackbody performance we biased the detector with 48V from shielded alkaline batteries, bypassed a 2uF capacitor and terminated through a 1M resistor.

The detector is cooled to about -25C thermoelectrically and is AC soupled via a low leakage 2uF capacitor to a discrete JFET (U430) differential preamp with an input noise voltage of 10nV/Hz at 1Khz. The preamp drives a low noise op-amp (OP05) for a final gain of 32.1dB, measured from the detector output to the amplifier output.

The amplified and detected signal is processed by a lock-in amplifier deriving its timing from an inductive pick-up on the tuning fork modulator. The lock-in phase-frequency filters the signal with an effective bandwidth of .025 Hz and can perform signal averaging by adjusting the time constant on the output filter.

A long time constant, however, limits the real time usefulness of the device, since at least half a time constant is necessary to estimate slowly varing signals. A 1s time constant requiring a S/N ratio about 3 times higher is probably the best usable for manual scanning and recording, although chart recording and motor scanning could use more like 1ms time constant.

FLOURIDE GLASSES

Fluoride glasses of the type drawn into fibers were first made in 1974. As materials they are formed from at least ternary mixtures of fluorides usually containing zirconium fluoride. The glass region of the possible mixtures is small and crystallization, particularly during fiber drawing, is common. Such crystallization weakens the fiber, initiating an unwanted cleave.

The major advantage of these glasses is that they do not need to contain oxygen derivatives such as OH which have strong near and mid-infrared absorption bands, while being easier to work with than true salt crystals by virtue of their flexibility. The use of coatings on the fiber has greatly improved the durability and flexibility of ZrF even during the course of this work. Despite the difficulty in making a high quality product, halide glasses are inherently optically superior to silicas with theoretical losses of .001 dB/Km and transmissions out to 5um for true glass chlorides, out to 11um for poly-crystalline chlorides.

The ZrF fiber available now is brittle and water soluble. We encased it in a protective sleeve of silica tubing for strength. The silica is also transparent to the irraditing laser and of low thermal conductivity preventing a disturbance of the steady state thermal conditions in the fiber vicinity.

ANALYTIC SIMULATION

In order to predict the sensitivity of the fiber radiometer to various parameters and to specify component requirements, we use an IBM PC/AT to model the system components and the radiating tissue. Since bandwidth is narrow and limited by the fiber, we use a photon detector for maximum detectivity. All optical signals are therefore measured in photons or quanta (q). The radiated sterances (q sec m sr) of the target tissue can be integrated over known wavelength bandwidth at given temperature and emmissivity using the Planck blackbody formula.

Knowing the radiation sterance, L, the photon flux for an arbitrary geometry can be calculated:

$$\phi = L A \Omega$$

The fiber has a numerical aperture of 0.2 and core diameter of 150um which define the entering flux measured in q/sec. The exiting flux minus the dB losses of two fresnel reflections and lumped scattering/absorption losses. Because of the need to place the chopper, filter and aperture between the fiber and the detector substrate, the usable NA is only .125 for a 1 x 1mm detector. This geometry was modeled as a loss of 4.1dB incurred by exiting photons which never reach the detector. Combining the input flux and the total attenuation of the optical system, the number of photons/sec reaching the detector can be predicted, giving a curve of output voltage vs temperature.

The ratio of the output signal in volts to the photon rate (no detector amplification), is the photon responsivity of the detector. This parameter is easier to use than the energy responsivity quoted on the manufacturer's data.

Predicting the photon flux using the computer, over the wavelength range of 1 to 4.4 Im, shows the responsivity is to be $3 \times 10"$ V/q/sec. This number allows a voltage vs temperature curve to be calculated for any input geometry.

A chopper temperature of 22C and emmisivity of 1 fit the data and are used as parameters for simulating the available signal through the 150um fiber. Signal levels in the mid 20C's are about 100nV, below the resolution of any instruments available to us; preamplification was therefore required.

We found that the longwavelength cutoff has a very great effect on the degree of signal present at the temperatures of interest, and signal at wavelengths shorter than 3um contributes only

insignificant radiation in the desired temperature range (80C).
In order to verify the performance of the system a series of measurements were made under controlled conditions. Fig. 1 shows the comparison of predicted vs/experimentally obtained signals from a calibrated balckbody source.

CONCLUSION

Experiments showed that ZrF fiber is too sensitive to water to be used unprotected within tissue. Heating a dry clay surface with argon irradation allowed a remote temperature measurement. Experimental results matched theoretical predictions quite closely. However, a similar measurement made by inserting the fiber into wet arterial plaque could not be made.

A probable explanation for this effect is the alteration of the index profile or scattering characteristics of the launching region, resulting in a loss of waveguiding properties of the ZrF fiber. Such a drawback can be overcome by using a protective window of sapphire, which is durable, waterproof and transmitting to over 5um in this thickness.

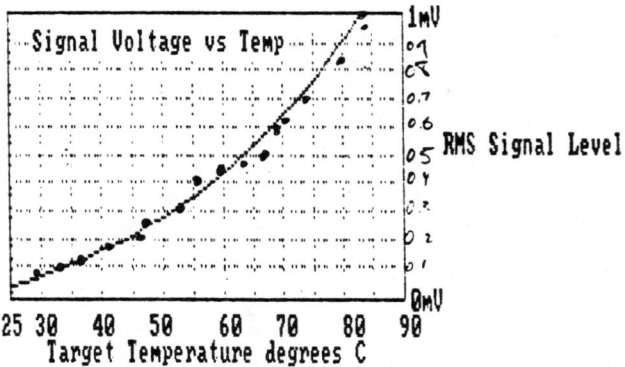

Fig. 1: Fiber radiometer signal vs temperature for:
  o   Experiment
  -   Theory

TISSUE DIAGNOSTICS USING LASER-INDUCED FLUORESCENCE TECHNIQUES

P.S. Andersson, J. Ankerst*, E. Kjellén*, S. Montán,
K. Svanberg* and S. Svanberg
Department of Physics, Lund Institute of Technology,
P.O. Box 118, S-221 00 Lund, Sweden

*Lund University Hospital, S-221 85 Lund, Sweden

ABSTRACT

The fluorescence emission from tissue that is irradiated with UV-light can be utilized for diagnostic purposes. The discrimination between tumors and normal tissue is of particular interest. We discuss natural tissue fluorescence as well as fluorescence due to injected hematoporphyrin agents. Results from studies on rats and mice are reported, as well as data from human biopsy samples. The development of clinical instrumentation for point measurements and imaging is discussed.

INTRODUCTION

Fluorescence monitoring of tissue is emerging as a promising technique for non-intrusive real-time diagnostics, e.g. for tumor localization. For some time the specific red fluorescence from previously injected tumor-seeking hematoporphyrin derivative (HPD) molecules has been used for such purposes. Recent progress in the field of tumor localization using HPD is described in the Proceedings of the First International Conference on the Applications of Photosensitization for Diagnosis and Treatment[1]. However, even without any agent injection tissue exhibits certain features in the fluorescence spectrum that can be used for tissue characterization. Several studies of this auto-fluorescence have been performed by our group[2-4] and others[5-6]. It seems that a combined use of auto-fluorescence and HPD features is particularly useful for tumor detection. We have shown that an enhanced contrast between tumor and surrounding muscle can be obtained for an experimental rat tumor system by dividing the background-free HPD fluorescence intensity by the blue tissue auto-fluorescence[7-9]. The tumor was found to be characterized by an increase in the HPD signal at the same time as the blue fluorescence decreased. Monitoring a ratio or any other dimensionless quantity also has important advantages, in particular in endoscopic (fiber-optic) applications:
1. Immunity to distance variations
2. Immunity to surface topography
3. Immunity to variations in excitation or detection efficiency
4. Immunity to wavelength-independent attenuation

Fluorescence measurements can be performed with a point monitoring device or with an imaging instrument. Recently part of the available instrumentation has been rewieved[10]. Point monitors can either be of filter type[11-13] or can employ optical multichannel analyzer techniques for full spectrum recovery[14]. The simplest type of fluorescence imaging is visual inspection of the UV-illuminated region. Electronic images can be obtained using intensified vidicon or diode matrix detectors[15], that detect the red fluorescence light in a selected passband. In one construction background can be intermittently subtracted by switching to a blue-transmitting filter[16]. Simultaneous monitoring in several selected bands and subsequent forming of an optimized dimensionless contrast function is possible using the computer-enhanced multi-color fluorescence imaging concept[9,3,17].

Fluorescence monitoring of tissue has mostly been focussed on cancer tumor detection. An emerging application is vessel characterization in connection with laser-based angioplasty[6]. In the present paper we will mostly discuss the former application but we will also present some results from our first studies related to the second field. Recent laboratory results from rats and mice are presented as well as some human biopsy specimen observations.

FLUORESCENCE MONITORING

Throughout this paper laser-induced fluorescence data are presented that have been obtained with an experimental set-up of the type shown in Fig. 1. In most of our studies an N_2 laser emitting light at 337 nm has been employed, but a XeCl (308 nm) excimer laser

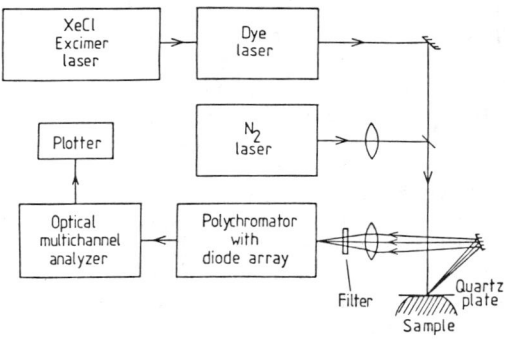

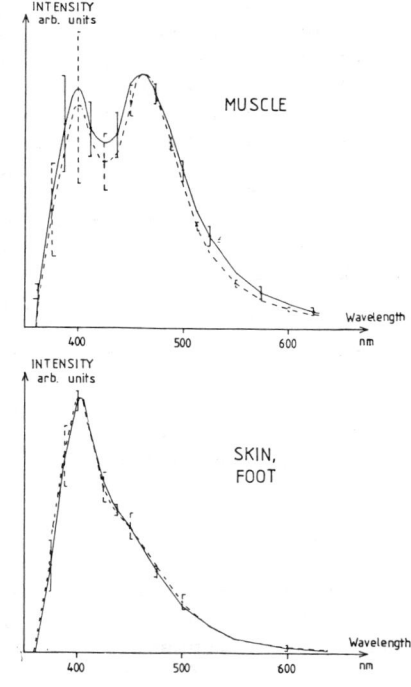

Fig. 1. Experimental set-up used in the studies of tissue fluorescence.

Fig. 2. Auto-fluorescence spectra of muscle fascia and skin of mouse. The full lines correspond to live animals and the dashed ones to dead animals, about 4 hours after sacrifice. Error bars indicate one standard deviation. (Ref. 4).

and an excimer-pumped dye laser have also been utilized. The radiation is directed onto the sample by means of mirrors. Fluorescent light is collected and directed to an optical multichannel analyzer system, that captures the entire fluorescence light distribution for every laser pulse. Spectra can be stored on floppy disks and be printed out on paper. Fluorescence intensities are measured in terms of a standard.

Extensive data on the fluorescence of different tissues of rat at various times after the injection of HPD have been presented in our previous papers[7,8]. In order to ascertain the validity of the approach used we have compared the fluorescence properties of live tissue and the tissue of animals that were sacrificed a few hours earlier[4]. Clearly, it is much easier to perform systematic fluorescence studies on sacrificed animals. In Fig. 2 comparative data for nude mice, first anesthetized with chloral hydrate and then sacrificed are shown. As can be seen the spectral features remain basically the same after sacrifice.

HPD molecules have their strongest absorption in the Soret band around 405 nm. For this reason violet excitation has normally been used to induce fluorescence. However, the contrast attainable between tumor and surrounding normal tissue for a certain injected dose of HPD is even more important than a strong fluorescence intensity. To enhance the contrast we also use the blue tissue auto-fluorescence, which also depends strongly on the excitation wavelength. We have performed extensive studies on induced tumors in Wistar/Furth rats using excitation wavelengths ranging from 308 to 405 nm[18]. In order to assess the contrast, scans extending from normal tissue into the tumor were performed. As an example such a scan for 337 nm excitation is shown in Fig. 3. In the inserted spectrum the relevant signal levels are denoted. In this scan it can be seen that by monitoring the characteristic HPD fluorescence intensity at 630 nm (A') low contrast is achieved. By subtracting the background and plotting the A level the situation is much improved. Rather than displaying a general fall-off in the blue fluorescence intensity (B) for the tumor the intensity falls off in particular at the tumor edges. In an A/B representation a high contrast is achieved and the tumor edges are strongly enhanced. The manual scan indicates what would be achievable with an imaging system. The full material is now being evaluated. In these studies we noted that a strong exposure to laser light can lead to sample bleaching thus changing the fluorescence characteristics. We have noted that an additional peak, at about 650 nm, can be induced in this way. Such a peak has previously been observed and utilized in human tumor detection[19].

An important aspect of tissue fluorescence characteristics is possible guidance in asssessing radicality in surgical tumor resection. Here the real-time capability of the fluorescence technique is of particular interest. We have studied a rat brain tumor system and find very encouraging results[20]. In Fig. 4 the results from a scan across the tumor are shown. 337 nm excitation is used in the measurement on rats that were injected with Photofrin II three days earlier. We again note, that background-free HPD monitoring strongly enhances the contrast, which is still further enhanced by normalizing to the

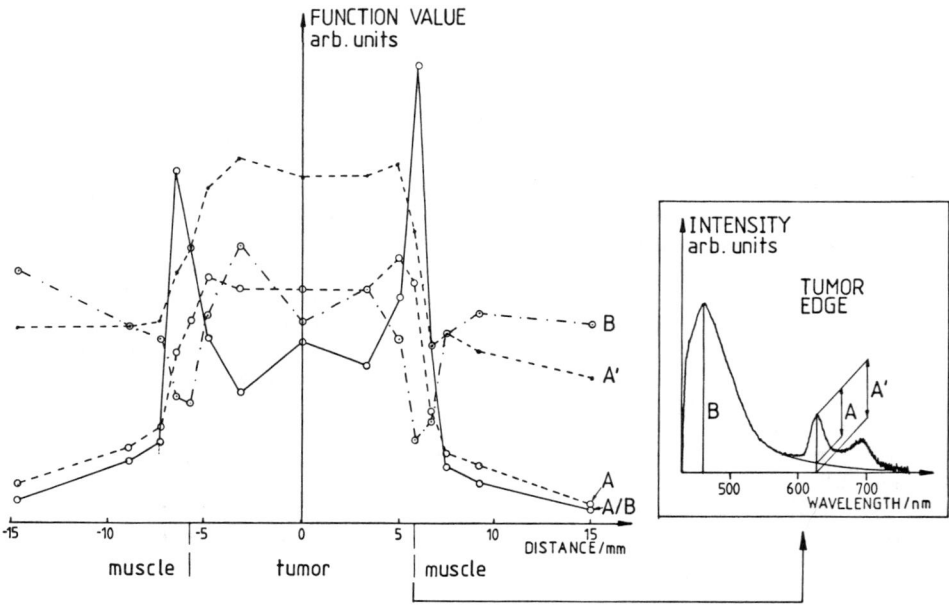

Fig. 3. Fluorescence results from a scan across a rat tumor in muscle.

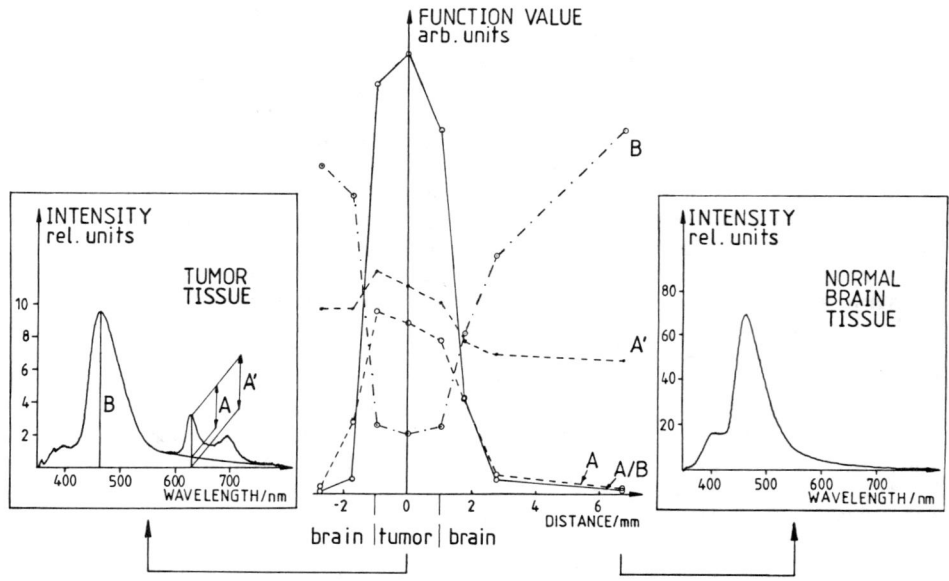

Fig. 4. Results from a scan across a tumor in rat brain. (Ref. 20).

blue fluorescence, which is strongly reduced for the tumor tissue. In the A/B representation a contrast of about 80 is achieved. The data in the figure were obtained using the low dose of 1 mg/kg bodyweight of the drug.

Recently the photodynamically more potent drug Photofrin II was introduced. This agent is enriched in DHE (dihematoporphyrin ether), which has been shown to be the most active porphyrin component in the previously used deriviative prepared by a modified Lipson procedure (Photofrin). It is well known that porphyrin dimers fluoresce much more weakly than monomers. Studies on saline solutions of Photofrin and Photofrin II solutions show[21], that the fluorescence is basically due to the monomeric "impurity" hematoporphyrin HP, and that the more pure substance Photofrin II exhibits a 3 times lower characteristic red fluorescence. However, in tissue strong transformational processes occur between the porphyrins. In order to evaluate the relative merits of the two agents with regard to tumor localization using fluorescence we have performed a comparative study on Wistar/Furth rats with inoculated tumors[22]. Groups of rats were injected with the same dose of the two agents in a blind test. In tissue Photofrin II gives rise to stronger specific fluorescence than Photofrin for the same injected drug concentration. However, since the more pure substance is injected at a reduced concentration (for cost and skin photosensitization reasons) it is not clear that the pure drug has an advantage over the previous one for the detection of tumors. We also observe a tendency towards reduced contrast for DHE since the non-tumor tissues seem to yield a relatively higher specific red fluorescence than for HPD. Clearly, these indications have to be further studied for definite assessment.

In Fig. 5 some results from our first experiments on fluorescence characterization of vessel status are given[23]. Different samples from a newly desceased 85 year-old person were studied using 337 nm excitation. Clear spectral differences between normal artery wall and wall modified by atherosclerotic plaque were found including the differences in the 550-600 nm region previously reported by Kittrell et al.[6]. Radiation at 337 nm has a much lower penetration depth than at 480 nm, the wavelength used in Ref. 6., and this has to be considered when interpreting the data. We plan to perform more extensive studies of this kind at different excitation wavelengths.

DISCUSSION

Fluorescence diagnostics of tissue has an interesting potential as a clinical aid both for localizing small occult cancer tumors and for ensuring radicality in surgical tumor resection. The full utilization of the available spectral information is important, particularly in the battle to keep the concentration of agents also causing ambient light hypersensitization at the lowest possible level. Tissue auto-fluorescence also provides interesting possibilities outside the tumor localization area. The development of powerful equipment, in particular with imaging capability, using extended spectral pattern recognition approaches seems to be a particular challenge.

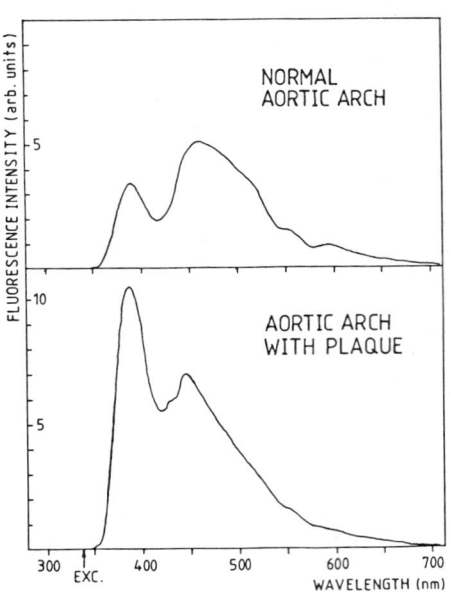

Fig. 5. Comparison between the fluorescence spectra from normal and damaged human aortic arch tissue. For 337 nm excitation, plaque was found to be characterized by a sharp peak at 450 nm and a reduced structure in the 550-600 nm region. (Ref. 23).

ACKNOWLEDGMENTS

The authors gratefully acknowledge valuable discussions with and encouragement from the other members of the Lund HPD Group. This work was supported by the Swedish Cancer Foundation (RmC) and the Swedish Board for Technical Developments (STUF).

REFERENCES

1. Y. Hayata (ed)., Proceedings of the First International Conference on the Applications of Photosensitization for Diagnosis and Treatment, Tokyo, April 30 - May 2, 1986, to appear.
2. S. Montán, Diploma Paper, Lund Reports on Atomic Physics LRAP-17, 1983.
3. J. Ankerst, S. Montán, E. Sjöblom, K. Svanberg and S. Svanberg, L.I.A. ICALEO 84, 43, 52 (1984).
4. P.S. Andersson, E. Kjellén, S. Montán, K. Svanberg and S.Svanberg, to appear.
5. Yanming Ye, Yuanlong Yang, Yufen Li and Fuming Li, CLEO'85 Technical Digest, p. 84.
6. C. Kittrell, R.L. Willett, C. de los Santos-Pacheo, N.B. Ratliff, J.R. Kramer, E.G. Malk, and M.S. Feld, Appl. Opt. 24, 2280 (1985).
7. J. Ankerst, S. Montán, K. Svanberg and S. Svanberg, Appl. Spectr. 38, 890 (1984).
8. K. Svanberg, E. Kjellén, J. Ankerst, S. Montán, E. Sjöblom, and S. Svanberg, Cancer Res. 46, 3803 (1986).

9. S. Montán, K. Svanberg, and S. Svanberg, Opt. Lett. 10, 56 (1985).
10. H. Kato and D. Cortese, Clinics in Chest Medicine 6,237 (1985).
11. J.H. Kinsey and D.A. Cortese, Revs. Sci. Instr. 51, 1403 (1980).
12. P.S. Andersson, S.E. Karlsson, S. Montán, T. Fersson, S. Svanberg and S. Tapper, to appear.
13. A.E. Profio, D.R. Doiron and J. Sarnaik, Med. Phys. 11, 516 (1984).
14. K. Aizawa et al. in Porphyrin Localization and Treatment of Tumours, (Alan R. Liss 1984) p. 227.
15. A.E. Profio, D.R. Doiron, O.J. Balchum and G.C. Huth, Med. Phys. 10, 35 (1983).
16. A.E. Profio, M.J. Carvlin, J. Sarnaik and L.R. Wudl, in Porphyrin in Tumor Phototherapy, Eds. A. Andreoni and R. Cubeddu (Plenum 1984) p. 321.
17. S. Montán and S. Svanberg, Patent pending.
18. P.S. Andersson, J. Ankerst, S. Montán, K. Svanberg and S. Svanberg, to appear.
19. M. Yamashita, M. Nomura, S. Kobayashi, T. Sato and K. Aizawa, IEEE J. Quant. Electr. QE-20, 1363 (1984).
20. P.S. Andersson, E. Kjellén, L.G. Salford, K. Svanberg and S. Svanberg, to appear.
21. R. Pottier, J.P. Laplante, Y.-F. Chow, and J. Kennedy, Can. J. Chem. 63, 1463 (1985).
22. P.S. Andersson, J. Ankerst, S. Montán, K. Svanberg and S. Svanberg, to appear.
23. P.S. Andersson, A. Gustafson, U. Stenram, K. Svanberg and S. Svanberg, unpublished results.

MECHANISTIC AND DIAGNOSTIC ASPECTS OF PHOTODYNAMIC ENHANCEMENT AND STONE FRAGMENTATION

David I. Rosen, Steven J. Davis, Arthur A. Boni, and John P. Campbell
Physical Sciences Inc., P.O. Box 3100, Andover, MA 01810

ABSTRACT

Metastable oxygen molecules in the singlet delta state $O_2(^1\Delta)$, are believed to be the active species that are produced upon laser irradiation of hematoporphyrin derivative (HPD), attached to tumors, and are responsible for cancer cell destruction. Although the presence of $O_2(^1\Delta)$ has been inferred by indirect chemical measurements, it has not been detected in real time. In this work, we review the optical characteristics of $O_2(^1\Delta)$ and describe an optical diagnostic procedure for its real-time detection during laser irradiation. Additionally, recent work has been presented whereby repetitively pulsed dye laser radiation can be delivered through a fiber optic probe to induce fragmentation of kidney stones and gallstones. In this paper, we review this work and describe experiments whose objective is to detect the fragmentation process in situ by acoustic emission.

PHOTODYNAMIC TREATMENT OF TUMOR CELLS

Photodynamic therapy is a rapidly developing branch of oncology. The technique is based upon the observation that certain compounds, such as porphyrins are preferentially accumulated and retained in malignant tumors when introduced through intravenous injection. Profio[1] has measured the fluorescence spectrum of hematoporphyrin derivative (HPD), a commonly used material that is derived from treatment of hemoglobin with acetic and sulfuric acids. It has been shown that HPD fluoresces in the red when it is excited by shorter wavelengths, cf. Figure 1 taken from Ref. 1. This fluorescent emission can be used to detect selectively the presence of tumors that have been treated with HPD. However, more significantly, it has been shown that upon excitation with photons from visible wavelength dye

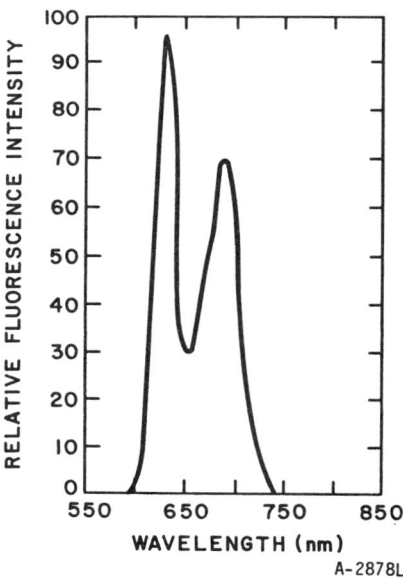

Fig 1. Fluorescence emission spectrum of hemotoporphyrin (Ref. 1)

lasers, the porphyrins initiate a reaction that selectively kills the
malignant cells to which they are attached. The apparent cytototoxic
agent that is responsible for cellular destruction is singlet oxygen
molecules, $O_2(^1\Delta)$.[2-8]

The accumulation of singlet oxygen following irradiation has
been detected by indirect analytical techniques that require several
hours to complete. It would be of great benefit to be able to
monitor singlet oxygen in real time simultaneously with the irradiation treatment. The ability to detect and correlate the amount of
singlet oxygen produced to the efficiency of treatment could lead to
the development of improved clinical methods and materials.

In a previous paper, we reviewed the background of photo-induced
processes and the specific role of singlet oxygen in the destruction
of cancer cells.[9] In this paper we briefly summarize the spectroscopic features of oxygen. These suggest the possibility of real-
time optical detection of $O_2(^1\Delta)$ as a potential clinical diagnostic
for its presence during cancer cell treatment.

The energy levels of the lower lying state of O_2 are shown in
Figure 2. The ground state of O_2 is $X(^3\Sigma_g)$. The metastable $O_2(a^1\Delta)$
radiates to the X state emitting at 1270 nm. The next highest
excited state, $b^1\Sigma$, emits to X at 762 nm. In addition to these O_2
emission bands, two $O_2(^1\Delta)$ molecules each in their lowest vibrational
levels can simultaneously emit one photon at 634 or 703 nm. This is
known as dimole emission and is easily seen visually in many situations. We also note that if one of the $O_2(^1\Delta)$ molecules is in its

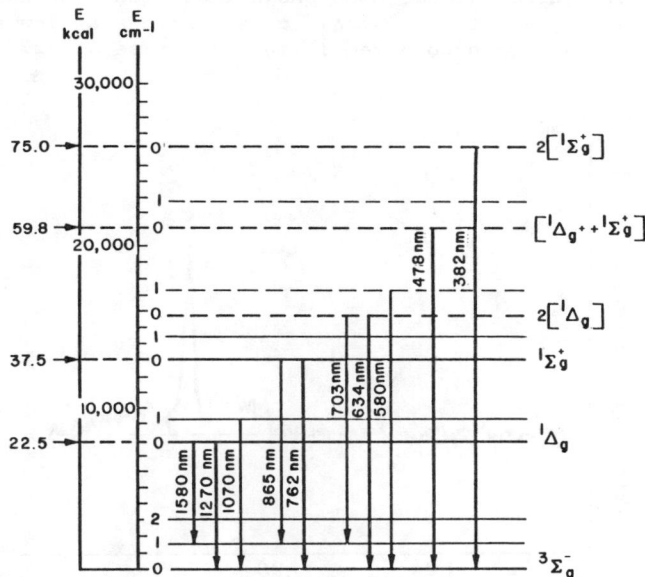

Fig. 2. Electronic and vibrational levels of the oxygen molecules.
Dotted lines correspond to the levels of molecular pairs from which
dimole transitions originate.

first excited vibrational level, then dimole emission at 580 nm is observed. The intensity of dimole emission varies as the square of the $O_2(^1\Delta)$ concentration.

When irradiated with visible light, HPD emits characteristic bands near 630 and 700 nm. Although some shifts due to solvent species and concentration have been reported, these two bands are typically observed. When HPD has been absorbed into cancerous tissue, the same fluorescence bands are observed. An HPD sample spectrum is shown in Figure 1, taken from Profio.[1] Interestingly and perhaps significantly is the fact that these two bands are very close to the positions of two of the O_2 dimole bands described earlier (634 and 703 nm). It is genrally believed that when HPD is attached to an O_2 source such as living cells that the visible irradiation leads to $O_2(^1\Delta)$ production via an intersystem crossing. It is possible that the $O_2(^1\Delta)$ can then emit via the dimole mechanism. Thus the HPD emissions from tumors at 630 and 700 nm could contain contributions from $O_2(^1\Delta)$ dimole emission, Recently, 580 nm emission has been observed during photochemotherapy by Andreoni and Cubeddu.[10] We previously pointed out that there are also O_2 dimole bands at 580 nm for vibrationally excited oxygen molecules. However, the electronic transition at 1270 nm is detectable and could be used both to identify and monitor the presence and concentration of singlet delta oxygen. Previous measurements by Davis et al.[11] taken in a discharge tube show a typical gas phase spectrum of the $O_2(^1\Delta-X^3\Sigma)$ transition, cf. Figure 3. There, it has been shown that even though this is a relatively weak transition, $O_2(^1\Delta)$ concentrations as low as 5×10^{12} cm^{-3} have been observed in the gas phase at 1270 nm.

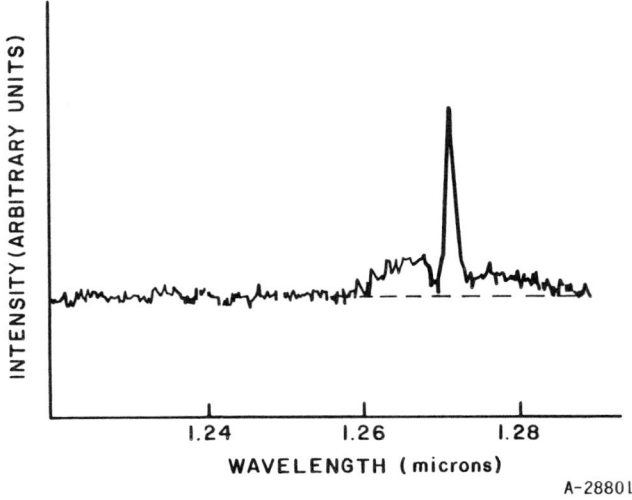

Fig. 3. Gas phase spectrum of $O_2(^1\Delta-^3\Sigma)$ emission from discharged O_2 at 1 torr (Ref. 11)

Measurements are currently under way in our laboratory using both pulsed and CW dye lasers in conjunction with appropriate detection and signal conditioning equipment to determine the feasibility of optical detection of singlet oxygen as a real time monitor of its presence during cancer cell treatment. Measurements in both the visible and near infrared are being made to study the above-mentioned transitions. Results from these studies will be reported in a subsequent publication.

IN SITU LASER INDUCED STONE FRAGMENTATION

Fragmentation of stones in the ureter[12] and in the common bile duct[13] is an established technique. However, the first generation fragmentation devices -ultrasonic and electrohydraulic probes - were large and rigid and relatively inflexible. The electrohydraulic probe causes severe damage to tissue when in close contact.[14] The laser was investigated because of the possibility of delivering low energy through fine, flexible quartz fibers. Experiments with the pulsed dye laser[15] have shown that a 1 μs laser pulse of 30 mJ in a 200 μm diameter fiber at 504 nm fragments 90 percent of upper urinary tract calculi and all biliary calculi. Experiments conducted in the pig ureter, common bile duct, and also in the rabbit bladder have further shown that fragmentation can be achieved without significant injury to the surrounding tissue even when the stone is impacted.[16]

In a recent publication, we reviewed work in this area and described mechanisms by which stone fragmentation occur.[9] A postulated mechanism or synergistic combination of two mechanisms for stone fracture has been proposed. The laser radiation is coupled into the stone via scattering and/or absorption of the visible radiation at the surface of the stone. At the intensities involved, it is possible that sufficient fluence may be accumulated to initiate and sustain electron cascade ionization, and to the formation of a plasma at the surface. The plasma formation process can occur in either vaporized stone or in the air adjacent to the stone. The presence of an ionized layer at the surface can then result in (additional) vaporization of the surface layer (as observed optically), and can also directly produce shock-induced fragmentation of the stone. The plasma pressure, which may exceed 10 to 100 atm, can be sufficient to induce an impulsive stress wave in the stone and lead to the existence of both compressive and tensile stresses. If this is the case, then the presence of the confining medium (water) can serve to amplify the magnitude of the stress wave. At laser intensities above the plasma formation threshold, laser-supported absorption (LSA) waves may be formed and propagate into the laser beam supported by the absorption of incoming laser radiation.

We also suggested a second independent, but potentially synergistic route to stone fragmentation. The energy may be absorbed at the surface of the stone, but be conducted into the stone prior to (as well as subsequent) to any plasma processes. This energy transport may result in vaporization of the water contained in the pores of the stone. As a result of vaporization of water, pressures of order 1000 atm may be generated internally (within the pore structure). Since relief of this pressure is limited by the resistance of the pore structure of the stone, the pressure buildup may fragment the stone directly. Additionally, and synergistically, the vaporization of water may be augmented by the presence of surface plasmas, thus resulting in stone fragmentation by the two mechanisms in concert. Sufficient data do not yet exist to indicate under which conditions of irradiation each of these processes is operative. However, the observations indicate that the presence of a surface plasma always occurs simultaneously with stone fragmentation, as does the emission of a strong acoustic signal. Therefore, the porosity-induced fragmentation may be a secondary process, or may occur simultaneously with plasma formation.

In Ref. 9, we correlated the fragmentation thresholds observed by Watson with thresholds for surface plasma formation observed previously. We found good correlation, thus suggesting that surface plasma formation is a precursor to stone fragmentation.

The laser has the advantage of producing smaller fragments than alternative fragmentation techniques. However damage was produced by instrumentation of the ureter and was found to a lesser or greater degree in all the ureters examined.[16] Damage correlated strongly with the size of the instrument inserted. When the laser was aimed directly at tissue it took a minimum of 50 pulses in precisely the same spot to penetrate the wall of the ureter and then the hole was a minute puncture which did not lead to extravasation. Since many calculi could be fragmented entirely in 100 to 200 pulses it is clear that the complications of instrumentation, i.e., the ureteroscope, outweigh the risks of direct damage by the laser on tissue. It was therefore proposed that the attendant acoustic signature[17] be used for monitoring the effect of the laser on calculi and for discriminating between tissue and stone as an alternative to visual monitoring by endoscopy. This would result in a small instrument for minimally-invasive therapy in the body which would increase its clinical usefulness in two common and serious situations, stones in the urinary tract and common bile duct.

The research that we are performing and describe herein is intended to develop an acoustic probe as a <u>real time</u>, <u>in-situ</u> monitor for laser fragmentation of kidney stones and/or gallstones. If successful, such a device would be invaluable for guiding the clinical procedure. In addition to providing a nearly instantaneous non-visual monitor of the status of stone fragmentation, the observed acoustic signature will also indicate when the laser delivery fiber is optimally targeted on the stone and help guide the optimum tuning of laser power level. Thus, the acoustic probe has the potential for greatly minimizing, if not eliminating, the need for direct visual endoscopic access which can certainly be problematic in the upper ureter and biliary duct. Once developed, the acoustic probe could be applied to the monitoring of virtually any clinical procedure involving the pulsed fragmentation and/or ablation of unwanted deposits within the body.

The need for improved monitoring devices for laser lithotripsy is clearly becoming more urgent. Clinical trials of the laser fragmentation of kidney stones have recently been completed[18,19] (with more than 75 patients already treated), and FDA approval is expected within the year. Clinical studies on gallstones remain to be done.

The specific objectives of our research to date were to investigate the extent to which the characteristic acoustic signatures generated[17] could be used as a non-visual diagnostic for guiding the laser fragmentation procedure. The approach taken for achieving these goals involved performing a series of well-controlled laboratory experiments on a variety of kidney and gallstone samples (as well as on selected "soft tissue" models) irradiated in-vitro by a repetitively-pulsed dye laser delivered via an optical fiber. The resulting acoustic signatures were characterized using a high speed hydrophone situated in the vicinity of the irradiated sample. In addition to recording the acoustic signature, sample response was further characterized by visual observations and video and still photography. The principal parameters varied in the experiment were the type of stone or tissue sample irradiated, the fiber to sample stand-off distance, and the magnitude of the laser pulse energy delivered. The recorded acoustic signals were analyzed in terms of frequency content by digitizing the waveforms and subjecting the digitized waveform to FFT (Fast Fourier Transform) analysis.

Kidney stone and gallstone samples were supplied to us through Massachusetts General Hospital from calculi which were salvaged from surgical procedures. The urinary calculi, as sent to us, were broadly classified as either calcium oxalate, calcium phosphate, or uric acid stones. To the best of our knowledge, all the calcium oxalate samples were of the dihydrate type. The two different biliary stone samples we received were broadly classified as a pigmented and a cholesterol type stone.

A schematic diagram of the apparatus used to measure the acoustic emissions accompanying stone ablation is shown in Figure 4. The laser source was a linear flashlamp-pumped dye laser (Model LFDL-2, Candela Corporation, Natick, MA). This laser typically produced pulse widths of ~1.4 μs (full width at half maximum) and could be operated up to pulse repetition rates as high as 20 Hz, although the present experiments were all performed using repetition rates <5 Hz. The laser was operated at 504 nm wavelength with a maximum output pulse energy of approximately 200 mJ.

The output of the laser was focused into a 200μ core diameter step-index quartz fiber (Spectran Corporation, Sturbridge, MA) which was in turn fed into the testing tank. The fiber was held and directed onto the sample being irradiated by way of a glass capillary tube attached to a multi-axis positioning device. In typical operation the distal end of the fiber was placed in direct contact with the sample being irradiated. In practice, the maximum fraction of the pulse energy that could be coupled through the fiber was about 50 percent.

For the majority of the experiments, test samples were held securely inside a short, vertical section of thin-walled tygon tubing (see Figure 4). A limited number of tests were also performed, however, with the stone samples held inside excised sections of cow face veins. There was no discernible difference observed between the

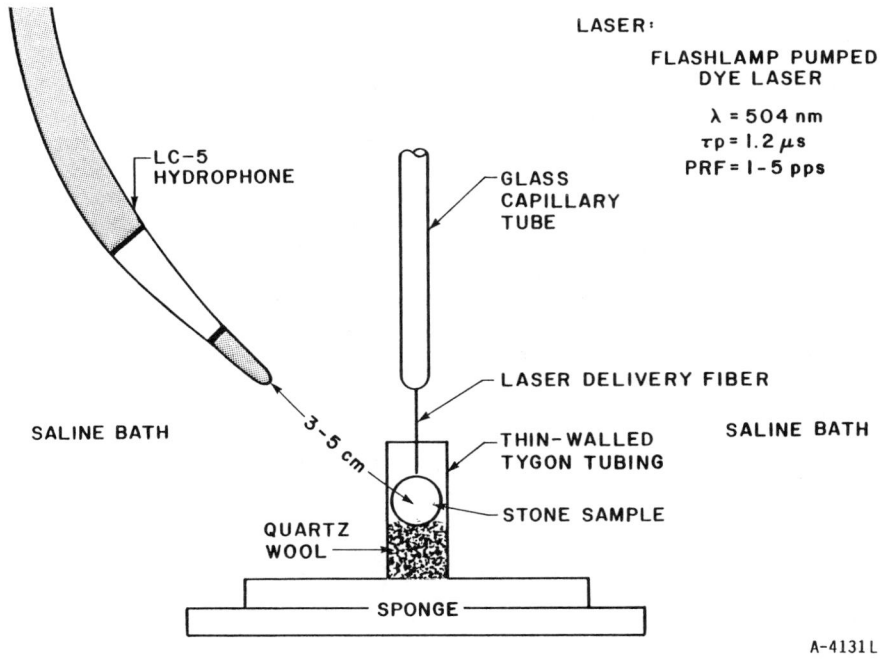

Fig. 4. Experimental setup for measuring in-vitro acoustic emissions from laser-irradiated stones

acoustic signals measured from stones irradiated in the thin walled tygon tubing and similarly irradiated stones contained in a section of cow vein. (This suggests that reflections from or attenuation in the wall of the containing tubing did not play a significant role in effecting the observed acoustic signature.)

The acoustic emissions were measured using a high frequency (useful response up to approximately 500 kHz) hydrophone (Celesco Model LC-5) which for most of the experiments was located approximately 3 cm from the irradiated stone surface. The diameter of the sensing tip of this hydrophone is 3 mm. The output of the hydrophone was typically fed into a high speed oscilloscope from which the waveforms were recorded photographically and later digitized with a commercial tracing digitizer. The only exception was for a period of about one week when a LeCroy high speed (150 MHz) digital data acquisition system was made available to us. Frequency analysis of the digitized waveforms was performed using an IBM PC and a standard software routine available for performing Fast Fourier Transform (FFT) analysis.

One of the first issues addressed in the in-vitro experiments was to investigate the correlation of the observed acoustic signal with the achievement of stone fragmentation. In particular we wished to evaluate the extent to which the strength of the acoustic signal generated correlated with the laser energy delivered to the stone and with the corresponding achievement of stone fragmentation.

Photographs of a calcium oxalate stone have been obtained after irradiation by a selected number of pulses at a given laser energy. Beneath each photo the corresponding acoustic signal observed as shown in Figure 5a-c. As can be seen, while there is a clearly detectable acoustic signal observed with pulse energies as small as 10 mJ per pulse, the acoustic signal really only increases dramatically to truly robust levels when the laser energy is above the threshold for fragmentation. This is demonstrated by comparing the results in frames (b) and (c) where it is seen that by just increasing the laser pulse energy from 10 mJ to 20 mJ the peak acoustic signal increased from only 25 to 30 mV to more than 400 mV. Furthermore, as shown by the photos, this very dramatic increase in acoustic signal appears to correspond with the onset of stone fragmentation.

Similar results were observed for all the stones tested during our experiments. These included about a dozen each of calcium oxalate dihydrate and calcium phosphate stones, and two uric acid stones. In short, stone fragmentation was only achieved in the presence of a robust acoustic signal, which for the conditions of these experiments meant peak acoustic signals > 400 mV, corresponding to an acoustic level > 4×10^5 dynes/cm^2 when $\sim$3 cm from the stone.

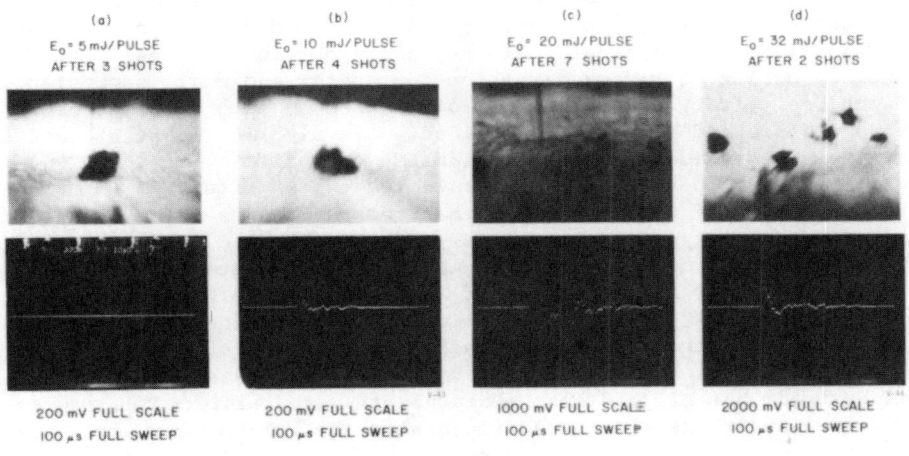

Fig. 5. Correlation of stone fragmentation with observed acoustic signal (Ca oxalate stone)

These findings demonstrate that the acoustic signal level is an extremely useful indicator for guiding the laser fragmentation procedure, with the absence of a robust acoustic signal generally indicating one of two things: either that the laser energy delivered by the fiber must be increased or, more typically, that the laser fiber is no longer adequately contacting the stone (and therefore requires repositioning).

Even in the present in-vitro experiments, the acoustic emission amplitude quickly became our preferred first level indicator of adequately pinpointing the laser fiber. This was especially true when the surrounding field became obscured by dust and debris during extended laser exposures. During such periods the stone or stone fragments could frequently shift slightly beneath the fiber causing poor direct contact, yet this situation would not be visually apparent. A more reliable indication was the loss of a robust acoustic signal.

As was mentioned earlier, the other important goal of our investigations was to evaluate the extent to which the acoustic signatures generated could be used to distinguish the irradiation of a stone from the inadvertent irradiation of any soft tissue surrounding it. Most importantly, we wished to establish the extent to which the acoustic emission from a stone can be distinguished from that which arises when the wall of ureter (or biliary duct) is irradiated. Furthermore, as recommended by Watson,[20] we also sought to investigate whether the acoustic emission from a stone could be distinguished from that arising from the irradiation of a blood clot.

In an attempt to address these issues simply in our laboratory, we decided to simulate the ureter and the blood clot with some rather simple tissue models that could be easily obtained. These were freshly excised cow face veins for the ureter and small segments of fresh beef liver for the blood clots. The cow face veins as used were drained and rinsed of any blood. Clearly, an important task in any future studies will be to establish the validity of these simulations by performing more realistic experiments.

The results obtained are summarized in Figures 6 and 7 in which we compare the typical acoustic signatures observed for a calcium oxalate stone, the inner wall of the cow vein (ureter model), and a small segment of fresh beef liver (blood clot model). In Figure 6 we show the directly measured acoustic signal amplitudes versus time and in Figure 7 we show the corresponding acoustic power density spectra (frequency distributions) obtained from FFT analysis. It should be emphasized that all of these measurements were performed as nearly as possible under identical conditions with the only differences being the type of sample material irradiated. The beef liver samples were cut to similar dimensions as the stone samples tested and each was irradiated while contained inside a short vertical section of the cow vein. The laser pulse energy and the location of the hydrophone relative to the test sample were kept fixed throughout this series of tests.

Turning our attention first to Figure 6, we can note several significant features of the results. Perhaps the most significant is the very large difference between the acoustic signal amplitude from

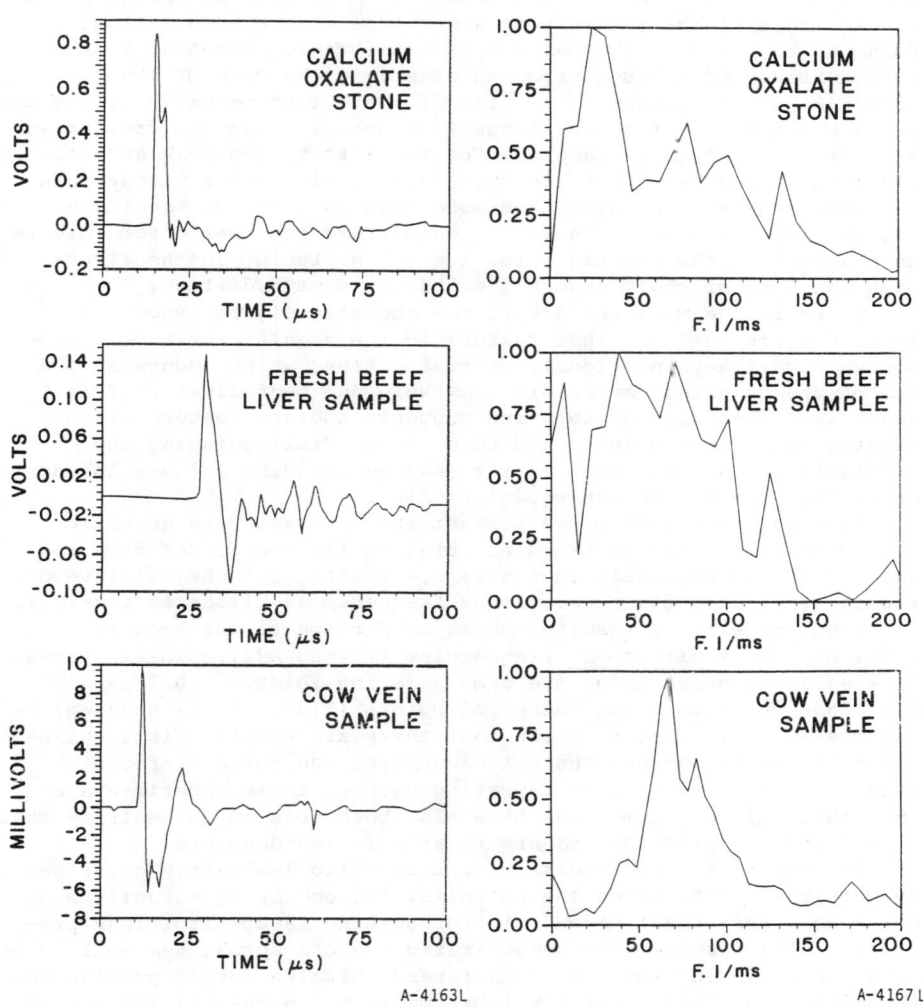

Fig. 6. Observed acoustic signal amplitudes versus time for three different materials (E_o = 32 mJ/pulse

Fig. 7. Observed acoustic power density spectra (frequency distributions) for three different materials

the irradiated stone and from the cow vein. While the peak amplitude from the stone is over 800 mV, the corresponding signal from the cow vein is only 10 mV - nearly two orders of magnitude lower. This is a direct result of the rather poor absorption of the vein wall at 504 nm as compared to the calcium oxalate stone. (Another reason is the occurrence of a laser-produced plasma in the case of the stone.[21]) It is further interesting to note that compared to the cow vein a considerably stronger acoustic signal is observed from irradiation of the beef liver sample. For the latter, the peak acoustic signal amplitude was typically only four to five times weaker than that observed from the stone and more than an order of magnitude stronger than from the vein wall. We believe the reason for this is the presence of the red pigments, i.e., hemoglobin, in the liver which can more strongly absorb the 504 nm laser radiation.

There is one more feature of the acoustic signals shown in Figure 6 worth noting. That feature is the significantly more pronounced negative-going signal, or rarefaction, which occurs in the case of the compliant materials (cow vein and beef liver). This result was quite reproducible and suggests another feature of the acoustic signatures which could be used for distinguishing the irradiation of a stone from soft tissue surrounding it. We intend to pursue this further in subsequent studies.

Finally, in Figure 6, we present the corresponding acoustic power density spectra or frequency distributions obtained from FFT analysis of the waveforms in Figure 5. Plotted are the relative distributions of acoustic power versus frequency in kilohertz ($(ms)^{-1}$). Note that for all the results the major portion of the acoustic signal occurs at <u>ultrasonic</u> frequencies (f > 20 kHz). In all cases FFT's were performed using the same sampling interval (0.2 µs per point) and the same total number of points (512). It is apparent from the acoustic spectra shown that there are clearly distinguishable differences between the emissions from the three different materials. Furthermore, by repeating each of these experiments at least three to four times, we have also been able to demonstrate that these characteristic spectra are remarkably reproducible.

To summarize, the results of our in-vitro laboratory experiments have clearly demonstrated the potential for employing acoustic monitoring as a non-visual diagnostic for guiding laser litotripsy procedures. The results have demonstrated the conceptual feasibility of using the attendant acoustic signatures generated for 1) guiding and pinpointing the laser delivery fiber on to the stone; 2) indicating an appropriate choice of laser power level; and 3) distinguishing the irradiation of stone from the inadvertent irradiation of soft tissue surrounding it.

The major research hurdles remaining are: 1) to perform appropriate in-vivo experiments to further verify these conclusions; 2) to devise a compact acoustic monitoring probe that is minimally invasive; and 3) to develop an appropriate signal acquisition and analysis scheme (and associated hardware) that will automatically provide, in real time, reliable and straightforward information for guiding the clinical procedure.

REFERENCES

1. A. Edward Profio, IEEE J. Quant. Elec. QE20, 1502 (1984).
2. L. Tomio, F. Calzavora, and G. Jori, Laser Photobiology and Photomedicine (Page 117), Edited by S. Martellucci and A.N. Chester, Plenum Press, New York (1985).
3. Kenneth R. Weishaupt, Charles J. Gomer, and Thomas J. Dougherty, Cancer Research, 36, 2326 (1976).
4. C.J. Gomer and T.J. Dougherty, Cancer Research, 39, 146 (1979).
5. T.J. Dougherty, R.E. Thoma, D.G. Boyle, and K.R. Weishaupt, Cancer Research, 41, 401 (1981).
6. R.L. Lipson, E.J. Balder, and A.M. Olsen, J. National Cancer Inst., 26, 1 (1961).
7. D. Kessel and T. Chow, Cancer Research, 43, 1994 (1983).
8. F. Docchio, R. Ramponi, C.A. Sacchi, G. Boltiroli, and I. Freitas, Laser Photobiology and Photomedicine (Page 83), Edited by S. Martellucci and A. N. Chester, Plenum Press, New York (1985).
9. Arthur A. Boni, David, I. Rosen, Steven J. Davis, and Leslie J. Popper, "High Power Laser Applications to Medicine," To be published in J. Quant. Spec. and Rad. Trsf. (1987).
10. A. Andreoni and R. Cubeddu, Laser Photobiology and Photomedicine (p. 109), Edited by S. Martellucci and A.N. Chester, Plenum Press, New York (1985).
11. P. Alsin, G. Simmons, and S.J. Davis, To be published in Chem. Phys. Lett.
12a. Reuter and Kern, "Electronic Lithotripsy of Ureteral Calculi," J. Urol., 110, 181-183 (1973).
  b. Goodfriend, "Ultrasonic and Electrohydraulic Lithotripsy of Ureteral Calculi," Urology, 23, 5-8 (1984).
  c. Hoffman, Bagli, Schoenberg, Lyon, "Transurethral Removal of Large Ureteral and Renal Pelvic Calculi using Ureteroscopic and Ultrasonic Lithotripsy," J. Urol., 130, 31-34, (1983).
13. Lear, Ring, Macoviak, and Baum, "Percutaneous Transhepatic Electrohydraulic Lithotripsy," Radiology, 150, 589-590, (1984).
14. Tidd, Wright, Oliver, Wallace, and Porteous, "Hazard to Bladder and Intestinal Tissues from Intravesical Underwater Electrical Discharges from a Surgecal Lithoclast," Urol. Res., 4, 49-54, (1976).
15. Watson, Whickham, Mills, Swain, Brown, and Salmon, "Laser Fragmentation of Renal Calculi," (Abstract) Lasers in Surgery and Medicine, 3, 115, (1983).
16a. Schuster and Schaeffer, "Economic Impact of Kidney Stones in White Male Adults," Urology, 24, 327-331, (1984).
  b. Sierakowski, Finlayson, Landes, Finlayson, and Sierakowski, "Frequency of Urolithiasis in Hospital Discharged Diagnoses in the United States," Investigative Urology, 15, 438-441, (1978).
  c. "Renal Stone Epidemiology - A 25 Year Study in Rochester, Minnesota," Kidney Int'l, 16, 624-631, (1979).
17. G. Watson, "Development of a Laser System for Fragmentation of Urinary Tract Calculi," Thesis, to be published.

18. Stephen P. Drethler, "Laser Fragmentation of Stones: In Patients," American Society for Laser Medicine and Surgery Conference, May 1986, Boston, MA.
19. S.P. Dretler, G.M. Watson, S. Murray, and J.A. Parrish, "Pulsed Dye Laser Fragmentation of Ureteral Calculi: Initial Clinical Experience," to be published, The Journal of Urology, February 1987.
20. G. Watson, Private communication.
21. P. Teng, R.R. Anderson, T.F. Deutsch, G.M. Watson, and J.A. Parrish, "Laser Fragmentation of Stones: In-Vitro Measurements (Mechanisms)," American Society for Laser Medicine and Surgery Conference, May 1986, Boston, MA.

ARTERIAL ANEURYSM MODEL USING LASER ENERGY

Joseph LoCicero, III, M.D., Renee S. Hartz, M.D.
Walter J. McCarthy, M.D., and Shou-Ren Shih, M.D.

Northwestern University Medical School, Chicago, IL

ABSTRACT

To elucidate the mechanisms of aneurysm production, controlled injuries of arterial intima were created with the pulsed CO_2 laser beam. Carotid and femoral arteries in rabbits were injured and were re-operated eight to 11 weeks after injuries. One carotid artery developed an aneurysm and eight femoral arteries developed aneurysms. Histologic examination revealed extensive medial necrosis with fragmentation of internal elastic lamina in the area of these aneurysms.

INTRODUCTION

Previous work from our laboratory and by other investigators suggest that vascular anastomoses can be successfully performed with various laser wavelengths.[1,2,3] These anastomoses are as strong as conventionally sutured anastomoses within one week of performance and have been demonstrated to exhibit growth with the surrounding vessel.[4] The occurrence of anastomotic aneurysms, however, is of concern and prompted this investigation.[5] The goal of this work is to determine the affects of a precisely quantitated intimal injury using a very low power CO_2 laser guided by an operating microscope.

METHODS

Ten New Zealand white rabbits were anesthetized and, using sterile technique, both carotid and femoral arteries were exposed and measured against the grid under the operating microscope. Each vessel was mobilized and occluded. Anterior longitudinal arteriotomies 1 cm in length were performed and the arterial lumen irrigated. The vessel edges were retracted and the posterior intima of the left carotid and right femoral were exposed to CO_2 laser energy. A spot size of 50 μm at a focal length of 350 mm was used to produce the injuries. Forty to 50 mW of energy was delivered over 0.5 sec. There were ten rows of four circular spots for a total injury area of 0.9 cm^2. Visual changes in the intima were regarded as the end-point of laser injury and the damage produced by each spot overlapped those surrounding it. The calculated power density was 270W/cm^2. For sham vessels, traction was maintained on the stay sutures for a duration similar to that of the laser-treated vessels.

Eight to 11 weeks later, the animals were anesthetized and vessels exposed and re-measured. They were fixed using buffered gluteraldehyde solution and examined histologically.

RESULTS

The mean initial diameter of the femoral arteries was 1.84 mm. This was significantly smaller than that of the carotid arteries which measured 2.51 mm. The mean diameter increase in the sham carotid group was not significantly different from the laser carotid group. For the femoral vessels, however, the mean diameter increase in the sham vessels was 0.54 mm as opposed to 1.35 mm in the laser group ($p < 0.001$).

Aneurysm formation was defined as an increase in the diameter of the vessel greater than or equal to 1.5 times the initial diameter and occurred as follows: none in the carotid sham group, one in the carotid laser group, two in the femoral sham group, and eight in the femoral laser group (see Table). Grossly, the aneurysm were fusiform and occurred at the side of the vessel injured with the laser. Histologic examination of the vessels demonstrated little change in the adventitia and complete luminal coverage with endothelial-like cells at reoperation. Extensive medial necrosis with pyknosis of the cells and abundant amorphous debris were present in the area of the aneurysms. There was extensive fragmentation of the internal elastic lamina which was not totally remodeled by 11 weeks.

DISCUSSION

The disturbing occurrence of anastomotic aneurysms after laser-assisted arterial anastomosis requires further elucidation before clinical application of this new technology. To date, lasers used for these investigations have been the continuous wavelength variety (heat-producing) and the mechanism of vascular fusion is thought to be the adherence of coagulated proteins in the vessel wall.[1,2] It is speculated that aneurysm formation is related to arterial thermal energy adjacent to the anastomosis. Rabbit arteries chosen for this study ranged from 1.2 to 3.6 mm in diameter with femoral arteries significantly smaller than the carotids. The medial layer of the femoral arteries was thus correspondingly thinner. The higher incidence of aneurysm formation in the femoral group may therefore be explained either by the relatively larger area of injury (all vessels were exposed to the same size grid of laser energy) or a greater proportion of medial injury distribution.

Although microscopic examination demonstrated that the aneurysms involved all layers of the vessel wall, the intima and adventitia were relatively spared compared to the tunica media.

There was extensive coagulation necrosis of the mecia with pyknosis of the cells and amorphous debris which remained even weeks after injury. This layer was extremely atteruated and the internal elastic lamina was fragmented.

This work involved arterial injury with only one laser wavelength and produced a predictable number of fusiform aneurysms. Conceivably, other continuous wavelength lasers will cause less medial damage and, therefore, less aneurysm formation. Hopefully, further experimentation with various wavelengths will continue to elucidate this problem prior to clinical trials of laser-assisted vascular anastomosis.

TABLE

FINDINGS AT RE-OPERATION

	Mean Diam. Change (mm)	p Value*	% Aneurysms**
Carotid Sham	0.15		0
		NS	
Carotid Laser	0.38		10
Femoral Sham	0.54		20
		p<.001	
Femoral Laser	1.38		80

*Paired T-Test

**Defined as increase in diameter greater than or equal to 1.5 times initial diameter

REFERENCES

1. O. M. Gomes, R. Macruz and E. Armelin, **Tex Heart Inst J 10:145, 1983.**
2. R. S. Hartz, J. LoCicero, and W. J. McCarthy, **Congress on Laser Neurosurgery III. (Northwestern University, Chicago, 1984), p. 97.**
3. K. K. Jain, **Vasc Surg 17:240, 1983.**
4. R. S. Hartz, J. LoCicero, S. R. Shih, W. J. McCarthy, and L. L. Michaelis, **Surg Form 36:457, 1985.**
5. M. R. Quigley, J. E. Bailes, H. C. Kwaan, L. J. Cerullo, J. T. Brown, and C. Lastre, **Lasers Surg Med 5:357, 1985.**

AIP Conference Proceedings

		L.C. Number	ISBN
No. 1	Feedback and Dynamic Control of Plasmas – 1970	70-141596	0-88318-100-2
No. 2	Particles and Fields – 1971 (Rochester)	71-184662	0-88318-101-0
No. 3	Thermal Expansion – 1971 (Corning)	72-76970	0-88318-102-9
No. 4	Superconductivity in d- and f-Band Metals (Rochester, 1971)	74-18879	0-88318-103-7
No. 5	Magnetism and Magnetic Materials – 1971 (2 parts) (Chicago)	59-2468	0-88318-104-5
No. 6	Particle Physics (Irvine, 1971)	72-81239	0-88318-105-3
No. 7	Exploring the History of Nuclear Physics – 1972	72-81883	0-88318-106-1
No. 8	Experimental Meson Spectroscopy –1972	72-88226	0-88318-107-X
No. 9	Cyclotrons – 1972 (Vancouver)	72-92798	0-88318-108-8
No. 10	Magnetism and Magnetic Materials – 1972	72-623469	0-88318-109-6
No. 11	Transport Phenomena – 1973 (Brown University Conference)	73-80682	0-88318-110-X
No. 12	Experiments on High Energy Particle Collisions – 1973 (Vanderbilt Conference)	73-81705	0-88318-111-8
No. 13	π-π Scattering – 1973 (Tallahassee Conference)	73-81704	0-88318-112-6
No. 14	Particles and Fields – 1973 (APS/DPF Berkeley)	73-91923	0-88318-113-4
No. 15	High Energy Collisions – 1973 (Stony Brook)	73-92324	0-88318-114-2
No. 16	Causality and Physical Theories (Wayne State University, 1973)	73-93420	0-88318-115-0
No. 17	Thermal Expansion – 1973 (Lake of the Ozarks)	73-94415	0-88318-116-9
No. 18	Magnetism and Magnetic Materials – 1973 (2 parts) (Boston)	59-2468	0-88318-117-7
No. 19	Physics and the Energy Problem – 1974 (APS Chicago)	73-94416	0-88318-118-5
No. 20	Tetrahedrally Bonded Amorphous Semiconductors (Yorktown Heights, 1974)	74-80145	0-88318-119-3
No. 21	Experimental Meson Spectroscopy – 1974 (Boston)	74-82628	0-88318-120-7
No. 22	Neutrinos – 1974 (Philadelphia)	74-82413	0-88318-121-5
No. 23	Particles and Fields – 1974 (APS/DPF Williamsburg)	74-27575	0-88318-122-3
No. 24	Magnetism and Magnetic Materials – 1974 (20th Annual Conference, San Francisco)	75-2647	0-88318-123-1
No. 25	Efficient Use of Energy (The APS Studies on the Technical Aspects of the More Efficient Use of Energy)	75-18227	0-88318-124-X

No. 26	High-Energy Physics and Nuclear Structure – 1975 (Santa Fe and Los Alamos)	75-26411	0-88318-125-8
No. 27	Topics in Statistical Mechanics and Biophysics: A Memorial to Julius L. Jackson (Wayne State University, 1975)	75-36309	0-88318-126-6
No. 28	Physics and Our World: A Symposium in Honor of Victor F. Weisskopf (M.I.T., 1974)	76-7207	0-88318-127-4
No. 29	Magnetism and Magnetic Materials – 1975 (21st Annual Conference, Philadelphia)	76-10931	0-88318-128-2
No. 30	Particle Searches and Discoveries – 1976 (Vanderbilt Conference)	76-19949	0-88318-129-0
No. 31	Structure and Excitations of Amorphous Solids (Williamsburg, VA, 1976)	76-22279	0-88318-130-4
No. 32	Materials Technology – 1976 (APS New York Meeting)	76-27967	0-88318-131-2
No. 33	Meson-Nuclear Physics – 1976 (Carnegie-Mellon Conference)	76-26811	0-88318-132-0
No. 34	Magnetism and Magnetic Materials – 1976 (Joint MMM-Intermag Conference, Pittsburgh)	76-47106	0-88318-133-9
No. 35	High Energy Physics with Polarized Beams and Targets (Argonne, 1976)	76-50181	0-88318-134-7
No. 36	Momentum Wave Functions – 1976 (Indiana University)	77-82145	0-88318-135-5
No. 37	Weak Interaction Physics – 1977 (Indiana University)	77-83344	0-88318-136-3
No. 38	Workshop on New Directions in Mossbauer Spectroscopy (Argonne, 1977)	77-90635	0-88318-137-1
No. 39	Physics Careers, Employment and Education (Penn State, 1977)	77-94053	0-88318-138-X
No. 40	Electrical Transport and Optical Properties of Inhomogeneous Media (Ohio State University, 1977)	78-54319	0-88318-139-8
No. 41	Nucleon-Nucleon Interactions – 1977 (Vancouver)	78-54249	0-88318-140-1
No. 42	Higher Energy Polarized Proton Beams (Ann Arbor, 1977)	78-55682	0-88318-141-X
No. 43	Particles and Fields – 1977 (APS/DPF, Argonne)	78-55683	0-88318-142-8
No. 44	Future Trends in Superconductive Electronics (Charlottesville, 1978)	77-9240	0-88318-143-6
No. 45	New Results in High Energy Physics – 1978 (Vanderbilt Conference)	78-67196	0-88318-144-4
No. 46	Topics in Nonlinear Dynamics (La Jolla Institute)	78-57870	0-88318-145-2
No. 47	Clustering Aspects of Nuclear Structure and Nuclear Reactions (Winnepeg, 1978)	78-64942	0-88318-146-0
No. 48	Current Trends in the Theory of Fields (Tallahassee, 1978)	78-72948	0-88318-147-9

No. 49	Cosmic Rays and Particle Physics – 1978 (Bartol Conference)	79-50439	0-88318-148-7
No. 50	Laser-Solid Interactions and Laser Processing – 1978 (Boston)	79-51534	0-88318-149-5
No. 51	High Energy Physics with Polarized Beams and Polarized Targets (Argonne, 1978)	79-64565	0-88318-150-9
No. 52	Long-Distance Neutrino Detection – 1978 (C.L. Cowan Memorial Symposium)	79-52078	0-88318-151-7
No. 53	Modulated Structures – 1979 (Kailua Kona, Hawaii)	79-53846	0-88318-152-5
No. 54	Meson-Nuclear Physics – 1979 (Houston)	79-53978	0-88318-153-3
No. 55	Quantum Chromodynamics (La Jolla, 1978)	79-54969	0-88318-154-1
No. 56	Particle Acceleration Mechanisms in Astrophysics (La Jolla, 1979)	79-55844	0-88318-155-X
No. 57	Nonlinear Dynamics and the Beam-Beam Interaction (Brookhaven, 1979)	79-57341	0-88318-156-8
No. 58	Inhomogeneous Superconductors – 1979 (Berkeley Springs, W.V.)	79-57620	0-88318-157-6
No. 59	Particles and Fields – 1979 (APS/DPF Montreal)	80-66631	0-88318-158-4
No. 60	History of the ZGS (Argonne, 1979)	80-67694	0-88318-159-2
No. 61	Aspects of the Kinetics and Dynamics of Surface Reactions (La Jolla Institute, 1979)	80-68004	0-88318-160-6
No. 62	High Energy e^+e^- Interactions (Vanderbilt, 1980)	80-53377	0-88318-161-4
No. 63	Supernovae Spectra (La Jolla, 1980)	80-70019	0-88318-162-2
No. 64	Laboratory EXAFS Facilities – 1980 (Univ. of Washington)	80-70579	0-88318-163-0
No. 65	Optics in Four Dimensions – 1980 (ICO, Ensenada)	80-70771	0-88318-164-9
No. 66	Physics in the Automotive Industry – 1980 (APS/AAPT Topical Conference)	80-70987	0-88318-165-7
No. 67	Experimental Meson Spectroscopy – 1980 (Sixth International Conference, Brookhaven)	80-71123	0-88318-166-5
No. 68	High Energy Physics – 1980 (XX International Conference, Madison)	81-65032	0-88318-167-3
No. 69	Polarization Phenomena in Nuclear Physics – 1980 (Fifth International Symposium, Santa Fe)	81-65107	0-88318-168-1
No. 70	Chemistry and Physics of Coal Utilization – 1980 (APS, Morgantown)	81-65106	0-88318-169-X
No. 71	Group Theory and its Applications in Physics – 1980 (Latin American School of Physics, Mexico City)	81-66132	0-88318-170-3
No. 72	Weak Interactions as a Probe of Unification (Virginia Polytechnic Institute – 1980)	81-67184	0-88318-171-1
No. 73	Tetrahedrally Bonded Amorphous Semiconductors (Carefree, Arizona, 1981)	81-67419	0-88318-172-X

No. 74	Perturbative Quantum Chromodynamics (Tallahassee, 1981)	81-70372	0-88318-173-8
No. 75	Low Energy X-Ray Diagnostics – 1981 (Monterey)	81-69841	0-88318-174-6
No. 76	Nonlinear Properties of Internal Waves (La Jolla Institute, 1981)	81-71062	0-88318-175-4
No. 77	Gamma Ray Transients and Related Astrophysical Phenomena (La Jolla Institute, 1981)	81-71543	0-88318-176-2
No. 78	Shock Waves in Condensed Matter – 1981 (Menlo Park)	82-70014	0-88318-177-0
No. 79	Pion Production and Absorption in Nuclei – 1981 (Indiana University Cyclotron Facility)	82-70678	0-88318-178-9
No. 80	Polarized Proton Ion Sources (Ann Arbor, 1981)	82-71025	0-88318-179-7
No. 81	Particles and Fields –1981: Testing the Standard Model (APS/DPF, Santa Cruz)	82-71156	0-88318-180-0
No. 82	Interpretation of Climate and Photochemical Models, Ozone and Temperature Measurements (La Jolla Institute, 1981)	82-71345	0-88318-181-9
No. 83	The Galactic Center (Cal. Inst. of Tech., 1982)	82-71635	0-88318-182-7
No. 84	Physics in the Steel Industry (APS/AISI, Lehigh University, 1981)	82-72033	0-88318-183-5
No. 85	Proton-Antiproton Collider Physics –1981 (Madison, Wisconsin)	82-72141	0-88318-184-3
No. 86	Momentum Wave Functions – 1982 (Adelaide, Australia)	82-72375	0-88318-185-1
No. 87	Physics of High Energy Particle Accelerators (Fermilab Summer School, 1981)	82-72421	0-88318-186-X
No. 88	Mathematical Methods in Hydrodynamics and Integrability in Dynamical Systems (La Jolla Institute, 1981)	82-72462	0-88318-187-8
No. 89	Neutron Scattering – 1981 (Argonne National Laboratory)	82-73094	0-88318-188-6
No. 90	Laser Techniques for Extreme Ultraviolt Spectroscopy (Boulder, 1982)	82-73205	0-88318-189-4
No. 91	Laser Acceleration of Particles (Los Alamos, 1982)	82-73361	0-88318-190-8
No. 92	The State of Particle Accelerators and High Energy Physics (Fermilab, 1981)	82-73861	0-88318-191-6
No. 93	Novel Results in Particle Physics (Vanderbilt, 1982)	82-73954	0-88318-192-4
No. 94	X-Ray and Atomic Inner-Shell Physics – 1982 (International Conference, U. of Oregon)	82-74075	0-88318-193-2
No. 95	High Energy Spin Physics – 1982 (Brookhaven National Laboratory)	83-70154	0-88318-194-0
No. 96	Science Underground (Los Alamos, 1982)	83-70377	0-88318-195-9

No. 97	The Interaction Between Medium Energy Nucleons in Nuclei – 1982 (Indiana University)	83-70649	0-88318-196-7
No. 98	Particles and Fields – 1982 (APS/DPF University of Maryland)	83-70807	0-88318-197-5
No. 99	Neutrino Mass and Gauge Structure of Weak Interactions (Telemark, 1982)	83-71072	0-88318-198-3
No. 100	Excimer Lasers – 1983 (OSA, Lake Tahoe, Nevada)	83-71437	0-88318-199-1
No. 101	Positron-Electron Pairs in Astrophysics (Goddard Space Flight Center, 1983)	83-71926	0-88318-200-9
No. 102	Intense Medium Energy Sources of Strangeness (UC-Sant Cruz, 1983)	83-72261	0-88318-201-7
No. 103	Quantum Fluids and Solids – 1983 (Sanibel Island, Florida)	83-72440	0-88318-202-5
No. 104	Physics, Technology and the Nuclear Arms Race (APS Baltimore –1983)	83-72533	0-88318-203-3
No. 105	Physics of High Energy Particle Accelerators (SLAC Summer School, 1982)	83-72986	0-88318-304-8
No. 106	Predictability of Fluid Motions (La Jolla Institute, 1983)	83-73641	0-88318-305-6
No. 107	Physics and Chemistry of Porous Media (Schlumberger-Doll Research, 1983)	83-73640	0-88318-306-4
No. 108	The Time Projection Chamber (TRIUMF, Vancouver, 1983)	83-83445	0-88318-307-2
No. 109	Random Walks and Their Applications in the Physical and Biological Sciences (NBS/La Jolla Institute, 1982)	84-70208	0-88318-308-0
No. 110	Hadron Substructure in Nuclear Physics (Indiana University, 1983)	84-70165	0-88318-309-9
No. 111	Production and Neutralization of Negative Ions and Beams (3rd Int'l Symposium, Brookhaven, 1983)	84-70379	0-88318-310-2
No. 112	Particles and Fields – 1983 (APS/DPF, Blacksburg, VA)	84-70378	0-88318-311-0
No. 113	Experimental Meson Spectroscopy – 1983 (Seventh International Conference, Brookhaven)	84-70910	0-88318-312-9
No. 114	Low Energy Tests of Conservation Laws in Particle Physics (Blacksburg, VA, 1983)	84-71157	0-88318-313-7
No. 115	High Energy Transients in Astrophysics (Santa Cruz, CA, 1983)	84-71205	0-88318-314-5
No. 116	Problems in Unification and Supergravity (La Jolla Institute, 1983)	84-71246	0-88318-315-3
No. 117	Polarized Proton Ion Sources (TRIUMF, Vancouver, 1983)	84-71235	0-88318-316-1

No. 118	Free Electron Generation of Extreme Ultraviolet Coherent Radiation (Brookhaven/OSA, 1983)	84-71539	0-88318-317-X
No. 119	Laser Techniques in the Extreme Ultraviolet (OSA, Boulder, Colorado, 1984)	84-72128	0-88318-318-8
No. 120	Optical Effects in Amorphous Semiconductors (Snowbird, Utah, 1984)	84-72419	0-88318-319-6
No. 121	High Energy e^+e^- Interactions (Vanderbilt, 1984)	84-72632	0-88318-320-X
No. 122	The Physics of VLSI (Xerox, Palo Alto, 1984)	84-72729	0-88318-321-8
No. 123	Intersections Between Particle and Nuclear Physics (Steamboat Springs, 1984)	84-72790	0-88318-322-6
No. 124	Neutron-Nucleus Collisions – A Probe of Nuclear Structure (Burr Oak State Park - 1984)	84-73216	0-88318-323-4
No. 125	Capture Gamma-Ray Spectroscopy and Related Topics – 1984 (Internat. Symposium, Knoxville)	84-73303	0-88318-324-2
No. 126	Solar Neutrinos and Neutrino Astronomy (Homestake, 1984)	84-63143	0-88318-325-0
No. 127	Physics of High Energy Particle Accelerators (BNL/SUNY Summer School, 1983)	85-70057	0-88318-326-9
No. 128	Nuclear Physics with Stored, Cooled Beams (McCormick's Creek State Park, Indiana, 1984)	85-71167	0-88318-327-7
No. 129	Radiofrequency Plasma Heating (Sixth Topical Conference, Callaway Gardens, GA, 1985)	85-48027	0-88318-328-5
No. 130	Laser Acceleration of Particles (Malibu, California, 1985)	85-48028	0-88318-329-3
No. 131	Workshop on Polarized ^{3}He Beams and Targets (Princeton, New Jersey, 1984)	85-48026	0-88318-330-7
No. 132	Hadron Spectroscopy–1985 (International Conference, Univ. of Maryland)	85-72537	0-88318-331-5
No. 133	Hadronic Probes and Nuclear Interactions (Arizona State University, 1985)	85-72638	0-88318-332-3
No. 134	The State of High Energy Physics (BNL/SUNY Summer School, 1983)	85-73170	0-88318-333-1
No. 135	Energy Sources: Conservation and Renewables (APS, Washington, DC, 1985)	85-73019	0-88318-334-X
No. 136	Atomic Theory Workshop on Relativistic and QED Effects in Heavy Atoms	85-73790	0-88318-335-8
No. 137	Polymer-Flow Interaction (La Jolla Institute, 1985)	85-73915	0-88318-336-6
No. 138	Frontiers in Electronic Materials and Processing (Houston, TX, 1985)	86-70108	0-88318-337-4
No. 139	High-Current, High-Brightness, and High-Duty Factor Ion Injectors (La Jolla Institute, 1985)	86-70245	0-88318-338-2

No. 140	Boron-Rich Solids (Albuquerque, NM, 1985)	86-70246	0-88318-339-0
No. 141	Gamma-Ray Bursts (Stanford, CA, 1984)	86-70761	0-88318-340-4
No. 142	Nuclear Structure at High Spin, Excitation, and Momentum Transfer (Indiana University, 1985)	86-70837	0-88318-341-2
No. 143	Mexican School of Particles and Fields (Oaxtepec, México, 1984)	86-81187	0-88318-342-0
No. 144	Magnetospheric Phenomena in Astrophysics (Los Alamos, 1984)	86-71149	0-88318-343-9
No. 145	Polarized Beams at SSC & Polarized Antiprotons (Ann Arbor, MI & Bodega Bay, CA, 1985)	86-71343	0-88318-344-7
No. 146	Advances in Laser Science–I (Dallas, TX, 1985)	86-71536	0-88318-345-5
No. 147	Short Wavelength Coherent Radiation: Generation and Applications (Monterey, CA, 1986)	86-71674	0-88318-346-3
No. 148	Space Colonization: Technology and The Liberal Arts (Geneva, NY, 1985)	86-71675	0-88318-347-1
No. 149	Physics and Chemistry of Protective Coatings (Universal City, CA, 1985)	86-72019	0-88318-348-X
No. 150	Intersections Between Particle and Nuclear Physics (Lake Louise, Canada, 1986)	86-72018	0-88318-349-8
No. 151	Neural Networks for Computing (Snowbird, UT, 1986)	86-72431	0-88318-351-X
No. 152	Heavy Ion Inertial Fusion (Washington, DC, 1986)	86-73135	0-88318-352-8
No. 153	Physics of Particle Accelerators (SLAC Summer School, 1985) (Fermilab Summer School, 1984)	87-70103	0-88318-353-6
No. 154	Physics and Chemistry of Porous Media—II (Ridge Field, CT, 1986)	83-73640	0-88318-354-4
No. 155	The Galactic Center: Proceedings of the Symposium Honoring C. H. Townes (Berkeley, CA, 1986)	86-73186	0-88318-355-2
No. 156	Advanced Accelerator Concepts (Madison, WI, 1986)	87-70635	0-88318-358-0
No. 157	Stability of Amorphous Silicon Alloy Materials and Devices (Palo Alto, CA, 1987)	87-70990	0-88318-359-9

No. 158	Production and Neutralization of Negative Ions and Beams (Brookhaven, NY, 1986)	87-71695	0-88318-358-7
No. 159	Applications of Radio-Frequency Power to Plasma: Seventh Topical Conference (Kissimmee, FL, 1987)	87-71812	0-88318-359-5